Undergraduate Texts in Mathematics

Robert L. Wilson

Much Ado About Calculus

A Modern Treatment with Applications
Prepared for Use with the Computer

Springer-Verlag
New York   Heidelberg   Berlin

Robert L. Wilson
Department of Mathematical Sciences
Ohio Wesleyan University
Delaware, Ohio 43105
USA

Editorial Board

P. R. Halmos
Managing Editor
Indiana University
Department of Mathematics
Bloomington, Indiana 47401
USA

F. W. Gehring
University of Michigan
Department of Mathematics
Ann Arbor, Michigan 48104
USA

AMS Subject Classification: 26-01, 68-01

With 145 Figures

Library of Congress Cataloging in Publication Data

Wilson, Robert Lee, 1917–
    Much ado about calculus.

    (Undergraduate texts in mathematics)
    Includes index.
    1. Calculus. I. Title.
QA303.W52   515   79-987

9 8 7 6 5 4 3 2 1

ISBN 0-387-90347-X   Springer-Verlag New York
ISBN 3-540-90347-X   Springer-Verlag Berlin Heidelberg

In memory of

William Harold Wilson

teacher of mathematics, researcher, and gentleman

Preface

The calculus has been one of the areas of mathematics with a large number of significant applications since its formal development in the seventeenth century. With the recent development of the digital computer, the range of applications of mathematics, including the calculus, has increased greatly and now includes many disciplines that were formerly thought to be non-quantitative. Some of the more traditional applications have been altered, by the presence of a computer, to an extent such that many problems hitherto felt to be intractable are now solvable.

This book has been written as a reaction to events that have altered the applications of the calculus. *The use of the computer is made possible at an early point, although the extent to which the computer is used in the course is subject to the decision of the instructor.* Some less traditional applications are included in order to provide some insight into the breadth of problems that are now susceptible to mathematical solution. The Stieltjes integral is introduced to provide for easier transition from the stated problem to its mathematical formulation, and also to permit the use of functions like step functions in later courses (such as statistics) with relative ease. The course is designed to include all the background material ordinarily associated with the first course in the calculus, but it is also designed with the user in mind. Thus, those topics that are felt to be most needed by the student who will take only one term of calculus are introduced early, so that such a student may be sure to have such materials before leaving the course. The exponential is a case in point. The development is done with one eye on rigor in order that students may see where the results originate. The amount of rigor introduced in a particular classroom is, again, up to the instructor, but the author and his colleagues have felt for some time that rigor should not be avoided. It is to be anticipated that the students we teach today will be using the calculus in ways we cannot anticipate. It will be up to them to know a sufficient amount of rigor to be able to determine whether the application they propose is valid or not. Only with an understanding of the underlying theorems can one make such a decision.

The text is designed to give maximum flexibility to the instructor. The first chapter includes those items that are needed throughout the remainder of the book. The instructor can omit such of these items as he/she may feel are not needed for a particular class. If more material of this variety is required, there are appendices on trigonometry and analytic geometry, each with exercises for the student. The programming languages are contained in the appendices. If some language other than BASIC or FORTRAN is to be used, there is nothing in the text, other than the abscence of a suitable appendix, to cause any problem.

This material has been taught in the classrooms of Ohio Wesleyan University since 1971. Prior to that time other material relating the computer and the calculus was used and found to place too great an emphasis on the computer. This book was designed to permit such emphasis as an instructor may desire, but also to permit almost total disregard of the computer if this be the desire of the instructor. The course at Ohio Wesleyan University has usually spent one or two days early in the course, usually in the first week, to introduce sufficient (and only sufficient) programming to handle a very simple problem, such as adding all of the odd integers less than 100,000. This serves to give all students some familiarity with a computer and with the concept of writing a program. During the second week the students will be asked to write a second program, perhaps one to solve an equation using the method of bisection. By the third or fourth week they may be asked to write a program which approximates the volume of a sphere by what amounts to integration. At this point the students should be able to do any programming required in the course without further training. If desired, this pace could be slowed to a considerable extent. It is possible, of course, to require a prerequisite or corequisite course in programming, but this has not appeared to be necessary. With this approach, the computer takes little extra time, and its judicious use will emphasize concepts at a later point to the extent that this time may well be regained.

The introduction of the Stieltjes integral may be viewed with some skepticism. It has been observed in the classroom that students who are being introduced to the calculus for the first time find it no more difficult to handle the Riemann–Stieltjes sum and integral than the Riemann sum and integral. In fact, the presence of the function $g(x)$ tends to make more specific the role that is sometimes lost with the simpler identity function. This is particularly true in understanding the proof of the fundamental theorem. It is also true that problems involving the determination of volume of revolution by the method of cylindrical shells is made easier to comprehend, for the function $g(x) = \pi x^2$ appears naturally and does not require elaborate explanation as to why the inside of the rectangle travels through a smaller circumference than does the outside. The Stieltjes integral is also helpful in such areas as substitution theorems and in many proofs, proofs that are made much shorter and more direct by the presence of the more general function $g(x)$.

In order to emphasize the role of the fundamental theorem, much use is made of this important result in using not only the anti-derivative, but also the anti-integral, to obtain additional formal results. This has proven to be a significant aid in obtaining a true understanding, of not only the meaning of the integral and the derivative, but also their interrelationship. In addition to achieving a deeper understanding, this also reduces the amount of time required to develop some of the usual formulas for differentiation and integration, thus permitting time for other topics.

In deference to the increasing use of the computer, chapters are included on interpolation and regression and on numerical methods. The chapter on interpolation and regression is of particular interest to those persons who do not have the large number of theoretically derived formulas from which to work. Such materials would be needed in most of the social sciences and in many of the life sciences. The physicist, on the other hand, is much more apt to have formulas that have been derived theoretically. Therefore, the inclusion of this chapter may depend on the particular class involved. Numerical methods are becoming increasingly important. It would be neither desirable nor possible to include a course in numerical analysis, but some of the more basic concepts of the subject will be needed by students who do not have time in their programs for a numerical analysis course. The inclusion of this chapter is, therefore, dependent on the particular class and the goals that the students may have.

Sufficient information on partial derivatives and iterated integrals is included to satify the needs of those students who may not continue through a course in linear algebra and multivariable calculus. Such students may have to use an occasional partial derivative, be concerned with extreme values, or be required to handle a multiple integral. However, such a student will probably need nothing more than some mechanical ability and an intuitive understanding. This has been provided. Students should be warned that if they will be making use of a great amount of analysis, they should continue with a second course. This represents no change from the situation occurring with the majority of courses designed for one year.

Once a text has been selected, the obvious choice for a syllabus is one that starts at the beginning and proceeds from chapter to chapter. More often than not, some modification of this procedure is desired. This book is intended as a first course in calculus, and beginning with Chapter II we will assume that certain basic mathematical information is familiar to the student. The specific information required is included in Chapter I and in Appendices A and B. The use of this material can be altered to fit a particular class, as determined by the instructor. Some may wish to go through the material in detail and others may feel that some, or perhaps all, of this material can be omitted. Chapters II and III introduce the Riemann–Stieltjes integral through the use of summations. While limits lurk in the background, the obvious mathematical material that is required includes only the least-upper-bound axiom. There is little that can be omitted in these two chapters, although the proofs can be de-emphasized if this is desired. Chapter IV introduces the derivative. Except for a few references to earlier material, there is no logical reason why this Chapter could not precede Chapters II and III. The fundamental theorem with its relation between the integral and the derivative is introduced in Chapter V, and the concept of inverse operations can hardly be brought in before this point.

Chapter VI deals primarily with techniques for differentiation and integration. Thus, this is a chapter which the majority of students should reach as soon as is feasible, since it will sharpen their skills in formal integration and differentiation. Chapter VII returns to the more rigorous aspects of the calculus and completes the work on limits and continuity which is taken somewhat intuitively in the earlier portion of the book. Applications of limits such as l'Hôpital's rule and improper integrals are included in this chapter. Chapter VIII discusses interpolation and regression techniques for finding functions that approximate given sets of data. There is no reason why this chapter cannot follow Chapter VI or, for that matter, Chapter IV. If it is included without Chapter VI, some care would be necessary to insure that only problems involving known techniques are used. The partial differentiation included in this chapter is intended to provide sufficient information to assist those who may wish to use partial differentiation in other courses before they have an opportunity to take further courses in mathematics.

Chapter XI discusses infinite series. This material requires work in limits and various techniques of differentiation and integration. It is doubtful whether this should be undertaken without having covered the material in Chapters VI and VII. As with the material on partial differentiation, the work on iterated integrals is intended to provide sufficient information to permit such use as may be required by persons who are not able to take further courses in mathematical analysis. The introduction of Fourier series provides additional background for those who may need this material before they would have it in other courses. The trigonometric series can be omitted without doing damage to the further work in the text. Chapter X deals with numerical techniques of differentiation and integration. It also provides an introduction to the numerical solution of linear differential equations. In view of the increasing breadth of applications, it is probable that students will come up against material that could hardly be handled in a formal way, as well as material that involves working with sets of data obtained from other computations. If this be the case, some insight into methods of handling such problems on a digital computer is highly desirable. This chapter should not be attempted if the students are not familiar with Taylor's series with error terms.

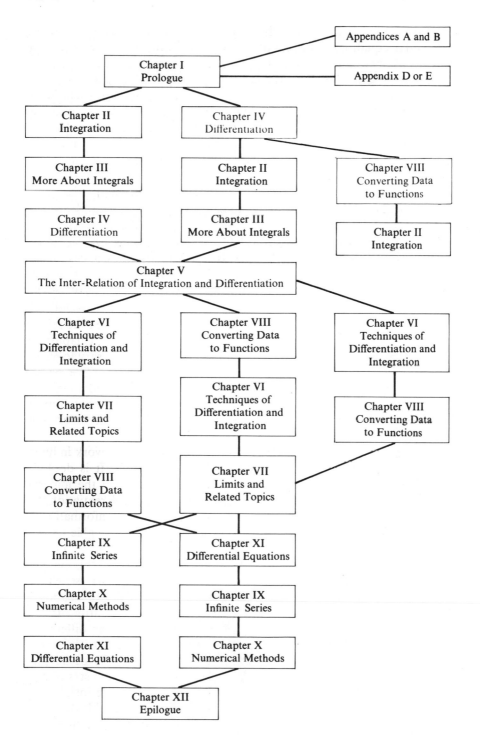

Chapter XI provides an introduction to the solution of the more common differential equations. This chapter would probably be more useful for students of physics and engineering, although differential equations appear in an increasing number of fields these days. Finally, Chapter XII is intended merely to inform students of material that might be useful to them in the future. Several students in any class will ask, or might like to ask, questions concerning further mathematics they might study, with a concern for the content and the areas of application appropriate to the material involved. It is suggested that students be reminded of the existence of this chapter near the end of the course, and then be invited to ask the instructor for more detailed information about local offerings if there is sufficient interest.

In particular, the flowchart shows some of the tracks that can be taught from this book.

With this book, as with any effort of this kind, there are many people involved. It would be impossible to mention all those who have had a part, but I would like to single out a few. I would like to give particular thanks to my colleagues at Ohio Wesleyan University and those who have born with me during the period of preliminary editions. Their suggestions and comments have been most helpful and appreciated. Professor S. B. Jackson read the manuscript thoroughly and made many helpful suggestions for which I am appreciative. Mrs. Marilyn Cryder has typed most of the manuscript for the several editions, and the remaining manuscript was typed by Mrs. Shirley Keller. They have put in many patient hours. The editorial and production staffs of Springer-Verlag have been most helpful and understanding. Finally, but most important, I would like to express my great appreciation to my wife, Anna Katherine, and to my family for their understanding and patience during the many occasions when the author was not available for family affairs. Without their support, an effort such as this could never have come to fruition.

Robert L. Wilson

Delaware, Ohio
September 1978

Contents

Chapter IV

Differentiation   205

Chapter V

The Interrelation of Intergration and Differentiation   245

Chapter VI

Techniques of Differentiation and Integration   330

Chapter VII

Limits and Related Topics   399

Chapter VIII

Converting Data to Functions   443

Chapter IX

Infinite Series   494

Chapter X

Numerical Methods   563

Chapter XI

Differential Equations   601

CHAPTER I

Prologue

I.1 A Preview

We are about to embark on an investigation of the calculus, a branch of mathematics about which you have heard a great deal, but concerning which you may possibly have some misgivings. The calculus deals with limiting processes, and therefore can be quite different in some respects from the algebra, geometry, and trigonometry with which you have had previous contact. When you try to solve problems which involve an infinite number of items, or which use limiting processes, strange things can happen. This doesn't mean that they always give unexpected results, but rather that you have to be very careful that unexpected results do not slip by. In order to illustrate this, let us consider an example.

EXAMPLE. The properties of the real numbers by which one can group terms and change the order of addition without changing the results have been used throughout most, if not all, of your mathematics. Let us see what happens when we apply these to the expression

$$x = 1 - \tfrac{1}{2} + \tfrac{1}{3} - \tfrac{1}{4} + \tfrac{1}{5} - \tfrac{1}{6} + \tfrac{1}{7} - \tfrac{1}{8} + \tfrac{1}{9} \cdots \qquad \text{(I.1.1)}$$

where the three dots are the mathematical equivalent of "etc." and indicate that we should continue to write terms using the indicated procedure without stopping. We can re-write (I.1.1) as

$$x = (1 - \tfrac{1}{2}) + (\tfrac{1}{3} - \tfrac{1}{4}) + (\tfrac{1}{5} - \tfrac{1}{6}) + (\tfrac{1}{7} - \tfrac{1}{8}) + \cdots \qquad \text{(I.1.2)}$$

in which case x is a sum of positive terms, since the number in each of the indicated subtractions is positive. We can also write

$$x = 1 - (\tfrac{1}{2} - \tfrac{1}{3}) - (\tfrac{1}{4} - \tfrac{1}{5}) - (\tfrac{1}{6} - \tfrac{1}{7}) - (\tfrac{1}{8} - \tfrac{1}{9}) - \cdots \qquad \text{(I.1.3)}$$

1

in which case x is the result of subtracting positive terms from one. Therefore, x must be a number less than one. From (I.1.2) and (I.1.3) it is apparent that x must be a number between zero and one (and this is correct). Now, let us multiply both sides of (I.1.1) by two. We have (I.1.4)

$$2x = 2 - 1 + \tfrac{2}{3} - \tfrac{1}{2} + \tfrac{2}{5} - \tfrac{1}{3} + \tfrac{2}{7} - \tfrac{1}{4} + \tfrac{2}{9} - \tfrac{1}{5} + \cdots. \qquad (I.1.4)$$

If we rearrange the terms in (I.1.4) and group those with equal denominators, we obtain

$$2x = (2 - 1) - \tfrac{1}{2} + (\tfrac{2}{3} - \tfrac{1}{3}) - \tfrac{1}{4} + (\tfrac{2}{5} - \tfrac{1}{5}) - \tfrac{1}{6} + \cdots. \qquad (I.1.5)$$

Upon simplifying the expressions in parentheses we have

$$2x = 1 - \tfrac{1}{2} + \tfrac{1}{3} - \tfrac{1}{4} + \tfrac{1}{5} - \tfrac{1}{6} + \cdots. \qquad (I.1.6)$$

From (I.1.1) and (I.1.6) we observe that we have $2x = x$. However, we have shown that $x \neq 0$, and hence division by x is possible, giving us the strange result $2 = 1$.

Rather clearly something seems to have gone wrong, despite the fact that we were performing only familiar operations and in a manner which we would expect to produce correct results. The difficulty here is a subtle one, and one which we shall not attempt to explain at this point, other than to say that this result could not have happened if we had only a finite number of terms. This result does demonstrate, however, why it is necessary to be very careful when we are considering any situation involving an infinite number of terms. (The reason for this strange occurrence will be cleared up in Chapter IX.)

Since the concepts that we will be dealing with must be handled with a great deal of care, we shall consider in this chapter some fundamentals which will be of assistance later on. *It cannot be emphasized too strongly that your primary concern in this chapter and throughout the calculus should be that of understanding the concepts with which we are working rather than memorizing any ordered set of words which come under the label of definition or theorem.* This will require attention to the fundamentals which underly the concepts. In proofs and derivations your concern should be focused on learning how to prove theorems and how to derive results. A computer can handle a formula with greater speed and greater accuracy than can a human being, but the computer is not able to determine which method can best be used in a given situation, nor how a particular result should be pursued. It is also true that with very few exceptions it is possible to prove theorems in a variety of ways and to solve problems by diverse methods. Therefore, there is no reason why your proof or solution must follow the same pattern you may find in this book, or which you find given by your instructor, or your classmates. The only requirement is that the proof be logically complete and correct or that the solution be one that can be defended if challenged.

One final bit of advice concerning proofs and solutions is in order. First read the theorem to be proven or the problem to be solved and be certain that you understand it. Second think of some procedure which would seem to provide a logical path from that which is given to that which is desired. Only after you have completed these two steps should you start writing down the proof or solution. By the time you have completed the first two steps you will have a direction in which to proceed, and this is important. As a part of the two steps given, if it is possible to draw a picture or sketch which is meaningful in interpreting the theorem or problem, by all means draw it—draw it large enough that you can label it clearly, and use this to aid your intuition and your logic in arriving at the proof or solution.

EXERCISE

1. Explain in detail why (I.1.2) shows that x must be positive.

2. Explain in detail why (I.1.3) shows that x must be less than one.

3. Why is it important in obtaining the conclusion "$2 = 1$" that x is a number other than zero?

4. Use an argument similar to that given in this section to show that x is greater than $1/2$ and less than $5/6$.

5. In what way would the argument of this section be made impossible if (I.1.1) had twenty terms instead of an infinite number of terms?

I.2 Some Properties of Numbers

Since the word *calculus* is derived from the same root[1] as the word *calculate*, it is not unreasonable to expect that we will be working with numbers, either explicitly as constants or perhaps a bit less obviously through our use of variables which represent numbers. You have worked with numbers for many years, and are familiar with many types of numbers, such as the natural numbers (the ones used for counting), the integers (which include the natural numbers, zero, and the negatives of the natural numbers), and the real numbers, among others. We shall not attempt in this section to go back and investigate all of the facts that you have been told concerning numbers. On the other hand, the example given in Section 1 indicates that some of the things that we have been in the habit of doing must be done with caution, particularly if we are dealing with an unlimited set of numbers in a given computation. We intend to state those assumptions (or in more sophisticated terms, those postulates) which we assume concerning the numbers we use.

---

[1] Both words are derived from *calculi* meaning *pebbles*. These pebbles would have been used on sand or strung on wires for counting purposes, somewhat in the manner of the abacus of today.

We will do this in a definition of a system that mathematicians call a *number field*.

Before stating the definition, it would be well to mention briefly some of the terms that are frequently used in connection with numbers. This should serve to give some background for the more formal definition. We will be dealing with a set (either finite or infinite in number) of elements which we usually call *numbers*. We have an equivalence relation, which we denoted by "=" and this is defined in such a way that we can say "$x = y$" if and only if x and y are the names (perhaps different ones) for the same number. Thus, to say "$a + b = c$" requires that the number represented by "$a + b$" is the same number as the one represented by "c". There is nothing in this explanation that differs from your previous experience, but it is well to set it forth so that there is no misconception concerning the meaning of the symbols that we use. We also have in most situations involving numbers two "binary" operations. A *binary operation* is one that requires that two (not necessarily distinct) numbers be given and which then specifies a unique third number (again not necessarily distinct from the first two). Thus, addition requires that we have two numbers, such as 2 and 3, and then produces a unique third number, in this case 5. If we wish to add more than two numbers, we employ the associative property and effectively add two at one time and then add the third to the sum of the first two. Thus, if we wish to add 2, 3, and 7, we should have $(2 + 3) + 7 = 5 + 7 = 12$. The fact that we can add in any order is covered by the commutative and associative properties of numbers. Note that in the definition these are stated for computations involving two or three numbers. (It is the use of these operations with an infinite set of numbers that causes trouble in the example of Section 1.) Finally, before stating the definition, it is well to mention that the operations of subtraction and division are given implicitly through the requirement that we have an inverse for addition and multiplication, respectively. Thus, we would define subtraction to be the operation of adding the additive inverse (or negative). This is equivalent to the subtraction that you have used in the past, but it is much easier to use the definition in this form. (For the purist we would note that we are aware of a certain redundancy in this definition, but this will assist our discussion here.)

Definition. A *number field* is a set of at least two elements (called numbers), denoted by $a, b, c, \ldots$, an equivalence relation "=", and two binary operations called addition $(+)$ and multiplication $(\cdot)$ such that for *any* three numbers a, b, and c the following assumptions hold:

1. *Closure*:	$a + b$ is a unique number;	$a \cdot b$ is a unique number	
2. *Commutativity*:	$a + b = b + a$;	$a \cdot b = b \cdot a$	
3. *Associativity*:	$(a + b) + c = a + (b + c)$;	$(a \cdot b) \cdot c = a \cdot (b \cdot c)$	
4. *Identity*:	there is a unique number 0 such that $a + 0 = 0 + a = a$	there is a unique number 1 such that $a \cdot 1 = 1 \cdot a = a$	

5. *Inverse*:   for any number a there is   for any number $a \neq 0$
a unique number ^-a such   there is a unique number
that $a + {}^-a = {}^-a + a = 0$; a^{-1} such that
$$a \cdot a^{-1} = a^{-1} \cdot a = 1;$$

6. *Distributive*: $a \cdot (b + c) = a \cdot b + a \cdot c$

Note that property 5 provides us with a *negative*, (^-a), and a *reciprocal*, (a^{-1}), the latter whenever we have a *non-zero number*, and these two assumptions give us the ability to subtract and to divide. We will define $(a - b)$ as meaning $a + (^-b)$ and $(a \div b)$ as meaning $a \cdot (b^{-1})$. This very definitely excludes division by zero as a result of property 5. Note also that the sign in (^-b) applies to a single number. This use of the negative sign illustrates a *unary operator*, or one that operates on a single number.

There are several sets of numbers that satisfy these properties. We will customarily be dealing with the real numbers, and the set of all real numbers constitute a number field. On the other hand, since the reciprocal of 2 is $1/2$, and $1/2$ is not an integer, the set of all integers do *not* form a number field, for this set does not fulfill the assumption which says that *every* non-zero number in the set has a reciprocal which is also in the set. The integers do satisfy all of the other assumptions in our list, however. It is worth noting that the set of all polynomials with real coefficients satisfy exactly the same set of assumptions which the integers satisfy. We will have occasion to speak of the *rational numbers*. This is the set of numbers, each of which is capable of being expressed as the *ratio* of two integers (note the first five letters of the word *rational*). These numbers also constitute a number field. There are many other fields, and we will be looking at one additional field in this chapter.

The *real numbers* can be further distinguished by including some additional properties concerning ordering of the numbers, but we will reserve these for Section 3 of this chapter and for Section 5 of Chapter II.

We have given a definition of a number field which includes many properties of the two operations $(+)$ and $(\cdot)$. In point of fact, these postulates (or properties) which we have assumed are more far-reaching than is apparent from reading them. For instance, these postulates cannot be true unless the product $(^-a)(^-b)$ has the same value as the product (ab), where we have implied the product by juxtaposition (or writing the two numbers without any intervening symbols). (See Exercise I.2.15). In order to indicate the additional properties which we have assumed, we shall *prove theorems* which are merely statements that are true because we have made these initial assumptions. The theorems are the consequences of the definitions and the postulates. We will have occasion rather frequently to prove theorems (or derive results).

[Almost any theorem can be proven in a great variety of ways. As we stated earlier, the proofs that you see in books and those presented in classrooms and in lectures usually appear to be very straightforward, to economize on the number of steps required, and to have an air of completeness and finality

about them. In the majority of cases, including most of the proofs presented
in this book, the first version of the proof was very awkward, and while it did
prove what it set out to prove, it did not do it in a very nice, neat way. By going
back over the work, it was possible to find shorter proofs, neater proofs, and in
general to produce this nice finished air. Anyone trying to prove any result for
the first time, even though he may have had prior experience in proving things,
should not be surprised if the proof can be made shorter. If, however, your
succession of statements constitute a proof, you have succeeded in your
assignment, and you should not be concerned if your particular proof differs
from another proof of the same result. The only note of caution is to be certain
that you start with assumptions which are permissible and using only definitions, postulates (or assumptions), and previously proven results (or theorems)
you are able to arrive at the conclusion using laws of logic which can be
supported at each step.]

As an illustration of a proof in which all reasons are given, we will consider
the following

EXAMPLE 2.1. Prove the theorem: If x is any number in a number field and 0
is the additive identity, then $x \cdot 0 = 0$.

PROOF

Statement	Reason
1. There is a number 0 in the field such that $1 + 0 = 1$	1. The property of additive identity from the field definition.
2. For any number x in the field, $x \cdot (1 + 0) = x \cdot 1$	2. The product of two numbers is unique.
3. $x \cdot (1 + 0) = x \cdot 1 + x \cdot 0$	3. The distributive property of the field.
4. $x \cdot 1 + x \cdot 0 = x \cdot 1$	4. Two things equal to the same things are equal to each other.
5. There is a unique number $^-(x \cdot 1)$	5. The additive inverse property from the field.
6. $^-(x \cdot 1) + [(x \cdot 1) + (x \cdot 0)] = {}^-(x \cdot 1) + (x \cdot 1)$	6. The result of addition is unique.
7. $[^-(x \cdot 1) + (x \cdot 1)] + (x \cdot 0) = {}^-(x \cdot 1) + (x \cdot 1)$	7. The associative property of addition.
8. $0 + (x \cdot 0) = 0$	8. The property of the additive inverse (or negative).
9. $(x \cdot 0) = 0$	9. Property of additive identity (zero). ☐

As an example of a slightly different type of proof, known as an *indirect
method of proof* we prove another theorem.

EXAMPLE 2.2. Prove: The numbers 0 and 1 are distinct. (That is, $0 \neq 1$.)

PROOF

Statement	Reason
1. Either $0 = 1$ or $0 \neq 1$	1. The determinative property of equivalence.
2. Let x be a number such that $x \neq 0$	2. A number field has at least two elements.
3. If $1 = 0$, then $x \cdot 1 = x \cdot 0$	3. $1 = 0$ is equivalent to stating that 1 and 0 are different names for the same number and multiplication gives a unique result.
4. $x \cdot 1 = x$	4. Identity property of multiplication.
5. $x \cdot 0 = 0$	5. Example 2.1.
6. $x = 0$	6. Replacement in 3 using 4 and 5.
7. Statements 2 and 6 are contradictory	7. By the determinative property of equivalence.
8. Assumption $1 = 0$ is false and $0 \neq 1$ must be true	8. Since resulting conclusion is contradictory. $\quad\square$

The examples given are illustrations and it is not implied that you would have given the same proofs. In fact, it is worth noting that it is often more difficult to find a method of proof for an *obvious* result than for one that appears less likely to be true. The important point is that you reason carefully and that you be able to give a reason for the validity of each step.

EXERCISE

1. Classify each of the following as an integer, a rational number, a real number or a specific combination of these types: 2, π, $\sqrt{3}$, 4.32, 1.732, 0, -5, $-3/7$, $100/9$, 3.1416, $22/7$.

2. Show that any terminating decimal (one with a finite number of decimal places) is a rational number.

3. Show that the set of real numbers form a number field. (That is, show that they satisfy all of the postulates of a number field.)

4. Show that the rational numbers form a number field.

5. Prove that 0 does not have a multiplicative inverse. (Hint: in statements such as the one you are to prove here, there are just two possible cases and it is frequently easier to use the indirect method.)

6. Prove that if a and b are real numbers and $a \cdot b = 0$, then either $a = 0$ or $b = 0$. How is this conclusion used in the solution of quadratic equations by factoring?

7. Find all real numbers, x, for which $x^3 - 6x^2 + 11x - 6 = (x - 1)(x - 2)(x - 3)$ $= 0$. Give a reason for each step in your solution. [Hint: apply Exercise 6.]

8. Find all real numbers satisfying each of the following equations. Give a reason for each step in your solution. [Hint: Apply Exercise 6.]

   (a) $x^2 + 3x = 0$
   (b) $x^2 + 3x + 2 = 0$
   (c) $2x^2 + 3x + 1 = 0$
   (d) $x^3 + 6x^2 + 11x + 6 = (x + 1)(x + 2)(x + 3) = 0$
   (e) $x^2 = 9$
   (f) $x^2 + 4x + 2 = (x + 2 + \sqrt{2})(x + 2 - \sqrt{2}) = 0$

9. If a/b is a quotient of real numbers which is defined, and if $a/b = 0$, prove that $a = 0$.

10. Prove that if a and b are real numbers then $(^-a) \cdot b = {}^-(a \cdot b)$.

11. Prove that if a is a real number, then $^-(^-a) = a$.

12. If x and y are real numbers, prove:

   (a) $x - (^-y) = x + y$
   (b) $^-(x + y) = (^-x) + (^-y)$

13. If x and y are real numbers, prove:

   (a) $(x^{-1})^{-1} = x$
   (b) $(x \cdot y)^{-1} = x^{-1} \cdot y^{-1}$

14. If x, y, and z are real numbers, prove:

   (a) If $x + z = y + z$, then $x = y$.
   (b) If $x \cdot z = y \cdot z$, and if $z \neq 0$, then $x = y$.

15. You have been asked to prove that if a and b are real numbers then $(^-a) \cdot (^-b) = a \cdot b$. You have the following set of statements:

   1. $[(^-a) \cdot (^-b) + (^-a) \cdot b] + a \cdot b = (^-a) \cdot (^-b) + [(^-a) \cdot b + a \cdot b]$
   2. $(^-a) \cdot [(^-b) + b] + a \cdot b = (^-a) \cdot (^-b) + [(^-a) + a] \cdot b$
   3. $(^-a) \cdot 0 + a \cdot b = (^-a) \cdot (^-b) + 0 \cdot b$
   4. $0 + a \cdot b = (^-a) \cdot (^-b) + 0$
   5. $a \cdot b = (^-a) \cdot (^-b)$

   For each statement give a reason which is correct and which follows either from the hypothesis or from previous statements. Does this constitute a proof of the desired conclusion? Does this prove that the assumption of the postulates for a number field *requires* that the product of two negatives be a positive? Why?

16. You find a piece of paper left by your roommate with the following outline of a proof:

    1. Let x and y be real numbers such that $x = y$
    2. $x \cdot x = x \cdot y$
    3. $x^2 - y^2 = xy - y^2$
    4. $(x + y)(x - y) = y(x - y)$
    5. $x + y = y$
    6. $x + x = x$
    7. $2x = x$
    8. $2 = 1$

    Give reasons where possible, and determine whether the reasons that your roomate had in mind are correct or not. (In other words, if the proof is not correct, where does it break down?) We seem to be forever proving that $2 = 1$. Show that if this proof were correct, then it would follow that $8 = 4$, and that $13 = 9$, and in general that each real number would be equal to every other real number. What do these latter proofs tell us about a proof as a logical exercise versus the truth of the conclusion?

M17. If we consider a number system consisting of three numbers designated by the symbols &, +, and Δ, and if in this system addition and multiplication are defined by the following tables:

Addition	&	+	Δ
&	&	+	Δ
+	+	Δ	&
Δ	Δ	&	+

Multiplication	&	+	Δ
&	&	&	&
+	&	+	Δ
Δ	&	Δ	+

Which of the properties of the definition of a field are satisfied? What symbol would represent the zero and what symbol would represent the one?

M18. If we consider a number system consisting of two numbers designated by the symbols Δ and $, and if addition and multiplication are defined by the tables:

Addition	Δ	$
Δ	Δ	$
$	$	Δ

Multiplication	Δ	$
Δ	Δ	Δ
$	Δ	$

do we have a field? If so, what represents zero and what represents one?

M19. If we were to replace the tables of Exercise 18 by

Addition	Δ	$
Δ	Δ	$
$	$	$

Multiplication	Δ	$
Δ	Δ	Δ
$	Δ	$

would we have a field? (This system is closely related to the one on which computer logic is based.)

I.3   Order Properties and Inequalities

As we have seen, the real numbers form a number field. However, the real numbers, and several other number fields, have additional properties which are very useful. One such property is that of *order*, that is given two numbers in the field they must be equal or one is larger than the other. This is not a property possessed by all fields. In the development of further results from this basic property of order, it will be easier to state this basic property in a form which does not use the terms *greater than* or *larger*. We note that in the case of the real numbers there is a proper subset, that is a subset which does not include all of the real numbers, such that both addition and multiplication are closed in that subset. The subset in this case is the subset of positive real numbers. Note that while the sum and product of positive numbers provide us again with positive numbers, the sum and product of negative numbers would not provide us with negative numbers in all cases. Using this information, we are now ready to state the order property.

Order property. A number field is an *ordered field* if and only if there is a proper subset of the field called the *positive numbers*, which we shall designate by P, such that

1. *Closure*: If a and b are two numbers in P, then $a + b$ and $a \cdot b$ are in P,
2. *Trichotomy*: For each number, a, in the field exactly one of the following three statements is true:
   (i) a is in P,
   (ii) $-a$ is in P,
   (iii) $a = 0$.

A number in P is said to be a *positive number*. A number whose additive inverse is in P is said to be a *negative number*. Any number field for which the order properties hold is called an *ordered field*. The rational numbers and the real numbers both provide us with examples of ordered fields. For an example of a field that is not ordered, see Exercise 20 at the end of this section. In an ordered field we distinguish three classes of numbers, by the trichotomy (or three choice) property. These classes are called *positive* (including just those numbers in P), *negative* (those numbers such that their additive inverses are in P), and *zero* (this class contains only a single number). Since P is a proper subset, we note that the set of positive numbers and the set of negative numbers must be non-empty, and by the field properties we know that there must be a number zero, hence each of the three possibilities indicated in the trichotomy property is possible.

We now proceed to the definition of the phrase that we used intuitively above, namely *is greater than*.

Definition. If a and b are numbers in an ordered field, then *a is greater than b*, $(a > b)$, if $a - b$ is positive, and *a is less than b*, $(a < b)$, if $b - a$ is positive.

This definition makes it possible to put all statements concerning greater than and less than into the context of the properties which apply to the positive numbers. Thus $4 > -5$ since $4 - (^-5) = +9$ is positive. We will use this definition in the following theorems to obtain information concerning inequalities in a form which may be easier to apply. Note that in these theorems we will be giving only a succession of statements, and will expect you to provide the reasons. This is normal procedure in the majority of proofs in mathematics books, and you might as well get used to it now. It is suggested that you have a pad of paper available as you read this and that you reproduce the proof complete with reasons.

Theorem 3.1 (Alternate Statement of Order Property). *If a and b are two real numbers, then $a > b$, $a = b$, or $a < b$, and exactly one of these statements is true.*

PROOF. Let $c = a - b$. Now c is a real number, and hence c is positive, $c = 0$ or ^-c is positive, but only one of these is true. If c is positive, then $a > b$. If $c = 0$, then $a = b$, if ^-c is positive, let $d = b - a$, and then $d = ^-c$ is positive and hence $a < b$. Each of these arguments can be reversed, and hence the "exactly one of these statements is true" must hold. □

We have been using the phrase "c is positive" or "c is in P." Each of these is somewhat cumbersome, and it would be convenient to have an abbreviated phrase to indicate when either of these statements is true. The following theorem will be convenient.

Theorem 3.2. *If c is a positive real number, then $c > 0$ and conversely.*

PROOF. If c is a positive real number, then $c = c - 0$ is positive, and $c > 0$. If $c > 0$, then $c - 0 = c$ is positive. □

In similar fashion we can show that "c is negative" is equivalent to the relation $c < 0$.

The problem of addition and subtraction of negative numbers is easily taken care of since ^-a is the additive inverse of a. We should investigate the problem of multiplication, however. We have already shown (Exercise I.2.15) that the product of additive inverses is equal to the product of the numbers of which they were inverses. We now turn to the multiplication of inequalities, and we find that we have two cases.

Theorem 3.3. *If a, b, and c are real numbers such that $a > b$,*
i. *if $c > 0$ then $ac > bc$,*
ii. *if $c < 0$ then $ac < bc$.*

PROOF. Since $a > b$, $a - b$ is positive. If $c > 0$, c is positive and hence $(a - b)c = ac - bc$ is positive and $ac > bc$. If $c < 0$, then $(a - b)(^-c) = a(^-c) - b(^-c) \equiv (^-ac) - (^-bc) > 0$. Then $^-[(^-ac) - (^-bc)] = ^-[(^-ac) + bc] = ^-(^-ac) + ^-(bc) = ac - bc < 0$ or $ac < bc$. □

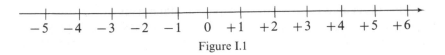

Figure I.1

This last result can be summed up by saying that if we multiply an inequality by a positive number, the products are unequal in the same order, but if we multiply by a negative number, the products are unequal in reverse order.

It is frequently helpful to visualize the integers, rational numbers, or real numbers with the use of a *number line*. Thus, in Figure I.1 we see a line, usually drawn horizontally, on which the larger numbers are to the right of the smaller numbers. Note that the number $a - b$ is the number of units measured from the point representing b to the point representing a, and $a - b$ will be positive if movement implied is from left to right whereas $a - b$ is negative for all cases in which the movement is from right to left. It is also worth noting that on this number line the integers occur with a distance of one unit between consecutive integers, whereas there is a one-to-one correspondence between the points on the number line and the real number. Since there are real numbers that are not rational, there are points on the number line for which no corresponding rational number exists, although there is a point on the number line for each rational number.

We have now considered all of the properties of the real field that we shall need to use with the exception of one which will be required in Section 5 of Chapter II and which will be given at that point. The fact that at least one more characterizing property is necessary should be apparent from the fact that both the real field and the rational field satisfy all of the properties that we have given so far, and there should be some property which distinguishes between these two fields.

EXERCISE

1. Show that $a > b$ implies that the point representing a on the number line is to the right of the point representing b. Show that this is the case regardless of whether a or b are both positive, both negative, or whether a is positive and b is negative. Show that the case in which a is negative and b is positive does not occur under the condition that $a > b$.

2. Which is the larger number in each of the following pairs? (Try this one *without* the use of a calculator). $(2.8, 29/11)$, $(3, -9)$, $(-0.47, -9/20)$, $(22/7, 355/113)$, $(22/7, \pi)$, $(-\sqrt{3}, -3/4)$, $(-3.2, -\sqrt{10})$, $(0.001, -1000)$, $(\sqrt{2}, \sqrt[3]{3})$.

3. Describe the set of numbers which satisfies the following relations:

   (a) $x > -2$,
   (b) $x \le -1.5$,
   (c) $x > \sqrt{2}$.

4. Put each of the following sets of numbers in ascending order.
   (a) $\{\sqrt{3}, -2, \pi, 22/7, 0, 2.8, -\sqrt{5/2}\}$,
   (b) $\{-3, 3.1, -2.9, 0.02, 1/25, 0\}$.

5. (a) Show that $(0.5)^2 < (0.5)$.
   (b) Describe carefully the set of all real numbers, x, for which $x^2 < x$.

6. If $a > b$, does it follow that $a^2 > b^2$? Prove your answer to be correct.

7. (a) Show that the truth of the statement $a^2 < a$ implies the truth of $a^3 < a^2$.
   (b) Show that there are some real numbers for which the converse of part (a) does not hold.

8. If $0 < a < 1$ and a is a real number, show that $0 < a^5 < a^4 < a^3 < a^2 < a < 1$.

9. (a) If you are given the fact that $3.1 < \sqrt{10} < 3.2$, what is the maximum amount by which $(3.1 + 3.2)/2 = 3.15$ could differ from $\sqrt{10}$?
   (b) If you are given the fact that $1.7320508 < \sqrt{3} < 1.7320509$, what is the maximum amount by which 1.73205085 would differ from $\sqrt{3}$?
   (c) If you are given the fact that $3.14 < \pi < 22/7$, what value could you find that would differ from π by the smallest possible amount, and what would be the maximum value of this difference?

10. Consider $3 > {}^-4$ and $2 > {}^-3$. Is it true that $(3)(2) > ({}^-4)({}^-3)$? Under what additional conditions will $a > b$ and $c > d$ assure that $ac > bd$?

11. If $b \neq 0$ and $d \neq 0$ and if a/b is defined to mean $a \cdot b^{-1}$, find a relation involving a, b, c, and d which is both necessary and sufficient to show that $a/b > c/d$. Prove that your relationship satisfies the above requirements.

12. Use the relationship derived in Exercise 11 to determine the truth of each of the following statements:
   (a) $20/29 < 0.7$,
   (b) $17/12 < \sqrt{2}$ [Hint: Find a relationship which involves only rationals.]

13. Prove that if c is a negative real number, than $c < 0$ and conversely.

14. Prove that if a, b, and c are real numbers such that $a < b$ and $b < c$, then $a < c$. (This is called the "transitive" property.)

15. (a) If a, b, and c are real number and $a < b$, prove that $a + c < b + c$.
   (b) If a, b, and c are real numbers and $a + c < b + c$, prove that $a < b$.

16. (a) If a and b are real numbers and $a < b$, show that there is a real number, c, such that $a < c < b$.
   (b) Repeat part (a) replacing the word "real" with the word "rational".
   (c) Show that the statement of part (a) is false if we replace "real numbers" with "integers".

17. (a) Show that $2 < 3$ implies that $-2 > -3$.
   (b) Show that $x < y$ implies $-x > -y$.
   (c) Show that $x - 2 > 3$ implies $2 - x < -3$.

M18. Is there a smallest real number, a, such that $a > 0$? Give a reason for your answer.

19. If we consider time, we note that two o'clock follows one o'clock and also two o'clock follows eleven o'clock. Is it possible to define $a > b$ when a and b are numbers to be read from the face of the clock?

M20. The system consisting of five elements, a, b, c, d, and e, in which addition and multiplication are defined by the following tables form a field.

Addition	a	b	c	d	e
a	a	b	c	d	e
b	b	c	d	e	a
c	c	d	e	a	b
d	d	e	a	b	c
e	e	a	b	c	d

Multiplication	a	b	c	d	e
a	a	a	a	a	a
b	a	b	c	d	e
c	a	c	e	b	d
d	a	d	b	e	c
e	a	e	d	c	b

The number a is the additive identity and b is the multiplicative identity. All of the requirements for the field definition are met. Show that there is no proper subset of this field which meets the requirements of P in our definition of positive numbers. This field is *not* an ordered field.

I.4   Complex Numbers

In the development of numbers it would seem natural that man would first learn to count, starting with the number one (probably counting on fingers) and then (if the climate were warm enough) on toes. The fact that the Mayans used a number system based on twenties would illustrate the latter observation. In some instances there is evidence that people even counted in terms of twos (perhaps because they were near-sighted and could only distinguish their two arms or fists). At some later point in history it was probably necessary to introduce negative integers when the first person ran into debt and found it necessary to borrow something. In other words, subtraction indicated the necessity for negative numbers of some variety unless subtraction were to be limited to certain types of number pairs. By the time man got around to paying off his debts, the amateur accountant was required to be aware of the existence of something equivalent to the number we call *zero*. In the next stage of development we can see our forefathers caught with the necessity for dividing some object and hence the concept of a fraction (or rational number) would come in. This was necessary if man was not to limit the numbers upon which he could perform that latest mathematical nightmare called *division*. Thus, it is reasonable, although we do not vouch for the fact, that this was precisely the order in which these concepts were developed, that the counting numbers were augmented throughout history as new situations developed for which the existing number system was not adequate.

At a later point in history man became interested in geometry and was concerned with finding the number of units of length in the diagonal of a square of which each side was one unit long. This required a completely new kind of number, one which we now call irrational, for it could not be expressed as a ratio of integers, no matter how hard one might try nor how ingenious the budding mathematicians might become. With the development of the real numbers, man had come to a plateau, or so he thought, and it was probably well for it took a great deal of effort to investigate the properties of the real numbers to be certain that everything was as it should be. That is not to say that the carpenter building the house was so concerned, but the mathematicians were concerned that they would know what these new numbers were all about. The fact that these numbers did not all come easily may be deduced from the fact that many early mathematicians denoted such numbers as (-2) by the name *fictitious* numbers. This name did not stick, but the term *negative* could be considered to be an adjective which is something less than complimentary. With the development of the real numbers, in answer to the need for being able to solve such equations as $x^2 - 2 = 0$, it seemed to many at the time that man had gone about as far as he could possibly go.

One of the difficulties at this stage of the development of numbers lay in the fact that if one took the square of any negative number, the result was a positive number. The nagging question arose in the mind of the very curious (those who were not very curious were able to push such questions aside) concerning the possibility of solving the equation $x^2 + 1 = 0$. This question became a matter of wider concern with the development of a formula for solving quadratic equations, and the consequent need to take a square root of a number which in many cases turned out to be negative. Consequently some brave soul (he must have been brave, for he most certainly had to endure the ridicule of his colleagues) suggested the postulation of a new number, usually designated i today, which would be the solution of this *unsolvable* equation $x^2 + 1 = 0$. This new number had been called *imaginary*, and we frequently call i the *imaginary unit*. (In parallel fashion we would denote the *real unit* by the number 1.) Since $i^2 + 1 = 0$ by the very definition of i, we know that $i^2 = -1$, and $i^4 = (i^2)^2 = (-1)^2 = +1$. In similar fashion we can deal with higher powers of i. This is a good start, but this does not take care of results such as those obtained by trying to solve $x^2 - 4x + 9 = 0$. One method of solution, using the old familiar factoring, would have us writing

$$\begin{aligned} x^2 - 4x + 9 &= (x^2 - 4x + 4) + 5 \\ &= (x - 2)^2 - (-1) \cdot 5 \\ &= (x - 2)^2 - (i^2) \cdot (\sqrt{5})^2 \quad \text{since } i^2 = -1 \\ &= [(x - 2) + i\sqrt{5}][(x - 2) - i\sqrt{5}]. \end{aligned}$$

We know (by Exercise I.2.6) that one of the two factors must be zero if the product is to be zero. Hence either

$$x - 2 + i\sqrt{5} = 0, \quad \text{whence } x = 2 - i\sqrt{5}$$

or

$$x - 2 - i\sqrt{5} = 0, \quad \text{and} \quad x = 2 + i\sqrt{5}.$$

In order to be certain that these are solutions of the equation $x^2 - 4x + 9 = 0$, we can replace x by $(2 + i\sqrt{5})$ and check to determine whether the resulting mathematical statement is correct. Then we can do the same with $(2 - i\sqrt{5})$. While this will use processes whose validity is established in definitions and theorems to follow, we will proceed to check $(2 - i\sqrt{5})$ both to illustrate procedures for checking complex roots and to illustrate the processes which we will later validate. Upon replacing x by $(2 - i\sqrt{5})$ in the given equation, we have

$$(2 - i\sqrt{5})^2 - 4(2 - i\sqrt{5}) + 9 = (4 - 4i\sqrt{5} + 5i^2) - (8 - 4i\sqrt{5}) + 9$$
$$= 4 - 4i\sqrt{5} - 5 - 8 + 4i\sqrt{5} + 9$$
$$= 0 + 0i$$
$$= 0.$$

Again we have used the fact that $i^2 = -1$. This shows that $x = (2 - i\sqrt{5})$ makes this equation a true statement. (It is always a good habit to check solutions whenever possible to do so.)

We are able to produce solutions for a much larger number of quadratic equations if we use complex numbers. The question arises, however, whether we can invoke a theorem which was established for real numbers, namely the one stating that $ab = 0$ if and only if either $a = 0$ or $b = 0$. Since this theorem depended only on the postulates for a number field, we can determine that this theorem will apply to complex numbers if the complex numbers obey the field postulates. Before doing this, however, we must be more precise in our definition of the complex numbers.

Definition 4.1. A number of the form $a + bi$ is a *complex number* if a and b are real numbers and i is a solution of the equation $i^2 + 1 = 0$. If $a = 0$, the number will be called a *pure imaginary* number.

In order to even consider whether the complex numbers form a field, it is also necessary to have a definition for equality of these numbers and a definition of what is meant by addition and multiplication of complex numbers. We will cover these in the next two definitions.

Definition 4.2. Two complex numbers, $a + bi$ and $c + di$, are said to be *equal* (that is, $a + bi = c + di$) if and only if $a = c$ and $b = d$.

Note that this reduces the determination of equality in the complex number system to that of determining equality in the real number system, something we have already considered.

Definition 4.3. If $a + bi$ and $c + di$ are two complex numbers, we define addition and multiplication of complex numbers by the relations:

$$\text{addition:} \quad (a + bi) + (c + di) = (a + c) + (b + d)i$$
$$\text{multiplication:} \quad (a + bi)(c + di) = (ac - bd) + (ad + bc)i$$

If the coefficient of i is zero in each of these numbers, we note that they have all of the attributes of the real numbers, and we can treat them as such. Hence, we can consider the real numbers as though they were a subset of the complex numbers. The definition of addition is a perfectly natural one. Multiplication is also natural if one considers

$$(a + bi)(c + di) = ac + adi + bic + bidi,$$

whence, if we assume the postulate that the real numbers commute with the imaginary unit, we have

$$(a + bi)(c + di) = (ac + bdi^2) + (ad + bc)i$$
$$= (ac - bd) + (ad + bc)i$$

as a result of the relation $i^2 = -1$.

We are now ready to consider whether the complex numbers form a field.

Theorem 4.1. *The complex numbers with the definition of equality given in Definition 4.2 and the definition of addition and multiplication given in Definition 4.3 form a number field.*

PROOF. The verification of each of the postulates for the field can be made directly, based upon the corresponding results for the real numbers. It should be noted that $0 + 0i$ is the additive identity and $1 + 0i$ is the multiplicative identity. The proof of the existence of an additive inverse of $(a + bi)$ merely requires the existence of a number $(x + yi)$ such that $(a + bi) + (x + yi) = 0 + 0i$, and it is easily shown that this requires that $^-(a + bi) = (^-a + ^-bi)$. The proof of the existence of a multiplicative inverse of $(a + bi)$ demands that we obtain a number $(x + yi)$ such that $(a + bi)(x + yi) = (ax - by) + (ay + bx)i = 1 + 0i$, or

$$ax - by = 1$$
$$bx + ay = 0.$$

Solving for x and y in terms of a and b, we obtain $x = a/(a^2 + b^2)$ and $y = -b/(a^2 + b^2)$. Thus, we have

$$(a + bi)^{-1} = \frac{a}{a^2 + b^2} + \frac{-bi}{a^2 + b^2} = \frac{a - bi}{a^2 + b^2}. \qquad \square$$

The numerator of this last expression, namely $(a - bi)$ is frequently referred to as the *complex conjugate* of $(a + bi)$. It is worth observing that $(a^2 + b^2)$ appears in the denominators, and therefore should not be zero,

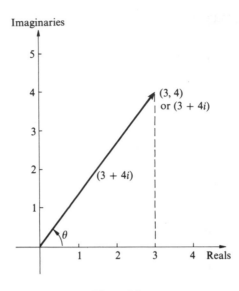

Figure I.2

since the division involved is real division. On the other hand, this merely requires that $a + bi \neq 0$, or that we do not attempt to obtain the multiplicative inverse of zero. Hence, the commandment "thou shalt not divide by zero," is just as necessary with complex numbers as with real numbers.

We pause for a moment to note that the use of i was introduced by Leonard Euler (1707–1748). The physicists frequently use j since they are in the habit of using i to represent the measure of current in an electrical circuit, but there is seldom confusion, and we will use the letter i to denote the imaginary unit. In an attempt to clarify the concept of a complex number, the Norwegian mathematician Wessel in 1799 represented the complex numbers graphically, using a horizontal axis to represent the real numbers and a vertical axis to represent the number of imaginary units. The work of Wessel seemed to have attracted very little attention, and the idea of a graphical representation was rediscovered in 1806 by Jean Robert Argand. This representation is now known under the name of the *Argand Diagram*. If we wish to graph the number $3 + 4i$, we can consider this as the point $(3, 4)$ in which the measurement along the axis of reals is of length 3 (the number of real units) and the measurement along the axis of imaginaries is 4 (the number of imaginary units). See Figure I.2. This uniquely determines a vector (or directed line segment joining the origin (or point $(0, 0)$) and the point $(3, 4)$. Observe that the axis of reals is in fact the number line that we had considered in the previous section, and for that matter the axis of imaginaries is also a copy of the number line placed in the vertical position. The vector of the preceding paragraph can now be described in its *polar* form, as was done by Roger Cotes in 1710, and later by Abraham de Moivre

(1730) and Euler (1743) by noting that the length of this vector is

$$5 = \sqrt{3^2 + 4^2},$$

and that the vector makes an angle θ with the positive real axis such that $\cos \theta = 3/5$ and $\sin \theta = 4/5$. Hence we can write this number as

$$5(\cos \theta + i \sin \theta).$$

[Since $\cos \theta = 0.6$ and $\sin \theta = 0.8$, we can determine from the tables in Appendix C that $\theta = 53°7'48''$. The angles $413°7'48''$ and $-306°52'12''$ would also work in this case. Later on (in Section III.5) we will find it profitable to use *radian measure* for angles. In radian measure we would have $\theta = 0.9273$, 7.2105, and -5.3559 for the angles given here in degrees.] The polar form can be easily obtained by plotting the point on the Argand diagram corresponding to the number and then noting the length of the vector involved and the angle that the vector makes with the positive real axis. *Note*: If you feel the need for either an introduction to trigonometry or a refresher in trigonometry, all of the material you will need is included in Appendix A. The definition of the sine and cosine and the other trigonometric functions are given there together with many relations involving these functions. The method for converting from degrees to radians and vice versa is also there.

EXAMPLE 4.1. Evaluate

i. $(3 - 4i) + (2 + 7i)$
ii. $(3 - 4i) - (2 + 7i)$
iii. $(3 - 4i)(2 + 7i)$
iv. $(3 - 4i) \div (2 + 7i)$

  Solution

(i) $(3 - 4i) + (2 + 7i) = [3 + (-4)i] + [2 + 7i]$
$$= (3 + 2) + (-4 + 7)i$$
$$= 5 + 3i$$

(ii) $(3 - 4i) - (2 + 7i) = a + bi$ is equivalent to the equation
$(2 + 7i) + (a + bi) = (3 - 4i)$. This gives
$(2 + a) + (+7 + b)i = 3 - 4i$

or
$$2 + a = 3$$
$$+7 + b = -4$$

whence
$$a + bi = 1 - 11i.$$

This can be done more quickly (but with less emphasis on the definition) by considering $(3 - 4i) - (2 + 7i) = (3 - 4i) + (-2 - 7i)$
$$= 1 - 11i.$$

(iii) $(3 - 4i)(2 + 7i) = [(3)(2) - (-4)(7)] + [(3)(7) + (-4)(2)]i$
$$= (6 + 28) + (21 - 8)i$$
$$= 34 + 13i$$

(iv) $(3 - 4i) \div (2 + 7i) = (a + bi)$ is equivalent to the equation
$(2 + 7i)(a + bi) = 3 - 4i$. This gives

$$2a - 7b = 3$$
$$7a + 2b = -4$$

whence

$$4a - 14b = 6$$
$$49a + 14b = -28$$

or

$$53a = -22$$

Since $a = -22/53$, $2b = -4 - 7(-22/53) = (-212 + 154)/53$
$$= -58/53$$
or $b = -29/53$.

Therefore $a + bi = -(22/53) - (29/53)i$.

This can be done more quickly (but with less emphasis on definitions)
by considering

$$(3 - 4i) \div (2 + 7i) = \frac{3 - 4i}{2 + 7i}$$

$$= \frac{3 - 4i}{2 + 7i} \cdot \frac{2 - 7i}{2 - 7i}$$

$$= \frac{(3 - 4i)(2 - 7i)}{(2 + 7i)(2 - 7i)}$$

$$= \frac{-22 - 29i}{53 \pm 0 \cdot i} = \frac{1}{53}(-22 - 29i).$$

In this last solution we noted a number in the denominator of the form
$(x + yi)$ and then multiplied both numerator and denominator by the complex
conjugate. Remember that the complex conjugate is obtained by replacing
the imaginary portion of the complex number by its negative. This assured
a real denominator since $(x + yi)(x - yi) = (x^2 + y^2) + 0i$.

EXAMPLE 4.2. Graph the complex number $(7 - 2i)$ and determine its polar
form (see Figure I.3).

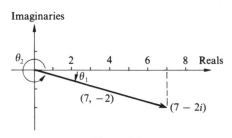

Figure I.3

Solution. The real component of this number is 7 and the imaginary component is $(-2i)$. We therefore plot the point $(7, -2)$ on the Argand diagram. (Note the number of real and imaginary units are 7 and (-2) respectively.) Either the point or the vector can be used as a representation of the number $(7 - 2i)$. The length of this vector is $\sqrt{7^2 + 2^2} = \sqrt{53}$. The vector can then be written in polar form as $\sqrt{53}\,(\cos \theta_1 + i \sin \theta_1)$ where θ_1 is the negative angle shown for which $\cos \theta_1 = 7/\sqrt{53}$ and $\sin \theta_1 = -2/\sqrt{53}$. Upon using a table of trigonometric functions (Appendix C) we find that $\theta_1 = -15°56'43''$ or $\theta_1 = -.278297$ radians. We could equally well have used the form

$$\sqrt{53}\,(\cos \theta_2 + i \sin \theta_2)$$

where θ_2 is a positive angle such that $\cos \theta_2 = 7/\sqrt{53}$ and $\sin \theta_2 = -2/\sqrt{53}$. Here we would have $\theta_2 = 344°3'17''$ or $\theta_2 = 6.00489$ radians. Note that in each case the angle would appear in the fourth quadrant

EXAMPLE 4.3. Graph the complex number whose polar form is

$$7\left(\cos \frac{5\pi}{6} + i \sin \frac{5\pi}{6}\right)$$

and express this number in *rectangular* form (see Figure I.4).

Solution. Draw a *ray* making an angle of $5\pi/6$ radians with the positive real axis. ($5\pi/6$ radians $= 5(180)/6 = 150$ degrees). On this ray mark off a vector 7 units long with the initial point at the origin. This vector represents the given number. The coordinates of the end point of this vector are

$$\left(7 \cos \frac{5\pi}{6}, 7 \sin \frac{5\pi}{6}\right) \quad \text{or} \quad \left(-\frac{7\sqrt{3}}{2}, \frac{7}{2}\right).$$

Thus this number can be described as $(-(7/2)\sqrt{3} + (7/2)i)$. This same result is obtained by writing

$$7\left(\cos \frac{5\pi}{6} + i \sin \frac{5\pi}{6}\right) = 7\left(-\frac{\sqrt{3}}{2} + i \cdot \frac{1}{2}\right) = \left(-\frac{7\sqrt{3}}{2} + \frac{7}{2}i\right).$$

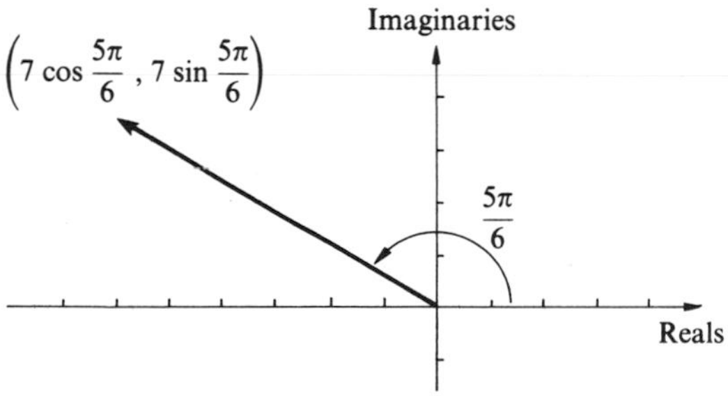

Figure I.4

The idea of imaginary or complex numbers seems strange at first to most people, but the extension of the number system to include these numbers is a natural extension in line with the earlier extensions from natural number to integer to rational to real. The complex numbers are a near necessity in problems such as those relating to alternating electrical currents.

EXERCISES

1. Evaluate:

    (a) $(2 + 3i) + (-1 + 2i)$
    (b) $(-3 - 4i) + (0 - i)$
    (c) $(1 - i) - (2 + 3i) + (3 - 4i)$
    (d) $(3 - 4i) + (3 - 4i) - (4 + 3i) + (4 + 3i)$

2. Evaluate:

    (a) $(3 - 2i)(4 + 7i)$
    (b) $(6 + 3i)(1 - 2i)$
    (c) $4(\cos 30° + i \sin 30°)(\cos 60° + i \sin 60°)$
    (d) $8(3 - 5i)$
    (e) $i(4 - i)$
    (f) $(1 + i)^4$

3. Evaluate:

    (a) $(3 - 2i)/(5i - 12)$
    (b) $(15 + 10i)/(3 - 4i)$
    (c) $(1/2)(2 + 3i)$
    (d) $(2 + i) + [(3 - 4i)(1 + 7i)]/[(1 - i)(2 + i)]$
    (e) $1/(\cos 30° + i \sin 30°)$

4. Solve the following equations and check your work.

    (a) $(4 + i) + x = 3 - 2i$
    (b) $(2 + 3i) - (4 + 7i)x = -27 + i$
    (c) $(2 + 3i) + (15 - 8i)x = 4 + 21i$
    (d) $(2 - 3i)x + (1 + i) = 6x - (4 - i)$
    (e) $(1 + i)x = (3 - 4i) + (1 - i)x$

5. Solve each of the following equations or show that no solution exists. Check each of the solutions you obtain.

    (a) $(2 - i)x + (4 - 4i) = ix$
    (b) $(3 + i) - (2 - i)x = (i - 2)$
    (c) $(1 + i)(2 - i)x = (2 + i)$

6. Solve the following quadratic equations and check all solutions:

    (a) $x^2 + 6x + 25 = 0$
    (b) $2x^2 + 7x = 0$
    (c) $3x^2 + 10 = 0$
    (d) $4x^2 + 5x + 6 = 0$
    (e) $x^2 + 6x + 15 = 0$

7. Plot on the Argand diagram and then convert to polar form each of the following:

(a) $(3 + 4i)$
(b) $-12 - 5i$
(c) $-6 - 6i$
(d) $4i$
(e) $3 + 5i$
(f) $(5 - 6i)/2$

8. Plot on the Argand diagram and then convert to rectangular form each of the following:
(a) $\cos 150° - i \sin 150°$
(b) $\cos(\pi/3)$
(c) $5(\cos \pi/4 + i \sin \pi/4)$
(d) $(\cos 3\pi/4 + i \sin 3\pi/4)/3$

9. Express each of the following in polar form:

(a) $8 - 15i$
(b) $2 + 2i\sqrt{3}$
(c) $-3 + 2i$
(d) $-12 + 5i$
(e) $-4 + 0i$
(f) $0 - 2i$

10. Show that each of the following statements is correct:

(a) $5(\cos \pi/3 + i \sin \pi/3) = 5(\cos 7\pi/3 + i \sin 7\pi/3)$
(b) $2(\cos \pi/4 + i \sin \pi/4) = 2(\cos 17\pi/4 + i \sin 17\pi/4)$
(c) $a\lfloor\cos \theta + i \sin \theta\rfloor = a\lfloor\cos(\theta + 2\pi n) + i \sin(\theta + 2\pi n)\rfloor$ for any integer n
(d) $5(\cos \pi/3 - i \sin \pi/3) = 5(\cos 5\pi/3 + i \sin 5\pi/3)$. Note that $\pi/3 + 5\pi/3 \doteq 2\pi$
(e) $a[\cos \theta - i \sin \theta] = a[\cos(2\pi - \theta) + i \sin(2\pi - \theta)]$

11. Show that $(\cos \theta + i \sin \theta)^2 = \cos 2\theta + i \sin 2\theta$.

12. Show that $(\cos A + i \sin A)(\cos B + i \sin B) = \cos(A + B) + i \sin(A + B)$.

13. Show that $(\cos A + i \sin A)^{-1} = \cos(-A) + i \sin(-A) = \cos A - i \sin A$.

14. Prove that the complex numbers satisfy the definition of a number field.

15. If a and b are real numbers, find the additive inverse of $a + bi$.

16. If a and b are real numbers and not both zero, find the multiplicative inverse of $a + bi$.

17. Show that $(-1 + i\sqrt{3})^3 = (-1 - i\sqrt{3})^3$. What other number has the same value for its cube?

18. Find a complex number z such that $z(x + iy)$ is a real number (that is the coefficient of i is zero).

19. Show that if a and b are two complex numbers, then the sum of their complex conjugates is the complex conjugate of their sum and the product of their complex conjugates is the complex conjugate of their product. [*Hint*: Let $a = r + si$ and $b = x + yi$ where r, s, x, and y are real numbers.]

20. What is the relation of the graph of a complex number and the graph of its complex conjugate on the Argand diagram?

21. If we consider the two forms of the complex number $(x + yi)$ and $r(\cos\theta + i\sin\theta)$, find x and y in terms of r and θ, and find r and θ in terms of x and y.

22. Plot any complex number on the Argand diagram, and then plot the result of multiplying that number by i. What is the relation between the two numbers geometrically? Does this depend on the particular complex number with which you started?

23. Plot $(2 + 3i)$ and $(-3 + 4i)$ on the Argand diagram. Also plot the sum and product of these numbers. What are the geometric relationships involved between the two given numbers and their sum? Between the two given numbers and their product? What is the relation between the lengths of the given vectors and the lengths of their sum and product? [*Hint*: for the sum consider the figure with vertices at the origin, at the end of the two given vectors and at the end of the sum. For the product compare the triangle with vertices at 0, 1, and $(2 + 3i)$ with the triangle having vertices 0, $(-3 + 4i)$, and the product.]

24. Plot $(5 - 2i)$ and $(2 + i)$ on the Argand diagram. Also plot the difference, $[(5 - 2i) - (2 + i)]$ and the quotient $(5 - 2i)/(2 + i)$. Answer questions similar to those of Exercise 23.

M25. Do you think it would be possible to find a proper subset of the complex numbers which would fit the requirements for the set P as used in defining the order relation? Give a reason for your conclusion.

I.5   Absolute Values and Intervals

We have seen that it is possible to associate any real or complex number with a point on the Argand diagram. We can use this diagram to associate with any real or complex number a unique, non-negative number by noting the distance between the point on the Argand diagram and the origin or zero point. Thus, in Figure I.5 we note the points associated with $+5$, -5, $3 + 4i$, $4 - 3i$, and $-5i$, and we observe that in each of these cases the distance between the point associated with the number and the origin is $+5$ units. In many cases we will be primarily concerned with this distance. In fact, this is of sufficient concern that we have given it a special name, as indicated in the following definition.

Definition. The *absolute value* of a number, z, whether real or complex, is the *positive distance* from the origin to the point which represents z on the Argand diagram. The absolute value of z is denoted by the symbol $|z|$.

Pursuant to the discussion above, we note that $|+5| = |-5| = |3 + 4i| = |4 - 3i| = |-5i| = +5$, as shown in Figure I.5. Other definitions of absolute

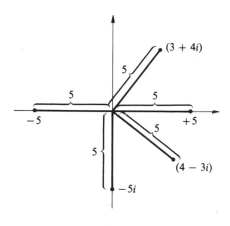

Figure I.5

value are frequently given, but we shall state these as theorems which are consequences of the definition given here.

Theorem 5.1. *If a is a real number and $a \geq 0$, then $|a| = a$. If a is a real number and $a < 0$, then $|a| = -a$. If a is a complex number and $a = x + iy$, then $|a| = (a^2 + y^2)^{1/2}$. [This last case covers the first two cases, for if a is real, then $y = 0$, and $|a| = (x^2)^{1/2}$.]*

In proving this theorem one only needs to consider each of the three cases and show that the conclusions of the theorem are the logical consequences of the hypotheses and the definition. It is suggested that you draw a diagram indicating all of the possible situations and use this to assist you in obtaining a proof.

We are frequently concerned not with the simple case of the absolute value of a single number, but rather with the absolute value of a mathematical expression. This expression may or may not include variables. In particular, we are often concerned with the absolute value of the difference of two numbers, that is with $|a - b|$.

We can represent the numbers a and b by the points A and B on the Argand diagram shown in Figure I.6, and also by the vectors indicated by a and b in this diagram. The distance $\overline{BA}$ is precisely the same as the distance $\overline{OP}$ since $\overline{BA}$ and $\overline{OP}$ are opposite sides of a parallelogram. Since the length of $\overline{OP}$ is $|a - b|$, the value $|a - b|$ is represented by the length of $\overline{BA}$. This relationship between absolute value and distance will be useful in the discussions to follow. [Statements concerning the direction of the vector $(a - b)$ would have to emphasize the distinction between $\overrightarrow{BA}$ and $\overrightarrow{AB}$, for the latter would represent the vector $(b - a)$. In the consideration of absolute value we do not have to be concerned with the direction, and therefore we can consider the length of either $\overline{AB}$ or $\overline{BA}$ in representing $|a - b|$.]

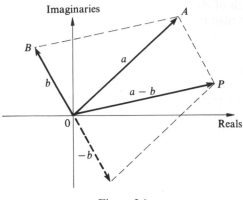

Figure I.6

Theorem 5.2 (Properties of Absolute Value). *If a and b are any two numbers, real or complex, then*

  i. $|a| \geq 0$, *and* $|a| = 0$ *if and only if* $a = 0$;
 ii. $|ab| = |a| \cdot |b|$;
iii. $|a| + |b| \geq |a + b|$.

PROOF. The proof of the first part follows from the definition of absolute value. To prove the second part note that every real or complex number, a, can be written in the form $a = r(\cos \theta + i \sin \theta)$ where r is a nonnegative real number. Then $|a| = r$. Furthermore, b can be written in the form $b = s(\cos \varphi + i \sin \varphi)$.

$$ab = r(\cos \theta + i \sin \theta) \cdot s(\cos \varphi + i \sin \varphi)$$
$$= rs[(\cos \theta \cos \varphi - \sin \theta \sin \varphi) + i(\cos \theta \sin \varphi + \sin \theta \cos \varphi)]$$
$$= rs[\cos(\theta + \varphi) + i \sin(\theta + \varphi)],$$

and $|ab| = rs = |a| \cdot |b|$. For the third part let A be the point corresponding to a (Figure I.7) and P be the point corresponding to $(a + b)$. Now by a theorem from geometry $\overline{OP}$ is no longer than the length of $\overline{OA}$ plus the length

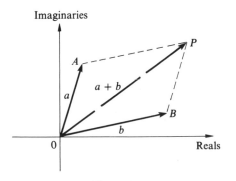

Figure I.7

of $\overline{AP}$. The length of $\overline{AP}$ is the same as the length of $\overline{OB} = |b|$. This proves the theorem and also indicates why the third part of this theorem is usually called the *triangle inequality*.   □

Note that it is sometimes more helpful to think of complex numbers in the *polar* form and sometimes more helpful in *rectangular* form. Since we have shown that for each expression in one form there is a corresponding equivalent expression in the other form, we have the freedom to use whichever is more convenient. (It is often helpful to go far enough into any portion of mathematics that we can express things in more than one way.) While we will normally be working with real numbers in this book, we will have occasion to work with complex numbers from time to time. It should be apparent from what we have said that these two number systems follow very much the same rules except that the operations for addition and multiplication are defined differently. It is also true that the complex numbers include a subset (which we frequently just call the real numbers) which behave exactly like the real numbers, namely the subset of numbers which are plotted on the axis of reals in the Argand diagram.

The determination of the *distance* between two numbers will be of great importance to us later on. For that reason, we will investigate this somewhat more fully in attempting to answer the question "By how much do two numbers differ?" For instance, we could pose the question "By how much do $(3 - 4i)$ and $(2 - 6i)$ differ?" One way of answering this question is by phrasing it in slightly different terms: "What is the distance between the location of $(3 - 4i)$ on the Argand diagram and the location of $(2 - 6i)$ on the same diagram?," or equivalently, "What is the absolute value of the expression $[(3 - 4i) - (2 - 6i)]$?" In order to be sure in your own mind that the last two questions are equivalent draw a graph and see what is meant by each question and determine for yourself whether they are equivalent. The practice of checking the meaning of each statement is a very good practice to follow in reading any book, and particularly a book involving mathematics. We will handle the algebraic (or arithmetic) part of the question here, but let you handle the geometric equivalent. If we consider the last question that we have posed, we have, since we are dealing with complex numbers,

$$|(3 - 4i) - (2 - 6i)| = |(3 - 2) + (-4 - (-6))i|$$
$$= |1 + 2i| = \sqrt{1 + 4} = \sqrt{5},$$

and hence we can state that these numbers differ by $\sqrt{5}$. Later on we shall be concerned with the notation of having one number sufficiently close to another, or having their difference sufficiently small to meet certain given conditions. Since our concern will be with the *size* of the differences, we shall be making liberal use of absolute values.

The difference of two numbers frequently arises in another way. We might be interested in knowing where to find all numbers which differ from 3

Imaginaries

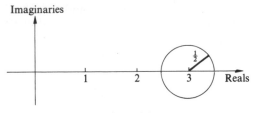

Figure I.8

by less than 0.5. Now we must ask another question before we can give an answer to this query. Do we intend to use all of the complex numbers which meet this requirement, or do we intend to restrict our attention to the real numbers? For the purposes of this book, we shall generally confine ourselves to the case in which we work with real numbers, but at some time you may have to consider the question in which the response is desired in terms of all of the complex numbers which would meet this requirement. We will consider both cases, and show their relation. It is intuitively reasonable that all of the points within 1/2 of a unit of the number (or point) + 3 are within a circle of radius 1/2 having the point + 3 as its center. This circle is shown in Figure I.8. Since any point, $(x + iy)$ is within this circle only if $\sqrt{(x - 3)^2 + (y - 0)^2}$ < 0.5, we see that $\sqrt{(x - 3)^2 + y^2} < 0.5$ or $|(x + iy) - 3| < 0.5$. This latter statement is certainly easier to write. Note that it means exactly the same thing as does the more complicated looking distance relation involving the radical sign. We have in reality considered both real and complex numbers, for if we are concerned with all possible complex numbers, we are ready and willing to use any numbers corresponding to points inside this circle, and if we wish to restrict our attention to real numbers, we are willing to consider precisely the points which are both within the circle and on the axis of reals. Since we have the possibility of using our order relations in the case of real numbers, we can word the solution for real numbers in a slightly different way. Thus, if x is real, and if $x \geq 3$, then x and $+3$ are closer together than the distance 0.5 provided $x - 3 < 0.5$. On the other hand, if $x < 3$ and x is nearer to $+3$ than 1/2, in order to use positive distance we would have to write $3 - x = -(x - 3) < 0.5$, but since $x - 3$ is negative, we know by Theorem 1 that $-(x - 3)$ can be written as $|x - 3|$. Therefore, we can cover both the case in which x is a number not less than three and the case in which x is a number less than 3 by the simple statement $|x - 3| < 0.5$. It is apparent that we can handle both the case of the complex numbers and the case of the real numbers by using the absolute value expression (that is $|x + iy - 3| <$ 0.5 or $|x - 3| < 0.5$) provided we indicate which system of numbers we wish to deal with.

In the case of real numbers, it is also possible to word the answer to this last question in a different manner. If we require that $-2.5 < x$, and at the same time $x < 3.5$, we would certainly have x closer to 3 than the distance 0.5 provided we restricted x to real numbers. Observe that these inequalities have not been defined if x is not a real number, and consequently we cannot interpret these inequalities in the case involving complex numbers. We frequently combine these two inequalities for the real numbers and express the resulting interval of the real axis by the relation $2.5 < x < 3.5$. Note that this is equivalent to the statement $-0.5 < (x - 3) < 0.5$.

In the example we have discussed, we have required that $|x - 3| < 0.5$ and this yields a circle or interval for which we want only the *inside* or *interior* points. On the other hand, we might have indicated that we wanted to consider $|x - 3| \leq 0.5$. In this case we must include the boundary of the circle or the end points of the interval in question. If we are dealing with the circle (or more properly the *disk*), we may talk about the interior or *open set*, if we consider $|x - 3| < 0.5$, and likewise the *closed set* if we mean $|x - 3| \leq 0.5$. We have the open or closed interval if we exclude or include the endpoints, respectively. If we include one but not both endpoints, we frequently refer to the *semi-open interval*. The disk refers to the complex plane, and the interval to the axis of reals, and hence to the real numbers.

Since these concepts are used very frequently, another notation has been developed for use with real numbers which frequently simplifies statements concerning intervals. If $a < b$, then we will write (a, b) to indicate the open interval $a < x < b$, and we will write $[a, b]$ to indicate the corresponding closed interval $a \leq x \leq b$. In similar fashion we would have $(a, b]$ if $a < x \leq b$, and we would have $[a, b)$ if $a \leq x < b$. You should note that we now have two possible interpretations for (a, b), one as the coordinates of a point and the other as an open interval. It is very seldom that you will not be able to tell which of the two meanings is to be assigned for the case in question from the context in which (a, b) appears. Since it is desirable to keep the symbolism used in mathematics (or any other discipline) as simple as possible, and there are a limited number of symbol combinations that suggest themselves, it is therefore not unexpected that some such combination should have to serve more than one purpose.

We will have many occasions to deal with inequalities. At times it will be more convenient to state these inequalities in terms of absolute value and at other times in terms of intervals such as $-0.5 < x - 3 < 0.5$ or $2.5 < x < 3.5$. The relation between these two equivalent forms is summarized in the following theorem.

Theorem 5.3. *If a and b are real numbers and x is to be a real number, the statement $|x - a| < b$ and $a - b < x < a + b$ are equivalent statements, and these statements have no values of x for which they are true statements if b is not positive. The statement $|x - a| > b$ is equivalent to the statement "either $x > a + b$ or $x < a - b$." Similar statements could be made if we were to use $\leq$ or $\geq$ in lieu of $<$ and $>$.*

The proof of this theorem can be written down easily, using our previous discussion. Note that the hypothesis of Theorem 5.3 applies only to real numbers, for while we can use inequalities in statements involving the absolute values of complex numbers, we have no way of defining inequality for complex numbers themselves in a way that would not yield contradictory results. (In the use of inequalities with absolute values of complex numbers we are really using inequalities only with real numbers, for the absolute value is always a real number.)

So far we have concerned ourselves with order relations involving rather simple expressions. Let us consider what would be meant by an expression such as $x^2 - 3x - 4 < 0$. This is equivalent to asking for those values of x for which the graph of $y = x^2 - 3x - 4$ would be below the x-axis, as shown in Figure I.9. We might consider factoring and obtaining $(x - 4)(x + 1) < 0$. This implies that for any real number x we must have the two numbers $(x - 4)$ and $(x + 1)$ of different sign in order that their product would be negative. Thus, we would either have $x - 4 < 0$ *and* $x + 1 > 0$ or we would have $x - 4 > 0$ *and* $x + 1 < 0$. If we take the first choice, then we must have $x < 4$ and also $x > -1$, or equivalently $-1 < x < 4$. If we take the second of these choices then we must have $x > 4$ and at the same time have $x < -1$, and no one has yet seen a number that fulfills both of these requirements at the same time. This is summarized in the table:

	First Choice	Second Choice
$x - 4$	negative; $x < 4$	positive; $x > 4$
$x + 1$	positive; $x > -1$	negative; $x < -1$
Summary	$-1 < x$ & $x < 4$ $-1 < x < 4$	$-1 > x$ & $x > 4$ it is not possible to fulfill both of these requirements at the same time.

Hence, the only real numbers that satisfy the inequality $x^2 - 3x - 4 < 0$ are those in the interval $-1 < x < 4$.

Note that this result describes exactly the interval on the x-axis for which the values of the quadratic expression are negative. We could have done this in a slightly different manner, for we could have written $x^2 - 3x - 4 = (x^2 - 3x + 9/4) - 25/4$ where we have carefully picked the 9/4 in order that the expression in the parentheses will be a perfect square. We then picked the 25/4 so that we would have $9/4 - 25/4 = -4$ in order to preserve the value of the original expression. Now we can write $x^2 - 3x - 4 = (x - 3/2)^2 - 25/4$, and this is negative provided $(x - 3/2)^2$ is smaller than 25/4 or equivalently $|x - 3/2| < 5/2$. By Theorem 5.3 we have $3/2 - 5/2 < x < 3/2 + 5/2$, but this is exactly the same as writing $-1 < x < 4$ or asking that x be in the interval $(-1, 4)$.

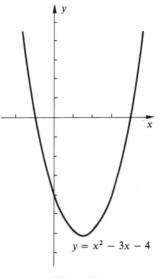

$$y = x^2 - 3x - 4$$

Figure I.9

You should be aware in each case that we have used our sense of logic and the meaning of the various symbols involved rather than trying to set up a formula. A formula might very well apply to only a small number of cases. If we were to pose the question of finding all values of x for which

$$|x^2 - 3x - 4| < 0,$$

it should be apparent at once from Theorem 5.2 that this is a very simple problem. The very first part of Theorem 5.2 tells us that there are no such values, since the absolute value of a real expression can never be negative. Note the reward here for being alert is the saving of algebra that might have given a meaningless answer. If you make use of a computer during this course, you will find that it is very easy to put a *formula* into a computer and come out with nonsense. In this case, for instance, the computer could give an incorrect answer unless you had been very careful to instruct it concerning all of the pitfalls in the problem. These instructions would have to be written, of course, in the computer language used, and should even be so detailed as to insure that the computer will halt if no solution is found (something that would seem obvious to a human being).

One final word is in order. We frequently find that inequalities or intervals are implied although there may be no specific statement that they are needed.

For instance, the formula for the height of an object in terms of time is usually meaningless except for the brief period of time after the object has started on its way and before it hits either the ground or another object. By the same token, any expression for population is only valid during a particular period—for instance in some more primitive country until some form of medical service is introduced which might cut the mortality rate.

EXERCISES

1. Plot each of the following on the Argand diagram and determine the absolute value of each:

$$2, 3, -2, \sqrt{2}, 3 - 2i, 4i - 8, 2i, -3i, i, 0, -\sqrt{25}, -i/2.$$

2. Plot each of the following on the Argand diagram and determine the absolute value of each:

$$-2 + i, 3 - i, 2i + 3, 1 - i, 1 + i, 5 + 12i,$$
$$15 - 8i, -4 + 3i, (2 - 3i) - (-1 + 2i).$$

3. Plot $4 - 5i$ on the Argand diagram. Draw a figure on this same diagram which includes all points representing numbers whose absolute value equals the absolute value of $4 - 5i$. How many numbers, $x + iy$, can you find having this absolute value for which both x and y are integers? Find at least one complex number and one real number having this absolute value for which the real component is irrational.

4. For each of the following pairs of points, plot the two points on the Argand diagram. Compare the vector joining the two points with the algebraic expression obtained by subtracting one of the numbers from the other. Does the order of subtraction affect the answer?

   (a) $3 - 4i, 2 - 6i$
   (b) $2i, 2$
   (c) $4 + 3i, 3 - 4i$
   (d) $0, i + 1$
   (e) $2 + 2i, -2 - 2i$
   (f) $2 + 3i, -2 + 3i.$

5. For each of the following inequalities find all values of z which satisfy the inequality if z is real. Do the same thing if z is complex.

   (a) $|z - 3| < 2$
   (b) $|z + 3| < 2$
   (c) $|z + 3| \geq 2$
   (d) $|2z - 5| \geq 3$
   (e) $|z - 4| < 0.01$

6. Express each of the following inequalities as an interval, that is as an expression of the form $a < x < b$ and as (a, b) if the interval is open or in an equivalent manner if the interval is closed:

   (a) $|x - 2| < 3$
   (b) $|x + 3| \leq 4$
   (c) $|2x - 4| < 3$
   (d) $|3x + 2| \leq 2$
   (e) $|2x - 7| < 1$
   (f) $|2x - 7| < -1.$

7. Express each of the following as an interval or union of intervals in the manner that seems most simple to you.

(a) $|x + 3| > 5$
(b) $|x - 2| > 0$
(c) $|x - 2| \geq 0$
(d) $|2x + 5| \geq 4$
(e) $|3x - 6| > -2$
(f) $|2x + 3| < 4$
(g) $2 < |3x - 2| < 7$
(h) $1 \leq |2x + 5| < 5$

8. Express each of the following as a single inequality using absolute values or explain why this is not possible.

(a) $-2 < x < 6$
(b) $3 < x < 5$
(c) $0 < x < 7$
(d) $-3 \leq x \leq -1$
(e) $-2 \leq x \leq 5$
(f) $-1 \leq x < 6$
(g) $x > 5$ and $x < 9$
(h) $x > 5$ or $x < 1$
(i) $x > 5$ and $x < 3$
(j) $x \neq 2$

9. Express each of the the following in polar form:

(a) $1/2 + \sqrt{3}i/2$
(b) $2 - 2i\sqrt{3}$
(c) $-3 + 3i$
(d) $4 - 3i$
(e) $13 + 12i$
(f) $-15 - 8i$
(g) $-3 + 2i$
(h) $14 + 0i$
(i) $-12 - 0i$
(j) $3i$
(k) $-2i$

10. Express each of the following in rectangular form:

(a) $\cos \pi/3 + i \sin \pi/3$
(b) $2(\cos \pi/6 + i \sin \pi/6)$
(c) $5(\cos 7\pi/6 + i \sin 7\pi/6)$
(d) $4(\cos 2 + i \sin 2)$ [Remember that we usually express angles as radians.]
(e) $3(\cos 0.5 + i \sin 0.5)$
(f) $\sqrt{3}(\cos 11\pi/3 + i \sin 11\pi/3)$
(g) $4[\cos(-0.2) + i \sin(-0.2)]$

11. Given that $|z| > 2$ and $|z| < 3$, find the best inequality you can under each of the following conditions:

(a) z is a real number.
(b) z is a complex number.   [*Hint*: use the Argand diagram.]
(c) z is a rational number.

12. In each of the following cases find the real values such that:

(a) $x^3 + 7x^2 \leq 0$
(b) $|4x + 8| < 3$
(c) $|x^2 - 2x| < 3$
(d) $x^2 - 4x - 5 < 0$
(e) $x^2 + 3x > 4$
(f) $|x + 2| + 3 > x$ [*Hint*: Consider this as two cases depending on whether $(x + 2)$ is negative or non-negative.]
(g) $|x + 2| = 3 + |x - 5|$
(h) $|x - 3| < 2 + |x + 1|$

13. Indicate graphically the location of all numbers for which $|z - 2| \leq 4$. How would the restriction that z is a real number affect your answer?

14. Prove that

$$[r(\cos \theta + i \sin \theta)][s(\cos \alpha + i \sin \alpha)] = (rs)[\cos(\theta + \alpha) + \sin(\theta + \alpha)].$$

15. Find the indicated result in each of the following cases:

(a) $[3(\cos \pi/4 + i \sin \pi/4)][4(\cos 2\pi/3 + i \sin 2\pi/3)]$
(b) $(\cos \pi/6 + i \sin \pi/6)^3$
(c) $[2(\cos \pi/4 + i \sin \pi/4)]^8$
(d) $(1 - i)^{11}$

16. Find all of the real values of x such that:

(a) $x^3 + 7x < 0$
(b) $|3x - 7| < 2$
(c) $2 < |x - 5| \leq 4$
(d) $|x + 3i| < 5$

17. Solve each of the following equations:

(a) $|x + 2| = 3 - |x - 5|$
(b) $|x - 5| = |x| + 5$
(c) $|2x + 1| = x^2 + 1$

18. Find a relation which gives all of the points in the Argand diagram that are closer to the point i than to the point $1 - i$.

19. Show that if $(a + bi)$ is a complex number and you consider all complex numbers such that $|(x + iy) - (a + ib)| = 5$, you have the circle of radius 5 with center at the point (a, b) in the complex plane. Use analytic geometry (See Appendix B) to show that the result of simplifying this expression involving absolute value is precisely the equation of the circle. Could you start with the equation of the circle and recover the relation involving absolute values? How would your work have been altered if the "$=$" sign in the relation involving absolute value had been "$<$"?

20. For what portion of the x–y plane is it true that $x \geq 0$ and $x > y$?

21. For what portion of the Argand diagram is it true that

$$|r(\cos \theta + i \sin \theta)| > 2 \quad \text{and} \quad \theta \geq \frac{\pi}{2}?$$

Is there any connection between your response and the fact that the complex field is not ordered?

22. If you started this particular month with a bank balance of $34.37, made deposits totaling $215.00, and wrote checks in the amount of $199.98, what would be the balance at the present time? What would be the result if you were to include all transactions by using their absolute value rather than noting whether you have debits or credits? Would it be necessary to know the exact amount of each transaction to determine either the correct result or the one involving absolute values?

23. An object is thrown up from the ground and its height above the ground at any time is given by the equation $h = 256t - 16t^2$ where t is the number of seconds it has been in the air and h is the height in feet. If you have an obstructed view so that you can see it only when it is at least 80 feet high, during what time interval are you able to see this object, assuming that the speed at which it is traveling is no hindrance to your perception?

M24. Give a formal proof of Theorem 5.1.

M25. Give a formal proof of Theorem 5.2.

I.6 Functions

The word *function* is used in many ways in the English language. We will focus on the use of this word as it would appear in the phrase "*item Y* is a *function* of *item X*." For example what you are wearing at the moment that you first read this is a function of the clothing you have available, the clothing you wish to save for some specific engagement in the near future, what seens to be comfortable at the present time, the weather, where you are, the time of day (or night), etc. In other words many facts can combine to form *item X* whereas *item Y* is a single entity. For our purposes we shall generally restrict the items X and Y to be elements of rather restricted sets, usually numbers. In recent years, however, the applications, particularly in the social sciences, have required us to consider functions for which the item X may include several thousand subitems as the set upon which the choice of an item Y is made. Having set a groundwork for the study of functions, we now proceed to the definition.

Definition 6.1. Let D be a set of elements (each element of which may be a set in its own right) and let S be another set of elements (not necessarily distinct from D). A "*function* from D to S" is a correspondence which determines for each element of D a unique element of S. The set D is called the *domain* of the function. The set R, which is a subset of S obtained by including precisely those elements of S which correspond to elements of D, is called the *range* of the function.

Notation. If d is an element of D in Definition 6.1 and r is the unique element of R corresponding to D under the function, and if we designate by f the particular function involved, we write $f(d) = r$, or we may also write

$$f:d \to r, \quad \text{or} \quad d \overset{f}{\to} r$$

We may also express the function as an *ordered pair* $(d, r) = (d, f(d))$ in which the first element of the ordered pair is an element of the domain and the second element is the corresponding element of the range.

These notations indicate some of the attributes of the concept of function, such as that of *correspondence* (or *mapping*), and you will find each to be convenient on occasion. We shall make greater use of the notations $f(d) = r$ and the ordered pairs in this book, but you will see the other notations from time to time. This concept can be further illustrated by the diagram given in Figure I.10. The circle labeled D is intended to convey the idea that the elements of the domain are included within the circle and can be thought of as points inside the circle. Similarly the points within the circle labeled R are intended to represent elements in the range. The arrows indicate that for each element in the domain there is a unique element in the range, and furthermore it is possible that one element in the range may be the function of more than one element in the domain, although the converse of this statement is not correct. Here we have pictured $f(d_1) = r_1$, while $f(d_2) = f(d_3) = r_2$.

EXAMPLE 6.1. Let D and S be the same set, namely the set of all integers. Let the correspondence be such that for any element of the domain, the uniquely determined element in the range is obtained by doubling the domain element. That is, $f(3) = (2)(3) = 6$, and $f(-5) = -10$. The domain would then consist of all integers, but the range would consist of the subset of S obtained by using the *images* of the domain elements, namely the even integers. Thus, R consists of the even integers, and R is the range. Note the use of the notation $f(d) = r$ in designating particular cases. It is much easier to write $f(5) = 10$ then to say "the value of $f(d)$ when $d = 5$ is given by 10." In this

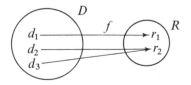

Figure I.10

case the function f denotes the process of doubling. Functions are not always this simple, but the concept remains valid, for the function can be thought of as a mechanism, formula, machine, or black box into which the domain element can be entered and the corresponding range value will appear. In mathematics we often have a formula, but in many cases we might have a situation in which it is not possible to express the entire relationship in a single formula.

EXAMPLE 6.2. If $f(x)$ is given by the formula $f(x) = x^3 - 3x - 5$, and if the domain is the set of all real numbers then it would appear that the set S would also be the set of all real numbers, for it would not be possible to put a real number in the function f and obtain anything other than a real number as a result. It is not obvious at first glance that the range includes all of the real numbers, for there might be some real numbers which cannot be obtained from this formula. The graph of $f(x) = x^3 - 3x - 5$, Figure I.11, would show us, however, that there are no real numbers which could not be an image of $f(x)$. We also can find particular values, such as $f(2) = 2^3 - 3(2) - 5 = 8 - 6 - 5 = -3$. More generally we can note that we would have $f(a) = a^3 - 3a - 5$ or

$$f(x^2 + 1) = (x^2 + 1)^3 - 3(x^2 + 1) - 5$$
$$= x^6 + 3x^4 + 3x^2 + 1 - 3x^2 - 3 - 5$$
$$= x^6 + 3x^4 - 7.$$

This latter case illustrates the fact that we can use any appropriate designation for a domain element, and provided we have a formula, we can obtain the corresponding element of the range. In this case it may well happen that

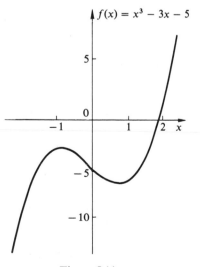

Figure I.11

we know the particular domain element to be the result of some other computation involving $x^2 + 1$, and consequently we have proceeded to use this designation in our evaluation of the function. For the values given here we could also write $(2, -3)$, $(a, a^3 - 3a - 5)$, and $(x^2 + 1, x^6 + 3x^4 - 7)$.

EXAMPLE 6.3. Let (x, y, z) be the coordinates of a point in space, and let t be the number of seconds since the first of the year. We then have for each point in space and for each time during the year a set of four numbers, (x, y, z, t), which uniquely designate the point and the time. At each point and time there is a unique temperature, which we can designate by the variable u. We can then write $T(x, y, z, t) = u$ to indicate that there is a function, which we have here called T, which identifies for each point and time the appropriate temperature. The fact that we would find it very difficult to write down a specific mathematical rule for calculating this temperature will in no way detract from the fact that this is a function. The fact that there is one and only one temperature for each point and each time is sufficient to fulfill our definition. Here the domain consists of all possible combinations of points and times, including times before this year if we allow t to be negative, and the range consists of all temperatures which have actually occurred. It is worth noting that we find some values for this function when we consult tables that give for a particular place (the point) and a particular time (perhaps every twelve hours) the temperature. Such tables are included in many newspapers under the description of the weather for the day.

EXAMPLE 6.4. A function may be defined as a set of ordered pairs, such as $(-2, 4)$, $(-1, 1)$, $(0, 0)$, $(1, 1)$, $(3, 9)$, and $(5, 25)$. In this case one can assume that the domain consists of the set of values $\{-2, -1, 0, 1, 3, 5\}$ and that the range consists of the values $\{0, 1, 4, 9, 25\}$. For each value in the domain we have a corresponding value in the range, but note that in two instances the same value of the range is obtained starting with different values from the domain. If one were to reverse the order of each pair, we would not have a function, for then the reversed pairs $(1, -1)$ and $(1, 1)$ would have a single element from the supposed domain indicating two distinct elements from the supposed range. This violates that portion of the definition of function which states that for each element in the domain there is a *unique* element in the range.

While the definition gives a very complete description of functions in general, it would be helpful at this point to mention several specific functions which we will find very useful in the work to follow. These functions include the constant function, the identity function, the square function, the square root function, and others whose definitions will be rather obvious once we have found the manner in which these names are given to the functions to which they refer.

Definition 6.2. A function, f, with domain, D, is said to be a *constant function* if the range consists of a single element. Thus, if r is the single element in the range, we have $f(d) = r$ for any element, d, selected from the domain.

EXAMPLE 6.5. The function $f(x)$ having for its domain the complex numbers and for its range the single number 7 is a constant function. In this case we know that $f(x) = 7$ for any value selected from the domain. In other words, we would have a few specific cases $f(3) = 7, f(21) = 7, f(-4) = 7, f(2 + 3i) = 7$, etc. This function could also be written

$$f: d \rightarrow 7 \quad \text{or written} \quad d \xrightarrow{f} 7$$

where d is any element taken from the domain.

EXAMPLE 6.6. As another example of a constant function, we might consider the absurd function having the set of all items in the Library of Congress for its domain and a printing press as the single element in the range. Then for any item in the Library of Congress, we would have the relation $f(d) =$ printing press where d is the item in the Library of Congress under consideration. (Of course it is apparent that the domain is not very well defined unless we restrict it to a particular moment in time since the contents of the Library of Congress are increasing steadily. Keeping track of the items there at that time would be quite a bookkeeping chore.) While it is true that such a function as the one described here would have little (if any) possible interest, it does illustrate both the fact that a constant function is one such that the *value* of the function is independent of the domain element selected, and that the domain and range need not be restricted to sets involving numbers.

Another function of great importance is the function for which the domain and the range are identical sets and the function assigns to each element the element identical to it—namely itself.

Definition 6.3. The *identity function* is the function, f, having a single set for the domain, D, and the range, R, and such that if x is any element in D, then $f(x) = x$.

EXAMPLE 6.7. If a function is established which relates people to people such that the domain and the range are the same set of people, then the function is the identity function if and only if each person is related only to himself or herself. The associated ordered pairs would appear as (*John, John*), (*Mary, Mary*), etc. Although the illustration is apparently anti-social, it does emphasize the nature of the identity function.

EXAMPLE 6.8. Let the domain and the range of a given function be the set of real numbers. The function is the identity function if we have $f(x) = x$ or if the ordered pairs are of the form (x, x). Thus, we might have $f(2) = 2$,

$f(-3/4) = -3/4$, or alternatively $(2, 2)$ and $(-3/4, -3/4)$. In this case in which we have an ordered field for both the domain and range we see that if $x > y$, then $f(x) > f(y)$.

We will not stop to list in great detail additional functions at this point, although as we mentioned before there are many, such as the square root function, the absolute value function, the square function, the sine function, the cosine function, and the exponential function. In each of these instances the nature of the function is rather clear from the name. The domain for most of these functions would be the set of real numbers, although the domain of the square root function would be limited to the non-negative real numbers if we desire that the range be real. In the case of the square function it would be possible to have a domain consisting of the complex numbers. A complete description of a function should indicate the domain in question, although in many instances it is possible to deduce the particular domain based upon the other characteristics or upon the origin of the rule for the function.

So far we have been talking about what functions are, and identifying certain particular functions which will occur frequently. Before considering operations which can be performed on functions, it would be well to note that it is not always necessary to be able to express a rule formally in order to have a function. Thus, for each individual in a community there is an amount of indebtedness, possibly zero dollars, for that individual. It might be extremely difficult to ascertain that amount for a given individual due to the use of several charge accounts, credit arrangements, etc. However, it seems fairly clear that such a function must exist (we might call it the indebtedness function) and that it would satisfy all of the requirements of our definition. There would be no rule in the nature of a formula, however, for determining the value of the function for a given individual.

We should also note that the function notation has a special meaning. The expression $f(x)$ does *not* indicate the product of f and x. Furthermore, it is seldom true that $f(x + y) = f(x) + f(y)$ or that $f(2x) = 2f(x)$. If you are ever in doubt concerning the freedom you have in this direction, you can usually check by using a function as simple as the square root function or the square function. The constant function and identity functions are *not* good functions to test in this way, for they obey many laws that functions in general do not obey.

If we have two functions, $f(x)$ and $g(x)$, with a common domain, and with a range in which it is possible to perform arithmetic operations, we can define arithmetic operations for the functions. Consider the following example

EXAMPLE 6.9. Let $f(x) = \sqrt{x}$ and $g(x) = \sqrt{25 - x^2}$. If these are functions with only real elements in their range, then the domain of f is the set of nonnegative real numbers and the domain of g is the set of real numbers such that their absolute value is no larger than five. If the set $[0, 5]$ is used as a domain for the two functions, they are both defined, and their values will always be real numbers. We can then define $f(x) + g(x)$ for the domain $[0, 5]$. Note that $f(x) + g(x)$ is a function of x and we write $(f + g)(x)$ to denote this function. We would have $(f + g)(4) = f(4) + g(4) = \sqrt{4} + \sqrt{25 - 4^2} = 2 + 3 = 5$. Thus, 5 is in the range of $(f + g)(x)$. Similarly, we can define a difference function $(f - g)(x) = f(x) - g(x)$, a product function $(fg)(x) = f(x)g(x)$, and a quotient function $(f/g)(x) = f(x)/g(x)$. The quotient function exists only for cases in which $g(x) \neq 0$, and consequently the domain of the quotient function may be smaller than the domain of the sum, difference, or product functions. In this case $(f - g)(4) = -1$, $(fg)(4) = 6$, and $(f/g)(4) = 2/3$. Note that $(f + g)(5) = \sqrt{5}$, and $(fg)(5) = 0$, but $(f/g)(5)$ fails to exist, and therefore 5 is not an element of the domain of the quotient function. Hence the domain for the quotient function is $[0, 5)$.

The observations of Example 6.9 can be restated in a definition as follows:

Definition 6.4. Let f and g be functions having a common domain and let their range be a subset of a system in which the arithmetic operations can take place. The sum, difference, product and quotient functions are defined as follows:

  i. $(f + g)(x) = f(x) + g(x)$;
 ii. $(f - g)(x) = f(x) - g(x)$;
iii. $(fg)(x) = f(x)g(x)$;
 iv. $(f/g)(x) = f(x)/g(x)$ when $g(x) \neq 0$. All values of x for which $g(x) = 0$ are excluded from the domain of this function.

It should be apparent that the definition of function requires that the domain and the range will each be a set. It should be rather natural to consider that the set which consistitutes the range of one function might well be a subset of the set which is the domain of a second function. If this is so, we can build a third function from the two given functions.

Definition 6.5. Let f be a function with domain D_1 and range R_1, and let g be a function with domain D_2 and range R_2, and further let R_1 be a subset of D_2. For any value, d_1 in D_1 we have $r_1 = f(d_1)$ and r_1 is in R_1 and therefore is in D_2. Consequently, there is an element $r_2 = g(r_1) = g(f(d_1))$. The function $h(x)$ defined such that $h(x) = g(f(x))$, that is such that $h(d_1) = g(f(d_1)) = g(r_1) = r_2$, is called the *composite function of f and g* or frequently *the composite function*. The notation $h = g(f)$ is sometimes used to designate this function. It should be noted that $g(f)$ is not the same as $f(g)$ in the majority of cases (that is composition of functions is not commutative).

This definition has several implications which should be noted. Since $g(r_1)$ must be defined here, it is apparent that r_1 must be an element of D_2. It is not necessary however, that every element of D_2 be an element of R_1.

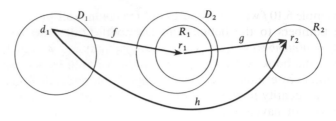

Figure I.12

The fact that $g(f)$ exists does not imply that $f(g)$ exists, for the range of g may not be included within the domain of f. We can use the mapping notation to write

$$d_1 \xrightarrow{f} d_2 \xrightarrow{g} r,$$

thus indicating the composite relationship in another manner. This can be further illustrated by the use of a diagram such as that in Figure I.12. It is noted that R_1 is a subset of D_2 by the fact that the circle for R_1 is contained within the circle for D_2. (It should be pointed out that this diagram is not meant to imply that D_2 is necessarily larger than R_1). The arrow from D_1 to R_2 represents the mapping of the function, h, described in Definition 6.5.

EXAMPLE 6.10. As an illustration of a composite function, consider the following. Let D_1 be the set of all book titles in a given library, and let D_2 be the set of all catalogue numbers associated with these books. Furthermore, let R be the set of all books in the library. If f is the function which assigns the catalogue numbers to the titles (sometimes called the cataloguer function) and if g is the function which locates the books corresponding to the numbers, then $h(title) = g(f(title)) = g(number) = book$. It would be meaningless to consider $f(g(number))$, for then we would have $f(book)$, and the domain of f consists of the titles, not the books. Symbolically we can write

$$title \xrightarrow{f} number \xrightarrow{g} book.$$

EXAMPLE 6.11. Let $f(x) = x + 3$ and $g(x) = x^2$ with each domain being the set of real numbers. We then have $h(x) = g(f(x)) = g(x + 3) = x^2 + 6x + 9$. If we use the domain element $x = 5$, we have $f(5) = 8$ and $g(f(5)) = g(8) = 64$. We would obtain the same results if we were to consider $h(5)$.

Note that $f(g(x)) = f(x^2) = x^2 + 3$, and this is not equal to $h(x) = g(f(x))$. This illustrates the fact that composition of functions is not commutative. It is therefore very important that you be sure of the order in which functions are to be used in composition.

If in Example 6.10 f were changed to become a function which assigns the catalogue numbers to the books (and not to the titles) we would have $g(f(book)) = g(number) = book$. If the cataloguing function is of any value to a library, the book referred to in $g(f(book))$ should be the same book which we obtain by evaluating $g(f(book))$, and hence the composite function should be the identity function. (If this is not the case, you may well have an excuse for not having finished a term paper.) Alternatively, we would have here $f(g(number)) = f(book) = number$, and the two occurrences of *number* should denote the same number. Note that in this case we must have the domain of f identical with the range of g and vice versa. This relationship is considered in the following definition.

Definition 6.6. If we have two sets, S and T, and if f is a function having domain S and range T whereas g is a function having domain T and range S and if furthermore $f(g)$ and $g(f)$ are *identity functions*, then g is the *inverse function* of f and f is the *inverse function* of g. The inverse function of f is frequently written as f^{-1}. This must not be confused with the reciprocal of f ($f^{-1} \neq 1/f$).

This definition suggests a theorem which will verify what may be rather obvious concerning the function we have defined to be the inverse function.

Theorem 1. *If f and g are inverse functions, one having domain S and range T and the other having domain T and range S, then f and g each define a one-to-one correspondence between the elements of S and the elements of T. Such a mapping can be illustrated by*

$$s \underset{g}{\overset{f}{\leftrightarrows}} t \quad \text{or} \quad s \underset{g}{\overset{f}{\leftrightarrow}} t.$$

PROOF. Let f have domain S and range T, and let s_1 be any element of S. Then we have $f(s_1) = t_1$. But since $g(f(s_1)) = s_1$ by the definition of an inverse function we have $g(t_1) = s_1$. Thus we have the unique correspondence of s_1 and t_1 relating each element of S to a single element of T. By considering elements of T, we can show that we also have a unique relationship of an element s_1 for each element t_1 of T. Thus, we have the one-to-one correspondence indicated in the theorem. □

To illustrate this concept, consider the following illustration. Insofar as is known each person having a right thumb has a right thumb print which is different from the right thumb print of any other individual. There are agencies which have made a table of the thumbprinting function such that for each person in the domain of right thumbed people there is established the unique value *thumb-print* (*person having a right thumb*). There are also agencies of identification establishing the inverse function, namely, *identification* (*right thumb-print*). These agencies have become so useful in some circles that the function and the inverse are known to have electronic connections via the various communication systems available. It has also been suggested that such sophisticated notions have increased the use of plastic surgery in isolated instances, but the author has done too little research in this area to be able to provide either positive or negative evidence of this conjecture. However, the thumb-print and identification functions are inverse functions.

In the more mathematical applications of these definitions, we note several rather interesting sidelights. Thus if we were to take the square function, $f(x) = x^2$, we would have $f(4) = 16$ and also $f(-4) = 16$. It is not possible, then, to have an inverse function. Where does this leave us? Stranded? No. Trust the mathematician to find a way to weasel out of this. He simply says that he will restrict the domain so that the domain is the set of non-negative real numbers, and then it would have been impossible to consider $f(-4)$ in the first place. This restriction of the domain actually defines a new function. This new function has for its domain the set of non-negative, real numbers, and on this domain the new function behaves exactly as the square function would have behaved on this domain. This is not the square function itself, though, for it has a different domain. You will observe that the square root function is now the inverse of this new function. We could, of course, have restricted our attention to the non-positive domain and then the function $g(x) = -\sqrt{x}$ would be the inverse function. Note that had we not restricted the domain of the square function, we could not have established a one-to-one correspondence between the domain and the range of the square function using the square root function.

EXAMPLE 6.12. Let the function f be defined by the relation

$$f(x) = \frac{x - 1}{2x + 3}.$$

If the domain is to contain no elements other than real numbers, what is the largest domain possible? Find the inverse function, if it exists, and then determine the range.

Solution. We note that $f(x)$ is a quotient function and hence $f(x)$ is defined for all real numbers except the one for which denominator function $(2x + 3) = 0$. Therefore the domain, D, includes all real numbers except $x = -1.5$. To obtain the inverse function, let $y = f(x)$ for any number x in D. Then we desire a function f^{-1} such that $f^{-1}(f(x)) = f^{-1}(y) = x$. This suggests that we solve the relation

$$y = \frac{x - 1}{2x + 3}$$

for x in terms of y. We have the successive steps

$$y(2x + 3) = x - 1$$
$$x(2y - 1) = -3y - 1$$

and if $2y - 1 \neq 0$, we obtain

$$x = \frac{3y + 1}{1 - 2y}.$$

This indicates that $f^{-1}(y) = (3y + 1)/(1 - 2y)$, and also that $y = 0.5$ is not in the domain of $f^{-1}(y)$, hence not in the range of $f(x)$. We can check our results by noting that

$$f^{-1}(f(x)) = f^{-1}\left(\frac{x - 1}{2x + 3}\right) = \frac{3\left(\dfrac{x - 1}{2x + 3}\right) + 1}{1 - 2\left(\dfrac{x - 1}{2x + 3}\right)}$$

$$= \frac{3(x - 1) + (2x + 3)}{(2x + 3) - 2(x - 1)} = \frac{5x}{5} = x.$$

Similarly $f(f^{-1}(y)) = y$. The fact that $y = 0.5$ is not in the range can be verified by attempting to find x such that

$$\frac{x - 1}{2x + 3} = \frac{1}{2}.$$

We have $2x - 2 = 2x + 3$ or zero multiplied by x equals five. Since zero multiplied by any number is zero, there is no solution possible in this case.

In attempting to find the inverse function by solving $f(x) = y$ for x in terms of y, it is possible that some operation may be required which gives more than one solution (such as taking the square root in which there are two choices from which a selection must be made). In such a case there will generally be no inverse function unless the domain of $f(x)$ was so restricted that only one choice was possible within the domain. Thus $f(x) = x^4$ has no inverse function unless the domain is restricted to a set for which no two elements have the same value for their fourth power. This could be done by establishing the domain as the set of non-negative real numbers or the non-positive real numbers, or the set consisting of the interval $[0, 2)$ together with the set of negative numbers which are not greater than -2. In similar manner, although the algebraic operations are not apparent, the inverse of $f(x) = \sin x$ would not exist unless the domain of $\sin x$ were restricted in such a way that no two domain elements have the same value for the sine.

We can also note that if we consider the functional notation of the ordered pair, the inverse function would merely reverse the order of the pairs. For example in the function $(0, 0)$, $(1, 1)$, $(2, 4)$, and $(3, 9)$, the inverse function would be $(0, 0)$, $(1, 1)$, $(4, 2)$, and $(9, 3)$. More generally the inverse of $(x, f(x))$ would be $(f(x), x)$. In this context our work in Example 6.12 provided us with the necessary information to be able to write $(f(x), x)$ as $(y, f^{-1}(y))$.

One final matter concerns us before concluding this very long section on functions. In Example 6.8 we noted that if the domain and range of the identity function is the real field, and if $x > y$, then we have $f(x) > f(y)$. Thus, $f(x)$ increases as x increases. It is natural to call this an increasing function. However, the question arises what one should call a function, such as the one which determines the amount of postage required for a letter as a function of the weight of the letter. In this case it is possible to have letters of two different weights requiring the same amount of postage. However, it is certainly true that if a letter weighs more, the postage will not be less. These situations are covered in the following definition.

Definition. 6.7. If f is a function from the real numbers to the real numbers, and if for $x > y$ it is true that $f(x) \geq f(y)$, f is said to be *monotonic increasing*. If further $f(x) > f(y)$ for $x > y$, then f is said to be *monotonic strictly increasing*. The term *increasing* is frequently used in place of *monotonic increasing* and the term *strictly increasing* in place of *monotonic strictly increasing*. Similar definitions hold for *monotonic decreasing, decreasing*, and for *monotonic strictly decreasing* and *strictly decreasing*. If a function is monotonic, then it is either monotonic increasing or monotonic decreasing.

In general one can expect that the height of an individual is an increasing function of age (although there are exceptions), but a calorie conscious industry has built its hopes on the fact that weight need not be such a monotonic function. Note that here we have numeric functions by implication, for we are thinking of age as the *number* of years (or months or days), of height as the *number* of inches or centimeters, and of weight as the *number* of pounds or kilograms. It is assumed that these numbers are real and not complex whence we are in an ordered field and it is possible to make such comparisons. In this last definition it was necessary to restrict our attention to domains and ranges in which the order properties hold. Any results requiring monotonic properties would not be meaningful in situations in which order is not defined.

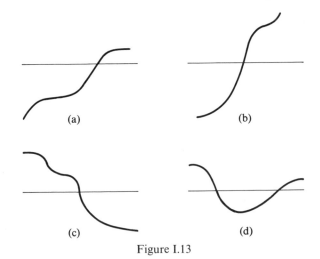

Figure I.13

These concepts are illustrated in Figure I.13. The functions graphed in Figure I.13a–I.13d represent respectively an increasing function, a strictly increasing function, a decreasing function, and a non-monotonic function. Note that parts of the graph in Figure I.13d are monotonic increasing and other parts are monotonic decreasing, but the function as a whole is not monotonic. You should observe that by our definition a constant function is both increasing and decreasing, but would be neither strictly increasing nor strictly decreasing.

EXERCISES

1. Which of the following sets of ordered pairs represent functions? Give a reason for your answer.

   (a) $(1, 2), (2, 3), (3,5), (4, 4), (-1, 0)$.
   (b) $(2, 3), (4, 1), (3, 3), (4, 5), (1, 1)$.
   (c) $(2, 3), (4, 1), (0, 3), (5, 4), (1, 1)$.

2. If $f(x) = x^2 - 3x$, find the value of $f(1), f(2), f(0), f(-3), f(a)$, and $f(x + 2)$.

3. If $g(x) = x/(x - 2)$, what real numbers cannot be in the domain of $g(x)$? For what values is $g(x)$ positive? Find the values of $g(0), g(2/3), g(\sqrt{2})$, and $g(y + 2)$.

4. If a function $f(x)$ is defined by the set of ordered pairs $(1, 1), (2, 3), (3, 6), (4,10), (5, 15), (6, 21)$, what is $f(3)$? $f(5)$? Is $f(0)$ defined? What is the domain of f? What is the range of f? Does this function have an inverse? If so, what is the inverse of $f(x)$?

5. Let the function f be defined by the relation $f(x) = x^2 - \sqrt{x}$. If the domain is a subset of the real numbers, what is the largest possible subset such that the range is contained in the set of real numbers? Evaluate $f(4), f(9), f(13)$. Is it true that $f(4) + f(9) = f(13)$? Does $2f(4) = f(8)$? Evaluate $f(0)$ and $f(1)$. Does $f(x)$ have an inverse?

6. If D is a set of three distinct objects and S is a set of four distinct objects, how many different functions can be defined having domain D and having the range included in S, perhaps including all of S? How many functions can be defined which possess an inverse function? What, if anything, must be done concerning the range of the function in order to have an inverse function?

7. If $f(x) = x^3 - 6x^2 + 11x - 4$, find $f(1), f(2), f(3)$. If you were only shown these latter results, what function might you think this would be? Can you explain why you might have made an incorrect assumption?

8. If $f(x) = x^2 - 3$ and $g(x) = (x + 3)/(2x - 4)$,

   (a) Find the value of $(f + g)(x)$ and $(f + g)(1)$. Check the latter value two ways (by using the sum function and by adding the two functions).
   (b) Find the value of $(f - g)(x)$ and $(f - g)(1)$. Check the latter value two ways.
   (c) Find the value of $(fg)(x)$ and $(fg)(1)$. Check the latter value two ways.
   (d) Find the value of $(f/g)(x)$ and $(f/g)(1)$. Check the latter value two ways.
   (e) Find the value of $f(g(x))$ and $f(g(1))$.
   (f) Find the value of $g(f(x))$ and $g(f(a))$.

9. Let $f(x) = x^2 + 3x$ and $g(x) = x - 1$. Evaluate $f(g(3))$ and $g(f(3))$. Let $h(x)$ be the composite function $f(g(x))$. Write out an expression for $h(x)$.

10. If $f(x) = x^2 - 3$ and $g(x) = 2x^2 - x + 3$, find $f(g(x))$, $g(f(x))$, $f(f(x))$, and $g(g(x))$.

11. Show that any polynomial can be formed from the identity function and the constant function with the four operations of algebra.

12. If we consider all real numbers for which the relation $f(x) = \sqrt{x - 2}$ is defined, is this a function? If so, what is the domain and what is the range? Does this function have an inverse function? If so, what is it?

13. If $f(x) = \sqrt{x - 5}$ and $g(x) = \sqrt{10 - x}$, what is their common domain, if the functions are to have real values? What is the value of

$$[f + g](x), \quad [f - g](x), \quad [fg](x) \quad \text{and} \quad \left[\frac{f}{g}\right](x)$$

for values of x in the common domain? Does each of these four functions obtained by arithmetic operations share the same domain? Evaluate each of these for $x = 6$ and $x = 9$. Also, evaluate these for $x = 7.5$. How would the common domain be affected if we were to permit $f(x)$ and $g(x)$ to assume complex values?

14. If the function $f(x)$ has a domain consisting of the five smallest positive integers and can be described by the number pairs $(1, 3)$, $(2, 2)$, $(3, 0)$, $(4, 5)$, $(5, 3)$, will this function have an inverse? If so, write it out. What would your answer have been if the last ordered pair had been $(5, 5)$?

15. Find a table giving the latest population figures for the 50 states and the District of Columbia. Do the pairs (state, population) form a function? Does this function have an inverse? Would it always follow that at any point in time this function would have an inverse?

16. Does a table of logarithms represent a function? If so, does this function have an inverse?

17. Does the table of sines for the angles in the first quadrant represent a function? If so, is this function increasing or decreasing?

18. If $f(x) = (x + 1)/(x - 2)$, what real number(s) cannot be in the domain of $f(x)$? If we assume that these numbers are not in the domain of $f(x)$, does $f(x)$ have an inverse? If so, find the inverse. [*Hint*: If there is an inverse, then we have $f(f^{-1}(x)) = x$.]

19. Are $f(x) = x^3$ and $g(x) = x^{1/3}$ inverse functions over the real numbers? Would they be inverse functions over the domain of complex numbers?

20. Does the function $f(x) = x^2$ have an inverse? If so, what is it? If not, would there be some domain over which the function could have an inverse? If so, state the domain and find the inverse.

21. Find the inverse of $g(x) = (3x + 5)/(4x - 7)$ or show that it does not exist.

22. If $g(x) = x + 2$, find $g^{-1}(x)$ and find $h(x) = 1/g(x)$.

23. In which intervals is $f(x) = x^2$ an increasing function and when is it a decreasing function?

24. If a function is monotonic (and of course with a real domain), does the function always have an inverse? Would the answer change if the function were known to be strictly monotonic? Would the answer depend on whether the function is increasing or decreasing? Give reasons.

25. If $f(x)$ and $g(x)$ are monotonic increasing functions, would either of the composite functions that can be formed from $f(x)$ and $g(x)$ be increasing? If so, which one(s).

26. If $f(x)$ and $g(x)$ are monotonic increasing functions, which ones of the four arithmetic operations will yield only an increasing function? What about the composite functions with regard to being increasing?

27. Using the definition, can you find one or more functions that are simultaneously increasing functions and decreasing functions at all points? Strictly increasing and strictly decreasing?

28. Michael Kassler in *A Sketch for the Use of Formalized Languages for the Assertion of Music*,[2] wrote "R is a single-valued function of D to E if and only if, for every element w of D, one and only one element z of E exists such that (w, z) is an element of R. R is the relational inverse of the relation R' of E to D if and only if R is the set of all elements (w, z) such that (z, w) is an element of R'. R is a one-to-one function of D to E if and only if R is a single-valued function of D to E and the relational inverse of R is a single-valued function of E to D. (In this last case, each of D and E is in one-to-one correspondence with each other.)" Sketch circles to indicate the sets D and E, and indicate the relationships described in this quotation.

29. State all of the information that is implied by the statement "$f(x)$ is a strictly decreasing function of x."

30. Let $f(x, y) = x^2 + y^2$, and $g(x) = \sqrt{x}$, with x and y both taken from the set of real numbers. What is the domain of f? Does $f(g(x))$ make sense here? Does $g(f(2, 3))$ have any interpretation? Does either of these functions have an inverse? If so, what is it?

I.7 Continuous Functions

We have seen many functions which have nice smooth graphs. These are functions having the real numbers, or a subset of the real numbers, as the domain and having a similar set for the range. The fact that we have used such functions for the majority of our examples to date does not mean that all functions whose domain and range consist of real numbers will be nice and smooth. Consider the *postage stamp function* or the *step function* as it is frequently called. The postage stamp function is given by $y = f(x)$ where

[2] Kassler, M. 1963. "A Sketch for the Use of Formalized Languages for the Assertion of Music." *Perspectives of New Music*. Vol. I, pp. 84–85.

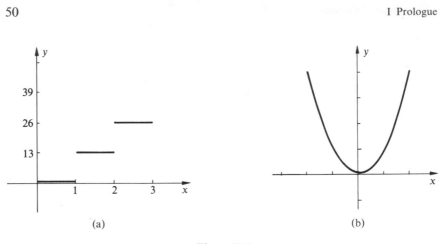

Figure I.14

x is the weight of a letter in ounces and y is the number of pennies in postage required to send the letter by first class mail. If x is in the interval $(0, 1]$, then $y = 13$ at the time of this writing. If x is in the interval $(1, 2]$, then $y = 26$, etc. The graph of this function is given in Figure I.14a. Note the reason for calling it a *step function*. This is an example of a function which is *discontinuous* at $x = 1$, $x = 2$, etc. It is easy to observe that there is no value of x for which $f(x) = 12$ in the case of this function. In fact the range consists of multiples of 13 and nothing else. On the other hand, the function $g(x) = x^2$, whose graph is given in Figure I.14b, has no apparent breaks, and for any positive value, say 12, it is possible to find a value of x such that $g(x) = 12$. In fact, we can find two values of x such that $g(x) = 12$, namely $\sqrt{12}$ and $(-\sqrt{12})$.

It is often necessary to distinguish between the case of the continuous function and the discontinuous function. One might do this by considering that in the case of the continuous function each point on the graph (or each value of the function) can be reached by approaching from either side in an orderly manner, or in other words approaching without having any break, either horizontally or vertically. If we were to consider as an example a point $x = c$, we would then require that there be a value $f(c)$. Moreover we would require that as x approaches c from either the positive or negative side the corresponding function values should approach $f(c)$ in a smooth manner. One method of expressing this relationship which has been used by mathematicians for more than a century involves thinking of a narrow horizontal band with the value $f(c)$ in the middle of the band and determining that there is a small interval about $x = c$ such that all corresponding functional values will be within the narrow band. This is illustrated in Figure I.15. While this may seem somewhat awkward, it does give a method which permits one to determine whether a given function does or does not fulfill the requirements necessary for continuity. We state this somewhat more abstractly in the following definition.

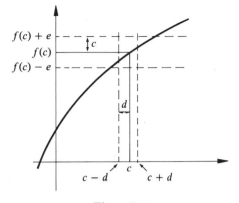

Figure I.15

Definition. A function $f(x)$ having both a domain and a range consisting of real numbers is said to be *continuous* over an interval (a, b) of its domain if for any point $x = c$ in (a, b) and for *any positive number, e*, (no matter how small e may be) it is possible to find an interval $(c - d, c + d)$, $d > 0$, such that the statement *x is in* $(c - d, c + d)$ will insure the truth of the statement $f(x)$ *has a value in the interval* $(f(c) - e, f(c) + e)$.

Note that in this definition the horizontal strip is the strip of width $2e$ which has $f(c) - e$ for its lower limit and $f(c) + e$ for its upper limit. The corresponding vertical strip goes from $c - d$ to $c + d$. This will be discussed at greater length in Chapter VII. However, this definition does make clear that as x gets closer to c, then $f(x)$ must get closer to $f(c)$ in a rather smooth fashion. On the other hand this situation does not occur for all points in the postage stamp function, for if we were to let $c = 2$, we would have $f(2) = 26$ while at the same time values of x arbitrarily close to 2 would produce the value of $f(x) = 39$ if x were only slightly larger than 2. More technically, if we were to pick a value of 1 for e, the statement $f(x)$ *is continuous* would require that there be some positive value of d such that when x is in the interval $(2 - d, 2 + d)$ we would have $f(x)$ in the interval $(26 - 1, 26 + 1)$. This is certainly not correct for all values of x that can be chosen in the interval $(2 - d, 2 + d)$ regardless of the value of d, as long as d *must* be positive as required by the definition.

Earlier in this section we hinted at a property of continuous functions which we shall need on several occasions. It is well to clearly identify such a property, for then we can refer to it by name when it is needed.

Intermediate value property of continuous functions. *If the function $f(x)$ is continuous over the real domain $[a, b]$, and if $f(a) = c$ while $f(b) = d$, and if furthermore y is any value between c and d, then there is some number x in the interval (a, b) such that $f(x) = y$.*

We can put the intermediate value property to work at once in finding the solution of equations.

EXAMPLE 7.1. Find the solution of $x^3 + x - 1 = 0$ which is between $x = 0$ and $x = 1$ with an error no greater than 0.05.

Solution. We should first check to see whether the function $f(x) = x^3 + x - 1$ is a continuous function. We will not go through the task of showing that at every point it does satisfy the requirements of the definition, but will rely on our intuition and past experience. In the case of polynomials this will be sufficient for our purposes here. We are working with the interval [0, 1], and therefore we will check $f(0)$ and $f(1)$. Since we have $f(0) = -1$ and $f(1) = +1$, and since $-1 < 0 < +1$, the intermediate value property assures us that there must be some value of x, $0 < x < 1$, such that $f(x) = 0$. (It is obvious that this is the value that we have been asked to find.) Now let us divide the interval into two intervals, and they might as well be of equal length. In other words, we will consider the interval [0, 0.5]. Since $f(0) = -1$ and $f(0.5) = -0.375$, we have no assurance that there is a value of x satisfying our requirements in the interval $0 < x < 0.5$. On the other hand, since $f(0.5) = -0.375$ and $f(1) = +1$, we do know that there is a value of x, $0.5 < x < 1$, such that $f(x) = 0$. We can now use the interval [0.5, 1] and again take the midpoint, repeating the entire procedure. It is easier to follow the reasoning if we make a short table, noting for each step the half of the previous interval which must contain the value of x that we are seeking.

Small x	$f(x)$	Large x	$f(x)$
0.0	-1.0	1.0	1.0
0.5	-0.375	1.0	1.0
0.5	-0.375	0.75	0.171875
0.625	-0.130859	0.75	0.171875
0.625	-0.130859	0.6875	0.012451

We now observe that the required value must be in the interval (0.625, 0.6875) and the midpoint of this interval is no further than 0.03125 from either end. Therefore, the midpoint value, $x = 0.65625$, gives a result which satisfies the demand of the problem as stated. Note that there are other values of x that would satisfy the requirements of the problem, but having one such value is sufficient. The intermediate value property is a basis for the solution by this method. We are assured that this property holds since $f(x)$ is continuous.

From time to time it is convenient to pick up additional items of frequently used mathematical notation. Mathematicians frequently use a *bracket function*, expressed as $f(x) = [x]$, where $[x]$ denotes the largest integer which is not larger than x. Thus $[2.3] = 2$, $[\pi] = 3$, $[5.9] = 5$, and

$[-3.4] = -4$. As a result of the manner in which this function is determined, it is also called the *greatest integer function*. You will find that computers will frequently convert real numbers to integers using this greatest integer function. Therefore, we have something to watch out for as a possible source of error in computer usage if we are mixing our usage of real numbers and integers in the same problem.

In this section we have given a definition of a continuous function without any elaboration and we have stated the intermediate value property without any proof. These will both be considered in more detail in Chapter VII, but the intuitive concepts herein presented will be sufficient at this point. It should be observed that polynomials are continuous, and the majority of functions appearing in applications also have the property of continuity. A graph, such as the ones discussed in the next section, will usually give sufficient insight to determine whether a function is continuous or not.

EXERCISES

1. Use the method of the Example of this section to find the decimal value of 5/13 to 4 decimal places. Check your result by division. [*Hint*: Find the value of x such that $f(x) = 13x - 5 = 0$. Start by finding an interval such that the function changes sign while going from one end of the interval to the other.]

2. Evaluate $\sqrt{37}$ with an error no greater than 0.05. [*Hint*: Find the solution of $x^2 - 37 = 0$.]

3. Evaluate $\sqrt[3]{20}$ with an error no greater than 0.1.

4. Given the postage stamp function described in this section, evaluate $f(3.4)$, $f(7.1)$, $f(0.9)$, and $f(3.001)$. Would $f(-3.2)$ make sense for the postage stamp function? Would this make sense if we were to consider the step function without regard to the use of this function for postal purposes?

5. Use the method described in this section to find a solution of $f(x) = x^3 + 2x - 4 = 0$ with an error no greater than 0.02. [There is only one real solution for this equation.]

6. Is the function $f(x) = x - [x]$ continuous? What is the range of this function?

7. Let $g(x) = [x^2]$. Is this function continuous? Give a reason for your answer. Evaluate $g(-0.75)$, and $g(2.3)$.

8. Show that the quotient function is always discontinuous when the denominator function is zero using the definition of this Section.

9. Show that the function $f(x) = x^2$ fulfills the requirements of a continuous function for the specific value of $c = 2$.

10. Show that the sum of two continuous functions having the same domain must be continuous.

P11. The energy of a given atom is always an integral number of a very small unit of energy known as a quantum. Show that the energy function for an atom is a discontinuous function of time provided the energy of the atom changes by more than one quantum during the time interval.

S12. The GNP (*Gross National Product*) is frequently given in millions of dollars. If the GNP is given in this manner as a function of time, show that the GNP must be a discontinuous function of time by our definition provided the GNP changes by at least one million dollars in the time interval under discussion.

M13. Show that a continuous function of a continuous function is continuous. [It is necessary to make certain assumptions concerning the range of one function and the domain of the other. What assumptions must you make?]

C14. Obtain the value of $\sqrt{12.5}$ with sufficient accuracy that the first six decimal places in the result are known to be correct.

C15. Find the value of $\sqrt[5]{10}$ correct to six decimal places.

I.8 Graphing

It is said that a picture is worth a thousand words. This is often just as true with relations in mathematics as with relations in other areas of human interest. The majority of mathematical pictures occur in the form of graphs. In fact, we have already mentioned graphs in connection with absolute values, complex numbers, and functions. We will pursue the matter of graphing somewhat further in this section. It should be pointed out that there is a much more comprehensive discussion of graphing in Appendix B, and this appendix should be consulted if the information on graphing in this Section is not adequate.

Our first graph will be the *number line*. This is frequently used to represent the real numbers or a subset thereof. An arbitrary point is selected to represent zero. This point is often called the *origin*. A line segment with one end at the origin will be selected to represent *one unit*. One direction will be chosen to indicate the positive direction. Once the choice of unit and direction have been made, the entire number line is determined as shown in Figure I.16. Any real number, x, may be represented by moving x units of distance from the origin in either the positive or negative direction depending upon the sign of x, thus locating the point corresponding to x. As we have already seen, complex numbers require two number lines, one to represent the real

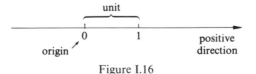

Figure I.16

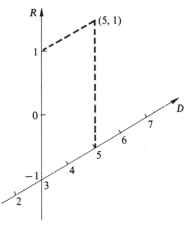

Figure I.17

component of the complex number and the other to represent the imaginary component. For the majority of our considerations in this book we shall be concerned with cases involving only the real numbers.

In order to represent a mathematical relation involving two sets of real numbers, we usually use two intersecting number lines. A function would be such a relation. In the case of a function, one line could represent the domain elements and the other line would include the range elements. Such a representation is shown in Figure I.17 in which the line labeled D represents the domain and the line labeled R represents the range. This graph appears somewhat strange, for we have made use of the arbitrariness which permits the two number lines to be other than perpendicular, which permits the origins to be at other than the point of intersection of the two lines, and which permits the selection of different lengths as units on the two lines. The one essential property is that the two lines intersect. In this case we could represent the point corresponding to the relation $f(5) = 1$ by locating the point on the domain line (D) corresponding to 5 and then through this point drawing a line parallel to the range number line. Similarly we would locate the point on the range line (R) corresponding to 1 and through this point draw a line parallel to the domain number line D. The intersection would then represent the relation $f(5) = 1$. Since we can represent this relation by the ordered pair (5, 1), we see the more familiar looking coordinates. A function or other relation would be represented by locating all of the points which satisfy the function or relation.

At this point a few comments are in order. This type of representation is restricted to those cases in which we are showing relationships between two sets involving only real numbers. The choice of angle formed by the intersection would ordinarily be a right angle unless there is some good reason for another selection such as the consideration of some crystals. The choice of right angles permits us to determine distances between two points by use of the Pythagorean Theorem rather than requiring the use of the law of cosines. The choice of point of intersection (whether the origin or not) and the length of the unit is usually dictated by the desire to make the portion of the graph in which we are interested as large as possible to facilitate easier use of this part of the graph. Thus, the location of this area of intersection on the graph and the size of this portion will play a part in the selection of the location of the origin and in the selection of the units.

Let us now consider the graphs of some representative functions.

EXAMPLE 8.1. Graph the constant function $f(x) = -4$.

Solution. In this case we find at once that any coordinates of the form $(x, -4)$ will represent a point determined by the function. Regardless of the value of x, we then have -4 as the value for y. Note the pictorial representation of the graph in Figure I.18, which illustrates exactly what we have been saying. (Parenthetically, it should be noted that any equation is a mathematical sentence with its subject, its verb, and its predicate, and like any other statement, such an equation has some fact to put forth. In this case the fact happens to be that regardless of the value of x, or the element selected from the domain set, the corresponding value in the range has the value -4, and the graph illustrates this with the horizontal line which has a y-value, or ordinate, of -4 for each point).

EXAMPLE 8.2. Graph the identity function.

Solution. This function is merely the function $f(x) = x$, and since the second number in the ordered pair of numbers is the function value, x, then the coordinates appear as (x, x) for any number x. This is equivalent to saying that we go as far to the right (or left) of the y-axis as we go up (or

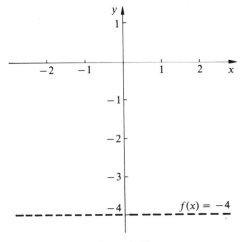

Figure I.18

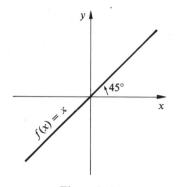

Figure I.19

down), or vice versa. This is illustrated in Figure I.19, and we see that we have the 45° line, or the line that makes an angle of 45° with the positive *x*-axis.

EXAMPLE 8.3. Graph the absolute value function, $f(x) = |x|$.

Solution. We remember that if $x \geq 0$, this is the same as the identity function. Hence, except for points to the left of the *y*-axis, we have the same graph that we had for the identity function. However, for the value of *x* such that $x < 0$, we have $f(x) = -x$, and this merely states that we will take the negative of the identity function, or we will measure the *y*-distances in the opposite direction from that which we would have used for the identity function, although we will use the same distances. This is an alternate way of saying that we wish to use a reflection in the *x*-axis of the identity function for those points to the left of the *y*-axis, or to consider the image we would have by observing the reflection of the negative portion of the identity function in this interval as it would be seen in a mirror which is located on the *x*-axis. The graph is illustrated in Figure I.20. Note that it is possible to obtain this from the identity function by stopping to think of the relation

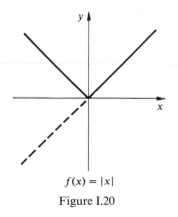

$f(x) = |x|$

Figure I.20

between the absolute value function and the identity function, noting carefully the portion of the domain in which one relation exists and the portion in which another relation exists. It is frequently possible to obtain graphs by this type of analysis rather easily, and it saves a great deal of rather dull work in the plotting of many individual points, with the consequent uncertainty about what happens to the graph *in between* the plotted points.

EXAMPLE 8.4. Graph the general linear function.

Solution. We frequently see this written as $ax + by = c$ or we see some relation which is algebraically equivalent to this. We will be concerned with this relation when a, b, and c are real constants. If $b = 0$ this reduces to the equation $ax = c$ and if further $a \neq 0$ we have $x = c/a$. This can be treated in a manner analogous to the treatment given the constant function and will produce a line parallel to the y-axis and at a distance c/a from the y-axis. (If both a and b are zero we have the equation $c = 0$ and this hardly merits a graph.) If $b \neq 0$, we can write $y = (c/b) - x(a/b)$, and we observe that the point $(0, c/b)$ is a point on the graph. This point is called the *y-intercept*. We further note that the value of y changes by a constant, $(-a/b)$, multiplied by the number of units by which x is changed. Hence, it is possible to draw the graph of this equation as illustrated in Figure I.21 where we have shown the y-intercept and the change in y corresponding to a unit change in x. This rate of change of y with respect to x is called the *slope* of the line. The slope is frequently denoted by the letter m, and consequently in this case we would have $m = -a/b$. Note the fact that if $a = 0$ we have a line with slope zero, that is a horizontal line.

EXAMPLE 8.5. Graph the relation $x^2 + y^2 = 16$.

Solution. We can analyze this relation by observing that if this is true, then either $\sqrt{x^2 + y^2} = 4$ or $\sqrt{x^2 + y^2} = -4$. Since the square root symbol is normally used to indicate only non-negative numbers (unless specifically noted otherwise), we can rule out the second of these options. The relation $\sqrt{x^2 + y^2} = 4$ is equivalent to the statement that the distance from the

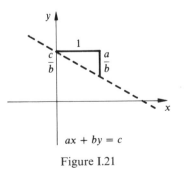

Figure I.21

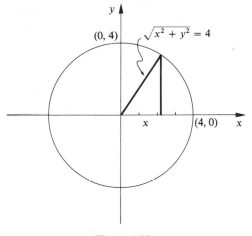

Figure I.22

origin to the point (x, y) is always 4. This can be seen in Figure I.22, for this relation is merely a restatement of the Pythagorean Theorem in this particular case. Consequently, the graph consists of just those points which are at a distance of 4 units from the origin, in other words a circle of radius 4 with center at the origin.

It is apparent that this is not a function for we have two values of y for each value of x selected from the interval $(-4, 4)$. On the other hand we can consider either the upper semicircle or the lower semicircle, and each of these would represent the graph of a function. Their respective equations would be $y = (16 - x^2)^{1/2}$ and $y = -(16 - x^2)^{1/2}$. Here we find an easy way to determine whether a graph represents a function by simply noting whether any value of x may give more than a single value of y.

It is worth noting that a careful analysis of the facts implied by a given mathematical statement (or equation) will often give sufficient information to at least help in drawing the graph, and this will cut back on the necessity for plotting many points and then wondering how the points should be connected to form the graph.

EXAMPLE 8.6. Graph $y = (x^2 - 4)/(x + 1)$ for values of x in the interval $[-4, 4]$.

Solution. Since $y = (x - 2)(x + 2)/(x + 1)$, we observe that $y = 0$ when $x = 2$, and when $x = -2$, and there is no value for y when $x = -1$ since then the denominator is zero. In fact, when x is near the value of -1, we have a denominator which is very near zero, and hence the quotient will be very large in absolute value. Therefore, not only will y fail to exist when $x = -1$, but when x is near the value of -1 we will have values of y which have large absolute values. We also note that when $x > 2$, the three factors $(x - 2)$, $(x + 2)$, and $(x + 1)$ are all positive and hence y must be positive.

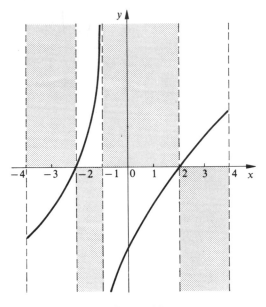

Figure I.23

We have indicated this fact in Figure I.23 by shading in the lower half of
the vertical strip to the right of $x = 2$ below the x-axis. This is done to indicate that the graph cannot appear in this area of the coordinate plane. In
similar manner, values of x in the interval $(-1, 2)$ will insure that y is negative, for $(x - 2)$ will be negative for these values while both $(x + 2)$ and
$(x + 1)$ will be positive. For this reason we have shaded the portion of the
strip between $x = -1$ and $x = 2$ above the x-axis to show that the curve
cannot appear there. This process can be continued to show that the values
of y must be positive if x is in $(-2, -1)$ and y must be negative if x is to the
left of $x = -2$. If we put all of this information together, noting where y
must be zero and where y must have values which will be far from the x-axis,
we can sketch the graph with essentially no plotting of individual points.
(Incidentally the line $x = -1$ which the graph approaches as y gets large
is called an *asymptote*.)

In general the type of analysis that we have used in the last three examples
will be of great assistance in sketching the graphs of equations. You should
carefully analyze the equation for any information which it may contain.
You will find that initially you may look for quite a while without seeing
much information which will be of value to you. However, if you are willing
to persist you will soon discover that you can see a great deal of information
and the matter of sketching curves will then become much easier. Since a
good sketch is often very helpful in setting up an application problem, this
ability to sketch graphs will be of great assistance later on.

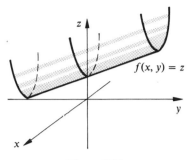

Figure I.24

In all that we have done in this section we have considered that the domain is the set of real numbers, and also that the range is the set of real numbers. From our earlier work with functions, we know that this need not be the case. Thus, the domain might well consist of points in a plane, that is ordered pairs of numbers of the form (x, y) and the range might still be the set of real numbers. If we have a case such as this, it is usual to think of a graph consisting of the plane, which constitutes the domain, and a number line (or axis) perpendicular to the plane, usually erected at the origin on the plane, which displays the values of the range. Thus, for each point in the plane, if the plane is horizontal, we have a height which determines a unique point on the graph. Joining these points will generally produce some type of surface. An illustration using the function which assigns to each point (x, y) the value $x^2 + y + 1$ is given in Figure I.24. Note that it is necessary here to attempt to portray a three dimensional figure in two dimensions. It is not difficult, if you think about it, to conceive that this can be done, but it may be difficult for anyone who is not a natural artist to do it to his satisfaction. Do not let that keep you from trying. While such functions occur frequently, we shall not have occasion to use many such graphs in this book.

As we noted in Section 7, one should not get the notion that all relations have nice smooth graphs of the type we have been drawing to date. Let us consider, for instance, the function such that if x is a rational number, $f(x) = x$, but if x is an irrational number, then $f(x) = 0$. Thus, all of the rational numbers would be represented by points on the 45° line, and all of the irrational numbers would have points on the x-axis. The graph would appear to consist then of two lines intersecting at a 45° angle, but in fact each line would have infinitely many holes. If you stop to check, $f(x)$ satisfies our definition of function, but the graph seems to be much more difficult to draw accurately. We shall consider this function again later, but you might try your hand at drawing the graph, just for fun. There is another problem which we might well face. What would happen if the domain consisted of points in a plane and the range also consisted of points in a plane, or if perhaps the domain consisted of all points in space and the range consisted of complex numbers? In these cases we can talk in geometric terms, if we like, but drawing the graph would be difficult, to say the least, since our experience has limited our perception to three dimensions. Fortunately, we will postpone the consideration of such problems until a future course, but it should be pointed out that the vast majority of problems in the real world have domains which involve a large number of factors (or variables), and the range may also involve many different items. Thus it would be impossible to draw graphs for such problems in the conventional sense. Also, if we were to consider a domain of complex numbers, we would remember that each complex number depends on two real numbers, and hence we would have certain immediate problems, for the complex numbers filled the plane using the Argand diagram.

Up to this point we have represented the domain and range by number lines. As a result we have constructed graphs using the *rectangular* or *Cartesian coordinate system*. The latter name gives credit to the French mathematician-philosopher *René DesCartes (1596–1650)* who first suggested such a coordinate system. He did this in an appendix to his most famous work on philosophy, *Discourse on Method*, published in 1637. It is not necessary, however, that we use this means of display. In another very useful coordinate system, called the *polar coordinate system*, we will let the domain elements be represented by the number of angular units, θ, which a *directed line* (or *vector*) has turned from the positive x-direction, and then we will let the corresponding range elements be denoted by the distance, r, of a point from the origin in the direction thus established. (We will usually use *radian* measure for the unit of angular measurement. The reason for this choice will become evident in Chapter III.)

EXAMPLE 8.7. Graph the function $f(\theta) = 3$ in polar coordinates.

Solution. Note that this is the constant function, but this time we have a circle of radius 3 since $r = 3$ for any value of θ. If $\theta > 2\pi$ we just keep on going and plot it on the second time around, so we have not really limited our domain. Similarly, if $\theta < 0$, we plot the angle in the negative or clockwise direction. This is shown in Figure I.25.

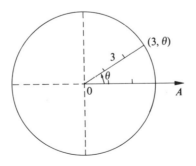

Figure I.25

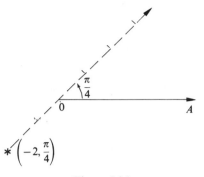

Figure I.26

In polar coordinates we use the term *pole* instead of origin. The pole has the interesting, and disturbing, property that the point in question is the pole any time the function has a value 0 regardless of the element of the domain which produces that value. Thus, many domain values may be plotted into a single point. While we are talking about terminology, the positive *x*-axis is often called the *polar axis*, and the value of the function is plotted along the *radius vector*. One more consideration arises from the fact that someone at some time started the custom in polar coordinates of putting the range element, namely *r*, first in the ordered pair, and the domain element, θ, second. Thus, we indicate the point (r, θ) when in fact we are usually considering that we have the relationship $r = f(\theta)$. Of course, it is true that we could have $\theta = f(r)$, or in the rectangular system we could have $x = f(y)$, but we rather seldom see these latter expressions in practice.

EXAMPLE 8.8. Plot the point $(-2, \pi/4)$ in polar coordinates.

Solution. To plot $(-2, \pi/4)$ we would first consider the vector making an angle of $\pi/4$ with the polar axis, as in Figure 1.26. We then proceed in the negative direction a distance of two units (since $r = -2$) to locate the required point.

Now we are ready to proceed to a somewhat more interesting illustration. In the following example we could plot many points, but it will be far easier to again analyze the function and determine from this analysis the general path which the curve must take.

EXAMPLE 8.9. Sketch the graph of $r = 2 + \sin \theta$ in polar coordinates.

Solution. Remember that the sine function starts with a value of zero when the angle is zero and then increases to the value of 1 by the time the angle is a right angle. The sine then decreases through zero at a straight angle and to -1 after three fourths of a revolution. Finally the sine returns to the value zero upon completing one revolution. This same performance is repeated

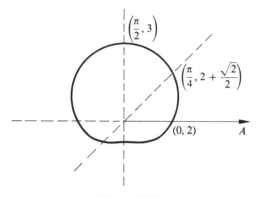

Figure I.27

with every complete revolution. In our case we have the distance r obtained by adding 2 to the value of the sine at the angle under consideration. Hence, r starts at 2 when $\theta = 0$ and increases to 3 when $\theta = \pi/2$. The value of r starts decreasing at this point and decreases through the value 2 at a straight angle and down to 1 when $\theta = 3\pi/4$. Finally the value of r starts to increase during the final fourth of the revolution. Since r is the distance from the pole along the radius vector for each angle, we are now ready to sketch the graph. We have indicated this process in Figure I.27 where we have shown representative radius vectors and corresponding distances. Note that the distance from the pole behaves in exactly the manner we have described above. You will usually find that this type of analysis is of great assistance in sketching a curve, although you may have to force yourself to analyze the relation in this way the first few times. Many people are so much in the habit of plotting points that it is difficult to break that habit, but the analysis will usually reduce the effort required to produce a reasonably good sketch.

Finally we should consider the relationship between the Cartesian or rectangular coordinate system and the polar coordinate system. This is usually done by thinking of superimposing the one coordinate system over the other with the pole and the origin being at the same point and with the polar axis coinciding with the positive x-axis. This is shown in Figure I.28. If we now take a representative point and let it be known by the coordinates (x, y) in the rectangular system and let the same point be known by (r, θ) in the polar system, we can draw the triangle indicated in Figure I.28. Since we have a nicely labeled right triangle, we observe at once that

$$x = r \cos \theta,$$
$$y = r \sin \theta.$$

We also observe that

$$r = \sqrt{(x^2 + y^2)},$$
$$\tan \theta = y/x.$$

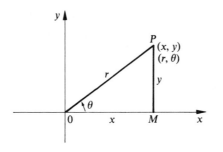

Figure I.28

In determining the value of θ from the values of x and y, we need to be careful to obtain the correct value of θ. If y is positive, θ will be in one of the first two quadrants. If y is negative, θ must be in the third or fourth quadrant.

EXERCISES

1. Plot and label each of the following points: $(3, 5)$, $(1, -2)$, $(-3, 2)$, $(-1, -3)$, $(4, 0)$, $(-3, 0)$, $(0, 2)$, $(0, -3)$, $(0, 0)$.

2. Sketch each of the following lines:

   (a) $2x - 3 = 0$
   (b) $3y - 2 = 0$
   (c) $x + \pi = 0$
   (d) $2y + 7 = 0$
   (e) $x = 0$
   (f) $y = 0$

3. Sketch each of the following lines and give the slope and y-intercept of each:

   (a) $2x + 3y = 6$
   (b) $5x - 3y = 7$
   (c) $4x = 2y - 3$
   (d) $3x + y = x + 3y - 7$
   (e) $2x + y = 2x - y$
   (f) $4x - 3y + 5 = 2x + 3y + 7$

4. Sketch each of the following pairs of lines and find the coordinates of the point at which the lines intersect:

   (a) $x + y = 4$
      $x - y = 6$
   (b) $2x + 3y + 4 = 0$
      $3x + 2y - 3 = 0$
   (c) $x = 3$
      $y = 4$
   (d) $2x = 3$
      $x - y = 3$

5. Sketch each of the following and approximate the value or values of x for which $y = 0$.

   (a) $y = x^2 - 3x - 5$
   (b) $y = 2x^2 - 5x + 2$
   (c) $x^2 - y^2 = 0$ [*Hint*: Factoring may help in *this* case.]
   (d) $x^2 + y^2 = 9$
   (e) $x^2 + 4x + y^2 - 6x = 23$ [*Hint*: Show that this is equivalent to $(x + 2)^2 + (y - 3)^2 = 36$.]
   (f) $x^2 - 4y^2 = 4$
   (g) $x^2 + 4y^2 = 4$

6. Find the equation of the line which passes through the pair of points in each case and sketch the graph.

   (a) $(2, 3)$ and $(-1, 7)$
   (b) $(3, 4)$ and $(-1, -2)$
   (c) $(-1, 3)$ and $(2, -1)$
   (d) $(0, 2)$ and $(3, 0)$
   (e) $(2, 3)$ and $(0, 0)$
   (f) $(-1, -3)$ and $(0, -5)$

7. In each of the following cases find the equation of the line and draw a sketch.

   (a) through $(2, 3)$ with a slope of 2
   (b) through $(-1, 3)$ with a slope of -3
   (c) through $(3, 0)$ with a slope of $-2/3$
   (d) through $(0, 4)$ with a slope of $3/4$
   (e) through $(0, 0)$ with a slope of $-7/3$

8. Sketch each of the following functions for values of x in the interval $[-5, 5]$:

   (a) $y = x^3 + 3x$
   (b) $f(x) = 1/(x^2 + 1)$
   (c) $f(x) = 2x/(x^2 + 1)$
   (d) $f(x) = x^4 - x^2$
   (e) $f(x) = 2^x$
   (f) $f(x) = (x - 1)(x + 4)/(x + 3)(x - 2)$

9. Sketch each of the following functions for values of x in the interval $[-3\pi, 3\pi]$.

   (a) $y = \sin x$
   (b) $f(x) = \cos x$
   (c) $f(x) = \tan x$
   (d) $f(x) = \cot x$
   (e) $y = \sec x$
   (f) $y = \csc x$

10. An object is thrown into the air and its height, h, above ground at any time, t, is given by the function $h = f(t) = 256t - 16t^2$. The height is measured in feet and the time is measured in seconds. Draw a graph showing the height at each time t, and determine the domain of this function which makes physical sense (that is for which the object is not below the surface). From your graph find the highest point reached and the time at which the object reached the highest point. Also find the total time the object was in the air.

11. The attractive force between two objects is inversely proportional to the square of the distance between them. That is, $F = k/r^2$ where k is a constant depending upon the masses of the objects, F is the force and r is the distance between the objects. Draw a graph showing the relation between the force and the distance for values of r which do not exceed 10. Let $k = 12$ for purposes of your graph.

12. Plot and label each of the following points in polar coordinates: $(2, \pi/2)$, $(-3, \pi/4)$, $(0, \pi)$, $(-1, 0)$, $(1, \pi)$, $(-1, 2\pi)$, $(3, -3\pi/4)$. Show clearly when (and if) two or more points coincide.

13. Sketch each of the following in polar coordinates:

   (a) $r = 4$
   (b) $r = -2$
   (c) $\theta = 0$
   (d) $\theta = 5\pi/6$
   (e) $r = \theta/\pi$ [Be sure to indicate what happens if θ is negative as well as what happens when θ is positive.]

14. Sketch each of the following in polar coordinates:

   (a) $r = 3 - 2\sin\theta$
   (b) $r = 2 + 2\cos\theta$
   (c) $r = 1 + 2\sin\theta$
   (d) $r\sin\theta = 2$
   (e) $r\cos\theta = 3$

15. Sketch each of the following:

   (a) $r = 4\sin\theta$
   (b) $r = 2\cos 2\theta$
   (c) $r = 3\sin 3\theta$
   (d) $r = -2\cos\theta$

16. Sketch each of the following in polar coordinates:

   (a) $r = 2\theta$
   (b) $r = 2^\theta$

17. Convert the equation $x^2 + y^2 - 2x = 0$ from rectangular to polar coordinates. Sketch the given curve on rectangular axes and the converted equation in the polar system and compare your results.

18. Convert the equation $r = \sin 2\theta$ to rectangular coordinates. Sketch the graph in one of the two coordinate systems and state why you chose the particular system you used. [*Hint*: $\sin 2\theta = 2\sin\theta\cos\theta$.]

19. Find the equation in rectangular coordinates of a line having slope m and y-intercept 3. (This equation should involve m.) Find the points at which this line intersects the curve $y = x^2 + 3x + 4$. For what values of m will the line have only one point in common with the parabola? For what values of m will the line have no points in common with the parabola? Draw a graph to explain your results.

I.9 Flowcharts

You have sometimes found yourself doodling when you should be listening to something, but then you rationalized that you were really mapping out in some graphic form the method by which you were going to accomplish that which was asked of you. We now wish to give you an alibi for such doodling but, as in every case, there is a catch. In this case the catch is that we are going to use somewhat standardized symbols for our doodling. These symbols are borrowed from those which apparently enjoy the greatest popularity at the moment among computer scientists. Our reason for using these is two-fold. Some of you may have access to computers, and in those cases it is helpful to start out using symbols that you may use later on. It is also true that these symbols will represent rather efficiently the items which we are most apt to consider in such a graphic representation. This particular form of doodling we will call *flowcharting* and oddly enough we will call the result a *flowchart*.

There are certain basic matters that we wish to consider in any analysis of a problem or task which is to be done. We must first consider the starting point, and at the end we must know where we are to terminate our efforts. While it would seem obvious that this is so, it may well be that our flowchart, when we get around to constructing it, is so complex that it is not apparent to anyone else, and perhaps not even to us, where we should begin and where we can finally call it quits. We must also be able to enter information into the problem and be able to extract information. In solving the problem we must be able to perform operations, and we must also be able at various points to make decisions concerning which path to follow as we proceed further. If the flowchart is very large, we may want to be able to extend it to a second (or third) page, and hence we would want the ability to show which parts from one page connect with parts from another page. It is not probable that you will be doing anything this complicated at an early stage, but we will include the appropriate symbols for reference purposes. Having indicated the types of items which must be considered, let us now consider the standardized symbols we will be using. These are shown in Figure I.29. In a complete flowchart these are joined by straight lines, and if the order of events would take us in any other direction than down or to the right, we put an arrow head on the line segment to indicate the direction of flow of the process involved. It never hurts to put in such arrow heads if you have any doubt about the clarity of the chart you have drawn.

We will illustrate the flowcharting technique with the following examples.

EXAMPLE 9.1. Draw a flowchart to indicate a potential path of your attempt to study this particular lesson.

Note that in Figure I.30 we have labeled the starting point. While the first question may have been more fundamental than locating the book

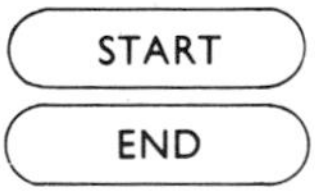

START
END — These symbols are used to indicate the beginning and the end of the process.

INPUT — This symbol, representing a data processing card, indicates that information is to be entered at this point.

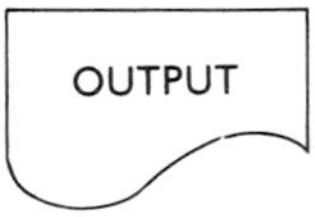

OUTPUT — This symbol, representing a piece of paper torn from a printer, is used to indicate that we are to get information or results from the process at this point.

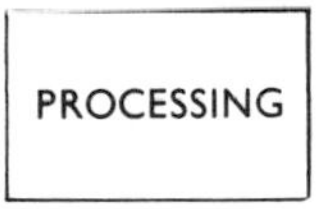

PROCESSING — This symbol, that is a rectangle, is used to indicate that some work, such as computation, analysis, logical thinking, etc., is to be done at this point.

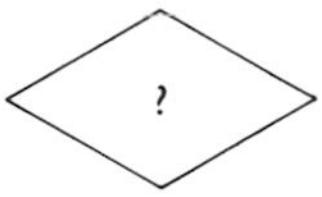

? — This symbol, the diamond, is used to indicate that a decision is to be made. The flow of the problem would enter at one of the vertices and the possible alternatives would exit at other vertices provided there are enough vertices, otherwise exits can be from any portion of the diamond.

TO — This symbol, called a connector, is used to indicate where one part of the flow chart is to connect logically with another. This is done by using matching symbols in connectors, one at each of the points which should be connected.

Figure I.29

(perhaps locating the assignment, checking with a fellow student, etc.), we have indicated the acquisition of the book as being the first item to be considered. The next step, of course, is to locate the correct page. (Perhaps we should have allowed for a false start, or for finding paper, etc.) After some period (we have indicated ten minutes) there is apt to be some type of interruption generated from without by a question or an invitation to get a snack, or perhaps a certain restlessness that accompanies the studying of mathematics, and at that time one is tempted to take inventory concerning whether a break can be justified. This is usually accomplished by checking the number of pages or problems yet to be done.

It is clear that this flowchart is only a simplified version of the one that should have been drawn. Can you improve on it? You might even allow for variable periods of study, such as starting with 5 minutes, working up to 15 minutes, and then finally getting down to a minute between successive checks of the time yet to be put in on this particular lesson. We might observe that many companies do, in fact, draw flowcharts of their activities in order to better coordinate and systematize their operations. Have you ever

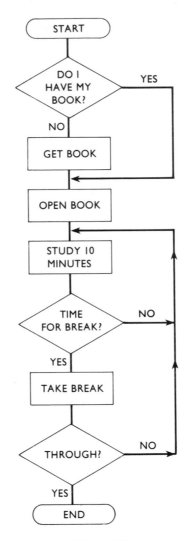

Figure I.30

encountered such a situation? Would your study schedule be improved if you were to draw a flowchart of your weekly intentions?

EXAMPLE 9.2. Draw a flowchart for finding $\sqrt{N}$ with an error no greater than one millionth for any positive number N.

Solution. In any problem such as this, it is first necessary to get the method to be used clearly in mind. Let us first try a particular problem using numbers. If we try to obtain $\sqrt{20}$, we might start by considering that $20 = (20)(20/20) = (20)(1)$, and then take the average of the two factors, 20 and 1. We will

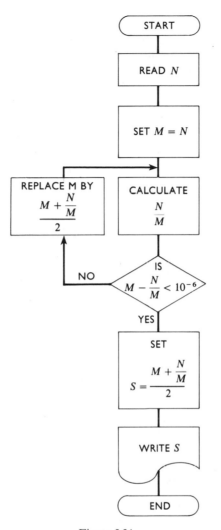

Figure I.31

rcpcat this process using $(20 + 1)/2 = 10.5$ for a factor. This would give $20 = (10.5)(20/10.5) = (10.5)(1.90476)$. If we proceed further, we will next use $(10.5 + 1.90476)/2 = 6.20238$. This process can be continued, getting closer to the correct value. When the two factors differ by less than 1 millionth, the result is certainly as accurate as required.

We will now let our specific 20 be replaced by N, and we will let our trial factor, first 20, then 10.5, etc., be denoted by M. The procedure we have followed then gives rise to the flowchart indicated in Figure I.31. The flowchart indicates the steps we used in our numeric solution and also indicates the order in which they must be performed. Furthermore, the flowchart indicates where the decision is to be made which determines whether we will try another step or whether we should quit. This type of illustration can be very helpful in clarifying the logic involved in solving many problems, whether they are strictly mathematical or not. It will be particularly helpful to use a flowchart if a problem is to be programmed for a computer, for the computer will follow the instructions given it and it is the responsibility of the person writing the program to have the correct logic.

We will use flowcharts from time to time throughout this text. They are intended merely to clarify some of the techniques or arguments that we employ. If you have access to a desk calculator or computer, the flowcharts can be used to map out the successive operations you will want to use. We will designate some problems as those particularly applicable to the use of a desk calculator or computer, and you will find it most helpful if you draw a flow chart of these problems prior to mapping out your set of instructions, or your program. It will also be helpful in many other instances to take advantage of this particular form of doodling, for such pictures help clarify the method of solution for the larger problems.

EXERCISES

1. Draw a flowchart to indicate the logic involved in proving that $(^-a)(^-b) = ab$.

2. Draw a flowchart to illustrate the procedures you use in planning your study session for the next evening on which you expect to study.

3. Use the flowchart given in Example 9.2 and following this flowchart find $\sqrt{30}$ correct to the number of decimal places indicated by your instructor.

4. What modifications would you have to make in the flowchart of Example 9.2 in order to permit N to be other than positive? (These modifications might include presentation of an error message, computation, or whatever else might be appropriate.)

5. Draw a flowchart to indicate the procedure for solving a quadratic equation.

6. Draw a flowchart for plotting a complex number on the Argand diagram and then putting the number in polar form.

7. Draw a flowchart for taking the words on this page and putting them in alphabetical order. (Ignore symbols and numbers.)

8. Draw a flowchart for finding all prime numbers less than 1000.

9. Draw a flowchart to determine whether a function has an inverse and to find the inverse function if it exists.

10. You are given a set of real numbers and are instructed to find the sum of all of the numbers which are greater than 5. Draw a flowchart indicating your method of solution.

11. Draw a flowchart for the solution of the equation $x^3 + 4x - 9 = 0$ using the intermediate value property.

I.10 Frequently Used Notation

As you have undoubtedly observed, the mathematical expression of a relation is usually much more concise than the expression in English. This brevity can be very helpful, but it also can be disturbing. It is essential that you be able to understand just what is meant by each mathematical expression if you are to be able to use that expression intelligently. On the other hand, the use of mathematical notation represents a type of shorthand which often aids the formulation of problems. In this section we wish to discuss certain notation that we will find very useful in the pages that follow.

The first notation we shall mention is represented by

$$\sum_{j=m}^{n} f(j).$$

In this notation we observe first the Greek letter capital sigma ($\sum$). This corresponds to the Latin letter (S), and stands for *sum*. The symbols m and n will refer to integers, and the variable j is the *index of summation*. This notation indicates that we are to take consecutive values of j starting with $j = m$ and stopping when we have used $j = n$, we are to calculate the values of the successive functions $f(j)$ for each of these values, and then we are to add these function values. From this it is apparent that $n \geq m$ and if $n = m$ there is to be a single term, namely $f(n)$. If f is the identity function and if m and n are respectively 4 and 7, we would have

$$\sum_{j=4}^{7} j = 4 + 5 + 6 + 7.$$

Of course there is no reason why we could not use some other index of summation, and hence

$$\sum_{k=1}^{25} k^2$$

indicates the sum of the squares of all positive integers up to and including 25. On the other hand

$$\sum_{n=17}^{17} \frac{n}{n+1} = \frac{17}{17+1}$$

since the index of summation, n, starts with $n = 17$ and also terminates with $n = 17$. As stated above, it would make no sense to start with 17 and terminate with 16, for we understand that n is to be incremented by one at each step and it would be impossible to reach 16 through incrementation if we were to start with 17.

We should pause to observe a rather obvious fact, but one which is sometimes misunderstood, namely

$$\sum_{j=m}^{n} 1 = n - m + 1.$$

This follows from the fact that $f(j)$ is the constant function $f(j) = 1$ and in adding the values of $f(j)$ we are in fact adding ones. If we look carefully we see that there are $(n - m + 1)$ values of j used, and hence this many ones added. In particular,

$$\sum_{j=3}^{7} 1 = 5.$$

Here we add a one for $j = 3$, for $j = 4$, for $j = 5$, for $j = 6$, and for $j = 7$, and the result will be five.

Having this notation, we will now prove some results which will aid us in the manipulation of the sigma notation.

Theorem 10.1. *If c is a constant, then*

$$\sum_{j=m}^{n} cf(j) = c \sum_{j=m}^{n} f(j).$$

PROOF. Since c is a constant, we can factor c out of each term of the summation by the distributive property. Note that since $f(j)$ is not generally a constant, the value of $f(j)$ would differ in the various terms involved and the distributive property would not apply to $f(j)$.   □

Theorem 10.2. *If p, q, and r are all integers and if $p \leq q < r$, then*

$$\sum_{j=p}^{q} f(j) + \sum_{j=q+1}^{r} f(j) = \sum_{j=p}^{r} f(j).$$

PROOF. Write out the terms involved and note that the second summation starts with the term that would have appeared immediately following the last term of the first summation.   □

Corollary 10.1

$$\sum_{j=m}^{n} f(j) + f(n + 1) = \sum_{j=m}^{n+1} f(j).$$

PROOF. Use Theorem 10.2 where $p = m$, $q = n$, and $r = n + 1$. In this case the second summation of Theorem 10.2 is just a single term.   □

Theorem 10.3

$$\sum_{j=m}^{n} [f(j + 1) - f(j)] = f(n + 1) - f(m).$$

PROOF.

$$\sum_{j=m}^{n} [f(j + 1) - f(j)] = [f(m + 1) - f(m)] + [f(m + 2) - f(m + 1)]$$

$$+ \cdots + [f(n) - f(n - 1)] + [f(n + 1) - f(n)].$$

From this it should be clear that for each positive term except $f(n + 1)$ there is a corresponding negative term to counteract it and for each negative term except $f(m)$ there is a corresponding positive term. Hence, the only two terms left are $f(n + 1)$ and the negative of $f(m)$. $\square$

Theorem 10.4

$$\sum_{j=m}^{n} [f(j) + g(j)] = \sum_{j=m}^{n} f(j) + \sum_{j=m}^{n} g(j).$$

OUTLINE OF PROOF. Write out the terms on each side of the equal sign and note that the associative and commutative properties of addition are sufficient to rearrange one side in order to obtain the other. $\square$

Another symbol that we use often is the *factorial* symbol. The factorial is indicated by the exclamation mark. Thus we have the definition

Definition 10.1. If n is a positive integer, then the product of all positive integers $1, 2, 3, \ldots, n$ is called n *factorial* and is written $n!$.

EXAMPLE 10.1. Evaluate $7!$.

Solution. $7! = (1)(2)(3)(4)(5)(6)(7) = 5040.$

Theorem 10.5. $(n!)(n + 1) = (n + 1)!$.

PROOF. Since $n!$ is the product of the first n positive integers, if we multiply $n!$ by $(n + 1)$ we have the product of the first $(n + 1)$ positive integers, and hence have $(n + 1)!$. $\square$

There are many formulas which we shall see that can be written more concisely with the factorial notation. However, it is often the case that there is one term (and possibly more) which would fit the formula if we could use $0!$ and if $0!$ were to have the value of one. While this seems somewhat odd to have $0! = 1$, there are in fact some very good reasons why this should be done. We shall see these later on when we study the *gamma function* in Chapter VII. In order to take care of the customary usage at this point, however, we will content ourselves with a definition.

Definition 10.2. $0! = 1$.

In this way we have the definition to which we can refer, and we will not have to concern ourselves with any further proofs.

In this section we have introduced two items of notation which have wide usage in mathematics. As in the case of all mathematical symbols you should make certain that you can translate each equality or inequality into words or ideas in order that you understand what the mathematical statement is trying to state. If you are careful to understand each mathematical statement, you will find that most of the things that we will be doing will be common sense.

EXERCISES

Note: The form $\sum_{k=1}^{n}$ is equivalent to the form $\sum_{k=1}^{n}$.

1. Write out and evaluate each of the following:

   (a) $\sum_{k=-2}^{5} k$
   (b) $\sum_{j=3}^{7} j^2$
   (c) $\sum_{k=4}^{10} 1$
   (d) $\sum_{t=-2}^{4} t^3$
   (e) $\sum_{j=2}^{6} j(j + 3)$
   (f) $\sum_{k=0}^{8} (-1)^k k$
   (g) $\sum_{j=0}^{6} (-1)^j j!$
   (h) $\sum_{k=0}^{6} 2^k$
   (i) $\sum_{k=2}^{7} \sqrt{k + 2}$

2. Express in sigma notation each of the following:

   (a) $4 + 9 + 16 + 25 + \cdots + 121 + 144$
   (b) $27 + 64 + 125 + 216 + 343$
   (c) $1 - 2 + 3 - 4 + 5 - \cdots + 99$
   (d) $2 + 6 + 18 + 54 + 162 + 486 + 1458$
   (e) $1 + 2 + 4 + 8 + 16 + 32 + 64 + 128 + 256$
   (f) $1 - 2 + 4 - 8 + 16 - 32 + 64 - 128 + 256$

3. Prove that each of the following is correct by quoting a theorem and also by writing out the terms.

   (a) $\sum_{k=1}^{4} k/(k + 1) + \sum_{j=5}^{8} j/(j + 1) = \sum_{k=1}^{8} k/(k + 1)$
   (b) $\sum_{k=3}^{7} k^2 + \sum_{k=8}^{10} k^2 = \sum_{k=3}^{10} k^2$
   (c) $\sum_{k=1}^{6} r^k + r^7 = \sum_{k=1}^{7} r^k$ (r is a constant.)
   (d) $\sum_{k=1}^{10} [(k + 1)^2 - k^2] = \sum_{k=1}^{10} [2k + 1] = 11^2 - 1^2 = 120$
   (e) $\sum_{k=0}^{6} [(k + 1)! - k!] = 7! - 0! = 5039$

4. (a) Show that $\sum_{j=1}^{6} j = \sum_{j=1}^{6} (6 - j + 1)$
   (b) Show that $\sum_{j=1}^{n} j = \sum_{j=1}^{n} (n - j + 1)$
   (c) Show that $\sum_{j=1}^{n} [j + (n - j + 1)] = n(n + 1)$
   (d) Show that $\sum_{j=1}^{n} j = n(n + 1)/2$

5. (a) Show that $\sum_{j=2}^{5} j(j + 1) = \sum_{j=3}^{6} (j - 1)j$
   (b) Show that $\sum_{j=m}^{n} j(j + 1) = \sum_{j=m+1}^{n+1} (j - 1)j$
   (c) Show that $\sum_{j=m}^{n} f(j) = \sum_{j=m-2}^{n-2} f(j + 2)$

6. Prove that $(x - y) \sum_{k=0}^{n-1} x^{(n-1)-k} y^k = \sum_{k=0}^{n-1} [x^{n-k} y^k - x^{(n-1)-k} y^{k+1}]$
$$= \sum_{k=0}^{n-1} x^{n-k} y^k - \sum_{k=1}^{n} x^{n-k} y^k$$
$$= x^n - y^n.$$

7. Prove that $(x + 1)^n - x^n = \sum_{k=0}^{n-1} (x + 1)^{(n-1)-k} x^k$.

8. (a) Prove if $x > 0$ then $\sum_{k=0}^{n-1} (x + 1)^{(n-1)-k} x^k < \sum_{k=0}^{n-1} (x + 1)^{n-1} = n(x + 1)^{n-1}$
   (b) Prove if $x > 0$ then $\sum_{k=0}^{n-1} (x + 1)^{(n-1)-k} x^k > \sum_{k=0}^{n-1} x^{n-1} = nx^{n-1}$
   (c) Prove if $x > 0$ then $n(x + 1)^{n-1} > (x + 1)^n - x^n > nx^{n-1}$

9. Find the value of each of the following.

   (a) $7!/(3!4!)$
   (b) $24!/22!$
   (c) $(12)(11!)$
   (d) $\sum_{k=0}^{5} ((-1)^k/k!)$

10. Show that there are 6! ways of arranging 6 people in 6 seats. [*Hint*: There are 6 places in which the first can be seated, 5 in which the second person can be seated after seating the first one, etc.]

11. (a) In how many ways could 30 persons be seated in 30 chairs?
    (b) If there are 100 drops of water in one cubic centimeter, and if we consider that the radius of the earth is 4000 miles with an atmosphere stretching out an additional 100 miles, show that if the earth and its atmosphere were entirely water there would be more ways of seating 30 persons in 30 chairs than there would be drops of water in this very wet world.

12. (a) Show that $n^n > n!$ for any integer $n > 1$.
    (b) Show that $3^6 > 6!$ but $3^7 < 7!$

13. (a) Show $6!/(2!4!) + 6!/(3!3!) = 7!/(3!4!)$
    (b) Show $11!/(4!7!) + 11!/(5!6!) = 12!/(5!7!)$
    (c) Prove $n!/(k!(n - k)!) + n!/((k + 1)!(n - k - 1)!)$
$$= (n + 1)!/((k + 1)!(n - k)!)$$

I.11 Mathematical Induction

There are many occasions in mathematics where it is necessary to know that a formula is correct for all positive integers, or for some infinite subset thereof. If it were necessary to use only a few values, these could be checked individually, but it is impossible to verify each case for an infinite set of values. Therefore, it is necessary to find some alternate method of proof. The usual method in this instance is known as *proof by mathematical induction* and it is this method which we will be investigating in this section. We will start with an Example.

EXAMPLE 11.1. Prove that for any positive integer, n, the sum of the first n even integers is given by $n(n + 1)$.

Solution. This is equivalent to asking us to show that for any positive integer, n, it is true that

$$\sum_{k=1}^{n} 2k = n(n + 1).$$

It is easy enough to check this result for $n = 1$, for we would then have

$$\sum_{k=1}^{1} 2k = 1(1 + 1) = 2$$

but the summation on the left side of this relation involves only the single term for which $k = 1$. In similar fashion we could check the result for $n = 2$, and we would have $2 + 4 = 2(2 + 1) = 6$. As we indicated in the first paragraph of this section, it would be impossible to continue this for all positive integers. Therefore, having established that this is true for $n = 1$ and $n = 2$, we will investigate whether the fact that it is true for some value of n assures us that it is also true for the next possible value of n. If this investigation is successful, we would then have the fact that since it is true for $n = 2$, it is also true for $n = 3$, but then it would be true for $n = 4$, and hence for $n = 5$, etc. We would thereby have accomplished our mission.

In order to carry out the suggested investigation, let us start with a value of n for which we know the result to be correct. This, of course, could be either 1 or 2 in this case. We then wish to investigate whether it is correct for $(n + 1)$. Using the Corollary of the last section we can write

$$\sum_{k=1}^{n+1} 2k = \sum_{k=1}^{n} 2k + 2(n + 1) = [n(n + 1)] + 2(n + 1) = (n + 1)(n + 2).$$

Our substitution of $n(n + 1)$ for $\sum_{k=1}^{n} 2k$ is permitted by our choice of n as a value for which these two expressions are equal. It is clear that

$$(n + 1)(n + 2) = (n + 1)[(n + 1) + 1]$$

would be the result obtained if in the original expression we were to replace n by $(n + 1)$. It is also clear that the correctness of the given expression for an integer n assures us that the expression is correct for $(n + 1)$. Hence we have done what we set out to do.

You would do well to read through this solution more than once, and to be sure that you can write it out, complete with the reasons for each step. The method of *mathematical induction* requires that we first establish the correctness of the desired result for some beginning value of n (in this case for $n = 1$). Then we further prove that if n is a value for which the statement is correct it must follow that the statement is also correct for $(n + 1)$. One of these parts without the other is not sufficient. This method of proof makes use of a set of integers starting with some initial value and then continuing with all subsequent integers. It is used sufficiently often that it merits a name of its own.

Definition 11.1. A set $S(N)$ is an *inductive set* provided N is an element of $S(N)$, and the statement "n is an element of $S(N)$" implies that $(n + 1)$ is also an element of $S(N)$.

The set of positive integers is the inductive set $S(1)$, and the set of non-negative integers is $S(0)$. Proof by mathematical induction, then, is a proof establishing the fact that a given statement is correct for some inductive set. Our proof in Example 1 involved showing that the *truth set* for the equation

$$\sum_{k=1}^{n} 2k = n(n + 1)$$

is the inductive set $S(1)$. We now proceed to another example.

EXAMPLE 11.2. Prove by mathematical induction that

$$\sum_{k=1}^{n} k(k + 1) = \frac{n(n + 1)(n + 2)}{3}$$

for all integers in $S(1)$, that is for all positive integers, n.

Solution. The first step in induction is to show that this statement is correct if $n = 1$. We do this by direct substitution. Thus

$$\sum_{k=1}^{1} k(k + 1) = 1(1 + 1) = (1)(2) = 2 = \frac{1(1 + 1)(1 + 2)}{3}.$$

We now proceed to the second step of induction. We assume that n is a number for which the statement is correct and then attempt to show that it must follow that the statement is also correct for the value $(n + 1)$. If we can do this, we have shown that the statement is in fact true for all values of n in $S(1)$, for we have included 1 and all integers which follow an included integer. Now to carry out the second step of the proof, assume that n is an integer for which this is correct, and hence

$$\sum_{k=1}^{n} k(k + 1) = \frac{n(n + 1)(n + 2)}{3}.$$

We wish to show that it follows that

$$\sum_{k=1}^{n+1} k(k + 1) = \frac{(n + 1)[(n + 1) + 1][(n + 1) + 2]}{3}$$

$$= \frac{(n + 1)(n + 2)(n + 3)}{3}.$$

But the Corollary of the last section states that

$$\sum_{k=1}^{n+1} k(k+1) = \sum_{k=1}^{n} k(k+1) + (n+1)[(n+1)+1]$$

$$= \frac{n(n+1)(n+2)}{3} + (n+1)(n+2)$$

$$= (n+1)(n+2)\left[\frac{n}{3}+1\right] = \frac{(n+1)(n+2)(n+3)}{3}$$

and this is just what we wished to demonstrate. You might observe that the algebra in this case is simplified by factoring out the common factors in the second line, hence making it unnecessary to multiply the factors together. The algebra can frequently be eased by noting items such as this.

As we have already noted, there are two parts to any proof using mathematical induction. It is essential that both parts be completed. For instance, it is true that

$$\sum_{k=1}^{n} k^3 = (2n-1)^2$$

for $n = 1$ and for $n = 2$, but it is also true that this is not correct for any other value of n. Consequently, the first step of the proof could be carried out but the second one would fail. On the other hand, the second part of the mathematical induction proof can be carried out for the relation

$$\sum_{k=1}^{n} k = \frac{n^2+n+1}{2}$$

but this statement is not correct for any value of n. Therefore, it would not be possible to carry out the first step of the proof.

It would be instructive to illustrate the fact that mathematical induction applies to problems involving inequalities in a manner very similar to that used for equalities.

EXAMPLE 11.3. Prove

$$\sum_{k=0}^{n-1} k^3 < \frac{n^4}{4}.$$

Solution. We note that the first value of n that fits in with our definition of a summation in this case is $n = 1$, and upon checking we see that with $n = 1$ this relation becomes $0 < 1/4$, a statement which is certainly correct. Thus, we observe that the first step in showing that this is correct for the inductive set $S(1)$ has been carried out.

For the second step we will assume that n is a value for which this is correct, and then see what happens for the value $(n + 1)$. Thus, we investigate

$$\sum_{k=0}^{n} k^3 = \sum_{k=0}^{n-1} k^3 + n^3 < \frac{n^4}{4} + n^3.$$

If we show that $n^4/4 + n^3 < (n + 1)^4/4$, we can then write

$$\sum_{k=0}^{n} k^3 < \frac{n^4}{4} + n^3 < \frac{(n + 1)^4}{4}$$

and we will have completed our proof. Note that $n^4/4 + n^3 < (n + 1)^4/4$ is equivalent to $(n + 1)^4 - n^4 > 4n^3$ and this is precisely the result that we had obtained in Exercise 8c of Section 10 with the substitutions $n = 4$ and $x = n$. Hence, we have shown the second part of the induction proof, and therefore the given result is correct. This example illustrates the fact that it is sometimes helpful to compare the result obtained by incrementing n with the result we desire in order to determine whether the desired result is correct. This particular result is illustrative of a situation in which we do not have the value of the summation, but we do have a *bound* for this value. In other words we know that the sum of terms which are the cubes of non-negative integers will be less than a number which we can easily calculate.

While we have started with the value $n = 1$ in each of the examples we have considered, there will be instances in which it will be necessary to start with some other starting value. Thus we might have started by showing that the statement is correct for $n = 5$. In this case, the inductive step would then have shown the statement to be correct for all numbers in $S(5)$. This would not state that the result is not correct for $n = 1, 2, 3,$ and 4, but it would not have implied that the result even had meaning for these values. If you were to try to prove the correctness of the statement

$$\sum_{k=2}^{n} \frac{1}{k(k - 1)} = \frac{n - 1}{n},$$

it should be apparent that $n = 1$ would be meaningless here. First of all you would start with the value $k = 2$ and end with the value $k = 1$ by incrementation, (a difficult feat), and secondly you would have a denominator of zero. Therefore, you would wish to verify that the truth set includes $S(2)$.

It would be in order to ask a question concerning what types of results should be proven by mathematical induction. Since this method clearly applies most easily to those cases in which the key numbers are integers, it would seem as a general rule that if we have a theorem involving all *real* numbers we would hardly expect to use mathematical induction, regardless of whether the letter n was used for the variable or not. It does not follow that if we are only using integers that we always use mathematical induction, but it would appear that if we could start at some point and work our way up, then induction would certainly be worth a try.

We can use a flowchart to illustrate the method of proof by mathematical induction. The first thing that we should consider is whether induction is appropriate, and if it is we must determine a starting value. If the relation is correct for the starting value, we proceed to the induction portion of the problem. While this flowchart is handy as an outline of the procedure which you should follow, it fails to indicate the care that you should take to insure that you make no mistakes in algebra. It also fails to indicate the use of the Corollary of the last section for those situations in which summations are used. In other words, you must do some thinking as you do your work. However, as you have already discovered, that is something that is required in doing anything in mathematics. This flowchart, as shown in Figure I.32,

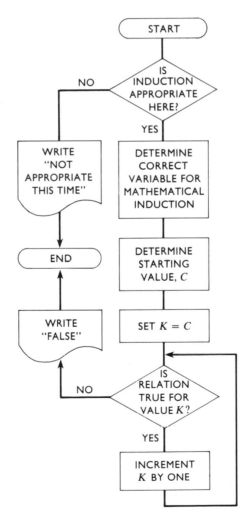

Figure I.32

would not be one you would wish to follow if the result is correct for all integers. Note that in this case there is no way of bringing your work to a halt. Do you think it would be possible to remedy this defect?

You will note that some of the exercises at the end of this section carry a notation that the results will be used at a later point. It is suggested that you pay particular attention to these exercises and note the results in order that you may return to them as needed in the next two chapters.

EXERCISES

1. Prove $\sum_{k=1}^{n} (2k - 1) = n^2$

2. Prove $\sum_{k=1}^{n} (4k - 3) = n(2n - 1)$

3. Prove $\sum_{k=2}^{n} 1/(k(k - 1)) = (n - 1)/n$

4. Prove $\sum_{k=1}^{n} j^2 = n(n + 1)(2n + 1)/6$

5. Prove $\sum_{k=1}^{n} j^3 = n^2(n + 1)^2/4$

6. Prove $\sum_{k=1}^{n} k(k - 1) = n(n^2 - 1)/3$

7. Prove $\sum_{k=1}^{n} (2k - 1)^2 = n(2n - 1)(2n + 1)/3$

8. Prove $\sum_{k=1}^{n} 1/(4k^2 - 1) = n/(2n + 1)$

9. Prove $\sum_{k=1}^{n} (1/(x + k)(x + k - 1)) = n/x(x + n)$ if $x > 0$

10. Prove $|\sum_{k=1}^{n} f(k)| \leq \sum_{k=1}^{n} |f(k)|$

11. Prove $\sum_{k=0}^{n-1} 3^k = (3^n - 1)/2$

12. Prove $\sum_{k=1}^{n} (0.1)^k = (1 - (0.1)^n)/9$

13. Prove $\sum_{k=1}^{n} ar^{j-1} = a(r^n - 1)/(r - 1)$, provided $r \neq 1$. [This result will be used later.]

14. Prove $\sum_{k=0}^{n} 2^j = 2^{n+1} - 1$

15. Prove $(\cos \theta + i \sin \theta)^n = \cos n\theta + i \sin n\theta$ for any positive integer n. This result is known as *DeMoivre's Theorem*. [This result will be used later.]

16. Use the result of Exercise 15 and the expansion of $(\cos \theta + i \sin \theta)^3$ to find formulas for $\cos 3\theta$ and $\sin 3\theta$.

17. Prove $\sum_{k=1}^{n} k^4 \geq n^5/5$.

18. Prove $\sum_{k=1}^{n} k^m > n^{m+1}/(m + 1)$.

19. Prove $\sum_{k=0}^{n-1} k^m < n^{m+1}/(m + 1)$.

20. Prove $\sum_{k=0}^{n-1} k^m < n^{m+1}/(m + 1) < \sum_{k=1}^{n} k^m$. [This result will be used later.]

21. Prove $(x + y)^n = \sum_{k=0}^{n} (n!/[(n - k)!k!])x^{n-k}y^k$. This result is known as the *binomial theorem*. [*Hint*: Use Example 13 of Section 10.]

22. Use the preceding result to expand $(2x - 3)^6$.

23. Use Exercise 21 to prove $\sum_{k=0}^{n} n!/((n - k)!k!) = 2^n$. [*Hint*: let x and y both have the value 1.]

24. Use Exercises 15 and 21 to evaluate $\cos 5\theta$ and $\sin 5\theta$.

CHAPTER II

Integration

II.1 Illustrative Problem

We will start this chapter with the consideration of a relatively simple problem in determining costs, and then consider what would happen if the problem were to be made more realistic. The results of this consideration will indicate many questions which must be answered, and will ultimately lead to the definition of one of the two basic concepts of the calculus.

Let us look in on the story of the Carefree Communication Company after they have perfected their new Smell-a-Phone for the aromatic touch. The Smell-a-Phone, henceforth referred to as SAP, went into production on January first of this year, and a careful record has been kept of all pertinent information. As a result there are numerous records of the cost of manufacture of each SAP on each day, and of the total number manufactured by the end of each day. Needless to say, these records are necessary not only for stockholders but also for tax purposes, negotiations, etc. Now the cost of manufacture will vary, for the cost of materials, the amount of overtime pay, and even the amount of plant overhead changes at surprisingly short intervals of time. Therefore it is not possible to compute manufacturing costs by taking the cost of manufacture of any single SAP and multiplying by the total number produced to date. However, the total cost of manufacture is required, and since no one who is not actively engaged in the manufacture of SAPs knows the multiplication tables, they have decided to hire you to compute the amount spent so far for the manufacture of these valuable instruments. When you agreed to take the job, you were given a table, the first portion of which has been reproduced in Table II.1, and you noted that for each day you have a function $C(t)$ which represents the unit cost of manufacture on the t-th day, and you also have the function $n(t)$ which indicates

Table II.1

t	$C(t)$	$n(t)$	t	$C(t)$	$n(t)$
1	8.94	6	11	9.24	105
2	8.95	13	12	9.23	119
3	9.02	21	13	9.26	131
4	9.08	30	14	9.25	145
5	9.07	41	15	9.31	156
6	9.19	51	16	9.43	169
7	9.19	62	17	9.39	183
8	9.21	72	18	9.51	198
9	9.22	84	19	9.52	213
10	9.20	94	20	9.51	229

the total number of SAPs that had been manufactured from the beginning of production to the end of the t-th day's work. Having been well trained, you start in a logical fashion by drawing a flow chart of the work to be performed. This flow chart is shown in Figure II.1. Needless to say, this was not the first flow chart, for you had started by putting in a computation block for each day, but you had soon observed that the chart could be refined.

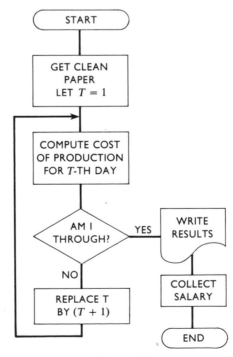

Figure II.1

Also you had become aware that many of these jobs had cycles or repetitions, and that you should take advantage of these by producing a loop in the flow chart.

At this point you start the computation, and following the directions given in the flow chart, you compute as indicated in the following expression:

$$
\begin{aligned}
C(1)n(1) &+ C(2)[n(2) - n(1)] \\
&+ C(3)[n(3) - n(2)] \\
&+ C(4)[n(4) - n(3)]
\end{aligned}
\tag{II.1.1}
$$

but then you stop, for you realize that you have a repetition and you could have written

$$
\text{Cost of Production} = \sum_{k=1}^{p} C(k)[n(k) - n(k - 1)],
\tag{II.1.2}
$$

provided, of course, you had agreed that $n(0)$, which had not previously seemed necessary, would have the value zero. Of course, p is the number of the present day of production, and you now have a formula for the cost of production to the present time. At this point it is a routine job to settle back, substitute the numbers from the table that has been provided you, and complete the calculation before collecting your salary. You are aware, of course, that we are obtaining the number of $SAPs$ manufactured on the k-th day by taking the difference $[n(k) - n(k - 1)]$, that is subtracting the number manufactured by the beginning of the k-th day, or the end of the $(k - 1)$-st day, from the number manufactured by the end of the k-th day. By the circumstances surrounding any decent manufacturing operation, this difference is bound to produce a non-negative result, and in fact it is presumed that it will be a positive result. (This is worth noting for later generalization of this illustration.)

Just as you begin to feel that you have accomplished that for which you were hired, some foreman breaks in to remind the Carefree head bookkeeper that in reality the $C(k)$ was never constant for an entire day, for the special Odor-metal that they were using is very expensive and hard to obtain, and they are seldom able to keep much inventory. The price seems to go up with each reorder, and therefore the $SAPs$ made later in a given day might well use some of a later, more expensive, shipment and would therefore cost more to produce. The foreman had a record of just when these different shipments came into use, and hence he could refine the table that you had been using with such painstaking care. Now instead of using k to represent the number of days since the start of the company it would become necessary to think of k (or some other variable) as representing a time at which the cost of manufacturing a SAP would change. In order to make this apparent you used the symbol x_k to represent the k-th time the manufacturing cost changed. Thus x_0 was the start of manufacturing $SAPs$, x_1 was the moment when the cost of production first changed, x_2 the instant of the second cost change, etc. Now you have $n(x_k)$ as representing the total number of SAPs manufactured at the time of the k-th change in the cost of manufacturing a single SAP. To further generalize the notation you decide to let t_k denote some moment of time in the interval when you have the k-th distinct manufacturing cost. Thus t_1 would be in the interval before x_1, or in the interval $[x_0, x_1)$, and t_2 would be in the interval $[x_1, x_2)$. Since the cost of manufacture is constant throughout the interval $[x_{k-1}, x_k)$, you do not really care which of the possible values of t_k you select from this interval. Now you proceed to write the expression analogous to (II.1.2) using this new notation, namely

$$\text{Cost of Production} = \sum_{k=1}^{N} C(t_k)[n(x_k) - n(x_{k-1})]. \qquad \text{(II.1.3)}$$

You note that $[n(x_k) - n(x_{n-1})]$ is the number of SAPs manufactured during the interval $[x_{k-1}, x_k)$, and thus you have added the total cost of manufacture for each of the time intervals involved, be they very short or very long. In this case the value N is the total number of such intervals since the start of manufacture, and if there have been many cost changes N can be rather large. At this point you pause to consider the situation and observe that $n(x)$ is an increasing function and the lengths of the intervals $[x_{k-1}, x_k)$ may differ.

If it were to happen that other persons came in with information indicating more frequent changes of production costs, you might find that the time intervals involved would become very short. In fact, a realistic view of the situation would suggest that the various incoming supplies would become mixed up and the cost of manufacture would appear to vary continuously. In this case it would be so inconvenient to use (II.1.3) with a very large number of small time intervals that one would be well advised to consider using some reasonably short time interval, say two minutes, and let t_k be some moment in the time interval such that $C(t_k)$ would be expected to give some kind of an average cost of production over that short interval. You realize, of course, that this would no longer give an exact amount, but it would be sufficiently close to the correct amount that it would probably serve for all practical purposes. It would be possible, of course, to select the highest cost during the given time interval and then to obtain an upper bound for the manufacturing cost. It would also be possible to obtain a lower bound by using the lowest cost in each time interval. The correct value would be somewhere between the upper and lower bounds, and if the two bounds are close to each other you may feel rather comfortable about your result. If they are far apart you will probably want to go back and repeat your calculations using a shorter time interval, say one minute or perhaps thirty seconds. It would seem intuitively clear that at some point the time interval would be sufficiently short that the two bounds would be within a dollar or less of each other, and this is close enough even for tax purposes.

EXERCISES

1. Using the data of Table II.1 of this section, calculate the cost of manufacturing the first 229 *SAP*s.

2. State clearly the relationship between the original choice of k in this section and the later choice of x_k and t_k. How do the various possible methods of selection of these values affect the calculation?

3. A certain student has observed that the amount learned on the k-th day of school is $p(k)$ percent of the number of pages read. If the number of pages read from the beginning of school through the k-th day is given by $n(k)$, find an expression for the total amount learned during the first 100 days of school. Ignore the fact that some material learned in the first few days may be forgotten by the 100-th day.

4. How would you change your answer to Exercise 3 if you were given the fact that the percent learned would vary not only with the day but with the time of day? Discuss a technique similar to that used in this section which would permit you to find an upper and lower bound for the amount learned.

5. Would the problem of the Carefree Communication Company be affected in any way if $C(x)$ and $n(x)$ were continuous? Would there be any affect if $C(x)$ and $n(x)$ were not monotonic? How would you interpret the fact that $C(x)$ might not be monotonic increasing? Could you interpret a situation in which $n(x)$ would not be an increasing function?

6. A solid object is made by gluing together several disks, each in the form of a right circular cylinder. The centers of the disks are along a common axis. If you think of the problem of finding the volume of this solid as a problem in which you proceed from one end to the other along this common axis with the volume of each disk being obtained by considering the length of travel through that particular disk multiplied by the cross sectional area, you will be able to express the volume as a sum in a manner similar to that of the problem of this section. Carry out this procedure for a solid consisting of 10 disks if the k-th disk has a radius of k and a thickness of $1/k$.

7. Consider a sphere of radius 6 as being approximated by a solid similar to that of Exercise 6 by taking a very large number of disks with centers along a diameter of the sphere, each of the disks being very thin. Would you be able to obtain a reasonable approximation of the volume of the sphere using the method of Exercise 6? If you have some calculating device or computer available, carry out this computation for disks with a thickness of 1 inch and for a thickness of 0.1 inch.

8. Give at least three additional examples of problems in which this type of computation could be used to give an approximation for a quantity which may be desired.

9. Why was it desirable to use the semi-open subintervals $[x_{k-1}, x_k)$ in the problem of this section? What affect would the inclusion of the end point, x_k, have had on the result? Would this inclusion of x_k have had a similar result in approximating the volume of the sphere in Exercise 7?

II.2   Partitions

In the consideration of the problem of Section II.1 we were given information
which made it desirable to break up the interval of time involved into subintervals. Thus, we considered the set of real numbers $\{x_0, x_1, x_2, \ldots, x_n\}$
to be the points of subdivision. Each successive pair of points determined one
of the sub-intervals of the interval $[x_0, x_n]$. We will now formalize this
procedure in a definition, with the understanding that we are concerned
here with real numbers only.

Definition 2.1. A *partition* of an interval $[a, b]$, $(a < b)$ is defined to be a set of
$(n + 1)$ points $\{x_k\}$, $(k = 0, 1, 2, \ldots, n)$, $(n \geq 1)$, such that $a = x_0 < x_1 <
x_2 < \cdots < x_n = b$. This partition is denoted by $P[a, b]$, and if there is no
doubt about the interval we will merely use the notation P.

It should be clear from the definition that there are a large number of
partitions for any interval on the real axis. (The fact that we have used
inequalities would limit us to the real numbers.) Thus, we might have
$P_1[2, 5] = \{2, 2.4, 4, 5\}$, and we might also have $P_2[2, 5] = \{2, 3, 5\}$. The
only points which are required are the two end points. We note that there
must be at least two points in the partition, namely the two end points. There
is no upper limit to the number of points since for any partition it is always
possible to add at least one additional point by subdividing one of the subintervals determined by the partition. Note that P_1 determines three subintervals, $[2, 2.4]$, $[2.4, 4]$, and $[4, 5]$, whereas P_2 determines the two
subintervals $[2, 3]$ and $[3, 5]$. Note that we have used closed subintervals
here in contrast to the semiopen subintervals in Section II.1. The distinction
usually has little effect on the result, although one might wish to make the
distinction in certain cases, as we did in Section II.1.

For a given problem there may appear to be a partition which occurs
naturally due to the nature of the situation from which it arises. The partition used in the problem in the last section, where the points of the partition
were determined by the time at which the production costs changed, was an
example. In the majority of cases, however, there is no such natural indication,
and for that reason the partition is to a very great extent an arbitrary matter
determined by the whim of the person creating the partition. This would
seem to provide a great deal of freedom in the manner in which a problem
could be approached, and it would appear that it would not be possible to
obtain a unique result. However, appearances here are quite deceptive.
Uniqueness follows as a result, although not an obvious one, of the following
definition.

Definition 2.2. If $P[a, b]$ is a partition of $[a, b]$ and if $P^*[a, b]$ is a second
partition of $[a, b]$ such that P is a proper subset of P^*, then P^* is a *refinement* of P.

Note that since P must be a proper subset of P^*, then P^* must have *at least one more point* than P, but must include all of the points of P. (This is nothing more than the definition of a proper subset.) Thus, if we were to consider the illustrations above, we might also consider

$$P_3 = \{2, 2.4, 3, 4, 4.6, 5\},$$

and we would observe that this is both a refinement of P_1 and P_2. Note that a partition is not a refinement of itself. Note also, that the intervals in a refinement are never longer than the intervals of the set of which we have the refinement.

The observations that we have made lead us to two important consequences of our definitions, one a further definition, followed by a theorem, and the other a theorem.

Definition 2.3. Let $P[a, b]$ be a partition of $[a, b]$, and let $d(P)$ be the largest difference of the form $(x_k - x_{k-1})$. We define $d(P)$ to be the *diameter* of the partition P.

Theorem 2.1. *Let $P^*[a, b]$ be a refinement of $P[a, b]$. Then $d(P^*) \leq d(P)$.*

OUTLINE OF PROOF. Consider the intervals of P and the intervals of P^*. Show that no interval of P^* is longer than the longest interval of P. □

If we were to consider a succession of refinements, and then the corresponding diameters, we would have a monotone decreasing sequence of diameters. Does this guarantee that the diameters of successive refinements will of necessity get smaller?

The other item coming from our considerations above was suggested by the partition, P_3, of the illustration used.

Theorem 2.2. *Let $P_1[a, b]$ and $P_2[a, b]$ be two partitions of the interval $[a, b]$. There exists a partition, $P[a, b]$, which is a refinement of both $P_1[a, b]$ and $P_2[a, b]$.*

PROOF. Let $P_1[a, b] = \{u_k\}$, $(k = 0, 1, 2, \ldots, n_1)$ and let $P_2[a, b] = \{v_j\}$, $(j = 0, 1, 2, \ldots, n_2)$. If P_1 is a subset of P_2, any refinement, P, of P_2 is also a refinement of P_1. If P_2 is a subset of P_1, then P can be any refinement of P_1. If P_1 is not a subset of P_2 and P_2 is not a subset of P_1, this indicates that P_1 has points that are not in P_2 and P_2 has points that are not in P_1. Hence $P_1 \cup P_2 = P$ provides a partition, P, which is both a refinement of P_1 and of P_2. (The method of proof is illustrated in Figure II.2). □

Corollary 2.1. *If $P_k[a, b]$, $(k = 1, 2, \ldots, n)$ are n partitions of $[a, b]$, then there exists a partition $P[a, b]$ which is simultaneously a refinement of each partition $P_k[a, b]$.*

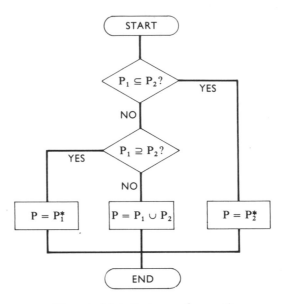

(The asterisk indicates a refinement)

Figure II.2

Corollary 2.2. *If $P_2[a, b]$ is a refinement of the partition $P_1[a, b]$, and if $P_3[a, b]$ is a refinement of $P_2[a, b]$, then P_3 is a refinement of P_1.*

Since we will generally be interested in successive refinements of partitions, it thus becomes clear that we are not likely to run into major difficulties, for if two persons start with different partitions of a given interval, there will always exist a common refinement.

In equation (II.1.3) we required two sets of points. The set $\{x_k\}$ constituted a partition of the interval involved. The set $\{t_k\}$ was related to this partition in that for each value of the subscript k we selected t_k in such a way that t_k was in the closed interval $[x_{k-1}, x_k]$. We will now dignify this set by means of a definition.

Definition 2.4. A set $T = \{t_k\}$, $(k = 1, 2, \ldots, n)$ is said to be an *evaluation set* of the partition $P[a, b]$ if t_k is in the closed interval $[x_{k-1}, x_k]$ for each value of k. We shall denote this set by $T(P)$ or $T(P[a, b])$, and if there is no confusion concerning which partition is involved we will denote the evaluation set merely by T.

It is clear that if we are given an interval, there are infinitely many partitions, and each partition will have infinitely many evaluation sets. If the diameter of the partition is small, the selection of each point of the single evaluation set is somewhat limited. In some cases we will find that we wish to make specific selections for the elements of the evaluation sets, and in other cases it will make little difference which evaluation set we will select for a given partition.

EXAMPLE 2.1. Let $f(x) = \sqrt{x}$, and $g(x) = x^2$. Write down a partition of $[1, 10]$ and an evaluation set for this partition. For your partition and evaluation set, write down the value of

$$\sum_{k=1}^{n} f(t_k)[g(x_k) - g(x_{k-1})]$$

where n is the number of intervals in your partition.

Solution. There are infinitely many partitions of $[1, 10]$, but we might use $\{1, 3, 7, 10\}$. Hence $n = 3$. There are 3 subintervals in this partition, namely $[1, 3]$, $[3, 7]$, and $[7, 10]$. The evaluation set must then have 3 points, one from each of the subintervals. We might use $T = \{2, 7, 7\}$. Note that 7 is a number in the second and in the third subinterval. Now

$$\sum_{k=1}^{3} f(t_k)[g(x_k) - g(x_{k-1})] = f(2)[g(3) - g(1)] + f(7)[g(7) - g(3)]$$

$$+ f(7)[g(10) - g(7)]$$

$$= \sqrt{2}(9 - 1) + \sqrt{7}(49 - 9) + \sqrt{7}(100 - 49).$$

At this point the problem is reduced to arithmetic, and you can complete the work.

After solving the problem, we observe that the arithmetic would have been easier for human solution if the values selected for the evaluation set had been squares of rational numbers. The set $\{1, 4, 9\}$ would have been an easier choice, leading to the result

$$1(9 - 1) + 2(49 - 9) + 3(100 - 49) = 8 + 80 + 153 = 241.$$

Another observation of this solution indicates the great amount of freedom we have in selecting not only the partition, but also the evaluation set. The resulting sum can have a variety of results, depending upon both the partition and the evaluation set selected. If we were to choose a refinement P^* of P such that the diameter of the refinement is no greater than 0.1 we would have much less variation in the possible values of the sum, both for P^* and for any refinement of P^*. This is a matter which we will pursue further in the next two Sections.

EXAMPLE 2.2. Find an approximation of the area bounded by $y = f(x) = x^2$, $y = 0$, $x = 1$, and $x = 4$. Also describe a method by which the area can be obtained more precisely.

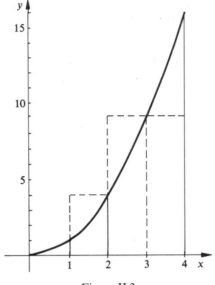

Figure II.3

Solution. This area is indicated in Figure II.3. A specific partition, $P[1, 4]$ is also indicated. We have chosen to use here the partition $P[1, 4] = \{1, 2, 4\}$, although we could equally well use any other partition that appealed to us. We have also selected an evaluation set, $T = \{2, 3\}$, to accompany this partition. In the figure we have indicated not only the partition, but also the evaluation set. Note that we have drawn horizontal lines through the points indicated by the evaluation set. We have also drawn in rectangles such that their width covers a subinterval determined by the partition and their height is determined by the height of the parabola at the corresponding evaluation point. The sum of the areas of these two rectangles approximates the area under the portion of the parabola in which we are interested. Since the width of the first rectangle is $[2 - 1]$ and the height is $f(2)$, and since the width of the second rectangle is $[4 - 2]$ and the height is $f(3)$, we have for our approximation of the desired area the expression

approx. area $= f(2)[2 - 1] + f(3)[4 - 2] = 4(1) + 9(2) = 22$

$$= \sum_{k=1}^{2} f(t_k)[g(x_k) - g(x_{k-1})]$$

where in the last expression we have denoted the points of the evaluation set by t_k, and we have let $g(x)$ be the identity function, $g(x) = x$, with partition points $\{x_k\}$, ($k = 0, 1,$ and 2).

With reference to the request for a method for improving our approximation, we note that we would only have to refine the partition used and then

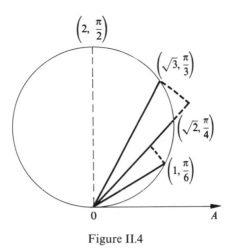

Figure II.4

make a corresponding modification of the evaluation set in order to obtain what appears to be a better approximation. This is a technique that we will be using in the next several Sections.

EXAMPLE 2.3. Sketch the graph of $r = 2 \sin \theta$ in polar coordinates, and then find an approximation for the area included inside this curve between the vectors determined by

$$\theta = \frac{\pi}{6} \quad \text{and} \quad \theta = \frac{\pi}{3}.$$

Solution. We note that as the angle θ increases from 0 to $\pi/2$, and thence to π that r increases from 0 to 2 and then decreases to 0. Hence, the general shape of the curve will be that which is indicated in Figure II.4. It is shown in Appendix B that this is a circle, but for this problem we do not need a precise graph in order to be able to determine an approximate area. We will now partition the interval

$$\left[\frac{\pi}{6}, \frac{\pi}{3} \right],$$

and we might well use the partition

$$\left\{ \frac{\pi}{6}, \frac{\pi}{4}, \frac{\pi}{3} \right\}.$$

We have drawn in the additional vector to indicate this partition in the graph. We next take an evaluation value in each subinterval of the partition. For these evaluation points we might select

$$\left\{ \frac{\pi}{6}, \frac{\pi}{3} \right\}.$$

If we now draw in a circular sector with center at the pole and with radius equal to the value of r corresponding to the evaluation point for each sub-interval of the partition, we have in this case two circular sectors as shown in Figure II.4. We can approximate the desired area by considering the sum of the areas of these two sectors. Since the area contained in the circle of radius $2 \sin (\pi/6)$ is given by

$$\text{area} = \pi \left[2 \sin \frac{\pi}{6} \right]^2$$

and the central angle involved is $((1/4)\pi - (1/6)\pi)$ radians, we have for the area of the first sector

$$\frac{\left(\frac{\pi}{4} - \frac{\pi}{6} \right)}{2\pi} \pi \left[2 \sin \frac{\pi}{6} \right]^2 .$$

In similar fashion the area of the second sector is

$$\frac{\left(\frac{\pi}{3} - \frac{\pi}{4} \right)}{2\pi} \pi \left[2 \sin \frac{\pi}{3} \right]^2 .$$

In each case the fraction indicates the proportion of the complete circle being used and

$$\pi [2 \sin t_k]^2$$

is the area of the complete circle having a radius $f(t_k) = 2 \sin t_k$ determined by the choice of evaluation point.

We can now express the approximate area required as

$$\text{approx. area} = \frac{\pi}{2\pi} \left(2 \sin \frac{\pi}{6} \right)^2 \left[\frac{\pi}{4} - \frac{\pi}{6} \right] + \frac{\pi}{2\pi} \left(2 \sin \frac{\pi}{3} \right)^2 \left[\frac{\pi}{3} - \frac{\pi}{4} \right]$$

$$= \frac{1}{2} (1)^2 \left(\frac{\pi}{12} \right) + \frac{1}{2} (\sqrt{3})^2 \left(\frac{\pi}{12} \right) = \frac{\pi}{6}.$$

Again we might observe that this falls into the same pattern that we observed before. We could write this expression as

$$\text{approx. area} = \sum_{k=1}^{2} \frac{1}{2} [f(t_k)]^2 [g(\theta_k) - g(\theta_{k-1})]$$

where t_k is a point of an evaluation set and $f(t_k)$ indicates the radius of the circular sector used in approximation, while $g(\theta) = \theta$ is the identity function and the difference indicates the size of the sector being used. A refinement of the partition with a corresponding change in the evaluation set would lead again to a more accurate approximation of the area involved.

While it is true that in the majority of cases we can use any partition, it is also true that it is frequently easier to use a partition in which the points are equally spaced throughout the given interval. Since we will find it convenient to use such a partition frequently, we will give it a name in the following definition.

Definition 2.5. A partition $P[a, b]$ is called *regular* if *all* subintervals have the same length.

From this definition it follows that a regular partition $P[a, b]$ having $(n + 1)$ points, and hence n subintervals, must have each subinterval of length $(b - a)/n$. In order to see this one must note that the length of the entire interval is $(b - a)$ and since there are n equal subintervals, each one must be of length $(b - a)/n$.

We now have all of the ingredients we need to discuss a summation, such as (II.1.3) more fully, and this we will do in the next section. We have considered sets without trying to give them specific origins or meanings in this section in order that we would not prejudice their use. Thus, the interval might be an interval of time, money, distance, amount of material to be mastered in this course, or any one of a number of other quantities. The partition may be one supplied by an outside source, such as nature, or may be one of our own choosing. The evaluation set may be prescribed, or may be selected as the set of points at which the task of evaluating a function, such as the $C(t)$ of (II.1.3), can be most easily carried out. Since we can start with two different partitions and end up with a common refinement, the particular starting point becomes a matter of less concern than the ultimate refinement with which we terminate the problem. This generality will be helpful in the sections that follow, and will also be helpful in permitting the application of these ideas to a broader selection of topics.

EXERCISES

1. Write down a partition of $[-1, 3]$ involving 7 points. What is the value of n for this partition? Write down another partition of this interval involving 5 points. Obtain three distinct partitions, each of which is a simultaneous refinement of your 7 point partition and your 5 point partition.

2. If $P[-1, 3]$ is a regular partition having 9 points, what are the points of this partition? What is the diameter of this partition? If $P*[-1, 3]$ is a refinement of $P[-1, 3]$, what is the minimum number of points in $P*$ if $P*$ is a regular partition?

3. If $P[0, 4] = \{0, 0.5, 2, 2.75, 4\}$, find the refinement, $P*$, of P having the least number of points such that $P*$ is regular.

4. Let $P[-4, -2]$ be a regular partition having 7 points. Let $T(P)$ be an evaluation set formed by taking the midpoints of the intervals of P. Give the points of P and of T.

5. Let $P_1[2, 5]$ be a regular partition with 5 points and $P_2[2, 5]$ be a regular partition having 7 points. Find a regular partition which is a common refinement of P_1 and P_2.

6. Let $P[0, 4]$ be a regular partition having 9 points. Show that the points, x_k, of the partition are given by the relation $x_k = 4k/8$ for the 9 values of k ($k = 0, 1, 2, 3, 4, 5, 6, 7, 8$).

7. Let $P[0, a]$ be a regular partition having $(n + 1)$ points. Show that $x_k = ka/n$ for each of the $(n + 1)$ values of k.

8. If $P[2, 5]$ is a regular partition having 7 points, show that $x_k = 2 + 3k/6$ for each of the 7 values of k.

9. If $P[a, b]$ is a regular partition having $(n + 1)$ points, or n subintervals, show that

$$x_k = a + \frac{k(b - a)}{n} = \frac{(n - k)a + kb}{n}$$

for each of the $(n + 1)$ values of k.

10. Give an illustration which verifies Corollary 2.1.

11. Give an illustration which verifies Corollary 2.2.

12. Give an illustration which verifies Theorem 2.1. Complete the proof of Theorem 2.1.

13. Write down a partition of $[2, 4]$ having at least 5 points, and then write down an evaluation set corresponding to your partition. If $f(x) = x^2$ and $g(x) = x^3$, find the value of $f(t_k)[g(x_k) - g(x_{k-1})]$ for each of the values of k.

14. Using the same partition that you used in Exercise 13 and the same functions $f(x)$ and $g(x)$, find the evaluation set $T(P)$ such that each of the terms $f(t_k)[g(x_k) - g(x_{k-1})]$ is as large as possible. Also find the evaluation set such that each of these terms is a small as possible.

15. Draw a graph of $f(x) = x^2$. Using the method of Example 2.2 of this section, find an approximation for the area under $y = f(x)$ over the interval $[0, 2]$ using $P[0, 2] = \{0, 0.4, 1, 1.2, 1.5, 2\}$ and using $T(P) = \{0.2, 0.8, 1.1, 1.5, 1.5\}$. Draw the rectangles on your graph corresponding to this partition and this evaluation set in a manner similar to that used in Figure II.3.

16. Using the graph and partition given in Exercise 15 find the evaluation set which gives the largest possible value for the approximation of the area under the curve $y = f(x)$ over the interval $[0, 2]$. Also find the evaluation set which would give the smallest possible value for this approximation. Using these evaluation sets find the upper bound and the lower bound for the number of square units of area. Does your answer to Exercise 15 fall within these bounds?

17. Show that the diameter of the partition in Exercise 15 is 0.6. Show that it is possible to refine this partition indefinitely without making the diameter smaller. Show that so long as the refinements considered do not decrease the diameter the upper bound for the area and the lower bound will differ by more than 0.5. Find a more precise value for this amount by which the upper and lower bounds of area must differ.

18. Let $f(x) = x^3$ and $g(x) = 1/x$. Using $P[1, 3] = \{1, 1.5, 2.3, 3\}$ and $T = \{1, 2, 3\}$, compute the value of

$$\frac{\sum_{k=1}^{3} f(t_k)[g(x_k) - g(x_{k-1})]}{g(3) - g(1)}.$$

Show that this value is between $f(1)$ and $f(3)$. How does this result relate to a concept of *average value*? How would it relate to *average value* if $g(x)$ were replaced with $g(x) = x$?

19. Draw a graph of the *cardioid* $r = f(\theta) = 1 + \cos\theta$. Following the method given in Example 3, use the partition $P[\pi/6, \pi/3] = \{\pi/6, 0.6, \pi/4, 1, \pi/3\}$ and the evaluation set $T(P) = \{0.55, 0.7, \pi/4, \pi/3\}$ to find an approximation for the area inside the curve between the radius vector corresponding to $\theta = \pi/6$ and the radius vector corresponding to $\theta = \pi/3$.

20. Find an approximation for the area within $r = \cos\theta$ and between the vectors $\theta = 0$ and $\theta = \pi/3$. Use an evluation set with three subintervals.

21. Using your partition of Exercise 20 find the evaluation set which would give the smallest possible approximation for area using this partition, and then evaluate this smallest possible approximation for area. Do the same thing for the largest possible approximation for area. Is your answer for Exercise 20 between the two approximations you have just obtained?

22. It is required to evaluate

$$\sum_{k=1}^{n} t_k[x_k - x_{k-1}]$$

where the values x_k are taken from the partition $P[-1, 5]$. If $d(P) < 0.1$, what is the maximum difference between the largest possible value for this sum and the smallest possible value. Note that the sum can vary depending on the choice of the partition and of the evaluation set.

23. If we were to consider the set of points on the circle $x^2 + y^2 = 4$ which are either on a positive axis or in the first quadrant, would it be possible to find a partition of this set? Could you find a refinement of this set? Would all of the results of this section hold in this case? [*Hint*: Could you put all of these points in any order from small to large?]

24. Would it be possible to find a partition of $[2 + 3i, -4i]$ in the Argand diagram? What points are permissible here? Does this differ significantly from the partitioning of a real interval? Give a reason for your answer.

25. In the preparation of income tax returns the government permits one to round numbers off to the nearest dollar. How does this relate to our use of partitions and evaluation sets? How much variation would you expect from the correct value if you were to use this method? If your time were worth the current minimum wage, would any possible gain to you be worth the time and effort involved in doing the additional arithmetic?

II.3 The Riemann–Stieltjes Sum

In the consideration of the cost of production problem of Section II.1 we first established (or had given by the terms of the problem) a partition and an evaluation set. We had a cost function provided which was bounded in that there was some finite cost which it could not exceed during the time interval in question. Furthermore, the counting function could only increase with time, and therefore was a monotonic function. This gave us a glimpse of a Riemann–Stieltjes sum. Since we will be using these sums very frequently, we will abbreviate this long title and merely call them RS sums. Since we will continue to use the adjective *bounded*, we will pause to define this term in a more specific manner before proceeding to a formal definition of an RS sum.

Definition 3.1. A function, $f(x)$, is said to be *bounded over an interval* $[a, b]$ provided there exists a positive number, M, such that $|f(x)| \leq M$ for all values of x in the interval $[a, b]$.

EXAMPLE 3.1. Is the function $f(x) = 1/x$ bounded over a domain consisting of all non-zero real numbers?

Solution. We need to determine whether there is a positive real constant, M, such that $|f(x)| \leq M$ for all values of x in the domain. Suppose there is such a value, M. Then if x is in the interval $(0, 1/M)$, we have $f(x) > M$. Hence, whatever value of M we might select, by virtue of the fact that it is a positive real constant, it must have a reciprocal which is nonzero and then there must be an open interval of values in the domain for which $f(x)$ exceeds M. Therefore, this function is not bounded over this domain. However, if we considered the domain to be all positive real numbers not less than 0.001, the resulting function would be bounded, for you could show rather easily that a value of $M = 1000$ would suffice in this case.

By limiting the domain it is often possible to remove points which would cause a function to be unbounded.

Definition 3.2. If $f(x)$ and $g(x)$ are bounded functions defined over the interval $[a, b]$, and if $P[a, b]$ is a partition of $[a, b]$ and T is an evaluation set of $P[a, b]$, then

$$S(P, T, f, g) = \sum_{k=1}^{n} f(t_k)[g(x_k) - g(x_{k-1})]$$

is a *Riemann–Stieltjes sum* (or an RS *sum*) of $f(x)$ with respect to $g(x)$ over the interval $[a, b]$ with partition P and evaluation set T.

Since we will be using RS sums through much of the remainder of this book, and since we will be using them to define the Riemann–Stieltjes integral in the next section, it is interesting to note that the early development of the calculus was very much from an intuitive point of view. *G. F. B. Riemann* (1826–1866) found in his work that he needed to obtain the integral of a function that had rather peculiar properties. In order to investigate this function and to be certain that the results were correct, he established the basis for the development of the integral, using what are called Riemann sums. These are just like the RS sums provided $g(x)$ is the identity function. The work of Riemann was fundamental in the development of the integral, and formed the basis for further development by many others, including the Dutch mathematician, *T. J. Stieltjes* (1856–1894). It is the modification of Riemann's work as done by Stieltjes that we are developing here.

EXAMPLE 3.2. Evaluate the RS sum of $f(x) = x^2 - x$ with respect to $g(x) = x^3 - 1$ over the interval $[-2, 3]$ with partition $P = \{-2, 0, 1, 2.5, 3\}$ and evaluation set $T = \{-1, 1, 2, 2.8\}$.

Solution. We note here that $f(x)$ and $g(x)$ are bounded in the prescribed interval. Hence, we have fulfilled the necessary requirements for an RS sum. Since P has five points, $n = 4$. Hence

$$S(P, T, f, g) = S(P, T, x^2 - x, x^3 - 1)$$

$$= \sum_{k=1}^{4} f(t_k)[g(x_k) - g(x_{k-1})]$$

$$= f(-1)[g(0) - g(-2)] + f(1)[g(1) - g(0)]$$
$$+ f(2)[g(2.5) - g(1)] + f(2.8)[g(3) - g(2.5)]$$

$$= (2)[(-1) - (-9)] + (0)[(0) - (-1)]$$
$$+ (2)[(14.625) - (0)] + (5.04)[(26) - (14.625)]$$

$$= 16 + 0 + 29.25 + 57.33 = 102.58.$$

From Example 3.2 we note that we can evaluate an RS sum by the following steps:

1. Check to make certain that the $f(x)$ and $g(x)$ are bounded over the interval in question.
2. Determine the partition to be used.
3. Determine the evaluation set to be used.
4. Evaluate $g(x)$ at the partition points and $f(x)$ at the evaluation points.
5. Calculate the separate terms in the summation and then calculate the sum.

Since the two functions and the interval will usually be apparent, there is no question concerning step (1). There are many partitions that are possible, and hence step (2) would suggest that there may be many possible values for the RS sum if only $f(x)$, $g(x)$, and the interval are given initially. Let us assume for the moment that we have selected a particular partition. We will consider the possible alternatives to this choice in the next section. Note that when we have selected a partition, we have determined via step (4) the value of each of the expressions $[g(x_k) - g(x_{k-1})]$ occurring in the summands. This indicates that the only terms that can affect the value of the sum at this point (having selected the partition) are the $f(t_k)$ which depend upon the evaluation set. Of course, the selection of a particular evaluation set will determine a specific RS sum, but we might wish to consider at this point the range of values which the RS sum might have for the given partition. Thus, we could attempt to find the largest possible RS sum given this partition, and we might also try to find the smallest such sum. This would give us an interval in which all other RS sums must occur. We will illustrate this in the following Example.

EXAMPLE 3.3. Find the largest and smallest possible values for the RS sums $S(P, T, x^2, x^3)$ if $P[1, 4] = \{1, 2, 4\}$.

Solution. Since we are given the partition in this case, we know that there are two subintervals and three points, and we also know that for an evaluation set $T = \{t_1, t_2\}$ we have

$$S(P, T, x^2, x^3) = t_1^2[2^3 - 1^3] + t_2^2[4^3 - 2^3] = 7t_1^2 + 56t_2^2.$$

If we wish to have the largest possible RS sum, we should select t_1 and t_2 so that we will make $7t_1^2 + 56t_2^2$ as large as possible. Now t_1 must be in the subinterval $[1, 2]$ and t_2 must be in the subinterval $[2, 4]$. Hence, the largest value for the sum will be obtained if we choose $t_1 = 2$ and $t_2 = 4$. In this case we would have $S(P, T, x^2, x^3) = 7(2^2) + 56(4^2) = 924$. In similar fashion, to find the smallest possible sum, noting that the partition is not altered, and hence we are still concerned with $7t_1^2 + 56t_2^2$, we would want to use the smallest possible value for t_1 and for t_2. Therefore, we would have $S(P, T, x^2, x^3) = 7(1^2) + 56(2^2) = 231$. We are now assured that any RS sum of the form $S(P, T, x^2, x^3)$ with the given partition would have a value in the interval $[231, 924]$. Needless to say we would not be happy to have such a range of values possible, but it would seem likely that we could have narrowed the range a great deal if we had started with a partition having a diameter smaller than 2. We could also write this result in the form of a three part inequality as $231 \le S(P, T, x^2, x^3) \le 924$.

We will frequently have occasion to obtain bounds of this nature, or at least to make use of the fact that such bounds exist. Therefore, as in many cases when we wish to make frequent use of a concept with a rather long description, we will define some shorter names for these two bounds.

Definition 3.3. If $f(x)$ and $g(x)$ are bounded functions defined over the interval $[a, b]$, and if $P[a, b]$ is a partition of $[a, b]$, then the largest possible RS sum considering all possible evaluation sets will be called the *upper RS sum* and will be denoted by $U(P, f, g)$. Similarly, the smallest possible RS sum considering all possible evaluation sets will be called the *lower RS sum* and will be denoted by $L(P, f, g)$.

Since the upper RS sum (sometimes called the *upper sum*) for a given $f(x)$, $g(x)$, and partition is obtained by using *the* evaluation set which gives the largest possible sum, the evaluation set has been essentially determined when we consider the upper sum. Therefore, it is no longer necessary to specify $T(P)$ in the expression $U(P, f, g)$. Of course, there may be more than one evaluation set which would give this largest value, but since the value is unaffected by the choice of evaluation set even if more than one such set exists, it is still not necessary to specify the evaluation set. A similar observation would hold for the lower sums. We now note that we can write

$$L(P, f, g) \le S(P, T, f, g) \le U(P, f, g). \qquad (\text{II.3.1})$$

While it is well to think about each case that arises, it is possible to make some generalizations. If $g(x)$ is an increasing function in the subinterval $[x_{k-1}, x_k]$, then we observe that $[g(x_k) - g(x_{k-1})]$ must necessarily be positive. In this instance we would wish to use the largest possible value of $f(x)$ over the interval $[x_{k-1}, x_k]$ in order to obtain the term to be used for the upper sum. If $f(x)$ is an increasing function in this subinterval, we would then use $f(x_k)$ as the value giving us the appropriate term for the upper sum. On the other hand, if $g(x)$ is decreasing the difference would be negative and we would wish to obtain the smallest possible value of $f(x)$ over the interval in order to have the term for the upper sum. In case $g(x)$ is not monotonic, one would have to simply compute the value of the difference to know which value of $f(x)$ to use. Furthermore, if $f(x)$ is not monotonic, then it is likely that the choice of x to obtain the appropriate $f(x)$ would not be at either end, and further investigation might be in order. Such a situation is considered in Example 3.4. There is the further possibility that $f(x)$ might be such a peculiar function that there is no *largest* value for the function on the interval. In such a situation, since $f(x)$ is a bounded function, there is an upper bound for the values of $f(x)$ on the subinterval, and we would then choose the smallest possible upper bound in order to determine the upper sum. This latter case is fortunately very rare, but it is worth mentioning that such cases can exist.

EXAMPLE 3.4. Evaluate $U(P, x^2 - x, x^3 - 1)$ and $L(P, x^2 - x, x^3 - 1)$ where $P[-2, 3] = \{-2, 0, 1, 2.5, 3\}$.

Solution. Since $g(x) = x^3 - 1$ is an increasing function throughout the interval $[-2, 3]$, we know that each of the differences $[g(x_k) - g(x_{k-1})]$ will be positive, and hence we want the maximum value that $f(x) = x^2 - x$ can have in each subinterval for the upper sum and we want the minimum value for the lower sum. We might help our consideration by drawing the graph of $y = f(x)$ over the domain $[-2, 3]$ as shown in Figure II.5. You will observe that over the interval $[-2, 0]$ $f(x)$ is decreasing, and hence the largest value of $f(x)$ occurs at $x = -2$ and the smallest at $x = 0$. For the subinterval $[0, 1]$ on the other hand we note that the largest value of $f(x)$

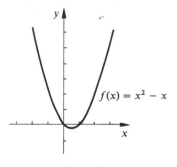

Figure II.5

is zero, and this is attained at both $x = 0$ and $x = 1$. The smallest value would appear to be about halfway between 0 and 1. We can show that this guess is correct by writing $f(x)$ as $f(x) = x^2 - x = (x - 0.5)^2 - 0.25$ and then note that $f(x)$ starts with the value (-0.25) and then adds to this value the square of $(x - 0.5)$. Since the square of a real number cannot be negative, all values of $f(x)$ must either be (-0.25) or be this value increased by some positive number. Hence the minimum value for $f(x)$ must be (-0.25) and must be attained when $(x - 0.5) = 0$ or when $x = 0.5$. For the subintervals $[1, 2.5]$ and $[2.5, 3]$ we see that $f(x)$ is an increasing function and therefore the largest value occurs at the right end of the subinterval and the smallest value occurs at the left end of the subinterval. Using all of this information we have

$$U(P, x^2 - x, x^3 - 1) = 6[(-1) - (9)] + 0[(0) - (-1)]$$
$$+ 3.75[(14.625) - (0)] + 6[(26) - (14.625)]$$
$$= 48 + 0 + 54.84375 + 68.25 = 171.09375.$$

Using the minimum values, we have for the lower sum

$$L(P, x^2 - x, x^3 - 1) = 0[(-1) - (-9)] + (-0.25)[(0) - (-1)]$$
$$+ 0[(14.625) - (0)] + 3.75[(26) - (14.625)]$$
$$= 0 - 0.25 + 0 + 42.65625 = 42.40625.$$

Note that we have used the evaluation set $\{-2, 0 \text{ (or 1)}, 2.5, 3\}$ to obtain the upper RS sum, and we have used the evaluation set $\{0, 0.5, 1, 2.5\}$ to obtain the lower RS sum. In the case of the upper RS sum we could use either of two values for t_2 in the second subinterval, and for the lower RS sum we found that one point in the evaluation set was not an end point of the corresponding subinterval. In the portion of the interval $[-2, 3]$ in which $f(x)$ was increasing, the point used to obtain the upper RS sum was the right hand end point, and in the portion in which $f(x)$ was decreasing we used the left hand end point. The situation was reversed for the case of the lower RS sum.

In general, if you are asked to evaluate an RS sum, you will need a partition of the given interval and also an evaluation set appropriate to the partition. If you are asked for an upper RS sum or a lower RS sum, you will need a partition, but you will also need to stop and consider the functions involved in order to assure yourself that you have added as much as possible in each of the summands of the summation, or you have added as little as possible. Remember that you are working with real numbers and negative numbers are smaller than positive numbers. A sketch of the functions involved may prove to be helpful, and may guide you into doing some algebra that will corroborate your guesses, just as we did in Example 3.4. Problems of the kind we have just considered do not lend themselves well to formulas or to a fixed set of rules which can be followed blindly. On the other hand, you will not find these problems difficult *if* you will stop at each step long enough to think through which values should be used in order to obtain the upper sum or the lower sum, as required.

EXERCISES

1. Let $P[2, 5] = \{2, 3, 4, 4.9, 5\}$, $f(x) = x^2$, and $g(x) = x^3$.

   (a) Evaluate $S(P, T, f, g)$ if the points of T are the midpoints of the subintervals determined by P. [That is $t_k = (x_{k-1} + x_k)/2$.]
   (b) Evaluate $S(P, T, f, g)$ if the points of T are left-hand endpoints of the subintervals determined by P. [That is $t_k = x_{k-1}$.]
   (c) Evaluate $S(P, T, f, g)$ if the points of T are the right-hand endpoints of the subintervals determined by P.
   (d) Evaluate $U(P, f, g)$.
   (e) Evaluate $L(P, f, g)$.
   (f) Determine whether the answers to parts (a), (b), and (c) are in the interval determined by the answers to parts (d) and (e).

2. Write down a partition of $[1, 4]$ having five points. If $f(x) = 1/x$ and $g(x) = x^2$, give answers to each of the six parts of Exercise 1.

3. Find a refinement, P^*, of your partition P of Exercise 2. Using the functions given in Exercise 2 find $U(P^*, f, g)$ and $L(P^*, f, g)$. Is the range for the possible RS sums larger using your refinement than for the original partition or is it smaller? Can you indicate why you think the change has gone in the direction you have observed? Does this agree with your intuition (if any) on this subject?

4. Compute $S(P, T, x^3 - 3x, x^2 + 1)$ if P is a regular partition of $[-2, 3]$ having 6 points and if T is the set of midpoints of the subintervals determined by P.

5. (a) Evaluate $U(P, x^2 - x, x^2)$ if P is a regular partition of $[-1, 2]$ having 4 points.
   (b) Evaluate $U(P, x^2 - x, x^2)$ if P is a regular partition of $[-1, 2]$ having 7 points.
   (c) Is the partition of part (b) a refinement of the partition of part (a)? Can you give a logical reason why your answer to the previous question should have been expected? (It is not sufficient to merely state that this is the way the results turned out.)

6. Repeat Exercise 5 using lower sums instead of upper sums.

7. Write down a partition $P_1[-1, 2]$ and a partition $P_2[2, 4]$, each having three points. Write down evaluation sets T_1 and T_2 corresponding to the two partitions. Let $f(x) = \sin(\pi x/6)$ and $g(x) = (x + 1)/2$.

   (a) Evaluate $S(P_1, T_1, f, g)$.
   (b) Evaluate $S(P_2, T_2, f, g)$.
   (c) Show that the partition $P = P_1 \cup P_2$ and the evaluation set $T = T_1 \cup T_2$ are partitions and evaluation sets for the interval $[-1, 4] = [-1, 2] \cup [2, 4]$.
   (d) Show that the sum of the answers in parts (a) and (b) is equal to $S(P, T, f, g)$.

8. Let $S_1 = S(P_1, T_1, f, g)$ be an RS sum over the interval $[a, b]$ and let $S_2 = S(P_2, T_2, f, g)$ be an RS sum over the interval $[b, c]$. Show that $S = S_1 + S_2$ is the RS sum $S(P, T, f, g)$ over the interval $[a, c]$ where $P = P_1 \cup P_2$ and $T = T_1 \cup T_2$.

9. (a) Using the information given in Example 3.4, find a refinement of the given partition such that $f(x)$ is monotonic in each subinterval of the refinement.
   (b) Evaluate the upper sum using this refinement of the partition.
   (c) Evaluate the lower sum using this refinement of the partition.
   (d) In what way does the evaluation of the upper and lower sums become easier if one uses a partition such that $f(x)$ is monotonic on each subinterval?

10. (a) Find a partition of $[-3, 4]$ such that $f(x) = \sin(\pi x/3)$ is monotonic on each subinterval.
    (b) If $g(x) = 1/(x^2 + 1)$ find the $U(P, f, g)$ and $L(P, f, g)$ if P is your partition of part (a) and $f(x)$ is the function of part (a).
    (c) Would it ease the computation if your partition included any additional points required to insure that $g(x)$ be monotonic on each subinterval?

11. (a) Let $f(x)$ be a constant function. Show that $S(P[a, b], T, f, g)$ is not dependent upon the choice of partition or evaluation set. In other words, all RS sums $S(P[a, b], T, f, g)$ will depend for their value only on the particular constant function $f(x)$, upon the function $g(x)$, and upon the interval $[a, b]$.
    (b) Let $g(x)$ be a constant function. Show that *all* RS sums $S(P, T, f, g)$ have the same value regardless of the choice of interval, partition, evaluation set, or function $f(x)$. What is this common value?

12. Sketch the graph of $f(x) = x^2 - x + 2$ over the interval $[-1, 3]$. Find a partition $P[-1, 3]$ having four points, and indicate these points on the x-axis of your sketch. Draw vertical lines through these points.

    (a) Set up an RS sum which approximates the area bounded by $f(x)$, the vertical lines $x = -1$ and $x = 3$, and the x-axis, and evaluate the sum.
    (b) Evaluate the upper sum and show that this is an upper bound for the area of part (a).
    (c) Evaluate the lower sum and show that this is a lower bound for the area of part (a).
    (d) Find the average of your answers in parts (b) and (c) and determine the maximum amount by which this average area could differ from the correct area. [*Hint*: The upper and lower sums represent extreme values for the area.]

13. Using the method suggested in Section II.2 and again in Exercise 12, find bounds for the area bounded by $y = \sqrt{x + 4}$, $x = -3$, $x = 0$, and the x-axis.

14. Using the method suggested in Example 2.3, find an approximation for the area within the curve $f(\theta) = r = 4 \sin 3\theta$, and between the radii $\theta = 0$ and $\theta = \pi/3$. Also find an upper bound for this area and a lower bound for this area.

15. Using the method of Exercise 14 find bounds for the area inside the smaller of the two loops of $f(\theta) - 2 - 4 \cos \theta$. [Hint: Sketch the curve and note the values of θ at which the smaller loop begins and at which it ends.]

16. (a) Sketch the circle $x^2 + y^2 = 16$.
    (b) Sketch in a sphere obtained by rotating this circle of part (a) about the x-axis.
    (c) Partition the interval [4, 4] using 9 points.
    (d) Sketch cross sections of the sphere of part (b) formed by plane cross sections passing through the partition points of part (c) such that the plane sections are perpendicular to the x-axis.
    (e) Replace each segment of the sphere formed by consecutive cross sections with a right circular cylinder whose radius is a radius of a cross section of the segment of the sphere at some point between the two sides of the segment.
    (f) Add up the volumes of the 8 cylinders of part (e) and show that you have an RS sum which approximate the volume of the sphere.
    (g) Find an upper and lower bound for the volume of the sphere using the partition of part (c).

C17. Repeat Exercise 16 using a regular partition of $[-4, 4]$ with n subintervals and find upper and lower bounds for each computation for $n = 25$, $n = 50$, $n = 75$, and $n = 100$. Also compute these bounds for $n = 250$, $n = 500$, $n = 750$, and $n = 1000$. Note whether the upper and lower bounds are approaching each other as the number of subintervals is increased.

C18. Repeat Exercise 12 using a regular partition having 100 subintervals.

C19. Repeat Exercise 14 using a regular partition having 100 subintervals.

M20. (a) A function $g(x)$ is said to satisfy the *Lipschitz condition* over an interval $[a, b]$ if there is a positive constant K such that for any subinterval $[x_{k-1}, x_k]$ of the given interval it is true that $|g(x_k) - g(x_{k-1})| \le K|x_k - x_{k-1}|$. Show that if P is a regular partition of $[a, b]$ and if both $f(x)$ and $g(x)$ are increasing functions satisfying the Lipschitz condition using the constant K, then

$$U(P, f, g) - L(P, f, g) \le \sum_{k=1}^{n} [f(x_k) - f(x_{k-1})][g(x_k) - g(x_{k-1})]$$

$$\le nK^2(x_k - x_{k-1})^2 = nK^2\left[\frac{b - a}{n}\right]^2 = \frac{K^2(b - a)^2}{n}.$$

(b) Show that under the conditions given in part (a) it would be possible to select a regular partition of $[a, b]$ such that the difference between the upper and lower sums can be made smaller than any given positive number.

(c) Show that under the conditions given in part (a) it is possible to find a regular partition of $[a, b]$ such that all RS sums have essentially the same value. By this we mean that all RS sums for the given partition would be within some arbitrarily small constant, such as one-millionth, of each other.

II.4 RS Sums and Refinements

The results of the last section, specifically those results summarized in (II.3.1), provide us with a means for obtaining an upper and a lower bound for all RS sums having the same functions and the same partition. It is clear that if the difference $[U(P, f, g) - L(P, f, g)]$ is small we could use any one of the RS sums and we would not be overly concerned with the particular choice of the evaluation set. On the other hand, the work that we have done thus far would indicate that if this difference is large we should consider refining the partition, for then this difference would probably become somewhat smaller. It therefore behooves us to investigate whether this is, in fact, correct. Consequently we shall investigate what results refinement of a partition would have on the corresponding RS sums, in particular on the upper and lower sums which serve to determine the range in which the other RS sums will fall.

Before we proceed further, there are two items that will simplify our development, and in the interests of simplicity we will look at these now. If we have a partition, $P[a, b]$, it is possible to obtain any refinement $P^*[a, b]$ by going through a succession of steps. each step including one more point in the partition. Thus we would have a sequence of refinements, each having an additional point, starting with the original partition and terminating with the refinement that we would ultimately like to have. Each one of these steps will be called a *one-refinement*. If $P^*[a, b]$ is a one-refinement of $P[a, b]$, P^* has all of the points of P and one additional point.

The second item that we wish to consider before proceeding concerns $g(x)$. While it is not essential for the definition of an RS sum that $g(x)$ be monotonic over the interval $[a, b]$, it will be convenient for our work in this Section if $g(x)$ is at least monotonic over each subinterval $[x_{k-1}, x_k]$. This is not difficult to achieve for unless $g(x)$ changes direction an infinite number of times in this subinterval we only need to include the points at which there is a change of direction in order that our partition will have this property. Therefore, we are not asking for too much when we ask that $g(x)$ be monotonic in each subinterval.

EXAMPLE 4.1. If $P[1, 4] = \{1, 3, 4\}$, and if $P^*[1, 4] = \{1, 2, 3, 4\}$ is a one-refinement of P formed by including the additional point $x = 2$, show that $U(P^*, x^2 - x, x^3) \le U(P, x^2 - x, x^3)$.

Solution. We observe that P has the subintervals $[1, 3]$ and $[3, 4]$ and that P^* has the subintervals $[1, 2]$, $[2, 3]$, and $[3, 4]$. They have the subinterval $[3, 4]$ in common. Now each of these subintervals will give rise to a term in the upper sum for each of the partitions. The term corresponding to $[3, 4]$ will appear in each of the sums, and therefore we can expect no difference in the value of the two upper sums with reference to this subinterval. The remaining part of $U(P, f, g)$ is given by $f(3)[g(3) - g(1)]$, since no other choice of evaluation point within this subinterval will make this term larger. On the other hand, the remainder of $U(P^*, f, g)$ will consist of two terms and these will be $f(2)[g(2) - g(1)] + f(3)[g(3) - g(2)]$. Thus, we need to compare the values of these two remainders (after omitting the term for the common interval $[3, 4]$). We see that $f(2) < f(3)$, and also that $[g(2) - g(1)]$ is positive. Consequently we have $f(2)[g(2) - g(1)] < f(3)[g(2) - g(1)]$. From this we have

$$f(2)[g(2) - g(1)] + f(3)[g(3) - g(2)]$$
$$< f(3)[g(2) - g(1)] + f(3)[g(3) - g(2)]$$
$$< f(3)[g(2) - g(1) + g(3) - g(2)]$$
$$< f(3)[g(3) - g(1)].$$

Therefore, the terms remaining after ignoring the common term have the relation indicated, that is the sum of the remaining terms of the upper sum using the refinement is less than the remaining term of the original partition. This proves the inequality $U(P^*, f, g) < U(P, f, g)$ to be correct.

Of course we could have been more direct and put in the numeric values at once. We will now proceed to do just that, but be certain to observe that we will be following similar steps and vindicating all of the statements that we have made so far. Using the values of the functions, we have

$$U(P, f, g) = f(3)[g(3) - g(1)] + f(4)[g(4) - g(3)]$$
$$= (6)[27 - 1] + (12)[64 - 27] = (6)(26) + (12)(37) = 600.$$

$$U(P^*, g, f) = f(2)[g(2) - g(1)] + f(3)[g(3) - g(2)] + f(4)[g(4) - g(3)]$$
$$= (2)[8 - 1] + (6)[27 - 8] + (12)[64 - 27]$$
$$= (2)(7) + (6)(19) + (12)(37) = 572.$$

While it is easy to observe that $572 < 600$, we should also pause to note that

$$(2)[8 - 1] + (6)[27 - 8] < (6)[8 - 1] + (6)[27 - 8]$$
$$= (6)[8 - 1 + 27 - 8] = (6)[27 - 1].$$

This latter series of inequalities and equalities parallels our earlier discussion. Be certain that you see how the two expressions are related.

In a manner similar to that used in this solution we could prove

$$L(P^*, x^2 - x, x^3) > L(P, x^2 - x, x^3)$$

using the partitions we have been discussing. Since the lower sums must be no greater than the upper sums provided we are using the same partitions for each, we could write $L(P, f, g) \leq L(P^*, f, g) \leq U(P^*, f, g) \leq U(P, f, g)$.

The result which we have just written as a result of our consideration in Example 4.1 is much more general than might appear at first glance. It is true that we may not always have the strict inequality, but if we were to replace the $(<)$ with $(\leq)$ the result would hold for all refinements provided, of course, the functions $f(x)$ and $g(x)$ were the same in the four summations involved (that is the two lower sums and the two upper sums). It is true that we have put the restriction on $g(x)$ that it would be monotonic within each subinterval. We will retain this restriction, but this is the only restriction we shall impose. In order to give this the importance it deserves, we will state this general result as a Theorem. (Note that $g(x)$ was monotonic over $[1, 4]$ in Example 4.1.)

Theorem 4.1. *If $f(x)$ and $g(x)$ are bounded functions over the interval $[a, b]$ and if $g(x)$ is monotonic over each subinterval of the partition $P[a, b]$ and $P^*[a, b]$ is a one-refinement of P, then $U(P^*, f, g) \le U(P, f, g)$.*

PROOF. Since P^* is a one-refinement of P, there is one additional point, x^*, in P^* that is not in P. Now suppose that x^* is in the particular subinterval $[x_{j-1}, x_j]$ so that $x_{j-1} < x^* < x_j$. We can write

$$U(P, f, g) = \sum_{k=1}^{j-1} f(t_k)[g(x_k) - g(x_{k-1})] + f(t_j)[g(x_j) - g(x_{j-1})]$$

$$+ \sum_{k=j+1}^{n} f(t_k)[g(x_k) - g(x_{k-1})]. \tag{II.4.1}$$

The first and third parts of the right hand side of (II.4.1) will not be changed by the one-refinement, and hence the first part will appear as the first part of $U(P^*, g, f)$ and the third part of (II.4.1) will appear as the final part of $U(P^*, f, g)$. However, the single term which constitutes the second part of (II.4.1) as it is written will be divided and will consist of two terms in $U(P^*, f, g)$. Since $g(x)$ is monotonic in the subinterval $[x_{j-1}, x_j]$, we might consider the case in which g is an increasing function and think of a sketch as shown in Figure II.6. (If $g(x)$ were decreasing we could use similar reasoning.) The value t_j must occur in one of the two subintervals $[x_{j-1}, x^*]$ and $[x^*, x_j]$ and it might occur at x^*. If t_j is in the second subinterval we will need an evaluation point in the first subinterval. Let us denote this evaluation

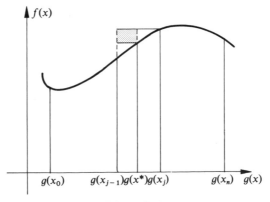

Figure II.6

point by t^*. Remember that these evaluation points are picked in such a way that we will obtain the upper sum over the refined partition. However, we know that $f(t_j) \geq f(t^*)$ if $g(x)$ is increasing for we have $f(t_j)[g(x_j) - g(x_{j-1})] \geq f(t^*)[g(x_j) - g(x_{j-1})]$ as required for the upper sum over the original partition. It follows that

$$f(t^*)[g(x^*) - g(x_{j-1})] + f(t_j)[g(x_j) - g(x^*)]$$
$$\leq f(t_j)[g(x^*) - g(x_{j-1})] + f(t_j)[g(x_j) - g(x^*)]$$
$$\leq f(t_j)[g(x^*) - g(x_{j-1}) + g(x_j) - g(x^*)]$$
$$\leq f(t_j)[g(x_j) - g(x_{j-1})].$$

This follows in precisely the same manner as did the argument in Example 4.1. Note that we can see each of these terms of the summand as an area in the graph of Figure II.6 in which the horizontal axis is the $g(x)$ axis and the vertical axis is the $f(x)$ axis. The shaded area represents that portion which is in the upper sum of the original partition but is missing from the upper sum of the one-refinement.

This essentially completes the proof, for the terms we have considered preceded by the first summation of (II.4.1) and followed by the last summation of the same expression will give the upper sum of the refinement if we include the two terms on the left of the above inequality, and will give the upper sum of the original partition if we use the single expression which appears as the end of the string of inequalities above. □

Corollary 4.1. *If P^* is any refinement of $P[a, b]$ and if $f(x)$ and $g(x)$ are bounded functions over $[a, b]$ and $g(x)$ is monotonic in each subinterval of P, then $U(P^*, f, g) \leq U(P, f, g)$.*

PROOF. Each time a point is added to P there is a one-refinement and Theorem 4.1 applies. Consequently the upper sum of the successive refinements cannot increase at any point. Since $g(x)$ is monotonic over each subinterval of P, it follows that it must be monotonic over each subinterval of the succession of refinements required to reach P^*. □

Theorem 4.2. *If $f(x)$ and $g(x)$ are bounded functions over the interval $[a, b]$ and if $g(x)$ is monotonic over each subinterval of the partition $P[a, b]$, and if $P^*[a, b]$ is a one-refinement of P, then $L(P^*, f, g) \geq L(P, f, g)$.*

PROOF. This proof is very similar to the proof of Theorem 4.1. □

Corollary 4.2. *If P^* is any refinement of $P[a, b]$ and if $f(x)$ and $g(x)$ are bounded functions over $[a, b]$ and $g(x)$ is monotonic in each subinterval of P, then $L(P^*, f, g) \geq L(P, f, g)$.*

Corollary 4.3. *If P^* is any refinement of $P[a, b]$ and if $f(x)$ and $g(x)$ are bounded functions over $[a, b]$, and if further $g(x)$ is monotonic in each subinterval of P, then $L(P, f, g) \leq L(P^*, f, g) \leq U(P^*, f, g) \leq U(P, f, g)$.*

We have used the requirement that $g(x)$ be monotonic in each subinterval of a given partition before the various theorems and corollaries will apply. As indicated in the beginning of this section, it is always possible to find such a partition provided $g(x)$ changes from increasing to decreasing and from decreasing to increasing only a finite number of times in the interval $[a, b]$. It is seldom that we will have to deal with a more complicated function. A somewhat more general result can be obtained, but we will not worry about that in this book. In fact, the situations which usually arise will have $g(x)$ monotonic throughout the interval or changing direction only once or twice. However, to show that this condition is necessary, let us consider the following Example.

EXAMPLE 4.2. Find the upper and lower sums if $f(x) = x$ and $g(x) = \sin(\pi x/2)$, and if $P[0, 4] = \{0, 2, 4\}$. Also show that there exist refinements of P for which the upper sum is larger and the lower sum is smaller than the corresponding upper and lower sum obtained using the partition, P.

Solution. Note that $g(x)$ is not monotonic in the subintervals $[0, 2]$ and $[2, 4]$. In fact we have $g(0) = g(2) = g(4) = 0$, and therefore all RS sums using the partition P have the value zero. In particular the upper and lower sums have the value zero.

If we let $P^*[0, 4] = \{0, 1, 2, 3, 4\}$, we then have the upper sum given by

$$U(P^*, f, g) = f(1)[g(1) - g(0)] + f(1)[g(2) - g(1)]$$
$$+ f(2)[g(3) - g(2)] + f(4)[g(4) - g(3)]$$
$$= (1)[1 - 0] + (1)[0 - 1] + (2)[(-1) - 0] + (4)[0 - (-1)]$$
$$= 1 + (-1) + (-2) + 4 = 2$$

and this is larger than $U(P, f, g)$. Similarly,

$$L(P^*, f, g) = f(0)[g(1) - g(0)] + f(2)[g(2) - g(1)]$$
$$+ f(3)[g(3) - g(2)] + f(3)[g(4) - g(3)]$$
$$= (0)[1 - 0] + (2)[0 - 1] + (3)[(-1) - 0] + (3)[0 - (-1)]$$
$$= 0 + (-2) + (-3) + 3 = -2$$

and this is smaller than $L(P, f, g)$.

From Example 4.2 it is apparent that we must be on the alert to insure that the partition we are using fulfils the necessary conditions if we wish to make use of the results of this section. We will assume in our discussions from this point on that we are dealing with partitions which meet these conditions, although in particular cases we will need to check to insure that the conditions are satisfied. Thus, we will assume in the remainder of this section that we are using only partitions which fulfil the condition that $g(x)$ is monotonic in any subinterval of the partition.

We wish to pause and take stock of the information that we have now acquired. All lower sums are bounded above by any upper sum, and con-

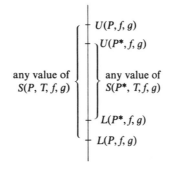

Figure II.7

versely all upper sums are bounded below by any lower sum. Furthermore, successive refinements of any given partition will cause the lower sums to either remain constant or to increase in value, whereas the upper sums will remain constant or decrease in value. Successive refinements then provide a sequence of numbers for the lower sums which is monotonic increasing, but bounded above, and also a sequence of numbers corresponding to the upper sums which is monotonic decreasing, but bounded below. If these upper and lower sums are not constant, the difference between the successive upper and lower sums will become smaller. This is illustrated schematically in Figure II.7. Since we could start with any number of different partitions, the question might arise whether the final results might depend upon the partition with which we started. However, since any finite number of partitions would have a common refinement, we need only include that refinement as one of the partitions in our succession of refinements to completely eliminate the influence of the choice of the starting partition.

We can summarize much of what has been said in this Section by giving an Example involving an application.

EXAMPLE 4.3. We are asked to find the total mass of a flywheel, and as is the case with many flywheels the wheel has been cast so that there is more weight near the rim than in the center. This particular flywheel has a radius of 10 centimeters, and varies in thickness so that the mass of a section r centimeters from the center is given by the relation $c(1 + r)$ grams per square centimeter. Find upper and lower bounds for this mass.

Solution. We have drawn a sketch of our wheel in Figure II.8. We have also shown a partition of the radius and have drawn in some concentric circles with centers at the center of the wheel and with radii equal to the distance from the center to partition points on the radius. Note that each ring so constructed consists of points all of which are at approximately the same distance from the center of the wheel, and therefore points for which we can use a common expression for the density per square centimeter. This permits us to use the expression $c(1 + r)$ for a given value of r and apply

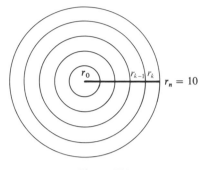

Figure II.8

it to all of the area with approximately that value of r, and hence with that density. Now the area of such a ring is given by the expression $[\pi r_k^2 - \pi r_{k-1}^2]$, since the area of the ring is the area that is inside the outer circle (of radius r_k) but outside the inner circle (of radius r_{k-1}). In the expression $c(1 + r)$, we could use any value of r in the interval $[r_{k-1}, r_k]$ as an evaluation point. However, if we wish to have upper and lower bounds for this mass we would use r_k for the upper bound, for this would give the greatest possible mass, and we would use r_{k-1} for the lower bound, for this would give us the smallest possible mass. Thus, we know that an upper bound for this mass would be given by $c(1 + r_k)[\pi r_k^2 - \pi r_{k-1}^2]$ and the lower bound for the mass of this ring would be given by $c(1 + r_{k-1})[\pi r_k^2 - \pi r_{k-1}^2]$. However, these values are only for the single ring in question. If we wish to have the bounds for the mass of the entire wheel, we could add up the lower bounds of the individual rings to obtain the lower bound of the mass of the wheel, and similarly we would add the upper bounds of the individual ring masses to obtain the upper bound for the mass of the wheel. Thus we have for the upper and lower bounds respectively

$$\sum_{k=1}^{n} c(1 + r_k)[\pi r_k^2 - \pi r_{k-1}^2]$$

and

$$\sum_{k=1}^{n} c(1 + r_{k-1})[\pi r_k^2 - \pi r_{k-1}^2].$$

It is now apparent that we have upper and lower RS sums, for these are merely $U(P, c(1 + r), \pi r^2)$ and $L(P, c(1 + r), \pi r^2)$. As so often happens, we did not start out looking for an RS sum, but through careful analysis of the problem we were able to write out an expression that would give us bounds for our answer and the bounds were in fact RS sums.

We could now take a specific partition of the interval $[0, 10]$ and do the necessary arithmetic to obtain upper and lower bounds for the mass of this wheel. However, we can save ourselves some arithmetic if we are willing to invest a bit of algebra. Let us take a regular partition of $[0, 10]$ having n

subintervals. In this case each subinterval will be of length $10/n$, and since $r_0 = 0$, we will have $r_1 = 1(10/n)$, $r_2 = 2(10/n)$, $r_3 = 3(10/n)$, and so on until we finally arrive at $r_n = n(10/n) = 10$. This is just the value that we should have for r_n. With a little thought we see that $r_k = k(10/n) = 10k/n$ and $r_{k-1} = (k-1)(10/n)$. Using these values our upper sum becomes

$$U(P, c(1+r), \pi r^2) = \sum_{k=1}^{n} c\left(1 + \frac{10k}{n}\right)\left[\pi\left(\frac{10k}{n}\right)^2 - \pi(k-1)^2\left(\frac{10}{n}\right)^2\right]$$

$$= \sum_{k=1}^{n} c\left(1 + \frac{10k}{n}\right)\pi\left(\frac{10}{n}\right)^2[k^2 - (k-1)^2]$$

$$= c\pi\left(\frac{100}{n^2}\right)\sum_{k=1}^{n}\left(1 + \frac{10k}{n}\right)[2k - 1]$$

$$= c\pi\left(\frac{100}{n^2}\right)\sum_{k=1}^{n}\left[\left(\frac{20}{n}\right)k^2 + \left(2 - \frac{10}{n}\right)k - 1\right]$$

$$= c\pi\left(\frac{100}{n^2}\right)\left[\left(\frac{20}{n}\right)\sum_{k=1}^{n}k^2 + \left(2 - \frac{10}{n}\right)\sum_{k=1}^{n}k - \sum_{k=1}^{n}1\right].$$

We have made liberal use of the theorems concerning summation that we derived in Section I.10. We have also made use of the fact that not only 10, but also n is a constant here, for we had to choose n before we had our partition in the first place. The three summations in the last line of our expression were evaluated in Section I.11 and we can substitute these values for the sums. Doing this, we get

$$U(P, c(1+r), \pi r^2) = c\pi\left(\frac{100}{n^2}\right)\left[\left(\frac{20}{n}\right)\frac{n(n+1)(2n+1)}{6}\right.$$

$$\left. + \left(2 - \frac{10}{n}\right)\left(\frac{n^2 + n}{2}\right) - n\right]$$

$$= c\pi\left(\frac{100}{n^2}\right)\left[\frac{20(2n^2 + 3n + 1)}{6} + n^2 + n - 5n - 5 - n\right]$$

$$= c\pi\left(\frac{100}{n^2}\right)\left[\frac{23n^2}{3} + 5n - \frac{5}{3}\right]$$

$$= \frac{2300\pi c}{3} + \frac{500\pi c}{n} - \frac{500\pi c}{3n^2}.$$

We now see that if $n = 1$, the upper bound would be $1100\pi c$. On the other hand if $n = 500$, the upper sum would be about $767.666\pi c$. If n were $1,000,000$ we would have the upper bound very close to $766.6667\pi c$. As the partition is refined the upper sum apparently does decrease, just as our theorems had predicted.

If we go through the same algebraic manipulation to find the lower sum, we would find that

$$L(P, c(1 + r), \pi r^2) = \frac{2300\pi c}{3} - \frac{500\pi c}{n} - \frac{500\pi c}{3n^2}.$$

This is obviously smaller, and subtraction tells us that the upper sum minus the lower sum is actually $(1000\pi/n)$. For very large values of n it is apparent that the upper and lower sums are very close together and we can find an approximation for the mass of the wheel which is as close to the actual mass as we may like.

We might take this solution one step further by noting that the number $2300\pi c/3$ is always between the upper and lower bounds, no matter how close these bounds are to each other, and hence it would seem reasonable to assume that $2300\pi c/3$ grams is the mass of the wheel.

This example illustrates several things. When we wish to evaluate some quantity it is frequently easier to think of taking small portions and finding the appropriate evaluation for each of these portions. We can then add the results to obtain the approximation for the entire quantity. If we give some thought to the problem at hand, there is frequently a method of determining the portions in such a way that the approximations are not difficult. This is apparent in our use of the concentric circles to obtain the rings in Example 4.3, for if the rings are narrow the density is nearly constant for each ring. Usually such a process will lead us to an RS sum. Notice that we did not start out to determine what should be the function $f(r)$ and what should be $g(r)$ in the last example, but rather we used common sense to set the problem up and the choice of f and g became obvious from the form of the solution. We also notice that by refining the partition, in this case making n greater, we were able to make the difference between the lower and upper bounds smaller, and hence obtain an ever better approximation. Finally, in each of these examples the function which played the role of $g(r)$, whether we knew in advance what the function would be or not, ultimately turned out to be monotonic in each subinterval. In the last case we observe that $g(r)$ was monotonic over the entire interval which certainly gives us the necessary assurance that we have monotonicity over the subintervals.

Exercises

Note. The adjective *marginal* is used in many places in economics. Consequently, you can expect to see this term in some of the Exercises which relate to the Social Sciences. While an alternate definition will be given in Chapter IV, we can assume for the present that *marginal cost at level n* would be the cost of increasing production from a level of n items to a level of $(n + 1)$ items. In other words, it is the cost of producing the $(n + 1)$-th item. Similarly, marginal revenue at level n would be the revenue from the

$(n + 1)$-th item. If the words *at level n* are omitted, it is understood that these words are implied.

1. Using the partition $P[-1, 3] = \{-1, -0.5, 2, 3\}$ determine the values of $U(P, x^2, x^3)$ and $L(P, x^2, x^3)$. Let P^* be a refinement of P obtained by including the points 0 and 1. Determine the upper and lower sums using P^*. Verify that Corollary 4.3 applies in this case. (Be sure to check that $g(x)$ fulfils the requirements of Corollary 4.3.)

2. In the interval $[-2, 2]$ $g(x) = x^3 - 3x$ has maximum values at $x = -1$ and $x = 2$ and $g(x)$ has minimum values at $x = -2$ and $x = 1$. Find a partition containing at least five points which will meet the conditions requisite for applying results of this section. Using this partition evaluate $U(P, x, x^3 - 3x)$ and $L(P, x, x^3 - 3x)$. Find a one-refinement of P and evaluate the upper and lower sums for the one-refinement. Verify that the results of this section hold in this case.

3. Follow the instructions of Exercise 2 after replacing $f(x) = x$ with $f(x) = x^2$.

4. Let $f(x)$ be a function defined over the interval $[0, 2]$ such that $f(x) = 1$ if x is a rational number and $f(x) = 0$ if x is an irrational number. Let $P[0, 2]$ be any partition of $[0, 2]$.

   (a) What is the largest value of $f(x)$ in each subinterval of P? [*Hint*: Since each subinterval has a non-zero length it must contain both rational and irrational points.]
   (b) What is the smallest value of $f(x)$ in each subinterval of P?
   (c) For your partition find the value of $U(P, f(x), 2 - x^2)$.
   (d) For your partition find the value of $L(P, f(x), 2 - x^2)$.
   (e) Does your result in parts (c) and (d) depend upon the partition?
   (f) In this case would there be any refinement such that the upper and lower sums would approach each other?

   While this may seem a somewhat bizarre example, it is unfortunately true that cases such as this do arise occasionally. Therefore, there will be some exercises of this type from time to time to illustrate the breadth of functions that can be handled with the information we are acquiring.

5. Let $g(x) = [x]$, the greatest integer function.

   (a) Show that $g(x)$ is an increasing function.
   (b) Find a partition $P[0, 3]$ having at least five points.
   (c) Evaluate $U(P, x^2, g(x))$ and $L(P, x^2, g(x))$.
   (d) Let P^* be a refinement of P having at least two additional points. Evaluate $U(P^*, x^2, g(x))$ and $L(P^*, x^2, g(x))$.
   (e) Does Corollary 4.3 apply in this case?
   (f) What is the maximum number of subintervals having a non-zero contribution to any RS sum in this case?

6. Let $P[1, 5]$ be a regular partition having three points. Let $f(x) = x^2$ and $g(x) = 1/x$.

   (a) Show that $g(x)$ is monotonic in this interval.

(b) Find the average of the upper and lower RS sums over this interval using the partition P.

(c) Find the RS sum $S(P, T, f, g)$ if the points of T are the midpoints of the sub-intervals of P. Compare this result with the result of part (b).

(d) Repeat parts (b) and (c) if P is a regular partition with five points.

(e) Repeat parts (b) and (c) if P is a regular partition with nine points.

(f) Show whether the conclusions of Corollary 4.3 apply with these functions and the given partitions.

7. (a) Show that if either $f(x)$ or $g(x)$ is constant, the upper sums are equal to the lower sums.

(b) Show that if $g(x)$ is a constant, the value does not depend upon either the interval involved or upon the function $f(x)$.

(c) If $g(x) = g_1(x) + g_2(x)$ is the sum of two functions, each being monotonic increasing over the interval $[a, b]$, show that

$$S(P, T, f, g) = S(P, T, f, g_1) + S(P, T, f, g_2)$$

for all RS sums, and hence for the upper and lower sums.

(d) If the functions of part (c) were both monotonic decreasing, show that the same result would hold.

8. Use the methods suggested in Section II.2 to set up an RS sum approximating the area bounded by the curve $y = 4 - x^2$ and the x-axis. Note that there is only one area whose boundaries are completely described in the preceding sentence. Use upper and lower sums to determine bounds for the actual area. How much error could you possibly have if you were to use the average of the upper and lower sums to represent the area? Refine your partition and notice whether you have reduced the margin of error.

9. Use the methods suggested in Section II.2 to set up an RS sum approximating the area inside the cardioid $r = 2 + 2\cos\theta$. Use upper and lower sums to find bounds for the actual area.

10. (a) Draw a sketch of a circular floor, 5 feet in radius.

(b) The weight on this floor is given by $\sqrt{r}$ pounds per square foot at a distance of r feet from the center. Draw in two circles with approximately equal radii and find the approximate weight per square foot of the flooring within the ring between the two circles.

(c) Find an expression for the number of square feet in the ring between the two circles.

(d) Find the approximate weight on the floor in this ring.

(e) By adding up the weights in the separate rings from the inside out to the edge of the floor, obtain an RS sum which approximates the total weight on the floor.

(f) Find upper and lower sums which give bounds for the weight on the floor.

(g) Refine the partition of the radius (put in more circles) and obtain bounds which are closer together.

11. Using the method suggested in Exercise 16 of Section II.3 find an upper and lower bound for the volume of a sphere of radius 5. Compute the value of the RS sum using the partition used for the preceding part of this problem, and using an evaluation set formed by using the midpoints of each subinterval. Is this result near the middle of the range formed by the upper and lower sums?

C12. Using a regular partition of $[0, 5]$ evaluate the upper sum, the RS sum using the midpoints of the subintervals for the evaluation points, and the lower sum for $10, 20, 30, \ldots, 100$ subintervals if $f(x) = 1/(x^2 + 1)$ and $g(x) = x^2$. List these in tabular form and notice the interval bounded by the upper and lower bounds for each of your partitions.

C13. Evaluate $U(P[1, 3], \sin x, 1/x)$ and $L(P[1, 3], \sin x, 1/x)$ for regular partitions having 20 subintervals, and then for 40 subintervals, and continuing by using an additional 20 subintervals each time until the range of possible values for the RS sums for the partitions you are working with is less than 0.001.

S14. The marginal cost in a certain plant in daily production is given by $M(n) = 100 - 0.02n + 0.00004n^2$ dollars per additional unit.

(a) Find the cost of increasing production from 400 units per day to 401 units per day.

(b) If demand is such that you wish to increase production from 400 units per day to 600 units per day, you would probably not want to calculate 200 terms and add them up. You might find an average cost for increasing from 400 to 425 by calculating the cost of going from 412 to 413 and then multiply this by 25. If you were to do this for increments of 25 you would only need to calculate eight terms. Carry this computation out and show that you have an RS sum.

(c) Using the partition suggested in part (b), find the upper limit to the increased cost and the lower limit for the increased cost of production due to the increase in output.

(d) Using the result of part (b) find the average cost per unit of increasing production from 400 units to 600 units per day.

S15. In a certain community the population doubles in slightly less than 35 years, and the population t years after 1975 is given by

$$P(t) = 100,000 + 2,000t + 20t^2 + 0.13t^3.$$

(a) What is the increase in population between 1982 and 1992?

(b) It is desired to set aside money for the construction of schools as they will be needed. If it is assumed that each additional person in the community will require an average of $\$300$ in 1975 money for school construction, the money to be available five years after the person entered the community, and if it further assumed that inflation will necessitate $(1.05)^t$ dollars t years after 1975 for each 1975 dollar, what is the approximate total number of dollars that will be needed for school construction between 1982 and 1992. [Remember that people added to the community in 1977 will require building funds in 1982, and the amount contributed in 1982 will have to be $300 multiplied by $(1.05)^7$ for each additional person.]

(c) The school authorities would like to know the upper bound for the money that will be required and the lower bound. Find these values for the school board.

S16. The use of land for solid waste disposal is currently a controversial subject. A particular city has a population given by $P(t) = 150,000(2^{t/50})$ where t is the number of years since 1970. Experience has shown that it takes about one acre per year for each 10,000 persons. The city manager is concerned with finding out for budgeting purposes the approximate increase in land required each year over that required for the preceding year for the period from December 31, 1980 through December 31, 1990. [Note that this does not require the total amount of land required, but does require that you find out for each year how much must be added, and then add the amounts for the ten year period.] Also find an upper and a lower bound for this amount of land.

P17. Find the mass of a wheel with radius 12 inches if the density at a point r inches from the center is given by $(1 + 3r)/400$ pounds per square inch of cross sectional area. Find upper and lower sums, and then find the correct value for this mass by the method of Example 4.3.

P18. The velocity of a certain object is given by the relation $v(t) = 100t - 100t^2$ feet per minute where t is measured in minutes.

   (a) Show that the object is at rest (velocity is zero) at $t = 0$ and $t = 1$.
   (b) Using the fact that the distance traveled in any period of 0.1 minute can be approximated by taking an average velocity during this time and multiplying that average velocity by the elapsed time of 0.1 minute, obtain the approximate distance the object has traveled during the minute.
   (c) Find an upper and a lower bound for the distance traveled.

P19. You are asked to calculate the approximate force on the face of a dam in the shape of a triangle, the triangle having a horizontal top 100 feet across, and being an isosceles triangle with a distance of 50 feet from top to bottom. The water level is at the top of the dam.

   (a) Make a rather large sketch of the face of the dam, and draw in several horizontal lines.
   (b) Note that between two successive horizontal lines, if they are drawn rather close to each other, the water depth would not vary a great deal from the top to the bottom. Using the fact that the force on the dam is the pressure of $62.4x$ pounds per square foot multiplied by the area involved, calculate the force on a horizontal segment of the dam. The value of x is the depth of the water at that point in feet, and an average depth could be used for this value.
   (c) Approximate the force on the dam by adding up the force on each of the horizontal segments assuming each horizontal strip is five feet from top to bottom.
   (d) Find an upper and lower bound for the force on the dam based upon the partition of the depth that you have used in parts (b) and (c).

P20. The physicist defines work to be the product of the force involved in moving an object multiplied by the distance the object moves. The force of attraction between a particle and a certain mass is given by the relation $F(x) = 1/x^2$ where x is the distance in feet from the object to the center of the mass.

   (a) Sketch the mass and the object as the object is to move away from the mass, and indicate the location of the object when it is 1 foot, 2 feet, 3 feet, and so on to 10 feet away from the mass.
   (b) Approximate the work done in moving each of the one foot distances sketched in part (a).

(c) Find an upper and lower bound for the work done in moving the object from a point one foot from the center of mass to a point ten feet from the center of mass.

(d) Refine your partition of part (a) and repeat part (c) until the difference between the upper and lower bounds is less than 0.5 foot pounds.

B21. Ordinarily a blood vessel has a circular cross section. Let the radius of the cross section be denoted by R. It is known that blood flows through such a blood vessel with the outer portion flowing more slowly due to friction with the walls of the blood vessel. In fact, the velocity of flow is given by the relation $v(r) = c(R^2 - r^2)$ where c is a constant determined in part by the viscosity of the blood and the velocity $v(r)$ is a function of the distance, r, of the particular portion of the cross section from the center of the cross section.

(a) Sketch the cross section, partitioning the radius and showing a ring between the circles of radii r_{k-1} and r_k.

(b) Write an expression for the volume of blood that flows through that ring in a single second if v is measured in centimeters per second and r is measured in centimeters.

(c) Given that $R = 0.1$ centimeters find an upper and lower bound for the amount of blood passing a given cross section of the blood vessel in one second. Note that this result will include the factor, c.

B22. One theory concerning the excitation of the retina of the eye by incident light indicates that the contribution of each square millimeter to the total excitation, $C(r)$, is given by the relation $C(r) = r^{-k}$ where r is the distance from the center of the retina measured in millimeters and k is a positive constant not greater than two. It is also assumed that the center of the retina does not obey this law. If the radius of the retina is 1.5 millimeters, and if the circular area in the center of radius 0.1 millimeters is inactive, find upper and lower bounds for the total excitation under the assumption that $k = 2$.

II.5 The Riemann–Stieltjes Integral

In the last section we learned that if $g(x)$ is monotonic in each subinterval of $P[a, b]$ then $U(P, f, g)$ is a decreasing function of the successive refinements of P, and similarly $L(P, f, g)$ is an increasing function. It is possible within the definition of these terms, of course, that neither the upper sum nor the lower sum changes as refinements are made, and hence the two sums may or may not be approaching each other. Example 4.3 would indicate that in at least some practical cases the two sums do approach each other, and do indicate a common result for the *ultimate* refinement. In this section it is our purpose to further investigate this particular situation.

In order to proceed we will need to state clearly one additional property of real numbers.

Postulate 5.1 (The Least Upper Bound Axiom). *If S is a finite or infinite set of real numbers and if S has an upper bound, there exists a smallest real number which is an upper bound of the set S. This number is called the* least upper bound *of S.*

This postulate distinguishes the rational numbers from the real numbers, for it is not true that there would be a rational number which is a least upper bound in every instance. If, for instance, we were to consider those positive rational numbers such that their square is less than two, we would find that if we had a rational upper bound we could always find a smaller one, and hence there would be no least upper bound. Just as we have this postulate which guarantees that there will be a least upper bound for a set which has an upper bound, if a set has a lower bound, there will be a real greatest (or largest) lower bound. We will be using these terms often enough that it will be convenient to abbreviate them, and we will frequently refer to the least upper bound as the *lub* and to the greatest lower bound as the *glb*.

EXAMPLE 5.1. Let S be the set of numbers consisting of $f(n) = n/(n + 1)$ for all positive integers n. Find the *lub* and the *glb*.

Solution. This is obviously an infinite set, and it may be instructive to write out the first few numbers in the set. Thus, we would have 1/2, 2/3, 3/4, 4/5, ... and we see that the numbers appear to be getting larger. Therefore we would expect that 1/2 would be the smallest number in the set, and regardless of how far we proceeded we would never get a smaller number. That makes the glb easy, for it is clear that 1/2 is a lower bound, and it is also clear that any larger number would not be a lower bound since 1/2 is itself in the set. As we proceed with this sequence of numbers, we see that they are getting closer to 1, albeit slowly. In fact, we could write $f(n) = 1 - 1/(n + 1)$ and we see that $f(n)$ will always be less than 1, but the difference will become progressively smaller, finally approaching a difference of zero. Thus, we have 1 as an upper bound, and we see that if we were to choose a number smaller than 1, say 0.999999, at some point $f(n)$ would be larger than this number. In particular $f(1,000,000)$ is larger than 0.999999. Hence 0.999999 could not be the lub since it is not an upper bound.

It may not have been obvious, but we made use of the least upper bound axiom earlier in this chapter. When we define the upper RS sum, we referred to the largest possible RS sum with the given partition. It is not at all obvious at first glance that such a largest sum may exist. However, under our assumption that $f(x)$ is bounded, and since each difference $[g(x_k) - g(x_{k-1})]$ is a constant, we see that the least upper bound axiom does in fact guarantee that there will be a least upper bound in each of the subintervals of the partition. If $f(x)$ is so bizarre that it does not actually attain its least upper bound in the subinterval, then we will use the lub instead, and this would give us the upper sum as required by the definition.

We have repeatedly commented on the fact that with appropriate conditions the lower sums can only increase under successive refinements and the upper sums can only decrease. However, the lower sums do have an upper bound, for any upper sum will serve as such an upper bound. Therefore, the least upper bound axiom guarantees that we have a least upper bound for the lower sums. By the same token we observe that any lower sum serves as a lower bound for the upper sums, and therefore the decreasing set of upper sums must have a greatest lower bound. These two numbers, the lub of the lower sums and the glb of the upper sums, play a very important role in the calculus. We therefore do the usual thing and give them special names as indicated in the following definition.

Definition 5.1. The *lower Riemann–Stieltjes integral of $f(x)$ with respect to $g(x)$ over the interval $[a, b]$* is the least upper bound of the set of *all* lower RS sums $L(P, f, g)$ over $[a, b]$, and is denoted by the symbol

$$\underline{\int_a^b} f(x)dg(x).$$

Definition 5.2. The *upper Riemann–Stieltjes integral of $f(x)$ with respect to $g(x)$ over the interval $[a, b]$* is the greatest lower bound of the set of *all* upper RS sums $U(P, f, g)$ over $[a, b]$, and is denoted by the symbol

$$\overline{\int_a^b} f(x)dg(x).$$

If you examine the notation we have just introduced you will observe that we have replaced the Greek sigma (denoting summation) by the *Old English long S*, and we have replaced the difference $[g(x_k) - g(x_{k-1})]$ by the symbol $dg(x)$. The latter will be discussed at much greater length later on, but suffice it to state at this point that this is part of the symbolism. It should not be unreasonable to think of the d for the moment as representing the difference of the two values of $g(x)$ taken at the end points of some very small interval that occurs in some ultimate refinement of the original partition. This interpretation is not necessary, and should *not* be taken literally, but it may help to provide a rationalization at this point. This notation was developed by *Gottfried Wilhelm Leibniz* (1646–1716), one of the co-developers of the calculus. [It is interesting to note that he had first used the notation *omn. f* involving an abbreviation for the Latin word for *all* thinking of the bringing together or *integration* of the values of $f(x)$.]

We can clarify the reasoning involved here by noting the diagram of Figure II.9. This diagram indicates that if we start with any lower and upper sum the successive refinements will cause the lower sums to either stay at the same level or increase and will cause the upper sums to either stay at the same level or decrease. In this diagram we have indicated the possibility

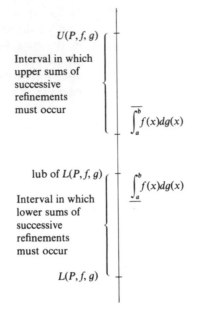

Figure II.9

that the lub of the lower sums may be less than the glb of the upper sums. It certainly could not be greater. It will frequently happen that the two have the same bounds, and in this case it becomes unnecessary to distinguish between the lower and the upper integral.

Definition 5.3. If the lower and upper RS integrals of $f(x)$ with respect to $g(x)$ over $[a, b]$ exist and if they are equal, then the common value is defined to be *the Riemann–Stieltjes integral of $f(x)$ with respect to $g(x)$ over the interval $[a, b]$*, and is denoted by the symbol

$$\int_a^b f(x)dg(x).$$

EXAMPLE 5.2. Evaluate the lower and upper sums $L(P, x, x^2)$ and $U(P, x, x^2)$ over the interval $[0, 2]$ and then find the values of the lower and upper RS integrals

$$\underline{\int_0^2} x\, dx^2 \quad \text{and} \quad \overline{\int_0^2} x\, dx^2.$$

Solution. Since we ultimately need the lub of *all* of the lower sums we might as well start with a partition that is easy to use. Our reasoning here stems from the fact that we will need to assure ourselves that we have an upper bound that cannot be exceeded by *any* lower sum, and therefore we would have to assure ourselves that we have effectively considered all possible partitions of $[0, 2]$. This means that we must use both the simple and the more complex partitions. However, it is more pleasant to contemplate starting with the easy partitions and then trying to determine whether we can use these to give us the information we need without having to use the more complicated ones. Probably the easiest one to use is the regular partition, and in this case we would have each subinterval of length $(2/n)$.

Further analysis in this case indicates that both x and x^2 are increasing functions over the entire interval $[0, 2]$, and therefore all of our theorems will hold. Furthermore, $[x_k^2 - x_{k-1}^2]$ will be positive in every case, and as a consequence we will use $x_k[x_k^2 - x_{k-1}^2]$ as the term in our summation for the upper sum and will use $x_{k-1}[x_k^2 - x_{k-1}^2]$ as the term in our summation for the lower sum. Hence, the upper sum will be

$$U(P, x, x^2) = \sum_{k=1}^{n} x_k[x_k^2 - x_{k-1}^2]$$

and the lower sum will be

$$L(P, x, x^2) = \sum_{k=1}^{n} x_{k-1}[x_k^2 - x_{k-1}^2].$$

Direct computation will give the following table for various values of n.

n	lower sum	upper sum
1	0.0	8.0
5	4.48	6.08
10	4.92	5.72
20	5.13	5.53
50	5.2528	5.4128
100	5.2932	5.3732
200	5.3133	5.3533
500	5.325328	5.341328
1000	5.329332	5.337332
10000	5.33293332	5.33373332
100000	5.3332933332	5.3333733332

We note that, as predicted, the lower sums are increasing and the upper sums are decreasing. Furthermore, again as predicted, the upper sums provide an upper bound for the lower sums and vice versa. In addition we see that these two sequences of numbers appear to be approaching the same number, namely 5.333333333. Thus, by direct computation, we would apparently have both the lower integral and the upper integral, and since they appear to be equal the integral exists and has the common value.

The actual arithmetic for a computation such as that carried out above is more than we would like to do on an every day basis, so it would be

worth investigating whether there may be a shorter way. Using the techniques of Example 4.3, we would note that using a regular partition of
$[0, 2]$, $x_k = (2k/n)$. Also, $x_{k-1} = 2(k - 1)/n$. With these values the upper
and lower sums become

$$U(P, x, x^2) = \sum_{k=1}^{n} \left(\frac{2k}{n}\right)\left[\left(\frac{2k}{n}\right)^2 - \frac{2^2(k - 1)^2}{n^2}\right] = \left(\frac{8}{n^3}\right) \sum_{k=1}^{n} k[2k - 1]$$

$$= \left(\frac{8}{n^3}\right)\left[\sum_{k=1}^{n} 2k^2 - \sum_{k=1}^{n} k\right]$$

$$= \frac{8}{n^3}\left[\frac{2n(n + 1)(2n + 1)}{6} - \frac{n^2 + n}{2}\right]$$

$$= \frac{16}{3} + \frac{4}{n} - \frac{4}{3n^2}.$$

Also

$$L(P, x, x^2) = \frac{16}{3} - \frac{4}{n} - \frac{4}{3n^2}$$

where the result for the lower sum is obtained in a manner completely
analogous to the method used for the upper sum. Both of these derivations
use previous results in a manner similar to that described in Example 4.3.
We could have derived these *formulas* and obtained our table above by
substitution, as you can find out by testing a few values. From these last
results we can see that as n becomes large the upper and lower sums both
approach $16/3$, and this agrees with our previous conjecture.

In order to insure that the integral exists, we might consider the difference
between the upper and the lower sum for each partition. In our algebraic
result we observe that $U(P, x, x^2) - L(P, x, x^2) = 8/n$. When $n = 100{,}000$,
the difference should be 0.00008, as indeed it is. As n increases further, the
difference diminishes. Note that both the lower integral and the upper
integral must be found in this very small gap between the lower and upper
sums. Since this gap can be made arbitrarily small by further refining the
partition, it is clear that there could not be two distinct values within this
gap (one for the upper integral and one for the lower integral). Therefore,
we have shown that only one value must exist for both integrals, and consequently there must be an RS integral in this case. Thus we have established
the fact that

$$\int_0^2 x \, dx^2 = \tfrac{16}{3}.$$

Before we leave this example we ought to consider one other point. You
will remember that we started with a discussion which attempted to rationalize our use of the regular partition. This, of course was the easiest path for
us algebraically, for we had a simple expression for the partition points, x_k.

However, since *no* lower sum could be larger than *any* upper sum, we see that no lower sum could be larger in this case than 16/3. Consequently, since we have already established the fact that the lub of the lower sum is at least 16/3, and we know that it can be no more than 16/3, we do not have to worry further about other possible partitions. By the same reasoning, we do not have to worry further about the glb of the upper sums, either. You will note that we have thus taken care of the concern that we might have had about which partition to start with. It is often the case that this approach will work in this fashion. We should not take it for granted, but we can start with this approach, and then check to insure that our results are comprehensive in the manner in which we handle things here.

You should follow the work of Example 5.2 carefully, and it might help to see whether you can do the algebra necessary to carry out the evaluation of the lower sum. This Example is a very good summary of much of the work that we have done to the present time, and it illustrates rather well how we can use the information we have derived in order to find the values of integrals. Note that the integral is a refinement of the concept of the sums, and therefore in the various exercises for which you have obtained approximations, you can now obtain precise values—provided, of course, the algebra can be handled with any degree of ease.

Examples 4.3 and 5.2 could be a bit misleading, for everything seems to work almost too well. In particular the upper and lower integrals are equal, and therefore the integral exists. This follows from the fact that the upper and lower sums are continually moving until they almost meet as we carry out further refinements. There are, however, instances in which matters do not work out quite so nicely. Let us consider the following example.

EXAMPLE 5.3. Let $f(x)$ be a function which has the value one if x is a rational number and has the value zero if x is an irrational number. Thus, this function must have either the value one or zero, and there is certainly nothing smooth about its performance. It would seem, then, from the very start that this function would not be amenable to much of anything. However, let us wait and see. We will consider $g(x) = x^2$, and ask for the integral over the interval [1, 3]. Remember, there are at least two integrals—an upper integral and a lower integral. If these two should happen to be equal we will have an integral.

Solution. Let $P[1, 3]$ be any partition of the interval [1, 3]. In each subinterval of the partition we must have both rational and irrational points. While this is probably intuitively obvious, it could be rigorously shown. We will not take the time to prove this fact here, however. In each subinterval of the partition there is a rational value of x and hence a point at which $f(x) = 1$ and thus a maximum value of $f(x)$ in the interval of one. On the other hand, there is also an irrational value of x, and therefore a point at which $f(x) = 0$ and this gives a minimum value of zero in each subinterval. Furthermore, $g(x)$ is monotone increasing in the interval $[1, 3]$, since for $1 \leq x \leq y \leq 3$, we would have $x^2 \leq y^2$. Therefore, we can form the RS sums

$$U(P, f, g) = \sum_{k=1}^{n} 1 \cdot [x_k^2 - (x_{k-1})^2]$$

and

$$L(P, f, g) = \sum_{k=1}^{n} 0 \cdot [x_k^2 - (x_{k-1})^2].$$

The value of n is merely the number of intervals in the partition. However, upon adding up the terms we find that

$$U(P, f, g) = (x_1^2 - x_0^2) + (x_2^2 - x_1^2) + \cdots$$
$$+ (x_n^2 - x_{n-1}^2) = x_n^2 - x_0^2 = 3^2 - 1^2 = 8,$$

and

$$L(P, f, g) = 0$$

regardless of the partition that is used. Refinements of P do not change this result. Therefore we can see that the glb of the upper sums must be 8 and the lub of the lower sum must be zero. We can write these results as

$$\overline{\int_1^3} f(x)dg(x) = \overline{\int_1^3} f(x)d(x^2) = 8$$

and

$$\underline{\int_1^3} f(x)d(x^2) = 0.$$

In this case the two integrals are not equal, and therefore there is no RS integral.

This is a somewhat unusual combination of functions, but note that we *can* obtain the upper and lower RS integrals. In fact, the functions $f(x)$ and $g(x)$, weird though they seemed, turned out to be very nice for they gave us no difficulty in obtaining the necessary glb and lub. As a matter of fact there are some instances in which functions almost as weird as these have applications, although we will not be seeing such applications in this book.

To make certain that the integration process is well understood, we will give one more example.

EXAMPLE 5.4. Evaluate

$$\int_0^a x \, dx \quad \text{if it exists.}$$

Solution. In this case we have $f(x) = g(x) = x$, the identity function, and we are using the interval $[0, a]$. While the functions involved are simple, be sure to observe the methods used. Needless to say, we must start with a partition of $[0, a]$. We will make use of three different evaluation sets. In order to obtain the upper sum, we will let $t_k = x_k$ in view of the fact that both f and g are increasing functions of x. For the lower sum we will use $t_k = x_{k-1}$. We will also make use of the midpoint of each interval in order to see what effect this might have on our sums, and in this case we will have $t_k = (x_{k-1} + x_k)/2$. The three RS sums in which we are interested then become

$$U(P, x, x) = \sum_{k=1}^{n} x_k[x_k - x_{k-1}]$$

$$L(P, x, x) = \sum_{k=1}^{n} x_{k-1}[x_k - x_{k-1}]$$

and

$$M(P, x, x) = \sum_{k=1}^{n} \frac{x_k + x_{k-1}}{2} [x_k - x_{k-1}]$$

$$= \frac{1}{2} \sum_{k=1}^{n} [x_k^2 - x_{k-1}^2].$$

Curiously enough the last of these sums is the easiest one to evaluate, for we observe that regardless of the partition used we have

$$M(P, x, x) = \frac{1}{2} [(x_1^2 - x_0^2) + (x_2^2 - x_1^2)$$

$$+ (x_3^2 - x_2^2) + \cdots + (x_n^2 - x_{n-1}^2)]$$

$$= \frac{1}{2} [x_n^2 - x_0^2] = \frac{a^2 - 0^2}{2} = \frac{a^2}{2}.$$

We have again made use of the fact that in the addition all terms appear both with a positive and a negative sign except for x_n^2 and x_0^2. For the evaluation of the upper and lower sums we must be more explicit with our partition. If we use a regular partition with n subintervals, we would have again $x_k = ak/n$ and $x_{k-1} = a(k-1)/n$. Thus, the sums in question become

$$U(P, x, x) = \sum_{k=1}^{n} \frac{ak}{n} \left[\frac{ak}{n} - \frac{a(k-1)}{n} \right] = \frac{a^2}{n^2} \sum_{k=1}^{n} k$$

$$= \frac{a^2}{n^2} \left[\frac{n^2 + n}{2} \right] = \frac{a^2}{2} + \frac{a^2}{2n}.$$

$$L(P, x, x) = \sum_{k=1}^{n} \frac{a(k-1)}{n} \left[\frac{ak}{n} - \frac{a(k-1)}{n} \right] = \frac{a^2}{n^2} \sum_{k=1}^{n} (k-1)$$

$$= \frac{a^2}{n^2} \left[\frac{n^2 + n}{2} - n \right] = \frac{a^2}{2} - \frac{a^2}{2n}.$$

Using reason similar to that we have used before, we see that as n increases the upper sum and the lower sum decrease and increase respectively to the same value we had previously obtained for the sum using midpoints. Furthermore, the difference between the upper sum and the lower sum will be a^2/n, and this difference decreases, giving rise to the same situation we had witnessed in Example 5.1. As a result of this development we see that

$$\overline{\int_0^a} x\, dx = \int_{\underline{0}}^a x\, dx = \frac{a^2}{2}$$

and since the upper and lower integrals are equal we have

$$\int_0^a x\, dx = \frac{a^2}{2}.$$

It should be pointed out that we seldom have the sum using the midpoint exactly equal to the correct result.

In each of the examples we have considered, we first consider a partition. We then have to consider the upper and lower sums and finally determine the glb of the upper sums and the lub of the lower sums. From this we are able to determine the upper integral and the lower integral and then determine whether the integral exists. It would seem from the last example that we might use the midpoint and then be through. However, it is not often that the midpoint used as the evaluation point will give precisely the correct result, and in fact we note that if we consider any sums other than the upper and lower sums, we fail to have any basis upon which to determine the upper and lower integrals. It is the comparison of these two integrals that lets us know whether the RS integral exists.

The procedure can be summarized in the flow chart of Figure II.10. Note the loop which occurs in attempting to insure that you have the upper and lower integrals. The determination required by this loop may well be the most difficult in the entire calculation. If you can show that with successive refinements you have $[U(P, f, g) - L(P, f, g)]$ getting arbitrarily close to zero, you can show that your results can not be altered by any other choice of partitions and successive refinements.

We have frequently used the fact that the RS sum

$$S(P, T, f, g) = \sum_{k-1}^{n} f(t_k)[g(x_k) - g(x_{k-1})]$$

over the interval $[a, b]$ is closely related to the RS integral

$$\int_a^b f(x)dg(x).$$

Note the similarities between the sum and the integral. We spent quite a bit of effort in the last section finding out how to handle problems of an applica-

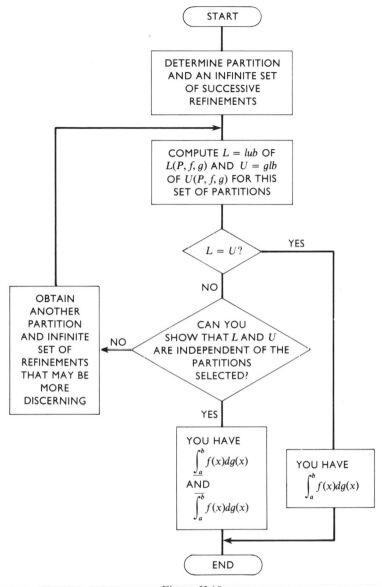

Figure II.10

tion nature, and how such problems often produce RS sums. We are now prepared to take the next step. Having set up the RS sum, we are able to write down the corresponding integral. If we were using upper sums, we would expect to find the upper integral, but we are more apt to have a case in which the integral exists and then we do not need to worry about whether we have an upper integral or a lower integral. Thus, the task of setting up an integral which yields a certain result is one that should give no difficulty.

First analyze the problem and set up the approximation of the result in terms of an RS sum, and then write down the RS integral. It only remains to evaluate the integral. We have shown how to do this in a few simple cases in this section. We will devote most of the next chapter to this problem, and will come back to it again in later chapters.

One final note is in order. If $g(x)$ in our definition is the identity function, $g(x) = x$, we obtain

$$\overline{\int_a^b} f(x)dx, \qquad \underline{\int_a^b} f(x)dx, \quad \text{and} \quad \int_a^b f(x)dx$$

as the upper integral, the lower integral and the integral respectively. These special cases are known as the *Riemann* integrals. Since they are special cases, then everything we develop for the Riemann–Stieltjes integral will also hold for the Riemann integral.

EXERCISES

1. Find the least upper bound of each of the following sets of numbers or show that the set is unbounded.

   (a) $4, 6, 7, 15/2, 31/4, 63/8, 127/16, \ldots, 2^{3-n}(2^n - 1), \ldots$
   (b) $f(n) = n!/n^n$ for $n = 1, 2, 3, 4, \ldots$
   (c) $f(n) = n!/10^n$ for $n = 1, 2, 3, 4, \ldots$
   (d) $1, -1/2, 1/4, -1/8, 1/16, -1/32, \ldots$
   (e) $f(n) = \sin n$ for $n = 1, 2, 3, 4, \ldots$ [Consider the angles to be measured in radians.]

2. (a) Find an algebraic expression for the upper RS sum and the lower RS sum $U(P, x^2, x)$ and $L(P, x^2, x)$ for a partition $P[0, 3]$ which is regular and has n subintervals.
   (b) Find $U(P, x^2, x) - L(P, x^2, x)$ and show whether this gets smaller as n increases.
   (c) Evaluate $\overline{\int_0^3} x^2 \, dx$ and $\underline{\int_0^3} x^2 \, dx$.
   (d) Does $\int_0^3 x^2 \, dx$ exist? If so, what is its value?
   (e) State whether you can be certain that your evaluation would not have been altered if you had used other than a regular partition. Be ready to explain the reasons for your answer in class, if asked.

3. Evaluate $\int_0^a x^2 \, dx$ if it exists. [Use the method of Exercise 2.]

4. (a) Let $P_1[0, 3]$ be a regular partition with $3n$ subintervals and let $P_2[0, 5]$ be a regular partition with $5n$ subintervals. Show that the points of P_1 are also points of P_2, and show that if we use the endpoints of the $2n$ subintervals which are in P_2 but not in P_1 we have a regular partition, P, of $[3, 5]$.
   (b) Using the information in part (a), show that $U(P, f, g) = U(P_2, f, g) - U(P_1, f, g)$ and similarly $L(P, f, g) = L(P_2, f, g) - L(P_1, f, g)$.

(c) Using the information from parts (a) and (b) show that

$$\int_3^5 x^2 \, dx = \int_0^5 x^2 \, dx - \int_0^3 x^2 \, dx.$$

(d) With the information of part (c) and with the result of Exercise 3 find the value of

$$\int_3^5 x^2 \, dx.$$

5. Evaluate $\int_1^4 dx$. [*Hint*: $f(x) = 1$ in this case.]

6. Evaluate $\int_2^5 x^3 \, d(1)$. [*Hint*: $g(x) = 1$ in this case.]

7. Evaluate $\int_a^b dg(x)$.

8. Let $f(x)$ be a function having the value one if x is a rational number and the value zero if x is an irrational number.

(a) Evaluate $\overline{\int_0^2} f(x)d(x^3)$ and $\underline{\int_0^2} f(x)d(x^3)$.

(b) Does $\int_0^2 f(x)d(x^3)$ exist? Give a reason for your answer.

9. Let $f(x)$ be a function such that $f(x) = x$ if x is a rational number and $f(x) = 0$ if x is an irrational number. [*Hint*: Example 5.3 may help.]

(a) Evaluate $\overline{\int_1^2} f(x)d(x^2)$ and $\underline{\int_1^2} f(x)d(x^2)$.

(b) Evaluate $\int_1^2 f(x)d(x^2)$ if it exists.

10. (a) Show that $d = U(P, f, g) - L(P, f, g) \geq 0$ for any choice of $P[a, b]$ and any choice of $f(x)$ and $g(x)$ subject to the condition that $g(x)$ is monotonic in each subinterval of P.

(b) Show that the set of numbers, d, is a decreasing function under successive refinements of P.

(c) Show that the set of possible values of d has a glb.

(d) Show that $\int_a^b f(x)dg(x)$ exists if and only if the glb of d is zero.

11. (a) A solid is formed by taking a right circular cone of radius 4 inches and altitude 4 inches from a cylinder of radius 4 inches and altitude 4 inches. The axes of the two solids are coincident and the base of the cone is identical with one base of the cylinder. Draw a sketch of the large cylinder. Draw a set of cylinders within the large cylinder, all of the smaller ones being concentric with the original cylinder.

(b) As a result of part (a) you have represented the portion of the cylinder remaining after the cone is removed by a set of nested solids, the cross section of each being a ring between two circles and the length being greater as the circular cross sections are larger. Consider one of these nested solids to be approximated by a solid similar to a piece of pipe. Find an expression for the volume of this solid.

(c) Using your result in part (b), set up an RS sum for the approximate volume of the solid described in part (a).

(d) Evaluate this integral, and thus find the volume of the solid described in part (a).

12. (a) Sketch the curve $y = x^2$ over the interval from $x = 0$ to $x = 3$.

   (b) Set up an RS sum which approximates the area bounded by $y = x^2$, $y = 0$ and $x = 3$.

   (c) Set up the RS integral which would give the number of square units of area in the area described in part (b).

   (d) Evaluate the integral and thus find the area.

13. (a) Sketch the curve $r = \theta$ in polar coordinates.

   (b) Set up an RS sum that approximates the area inside this curve between the vectors $\theta = 0$ and $\theta = \pi/2$.

   (c) Set up an integral that gives the value of this area.

   (d) Evaluate this integral and find the area.

C14. Use a regular partition of $[1, 3]$ with 25, 50, 75, 100 and so on to 1000 intervals to find the upper and lower RS sums if $f(x) = x^3$ and $g(x) = \sqrt{x}$. Use these results to estimate the value of

$$\overline{\int_1^3} x^3 \, d\sqrt{x} \quad \text{and} \quad \int_{\underline{1}}^3 x^3 \, d\sqrt{x}.$$

From the information you have, determine whether the integral exists, and if you think it does, estimate the value of the integral.

C15. Use a regular partition of $[-1, 2]$ with 20, 40, 60, and so on to 200 intervals to find the upper and lower RS sums if $f(x) = x^2$ and $g(x) = \sin x$. Use these results to estimate the value of

$$\overline{\int_{-1}^2} x^2 \, d(\sin x) \quad \text{and} \quad \int_{\underline{-1}}^2 x^2 \, d(\sin x).$$

From this information determine whether the integral exists, and if you think it does, estimate the value of the integral.

P16. A metal rod is one foot long, and the mass is distributed in such a way that the mass of the first x inches from one end is $(0.2x + 0.004x^2)$ pounds per inch of rod.

   (a) Sketch the rod and partition the length.

   (b) Find an expression for the mass of the portion of rod between two successive partition points.

   (c) The k-th *moment of mass* about a given axis is defined to be the product of the mass by the k-th power of the distance of that mass from the axis. Write down an expression for the second moment of the increment of mass between two partition points about the axis perpendicular to the rod at the end of the rod where $x = 0$.

   (d) By adding up the second moments for all of the subintervals of your partition, find an RS sum which approximates the second moment of the entire rod about the axis through the end of the rod at which $x = 0$.

   (e) Find an integral which would give the second moment of the rod about the axis through $x = 0$.

   (f) Find upper and lower bounds for the value of the integral in part (e). (The *second moment* is also known as *the moment of inertia*).

P17. Set up an integral to evaluate the mass of the wheel of Exercise II.4.17. Evaluate this integral to find the actual mass of this wheel.

P18. Set up an integral to evaluate the force on the face of the dam in Exercise II.4.19. Evaluate this integral to find the force on the face of the dam.

M19. Show that there is no rational number which can serve as the least upper bound of the set of positive numbers such that $x^2 < 2$. [*Hint*: If you assume that c is such a rational number, then either $c^2 > 2$ and there is a rational number smaller than c that would also be an upper bound or if $c^2 < 2$ then there is a rational number larger than c such that its square is less than two.]

M20. Let $f(x)$ be a function over the interval $[a, b]$ such that for any positive real number, e, it is possible to find a positive real number, d, such that if the diameter of the partition $[a, b]$ is no greater than d units, then the maximum and minimum values of $f(x)$ over each subinterval differ by less than e.

(a) Show that if $P[a, b]$ is a partition such that $d(P) < d$ then

$$U(P, f, g) - L(P, f, g) < e \sum_{k=1}^{n} |g(x_k) - g(x_{k-1})|.$$

(b) Show that under these conditions $\int_a^b f(x)dg(x)$ exists.
(c) If $g(x)$ is an increasing function, show that

$$U(P, f, g) - L(P, f, g) < e[g(b) - g(a)].$$

B21. Set up an integral for the evaluation of the amount of blood passing one point in one second in the blood vessel of Exercise II.4.21. Given the fact that

$$\int_0^{0.1} r^2 \, dr^2 = 5(10^{-5})$$

evaluate the integral (if possible) and determine exactly the rate of flow of blood in the blood vessel.

B22. It can be shown that under rather general conditions the concentration of a solute in a circular tube (such as a blood vessel) is closely approximated by $C(x) = C_0 + (C_i - C_0)2^{-ax}$ where x is the distance in centimeters from one end of the tube, C_0 g/cm^3 is the concentration outside of the tube, C_i is the concentration of the solute inside of the tube at the point where $x = 0$ and a is a positive constant. The decrease in concentration is due to the diffusion of some of the solute through the walls of the tube.

(a) Sketch the tube if it is 20 centimeters long, and indicate a partition of the length.
(b) Find an expression for the amount of solute in the tube between successive partition points. (Assume a tube radius of 2 millimeters.)
(c) Set up an integral which would give the amount of solute in the tube between $x = 0$ and $x = 20$. (This will involve the constants a, C_i, and C_0. Do not try to integrate this expression.)

S23. Set up an integral for the total increase in cost of production in Exercise II.4.14. Evaluate this integral and find the increase in cost of production.

S24. In a certain town the amount of electricity used per person per day is determined
to be $(3 + 0.1t)$ kilowatts where t is the number of years since January 1, 1975.
The population of the town is given by the expression $p(t) = 18,000 + 360t + 4t^2$.

(a) Find an expression for the daily use of electricity in the town t years after
January 1, 1975. [It is not essential that t be an integer.]
(b) Find an approximation of the total use of electricity in the town over a period
of time from t_{k-1} to t_k where the values of t_k are taken from a partition
$P[0, 10]$ of the period from January 1, 1975 to January 1, 1985.
(c) Set up an integral which would give the total use of electricity during the
decade from January 1, 1975 to January 1, 1985.
(d) Find the average daily use of electricity in this town during this period of time.
[Approximate this use using upper and lower bounds for the integrals.]
(Ignore leap years. This type of computation is useful in trying to project the
community requirements for some utility in order to permit an orderly
expansion of available services consistent with community needs.)

II.6   Some Facts Concerning Evaluation of Integrals

Now that we have defined upper integrals, lower integrals, and integrals,
the question must naturally arise concerning how we can find values for
such things. In the first place, it is not easy to contemplate having to set up
the RS sums, and it is even less appetizing to think of trying to find the lubs
and glbs and assuring ourselves both that an integral exists and that we have
a sufficiently good approximation of its value. We gave some illustrations
in the last section, but these were not necessarily reassuring. One was rather
bizarre, to say the least, and the others seemed a bit on the special side. As
is usually the case, however, if something has been mentioned in a book such
as this, there is some way of getting around the difficulties in at least a few
instances. It should be made clear at this point that the problem of evaluating
integrals in general is not a simple one, and we will ordinarily confine our
attention to the more tractable members of the set of all integrals. Fortunately, the world of applications, as well as that of more sophisticated mathematics, tends to use those integrals that are easier to handle. (This might
be due to the fact that these are the only ones that have been given serious
consideration, except for the rare cases in which applications provide no
other choice.)

There are basically two ways of evaluating integrals. The first method
makes use of the definition, utilizing the RS sums with a partition that has
a very small diameter, and for which we can expect little fluctuation between
the upper sums and the lower sums. Since the correct value is never greater
that any upper sum nor less than any lower sum, we have bounds for the
correct result. This method would not be very appealing if $|b - a|$ is large
and a small diameter is required for the partition, for we would then have a

very large number of summands. On the other hand, if we had a hand calculator, or better yet a computer, we could rather easily consider this method. The second method that we referred to above is that of finding some expressions or formulas which will permit us to evaluate integrals directly. For instance, we might use the results of the last section to start a table of values such as

$$\int_0^a x \, dx = \frac{a^2}{2}.$$

When these particular integrals arise again we would not have to repeat this work. This method will be employed in Chapter III. Suffice it to say that this can be done for only a restricted class of integrals, and therefore the method of obtaining formulas, while the more appealing of the two methods, is not of the general applicability that we would like.

We will consider the first of these two methods, the computational method, here. In order to be able to bring some order out of the possible chaos, we will restrict our attention to the case in which both $f(x)$ and $g(x)$ are monotonic, and for convenience we will consider that $g(x)$ is increasing. A little thought should indicate what needs to be altered if $g(x)$ is decreasing. If $g(x)$ is increasing part of the time and decreasing part of the time, one would use a partition including all points at which the direction of monotonicity changes. Since we are assuming that $g(x)$ is increasing, then the portion of each summand which is of the form $[g(x_k) - g(x_{k-1})]$ is non-negative. If $f(x)$ is also increasing, we would use $f(x_k)$ as the value for the upper sum and use $f(x_{k-1})$ as the value for the lower sum. Consequently, we would have

$$\left. \begin{array}{l} L(P, f, g) = \displaystyle\sum_{k=1}^{n} f(x_{k-1})[g(x_k) - g(x_{k-1})] \\[2mm] U(P, f, g) = \displaystyle\sum_{k=1}^{n} f(x_k)[g(x_k) - g(x_{k-1})]. \end{array} \right\} \qquad \text{(II.6.1)}$$

If $f(x)$ is a decreasing function, similar reasoning would produce

$$\left. \begin{array}{l} L(P, f, g) = \displaystyle\sum_{k=1}^{n} f(x_k)[g(x_k) - g(x_{k-1})] \\[2mm] U(P, f, g) = \displaystyle\sum_{k=1}^{n} f(x_{k-1})[g(x_k) - g(x_{k-1})] \end{array} \right\} \qquad \text{(II.6.2)}$$

It becomes a relatively simple matter to plan the computation of either of these pairs of sums, as indicated by the flow chart in Figure II.11. Such a flow chart can be rather easily transformed into a computer program. Note that it is possible to compute both the lower and upper sums in the one algorithm. However, we still have the question whether we will have the required result with sufficient accuracy. If $U - L$ is sufficiently small, then the average of U and L will be sufficiently close to the correct value of the integral that the

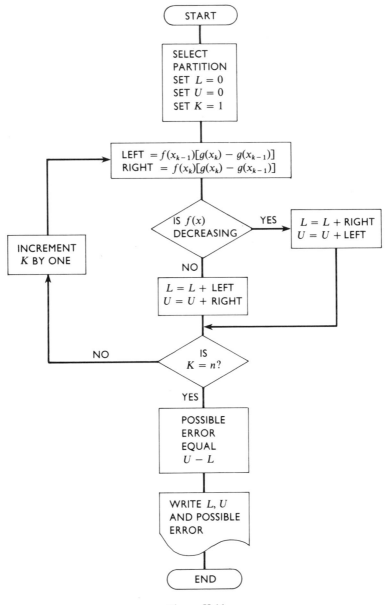

Figure II.11

result will satisfy our needs. On the other hand, if $U - L$ is too large we might have the correct result by taking the average, but on the other hand we might have far too great an error. It would be very helpful then to have some upper bound on the value of $U - L$ in order that we could have some assurance that our computation would not be in vain. Even when using a computer there is a great expenditure of resources and we should not undertake work which will not have a reasonable chance of giving a useful result.

In order to determine whether a given partition $P[a, b]$ might give a satisfactory result, we can start by determining the difference

$$U(P, f, g) - L(P, f, g) = \sum_{k=1}^{n} |f(x_k) - f(x_{k-1})| [g(x_k) - g(x_{k-1})].$$

$$(II.6.3)$$

This is obtained by combining the terms of the upper and lower sums of either (II.6.1) or (II.6.2) having $[g(x_k) - g(x_{k-1})]$ in common and observing that the absolute value $|f(x_k) - f(x_{k-1})|$ takes care of both the increasing and the decreasing case. As it stands (II.6.3) is much more complicated than we would like. It would help if either $|f(x_k) - f(x_{k-1})|$ or $[g(x_k) - g(x_{k-1})]$ were constant, for then we could factor out a common factor. It would be most unlikely if this were so, but we can make use of this idea by considering the inequality we would have if we were to replace each occurence of $|f(x_k) - f(x_{k-1})|$ by F where F is chosen to be the largest of the n values of the form $|f(x_k) - f(x_{k-1})|$. We would then have

$$U(P, f, g) - L(P, f, g) \le \sum_{k=1}^{n} F[g(x_k) - g(x_{k-1})]$$

$$\le F \sum_{k=1}^{n} [g(x_k) - g(x_{k-1})] = F[g(b) - g(a)].$$

$$(II.6.4)$$

In other words, $F[g(b) - g(a)]$ is an upper bound for the difference $U - L$. Now $[g(b) - g(a)]$ is a constant that is easily determined. Furthermore, by our assumptions this is positive. If D is an upper limit to a useful size for $U - L$, then if $F[g(b) - g(a)] \le D$ we are certain that $U - L \le D$, for we have

$$U(P, f, g) - L(P, f, g) \le F[g(b) - g(a)] \le D. \qquad (II.6.5)$$

Thus, we can test any partition before doing the computation to determine whether that partition is likely to meet our needs. It is true that this is an upper bound, and the partition might fail this test and yet be satisfactory. However, if the partition passes this test, it is certain to suffice.

We might go one step further and use the information in (II.6.5) to help us find a satisfactory partition. Consider the following Example.

EXAMPLE 6.1. Find an approximation for the value of $\int_2^3 (1/x)d(x^2)$ with an error no greater than 0.1.

Solution. In this case we might let $D = 0.1$. Since $g(x) = x^2$, we have $[g(3) - g(2)] = 3^2 - 2^2 = 5$. We then wish F to be determined such that $5F \leq 0.1$ in order to satisfy (II.6.3). Therefore, $F \leq 0.02$. This is equivalent to saying that we should have a partition such that $|(1/x_k) - (1/x_{k-1})| \leq 0.02$. Since $x_0 = 2$, we have $|(1/x_1) - (1/2)| \leq 0.02$. Since $x_1 > 2$, we know that $[(1/x_1) - (1/2)]$ is negative and hence using the absolute value we have $1/x_1 \geq (1/2) - 0.02 = 0.48$. Consequently, we have

$$x_1 \leq 1/0.48 = 2.0833333.$$

Since we now have a maximum value for x_1, we can select any value we wish to use in the interval [2, 2.0833] and then proceed to determine a value for x_2 in a similar manner. This process could be continued until we had a partition of the interval [2, 3] that would be guaranteed to give the required accuracy.

In this case, we have done about twice as much work as was really needed, for we could have observed that

$$U - \frac{U + L}{2} = \frac{U + L}{2} - L = \frac{U - L}{2}.$$

Therefore, if we were to use as an approximation for the integral the value $(U + L)/2$, we could have let D be bounded by $2(U - L)$ and then our error would have been at most $D/2$ or $(U - L)$. In this example we would then let $D = 0.2$ and then we would have $5F \leq 0.2$. This gives a bound for F of 0.04. Consequently our choice for x_1 would be limited by the relation $(1/x_1) \geq (1/2) - 0.04 = 0.46$, and we would have $x_1 \leq 2.1739$. If we continued the process, we would find that a partition consisting of the points $\{2, 2.16, 2.36, 2.60, 2.90, 3\}$ would guarantee a sufficiently accurate result. This, of course, is not the only partition that we might use, but it is a compromise involving as few points as possible and numbers that do not carry too many decimal places. Using these points and evaluating the lower sum we obtain the approximation 1.9147. Similarly the upper sum would give 2.0938. The average is 2.00425, and this is certainly within 0.1 of either the upper or lower sum. Hence, it is certain that it differs from the correct value of the given integral by less than 0.1. In fact, the value of the given integral is actually 2, and we see that we have a very good approximation to this value.

It is certainly clear that there is a considerable amount of work in the approximation of an integral by the method of Example 6.1, but if we need to find an approximation with a given error bound we now have the means for doing so. You will note that we cut our work by quite a bit by the device of using the average of the upper and lower sums. We could have obtained this average either by computing the upper and lower sums first or by using the relation

$$T(P, f, g) = \frac{[U(P, f, g) + L(P, f, g)]}{2}$$

$$= \sum_{k=1}^{n} \frac{f(x_k) + f(x_{k-1})}{2} [g(x_k) - g(x_{k-1})].$$

(II.6.6)

This method is frequently called the *trapezoidal method*, hence the letter T. The reason for this name is clear from Figure II.12, for we see that if we were to think of a curve plotted on the $f(x)$ axis versus the $g(x)$ axis, the upper sum contribution would be the rectangle with the upper top, the lower sum contribution the one with the lower top, and the trapezoidal sum contribution, being the average of the other two, is the trapezoid indicated in the Figure.

It is probably better to obtain the trapezoidal result by calculating the upper and lower sums and then averaging them, for in this way you are certain to have bounds for the correct value of the integral in question.

One final computational method worth considering is the one using the midpoint of the interval $[x_{k-1}, x_k]$ as the choice for t_k. Thus $t_k = (x_{k-1} + x_k)/2$. This will produce another RS sum. Since we are using the midpoint for our computation of the value of $f(x)$ and since we have assumed $f(x)$ to be a monotonic function, this would seem to be a better choice. In many cases it will be, but there is no guarantee that such will be the case. Furthermore, there are no bounds available if this is the only computation we perform. This sum, however, which we will denote by $M(P, f, g)$ (where we use M for midpoint) is one that we should not ignore as a possibility.

We can illustrate the midpoint rule with the following Example.

EXAMPLE 6.2. Approximate $\int_2^3 (1/x)d(x^2)$ using the partition $\{2, 2.16, 2.36, 2.60, 2.90, 3\}$.

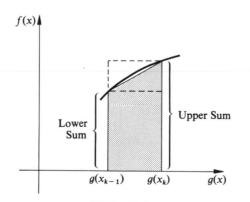

Figure II.12

Solution. We note that this is the same partition that we used in Example 6.1, but this time we do not have the assurance that our result will be within 0.1 of the correct result. It must certainly be in the interval [1.9147, 2.0938], but if we had not been told that the correct value is 2, we might have assumed that the correct value would be 1.92 and we could then have obtained 2.08 as some RS sum, clearly a result that would be in error by more than 0.1. This illustrates the possible danger of calculating an RS sum without having the bounds provided by the upper and lower sums.

However, the discussion of the preceding paragraph does not perform the calculation requested in this Example. Here we have

$$M(P, 1/x, x^2) = \left(\frac{1}{2.08}\right)[2.16^2 - 2^2] + \left(\frac{1}{2.26}\right)[2.36^2 - 2.16^2]$$

$$+ \left(\frac{1}{2.48}\right)[2.6^2 - 2.36^2]$$

$$+ \left(\frac{1}{2.75}\right)[2.9^2 - 2.6^2] + \left(\frac{1}{2.95}\right)[3^2 - 2.9^2]$$

$$= 0.32 + 4 + 0.48 + 0.6 + 0.2 = 2.0$$

and in this instance at least we have the exact value. It seldom turns out this well, but the midpoint rule is often a good compromise.

In summary, we will not use these computational methods in practice if we can find a formal means that will take care of the situation at hand (once we have developed some formal techniques), but there will continue to arise cases in which the formal techniques are not adequate and in which we will have to resort to numerical methods of some sort. We have indicated some methods here that will permit us to find approximations sufficiently good to meet any requirements. It would be well to close with a comment concerning the *bell curve* which is used extensively in statistics and which finds occasional use in determining grade distributions. The use of this curve for statistical purposes requires the evaluation of an integral for which no formula is known. The integral in question is

$$\int_a^b \frac{1}{\sqrt{2\pi}} e^{-x^2/2} \, dx,$$

where e is a constant and the approximate value of e is 2.71828. Here $g(x)$ is the identity function and

$$f(x) = \frac{1}{\sqrt{2\pi}} e^{-x^2/2}.$$

Since the value of this integral is so widely used, it is almost essential that we not only have the value determined for various intervals [a, b], but also that we have information concerning just how accurate are the values we have available to us. Therefore, we would need to evaluate the upper and lower sums in order to have bounds for the correct value of the integral.

EXERCISES

1. Use the methods of this Section to obtain a lower bound, an upper bound, and then use the trapezoidal rule and the midpoint rule to find the value of each of the following integrals, using a regular partition with 5 intervals in each case.

$$\int_0^1 3x^2\, dx, \qquad \int_0^2 3x^2\, dx, \qquad \int_0^3 3x^2\, dx, \qquad \int_0^4 3x^2\, dx.$$

Can you make a guess as to the formula that would explain each of these results?

2. Evaluate

$$\int_1^2 x\, d(x^3)$$

with an error no greater than 0.1. Find a partition which would give the value of this integral with an error no greater than 0.001.

3. Evaluate $\int_1^2 (1/x)dx$ with an error no greater than 0.1.

4. You need to evaluate each of the following integrals by the trapezoidal rule with an error no greater than 0.01. Find a partition that would suffice in each case. [It is sufficient to find the first 4 points and indicate the procedure required to obtain the remaining points.]

(a) $\int_0^3 x^2\, d(x^3)$

(b) $\int_0^4 \sqrt{16 - x^2}\, d(2\pi x)$

(c) $\int_1^3 2^x\, d(1/x)$

5. Show that $\int_{-1}^2 x^4\, d(x^2)$ exists. Note that neither function involved is an increasing function or a decreasing function throughout $[-1, 2]$.

6. Use the numerical methods that we have discussed to find bounds and an approximate value for

$$\int_{-4}^0 2^x\, dx \quad \text{and for} \quad \int_0^4 2^x\, dx$$

if you use four subintervals in each case. Which of the two results is more apt to have a smaller error?

7. (a) Sketch the graph of $y = 2^x$ over the interval $[-4, 4]$.
   (b) Put in the partition points of your partitions in Exercise 6.
   (c) Show that the trapezoidal rule would give values which are too large in your evaluation of the integrals in Exercise 6.
   (d) Indicate the midpoints of your intervals, and then draw lines indicating the rectangles that would be used in the evaluation of $M(P, 2^x, x)$.
   (e) In each subinterval draw a line through the point $(x_m, 2^{x_m})$ tangent to the curve $y = 2^x$. [x_m is the midpoint of the subinterval.]

(f) Show that the trapezoid formed in each subinterval in part (e) by the four lines $y = 0$, the tangent line and the lines $x = x_{k-1}$ and $x = x_k$ has the same area as the rectangle formed by the lines $y = 0$, $y = x_m$, $x = x_{k-1}$, and $x = x_k$. Use this result to show that in this case the midpoint rule would give a value too small for the value of the integral.

8. Use a partition with five intervals to obtain bounds for the RS integral

$$\int_{-1}^{2} x^2 \, dx.$$

Do the same thing with a partition having ten intervals. What values do you have with the trapezoidal rule and the midpoint rule in this case? [At least half of your computation for ten intervals should be complete when you have completed the five interval case.]

9. It is desired to find the value of

$$\sum_{k=0}^{50} 2^{-k}.$$

Someone has suggested approximating this by evaluating

$$\int_{0}^{50} 2^{-x} \, dx.$$

Can you find any connection between these two results? How accurate do you think the approximation would be? Would it be too large or too small?

10. Show that if either $f(x)$ or $g(x)$ is a constant all RS sums are equal and therefore any partition will serve to find the correct value of the integral.

11. If we restrict ourselves to the interval $[0, \pi/2]$ it is true that for values $x > y$ we have $[\sin x - \sin y] < [x - y]$. Find a partition which would permit you to evaluate $\int_0^{\pi/2} \sin x \, dx^2$ with an error no greater than 0.1.

12. Find a partition which would permit you to evaluate $\int_0^{\pi/2} x \, d(\sin x)$ with an error no greater than 0.1.

13. (a) Sketch the graph of $y = 4 - x^2$.
    (b) Using the fact that the area bounded by this curve and the x-axis is symmetric about the y-axis, find the area under this curve by finding the area to the right of the y-axis and doubling that area.
    (c) Find the area determined by the upper sum using a partition with only one subinterval. Sketch on your graph the outline of this rectangle.
    (d) Determine the ratio of the area of part (b) to the area of part (c). [This ratio was discovered analytically by Archimedes not later than 212 B.C.]

C14. (a) Sketch the graph of $y = (4 - x^2)^{1/2}$ over the interval $[0, 2]$.
    (b) Sketch in a regular partition having n subintervals for a small value of n, such as $n = 5$.
    (c) For one subinterval of your partition (not either end subinterval) sketch in the rectangles for the term in the upper sum, the lower sum, and the midpoint sum, and sketch in the trapezoid for use with the trapezoidal rule.

(d) From your sketch determine which of the various rules of evaluation are likely to give values larger than the area under the curve and which ones will give smaller values.

(e) Use the upper sum, the lower sum, the midpoint sum, and the trapezoidal rule to find the area under the curve for $n = 25, 50, 75, 100, \ldots, 500$, and then estimate the correct area.

(f) Use geometry to find this area and check your results.

CHAPTER III

More About Integrals

III.1 Some Integrals to Remember

While it may be necessary from time to time to use the computational methods of the last chapter in evaluating integrals, we would prefer other methods if they can be made available. It will be the purpose of this chapter to develop some alternatives to the use of RS sums for the evaluation of integrals. It might be worth listing the result of Example (II.5.4), for this could be the first of a list of *formulas* that would help.

$$\int_0^a x \, dx = \frac{a^2}{2}. \tag{III.1.1}$$

This result would permit us to give the value of any integral with $f(x) = g(x) = x$ over an interval having the origin as left-hand point. We now turn to some other integrals in order to add to our list of formulas.

EXAMPLE 1.1. If $f(x) = c$, the constant function, and if $g(x)$ is monotonic in each of the subintervals of some partition of $[a, b]$, show that

$$\int_a^b c \, dg(x) = c[g(b) - g(a)]. \tag{III.1.2}$$

Solution. If $P[a, b]$ is a partition of $[a, b]$, then any RS sum over this partition is given by the relation

$$S(P, T, c, g(x)) = \sum_{k=1}^n c[g(x_k) - g(x_{k-1})] = c \sum_{k=1}^n [g(x_k) - g(x_{k-1})]$$

$$= c[g(x_n) - g(x_0)] = c[g(b) - g(a)].$$

since $f(t_k) = c$ in every case, and since the summation

$$\sum_{k=1}^{n} [g(x_k) - g(x_{k-1})]$$

has the interesting property we have seen before in that each term except $g(x_n)$ and $g(x_0)$ appears twice, once with each sign. The only terms that do not "cancel" are the terms $g(x_n) = g(b)$ and $g(x_0) = g(a)$. In particular, since all RS sums have the same value, the upper and lower sums have this value. But the integral has a value not less than the lower sum nor more than the upper sum, and therefore the integral must have this same value. Note the generality of this result in that this holds over any interval and for any function $g(x)$ that can meet our requirements for integrability.

EXAMPLE 1.2. If $f(x)$ is a bounded function, show that

$$\int_{a}^{b} f(x)dc = 0. \tag{III.1.3}$$

Solution. In this case we have $g(x) = c$, a constant function. Therefore $[g(x_k) - g(x_{k-1})] = [c - c] = 0$ for each choice of k. As a result we have for any partition the fact that the RS sum must have the value zero since each term must be zero and the RS sum is merely the sum of n zeroes.

We next turn our attention to an example which is a generalization of the result expressed in (III.1.1).

EXAMPLE 1.3. Let m be a positive integer and $a > 0$. Find the value of $\int_0^a x^m \, dx$.

Solution. Since we have no prior formula to call upon, we must again resort to RS sums. Since it is the easiest one to use, let us try using again a regular partition with n subintervals. Since both $f(x) = x^m$ and $g(x) = x$ are increasing functions, we would have $t_k = x_k$ for the upper sum and $t_k = x_{k-1}$ for the lower sum. Again using the fact that $x_k = ak/n$ and since

$$[g(x_k) - g(x_{k-1})] = (ak/n) - [a(k-1)/n] = a/n,$$

we have

$$U(P, x^m, x) = \sum_{k=1}^{n} \left(\frac{ka}{n}\right)^m \left[\frac{a}{n}\right] = \left(\frac{a}{n}\right)^{m+1} \sum_{k=1}^{n} k^m.$$

In Exercise (I.11.20) it was shown that

$$\sum_{k=1}^{n} (k-1)^m = \sum_{k=0}^{n-1} k^m < \frac{n^{m+1}}{m+1} < \sum_{k=1}^{n} k^m. \tag{III.1.4}$$

From the right hand inequality in (III.1.4) we obtain

$$U(P, x^m, x) = \left(\frac{a}{n}\right)^{m+1} \left[\sum_{k=1}^{n} k^m\right] \geq \left(\frac{a}{n}\right)^{m+1} \left[\frac{n^{m+1}}{m+1}\right] = \frac{a^{m+1}}{m+1}.$$

Using the left hand inequality in (III.1.4) we have

$$L(P, x^m, x) = \sum_{k=1}^{n} [(k - 1)a/n]^m [a/n]$$

$$= (a/n)^{m+1} \sum_{k=1}^{n} (k - 1)^m = (a/n)^{m+1} \left[\sum_{k=0}^{n-1} k^m \right]$$

$$\leq (a/n)^{m+1} [n^{m+1}/(m + 1)] = \frac{a^{m+1}}{m + 1}.$$

Thus we have

$$L(P, x^m, x) \leq \frac{a^{m+1}}{m + 1} \leq U(P, x^m, x). \qquad \text{(III.1.5)}$$

It is also true that

$$U(P, x^m, x) - L(P, x^m, x) = \left(\frac{a}{n}\right)^{m+1} \left[\sum_{k=1}^{n} k^m - \sum_{k=0}^{n-1} k^m \right]$$

$$= \left(\frac{a}{n}\right)^{m+1} n^m = \frac{a^{m+1}}{n}.$$

Now, suppose that

$$\overline{\int_0^a} x^m \, dx \neq \underline{\int_0^a} x^m \, dx.$$

If they are not equal, then the upper integral must exceed the lower integral by some positive number, e. However, by (III.1.5) we can select a regular partition, P, with a value of n large enough that $a^{m+1}/n < e$ for any e which is positive. If we assume we have made use of this particular regular partition, we now have the string of inequalities

$$U(P, x^m, x) \geq \overline{\int_0^a} x^m \, dx > \underline{\int_0^a} x^m \, dx \geq L(P, x^m, x)$$

and this presents a contradiction for the two sums differ by less than e whereas the two integrals differ by more than e. This contradiction was brought about by the assumption that the upper and lower integrals were not equal. Hence, we must have the upper and lower integrals equal. Therefore

$$\int_0^a x^m \, dx$$

must exist. Since

$$U(P, f, g) \geq a^{m+1}/(m + 1) \geq L(P, f, g)$$

for any regular partition, P, and since the upper and lower sum differ by an arbitrarily small amount with the integral also included between the upper and lower sums, the assignment of any value other than $a^{m+1}/(m + 1)$ to the integral would lead to a contradiction. Therefore, the result we sought is

$$\int_0^a x^m\, dx = a^{m+1}/(m + 1). \tag{III.1.6}$$

We should observe that (III.1.1) is merely a special case of (III.1.6) in which $m = 1$. Hence, we have only developed three results so far that you need to remember for later use, and these are the results of (III.1.2), (III.1.3), and (III.1.6).

EXAMPLE 1.4. Find the value of

$$\int_0^4 x^6\, dx$$

Solution. Here we have (III.1.6) with $a = 4$ and $m = 6$. Hence the required value is $4^7/7$, or $16384/7$. Ordinarily we will be willing to accept $4^7/7$ rather than requiring the additional practice in multiplication. (Isn't this easier than developing the RS sums and trying to find the appropriate partitions to make it possible to find the lubs and glbs?)

We will consider just one more example in this Section. This is a slight generalization of Example (II.5.2), and you may wish to refer to the detailed description of the earlier solution.

EXAMPLE 1.5. Find the value of $\int_0^a x\, d(x^2)$.

Solution. Note that $f(x)$ is again an increasing function, and hence we can obtain upper and lower sums as before. Again using the regular partition we have

$$U(P, x, x^2) = \sum_{k=1}^n x_k[x_k^2 - x_{k-1}^2] = \sum_{k=1}^n (ka/n)[(ka/n)^2 - ((k-1)a/n)^2]$$

$$= (a/n)^3 \sum_{k=1}^n [2k^2 - k]$$

$$= (a/n)^3 \left[2\sum_{k=1}^n k^2 - \sum_{k=1}^n k \right]$$

$$= \left(\frac{a}{n}\right)^3 \left[\frac{2n(n+1)(2n+1)}{6} - \frac{n(n+1)}{2} \right]$$

$$= \left(\frac{a}{n}\right)^3 \left[\frac{4n^3}{6} + \frac{3n^2}{6} - \frac{n}{6} \right] = \frac{2a^3}{3}\left[1 + \frac{3}{4n} - \frac{1}{4n^2} \right]$$

$$= \frac{2a^3}{3} + \frac{a^3}{2n} - \frac{a^3}{6n^2}.$$

The results of the summations were obtained by mathematical induction. We now observe that if we make successive refinements, always using a regular partition, n increases and $U(P, x, x^2)$ approaches $(2a^3)/3$. A similar computation would show us that the lower sum $L(P, x, x^2)$ also approaches $(2a^3)/3$, and therefore we can conclude that

$$\int_0^a x \, dx^2 = \frac{2a^3}{3}. \qquad (\text{III.1.7})$$

You could check the difference between the upper sum and the lower sum and show that this is $(a^3)/n$, and hence that successive refinements will make this arbitrarily small. You will note that we now have available to us the same type of reasoning we used in Example 1.3 in order to assure ourselves that the integral does exist and that (III.1.7) gives us the correct result.

This result is a special case of a more general formula. You are asked to develop this generalization in Exercise 5. Note that (III.1.6) is also a special case of this new result.

EXERCISES

1. Evaluate:

   (a) $\int_0^5 3 \, dx$
   (b) $\int_0^5 x^3 \, dx$
   (c) $\int_{\pi/6}^{\pi/4} 7 \, d(\sin x)$
   (d) $\int_{-1}^2 x^0 \, d(x^4)$
   (e) $\int_2^8 x^3 \, d(3)$
   (f) $\int_4^9 \pi \, d(\sqrt{x})$
   (g) $\int_0^1 x^{14} \, dx$
   (h) $\int_{\pi/6}^{\pi} \sin x \, d(\pi)$

2. Evaluate:

   (a) $\int_0^1 x^3 \, dx$
   (b) $\int_0^2 x^3 \, dx$
   (c) $\int_0^3 x^3 \, dx$
   (d) $\int_0^4 x^3 \, dx$
   (e) $\int_0^5 x^3 \, dx$
   (f) $\int_0^6 x^3 \, dx$

3. (a) Sketch the curves $y = x$ and $y = x^3$.
   (b) Find a positive value of a such that

$$\int_0^a x \, dx = \int_0^a x^3 \, dx.$$

   (c) Using the value of a found in part (b), sketch $x = a$ on your graph.
   (d) Interpret the geometric meaning of the value of a using your graph.

4. If m and n are positive integers and if $m < n$

(a) Show that $\int_0^1 x^m \, dx > \int_0^1 x^n \, dx$

(b) Show that $\int_1^2 x^m \, dx < \int_1^2 x^n \, dx$

(c) Why is the inequality of part (a) reversed for part (b)?

5. (a) Use a regular partition of $[0, a]$ with p subintervals and give the upper and lower sums corresponding to

$$\int_0^a x^n \, dx^m.$$

(b) Show that the upper sum can be written as

$$U(P, x^n, x^m) = \frac{a^{n+m}}{p^{n+m}} \sum_{k=1}^{p} k^n [k^m - (k - 1)^m].$$

(c) Write the lower sum in a manner similar to that of part (b).

(d) Using the result from Exercise I.10.8 which states that

$$m(k - 1)^{m-1} < [k^m - (k - 1)^m] < mk^{m-1} \quad \text{show that}$$

$$U(P, x^n, x^m) < \frac{ma^{n+m}}{p^{n+m}} \sum_{k=1}^{p} k^{n+(m-1)} \quad \text{and}$$

$$L(P, x^n, x^m) > \frac{ma^{n+m}}{p^{n+m}} \sum_{k=1}^{p} (k - 1)^{n+(m-1)} = \frac{ma^{n+m}}{p^{n+m}} \sum_{k=0}^{p-1} k^{n+m-1}.$$

(e) Using a result similar to (III.1.4) together with the inequalities of part (d), show that

$$L(P, x^n, x^m) < U(P, x^n, x^m) \quad \text{and}$$

$$U(P, x^n, x^m) - L(P, x^n, x^m) < \frac{ma^{n+m}}{p}$$

(f) Show that $\int_0^a x^n \, dx^m$ exists and

$$\int_0^a x^n \, dx^m = \frac{ma^{n+m}}{n + m}. \qquad (III.1.8)$$

6. Use (III.1.8) to evaluate each of the following:

(a) $\int_0^2 x^3 \, dx^2$

(b) $\int_0^3 x^2 \, dx^3$

(c) $\int_0^1 x^4 \, dx^5$

(d) $\int_0^{\sqrt{2}} x^3 \, dx^4$

(e) $\int_0^{\sqrt{3}} x^2 \, dx^3$

(f) $\int_0^2 x^4 \, dx^5$

7. Use (III.1.8) to evaluate each of the following:

(a) $\int_0^2 x^0 \, dx^6$

(b) $\int_0^2 x \, dx^5$

(c) $\int_0^2 x^2 \, dx^4$

(d) $\int_0^2 x^3 \, dx^3$

(e) $\int_0^2 x^4 \, dx^2$

(f) $\int_0^2 x^5 \, dx$

(g) $\int_0^2 x^6 \, dx^0$

8. Evaluate each of the following and verify whether the result given by (III.1.8) agrees with results you would have obtained for the following integrals using methods derived in this section.

   (a) $\int_0^a x^n \, dx$

   (b) $\int_0^a x \, dx^2$

   (c) $\int_0^a dx^n = \int_0^a x^0 \, dx^n$

   (d) $\int_0^a x^n \, d(1) = \int_0^a x^n \, dx^0$

9. (a) Sketch the general shape of $y = x^n$ for an integer $n > 1$.

   (b) Set up the integral for the area bounded by $y = x^n$, $y = 0$ and $x = a$, where a is some positive real number.

   (c) Find the number of square units of area in the figure defined in part (b).

   (d) Find the area in the rectangle bounded by $y = 0$, $x = a$, $y = a^n$, and $x = 0$.

   (e) Show that the area of part (b) is included in the area of part (d), and show that the ratio of the portion of the rectangle which is not in part (b) to the area in part (b) is equal to n.

   (f) Show that the area in part (b) is $[1/(n + 1)]$ times the area of the rectangle in part (d).

10. (a) Show that $\int_0^a x^n \, dx^n = (a^n)^2/2$ for any positive integer n.

    (b) Show that $\int_0^a x^{kn} \, dx^n = (a^n)^{k+1}/(k + 1)$ for any positive integers k and n.

    (c) Evaluate $\int_0^{\sqrt[3]{3}} x^9 \, dx^3$.

III.2 Theorems Concerning Integrals

In Section III.1 we developed formulas which are of great assistance to us in the evaluation of a limited variety of integrals. It would be very helpful if we could find some means for expanding the utility of these formulas. It would be nice, for instance, to be able to use our earlier results to evaluate

$$\int_0^6 (x^7 - 9x^4 + 8)dx.$$

Since this integral has three terms, each of which is of the type we considered in the last section, it would not seem impossible that we might build upon our earlier work. It will be the purpose of this section and the next to develop theorems which will permit this expansion of the usage of the earlier results.

In the majority of theorems the proofs are rather straightforward and only the outline of the proof will be given. In some instances only a suggestion will be offered. You should be aware, of course, that in cases where the proof is not given in detail it is up to you to supply the missing steps. You are also aware that you should not use any result which has not been proven as a basis for a step or a statement in any proof or solution. Since we do not have a large backlog of proven theorems concerning integrals it is not unexpected that for the first proofs, at least, we will have to resort to the RS sums upon which the integrals are based. It is worth mentioning that it is frequently helpful to work through an example of the statement of the theorem, for that may give clues concerning a possible method of proof.

Theorem 2.1. *If*

$$\int_a^b f_1(x)dg(x) \quad and \quad \int_a^b f_2(x)dg(x)$$

exist, then

$$\int_a^b [f_1(x) + f_2(x)]dg(x) = \int_a^b f_1(x)dg(x) + \int_a^b f_2(x)dg(x).$$

OUTLINE OF PROOF. If we select any partition $P[a, b]$ and examine the upper sums $U(P, f_1 + f_2, g)$, $U(P, f_1, g)$ and $U(P, f_2, g)$ we observe that in any subinterval it is possible, if not probable, that the evaluation point used for $f_1(x)$ will not be the same evaluation point as that used for $f_2(x)$. However, since $[g(x_k) - g(x_{k-1})]$ will have the same sign in each of the three upper sums, we would need either the largest value of f_1 and the largest value of f_2, or else we would need the smallest values in the subinterval. If the evaluation points for the two functions f_1 and f_2 differ within the subinterval, then it can be shown that

$$[f_1(r_k) + f_2(r_k)][g(x_k) - g(x_{k-1})] \le f_1(s_k)[g(x_k) \\ - g(x_{k-1})] + f_2(t_k)[g(x_k) - g(x_{k-1})]$$

where r_k is the evaluation point used for the integral involving $f_1 + f_2$, s_k is the evaluation point used for the integral involving only f_1, and t_k is the evaluation point used for the integral involving only f_2. From this we can deduce the fact that

$$\overline{\int_a^b} [f_1(x) + f_2(x)]dg(x) \le \overline{\int_a^b} f_1(x)dg(x) + \overline{\int_a^b} f_2(x)dg(x).$$

By similar reasoning we can obtain a corresponding relation for the lower integrals. Hence, we have

$$\underline{\int_a^b} f_1(x)dg(x) + \underline{\int_a^b} f_2(x)dg(x) \le \underline{\int_a^b} [f_1(x) + f_2(x)]dg(x)$$

$$\le \overline{\int_a^b} [f_1(x) + f_2(x)]dg(x) \qquad \text{(III.2.1)}$$

$$\le \overline{\int_a^b} f_1(x)dg(x) + \overline{\int_a^b} f_2(x)dg(x).$$

Since the two integrals $\int_a^b f_1(x)dg(x)$ and $\int_a^b f_2(x)dg(x)$ exist, the first and last expressions of (III.2.1) are equal, and hence each of the inequalities of this expression can be replaced by an equality.   □

Theorem 2.2. *If the integrals involved exist, and if c is a constant, then*

$$\int_a^b cf(x)dg(x) = \int_a^b f(x)d[cg(x)] = c\int_a^b f(x)dg(x).$$

OUTLINE OF PROOF. Show that

$$S(P, T, cf, g) = S(P, T, f, cg) = cS(P, T, f, g)$$

for any RS sum $S(P, T, f, g)$. The Theorem follows from this fact and our previous Theorems.   □

Theorem 2.2 permits us to move a factor outside of the integral provided the factor is a constant. It cannot be emphasized too greatly that it is *not* possible in general to move a factor outside of the integral if the factor is not a constant.

With these two theorems and the results of the last section we can now handle the integral that was proposed in the first paragraph of this section.

EXAMPLE 2.1. Evaluate

$$\int_0^6 (x^7 - 9x^4 + 8)dx.$$

Solution. By Theorem 2.1 we have

$$\int_0^6 (x^7 - 9x^4 + 8)dx = \int_0^6 x^7\, dx + \int_0^6 (-9)x^4\, dx + \int_0^6 8\, dx.$$

Theorem 2.2 permits us to write

$$\int_0^6 x^7\, dx + \int_0^6 (-9)x^4\, dx + \int_0^6 8\, dx$$

$$= \int_0^6 x^7\, dx + (-9)\int_0^6 x^4\, dx + 8\int_0^6 dx.$$

Combining these two equations and using the results of the last section, we now have $\int_0^6 (x^7 - 9x^4 + 8)dx = (6^8/8) - 9(6^5/5) + 8(6)$. From this point on we have just arithmetic, and hence we have been able to evaluate the integral in question.

This demonstrates that theorems can be very helpful at times.

Theorem 2.3. *If all the integrals involved exist, then*

$$\int_a^b f(x)d[g_1(x) + g_2(x)] = \int_a^b f(x)dg_1(x) + \int_a^b f(x)dg_2(x).$$

OUTLINE OF PROOF. This proof is somewhat similar to the proof of Theorem 2.2. The biggest pitfall would be that of ignoring the situation in which g_1 might be increasing in a subinterval and g_2 might be decreasing in the same interval. In this case we would have for the upper sum

$$f(r_k)[(g_1(x_k) + g_2(x_k)) - (g_1(x_{k-1}) + g_2(x_{k-1}))]$$
$$\leq f(s_k)[g_1(x_k) - g_1(x_{k-1})] + f(t_k)[g_2(x_k) - g_2(x_{k-1})].$$

Where r_k, s_k, and t_k are evaluation points for the respective upper sums. $\square$

We can combine the results of Theorems 1, 2, and 3 in the following corollary.

Corollary 2.1. *If c_1, c_2, c_3 and c_4 are constants and if the integrals in the following expression exist, then*

$$\int_a^b [c_1 f_1(x) + c_2 f_2(x)]d[c_3 g_1(x) + c_4 g_2(x)]$$

$$= c_1 c_3 \int_a^b f_1(x)dg_1(x) + c_1 c_4 \int_a^b f_1(x)dg_2(x)$$

$$+ c_2 c_3 \int_a^b f_2(x)dg_1(x) + c_2 c_4 \int_a^b f_2(x)dg_2(x).$$

While we have considered several theorems which apply over a general interval $[a, b]$, the majority of the results already obtained which help us evaluate integrals apply only over an interval $[0, a]$ starting at the origin and assuming $a > 0$. It is not to be expected that such an interval will suffice for all of our needs, and consequently we will consider some theorems which address this problem.

Theorem 2.4. *If $a < b < c$ and if all of the integrals involved exist, then*

$$\int_a^b f(x)dg(x) + \int_b^c f(x)dg(x) = \int_a^c f(x)dg(x).$$

OUTLINE OF PROOF. Start with the RS sums $S(P_1[a, b], T_1, f, g)$ and $S(P_2[b, c], T_2, f, g)$. Show that $P = P_1 \cup P_2$ is a partition of $[a, c]$ and $T = T_1 \cup T_2$ is a corresponding evaluation set. Then show that $S(P_1, T_1, f, g) + S(P_2, T_2, f, g) = S(P, T, f, g)$. From this obtain the necessary upper and lower sums and show that the conclusion of the theorem holds. $\square$

This last result has rather interesting implications, for it permits us to evaluate the integral over $[a, b]$ and over $[b, c]$ and then know that the sum of the results is the result we would have obtained by evaluating the integral over $[a, c]$. Of course, it would also be possible to evaluate the integral over $[a, c]$ and then subtract either the integral over $[a, b]$ to obtain the integral over $[b, c]$ or to subtract the integral over $[b, c]$ to obtain the integral over $[a, b]$. By our definitions, of course, we have assumed that $a < b < c$. It would be nice to be able to relax this last restriction in Theorem 2.4. For instance, we might wish to let $b = c$. Then, of course, by mere substitution we would have

$$\int_a^b f(x)dg(x) + \int_b^b f(x)dg(x) = \int_a^b f(x)dg(x)$$

This could not be correct unless we were to have $\int_b^b f(x)dg(x) = 0$. While it seems reasonable that it would be true that the RS sum in which the domain starts and stops at the same point should have a zero value, we will be on safer ground if we define this to be true as follows.

Definition 2.1. For any $f(x)$ and $g(x)$ it is true that $\int_a^a f(x)dg(x) = 0$.

We have solved one problem by this definition and made Theorem 2.4 one of slightly larger application. However it would be nice to be able to remove all restrictions on the order of a, b, and c. If we were to do this, we would then be able to write, letting $c = a$,

$$\int_a^b f(x)dg(x) + \int_b^a f(x)dg(x) = \int_a^a f(x)dg(x).$$

By our definition the right hand side is zero, and this can only be true provided

$$\int_b^a f(x)dg(x) = -\int_a^b f(x)dg(x).$$

Since we have not covered the case in which the interval of integration goes from the larger to the smaller value, we will use the definition route again.

Definition 2.2. If the integral $\int_a^b f(x)dg(x)$, $b > a$, exists, then the integral $\int_b^a f(x)dg(x)$ is defined by the relation $\int_b^a f(x)dg(x) = -\int_a^b f(x)dg(x)$.

This use of definition is not new, for you will remember that we had a definition of a^n for all cases in which n is a positive integer early in our study of algebra, and we then found certain laws which the exponents obeyed. In order to make these laws apply to a larger class of exponents, we defined a^{-n} to be $1/a^n$ and we also defined $a^0 = 1$. In similar fashion we define $0! = 1$. The results defined above will help, for we can now extend our earlier results to all intervals $[a, b]$. Thus

$$\int_a^b f(x)dg(x) = \int_a^0 f(x)dg(x) + \int_0^b f(x)dg(x) = \int_0^b f(x)dg(x) - \int_0^a f(x)dg(x).$$

In particular we can write

$$\int_a^b x^n \, dx = \int_0^b x^n \, dx - \int_0^a x^n \, dx = \frac{b^{n+1}}{(n+1)} - \frac{a^{n+1}}{(n+1)}$$

$$= \frac{(b^{n+1} - a^{n+1})}{(n+1)}. \qquad \text{(III.2.2)}$$

EXAMPLE 2.2. Evaluate

$$\int_1^2 (3x^3 - x^2 + 2) d(x^3 - 3x).$$

Solution. Using the corollary we can write this as

$$3 \int_1^2 x^3 \, d(x^3) - \int_1^2 x^2 \, d(x^3) + 2 \int_1^2 d(x^3) - 9 \int_1^2 x^3 \, dx$$

$$+ 3 \int_1^2 x^2 \, dx - 6 \int_1^2 dx.$$

Now with the aid of Exercise (III.1.5) and the reasoning of Equation (III.2.2) we have

$$3 \left[\frac{3(2)^6}{6} - \frac{3(1)^6}{6} \right] - \left[\frac{3(2)^5}{5} - \frac{3(1)^5}{5} \right] + 2[(2)^3 - (1)^3]$$

$$- 9 \left[\frac{(2)^4}{4} - \frac{(1)^4}{4} \right] + 3 \left[\frac{(2)^3}{3} - \frac{(1)^3}{3} \right] - 6[(2) - (1)]$$

$$= 3 \left[\frac{3}{6}(63) \right] - \left[\frac{3}{5}(31) \right] + 2[(7)] - 9 \left[\frac{1}{4}(15) \right] + 3 \left[\frac{1}{3}(7) \right] - 6[(1)] = 57.15.$$

EXAMPLE 2.3. If $f(x)$ is a function such that $f(x) = x^2$ for values of $x \le 3$ and $f(x) = 9$ for values of $x > 3$, find the value of the integral

$$\int_1^6 f(x) dx^2.$$

Solution. In this case we have two definitions for $f(x)$, one valid in the interval to the left of and including $x = 3$ and the other definition valid in the interval from $x = 3$ extending toward the right. Since we are concerned with the interval $[1, 6]$, it is clear that we will have to use both definitions. We can do this by noting that we can represent $[1, 6]$ as the union of $[1, 3]$ and $[3, 6]$. Since it is possible to express the integral over $[1, 6]$ as the sum of the integrals over $[1, 3]$ and $[3, 6]$, we can then write

$$\int_1^6 f(x) dx^2 = \int_1^3 f(x) dx^2 + \int_3^6 f(x) dx^2 = \int_1^3 x^2 \, dx^2 + \int_3^6 9 \, dx^2$$

$$= \frac{2(3^4 - 1^4)}{4} + 9(6^2 - 3^2) = \frac{2(80)}{4} + 9(27) = 283.$$

Note that in this example we have considered the original integral as the sum of two separate integrals. When there are two or more definitions of a function, each depending upon the portion of the domain under consideration, the method shown in Example 2.3 provides us with a means of evaluating the integral. This is another illustration of the versatility provided us by the theorems we have proven up to this point.

Having considered several theorems and definitions which involve equalities, we will now consider one involving inequalities.

Theorem 2.5. *If $f_1(x) \le f_2(x)$ for every x in the interval $[a, b]$, and if $g(x)$ is monotonic increasing throughout $[a, b]$, then*

$$\int_a^b f_1(x)dg(x) \le \int_a^b f_2(x)dg(x).$$

OUTLINE OF PROOF. For any partition $P[a, b]$ and evaluation set T over that partition, we have $S(P, T, f_1, g) \le S(P, T, f_2, g)$. This follows from the fact that $f_1(t_k)[g(x_k) - g(x_{k-1})] \le f_2(t_k)[g(x_k) - g(x_{x-1})]$ for each t_k and $[x_{k-1}, x_k]$ since $[g(x_k) - g(x_{k-1})] \ge 0$. $\square$

EXAMPLE 2.4. Find upper and lower bounds for the value of the integral $\int_{\pi/6}^{\pi/4} \tan x \, dx^2$.

Solution. Since $\tan x \ge 1/\sqrt{3}$ and $\tan x \le 1$ over the interval $[\pi/6, \pi/4]$, and since x^2 is increasing over this interval, the theorem applies.

Therefore, we have $\int_{\pi/6}^{\pi/4} (1/\sqrt{3})dx^2 \le \int_{\pi/6}^{\pi/4} \tan x \, dx^2 \le \int_{\pi/6}^{\pi/4} 1 \, dx^2$ or $(1/\sqrt{3})[(\pi^2/16) - (\pi^2/36)] \le \int_{\pi/6}^{\pi/4} \tan x \, dx \le 1[(\pi^2/16) - (\pi^2/36)]$.   Thus $5\pi^2/(144\sqrt{3}) \le \int_{\pi/6}^{\pi/4} \tan x \, dx^2 \le 5\pi^2/144$ or $0.19785 \le \int_{\pi/6}^{\pi/4} \tan x \, dx^2 \le 0.34269$.

EXERCISES

1. Evaluate each of the following:

   (a) $\int_1^2 4x^3 \, dx$
   (b) $\int_1^3 5x^2 \, dx^2$
   (c) $\int_1^2 2x^3 \, dx^4$
   (d) $\int_2^3 5x \, dx^4$
   (e) $\int_0^{\pi/2} 2 \, d(\cos x)$
   (f) $\int_\pi^{2\pi} 7 \sin x^2 \, d(\pi)$
   (g) $\int_0^2 3(x^2)^3 \, dx$
   (h) $\int_{0.1}^1 (2x^3)^2 \, dx^2$

2. Evaluate each of the following:

   (a) $\int_2^5 (x + 2)dx$
   (b) $\int_1^4 (x^2 - 3x + 1)d(x + 1)$

(c) $\int_2^1 (x + 2)^2 \, dx^2$

(d) $\int_1^1 (3x^2 - 4x)^2 \, d(x^2 - 5)$

(e) $\int_0^2 (2x - 1)^3 \, dx$

(f) $\int_3^1 (x^2 + x)d(x^2 - x)$

(g) $\int_1^2 (x^2 + x + 1)^2 \, d(2x - 3)$

(h) $\int_2^3 x \, d(x + 1)^2$

3. Evaluate:

(a) $\int_1^3 (4x^2 - 7x^5 + 2)dx$

(b) $\int_1^2 (x^3 + 4x)dx$

(c) $\int_a^0 (3x + 7x^3 - 9x^5)dx$

(d) $\int_0^2 (x^2 - 3x)d(7x)$

(e) $\int_2^4 (x - x^2)d(x + x^2)$

(f) $\int_4^2 4x^2 \, dx$

4. Find the value of $\int_1^2 (\sum_{k=0}^7 x^k)d(5x - 7)$.

5. If $f(x) = x^2$ when x is in the interval $[1, 2]$ and $f(x) = x^3 - x^2$ when x is in the interval $[2, 4]$, find the value of $\int_1^4 f(x)d(5x^2)$.

6. Evaluate $\int_1^5 f(x)dx^3$ if $f(x) = 4 - x^2$ when x is in the interval $[1, 3]$ and $f(x) = x - 8$ when x is in the interval $[3, 5]$.

7. (a) Evaluate $\int_0^4 (x^2 - 4)dx$.

   (b) Sketch the graph of $y = x^2 - 4$ over the interval $[0, 4]$.

   (c) Note the portion of the graph of part (b) that is below the x-axis and the portion that is above. Sketch the area bounded by $x = 0$, $x = 4$, and the curve $y = x^2 - 4$.

   (d) Calculate the area of part (c) which is below the x-axis and the area which is above the x-axis. Note that the integral determining the area below the x-axis has a negative value.

   (e) Interpret the integral of part (a) as an area, noting whether the two areas of part (c) are added together taking into account the signs of the two areas (the area above being positive and the area below being negative) or whether you have the sums of the absolute values of the two areas.

   (f) Show how you could obtain the sum of the absolute values by breaking the interval $[0, 4]$ into two parts and using a different $f(x)$ for each part.

8. Show that $\int_0^4 (x - 2)dx = 0$ and interpret this geometrically in terms of area under the curve $f(x) = x - 2$ using the general approach indicated in Exercise 7.

9. The triangle with vertices $(0, 0)$, $(5, 0)$, and $(5, 3)$ is rotated about the x-axis to form a cone. If you partition the x-axis and then take cross sections perpendicular to the x-axis through the partition points, you have the cone as being formed by a series of disks with circular cross sections.

   (a) Sketch this figure.

   (b) Set up the RS sum for the volume of the cone by adding the volumes of the disks.

(c) Set up the integral corresponding to your RS sum.

(d) Evaluate the integral, and compare your result with the result you would have found if you had used the formula for the volume of a cone.

10. Repeat the work of Exercise 9 if the vertices of the original triangle are $(0, 0)$, $(h, 0)$, and (h, r). Show that this gives the formula for the volume of the cone.

11. The quarter circle $y = (r^2 - x^2)^{1/2}$ in the first quadrant is rotated about the x-axis to form a hemisphere. Sketch this hemisphere. Set up the integral for the volume of this hemisphere and then find the volume. Show that your result gives the formula for the volume of a hemisphere of radius r.

12. Find bounds for the value of each of the following integrals.

(a) $\int_0^3 2^{-x} \, dx$

(b) $\int_0^{\pi/3} \sin x \, dx^2$

(c) $\int_0^{\pi/3} \sec x \, dx$

(d) $\int_1^{10} \log_{10} x \, dx$

(e) $\int_3^4 \sqrt{25 - x^2} \, d\sqrt{x - 3}$

(f) $\int_1^4 \cos^2 x \, dx^3$

13. Show that $|\int_a^b (\sin x + \cos x) dg(x)| < \sqrt{2}[g(a) - g(b)]$ if $a < b$ and $g(x)$ is decreasing in $[a, b]$. [*Hint*: Show that $\sin x + \cos x = \sqrt{2} \cos(x - \frac{1}{4}\pi)$.]

14. Let $g(x) = [x]$ where $[x]$ is the bracket function, and as we mentioned earlier is defined to be the largest integer which is not greater than x.

(a) Set up the RS sum $S(P[1, 5], T, 5, g(x))$ for some partition P of $[1, 5]$.

(b) Observe that if the diameter of P is less than one there are some terms in the RS sum of part (a) which have the value zero.

(c) Evaluate the integral

$$\int_1^5 5 \, dg(x) = \int_1^5 5 \, d[x].$$

(Integrals of this type are useful in statistics and in other applications.)

15. Evaluate the integral

$$\int_1^5 x^2 \, d([2x]).$$

[Note that you have the greatest integer function of $(2x)$ and not twice the greatest integer function of x.]

P16. You are given a circular disk with a radius of 5 inches such that for each portion of the disk having a square inch of surface the mass is 0.5 pounds. Find the second moment of this disk about an axis through the center of the disk and perpendicular to the disk.

P17. Find the distance traveled by an object in the first 10 seconds of movement if its velocity is given by the relation $v(t) = 16t^2 + 32t$.

B18. Evaluate the integral involved in Exercise (II.4.21) and thus evaluate the amount of blood flowing past a point in the blood vessel in one second.

S19. If the cost of the first 100 items of weekly production in a given plant is \$15,000 and if the marginal cost is given by the relation $M(n) = 4 + 10n - 0.000001n^2 + 0.000000003n^3$, find the cost of producing 300 items in one week. $M(n)$ is the increase in the number of dollars required to increase weekly production from n items to $(n + 1)$ items.

M20. Prove Theorem 2.1.

M21. Prove Theorem 2.2.

M22. Prove Theorem 2.3.

M23. Prove Theorem 2.4.

M24. Prove Theorem 2.5.

III.3   Some Additional Theorems

In Section III.2 we discussed certain theorems which assist us in evaluating integrals. These theorems express integrals with many terms as a combination of integrals of a simple nature. This capability is not always sufficient to permit us to evaluate an integral, however. If, for instance, we were faced with the integral

$$\int_{\pi/6}^{\pi/2} \sin^2 x \, d(\sin x)$$

we would be lost, for this does not resemble anything we have seen so far. In this section our purpose will be to obtain some additional theorems which among other things will help us evaluate integrals of this type.

A very important Theorem, called the *integral theorem of the mean* or *the mean value theorem for integrals* follows quickly from Theorem (III.2.5). The word *mean* is used here, as in all of mathematics, as it comes to us from the older English with the meaning of middle, or a value in the middle. You should see why we have chosen this name when you read the Theorem.

Theorem 3.1. *If $f(x)$ is a bounded, continuous function and $g(x)$ is monotonic and bounded over the interval $[a, b]$ then there is a number c such that $a < c < b$ and $\int_a^b f(x)dg(x) = f(c)[g(b) - g(a)]$.*

OUTLINE OF PROOF. Since $g(x)$ is monotonic, we can assume that it is increasing. (A similar argument would hold if it were decreasing.) Since $f(x)$ is bounded, it must have a lub, M, and a glb, m. on $[a, b]$. Furthermore, there must be values x_1 and x_2 in $[a, b]$ such that $f(x_1) = M$ and $f(x_2) = m$. This will be shown in Chapter VII, but we will assume it for the present. (It is not an unreasonable assumption, although it is one that should be questioned.) Since $m \le f(x) \le M$ for all values in $[a, b]$, it follows from

Theorem III.2.5 that

$$\int_a^b m \, dg(x) = m[g(b) - g(a)] \le \int_a^b f(x) dg(x)$$

$$\le \int_a^b M \, dg(x) = M[g(b) - g(a)].$$

Thus,

$$\int_a^b f(x) dg(x)$$

has a value between $m[g(b) - g(a)]$ and $M[g(b) - g(a)]$. Consequently this integral must be equal to $C[g(b) - g(a)]$ for some value of C between m and M. However, by the intermediate value property of continuous functions, $f(x)$ must assume the value C at some point x between x_1 and x_2. Therefore, $f(x) = C$ at a point in the interval $[a, b]$. Let us denote this point by $x = c$. Then $f(c) = C$, and we have

$$\int_a^b f(x) dg(x) = f(c)[g(b) - g(a)]$$

as required.   □

EXAMPLE 3.1. Find a value c in $(1, 3)$ such that

$$\int_1^3 f(x) dg(x) = f(c)[g(3) - g(1)]$$

if $f(x) = x^2 + 2x - 5$ and $g(x) = x^2$.

Solution.

$$\int_1^3 f(x) dg(x) = \int_1^3 (x^2 + 2x - 5) d(x^2)$$

$$= \int_1^3 x^2 \, d(x^2) + 2 \int_1^3 x \, d(x^2) - 5 \int_1^3 d(x^2)$$

$$= \frac{2}{4}[(3)^4 - (1)^4] + 2\left(\frac{2}{3}\right)[(3)^3 - (1)^3] - 5[(3)^2 - (1)^2]$$

$$= 40 + \frac{104}{3} - 40 = \frac{104}{3}.$$

Also

$$f(c)[g(3) - g(1)] = (c^2 + 2c - 5)[(3)^2 - (1)^2] = 8c^2 + 16c - 40.$$

These two results will be equal if

$$8c^2 + 16c - 40 = \frac{104}{3} \quad \text{or} \quad 3c^2 + 6c - 28 = 0.$$

Hence $c = -1 + \frac{1}{3}\sqrt{93}$ *or* $-1 - \frac{1}{3}\sqrt{93}$. It is clear that the latter value is not in (1, 3) but a bit of computation shows that $-1 + \frac{1}{3}\sqrt{93}$ is approximately $-1 + \frac{1}{3}(9.64)$ or 2.21 and this is the value requested.

It is always necessary in a problem of this variety to check your results, for it is required that the result in this case be in the interval (1, 3), and any result not in this interval fails to provide a usable answer. We should also note that we have required that the value be in (1, 3) and not in [1, 3]. This is possible since Theorem 1 guarantees that there will be a suitable value *between* the two end points. Of course, we should also note that in this case $g(x)$ is increasing, and therefore monotonic throughout the interval.

Theorem 3.1 will be used most frequently to assure us that a value, c, exists without requiring that we obtain a specific value. However, there are occasions when such a value is required. In all cases be sure that you check whether $g(x)$ is monotonic, for the Theorem does not apply otherwise. Also be certain to check that your value of c is in the required open interval.

It is often very helpful in evaluating integrals to be able to make a substitution for some function which occurs more than once in the integrand. This would help, for instance, in the integral given at the beginning of this section. It is true that such a substitution should not change the value of the integral, but it is usually much easier to see how to handle a problem if the problem looks somewhat less complicated. The following Theorem will help us in this regard.

Theorem 3.2. *If $g(x)$ is a monotonic, continuous, bounded function on the interval $[a, b]$ and if the integral*

$$\int_a^b f(g(x))dg(x)$$

exists, then

$$\int_a^b f(g(x))dg(x) = \int_{g(a)}^{g(b)} f(y)dy.$$

PROOF. Since the integral

$$\int_a^b f(g(x))dg(x)$$

exists, we can content ourselves with using either the upper or lower RS. sum, but without the requirement that we consider both. Therefore, this integral will be considered here as the lub of the lower RS sum. Let us suppose that T is the evaluation set for the partition P such that

$$L(P, f, g) = \sum_{k=1}^n f(g(t_k))[g(x_k) - g(x_{k-1})].$$

If we let $y = g(x)$, then the set of points P_y such that $y_k = g(x_k)$ will give us a partition of $[g(a), g(b)]$ provided we eliminate any points which are repeated. We can now establish an evaluation set T_y by using the points $s_k = g(t_k)$. The fact that $g(x)$ is monotonic assures us that s_k is in the interval $[y_{k-1}, y_k]$. It follows at once that

$$\sum_{k=1}^{n} f(g(t_k))[g(x_k) - g(x_{k-1})] = \sum_{k=1}^{n} f(s_k)[y_k - y_{k-1}].$$

Therefore, for each lower sum corresponding to the integral

$$\int_a^b f(g(x))dg(x)$$

we have a corresponding lower sum with equal value associated with the integral

$$\int_{g(a)}^{g(b)} f(y)dy.$$

Note that we have determined the limits of integration in the latter integral by noting that the lower limit $x = a$ indicates a corresponding lower limit $y = g(a)$ and similarly for the upper limit. Thus, we are assured that the partition $P[a, b]$ is equivalent to the partition $P_y[g(a), g(b)]$.

Since the lower sums of the first integral each give rise to lower sums with equal values for the second integral, we would apparently be able to conclude that the two sets of lower sums have the same lub. However, it might be possible for an additional lower sum to creep in for the integral

$$\int_{g(a)}^{g(b)} f(y)dy$$

which would alter the lub, and consequently we must be sure that we can go in the other direction. However, since we have assumed that $g(x)$ is a continuous, monotonic function it follows if x is in $[a, b]$ then $g(x)$ is in the interval $[g(a), g(b)]$. By the intermediate value property, we know that for any value of y in $[g(a), g(b)]$ there is a value of x in $[a, b]$ such that $g(x) = y$. Hence, if we were to consider the lower RS sum which is associated with

$$\int_{g(a)}^{g(b)} f(y)dy$$

we could then make the reverse substitution and obtain the corresponding lower RS sum associated with

$$\int_a^b f(g(x))dg(x).$$

Therefore, we have for each lower RS sum for one of the two integrals a corresponding lower RS sum for the other one which has the same value.

Consequently, any upper bound for one set of lower sums is also an upper bound for the other. This implies that the lubs are also equal, and hence the integrals are equal.   □

EXAMPLE 3.1. Evaluate

$$\int_{\pi/6}^{\pi/2} \sin^2 x \, d(\sin x).$$

Solution. We note that $g(x) = \sin x$ is monotonic, continuous, and bounded over the interval $[\pi/6, \pi/2]$, and hence Theorem 3.2 applies. Since $\sin(\pi/6) = 1/2$ and $\sin(\pi/2) = 1$, our new integral will apply over the interval $[1/2, 1]$, and, of course, we will have $y = g(x) = \sin x$. Consequently,

$$\int_{\pi/6}^{\pi/2} \sin^2 x \, d(\sin x) = \int_{1/2}^{1} y^2 \, dy = \frac{1^3 - (1/2)^3}{3} = \frac{7}{24}.$$

It is well to be certain that Theorem 3.2 applies before trying to make use of it, for it is possible that a situation might occur in which some of the hypotheses fail to hold, and hence the theorem might not apply. We will broaden the coverage of this theorem in some of the remaining theorems and corollaries of this section, but there will still remain some limitations.

From Example 3.1 it is apparent that this substitution concept is going to be of great assistance to us. However, it often happens that we may not have f as a function involving g, but rather that f and g may both be composite functions involving a common function, say h. To cover this situation, we have the following theorem.

Theorem 3.3. *If $h(x)$ is bounded and monotonic on the interval $[a, b]$ and $g(x)$ is continuous, bounded, and strictly monotonic on the interval $[h(a), h(b)]$, and if the integral*

$$\int_a^b f(h(x))dg(h(x))$$

exists, then

$$\int_a^b f(h(x))dg(h(x)) = \int_{h(a)}^{h(b)} f(y)dg(y).$$

PROOF. Let $g(h(x)) = k(x)$ and let $g^{-1}(x)$ be the inverse function of $g(x)$ on the interval $[h(a), h(b)]$. The inverse exists since $g(x)$ is strictly monotonic. We know that $h(x) = g^{-1}(g(h(x))) = g^{-1}(k(x))$, and therefore $f(h(x)) = f(g^{-1}(k(x)))$. It is also true that $g^{-1}(g(y)) = y$. With this information, we have

$$\int_a^b f(h(x))dg(h(x)) = \int_a^b f(g^{-1}(k(x)))dk(x) = \int_{k(a)}^{k(b)} f(g^{-1}(z))dz$$

from Theorem 3.2. On the other hand, we also know that

$$\int_{h(a)}^{h(b)} f(y)dg(y) = \int_{h(a)}^{h(b)} f(g^{-1}(g(y)))dg(y)$$

$$= \int_{g(h(a))}^{g(h(b))} f(g^{-1}(z))\, dz = \int_{k(a)}^{k(b)} f(g^{-1}(z))dz.$$

Therefore

$$\int_{a}^{b} f(h(x))dg(h(x)) = \int_{h(a)}^{h(b)} f(y)dg(y)$$

since these two integrals are equal to the same integral. $\square$

Corollary 3.1. *If $h(x)$ is bounded and monotonic on the interval $[a, b]$ and $g(x)$ is continuous, bounded, and monotonic on the interval $[h(a), h(b)]$, and if the integral*

$$\int_{a}^{b} f(h(x))dg(h(x))$$

exists, then

$$\int_{a}^{b} f(h(x))dg(h(x)) = \int_{h(a)}^{h(b)} f(y)dg(y).$$

OUTLINE OF PROOF. This Corollary differs from Theorem 3.3 in that $g(x)$ is required to be monotonic, but not required to be strictly monotonic. Partition $[a, b]$ such that on each subinterval $[x_{k-1}, x_k]$ either $g(h(x))$ is strictly monotonic or else $g(h(x))$ is constant. This is possible since $g(x)$ is monotonic on $[h(a), h(b)]$. Since the integral over an interval can be obtained by taking the sum of the separate integrals over the subintervals of a partition of the given interval, we have

$$\int_{a}^{b} f(h(x))dg(h(x)) = \sum_{k=1}^{n} \int_{x_{k-1}}^{x_k} f(h(x))dg(h(x)).$$

However, the integrals in the summation either satisfy the requirements of Theorem 3.3 or else $g(x)$ is a constant in the subinterval and hence the integral has a zero value. In the former case the substitution indicated by the theorem is valid, and in the latter case the substitution would continue to give a zero value. Thus, the results of the theorem apply with the removal of the adjective *strictly* from the assumptions required in the theorem. $\square$

EXAMPLE 3.2. Evaluate

$$\int_{1}^{8} x^{4/3}\, dx^{5/3}.$$

Solution. If we let $h(x) = x^{1/3}$, we note that $x^{4/3} = (x^{1/3})^4 = [h(x)]^4$. In similar fashion, $g(x) = x^{5/3} = [h(x)]^5$. Furthermore, the hypotheses of the corollary are satisfied. Since $h(1) = 1^{1/3} = 1$ and $h(8) = 8^{1/3} = 2$, the interval of integration after substitution will be $[1, 2]$. Therefore, we have

$$\int_1^8 x^{4/3} \, dx^{5/3} = \int_1^2 y^4 \, dy^5 = \frac{5[2^9 - 1^9]}{9} = \frac{5(511)}{9} = \frac{2555}{9}.$$

Corollary 3.2. *Let $h(x)$ be a bounded function with a finite number of maxima and minima over the interval $[a, b]$, and let M be the maximum value and m the minimum value of h over $[a, b]$. If $g(x)$ is continuous, bounded, and has a finite number of maxima and minima in the interval $[m, M]$ then*

$$\int_a^b f(h(x)) dg(h(x)) = \int_{h(a)}^{h(b)} f(y) dg(y).$$

PROOF. Let $P = \{x_1, x_2, x_3, \ldots, x_{n-1}, x_n\}$ be the set of points which include all of the points at which $h(x)$ has a maximum or a minimum and the points at which $g(h(x))$ has a maximum or a minimum. In any interval $[x_{k-1}, x_k]$, $h(x)$ is monotonic and $g(h(x))$ is also monotonic. If we let $x_0 = a$ and $x_n = b$, we can then write

$$\int_a^b f(h(x)) dg(h(x)) = \sum_{k=1}^n \int_{x_{k-1}}^{x_k} f(h(x)) dg(h(x)) = \sum_{k=1}^n \int_{h(x_{k-1})}^{h(x_k)} f(y) dg(y)$$

$$= \int_{h(a)}^{h(b)} f(y) dg(y).$$

Note that we have satisfied the requirements on g in each of the subintervals $[x_{k-1}, x_k]$. (We might note the fact that if $g(x)$ is continuous on a closed interval, it must be bounded. This will be shown in Chapter VII.) □

We have made heavy use of the results of Example (III.1.3) and Exercise (III.1.5). However, these two results have limited our integration of power functions to intervals having only non-negative numbers. This prevents our using these results in many instances when they would otherwise be helpful. It is time to consider removing these restrictions.

Corollary 3.3. *If a is a negative number then*

$$\int_a^0 x^m \, dx^n = -\frac{na^{m+n}}{m+n}$$

PROOF. Since $f(x) = f(-(-x))$ and $g(x) = g(-(-x))$, we can make the substitution $y = (-x)$, and the limits of integration then become $(-a)$ and $(-0) = 0$ respectively. Also, we would have

$$\int_a^0 x^m \, dx^n = \int_{-a}^0 (-y)^m \, d(-y)^n = c \int_{-a}^0 y^m \, dy^n = -c \int_0^{-a} y^m \, dy^n$$

where $c = (+1)$ if $(m + n)$ is an even integer and $c = (-1)$ if $(m + n)$ is an odd integer. However,

$$-c \int_0^{-a} y^m \, dy^n = -c \frac{n(-a)^{m+n}}{m + n} = -\frac{na^{m+n}}{m + n}$$

by again factoring out the negative signs and noting that there are an even number of them if $(m + n)$ is even and an odd number otherwise. This gives us the proof. $\square$

Corollary 3.4. *If a and b are any real numbers such that $a < b$, then*

$$\int_a^b x^m \, dx^n = \frac{n(b^{m+n} - a^{m+n})}{m + n}.$$

PROOF. Either $0 < a < b$ in which case the given integral is the difference of two integrals as shown in (III.2.2), or $a < 0 < b$ in which case we have the sum over the two interval $[a, 0]$ and $[0, b]$, or we have $a < b < 0$ in which case we have the difference obtained by subtracting the integral over $[a, 0]$ from the integral over $[b, 0]$. $\square$

EXAMPLE 3.3. Evaluate $\int_{-2}^2 x^3 \, dx^2$ and show that Theorem 3.1 (the integral theorem of the mean) does not apply in this case.

Solution. Using the results of Corollary 3.4 we have

$$\int_{-2}^2 x^3 \, dx^2 = \frac{2[(2)^5 - (-2)^5]}{5} = \frac{2[32 - (-32)]}{5} = \frac{128}{5}.$$

However, if we were to try to use the integral theorem of the mean, we would be seeking a value c such that

$$c^3[(2)^2 - (-2)^2] = \int_{-2}^2 x^3 \, dx^2 = \frac{128}{5}.$$

No such c exists, for this would require us to have a c such that zero multiplied by c^3, obviously a zero result, would equal a non-zero number. In this case the integral theorem of the mean does not fail, but we did not satisfy all of the hypotheses. You will remember that we required that $g(x)$ be monotonic in the interval of integration, but $g(x)$ is not monotonic in the interval $[-2, 2]$. Hence the theorem is intact, but it does pay to check hypotheses before attempting to use the theorem.

Corollary 3.5. *If*

$$\int_a^b f(x + c) \, dx$$

exists, then

$$\int_a^b f(x + c) \, dx = \int_{a+c}^{b+c} f(x) \, dx.$$

PROOF. Observe that

$$\int_a^b f(x+c)dx = \int_a^b f(x+c)d[(x+c)-c]$$

$$= \int_a^b f(x+c)d(x+c) - \int_a^b f(x+c)dc.$$

However, the second integral in the last line has a value of zero since $g(x)$ is a constant, and by Corollary 3.2

$$\int_a^b f(x+c)dx = \int_a^b f(x+c)d(x+c) = \int_{a+c}^{b+c} f(y)dy. \qquad \square$$

Corollary 3.6. *If the integral $\int_{ca}^{cb} f(x)dx$ exists then $\int_a^b f(cx)dx = (1/c)\int_{ca}^{cb} f(x)dx$.*

PROOF.

$$\int_a^b f(cx)dx = \int_a^b f(cx)d\left[\frac{1}{c}(cx)\right] = (1/c)\int_a^b f(cx)d(cx)$$

since $(1/c)$ is a constant and can therefore be factored out of the integral. But now we can apply Theorem 3.3, and we have

$$\int_a^b f(cx)dx = \frac{1}{c}\int_a^b f(cx)d(cx) = \frac{1}{c}\int_{ca}^{cb} f(y)dy = \frac{1}{c}\int_{ca}^{cb} f(x)dx.$$

The last step is possible in view of the fact that the particular letter used to represent the variable of integration is immaterial, for a change of letter only involves a change of letter throughout the RS sums. However, the letters used to represent the variable are representing the same numbers from the same partitions. $\qquad \square$

EXAMPLE 3.4. If you are told that $\int_0^{\pi/2} \sin x \, dx = 1$, find the value of $\int_0^{\pi/4} \sin 2x \, dx$.

Solution. By Corollary 3.6 we know that

$$\int_0^{\pi/4} \sin 2x \, dx = \frac{1}{2}\int_{2(0)}^{2(\pi/4)} \sin x \, dx = \frac{1}{2}(1) = \frac{1}{2}.$$

Corollary 3.7. *If m and n are positive, rational numbers and if a^m, a^n, b^m, and b^n are real, then*

$$\int_a^b x^m \, d(x^n) = \frac{n(b^{m+n} - a^{m+n})}{m+n}.$$

OUTLINE OF PROOF. If m or n have odd numerators and even denominators, for example $3/2$, then negative values would yield imaginary results. Thus $(-4)^{3/2}$ requires the computation of the square root of a negative number.

This is imaginary. Since we have developed our results only for the case in which all of the values involved are real, we will exclude the imaginary possibility from our consideration here. This explains the restriction on a^m, a^n, b^m, and b^n in the corollary.

Let r be the least common denominator of the two rational numbers m and n, and then $m = p/r$ and $n = q/r$ where p, q, and r are positive integers. If $h(x) = x^{1/r}$, then $x^m = x^{p/r} = [h(x)]^p$ and $x^n = x^{q/r} = [h(x)]^q$. Therefore, we have

$$\int_a^b x^m \, d(x^n) = \int_a^b [h(x)]^p \, d([h(x)]^q) = \int_{h(a)}^{h(b)} y^p \, d(y^q)$$

$$= \frac{q([h(b)]^{p+q} - [h(a)]^{p+q})}{p + q}$$

$$= \frac{(q/r)[b^{(p+q)/r} - a^{(p+q)/r}]}{(p + q)/r}$$

$$= \frac{n(b^{m+n} - a^{m+n})}{m + n}. \qquad \square$$

This corollary extends the result of Corollary 3.4, for we can now apply this result when m and n are positive rational numbers and we do not require that they be integers.

EXAMPLE 3.5. Evaluate $\int_{-1}^8 x^{5/3} \, dx^{2/3}$.

Solution. Since both 5/3 and 2/3 are rational, we can apply Corollary 3.7. Thus we have

$$\int_{-1}^8 x^{5/3} \, dx^{2/3} = \frac{(2/3)[8^{(5/3)+(2/3)} - (-1)^{(5/3)+(2/3)}]}{(5/3) + (2/3)}$$

$$= \frac{(2/3)[8^{7/3} - (-1)^{7/3}]}{(7/3)}$$

$$= \frac{(2)[2^7 - (-1)^7]}{7}$$

$$= \frac{(2)[128 - (-1)]}{7} = \frac{2(129)}{7} = \frac{258}{7}.$$

We will close this section with what amounts to an application. You will remember that when we wished to obtain an average of a set of values, we added them up and divided by the number of values. In some instances a value might occur more than once, and then it was necessary to count each occurrence of that value. Equivalently we could add the products obtained by multiplying each value by its respective *weighting function* (or number of occurrences) and then divide by the sum of the weighting functions. This can also be done when the weighting function indicates the relative importance of the value. Thus, if in your class you had three hour tests and a final exam, and the final is to count as the equivalent of two hour tests, you would obtain your average by computing

$$\frac{T_1 + T_2 + T_3 + 2E}{1 + 1 + 1 + 2} = \frac{1}{5}(T_1 + T_2 + T_3 + 2E).$$

In the case of attempting to find the average value of a function, we do the same thing. In this case we will be trying to find the average value of $f(x)$ with respect to a weighting function $g(x)$. If we think of the summations, we will have to consider first the interval over which we are interested. Let us assume it to be $[a, b]$, and then we can assume some suitable partition of the interval, $P[a, b]$. If we use an evaluation set, T, we will obtain

$$\overline{f(x)_g} \doteq \frac{\sum\limits_{k=1}^{n} f(t_k)[g(x_k) - g(x_{k-1})]}{\sum\limits_{k=1}^{n} [g(x_k) - g(x_{k-1})]}$$

where we have indicated the average or *mean value* by placing a bar over the function, using a subscript g to indicate that the mean value is obtained with respect to the weighting function $g(x)$. We have also indicated that this is an approximation by placing the dot over the " $=$ ", since we have some doubts concerning whether we selected the correct partition and evaluation set. It should not come as a surprise that we will proceed from the two RS sums that we have here, one in the numerator and the other in the denominator, to integrals by successive refinements. Thus, we have the following definition.

Definition 3.3. The mean value of $f(x)$ with respect to $g(x)$ over $[a, b]$ is

$$\overline{f(x)_g} = \frac{\int_a^b f(x)dg(x)}{\int_a^b dg(x)} = \frac{1}{g(b) - g(a)} \int_a^b f(x)dg(x).$$

The mean value has many interpretations, including the usual ones in statistics, but also including centroid (essentially center of gravity) among others.

EXAMPLE 3.4. Find the mean value of x^3 with respect to x^2 over $[1, 4]$.

Solution. The mean value is given by the relation

$$\text{Mean value} = \frac{1}{(4)^2 - (1)^2} \int_1^4 x^3 \, d(x^2) = \frac{1}{15}\left[\frac{2}{5}(4^5) - \frac{2}{5}(1^5)\right]$$

$$= \frac{2}{75}[1023] = \frac{682}{25} = 27.28.$$

Note that the mean value is between the minimum and maximum value assumed by x^3 in $[1, 4]$.

EXERCISES

1. Find a value c which satisfies the integral theorem of the mean or else show that such a value fails to exist.

   (a) $\int_2^4 x \, dx$
   (b) $\int_0^3 x^3 \, dx$
   (c) $\int_1^3 (x^2 - 2x + 2)dx$
   (d) $\int_{-2}^2 (x - 3)^2 dx$
   (e) $\int_{-2}^2 x \, dx^2$
   (f) $\int_{-1}^2 x^3 \, dx^2$
   (g) $\int_0^{\pi/2} 4 \, d(\sin x)$
   (h) $\int_{-1}^2 (x^2 + \sin x)d(4)$

2. Evaluate:

   (a) $\int_1^2 (3x - 2)^2 dx$
   (b) $\int_{-2}^0 (x + 2)^2 d(x + 3)$
   (c) $\int_{-2}^5 (x + 3)^{2/3} d(x + 3)^{4/3}$
   (d) $\int_{-2}^1 \sqrt{x + 3} \, dx$
   (e) $\int_{-1}^1 (2x^3 - 3x^2 + 4x - 17)dx^2$
   (f) $\int_0^2 (x^2 - 5)d(x^2 - x + 3)$
   (g) $\int_{-1}^2 x^2 \, d(x - 3)^3$
   (h) $\int_1^{64} \sqrt[3]{x^2} \, d\sqrt{x^3}$

3. Evaluate:

   (a) $\int_{-\pi/6}^{\pi/2} \sin^2 x \, d(\sin^3 x)$
   (b) $\int_0^{\pi/3} \cos^3 x \, d(\cos x)$
   (c) $\int_0^{\pi/4} \sqrt{\tan^3 x} \, d(\sec^2 x)$
       [*Hint*: $\sec^2 x = 1 + \tan^2 x.$]
   (d) $\int_{-2}^2 \sqrt{x^2 + 5} \, dx^2$
   (e) $\int_{-1}^2 2^{3x} \, d(2^{x/2})$ [*Hint*: $2^{3x} = (2^x)^3$]
   (f) $\int_0^{\pi/2} \cos^4 x \, d(\sin x)$
       [*Hint*: $\cos^2 x = 1 - \sin^2 x.$]
   (g) $\int_0^3 (x^3 + 27)^{1/3} \, dx^3$

4. Evaluate:

(a) $\int_{-2}^{1} x(x + 1)d(x + 2)$

(b) $\int_{-1}^{32} x^{2/5} dx^{4(5}$

(c) $\int_{0}^{3} (4 - y)^{1/2} dy^2$
[Hint: $y^2 = \{4 - (4 - y)\}^2$]

(d) $\int_{1}^{2} (4/x)d(1/x^3)$

(e) $\int_{3}^{5} (x^2 - 9)^{1/2} dx^2$

(f) $\int_{1}^{3} (2x^4)dx$

5. Evaluate:

(a) $\int_{0}^{64} \sqrt[3]{x} \, d\sqrt{x}$

(b) $\int_{0}^{3} (x + 1)^{3/2} \, dx$

(c) $\int_{0}^{\pi/3} (\cos x)^{2/3} d(\sqrt{\cos x})$

(d) $\int_{0}^{\pi/4} \sec^4 x \, d(\tan^2 x)$
[Hint: $\sec^2 x = \tan^2 x + 1$.]

(e) $\int_{-1}^{1} \sqrt{x^2 + 3} \, d(x^2)$

(f) $\int_{0}^{2} \sqrt{x^2 + 3} \, d(x^2)$

(g) $\int_{1}^{3} (3x)^2 dx$

6. Find the mean value of $f(x)$ with respect to $g(x)$ over the interval $[a, b]$ in each of the following cases.

(a) $f(x) = x$, $g(x) = x^2$, over $[1, 4]$.

(b) $f(x) = x^3$, $g(x) = x$, over $[-2, 2]$.

(c) $f(x) = x$, $g(x) = x$, over $[0, 4]$.

(d) $f(x) = x^{0.4}$, $g(x) = x^{0.6}$, over $[1, 5]$.

7. (a) Find the area bounded by $f(x) = \sqrt{x}$, $x = 1$, $x = 4$ and the x-axis.

(b) Find the mean value of $f(x) = \sqrt{x}$ with respect to $g(x) = x$ over the interval $[1, 4]$.

(c) Sketch the graph of $y = \sqrt{x}$ over the interval $[1, 4]$, and then draw the line $y = m$ where m is the mean value found in part (b).

(d) Compare the area of part (a) with the area of the rectangle bounded by $x = 1$, $x = 4$, $y = 0$, and $y = m$ where m is the mean value found in part (b).

8. Find both upper and lower bounds for each of the following integrals.

(a) $\int_{\pi/6}^{\pi/2} \sin x \, dx$

(b) $\int_{0}^{\sqrt{8}} \sqrt[3]{x^4 + 1} \, d(x - 2)$

(c) $\int_{0}^{\pi/2} \sin^2 x \, dx$

(d) $\int_{0}^{4} \sin^3 x \, d(x\sqrt{x})$

(e) $\int_{0}^{\pi/4} \tan x \, dx^2$

(f) $\int_{1}^{10} \log_{10} x \, dx$

(g) $\int_{0}^{\pi/3} \cos^3 x \, dx^2$

(h) $\int_{-\pi/6}^{\pi/6} x^3 \, d(\sin x)$

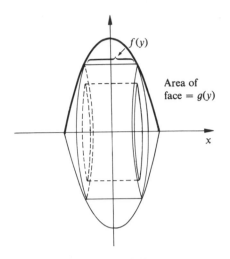

Figure III.1

9. (a) Prove $\int_1^b (1/x)dx = \int_1^b (1/ax)d(ax)$.
   (b) Prove $\int_1^b (1/ax)d(ax) = \int_a^{ab} (1/y)dy = \int_a^{ab} (1/x)dx$
   (c) Prove $\int_1^b (1/x)dx = \int_a^{ab} (1/x)dx$
   (d) Prove $\int_1^a (1/x)dx + \int_1^b (1/x)dx = \int_1^a (1/x)dx + \int_a^{ab} (1/x)dx = \int_1^{ab} (1/x)dx$
   (e) If $h(a) = \int_1^a (1/x)dx$, then part (d) states $h(a) + h(b) = h(ab)$. Can you think
       of any other function that behaves in this way (that is such that the sum of
       the functions of a and b give the function for the product ab)?

10. The portion of the curve $y = 4 - x^2$ which is above the x-axis is revolved about
    the x-axis to form the surface of a solid of revolution as shown in Figure III.1.

    (a) Partition the interval $[0, 4]$ of the y-axis, and sketch the right circular
        cylinders having the x-axis as the axis of the cylinder and with a radius equal
        to the value of y_k for each partition point.
    (b) If the cylinders are just long enough to fit inside the surface of revolution,
        show that the approximate volume which is inside one of these cylinders but
        outside the next smaller one is given by $2(4 - t_k)^{1/2}[\pi y_k^2 - \pi y_{k-1}^2]$.
    (c) Add the volumes of the type given in part (b) to find an RS sum which ap-
        proximates the volume of the solid of revolution.
    (d) Set up the corresponding RS integral and find the volume of this solid of
        revolution.
    (e) Partition the x-axis and set up an RS sum and then an RS integral to find
        this same volume. Compare your two results.

11. The area bounded by $y = x^2$, $y = 0$, and $x = 5$ is rotated about the y-axis. Use
    the method of Exercise 10 (that is the method of using *cylindrical shells*) to find
    the volume of the resulting solid.

12. Find the area bounded by the spiral $r = \theta/2$, $\theta = \pi/6$, and $\theta = \pi/2$.

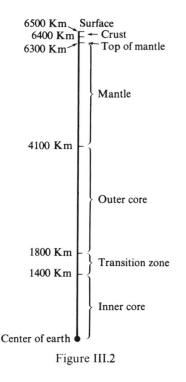

Figure III.2

P13. The inner core of the earth as shown in Figure III.2, comprises a sphere in the center of the earth of radius 1400 km. The density at the center is 16.5 and the density at the outer layer of this inner core is 14.0. (If the density is 5, each cubic centimeter of volume weighs 5 g.)

   (a) If we assume that the density is a linear function of the radius, find the equation relating density and radius.
   (b) Find the mass of the inner core.
   (c) Given that the outer core of the earth is bounded by spheres of radii 1800 km and 4100 km and the information that the density at 1800 km is 11.8 whereas the density at 4100 km is 9.5, find the mass of the outer core. Assume that the density is a linear function of radius.
   (d) Find the mass of the lower mantle under the assumption that the density is a linear function of radius and given that the lower mantle has spheres of radii 4100 km and 6300 km as boundaries. The density is 5.5 at the inside of the lower mantle and is 4.20 at the outside of this mantle.

P14. The portion of the earth's crust called the "simatic layer" is that portion which starts about 5 km below the earth's surface and goes down to a depth of 60 km. If the density of this layer is about 3.00 and the earth is 6500 km in radius, find the mass of this layer.

M15. If the conditions for the integral theorem of the mean are satisfied, show that the mean value of $f(x)$ with respect to $g(x)$ over the interval $[a, b]$ is exactly $f(c)$ where c is the value guaranteed by the integral theorem of the mean.

BS16. The number of organisms in a certain culture t hours after the culture was started is given by the relation $p(t) = 100 + (t/2) + (t^2/8) + (t^3/48)$. Find the average (or mean) value of the number of organisms in the culture between $t = 0$ and $t = 3$. (If this same question were to be asked concerning the number of people in a given area, the relation would be approximately the same if the unit of time were about 25 years.)

M17. (a) Evaluate $\int_a^b (cx)^n dx$ using Corollary 3.6.

    (b) Evaluate $\int_a^b (cx)^n dx = c^n \int_a^b x^n \, dx$ using Corollary 3.7.

    (c) Show that the results of parts (a) and (b) are equal.

P18. (a) Find the first moment of the area bounded by $x = 0$, $y = 0$, and $f(x) = 4 - x^2$ about the y-axis.

    (b) Find the mean value of the first moment of the area of part (a) with respect to the area. This value is the x-coordinate of the centroid (or center of gravity) of the area of part (a).

    (c) Find the second moment (moment of inertia) of the area about the y-axis.

P19. (a) Find the area of the triangle having vertices $(0, 0)$, $(4, 0)$, and $(4, 3)$.

    (b) Find the first moment of this area about the y-axis. Note that it is necessary to use a partition of the x-axis in this case.

    (c) Find the x-coordinate of the centroid of the triangle of part (a).

    (d) Find the y-coordinate of the centroid of the triangle of part (a). In this case it will be necessary to partition the y-axis.

    (e) Show that the centroid is located at the point where the medians of the triangle intersect, and that the centroid divides each median in such a way that one part is twice the length of the other.

III.4 The Exponential Function

You have often heard it said that something is growing *exponentially*. Have you ever questioned precisely what is meant by this, other than the fact that the context usually indicates that it is growing very fast? Let us consider the function $f(x) = 2^x$ as shown in Figure III.3. We see that $f(0) = 1$, $f(1) = 2, f(2) = 4, f(3) = 8, \ldots, f(20) = 1{,}048{,}576$, etc. This would indicate rapid growth. Of course, it is possible to use other values for x, too, for we could write $f(-4) = 1/16$, $f(1/2) = \sqrt{2}$, etc. We note at once that $f(x)$ is monotonic increasing. If we consider $x > y$ and then consider $f(x)$ and $f(y)$, we have $f(x) - f(y) = 2^x - 2^y = 2^y[2^{x-y} - 1]$. But $2^y > 0$, and since $x > y$, $x - y > 0$ and $2^{x-y} > 1$. Therefore $[2^{x-y} - 1] > 0$, whence $f(x) - f(y) > 0$ or $f(x) > f(y)$. Note that the same result could have been obtained if we had replaced 2 by any number $a > 1$. The function $g(x) = a^x$ is called the *exponential function*. This is a very important function in many applications, particularly those having to do with problems of growth or decay, such as population growth, economic growth, radioactive decay, etc. We will consider the problem of integrating the exponential function in this section.

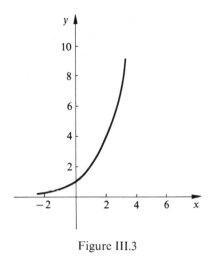

Figure III.3

We must first assure ourselves that the integral in which we are interested actually exists, for the exponential function differs from the kind of function we have considered in the past.

Theorem 4.1. *If $a > 1$ and $c > b$, then the integral $\int_b^c a^x \, dx$ exists.*

PROOF. Let $P[b, c]$ be a regular partition with n subintervals. In this case we have $x_k = b + k(c - b)/n$, and we see that $x_k - x_{k-1} = (c - b)/n$. Now

$$U(P, a^x, x) - L(P, a^x, x) = \sum_{k=1}^{n} \frac{x^{b + k(c - b)/n}(c - b)}{n}$$

$$- \sum_{k=1}^{n} \frac{x^{b + (k - 1)(c - b)/n}(c - b)}{n}$$

$$= \frac{a^c(c - b) - a^b(c - b)}{n} = \frac{(a^c - a^b)(c - b)}{n}.$$

However, this last expression has a constant numerator, and as we increase the number of subintervals we increase the denominator. Thus, the fraction can be made as small as we like by making n sufficiently large. Therefore, the upper and lower sums become arbitrarily close to each other by taking a regular partition with a sufficient number of subintervals. Thus, by arguments we have used before, we see that the integral must exist.   □

Our next task will be that of finding the value of the integral $\int_0^b a^x \, dx$ given that $a > 1$ and $b > 0$. Of course this integral considers only the interval $[0, b]$, but we will worry about taking a more flexible approach to the interval later on. For the moment this choice will greatly simplify the algebra in our proof. Since we have no formulas to which we can refer for this integral, we will have to go back to the RS sums. The proof of Theorem 4.1 demonstrates that we can make use of regular partitions (always somewhat easier to use) and be assured that either the upper or lower sum can be brought arbitrarily close to the value of the integral itself by requiring a suitably large number of subintervals. With this in mind we will again follow the procedure of obtaining a formula for the value of the sum based upon n and determine what would happen if n is made very large. In this case it will be slightly more convenient to use the lower sum.

Since we know that under the given conditions both $f(x) = a^x$ and $g(x) = x$ are increasing functions, we will wish to use the value x_{k-1} as the evaluation point for the k-th subinterval. Furthermore, we know from past experience that if $P[0, b]$ is a regular partition with n subintervals then $x_k = (kb)/n$. Hence, we have

$$L(P, a^x, x) = \sum_{k=1}^{n} a^{(k-1)b/n} \left[\frac{kb}{n} - \frac{k(b-1)}{n} \right] \qquad \text{(III.4.1)}$$

$$= \frac{b}{n} \sum_{k=1}^{n} a^{(k-1)b/n}.$$

However,

$$\sum_{k=1}^{n} a^{(k-1)b/n} = a^0 + a^{b/n} + a^{2b/n} + \cdots + a^{(n-1)b/n}$$

$$= [a^{b/n}]^0 + [a^{b/n}]^1 + [a^{b/n}]^2 + \cdots + [a^{b/n}]^{n-1}$$

$$= \frac{a^{nb/n} - 1}{a^{b/n} - 1} = \frac{a^b - 1}{a^{b/n} - 1}.$$

This result stems from the fact that we have a geometric progression with the first term being $a^0 = 1$ and the ratio being $a^{b/n}$. By inserting this result in (III.4.1), we have

$$L(P, a^x, x) = \frac{b}{n} \frac{a^b - 1}{a^{b/n} - 1} = \frac{a^b - 1}{(a^{b/n} - 1)/[(b/n) - 0]} \qquad \text{(III.4.2)}$$

The final expression in (III.4.2) has been written in such a way that the numerator, at least, does not involve n, and is therefore completely determined regardless of the regular partition used. A close look at the denominator should bring back memories of the fraction $(y_2 - y_1)/(x_2 - x_1)$ which gives the slope of the line segment joining (x_1, y_1) and (x_2, y_2). This indicates that the denominator of (III.4.2), that is

$$\frac{a^{b/n} - a^0}{(b/n) - 0},$$

is the slope of the line segment joining $(0, a^0)$ and $(b/n, a^{b/n})$. Both of these points are points of the curve $f(x) = a^x$ as shown in Figure III.4. Since we are ultimately interested in what happens as this partition is refined without stopping, we are concerned with this slope as (b/n) becomes smaller. This

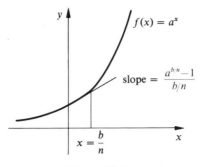

$f(x) = a^x$

$\text{slope} = \dfrac{a^{b/n}-1}{b/n}$

$x = \dfrac{b}{n}$

Figure III.4

follows, of course, since n becomes larger. We are not prepared to consider every detail of this process at this point, but we can follow very closely in an intuitive manner. We should also observe that since we must divide by the *ultimate* value of this slope, that is the value that this slope will approach as we continue to refine the partition, it would be convenient if this ultimate value were one. Thus, we would like to select a value for a such that as (b/n) gets closer and closer to zero the value of

$$\frac{a^{b/n} - 1}{(b/n)}$$

gets correspondingly closer to one.

Since the fraction (b/n) occurs so often, we will simplify our writing by replacing this value by h. Thus, we will examine $(a^h - 1)/h$ as h becomes progressively closer to zero. If we ask that this fraction have the value one, we would then have $(a^h - 1)/h = 1$ or $a^h = 1 + h$. From this we have

$$a = (1 + h)^{1/h}. \tag{III.4.3}$$

Of course we desire that this be the case for very small values of h, therefore we must regard this equation as an approximation, but we will proceed nevertheless. We can see the value obtained for a using this expression and various values of h in the following table.

Table III.1

h	$a = (1 + h)^{1/h}$
0.1	2.593742
0.01	2.704814
0.001	2.716924
0.0001	2.718146
0.00001	2.718255

It would seem that the value of a is approaching some limit as h gets smaller.

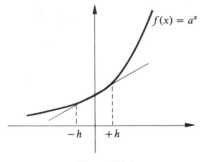

Figure III.5

If we again draw a graph of $f(x) = a^x$ and this time determine the points corresponding to both $x = -h$ and $x = h$, we would have the situation shown in Figure III.5. It would seem that the slope over the interval $[0, h]$ would be larger than the slope over the interval $[-h, 0]$ and perhaps we could improve our choice of a value for a if we were to average these two slopes and ask that the average be one. The fact that the slope to the right of the axis is actually larger than the slope to the left can be shown by the relation

$$\frac{a^h - 1}{h} = a^h \frac{1 - a^{-h}}{h} = a^h \frac{a^{-h} - 1}{-h}$$

when we realize that $a^h > 1$ if $h > 0$, $(a^h - 1)/h$ is the slope to the right of the y-axis, and $(a^{-h} - 1)/(-h)$ is the slope to the left of the y-axis. Now to pursue this idea of setting the average equal to the number one, we would have

$$\frac{(a^h - 1)/h + (a^{-h} - 1)/-h}{2} = \frac{a^h - a^{-h}}{2h} = 1$$

or

$$a^h - a^{-h} = 2h.$$

If we multiply both sides of this equation by a^h and move all terms to the left side, we obtain

$$a^{2h} - 2ha^h - 1 = (a^h)^2 - 2h(a^h) - 1 = 0.$$

This is obviously a quadratic equation in the variable (a^h), and hence we have

$$a^h = \frac{2h \pm \sqrt{4h^2 + 4}}{2} = h \pm \sqrt{h^2 + 1}.$$

This is equivalent to

$$a = (\sqrt{h^2 + 1} + h)^{1/h}. \qquad (\text{III.4.4})$$

We can use this result to obtain a table similar to Table III.1.

Table III.2

h	$a = (\sqrt{h^2 + 1} + h)^{1/h}$
0.1	2.713775362
0.01	2.718236596
0.001	2.718281285
0.0001	2.718281828 ·

These values are, in fact, much better than those obtained using (III.4.3). It has been shown that the limiting value of a as h becomes smaller would actually be 2.718281828459045.... This number is one that is very useful in a great many applications of mathematics, and consequently it is given a name. It is called e. It has been computed to more than 100,000 decimal places. The value given here is more than sufficient for the majority of applications, however.

Since we have now selected a specific value for the a of the function $f(x)$, namely e, we have

$$\int_0^b e^x \, dx = \frac{e^b - 1}{1} = e^b - 1. \tag{III.4.5}$$

We have used the fact that if we let $a = e$ the denominator of (III.4.2) will approach one with further refinements of the partition.

EXAMPLE 4.1. Evaluate $\int_0^3 e^x \, dx$.

Solution. Using (III.4.5) we have $\int_0^3 e^x \, dx = e^3 - 1 = 20.0855 - 1 = 19.855$. The values of e^x are given in Table 2 of Appendix C. Many of the hand-held calculators also have a function key which will permit computation of the powers of e.

We have developed the integral of the exponential function for a rather special case, for we have required that we have a power of e and that the interval be an interval starting at $x = 0$ and proceeding to the right. In the first case we can gain relief by noting that for any value of $a > 0$ there must be some value of c such that $a = e^c$. This would seem reasonable if we look at the graph of e^x for we only need to show that there is a value of x such that $e^x = a$ as shown in Figure III.6. It can be shown that $f(x) = e^x$ is a continuous function, and since it is possible to find positive exponents sufficiently large that the exponential function will exceed any given value of a and it is also possible to find negative exponents which give values of the exponential function sufficiently close to zero that the value would be less than a, the intermediate value theorem assures us that the value c referred to above must exist. Since $e^c = a$, c is the logarithm of a to the base e by definition of the logarithm. We can write this $c = \log_e a$. In view of the fact that logarithms

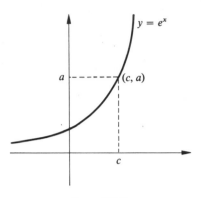

Figure III.6

to the base e, often called *natural logarithms*, occur so frequently, we will introduce the special notation "ln a" to indicate the natural logarithm.

We have now reduced the problem of integrating the more general exponential function to the problem of finding the integral of

$$\int_0^b a^x \, dx = \int_0^b (e^c)^x \, dx = \int_0^b e^{cx} \, dx.$$

Corollary 3.6 of the last section tells us that this last value is given by the relation

$$\int_0^b a^x \, dx = \int_0^b e^{cx} \, dx = \frac{1}{c} \int_0^{cb} e^x \, dx = \frac{e^{cb} - 1}{c} = \frac{a^b - 1}{\ln a}. \qquad \text{(III.4.6)}$$

Since $\ln e = 1$ we see that (III.4.6) includes the result of (III.4.5). Since formula (III.4.5) is probably easier to remember you may wish to restrict your computation to this formula in the manner indicated in the next example.

EXAMPLE 4.2. Evaluate $\int_0^3 2^x \, dx$.

Solution. Since $\ln 2 = 0.6931$, we know that $2 = e^{0.6931}$. Therefore we wish to find the value of

$$\int_0^3 2^x \, dx = \int_0^3 e^{0.6931x} \, dx = \frac{1}{0.6931} [e^{(0.6931)(3)} - 1]$$

$$= 1.4428(2^3 - 1) = 10.0996.$$

We now take a look at the possibility of integrating over some interval other than $[0, b]$.

Theorem 4.1.

$$\int_a^b e^x \, dx = e^b - e^a.$$

PROOF.

$$\int_a^b e^x \, dx = \int_a^b e^a e^{x-a} \, dx = e^a \int_a^b e^{x-a} \, dx = e^a \int_0^{b-a} e^x \, dx = e^a(e^{b-a} - 1)$$

$$= e^a e^{b-a} - e^a = e^b - e^a.$$

Here we have used the fact that e^a is a constant and can therefore be factored out of the integrand. We have also used Corollary 3.5 of the last section. Note that this theorem removes all restrictions on the interval. It is also true that this includes the previous result, for there is no reason why a could not have the value zero.   □

EXAMPLE 4.3. Evaluate $\int_{-1}^2 3^{2x} \, dx$.

Solution. Since $\log_e 3 = \ln 3 = 1.0986$, we have

$$\int_{-1}^2 3^{2x} \, dx = \int_{-1}^2 e^{(1.0986)(2x)} \, dx = \frac{e^{(1.0986)(2)(2)} - e^{(1.0986)(2)(-1)}}{(1.0986)(2)}$$

$$= \frac{3^{(2)(2)} - 3^{(2)(-1)}}{2.1972} = \frac{81 - (\frac{1}{9})}{2.1972} = 36.815.$$

We conclude this section with a summary of the results of the section and with one additional note on notation.

Notation. The functional notation $\exp(x)$ is often used to denote the exponential function. Thus, by definition $\exp(x) = e^x$.

The results of the work in this section can be summed up in the single formula

$$\int_b^c a^x \, dx = \frac{1}{\ln a} (a^c - a^b). \tag{III.4.7}$$

Since $\ln e = 1$, this reduces to the simpler formula

$$\int_b^c \exp(x) dx = \exp(c) - \exp(b) \tag{III.4.8}$$

if $a = e$.

EXERCISES

The following table may be helpful. A more complete table can be found in Appendix C.

c	2	3	4	5	6	10
$\ln c$	0.6931	1.0986	1.3863	1.6094	1.7918	2.3026

1. Evaluate each of the following:

(a) $\int_{-1}^1 e^x \, dx$

(b) $\int_0^2 e^{2x}\, dx$

(c) $\int_1^3 e^{-x}\, dx$

(d) $\int_1^3 e^x\, dx$

(e) $\int_{-2}^2 3^x\, dx$

(f) $\int_0^2 4^{-x}\, dx$

(g) $\int_{-2}^2 3^{-2x}\, dx$

(h) $\int_0^2 2^{x+1}\, dx$

2. Evaluate each of the following:

(a) $\int_{-1}^1 (e^x - 1)^3 dx$

(b) $\int_1^2 e^{(x^2)}\, dx^2$

(c) $\int_{-1}^2 (x^2 - 2^x)dx$

(d) $\int_{-1}^1 (e^x + e^{-x})^2 dx$

(e) $\int_{-2}^2 (1/2^x)dx$

(f) $\int_{-\pi/6}^{\pi/3} \sin x\, d(\sin x)$

(g) $\int_0^{\pi/4} 3^{\tan x}\, d(\tan x)$

(h) $\int_0^{\ln 2} e^{(e^x)}\, d(e^x)$

3. Evaluate each of the following:

(a) $\int_{-1}^2 \exp(x)dx$

(b) $\int_{-1}^1 [1 - \exp(x)]^3 dx$

(c) $\int_{-2}^2 (2^x - e^x)dx$

(d) $\int_1^3 \exp(3x)dx$

(e) $\int_1^3 \exp(x)d(3x)$

(f) $\int_1^4 \exp(\sqrt{x})d(\sqrt{x})$

4. Let $F(x)$ be defined as indicated in each of the following parts. The domain in each case consists of all real numbers x such that $x > 1$. In each case determine whether $F(x)$ is an increasing or a decreasing function, or whether it is non-monotonic.

(a) $F(x) = \int_0^3 x^y\, dy$. [Hint: Remember that x is a constant insofar as the integration is concerned.]

(b) $F(x) = \int_{-3}^0 x^y\, dy$

(c) $F(x) = \int_0^3 x^{-y}\, dy$

(d) $F(x) = \int_{-3}^0 x^{-y}\, dy$

(e) $F(x) = \int_0^x e^y\, dy$

(f) $F(x) = \int_{-x}^0 e^y\, dy$

(g) $F(x) = \int_0^x e^{-y}\, dy$

(h) $F(x) = \int_{-x}^0 e^{-y}\, dy$

5. The *hyperbolic sine* (sinh x) and *hyperbolic cosine* (cosh x) are defined by the relations sinh $x = (\exp(x) - \exp(-x))/2$ and cosh $x = (\exp(x) + \exp(-x))/2$.

(a) Evaluate $\int_0^a \sinh x\, dx$. (Express the result in terms of hyperbolic functions.)

(b) Evaluate $\int_0^a \cosh x\, dx$. (Express the results in terms of hyperbolic functions.)

(c) Show that $\cosh^2 x - \sinh^2 x = 1$.

(d) Show that $\exp(x) = \sinh x + \cosh x$.

(e) Evaluate $\int_0^2 (\sinh x)(\cosh x)dx$.

6. Find the area bounded by $y = x^2$, $y = 2^x$, and $x = 0$. Note that $x^2 = 2^x$ if $x = 2$. (Show that there is no intersection of these curves in the first quadrant to the left of $(2, 4)$.)

7. Find the area inside the logarithmic spiral $r = e^\theta$ between the radii $\theta = -\pi/6$ and $\theta = \pi/4$. [The spiral of the snail's shell is given by this equation.]

8. Find the mean value of the function $f(x) = 2^x$ with respect to $g(x) = x$ over the interval $[-2, 2]$. Sketch the graph of $y = f(x)$ and show this mean value on the graph. Determine whether your result seems reasonable.

9. (a) Show that $e^{-x} < 1$ if $x > 0$. Then show $\int_0^x e^{-t} dt \le x$ whence $e^{-x} \ge 1 - x$.

(b) Continue with the same line of reasoning to show $1 - x + (x^2/2) \ge e^{-x} \ge 1 - x$ and then

$$\sum_{k=0}^{2} \frac{(-x)^k}{k!} = 1 - x + \frac{x^2}{2} \ge e^{-x} \ge 1 - x + \frac{x^2}{2} - \frac{x^3}{6} = \sum_{k=0}^{3} \frac{(-x)^k}{k!}$$

for $x > 0$.

(c) Use mathematical induction to show

$$\sum_{k=0}^{2n} \frac{(-x)^k}{k!} \ge e^{-x} \ge \sum_{k=0}^{2n+1} \frac{(-x)^k}{k!} \quad \text{for } x > 0.$$

(d) Use the first eight terms to show that e is approximately $280/103$. (Use the terms including the one that has 7! in the denominator.) What is the maximum error in e^{-x} using the first 8 terms? (Hint: What would be added to change your result to an upper bound?)

C10. Write a program to use the results of Exercise 9 and obtain a table of values of e^{-x} for $x = 0, 0.02, 0.04, \ldots, 1.0$ correct to 6 decimal places.

C11. (a) Draw a flowchart and write a program to evaluate $f(a) = a^3 - 1 - \int_0^3 a^x dx$ using upper sums and using lower sums for the evaluation of the integral.

(b) Use your program to show that $f(2) < 0$ and $f(3) > 0$. [Determine whether you should use upper or lower sums in each case to be sure that your result is correct. You might stop to think that if you have the smallest possible approximation for the integral and the function has a negative value, it would certainly have a negative value for the correct value of the function.]

(c) Use the bisection method based on the intermediate value property for $f(a)$ to determine the value of a for which $f(a) = 0$ to six decimal places.

(d) Interpret the results that you have obtained in this Exercise.

BP12. The emission rate of a radioactive material decays in an exponential manner. Hence, if the emission rate on a given day is R milliroentgens per day, the emission rate t days later will be $r(t) = Re^{ct}$ milliroentgens per day for some constant c depending upon the particular substance. It is known that *radioactive iodine* (atomic weight 131) has a half-life of 8 days. Therefore, $r(8) = 0.5r(0)$. This short half-life is one of the reasons why this isotope of iodine is useful in curing certain types of cancer, such as cancer of the goiter.

(a) Given that the half-life of radioactive iodine is 8 days, find the value of c in the formula for $r(t)$.

(b) If the total emission from a certain amount of radioactive iodine during an eight day period is 10 milliroentgens, find the value of R, and hence the formula for $r(t)$ in this case. [*Hint*: The total emission is the sum of the emissions over relatively short intervals of time, and this would lead to an RS sum.]

(c) Find the total emission from this amount of radioactive iodine during the period from $t = 72$ to $t = 80$.

(d) Find the total emission from this amount of radioactive iodine during the period from $t = 0$ to $t = 80$. What percentage of this emission occurs in the first 8 days?

BP13. A salt will dissolve in a solution at the rate of ae^{-bt} grams per minute at a time t minutes after the process starts. The constants a and b are determined by the amount of the salt involved in relation to the amount of solution and also related to the particular salt and solution. In a given situation we have $a = 5$ and $b = 0.25$.

(a) Find the amount of salt dissolved in the first ten minutes.

(b) Find the amount of salt dissolved in the first 100 minutes.

(c) Find the amount of salt dissolved in the first day.

(d) What is the maximum amount of salt that would be dissolved assuming the process were to continue without disturbance?

(e) By what time will 95% of the maximum amount have been dissolved?

S14. If the population of a community at a given time is $P(0)$ persons, the population t years later will be closely approximated by the function $P(t) = P(0)e^{rt}$ where r is a constant determined by the rate of growth. This relation ignores sudden unusual changes such as might be caused by the establishment of a major industry with many potential jobs or perhaps by a devastating epidemic of influenza. However, it provides us with a good general formula.

(a) A certain community has 100,000 people in 1960 and 128,400 people in 1970. Find the rate of growth, r, for this community.

(b) If each person in this community uses 10 kilowatts of electricity per day, find the total amount of electricity that will be used in this community during the decade from 1980 to 1990.

(c) Find the average daily electricity usage in this community during the period from 1980 to 1990.

S15. A company starts manufacturing a new product, and since people have to learn the new processes the number of items manufactured per day can be expected to increase as time goes on. The cost of the items can also be expected to increase with inflation. If the number of items manufactured from the beginning of operation through the t-th month is given by $N(t) = 1000t - 400e^{-0.2t}$ and the cost is given by $C(t) = 60 + 3e^{0.005t}$, find the total cost of production for each of the first five years of operation. [*Hint*: $e^{-0.2} = (e^{0.005})^{-40}$.]

M16. (a) Expand $(1 + h)^{1/h}$ using the binomial expansion. Consider that h is so chosen that $1/h$ is an integer.

   (b) Using your expansion of part (a). show that if h is very small then $(1 + h)^{1/h} \doteq \sum_{k=0}^{n} (1/k!)$ where the symbol $\doteq$ indicates "approximately equal" and $n = 1/h$.

   (c) Find the value of the summation of part (b) if $n = 8$ and then compare this value with the value of e.

   (d) Repeat part (c) for $n = 10$.

M17. (a) If you have \$1.00 and invest it at an annual rate of r percent interest compounded n times per year, show that at the end of the year you would have $(1 + r/n)^n$ dollars on deposit.

   (b) Show that this amount could also be written $[(1 + r/n)^{n/r}]^r = [(1 + h)^{1/h}]^r$ if we replace r/n by h.

   (c) Using the results of Exercise 16, show that if n is very large (that is we are compounding a very large number of times per year) this value can be approximated by e^r.

   (d) Show that if n is very large, the amount on deposit after t years will be given by e^{rt} dollars.

M18. (a) Sketch the curve $y = e^x$.

   (b) Find the area bounded by $y = e^x$, $y = 0$, $x = -10^6$ and $x = a$.

   (c) Show that the area you have found in part (b) is essentially the same as the height of the curve at the right hand end.

III.5 Trigonometric Functions

We have considered the integral of the power function x^n, and the integral of the exponential function, a^x. Another class of functions which appear frequently includes the trigonometric functions. Any problem involving periodic change is apt to use trigonometric functions. We will also find that the evaluation of many integrals of an algebraic nature can be facilitated through the use of trigonometric functions (see Chapter VI). The trigonometric functions are described in detail in Appendix A. Our concern here will concentrate on the integration of the sine and cosine, probably the two most used of the six trigonometric functions. While the tangent is also very much used, we shall postpone consideration of the tangent until Chapter VI, since it involves material which we have not yet derived. The present section can be considered in some ways to be a review section, for we will be calling on many items we discussed previously but which appear to have been forgotten to this point. As an interesting exercise you might thumb back through this book to see how many things we have derived that have not yet been used, and then see how many of these things are mentioned in this section. (That is one of the difficulties with mathematics—seemingly unimportant bits of information turn out to be used at some later

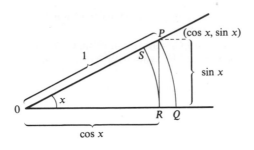

Figure III.7

time and there is no leeway for those of us who have somewhat less than perfect memories!)

In order to keep the flow of thought uninterrupted when we get to the derivation of the integrals of the sine and cosine, we will introduce a theorem at this point which has a conclusion that is necessary later on, although on the surface of things it does not appear to relate to the integral.

Theorem 5.1. *Let U be the number of units of angular measure in one complete revolution. Let x be the number of units in a positive angle no larger than a right angle. As the size of the angle shrinks toward zero, the value of* (sin x/x) *approaches* (2π/U). *For any positive number c there are sufficiently small positive values of x that* |sin x/x − 2π/U| < c.

PROOF. Let ∠POQ of Figure III.7 be an angle containing x angular units, and let OP be a segment of length one. The coordinates of P are (cos x, sin x). The area of sector ORS is to the area of the circle of radius (cos x) as x is to U since the sector occupies x/U times the area of the circle. Therefore, the area of ORS is $(x/U)[\pi(\cos x)^2]$. In similar fashion the area of sector OPQ is $(x/U)[\pi(1)^2]$. The area of the triangle OPR is $\frac{1}{2}(\sin x)(\cos x)$. Hence, we have the relationship

$$\frac{x}{U}(\pi 1^2) \geq \frac{1}{2}(\sin x)(\cos x) \geq \frac{x}{U}(\pi \cos^2 x). \qquad (III.5.1)$$

In this expression we have respectively the areas of the sector OPQ, the triangle OPR, and the sector ORS. The inequalities are established by the inclusion of one area within another. We seek (sin x)/x. We can obtain this function from (III.5.1) by multiplying each term by 2/(x cos x). By the conditions we have imposed, we see that 2/(x cos x) is a positive number, and we can multiply all terms by this number without disturbing the inequalities. It follows that

$$\frac{2\pi}{U}\frac{1}{\cos x} \geq \frac{\sin x}{x} \geq \frac{2\pi}{U}\cos x. \qquad (III.5.2)$$

We note that the difference between the first term and the third term can be expressed as

$$\frac{2\pi}{U}\frac{1}{\cos x} - \frac{2\pi}{U}\cos x = \frac{2\pi}{U}\left(\frac{1 - \cos^2 x}{\cos x}\right) = \frac{2\pi}{U}\frac{\sin^2 x}{\cos x}.$$

But we can establish the fact that

$$\sin x < \frac{2\pi x}{U} \tag{III.5.3}$$

since $\overline{PR} = \sin x$ and $\overline{PQ} = (x/U)(2\pi)$. Consequently, we can write

$$\frac{2\pi}{U}\frac{1}{\cos x} - \frac{2\pi}{U}\cos x = \frac{2\pi}{U}\frac{\sin^2 x}{\cos x} < \frac{2\pi}{U}\frac{(2\pi x)^2}{U^2}\frac{1}{\cos x} = \frac{8\pi^3}{U^3}\frac{x^2}{\cos x}.$$

We are interested in the case in which x is small. If x is less than 0.14 radians (about 8 degrees), $\cos x > 0.99$. Therefore, if we restrict x to values which are less than 0.14, we know that the difference obtained by subtracting the third term of (III.5.2) from the first term is less than

$$\frac{8\pi^3}{0.99U^3}x^2.$$

It is clear that for any positive number c we can find a value of x sufficiently small that

$$\frac{8\pi^3}{0.99U^3}x^2 < c. \tag{III.5.4}$$

We only need to take the smaller of the values 0.14 and the value of x used to satisfy this last inequality, and we have a value of x which satisfies (III.5.4).

In view of the fact that the first and third terms of (III.5.2) are closer together than some given positive integer, c, it follows that $(\sin x)/x$ is closer to either $(2\pi/U)[1/(\cos x)]$ or $(2\pi/U)[\cos x]$ than c, for $(\sin x)/x$ is in the middle in (III.5.2). Furthermore, $2\pi/U$ is also between the first and third terms of (III.5.2). Therefore $\sin x/x$ and $2\pi/U$ must be closer together than c. In other words $|(\sin x)/x - 2\pi/U| < c$.   □

We were more restrictive than necessary in stating Theorem 5.1. If x is a negative angle, but $|x| < \pi/2$ then $\cos x$ is positive and $(\sin x)/x$ is positive. Furthermore, (III.5.2) still holds. Hence, the same argument could be used with only the modification of including a few absolute value signs and the same result would follow.

EXAMPLE 5.1. Make a table of values for $\sin x/x$ both for degree measure and for radian measure as x diminishes from $x = 1$ radian to $x = 0.01$ radian.

Solution. For degrees we have $U = 360$, and hence the anticipated results will have $(\sin x)/x$ approaching $2\pi/360 = 0.017453$. For radians we will have $U = 2\pi$ and consequently we expect $(\sin x)/x$ to approach $2\pi/2\pi = 1$.

x (radians)	$(\sin x)/x$ (radians)	x (degrees)	$(\sin x)/x$ (degrees)
1.00	0.84147	57.29578	0.018686
0.50	0.95885	28.64789	0.016735
0.20	0.99335	11.45926	0.017337
0.10	0.99833	5.72957	0.017424
0.05	0.99958	2.86479	0.017446
0.01	0.99998	0.57296	0.017453

We note here that the predicted results appear to be verified by the result of the computations.

A second result that will be needed can now be derived. We need to know the value $(1 - \cos x)/x$ will assume as x approaches the value zero.

Theorem 5.2. *If an angle of x units is nearing the zero angle, then $|(1 - \cos x)/x|$ will be nearing the value zero. The smaller the value of x, the smaller the value of $|(1 - \cos x)/x|$.*

PROOF. Algebra and trigonometry provide the following derivation:

$$\left| \frac{1 - \cos x}{x} \right| = \left| \frac{1 - \cos^2 x}{x(1 + \cos x)} \right| = \left| \frac{\sin x}{x} \right| \cdot \left| \frac{\sin x}{1 + \cos x} \right|.$$

By Theorem 5.1, we can select any positive number c and then be sure that x can be made small enough that $|(\sin x)/x| < 2\pi/U + c$. For values of x less than $U/4$ we know that $\cos x > 0$, and consequently $(1 + \cos x) > 1$. Therefore, $|(\sin x)/(1 + \cos x)| < |\sin x|$. We can now state the inequality

$$\frac{1 - \cos x}{x} < \left(\frac{2\pi}{U} + c \right) \sin x < \left(\frac{2\pi}{U} + c \right) \frac{2\pi x}{U} = \left(\frac{2\pi}{U} + c \right) \left(\frac{2\pi}{U} \right) x.$$

Here we have made use of (III.5.3). Since the coefficient of x in the last term is bounded, we can always find a value of x sufficiently small that this last expression is less than any positive number we might care to think of. Thus, we are assured that $(1 - \cos x)/x$ gets closer to zero as x gets closer to zero. $\square$

EXAMPLE 5.2. Make a table similar to the table in Example 5.1 illustrating the behavior of $(1 - \cos x)/x$.

Solution. In this case the result should be approaching zero regardless of the unit used for measuring the value of x.

x (radians)	(1 − cos x)/x (radians)	x (degrees)	(1 − cos x)/x (degrees)
1.00	0.45970	57.29578	0.008023
0.50	0.24483	28.64789	0.004273
0.10	0.04996	5.72957	0.000872
0.05	0.02499	2.86479	0.000436
0.01	0.00500	0.57295	0.000087

Again we can see that the predicted results appear to be forthcoming.

Now we turn to the problem of obtaining integrals for the sine and cosine, the problem we posed at the beginning of this section. In this derivation we will make the assumption that we can evaluate an RS sum in which $f(x)$ is complex provided the values of the real parts at successive evaluation points can be brought sufficiently near each other, and provided the same property would hold for the imaginary parts of the complex values. We will also assume that we can still move a constant factor of the integrand outside the integral as a coefficient of the integral. Both of these assumptions are shown to be valid in later work in mathematics.

We wish to obtain the values of the two integrals

$$\int_0^b \sin x \, dx \quad \text{and} \quad \int_0^b \cos x \, dx.$$

We will obtain these in one result by starting with the integral

$$\int_0^b (\cos x + i \sin x) dx.$$

By the Corollary of Theorem 2.2 of this chapter, this can be written

$$\int_0^b (\cos x + i \sin x) dx = \int_0^b \cos x \, dx + i \int_0^b \sin x \, dx. \qquad \text{(III.5.5)}$$

Since we are using complex quantities here, the rules for complex numbers will hold. Hence the real portion of the result must be equal to $\int_0^b \cos x \, dx$ and the coefficient of the imaginary unit must be equal to $\int_0^b \sin x \, dx$.

We are again in the position of having to use RS sums, for we have no other recourse. It would seem advisable to use a regular partition of $[0, b]$, for the regular partition has proven to be somewhat easier to handle in the past. There is the question whether this partition will give us the value we desire, but we can assure ourselves that it will if we can show that with successive refinements we can make $U - L$ smaller than any given positive number.

Now if we take a regular partition of $[0, b]$ with n subintervals, each sub-interval is of length (b/n). Therefore we will have

$U(P, \cos x + i \sin x, x) - L(P, \cos x + i \sin x, x)$

$$= \left[\sum_{k=1}^{n} (\cos u_k + i \sin u_k)\left(\frac{b}{n}\right) - \sum_{k=1}^{n} (\cos l_k + i \sin l_k)\left(\frac{b}{n}\right) \right]$$

$$= \frac{b}{n} \left[\sum_{k=1}^{n} (\cos u_k - \cos l_k) + i \sum_{k=1}^{n} (\sin u_k - \sin l_k) \right].$$

Here we have used u_k as the evaluation point associated with the upper sum and l_k as the evaluation point associated with the lower sum. In no quadrant can either

$$\sum_{k=1}^{m} (\cos u_k - \cos l_k) \quad \text{or} \quad \sum_{k=1}^{m} (\sin u_k - \sin l_k)$$

exceed one, and hence we know that $U - L < Q(1 + i)b/m$ where Q is the number of quadrants in which at least part of the angle of magnitude b can be found. (We have used m to indicate the number of terms in a quadrant.) For instance, if b is 500 degrees, then $Q = 6$ for b would extend through a complete revolution of 4 quadrants and would go on through the first and part of the second quadrant for a second time. Since $Q(1 + i)b$ is a constant, it follows that upon dividing it by a very large value of m we have a very small difference $U - L$. Consequently, we can conclude that the integral exists and we could evaluate this integral by using any RS sum and a regular partition, for any such sum must be either the upper sum, the lower sum, or must have some value in between.

We will use the left hand points of the partition as evaluation points, and hence $t_k = x_{k-1} = b(k - 1)/n$. Thus we have the sum

$$S(P, T, \cos x + i \sin x, x) = \sum_{k=1}^{n} \left[\cos \frac{b(k - 1)}{n} + i \sin \frac{b(k - 1)}{n} \right]\left(\frac{b}{n}\right)$$

$$= \frac{b}{n} \sum_{k=0}^{n-1} \left[\cos \frac{bk}{n} + i \sin \frac{bk}{n} \right]$$

$$= \frac{b}{n} \sum_{k=0}^{n-1} \left[\cos \frac{b}{n} + i \sin \frac{b}{n} \right]^k,$$

where the last step was obtained by using DeMoivre's Theorem. However, the last expression is a geometric progression. We can handle this as we handled the geometric progression in the development of the integral for the exponential function. Note the ratio here is $[(\cos b/n) + i(\sin b/n)]$.

Thus, we obtain

$$S(P, T, \cos x + i \sin x, x) = \frac{b}{n} \sum_{k=0}^{n-1} \left[\cos \frac{b}{n} + i \sin \frac{b}{n} \right]^k$$

$$= \frac{b}{n} \frac{\left(\cos \dfrac{b}{n} + i \sin \dfrac{b}{n} \right)^n - 1}{\left(\cos \dfrac{b}{n} + i \sin \dfrac{b}{n} \right) - 1}$$

$$= \frac{(\cos b + i \sin b) - 1}{\dfrac{\cos(b/n) - 1}{(b/n)} + i \dfrac{\sin(b/n)}{(b/n)}}.$$

As we consider successive refinements, we have (b/n) approaching zero, and hence the two fractions in the denominator of the last expression conform to the fractions appearing in Theorems 5.1 and 5.2. As n increases we have the first fraction of the denominator approaching zero and the second fraction approaching $(2\pi/U)$. Hence the RS sum approaches the value

$$\frac{(\cos b - 1) + i \sin b}{0 + i\left(\dfrac{2\pi}{U}\right)} = \frac{U}{2\pi i} [(\cos b - 1) + i \sin b]$$

$$= \frac{Ui}{2\pi i^2} [(\cos b - 1) + i \sin b]$$

$$= \frac{Ui}{-2\pi} [(\cos b - 1) + i \sin b]$$

$$= i\left[\frac{U}{2\pi} (1 - \cos b) \right] + \frac{U}{2\pi} \sin b.$$

This gives us for the integral the value

$$\int_0^b (\cos x + i \sin x)dx = \int_0^b \cos x \, dx + i \int_0^b \sin x \, dx$$

$$= \frac{U}{2\pi} \sin b + i\left[\frac{U}{2\pi} (1 - \cos b) \right].$$

By equating the real and imaginary parts of these expressions we have

$$\int_0^b \cos x \, dx = \frac{U}{2\pi} \sin b \qquad\qquad\qquad (III.5.6)$$

and

$$\int_0^b \sin x \, dx = \frac{U}{2\pi} (1 - \cos b). \qquad\qquad\qquad (III.5.7)$$

EXAMPLE 5.3. Evaluate $\int_0^{60°} \sin x \, dx$ and $\int_0^{30°} \cos x \, dx$ where the unit of angular measure is the degree.

Solution. Here $U = 360$, and consequently our formulas give us

$$\int_0^{60°} \sin x \, dx = \frac{360}{2\pi} (1 - \cos 60°) = \frac{180}{\pi} \left(1 - \frac{1}{2}\right) = 28.6479$$

and

$$\int_0^{30°} \cos x \, dx = \frac{360}{2\pi} \sin 30° = \frac{180}{\pi} \left(\frac{1}{2}\right) = 28.6479.$$

EXAMPLE 5.4. Evaluate $\int_0^{0.4} \sin x \, dx$ and $\int_0^{0.6} \cos x \, dx$ where the unit of angular measure is the radian.

Solution. Here $U = 2\pi$, and consequently our formulas give us

$$\int_0^{0.4} \sin x \, dx = \frac{2\pi}{2\pi} (1 - \cos 0.4) = (1 - 0.92106) = 0.07894.$$

and

$$\int_0^{0.6} \cos x \, dx = \frac{2\pi}{2\pi} \sin 0.6 = 0.56464.$$

We obtained the values of the sine and cosine in this last case by means of Table 1 of Appendix C. We note that the fraction $(2\pi/U)$ is not troublesome in this example, for it has the value one.

It is clear from these examples that the formulas for the integration of the sine and cosine would be somewhat simpler to apply if we were to use radian measure, for then the fraction $(2\pi/U)$ would have the value one. It is for this reason that radian measure is used in much of mathematics. From this point on we will agree that we will consider all angles to be measured in radians unless specifically stated otherwise. With this convention we have the formulas

$$\int_0^b \sin x \, dx = 1 - \cos b$$

$$\int_0^b \cos x \, dx = \sin b.$$

(III.5.8)

It is unrealistic to assume that we will always wish to use the interval $[0, b]$ starting at the origin as the interval of integration. We can avoid this requirement by the use of the following theorem.

Theorem 5.3. *If $a < b$, then $\int_a^b \sin x \, dx = \cos a - \cos b$ and $\int_a^b \cos x \, dx = \sin b - \sin a$.*

PROOF.

$$\int_a^b \sin x \, dx = \int_a^b \sin[a + (x - a)]dx$$

$$= \int_a^b [\sin a \cos(x - a) + \cos a \sin(x - a)]dx$$

$$= \sin a \int_a^b \cos(x - a)dx + \cos a \int_a^b \sin(x - a)dx$$

$$= \sin a \int_0^{b-a} \cos x \, dx + \cos a \int_0^{b-a} \sin x \, dx$$

$$= \sin a \sin(b - a) + \cos a[1 - \cos(b - a)]$$

$$= \cos a - [\cos a \cos(b - a) - \sin a \sin(b - a)]$$

$$= \cos a - \cos[a + (b - a)] = \cos a - \cos b.$$

The other half of the theorem can be proved in a similar manner.   □

EXAMPLE 5.5. Evaluate $\int_{\pi/4}^{\pi/3} (3 \sin x - 4 \cos x)dx$.

Solution.

$$\int_{\pi/4}^{\pi/3} (3 \sin x - 4 \cos x)dx = 3\left(\cos \frac{\pi}{4} - \cos \frac{\pi}{3}\right) - 4\left(\sin \frac{\pi}{3} - \sin \frac{\pi}{4}\right)$$

$$= 3\left(\frac{\sqrt{2}}{2} - \frac{1}{2}\right) - 4\left(\frac{\sqrt{3}}{2} - \frac{\sqrt{2}}{2}\right)$$

$$= \frac{7\sqrt{2}}{2} - 2\sqrt{3} - \frac{3}{2}.$$

As a final bit of generalization, we can again invoke Corollary 3.6 and obtain the following corollary.

Corollary 5.1. *If $c \neq 0$, $\int_a^b \sin cx \, dx = (1/c)(\cos ca - \cos cb)$ and $\int_a^b \cos cx \, dx = (1/c)(\sin cb - \sin ca)$.*

OUTLINE OF PROOF. This follows immediately from Theorem 5.3 and Corollary III.3.6. In case $a > b$, we would also invoke Definition III.2.2.   □

EXAMPLE 5.6. Evaluate $\int_0^{\pi/6} (2 \cos 3x - 4 \sin 2x)dx$.

Solution. Using Corollary 5.1, we will have:

$$\int_0^{\pi/6} (2\cos 3x - 4\sin 2x)dx = 2\int_0^{\pi/6}\cos 3x\,dx - 4\int_0^{\pi/6}\sin 2x\,dx$$

$$= 2\left[\left(\frac{1}{3}\right)\left(\sin\frac{3\pi}{6}\right)\right] - 4\left[\left(\frac{1}{2}\right)\left(1 - \cos\frac{2\pi}{6}\right)\right]$$

$$= \left(\frac{2}{3}\right)\left(\sin\frac{\pi}{2}\right) - 2 + 2\cos\frac{\pi}{3}$$

$$= \left(\frac{2}{3}\right) - 2 + 1 = \left(-\frac{1}{3}\right).$$

EXERCISES

1. Evaluate each of the following:

   (a) $\int_0^{\pi/2}\sin x\,dx$

   (b) $\int_0^{\pi/2}\cos x\,dx$

   (c) $\int_0^{\pi/20} 2\sin 5x\,dx$

   (d) $\int_0^{\pi/3}(\sin 2x - \cos x)dx$

   (e) $\int_0^{\pi/8}\cos 4x\,dx$

   (f) $\int_0^{\pi/9}\sin 3x\,dx$

   (g) $\int_0^{\pi/6} 3\cos 2x\,dx$

   (h) $\int_0^{\pi/3}(x^2 - 4\cos x)dx$

2. Evaluate each of the following:

   (a) $\int_{1/6}^{1/2}\sin \pi x\,dx$

   (b) $\int_{-\pi/6}^{\pi/6} 4\,d(\sin x)$

   (c) $\int_{-1/6}^{1/3} 3\cos \pi x\,dx$

   (d) $\int_1^2\cos 5x\,d(3)$

   (e) $\int_{-1/6}^{1/4} 3\sin \pi x\,dx$

   (f) $\int_{-\pi/3}^{\pi/6} 4\cos 2x\,dx$

   (g) $\int_{-\pi/4}^{\pi/2}(3\cos 2x - \sin 2x)dx$

   (h) $\int_1^2 (x^2 - 3x + \sin 2x)dx$

3. Evaluate each of the following:

   (a) $\int_0^{\pi/3}(\cos x + 4)d(\cos x)$

   (b) $\int_0^{\pi/3}(\cos x + 4)d(\cos x + 4)$

   (c) $\int_0^{\pi/3}(\cos x + 4)dx$

   (d) $\int_{\pi/3}^{\pi/2} e^{\sin x}\,d(\sin x)$

   (e) $\int_0^1 \sin 2e^x\,d(e^x)$

(f) $\int_0^{-1} \sin(\pi x/2)dx$

(g) $\int_1^3 \cos(\pi x/6)dx$

(h) $\int_0^{0.5} \sin \pi x^2 \, dx^2$

4. Evaluate each of the following:

(a) $\int_{(1/2)(\ln \pi/6)}^{(1/2)(\ln \pi/2)} \cos(3e^{2x})d(e^{2x})$

(b) $\int_0^{\pi/4} \cos 2x \tan 2x \, dx$

(c) $\int_0^{\pi/6} \sin 3x \cot 3x \, dx$

(d) $\int_0^{\pi/6} e^{\sin 2x} \, d(\sin 2x)$

(e) $\int_0^{\pi/6} \sin x \cos x \, dx$

   [Hint: $\sin A \cos A = \frac{1}{2} \sin 2A$.]

(f) $\int_0^{\pi/6} \sin^2 x \, dx$ [Hint: $\sin^2 A = (1 - \cos 2A)/2$.]

(g) $\int_0^{\pi/6} \cos^2 2x \, dx$ [Hint: $\cos^2 A = (1 + \cos 2A)/2$.]

(h) $\int_0^{\pi/6} (\sin x + \cos x)^2 dx$

5. Evaluate each of the following.

(a) $\int_{-1}^{1} [e^{2x} + \cos(\pi x/3)]dx$

(b) $\int_{-1}^{3} \sin(\pi x/4)dx$

(c) $\int_{-1}^{4} d(\cos(\pi x/3))$

(d) $\int_0^{\pi/4} e^{\sin 2x} \, d(\sin 2x)$

(e) $\int_1^2 \sin x^2 \, dx^2$

(f) $\int_{-2}^{3} (x^3 + e^{-x} - \sin(\pi x/12) + \cos(\pi x/6))dx$

6. (a) Use the relation $\cos(5x) = \cos(4x + x) = \cos 4x \cos x - \sin 4x \sin x$ and $\cos 3x = \cos(4x - x) = \cos 4x \cos x + \sin 4x \sin x$ to show that $\cos 4x \cos x = \frac{1}{2}(\cos 3x + \cos 5x)$.

   (b) Use part (a) and evaluate $\int_{-\pi/6}^{\pi/4} \cos 4x \cos x \, dx$.

   (c) Use a technique similar to that of part (a) and then evaluate $\int_{-\pi/6}^{\pi/4} \sin 4x \sin x \, dx$.

7. (a) Find the area of the rectangle bounded by $x = 0$, $x = \pi/2$, $y = 0$, and $y = 2/\pi$.

   (b) Find the area of the region bounded by $y = \sin x$, $y = 0$, and $x = \pi/2$.

   (c) Sketch the regions of parts (a) and (b) on the same set of axes.

   (d) Explain the fact that the areas of these two regions are equal, using your sketch.

8. Find the area bounded by $r = (\sin \theta)^{1/2}$ and the radii $\theta = 0$ and $\theta = \pi/4$.

9. Find the mean value of $\sin x$ with respect to x over the interval $[0, \pi/2]$.

10. (a) Since $\cos x \leq 1$, $\int_0^x \cos t \, dt \leq \int_0^x 1 \, dt$ if $x > 0$. Use this to derive the relation $\sin x \leq x$ for $x > 0$.

   (b) Use the result of part (a) to derive the relation $1 - \cos x \leq x^2/2$ and then $1 - (x^2/2) \leq \cos x$.

   (c) Use the relations of parts (a) and (b) to obtain $x - (x^3/6) \leq \sin x \leq x$ (if $x > 0$).

(d) Continue with this process (using mathematical induction if appropriate) to obtain

$$1 - \frac{x^2}{2!} + \frac{x^4}{4!} - \frac{x^6}{6!} + \frac{x^8}{8!} - \frac{x^{10}}{10!} \le \cos x \le 1 - \frac{x^2}{2!} + \frac{x^4}{4!} - \frac{x^6}{6!} + \frac{x^8}{8!}$$

and

$$x - \frac{x^3}{3!} + \frac{x^5}{5!} - \frac{x^7}{7!} + \frac{x^9}{9!} - \frac{x^{11}}{11!} \le \sin x \le x - \frac{x^3}{3!} + \frac{x^5}{5!} - \frac{x^7}{7!} + \frac{x^9}{9!}$$

(if $x > 0$).

(e) Show that the successive lower bounds in these two inequalities form a monotone increasing sequence and the upper bounds form a monotone decreasing sequence provided x is not too large. (If $x = 10$, for instance, this monotonicity would not show up until we had passed the term having 10! in the denominator.)

C11. Using the relations derived in Exercise 10 write a program and obtain a table giving the values of the sines and cosines for $0°$, $1°$, $2°$, and so on to include $90°$. Note that the results of Exercise 10 only apply for radian measure, and therefore you will have to convert $1°$ to radian measure for computation, and similarly with the other values. If your facilities will permit, obtain the results to six decimal places, and then compare your table with printed tables of trigonometric functions.

12. In 1777 the *Comte de la Buffón* proposed a method for approximating the value of π through a process that involved tossing a needle of length l and letting it fall on a floor consisting of boards running lengthwise of the room, each board being of width l. He asserted that if one dropped this needle a very large number of times, it should land so that it covers a crack between two boards about $(2n/\pi)$ times out of each n drops of the needle for any integer n. He further stated that this approximation should be very close if n is very large.

(a) In Figure III.8, show that if the angle which the needle makes with the line $\overline{PQ}$ is likely to be any angle in the interval $[0, \pi/2]$ then the effective length of the needle parallel to PQ is $2l/\pi$. [*Hint*: Show that you have the mean value of $l \cos \theta$ with respect to all values of θ in the interval $[0, \pi/2]$.]

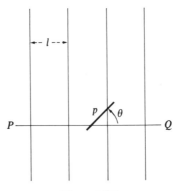

Figure III.8

(b) Using the information from part (a), show that the probability that the needle lands on a crack is given by $2/\pi$.
(c) Confirm the conjecture of the Comte de la Buffón.
(d) Try this experiment 200 times and determine whether your results appear to confirm this conjecture.
(e) If your computer has a random number generator, write a program which will carry out this experiment and determine the results after 5000 simulated tosses of the needle.

III.6 Summary

In this Chapter we have developed many results and established many Theorems which can be of assistance in evaluating integrals. This will be of aid in problems involving applications, however, only if an integral is involved in the application, and then only after the integral has been obtained. In order to obtain the integral in the first place, it is suggested that you should first of all think of a problem, such as one of those in the Exercises of this Chapter, as being approximately solved by adding up a large number of values, each of which describes a small portion of the total problem. In other words, set the problem up in such a way that you approximate the the solution through an RS sum. You can then determine the integral which would result from using the sum with continuing refinement of the interval involved. At this point and only at this point the results of this chapter can be used to evaluate the integral.

In the evaluation of integrals we must use the results obtained herein exactly as they are derived. If an integrand *almost* fits, that is not sufficient. This means, of course, that in some cases we may not find a corresponding result and may yet have to resort to RS sums. In the material that is ahead of you you will find more results which will assist in the evaluation of integrals, but it is probable that if you use a great deal of calculus in applications you will still have to resort to RS sums or some equivalent alternative from time to time. Unfortunately, there are integrals that occur in practice for which no formulas have been found, and for which there is an assurance that no formulas will be found.

In order to bring together the formulas which we have derived so far and the conditions under which these formulas are applicable, we give the following list.

1. $\int_a^b x^m \, dx^n = m(b^{m+n} - a^{m+n})/(m + n)$, m and n are non-negative, rational numbers.
2. $\int_p^q a^{cx} \, dx = (a^{cq} - a^{cp})/(c \ln a)$, $a > 1$ and $c \neq 0$.
2a. $\int_a^b e^{cx} \, dx = (1/c)(e^{cb} - e^{ca})$, $c \neq 0$.
3. $\int_a^b \sin cx \, dx = (1/c)(\cos ca - \cos cb)$, $c \neq 0$.
4. $\int_a^b \cos cx \, dx = (1/c)(\sin cb - \sin ca)$, $c \neq 0$.

Note that (2a) is a special case of (2). In general you will find that (2a) is sufficient for most purposes, although both are stated here for purposes of completeness.

In addition to the calculus, we have seen Exercises that cover a great many applications throughout Chapters II and III. It might be well to review these. Note that we have found two methods for finding volumes of revolution. We have also discussed moments, work, force, marginal cost, etc. Some of these concepts come up from time to time in problems, and any person concerned with applications will find it necessary on occasion to use the index and look up methods, to carefully analyze a problem, and to handle each problem as a unique challenge. You may find that some of these suggestions will be helpful in the exercises at the end of this section.

EXERCISES

1. Evaluate the following integrals:

   (a) $\int_1^3 x^3 \, dx$
   (b) $\int_{-4}^4 \frac{1}{2}x^7 \, dx$
   (c) $\int_{-2}^4 3x^6 \, d(\frac{1}{2}x)$
   (d) $\int_0^{\pi/4} \sin 2x \, dx$
   (e) $\int_1^3 e^{x/2} \, dx$
   (f) $\int_{-1}^3 2^x \, dx$
   (g) $\int_0^3 e^{-x} \, dx$
   (h) $\int_1^{3/2} (4x^2 - 7 + 3 \cos \pi x - 2e^{-x}) dx$

2. Evaluate the following integrals:

   (a) $\int_1^4 x^3 \, d(2 + \sqrt{x})$
   (b) $\int_1^2 e^{x^2} \, d(x^2)$
   (c) $\int_0^{\pi/2} e^{-\cos x} \, d(\cos x)$
   (d) $\int_1^4 \cos^3 2x \, d(\cos 2x)$
   (e) $\int_{-1}^2 3^{2x} \, dx$
   (f) $\int_1^4 x^{3/2} \, dx$
   (g) $\int_0^8 x^{4/3} \, dx$
   (h) $\int_4^9 x^2 \sqrt{x} \, dx$

3. Evaluate the following integrals:

   (a) $\int_{-1}^1 \sinh 2x \, dx$
   (b) $\int_{-1}^1 \cosh 3x \, dx$
   (c) $\int_1^2 \sinh x \cosh x \, dx$
   (d) $\int_{\pi/3}^{3\pi/4} \sin(2x - \pi) dx$
   (e) $\int_{-1}^3 (4/\exp(2x)) dx$

(f) $\int_1^2 (2/x^2)d(1/x)$
   [*Hint*: try a substitution.]
(g) $\int_{-2}^{-1} (e^x - e^{-x})^3 dx$
(h) $\int_{-\pi/6}^{\pi/6} \cos(3x + \pi/2)dx$

4. Evaluate the following integrals:

   (a) $\int_1^{32} x^{4.2} dx^{0.8}$
   (b) $\int_1^3 e^{x+2} \, dx$
   (c) $\int_0^2 [x^2 - \sin(2x - 4) + e^{2x+3}]dx$
   (d) $\int_0^{\pi/9} \sin^2 3x \, dx$
       [*Hint*: See hints in exercises for Section III.5.]
   (e) $\int_3^6 (x - 4)^6 dx$
   (f) $\int_1^2 3 \sin(x - 1)dx$
   (g) $\int_0^8 (x^{2/3} - 3)^2 d(x + 1)^2$
   (h) $\int_{\pi/12}^{\pi/2} \sin 2x \sin x \, dx$

5. (a) Sketch the curve $y = 2 \cosh x$ over the interval $[-3, 3]$.
   (b) Find the area bounded by $x = -3$, $x = 3$, $y = 2 \cosh x$, and $y = 0$.
   (c) Find the volume of the solid of revolution formed by revolving the area of part (b) about the x-axis.
   (d) Find the mean value of $2 \cosh x$ with respect to x over the interval $[-3, 3]$.

6. (a) Evaluate the integral $\int_{-4}^4 (x^2 - x - 6)dx$
   (b) Evaluate the integral $\int_{-4}^4 |(x^2 - x - 6)|dx$
   (c) Indicate the difference in geometric interpretation of these two integrals as areas.

7. Find the area bounded by $y = e^{-2x}$ and $x = 3$ which is in the first quadrant.

8. Find the area bounded by $y = 18 - x^2$, and $y = x^2$.

9. (a) Sketch the curve $y = x^{3/2}$ over the interval $[0, 4]$, and the lines $x = 0$, $x = 4$, $y = 0$, and $y = 8$. Note that the rectangle formed by the four lines is cut into a larger and smaller portion by the semi-cubical parabola.
   (b) Find the area of the larger portion and the smaller portion of the two areas formed in part (a).
   (c) Find the volume formed by revolving the larger portion about the x-axis and the volume formed by revolving the smaller portion about the x-axis.
   (d) Find the volume formed by revolving each of these two areas about the y-axis.
   (e) Find the volume formed by revolving each of these two areas about the line $x = 4$. [*Hint*: Sketch a partition and small solids, and then set up the RS sum and hence the integral in each case.]
   (f) Find the volume formed by revolving each of these two areas about the line $y = 8$.

10. Find the area bounded by the curve $r = e^\theta$ and $\theta = 0$ which has a central angle equal to a right angle. Note that this describes two areas. Find the area of each.

11. Find the area inside one loop of the curve $r = \sin 2\theta$.

12. Find the volume of a sphere of radius R. Note that this sphere can be thought of as the solid of revolution formed by revolving the circle $x^2 + y^2 = R^2$ about either the x-axis or the y-axis.

13. (a) Find the first moment of the area bounded by $y = x^2$, $x = 3$, and the x-axis about the y-axis.
    (b) Find the quotient obtained by dividing this first moment by the number of square units of area in the region described in part (a). This quotient gives the x-coordinate of the center of gravity of this area.
    (c) Find the first moment of this area about the x-axis.
    (d) Use your result in part (c) to find the y-coordinate of the center of gravity for this area.
    (e) Plot the point whose coordinates are the x-coordinate and the y-coordinate of the center of gravity respectively. Show that this location is a reasonable location for the center of gravity of this area.

14. Using the method of Exercise 13 find the coordinates of the center of gravity of the area bounded by $y^2 = 4x$ and $x = 4$ which is in the first quadrant.

P15. (a) Find the moment of inertia (the second moment) of the area of Exercise 13 about the x-axis.
    (b) Find the quotient of the moment of inertia divided by the area involved.
    (c) Find the square root of the quotient of part (b). This square root is called the *radius of gyration* of the area about the axis.

P16. Find the radius of gyration of the area of Exercise 14 about both the x-axis and the y-axis.

P17. A certain light bulb has a resistance of $R = 200$ ohms. The voltage on the circuit of which the bulb is a part is given by $E = 100 \sin(120\pi t)$ volts. Time is measured in seconds. The power consumed is given by $P = E^2/R$ watts. How many kilowatt-hours (1000 watts for one hour) are used by the bulb in the course of 3 hours? [*Hint*: Since the voltage is that of an alternating current at 60 cycles per second, the voltage varies. Hence, it is necessary to think of the power used in a short increment of time, and add up such individual power uses to find the total power used. This should involve an RS sum and then a corresponding integral.]

B18. A certain dose of medicine is taken, and it is being absorbed by the body at a rate given by $M(t) = 0.4e^{-0.5t}$ milligrams per hour at a time t hours after it was ingested.

    (a) How much is absorbed by the body in the first two hours after the medicine is taken?
    (b) If the medicinal effect is negligible when the absorption rate is 0.08 milligrams per hour, at what time does this occur? (At this time one would ordinarily take another dose.)
    (c) How much of this medication is absorbed before another dose is needed as determined by your response to part (b)?

B19. It is not infrequent that a medical prescription prescribes that two tablets be taken at once and then one tablet is to be taken at stated intervals thereafter. The time interval between doses is ideally the time such that one additional tablet will restore the absorption rate of the drug to that which the two tablets had given initially. The rate of absorption closely approximates $M(t) = De^{-rt}$.

(a) If the first dose is taken when $t = 0$, find the time for the second dose in terms of r if the second dose is to be given at such time that the one tablet will bring the absorption level back to that which prevailed at time $t = 0$.

(b) Find the number of milligrams absorbed between doses if $D = 0.5$ milligrams and $r = 0.3$.

(c) Find D if it is desired that 0.4 milligrams be absorbed in the first hour.

(d) For what value of r would it be desirable to take a dose of medicine under the conditions of this problem every four hours?

B20. Ecologists often use heat accumulation above a prescribed temperature threshold over a given period of time as a means of predicting temperature effects on biological processes. C. Y. Arnold has shown that a very good approximation to the number of degree days of heat accumulation can be obtained by considering the area which is under a sine curve and above the given threshold temperature. Suppose the given threshold temperature for a given day is 50°F, and the temperature t hours after sunrise is given by $T(t) = 50 + 25 \sin(\pi t/12)$ measured in degrees Fahrenheit. Find the effective heat accumulation for this day measured in degree-days. [*Hint*: a degree-day is the additional energy that would be obtained were the temperature to have been one degree higher for the entire 24 hour period. It is necessary to watch time units in this problem.]

S21. The amount of ore extracted from a certain mine is given by the equation $T(t) = 1000e^{-0.001t}$ tons per day where t is the number of working days since January 1, 1978. Find the amount of ore extracted during the year 1978 if it is assumed that there are 50 weeks of 5 days each during which work is in progress each year. (Note that the loss of two weeks would just about cover the standard vacation.) At some point in the life of a mine the behavior indicated by this equation would be reasonable, for the amount of ore remaining in the mine would become sufficiently small that it would be impossible to continue to mine at a constant rate without an expenditure that would make the mine uneconomical to operate.

S22. If the marginal cost of production of a certain item at a level of n items of production per day is given by the relation $MC(n) = 12 + 6n + (e^{-0.1n}/12)$, find the cost of producing 40 items per day. Find the average cost per item if the production rate is 40 items per day.

S23. The price at which a product sells determines for most items the approximate number that can be sold. The higher the price, the less the demand is likely to be. Consequently, if the equation relating price and the quantity demanded is graphed we can expect a curve which generally looks like the curve AB of Figure III.9. In similar manner, the greater the price the more items will probably be manufactured, and the equation relating price and supply will appear as the curve BC of the graph. If we envisage an economy with *pure competition*, it is reasonable to assume that the selling price will ultimately settle on the price indicated by the intersection of these two curves, for a lower price would yield an unfulfilled demand and a higher price would tend to leave items unsold. For any price used, the horizontal line representing that price will divide the area ABC into two parts. For this selling price the area ABD represents money at least part of which would have been spent by some consumers if the price had been higher. Consequently, this is called *consumer's surplus* since at the price represented by the horizontal line this

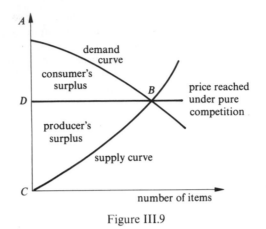

Figure III.9

area represents a gain for the purchaser. On the other hand, the area represented by the portion of *ABC* below the line involves money obtained by the supplier due to higher prices on goods he would have been willing to produce for less. This area represents a *producer's surplus*. In Figure III.9 the line *BD* represents the limiting price under pure competition, the area *ABD* represents the consumer's surplus and the area *BCD* represents the producer's surplus.

If the demand and supply curves are respectively $2x + 3y = 195$ and $20y = x^2$, find the price reached under pure competition. Find the consumer's surplus and the producer's surplus. (Assume price to be in dollars and the quantity of demand to be the number of individual items.)

S24. If the equation of the demand curve is $10y + 2e^{0.1x} = 90$, find the consumer's surplus if the price of equilibrium is $3. Do the same thing if the price is $1, $2, $4, and $5 respectively. For what interval would this demand curve have a sensible interpretation? [*Hint*: Look at the graph of this equation.]

S25. If the equations of the demand and supply curves are respectively $y = 440 - x^2 - 2x$ and $2y = 4 + (x + 4)^2$, find the price reached under pure competition. Also find the consumer's surplus and the producer's surplus for this price. Use RS sums (with a partition of the y-axis) to show that this is the price at which the sum of the two surpluses is a minimum.

S26. (a) The equation of a supply curve is given by $y = 8e^{0.2x}$. Find the producer's surplus if the price of equilibrium is $16.
   (b) Do the same if the price of pure equilibrium is $10.
   (c) Show why you could not have a price of equilibrium less than $8, and why $8 is most unlikely. [*Hint*: Examine the graph of this curve.]

CHAPTER IV

Differentiation

IV.1 Rates of Change

In our work so far we have developed the integral and discussed the process of integration at some length. The calculus is built around two concepts of which integration is only one. The other part of the calculus will investigate the rate at which a function changes. One of the important questions concerning any thing that is changing is "How fast is it changing?" This is just as true in matters of economic and social change as it is for change of location or change in the amount of your mathematical knowledge since you first opened this book. It is this question which we wish to address in this chapter. We shall find that it has very far reaching ramifications, and that there are many rather subtle matters that must be considered. We shall be more alert to the nature of the ideas involved, however, than the rigorous treatment of the subtleties in this chapter. We shall come back to consider the latter in more detail in Chapter VII.

In order to provide a brief insight into the manner of attack we will use an illustrative example. Let us consider the question of determining how fast a vehicle is going as it passes a given spot. We might even consider that the speedometer is not operating properly, and the driver has occasion to try to prove in court that he was not going as fast as he had been accused of traveling. It should be pointed out, however, that this latter case is unlikely here, for few would be so foresighted as we will demand if they then take chances with speed limits. We will assume that the aforementioned driver has available to him some very accurate timing equipment and measuring equipment, and that it is set up so that he can measure with very great accuracy the distance traveled in the first second after passing the spot in question. If he knows that he has traveled 66 feet in this one second period, he knows that he would travel sixty times as far in one minute, or a distance of 3960 feet (three-fourths of a mile), and this very astute driver would then be able to determine that if he were to maintain for one hour the same average speed he had maintained over that one second, he would have been traveling at 45 miles per hour. (Note that if and when the metric system is adopted, this example will have to be altered—one of the problems that progress always brings!) Now, we have already made certain implicit assumptions that are hardly justified. For instance, we have introduced a very big IF concerning the maintenance of the same *average* speed throughout the period of an hour. It is not at all certain that the speed was constant throughout the second, and hence we have had to use the word average here. Since some acceleration or deceleration is possible in one second, we would have done better to take a shorter time interval, say one third of a second. It is possible to raise the same argument here, though, for it does not follow that the same average speed is maintained throughout one third of a second, although this would seem more likely than the fact that it would be maintained during an entire second. With this accurate timing mechanism, however, there is no reason why we can't go further and consider a time interval of one tenth of a second, or one one-hundredth of a second, or one one-thousandth of a second, etc. As we consider shorter and shorter intervals, we are allowing less time for any change of speed, and hence are more likely to have the speed that the car had at the moment it passed the spot in question.

Now it should be very clear that we cannot hope to obtain the speed at the *moment* that it passes the spot, for this would imply in our rather imprecise use of the word moment that there is only one point involved and this is not enough information to establish a speed. Therefore, the best that we would seem to be able to do would be to list the average speeds over shorter and shorter time periods, each of which would include the instant at which we passed the spot in question, and then determine whether this set of numbers would approach some limiting value. We have already seen examples of limiting situations in our use of lubs and glbs to derive integral formulas. In these cases we were able to establish upper and lower bounds for the final result and then to make use of these bounds in establishing the desired limiting value. This matter of obtaining bounds might provide us with some assistance in finding limits in the future.

While we have selected a particular example involving speed, we could equally well have selected any number of illustrations from a wide variety of disciplines. These rates of change have such names as *velocity, marginal cost, rate of solution, slope*, etc., but they all share the same concept. Rather than having to concern ourselves with such a variety of entities we will call this limiting rate of change a *derivative*, and remember that this has application in a number of disciplines under names that seem appropriate to those disciplines. We shall mention some of these names in the application Exercises as we go along.

In talking about the rate of change of distance, we assumed that the problem of measuring speed is one that has occurred before. However, to make certain that the computations involved are well understood, let us take just a moment for review. If $s(t)$ denotes the distance from some starting point, at a time denoted by t, then the location at time $(t + \Delta t)$ will be given by $s(t + \Delta t)$. We have used Δt here to denote a difference in time, since Δ is the Greek equivalent of our letter D, the first letter of difference. The distance traveled during the period from t to $(t + \Delta t)$ is then $s(t + \Delta t) - s(t)$, or the odometer reading at the end of the time interval minus that at the beginning. (The odometer is that portion of the mileage measuring instrument which indicates how far the car has gone.) Since this distance was traversed in $\Delta t = (t + \Delta t) - t$ units of time, the average speed is then obtained by computing the value

$$\frac{s(t + \Delta t) - s(t)}{(t + \Delta t) - t} = \frac{s(t + \Delta t) - s(t)}{\Delta t}.$$

We list both of these, for they will give rise to two forms of the same expression, one of which will be of greater use in some instances and the other at other times. Our concern here is to try to find the value toward which this expression seems to be moving as Δt assumes smaller and smaller values. Note that this is very much like the problem we had with denominators occurring in the derivations of Sections 3.4 and 3.5. Since we will be using this expression rather frequently, it will be convenient to give it a name. Henceforth, we will refer to it by the name *differential quotient*.

Definition 1.1. If $f(x)$ is a function defined over the interval $[a, b]$ then the *differential quotient* of $f(x)$ with respect to x over the interval $[a, b]$ is given by $DQf(a, b) = [f(b) - f(a)]/(b - a)$.

We have used the notation $DQf(a, b)$ to indicate that we have the differential quotient of $f(x)$ and the interval involved is the interval $[a, b]$. In particular

$$DQf(x, x + \Delta x) = \frac{f(x + \Delta x) - f(x)}{(x + \Delta x) - x} = \frac{f(x + \Delta x) - f(x)}{\Delta x}$$

is the differential quotient over the interval from x to $x + \Delta x$. It is clear that the differential quotient represents an average rate of change over an interval, and specifically $DQf(a, b)$ represents the rate of change (or slope) from $(a, f(a))$ to $(b, f(b))$. Also $DQf(x, x + \Delta x)$ represents the rate of change (or slope) from $(x, f(x))$ to $(x + \Delta x, f(x + \Delta x))$.

EXAMPLE 1.1. If $f(x) = x^2 - e^x$, find $DQf(1, 2)$.

Solution.

$$DQf(1, 2) = \frac{f(2) - f(1)}{2 - 1} = \frac{(4 - e^2) - (1 - e)}{1} = 3 - e^2 + e.$$

EXAMPLE 1.2. Find $DQf(x, x + \Delta x)$ if $f(x) = 1/x^2$.

Solution.

$$DQf(x, x + \Delta x) = \frac{f(x + \Delta x) - f(x)}{(x + \Delta x) - x} = \frac{\frac{1}{(x + \Delta x)^2} - \frac{1}{x^2}}{\Delta x}$$

$$= \frac{x^2 - [x^2 + 2x(\Delta x) + (\Delta x)^2]}{(\Delta x)x^2(x + \Delta x)^2}$$

$$= \frac{(\Delta x)(-2x - \Delta x)}{(\Delta x)x^2(x + \Delta x)^2}.$$

This last result can be further simplified if $(\Delta x) \neq 0$. It would, of course, be meaningless if (Δx) were to be zero, for then we would have a fraction with both numerator and denominator zero. To complete the simplification under the assumption that $(\Delta x) \neq 0$, we would have

$$DQf(x, x + \Delta x) = -\frac{2x + \Delta x}{x^2(x + \Delta x)^2}.$$

EXERCISES

1. If $f(x) = x^2$ find the value of each of the following differential quotients.

  (a) $DQf(3, 6)$
  (b) $DQf(3, 5)$
  (c) $DQf(3, 4)$
  (d) $DQf(3, 3.1)$
  (e) $DQf(3, 3.01)$
  (f) $DQf(3, 3.001)$
  (g) $DQf(3, 3.0001)$
  (h) $DQf(3, 0)$
  (i) $DQf(3, 1)$
  (j) $DQf(3, 2)$
  (k) $DQf(3, 2.9)$
  (l) $DQf(3, 2.99)$
  (m) $DQf(3, 2.999)$
  (n) $DQf(3, 2.9999)$

2. Evaluate $DQf(x, x + \Delta x)$ for each of the following functions. Simplify each as far as possible under the assumption that $(\Delta x) \neq 0$. Show where you use this assumption in each case.

  (a) $f(x) = 1/x$
  (b) $f(x) = 3x - 2$
  (c) $f(x) = 4 - x + x^2$
  (d) $f(x) = \sqrt{x}$
  (e) $f(x) = x^3$
  (f) $f(x) = x^2 - x^{-1}$

(g) $f(x) = 12.3$
(h) $f(x) = 2x^{-1/2}$
(i) $f(x) = (2x - 3)^2$
(j) $f(x) = (2x - 3)^3$

3. Show that $DQf(a, b)$ does not depend on the values of a and b if $f(x)$ is a constant function.

4. Show that $DQf(a, b)$ does not depend on the values of a and b if $f(x)$ is any constant multiple of the identity function.

5. Show that $DQf(a, b) = DQf(b, a)$ for any function and for any values of a and b. Interpret this result on the graph of $y = f(x)$.

6. (a) Show that the differential quotient of a monotonic function cannot be positive over one interval and negative over another interval.
   (b) If the monotonic function is increasing, show that its differential quotient can never be negative.
   (c) If the monotonic function is strictly increasing, show that its differential quotient is always positive.
   (d) Repeat parts (b) and (c) for decreasing and strictly decreasing functions respectively with the appropriate algebraic signs.

7. If $f(x) = x^m$ use the results of Exercise 8 of Section I.10 to show that $ma^{m-1} \leq DQf(a, a + 1) \leq m(a + 1)^{m-1}$ for any positive integer m.

8. If $P[a, b]$ is a partition of the interval $[a, b]$, show that

$$\sum_{k=1}^{n} DQf(x_{k-1}, x_k)(x_k - x_{k-1}) = f(b) - f(a).$$

9. (a) Write an expression for $DQf(0, b/n)$ if $f(x) = a^x$.
   (b) Locate this result in the derivation of Section III.4.
   (c) Use the results of Section III.4 to determine the value of $DQf(0, b/n)$ when b/n is very close to zero.

10. (a) Write out $DQ \cos(0, b/n)$.
    (b) Locate this result in the derivation of Section III.5 and determine the value of this differential quotient when b/n is very close to zero.

11. (a) Write out $DQ \sin(0, b/n)$.
    (b) Locate this result in the derivation of Section III.5 and determine the value of this differential quotient when b/n is very close to zero. Note that this result depends upon the unit used for measuring the angle.

S12. If $C(n)$ is the cost of producing n items in one day, show that the marginal cost of increasing production from a level of n items per day is given by $DQf(n, n + 1)$.

B13. If the number of organisms in a given culture at a time t hours after the culture is started is given by $n(t) = 100e^{0.1t}$, find the average rate of change in the number of organisms during the first 100 hours. During the first 50 hours. During the first 20 hours. During the first 10 hours. During the first 5 hours. During the first 3 hours. During the first 2 hours. During the first hour. Does this sequence of numbers appear to be approaching some number? If so, what is the number that this appears to be approaching?

IV.2 The Derivative

In the preceding section we defined the differential quotient

$$DQf(x, x + \Delta x),$$

and we discussed some of the consequences of letting Δx get smaller and letting it eventually approach zero. It was carefully pointed out that it would make no sense to let Δx actually become zero, for then we would have a quotient with zero for both the numerator and the denominator. Such a quotient is undefined. In similar fashion if we consider the differential quotient to represent a slope, letting Δx be zero would be equivalent to asking for the slope of a line determined by a single point and it is obvious that a single point does not determine a unique line.

We will start this discussion by considering the function $f(x) = x^{1/2}$, and the differential quotient $DQf(4, 4 + \Delta x)$. The differential quotient gives the slope of the line joining the point $(4, f(4))$ and the point $(4 + \Delta x, f(4 + \Delta x))$. This is pictured in Figure IV.1. We will permit Δx to be either positive or negative, and will follow the behavior of this differential quotient as the value of Δx gets closer to zero.

As a first approximation we consider Δx to be -1 and then Δx to be $+1$, giving us the average rate of change of $f(x)$ in the interval one unit to the left of $x = 4$ and also in one unit to the right of $x = 4$. We have

$$DQf(4, 3) = \frac{\sqrt{3} - \sqrt{4}}{3 - 4} = \frac{1.7320508 - 2}{-1} = 0.2679492$$

and

$$DQf(4, 5) = \frac{\sqrt{5} - \sqrt{4}}{5 - 4} = \frac{2.2360680 - 2}{+1} = 0.2360680.$$

There results are rather far apart, but they indicate the manner in which such results must be calculated. We proceed to make a table starting with these values and using ever smaller values for Δx. This table is given in Table IV.1. It seems clear that the value of the differential quotient is approaching 0.25 as x approaches zero, and it does not depend on whether x

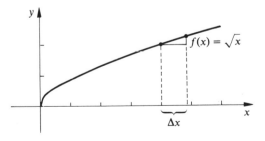

Figure IV.1

Table IV.1

Δx	$DQf(4, 4 - \Delta x)$	$DQf(4, 4 + \Delta x)$
1.0000	0.267949	0.236068
0.1000	0.251582	0.248457
0.0100	0.250156	0.249844
0.0010	0.250016	0.249984
0.0001	0.250002	0.249998

is negative or whether x is positive. At best, however, this deduction is intuitive, and it would be well to examine this situation a bit more thoroughly.

Using this same function, let us use a more formal approach to investigate the behavior of this differential quotient as Δx nears zero. If we write $DQf(4, 4 + \Delta x) = (\sqrt{4 + \Delta x} - \sqrt{4})/\Delta x$ we observe that as Δx gets smaller, both the numerator and denominator get smaller, and there is a real question concerning which will win the race toward zero. We know that we must refrain from letting Δx go all of the way to zero. Consequently, we will require that at no time shall $\Delta x = 0$. There are times when some algebraic legerdemain may be helpful, and here we will use some to rationalize the numerator! Consider the following

$$DQf(4, 4 + \Delta x) = \frac{\sqrt{4 + \Delta x} - \sqrt{4}}{\Delta x} \cdot \frac{\sqrt{4 + \Delta x} + \sqrt{4}}{\sqrt{4 + \Delta x} + \sqrt{4}}$$

$$= \frac{(4 + \Delta x) - 4}{\Delta x[\sqrt{4 + \Delta x} + \sqrt{4}]} = \frac{\Delta x}{\Delta x[\sqrt{4 + \Delta x} + \sqrt{4}]}. \quad \text{(IV.2.1)}$$

You will note that we have multiplied both numerator and denominator by an expression that is similar to the conjugate in complex numbers. It would appear reasonable at this point to divide the numerator and denominator of the last expression by Δx. This can be done since we have decreed that $\Delta x \neq 0$. However, we should note once more that this would not be possible if we were to permit Δx to be zero. Now we can write $DQf(4, 4 + \Delta x) = 1/(\sqrt{4 + \Delta x} + \sqrt{4})$ for those values of Δx other than zero. We are in no way limiting any non-zero values for Δx, no matter how small, and regardless of sign. If Δx is very small, then $\sqrt{4 + \Delta x}$ will have a value very close to 2. Thus $DQf(4, 4 + \Delta x)$ would be very close to 1/4. We would then say that the limit of $DQf(4, 4 + \Delta x)$ is 1/4 as Δx approaches zero. This agrees with our earlier supposition concerning the value this differential quotient should be approaching.

It is clear from the graph in Figure IV.1 that we are taking the slope as Δx gets smaller, and thus we are obtaining the slopes of lines which approach

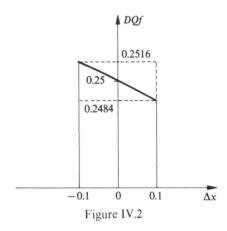

Figure IV.2

the limiting position given by the tangent line. Since this limit is 0.25 we can then say that the tangent line at the point $(4, f(4))$ has a slope of $1/4$.

This discussion raises several issues. We have taken the "limit" of the differential quotient as Δx approached (but did not reach) zero. We could word this in a reverse manner by indicating that if we wished to insure that $DQf(4, 4 + \Delta x)$ had values in the interval $(0.2484, 0.2516)$ we need only be certain that Δx has values in the interval $(-0.1, 0.1)$. This is shown in Figure IV.2. If we wished to insure that $DQf(4, 4 + \Delta x)$ has values restricted to the interval $(0.249, 0.251)$, we need only limit the values of Δx to the interval $(-0.063, 0.064)$. This suggests that if we wish to show that the value of the differential quotient is approaching 0.25 as Δx approaches zero, we might specify an open interval, E, which includes the value 0.25 and then see whether there is an interval, D, such that values of Δx in D other than $\Delta x = 0$ insure the fact that the differential quotient has a value in E. We must, of course, be able to do this for *any* open interval E which includes 0.25. We shall use this approach as a basis for a definition of the *limit* concept. We will specify that the set of points, D, be a *deleted neighborhood of* 0 and by this we will mean that D is an open interval with zero as an interior point but with the single point, $\Delta x = 0$, deleted. This terminology will permit us to avoid the repeated indication that zero is to be avoided from this point on.

Definition 2.1. The *limit of the function $f(x)$ as x approaches the value a is L* if and only if for *any* open interval E which includes L there is a corresponding deleted neighborhood D about the point a such that if x is an element of D, then $f(x)$ is an element of E. We denote this limit by $\lim_{x \to a} f(x) = L$.

We can now write for our earlier example $\lim_{\Delta x \to 0} DQf(4, 4 + \Delta x) = 0.25$. From our discussion above we see that we have satisfied the requirements of the definition. (The abbreviation *lim* obviously stands for *limit* and the notation "$\Delta x = 0$" indicates that the limit is taken as Δx approaches zero.)

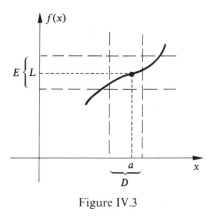

Figure IV.3

It is important to note in the definition of the limit that the choice of the open interval E is arbitrary but for the fact that it must include as an interior point the limit, L. As shown in Figure IV.3, we are taking an interval on the $f(x)$ axis which effectively determines a horizontal strip on the graph. We are now asking that there be a deleted neighborhood, D, about the point $x = a$ such that when x is selected as being a member of D, we are assured that $f(x)$ has a value in the horizontal strip. We can use D as the base for a vertical strip such that no point of the graph $y = f(x)$ which is in the strip D is outside the strip E.

We started our discussion by considering a differential quotient and then taking the limit of the differential quotient $DQf(x, x + \Delta x)$ as Δx approaches zero. This operation occurs frequently enough to merit a definition.

Definition 2.2. The *derivative of* $f(x)$ *at* $x = x_0$ is defined to be $D_x f(x_0) = \lim_{\Delta x \to 0} DQf(x_0, x_0 + \Delta x)$ provided the limit exists. If the limit fails to exist, $f(x)$ has no derivative at $x = x_0$.

The process of finding the derivative is called *differentiation*. The derivative has many aliases. It is also known as the slope in connection with the direction of a line, as velocity or acceleration in connection with the movement of a particle, as marginal cost in consideration of a problem in economics, etc. Thus, you may find that the derivative appears under other names in disciplines outside of mathematics, but the properties of the derivative hold for each of the aliases which the derivative has.

There are many notations for the derivative, also. We will use either $D_x f(x)$ or $f'(x)$ most frequently. The former denotes the derivative with respect to x in a very unmistakable manner. The notation $f'(x)$ is a modification of the notation introduced by *Sir Isaac Newton* (1642–1727). Newton had placed a dot over the function to indicate the derivative. Leibniz, on the other hand, noted that the differential quotient involved a change in f divided by a change in x, or in other words $\Delta f / \Delta x$. From this he derived the notation

df/dx to represent the derivative. We shall use $D_x f(x)$ and $f'(x)$, for they are less likely to cause one to inadvertently treat the derivative as a fraction. If we have $y = f(x)$, we may also refer to the derivative as y'.

Note that the definition starts with the differential quotient and then has us taking the limit. We can illustrate this process with the following Example.

EXAMPLE 2.1. Find $f'(x_0)$ if $f(x) = 2x^3$.

Solution.

$$DQf(x_0, x_0 + \Delta x) = \frac{f(x_0 + \Delta x) - f(x_0)}{\Delta x} = \frac{2(x_0 + \Delta x)^3 - 2x_0^3}{\Delta x}$$

$$= \frac{2[(x_0 + \Delta x) - x_0][(x_0 + \Delta x)^2 + (x_0 + \Delta x)x_0 + x_0^2]}{\Delta x}$$

$$= \frac{2\Delta x[3x_0^2 + 3x_0 \Delta x + (\Delta x)^2]}{\Delta x}$$

$$= 2[3x_0^2 + 3x_0 \Delta x + (\Delta x)^2] = 6x_0^2 + \Delta x(6x_0 + 2\Delta x)$$

provided we do *not* permit $\Delta x = 0$. We are now able to take the limit and thus obtain

$$f'(x_0) = D_x f(x)|_{x=x_0} = \lim_{\Delta x \to 0} DQf(x_0, x_0 + \Delta x)$$

$$= \lim_{\Delta x \to 0} [6x_0^2 + \Delta x(6x_0 + 2\Delta x)]$$

$$= 6x_0^2.$$

This follows since the definition of limit specifically precludes the possibility that $\Delta x = 0$, and for any interval E which includes the value $6x_0^2$, we can find a deleted neighborhood D of $\Delta x = 0$ such that if Δx is in D then the differential quotient is in E. Since the amount by which the differential quotient differs from $6x_0^2$ is $[\Delta x(6x_0 + 2\Delta x)]$, we observe first that we can find a deleted neighborhood D which allows so little variation for Δx that Δx can be kept as small as we may need. Since for such a neighborhood we are assured that there is some positive number M such that $|6x_0 + 2\Delta x| < M$, we only need to further insure that Δx is small enough that $M(\Delta x)$ is small enough that $6x_0^2 + M(\Delta x)$ is within E for both positive and negative values of (Δx).

Observe that we have very carefully considered what happens as Δx takes on values close to zero. One should guard against just substituting zero for Δx. While this appears to work in many instances there will be some instances in which it is not possible to obtain a result using this method, and it is a bad habit to start. (The difficulty with bad habits is that just about the time they become habits there is some instance in which the habit gives an incorrect answer. Hence, we emphasize here the desirability of correct thinking from the beginning.)

In looking at the above result, we have really done more than might appear on the surface. We have obtained $f'(x_0)$ for $f(x) = 2x^3$. But we have a general result, for we placed no restrictions on the choice of x_0. In fact, x_0 could have been any value of x. Therefore, our result $f'(x_0) = 6x_0^2$ would indicate that we would have the derivative if we would evaluate $6x^2$ at the point x_0. We do, indeed, have a derived function, and there is no reason why we should not drop the subscript and write $f'(x) = 6x^2$. In this case we didn't need to specify a specific point, and hence we could equally well write $D_x f(x) = D_x 2x^3 = 6x^2$. The subscript, x, placed on the D (for derivative) is to remind us that it was the variable x that we were considering as the variable in obtaining this particular derivative. This convention will be particularly helpful in case a function may have more than one literal expression with all but one being constant for the purpose of evaluating the derivative. Thus, we might have considered the function $g(x) = ax^3$ with the understanding that a is a constant coefficient and we might wish to consider the derivative with respect to the variable x. If $a = 2$ in this instance we would have the result considered above.

The process of obtaining the derivative is called *differentiation*, and we *differentiate* a function in order to obtain its derivative. Since the derivative is the rate of change, anything which is a rate of change is therefore essentially a derivative of something with respect to something. Note that the expression $(a^{b/n} - 1)/(b/n)$ was in fact a differential quotient. If $f(x) = a^x$, then $DQf(0, b/n) = (a^{b/n} - 1)/(b/n)$. When we tried to find the value of the differential quotient for a very small value of b/n we were approximating the derivative. If $a = e$ we have $f'(0) = 1$.

In order to evaluate a derivative using the definition the procedure is very straightforward. Evaluate the differential quotient, considering if you like that Δx can be small for all cases in which you will be interested, and then obtain the limit of this differential quotient as Δx approaches zero. This process is given in the flow chart in Figure IV.4. The main question concerns whether one can find the limit. Cases exist for which there is no limit. We shall see that in the cases in which we are interested we will usually be able to find the limit. (Could this have something to do with the fact that we are interested in these cases—namely that we can handle them?) Later we will develop formal means for finding derivatives, just as we did for finding integrals. For situations not covered by the formal results you can always resort to the definition, for the definition is the point from which all of these results start.

At this point we should comment on a matter of notation in Example 2.1. You will observe that we were very careful to use $(\Delta x)^2$ and not to use Δx^2. Since the Greek letter Δ denotes a difference or change, Δx^2 would denote a difference or change in x^2. Hence, we would have $\Delta x^2 = (x + \Delta x)^2 - x^2 = 2x\Delta x + (\Delta x)^2$. One of the difficulties experienced by many people in obtaining derivatives is in the use of algebra, and you will find it most helpful to be *very* careful to write algebraic expressions correctly. This includes the use of parentheses where they are needed.

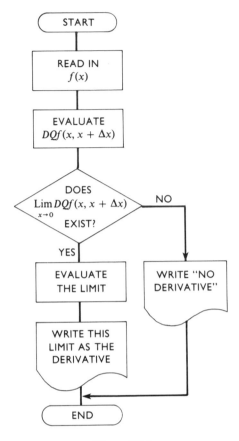

Figure IV.4

It should be emphasized that the derivative is a rate of change. It can appear under many different aliases depending upon the circumstances. For instance the derivative may appear as velocity or acceleration in physics; it may appear as a *marginal* construct, such as marginal cost, in economics; it can appear as slope in geometry; it can appear as rate of solution in chemistry; etc. If the concept involved is that of a rate of change, it is safe to conjecture that you have what the mathematician will call a derivative.

EXAMPLE 2.2. Find an expression for the slope of the line tangent to the curve $y = 1/x$ at a general point (x, y).

Solution. Let P be a point (x, y) on the curve whose equation is $y = 1/x$ as shown in Figure IV.5. Let Q be a second point on the curve with coordinates $(x + \Delta x, y + \Delta y)$. The slope of the line through P and Q is given by the relation $m = ((y + \Delta y) - y)/((x + \Delta x) - x) = \Delta y/\Delta x$. In this par-

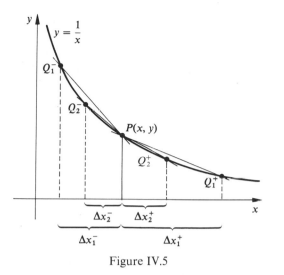

Figure IV.5

ticular case where $y = 1/x$ and $y + \Delta y = 1/(x + \Delta x)$ we now have

$$m = \dfrac{\dfrac{1}{x + \Delta x} - \dfrac{1}{x}}{x + \Delta x} = \dfrac{(x) - (x + \Delta x)}{(\Delta x)(x)(x + \Delta x)} = \dfrac{-1}{x(x + \Delta x)}.$$

Here we have used the fact that P and Q are distinct points, and therefore $\Delta x \neq 0$. This all looks very much like the differential quotient of Example 2.1. We illustrate this in Figure IV.5 by several choices for Q, labeling each one with a subscript, and further labeling those to the right of P with $(+)$ and those to the left with $(-)$. It seems probable that we will have the slope of the tangent line if we hold P fixed and let Q move toward P. It is clear that Q could move either from the right or from the left. It is equally clear that as Q moves toward P we have Δx approaching zero. Therefore, we again have the derivative. Thus, we would have in this case

$$m^* = \lim_{\Delta x \to 0} \frac{\Delta y}{\Delta x} = \lim_{\Delta x \to 0} \frac{-1}{x(x + \Delta x)} = \frac{-1}{x^2}.$$

We have denoted the limiting value of the slope by m^* to distinguish it from the value m used when we obtained the slope of the line PQ. Note that we really have the derivative, and thus $m^* = D_x y = y'$ where we have again used the prime ($'$) to denote the derivative. We can establish the fact that the limit involved is really $-1/x^2$ by again establishing the fact that the definition of limit is satisfied. While it is somewhat less straightforward than in Example 2.1, it is suggested that you try applying the definition to prove that this result is correct.

We have introduced two notations commonly used for the derivative. There are others. Leibniz introduced the notation df/dx or dy/dx where

$y = f(x)$. He was thinking in terms of the differential quotient which he wrote as $\Delta f/\Delta x$ or as $\Delta y/\Delta x$. He then considered that the lower case Latin letter, d, would indicate that he had taken the limit. While this notation is appealing and is widely used, it has the danger that one might consider that the derivative is a quotient, and one might try some algebra that is not warranted. (It is true that this can be considered as a quotient from a different point of view, but we will reserve this for a later portion of this book.) If x is a function of time, and hence $x = x(t)$, the derivative is sometimes written as $\dot{x}$ (that is with a dot over the x). This notation was due to Sir Isaac Newton. The Leibniz notation for the derivative is more widely used, but each of these notations can be found in mathematical literature.

As a final note in the section we should mention the problem of finding the derivative of a function $f(x)$ defined over the domain $[a, b]$ at one of the end points of the domain. If we consider the derivative at $x = a$, the definition of limit requires that we consider a deleted neighborhood of $x = a$, or equivalently that in considering $a + \Delta x$ the value Δx must be in a deleted neighborhood of zero. Since $f(x)$ is not defined for points to the left of $x = a$, we run into an impasse. We get around this by considering those points for which $f(x)$ is defined, that is the points to the right of $x = a$. Thus, our deleted neighborhood becomes an open interval (a, c) for some value c such that $c > a$. The resulting derivative, if the limit exists, is called *the right-hand derivative of $f(x)$* at $x = a$ and is denoted by $f'_+(a)$. We would handle the *left-hand derivative* in similar fashion and write $f'_-(b)$. We will have occasion to use these one-sided derivatives later on.

We can also find functions such that at certain points only one of the one-sided derivatives will exist. (It should be clear that if both one-sided derivatives exist and they are equal, we have the derivative. This is reminiscent of the fact that when the upper and lower integrals exist and are equal we have the integral.)

EXAMPLE 2.3. Find the one-sided derivatives of $f(x) = [x]$ at the point $(3, 3)$.

Solution. We first consider the right-hand derivative and have

$$f'_+(3) = \lim_{\Delta x \to 0^+} \frac{[3 + \Delta x] - [3]}{\Delta x} = \lim_{\Delta x \to 0^+} \frac{0}{\Delta x} = 0$$

provided we consider Δx to be in the open interval $(0, 1)$. Since we are not to permit Δx to be zero, and since Δx must approach zero, we are certain of having Δx in this interval if Δx is positive. For the right-hand derivative, we are approaching $x = 3$ from the right, and this assures us that Δx must be positive in this case. (We have used, of course, the fact that $[X] = 3$ provided $3 \leq X < 4$, for this comes from the definition of the greatest integer function.) We have also noted that Δx approaches zero from the right by our use of $\Delta x \to 0^+$ under the phrase *lim*. Here the right hand derivative exists and is equal to zero. On the other hand, if we examine the left-hand derivative, we know that we are approaching three from the left, and consequently Δx must be negative. Therefore we have

$$f'_-(3) = \lim_{\Delta x \to 0^-} \frac{[3 + \Delta x] - [3]}{\Delta x} = \lim_{\Delta x \to 0^-} \frac{-1}{\Delta x}$$

but the limit in this case fails to exist. We have considered here that if Δx is going to approach zero from the left, it must at some time arrive and then stay in the interval $-1 \le x < 0$, and in this interval we have $[3 + \Delta x] = 2$. Therefore, the numerator of this differential quotient becomes (-1) and then remains at this value. However, the denominator, while negative, is approaching zero. It should be intuitively clear that when one tries to divide (-1) by a number which is approaching zero, the quotient becomes larger without limit. Since any result we could obtain would be useless in view of the infinitely large size which would result, we can denote this by stating, as we have done, that the limit fails to exist. Note that here we have a function which at the point 3, at least, has a derivative on one side, in this case the right side, but no derivative if we approach from the other side. While such functions are not common, they do occur from time to time.

EXERCISES

1. Differentiate each of the following functions at the point indicated:

   (a) $f(x) = 3x^2 - x + 1$ at $x = 2$
   (b) $f(x) = 6/x$ at $x = 3$
   (c) $f(x) = 3x - x^{1/2}$ at $x = 4$
   (d) $f(x) = (2x - 3)^2$ at $x = 0$
   (e) $f(x) = 4x^3 - 2x$ at $x = 2$
   (f) $f(x) = \sqrt{2x}$ at $x = 8$
   (g) $f(x) = x^3 + x^2$ at $x = -2$
   (h) $f(x) = 2x^{-1/2} - \pi^3$ at $x = 4$

2. Find $f'(x)$ in each of the following cases:

   (a) $f(x) = x$
   (b) $f(x) = (x - 3)^3$
   (c) $f(x) = (3 - 2x)^3$
   (d) $f(x) = 4x^3 - x - e$
   (e) $f(x) = 4 - x^4$
   (f) $f(x) = x^3 - 1/x$
   (g) $f(x) = 1/2$
   (h) $f(x) = (x - 3)^{-1}$

3. (a) If $y = x^3 - 3x^2 - 9x + 7$, find y'.
   (b) Find all values for which $y' = 0$.
   (c) Find all points at which the graph of $y = x^3 - 3x^2 - 9x + 7$ has a horizontal tangent. (Be sure to give both x and y coordinates.)
   (d) Sketch the graph of this curve and show the horizontal tangents.

4. Repeat the work of Exercise 3 if $y = x^3$. Note that the horizontal tangent in this
case does not occur at a point which is either a maximum point of the graph nor a
minimum point of the graph.

5. (a) Solve the general linear equation $ax + by + c = 0$ for y.
   (b) Find $D_x y$.
   (c) Show that $D_x y$ is a constant, and is equal to the slope of the given line.

6. (a) Sketch the graph of $y = x^2 - 2x$.
   (b) Find the interval for which $D_x y > 0$ and the interval for which $D_x y < 0$.
   (c) Determine whether there is a relationship between the sign of the derivative
       and the direction of the slope of the curve. Explain any relationship you find.

7. (a) Sketch the graph of $y = 1/x$.
   (b) Find $D_x y$.
   (c) Show that the derivative of part (b) is either undefined or its sign is always
       negative.
   (d) Interpret the result of part (c) geometrically. Find the value of the slope of
       the curve at points near the y-axis, and use the derivative to help explain the
       behavior of this curve near the y-axis.

8. Find the derivative of x^k for $k = 0, 1, 2, 3$, and 4. Examine your results and see if
you can determine a formula that would appear to hold for all of these results.

9. Show that $\lim_{x \to 0} 3x = 0$ and $\lim_{x \to 0} x = 0$.
   (b) Find $\lim_{x \to 0} (3x/x)$.
   (c) Find $\lim_{x \to 0} (x/3x)$.
   (d) Discuss the reason for saying that zero divided by zero is undefined. You
       may want to use information other than the information which is included
       in parts (a), (b), and (c) of this Exercise.

10. (a) Show that the right-hand derivative of $f(x) = [x]$ exists for each value of x.
    (b) Show that the left-hand derivative of this function exists for each value of x
        which is not an integer.
    (c) Show that the greatest integer function has a derivative at every point
        except when x is an integer.

11. (a) Evaluate $F(x) = \int_2^x (3t + 2)dt$.
    (b) Find $F'(x)$.

12. (a) Evaluate $F(x) = \int_1^x (t^2 - 4t + 3t^{1/2})dt$.
    (b) Find $D_x F(x)$ and compare this with the integrand in part (a).

13. (a) Sketch the curve $f(x) = x^{1/2}$, and locate the point $(9, 3)$ on the curve.
    (b) Draw in the line which is tangent to the curve at $(9, 3)$, and find its slope.
    (c) Locate the point on this tangent line at which $x = 10$ and give the coordinates
        of this point.
    (d) Compare the y-coordinate of this point with $\sqrt{10}$, and explain their relative
        closeness using the graph.

14. (a) Sketch the curve $f(x) = 2x^2$, and sketch the line tangent to this curve at the point (3, 18).
   (b) Find the equation of the line tangent to the curve at (3, 18) and determine the y-value on this line corresponding to $x = 2.8$.
   (c) Compare the value found in part (b) with $f(2.8)$.

15. (a) Find an expression which would give the area under the curve $y = x^2 + 3x + 1$ to the right of the y-axis, and to the left of an arbitrary value $x = x_0$.
   (b) Let x_0 of part a be a variable so that you have a formula for the area from $x = 0$ up to and including $x = x_0$. Differentiate this formula with respect to x_0.
   (c) Compare the results you obtained in part (b) with the equation of the curve.

C16. If $f(x) = x^2$, write a program that will print out Δx, $DQf(2, 2 - \Delta x)$, and $DQf(2. 2 + \Delta x)$ for $\Delta x = 2^{-k}$ and $k = 0, 1, 2, 3, \ldots, 20$. Compare your result with $f'(2)$.

C17. If $f(x) = \sin x$, write a program that will print out Δx, $DQf(\pi/3, \pi/3 - \Delta x)$ and $DQf(\pi/3, \pi/3 + \Delta x)$ for $x = 2^{-k}$ with $k = 0, 1, 2, 3, \ldots, 20$.

SB18. It has been postulated that epidemics spread in a manner which is determined by the number who are currently infected. Let $p(t)$ be the proportion of the population currently infected. Then $[1 - p(t)]$ will be the proportion free of the infection. The rate of change of the proportion infected is assumed to be proportional to the product of the proportion infected and those not infected. Thus, the rate of change would be proportional to $p(t)[1 - p(t)]$.

   (a) Express this in terms of a derivative.
   (b) Show that the rate of change is very small when $p(t)$ is near 0 or near 1.

B19. If in Exercise 18 we let $x = p(t)$, then the rate of change of the proportion infected is given by the relation $kx(1 - x)$. It is apparent that x must be in the interval $[0, 1]$.

   (a) When is the rate of change increasing and when is it decreasing?
   (b) Can you deduce anything about the percentage which would give the maximum rate of change? If so, what?

S20. The cost of producing n items is given by $C(n) = 12000 + 34n + 0.02n^2$ dollars. Find the marginal cost when $n = 100$. Find the marginal cost when $n = 200$. Find the marginal cost when $n = n_0$.

SB21. The population of a certain community is given by the expression $P(t) = 25000 + 120t + 2t^2$ where t is the number of years since January 1, 1975.

   (a) Find the rate of growth of the population in this community on January 1, 1979.
   (b) Compare this with the average rate of growth of this population during the year 1978.

P22. A car travels with velocity $9t - 0.01t^3$ at a time t seconds after starting where the velocity is measured in feet per second.

   (a) Find the acceleration of the car at $t = 2$.
   (b) When is the car going faster and when is it slowing down?

P23. The distance an object falls in t seconds from a very high point is given approximately by the formula $s = 16t^2$. This formula assumes that the object is falling in a vacuum and it therefore ignores air resistance.

(a) Find the velocity of fall t seconds after release.
(b) Find the acceleration t seconds after release. [The acceleration is the rate at which velocity is changing.]
(c) If the object is dropped from a height of 10,000 feet, find the speed with which it hits the ground. [*Hint*: First find the number of seconds which it will fall before it hits the ground.]

P24. An object is thrown upward from the top of a building 512 feet tall. The height of the object at a time t seconds after it was thrown is given by $h(t) = 512 + 96t - 16t^2$.

(a) Find the velocity of this object t seconds after being thrown.
(b) When is the velocity zero?
(c) Find the highest point which this object attains on its trajectory.
(d) Find the speed with which this object hits the ground.

IV.3 Some Special Cases

In the last section we defined the derivative and gave some examples showing how one might determine the derivative. You will remember that in the case of the integral we derived certain special cases and then found that by means of certain theorems we could use these cases to obtain the integrals of more complicated functions. It should not be surprising, then, that the same thing is possible in the case of the derivative. Before starting on the theorems, we will develop derivatives for some special cases. We will then use these as the starting point for obtaining results with wider applications. The functions with which we will start are all algebraic in nature as opposed to the exponential and trigonometric functions we will consider later. In each case we will start with the definition of the derivative, both because this is the only basis for obtaining these derivatives at this time and because this will give an opportunity to become more familiar with the definition.

EXAMPLE 3.1. Find $D_x f(x)$ where $f(x)$ is a constant function.

Solution. Since $f(x)$ is a constant function, $f(x) = c$ where c is a constant. Consequently, $f(x + \Delta x) = c$, and we have

$$D_x c = \lim_{\Delta x \to 0} \frac{f(x + \Delta x) - f(x)}{\Delta x} = \lim_{\Delta x \to 0} \frac{c - c}{\Delta x} = \lim_{\Delta x \to 0} \frac{0}{\Delta x} = 0$$

since zero divided by any non-zero number is zero, and since Δx is of necessity non-zero here. Since the value of the differential quotient in this case is always zero, our limit is rather easy to take. Given any interval, E, which includes zero it makes no difference which deleted neighborhood, D, of zero we use for our choice of Δx. The differential quotient in this Example will always have a value in E, namely zero, regardless of the choice of Δx.

Result. $D_x c = 0$ for any constant c.

This result should have been anticipated, for if we have a function whose value is unable to change, the *rate* at which it changes must be zero. We pause to note here that a positive rate of change means that the function is increasing and a negative rate of change that it is decreasing. Neither of these could apply in the case of a constant function, and since the only other choice, given the order relation for real numbers, is zero, this result should not be at all unexpected.

EXAMPLE 3.2. Find $D_x f(x)$ where $f(x)$ is the identity function.

Solution. Since $f(x)$ is the identity function, $f(x) = x$. The difficulty in this case lies in the fact that things are almost too nice. We have at once that $f(x + \Delta x) = x + \Delta x$, and consequently the differential quotient is $DQf(x, x + \Delta x) = [(x + \Delta x) - x]/\Delta x = \Delta x/\Delta x = 1$. To complete the computation of the derivative we take the limit and have

$$\lim_{\Delta x \to 0} DQf(x, x + \Delta x) = \lim_{\Delta x \to 0} 1 = 1$$

since if we take any interval, E, which includes 1 as an interior point, then for any Δx whatsoever the differential quotient has the value 1. This result is not surprising, for by the very nature of the identity function, the function should be increasing at the same rate that the domain element is increasing.

Result. $D_x x = 1$.

So far we have discussed two of the very basic functions, and these results will be used from this point on. While it is possible to re-derive them each time you encounter them, it is probably easier to remember them, for the results are not difficult to remember. In Section IV.5 we will find one result which will encompass both of these, so the requirement for knowing these individually is rather short-lived. However, they are used often enough that you may still wish to retain them as special cases. We now proceed to other cases that will occur from time to time.

EXAMPLE 3.3. Evaluate $D_x x^2$.

Solution. The differential quotient is given by

$$DQf(x, x + \Delta x) = \frac{(x + \Delta x)^2 - x^2}{\Delta x} = \frac{[(x + \Delta x) - x][(x + \Delta x) + x]}{\Delta x}$$

$$= \frac{\Delta x[2x + \Delta x]}{\Delta x}.$$

Since $\Delta x \neq 0$, we have the fact that this differential quotient is equal to $(2x + \Delta x)$. We take the limit and have

$$\lim_{\Delta x \to 0} (2x + \Delta x) = 2x$$

Result. $D_x x^2 = 2x$.

If in this example we take an open interval, E, with $2x$ as an interior point, there must be points on either side of $2x$. Let D be the set of points which are obtained by subtracting $2x$ from each point of E, and then remove the zero from D. We leave it for you to show that if Δx is in the deleted neighborhood of zero given by D then $\Delta x \neq 0$, and furthermore $\Delta x + 2x$ is in E.

EXAMPLE 3.4. Evaluate $D_x(1/x)$.

Solution.

$$DQf(x, x + \Delta x) = \frac{\dfrac{1}{x + \Delta x} - \dfrac{1}{x}}{\Delta x} = \frac{x - (x + \Delta x)}{(\Delta x)(x)(x + \Delta x)} = \frac{-1}{x(x + \Delta x)}.$$

Again we have used the fact that $\Delta x \neq 0$. (If we seem to keep emphasizing this point, it is because we wish to point out very emphatically that the algebra we are using to simplify the fractions which arise would not be legitimate if Δx were zero.) As we approach the limit in this case, it seems reasonable that the denominator would approach the value x^2, but it is not quite as easy to find the set, D, given the set E. Such sets do exist, and we leave it to you to find them. We would like to give the hint, however, that the relative size of the set D will depend not only on the size of E, but also on the value of x selected for the particular problem at hand. We have, once you have shown that this guess for the limit is correct, the following result.

Result. $D_x(1/x) = -1/x^2$.

EXAMPLE 3.5. Find $f'(x)$ if $f(x) = \sqrt{x}$.

Solution. This is very similar to the first example of the last section, with the exception that we will obtain the derivative for a more general value than $x = 4$. The procedure is the same. Hence we will proceed without further comment.

$$DQf(x, x + \Delta x) = \frac{\sqrt{x + \Delta x} - \sqrt{x}}{\Delta x} = \frac{(x + \Delta x) - x}{\Delta x[\sqrt{x + \Delta x} + \sqrt{x}]}$$

$$= \frac{1}{\sqrt{x + \Delta x} + \sqrt{x}}.$$

By reasoning not unlike that used in the earlier case and with the type of reasoning required in Example 3.4 relative to the sets E and D, we find

Result.

$$f'(x) = \lim_{\Delta x \to 0} DQf(x, x + \Delta x) = \frac{1}{2\sqrt{x}}.$$

This bears out our former result, for when $x = 4, f'(x) = 1/4$.

Note that all of the Examples given follow the same pattern. The development of the differential quotient is not difficult in any of these cases, although the simplification may require some type of trickery as in the case of the square root function. Before you take the limit you prefer not to have a Δx in the denominator as a factor, if you can avoid it, and you therefore need to look for methods by which you can divide out the Δx which occurs there naturally. The second part of obtaining the derivative is that of obtaining the limit. This can usually be done by making an educated guess and then determining whether you are right by trying to establish the fact that for *any* set E given pursuant to the requirements of the definition of a limit there is a corresponding deleted neighborhood D. Note that the neighborhood D can be any set whatsoever that fulfills the requirement of the definition. It does not have to be the largest possible neighborhood.

EXERCISES

1. Differentiate:

   (a) $3x^2 - 5x + 2$
   (b) $4 - x^{-1} + x^{1/2}$
   (c) $(2x - 3)^2$
   (d) $(4x^3 - 3x + 7)/x$
   (e) $(x/2 - 1/3)^2$
   (f) $(24 + 3\sqrt{x})^2$
   (g) $x^3 - 3x^2 + 7x - 4/x$
   (h) $(x + 1/\sqrt{x})^2$

2. Differentiate:

   (a) $(2x - 3/x)^2$
   (b) $x(2x - 3/x)^2$
   (c) $x^2(3 + 1/x)$
   (d) $(\sqrt{x} - 1/\sqrt{x})^2$
   (e) $x/2 - x/3 - x/6 + \pi^3$
   (f) $x^2(3 + 1/x)^2$

3. (a) Sketch $f(x) = x^2$ and $g(x) = \sqrt{x}$ on the same set of axes.
   (b) Find the slope of $f(x) = x^2$ at the point (a, a^2).
   (c) Find the slope of $g(x) = \sqrt{x}$ at the point (a^2, a).

(d) Show that the product of the slopes of parts (b) and (c) is always 1 regardless of the value of a.

(e) For some value a, plot (a, a^2) and (a^2, a) on your graph. Show that these two points are symmetric with respect to the line $y = x$. Use this sketch to explain the result of part (d).

4. (a) For which values of x is $D_x x^2 > D_x \sqrt{x}$?

(b) For which values of x is $D_x x^2 < D_x \sqrt{x}$?

(c) Use a sketch of $y = x^2$ and $y = \sqrt{x}$ on the same set of axes to illustrate graphically your answers for parts (a) and (b).

5. (a) Show that the slope of $y = 1/x$ is always negative.

(b) Show graphically what happens to the curve and the slope of the curve $y = 1/x$ when x is very small.

(c) A line which is tangent to the curve $y = 1/x$ at the point $(a, 1/a)$ will cross both the x and y-axes. Find the values of a such that the x and y-intercepts of the tangent at $(a, 1/a)$ are equal.

6. Find the equation of the line tangent to each of the following curves at the indicated point. [*Hint*: You have a point and a slope in each case.]

(a) $y = x^2 - 3x + 2$ at $(2, 0)$.

(b) $y = \sqrt{x}$ at $(4, 2)$

(c) $y = x^2 + 1/x$ at $x = (1, 2)$

(d) $y = 2x - 3$ at $(4, 5)$

(e) $y = 7$ at $(-1, 7)$

7. Find all of the points (if any) at which $f(x) = x^3 - 3x$ has the following slopes. Illustrate each answer on a sketch of this curve.

(a) 0

(b) 9

(c) -3

(d) -4

8. If f and g are two functions such that $f(x) = g(x + a)$ for each value of x, use the Definition of the derivative to prove that $f'(b) = g'(b + a)$ for any constant b. It is assumed here that a is a constant.

S9. If Q is a demand function relating the quantity $Q(p)$ and the price, p, the *elasticity of demand* is defined to be the ratio of the percentage change in quantity demanded to the percentage change in price. This is given by the equation

$$E(p) = \frac{\dfrac{-D_p Q(p)}{Q(p)}}{\dfrac{D_p p}{p}} = -p \frac{Q'(p)}{Q(p)}.$$

The negative sign is inserted in order to make the elasticity non-negative.

(a) Show why $Q'(p)$ is normally a negative quantity.

(b) Show that the simplification from the first fraction to the second fraction in the elasticity function is correct.

(c) If $Q(p) = c/p$ where c is a constant, show that the elasticity is one.

P10. If a body falls from rest in a vacuum, it falls $s(t) = \frac{1}{2}gt^2$ feet in t seconds.

    (a) Find the velocity of the body after t seconds of fall.

    (b) Find the acceleration of the body after t seconds of fall.

    (c) The force which causes a body to move is the mass of the body multiplied by the acceleration which the force gives to the body. Find the force of gravity on an object weighing ten pounds.

    (d) In the case of a body falling in a vacuum, does the force depend upon the time during which the body has been falling?

B11. For a certain individual the number of units of excitation of the retina of the eye is given by $C(r) = kr^{-2}$ where k is a constant determined by the intensity of the incoming light.

    (a) Find the rate at which the excitation is changing with respect to radius as one moves away from the center of the retina.

    (b) Is the excitation increasing or decreasing as the point in question is moved further from the center of the retina? Relate this to the sign of the derivative in part (a).

M12. Let $F(x) = \int_a^x f(t)dg(t)$.

    (a) Show that $F(x)f(x) \geq 0$ if $g(x)$ is increasing throughout the interval $[a, b]$ and if x is in this interval. [You may wish to use RS sums here.]

    (b) Show that $F(x)f(x) \leq 0$ if $g(x)$ is decreasing throughout the interval $[a, b]$ and if x is in this interval.

    (c) Under what conditions would equality hold in parts (a) and (b)?

IV.4 Some Theorems Concerning Derivatives

It would appear that one could find all of the derivatives we might need by a direct application of the definition, but it would seem unreasonable to contemplate actually carrying out such computations in the case of more complicated functions. For this reason, we will pause to see what relationships we may be able to develop which might permit us to differentiate functions such as $f(x) = 4x^2 - 3x + 7(x - 2)^2/x + \sqrt{24x}$ by making use of the results of the last section. Whereas we were only able to find formulas for addition and subtraction in the case of the integral, we will find that the logic of mathematics treats us somewhat more kindly at this point. We will be able to find results relating to each of the four basic operations of algebra.

Theorem 4.1. *If $D_x f(x)$ and $D_x g(x)$ both exist, then $D_x[f(x) + g(x)] = D_x f(x) + D_x g(x)$.*

PROOF.

$$DQ[f + g](x, x + \Delta x) = \frac{[f(x + \Delta x) + g(x + \Delta x)] - [f(x) + g(x)]}{\Delta x}$$

$$= \frac{[f(x + \Delta x) - f(x)] + [g(x + \Delta x) - g(x)]}{\Delta x}$$

$$= \frac{f(x + \Delta x) - f(x)}{\Delta x} + \frac{g(x + \Delta x) - g(x)}{\Delta x}$$

$$= DQf(x, x + \Delta x) + DQg(x, x + \Delta x).$$

In order to establish the fact that $\lim_{\Delta x \to 0} DQ[f + g](x, x + \Delta x)$ exists, we must start by taking an interval E which includes the proposed limit, namely $f'(x) + g'(x)$. Let e be some positive constant such that the interval $(f'(x) + g'(x) - e, f'(x) + g'(x) + e)$ is entirely within the interval E. We can now take two intervals, $E_f = (f'(x) - e/2, f'(x) + e/2)$ and $E_g = (g'(x) - e/2, g'(x) + e/2)$ and determine deleted neighborhoods D_f and D_g such that if Δx is in D_f then $DQf(x, x + \Delta x)$ is in E_f and if Δx is in D_g then $DQg(x, x + \Delta x)$ is in E_g. Let the deleted neighborhood D be the intersection of D_f and D_g. Now if we take any Δx in D, we know that it is in both D_f and D_g. Consequently we know that $DQf(x, x + \Delta x)$ is in E_f and $DQg(x, x + \Delta x)$ is in E_g. However, the sum of any value in E_f and any value in E_g is a value in E. Therefore if Δx is in D we can be sure that $DQ[f + g](x, x + \Delta x)$ is in E. This establishes the fact that

$$\lim_{\Delta x \to 0} DQ[f + g](x, x + \Delta x) = \lim_{\Delta x \to 0} DQf(x, x + \Delta x)$$
$$+ \lim_{\Delta x \to 0} DQg(x, x + \Delta x)$$

and consequently $D_x[f(x) + g(x)] = D_x f(x) + D_x g(x)$. (It is worth noting that since both D_f and D_g are deleted neighborhoods, D is a deleted neighborhood, and again we have ruled out $\Delta x = 0$.)   □

The fact that the derivative of the sum of two functions is the sum of the derivatives is not surprising. We will now take a look at the product. Leibniz found this a difficult problem, for he tried for some time to show that the derivative of a product is the product of the derivatives. Of course, this is the result that seems natural at first glance. However, if we consider the simple example of $f(x) = x^2 = (x)(x)$, we find that this does not work. We would have on the one hand $D_x(x^2) = 2x$. Using the factored form we would have $(D_x x)(D_x x) = (1)(1) = 1$. It is obvious that these two results are not equal for all values of x. After much searching Leibniz found the correct rule for the product. We have it in the following theorem.

Theorem 4.2. *If $D_x f(x)$ and $D_x g(x)$ both exist, $D_x[f(x)g(x)] = f(x)D_x g(x) + g(x)D_x f(x)$.*

PROOF.

$$DQ[fg](x, x + \Delta x) = \frac{f(x + \Delta x)g(x + \Delta x) - f(x)g(x)}{\Delta x}.$$

It is not immediately obvious that one can relate this expression to the respective differential quotients as we did in Theorem 4.1, for there does not appear off hand to be any connection. However, one might try (and we would hardly suggest it if it would not work) finding terms which either involve both $f(x + \Delta x)$ and $g(x)$ or involve $f(x)$ and $g(x + \Delta x)$ such that we could use the distributive property to obtain at least one differential quotient along the way. In fact we can use either of these choices, and the selection is an arbitrary one. We will use the choice which involves the factor $f(x + \Delta x)$, and therefore will want to have

$$f(x + \Delta x)[g(x + \Delta x) - g(x)] = f(x + \Delta x)g(x + \Delta x) - f(x + \Delta x)g(x).$$

We note that the first of these two terms is already on hand, but we do not have a term in our differential quotient which resembles $f(x + \Delta x)g(x)$. Since we are not permitted to change any values, we must insert a compensating term, and thus we would have

$DQ[fg](x + \Delta x)$

$$= \frac{f(x + \Delta x)g(x + \Delta x) - f(x + \Delta x)g(x) + f(x + \Delta x)g(x) - f(x)g(x)}{\Delta x}$$

$$- f(x + \Delta x)\frac{g(x + \Delta x) - g(x)}{\Delta x} + g(x)\frac{f(x + \Delta x) - f(x)}{\Delta x}$$

$$= f(x + \Delta x)DQg(x, x + \Delta x) + g(x)DQf(x, x + \Delta x).$$

Having gone this far, the next step is that of taking the limit as Δx approaches zero. The two differential quotients on the right side present no problem by virtue of our assumption that the derivatives exist. The factor $g(x)$ is not affected in any way by Δx, and therefore presents no difficulty. Since we have assumed that $D_x f(x)$ exists, we know that for any interval E including the value $f'(x)$ a deleted neighborhood D can be found such that when Δx is in D, $(f(x + \Delta x) - f(x))/\Delta x = s$ is some number in E. This assures us that for such a value of Δx, we have $f(x + \Delta x) = f(x) + s\Delta x$, where s is in E. Consequently, as Δx approaches zero, it follows that $f(x + \Delta x)$ approaches $f(x)$, for $s\Delta x$ will approach zero. (Note that this appears to insure that if $f(x)$ has a derivative at a point, then $f(x)$ is continuous at that point. We will have more to say about this in Chapter VII.) Using this information concerning the behavior of $f(x + \Delta x)$ and the other information we have available about limits, we are now in a position to state that

$$\lim_{\Delta x \to 0} DQ[fg](x, x + \Delta x) = f(x)D_x g(x) + g(x)D_x f(x)$$

$$= f(x)g'(x) + g(x)f'(x).$$

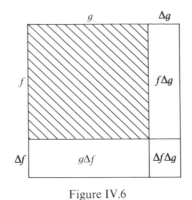

Figure IV.6

The reader is encouraged to fill in any details that are missing in showing that this last statement follows as a result of a proper use of the definition of limit. A more thorough discussion of the intricacies involved will be given in Chapter VII.   □

This result should not be too surprising if we think of it in terms of Figure IV.6. Note that an increase in $f(x)$ causes a change Δf and an increase in $g(x)$ causes a change Δg. The area calculated by the product $f(x)g(x)$ is then increased by the three areas $f(x)\Delta g$, $g(x)\Delta f$, and $\Delta f \Delta g$. If the change is very small, then $\Delta f \Delta g$ will be very small, for it is the product of two small numbers. If we take the limit and let Δx approach zero, then we can expect Δf to approach zero and Δg to approach zero. Thus, we have the fact that $\Delta f \Delta g$ rapidly diminishes in importance, and the consequent change in area is effectively given by $f(x)\Delta g + g(x)\Delta f$. We have only to divide by Δx to obtain the corresponding rates of change, and this gives us in an intuitive manner the result we just obtained.

We can now return to the example which caused trouble when we tried to use an incorrect formula for differentiating a product. If we consider

$$D_x(x^2) = D_x[(x)(x)] = xD_x x + xD_x x = x(1) + x(1) = 2x,$$

we see that this formula gives us a result, at least in this case, which agrees with our earlier derivations.

We now turn to the quotient. After our work with the product we would not anticipate anything simple here, and we will not be disappointed. We have the following theorem.

Theorem 4.3. *If $D_x f(x)$ and $D_x g(x)$ both exist, and if there is an interval (a, b) in which $g(x)$ is never equal to zero, then in that interval*

$$D_x \left[\frac{f(x)}{g(x)} \right] = \frac{g(x)D_x f(x) - f(x)D_x g(x)}{[g(x)]^2}.$$

PROOF.

$$DQ\left[\frac{f}{g}\right](x, x + \Delta x) = \frac{\dfrac{f(x + \Delta x)}{g(x + \Delta x)} - \dfrac{f(x)}{g(x)}}{\Delta x}$$

$$= \frac{g(x)f(x + \Delta x) - f(x)g(x + \Delta x)}{\Delta x g(x)g(x + \Delta x)}.$$

Now we have a dilemma not dissimilar to that which we had in the proof of Theorem 4.2. If an idea works once, it is certainly worth trying again, although you might observe that this time the product to be inserted and then deleted by addition and subtraction respectively should be a different one. While it would be possible to use $f(x + \Delta x)g(x + \Delta x)$, it will be slightly more convenient to use $f(x)g(x)$. Hence,

$$DQ\left[\frac{f}{g}\right](x, x + \Delta x)$$

$$= \frac{g(x)f(x + \Delta x) - g(x)f(x) + f(x)g(x) - f(x)g(x + \Delta x)}{\Delta x g(x)g(x + \Delta x)}$$

$$= \frac{g(x)\dfrac{f(x + \Delta x) - f(x)}{\Delta x} - f(x)\dfrac{g(x + \Delta x) - g(x)}{\Delta x}}{g(x)g(x + \Delta x)}$$

$$= \frac{g(x)DQf(x, x + \Delta x) - f(x)DQg(x, x + \Delta x)}{g(x)g(x + \Delta x)}$$

where we have factored out $[-f(x)]$ from the second half of the numerator in order to obtain the desired $DQg(x, x + \Delta x)$. Now we are ready to take limits. Using the discussion of the last proof we see that $g(x + \Delta x)$ must approach $g(x)$. Since we have assumed in the hypothesis that for any x in (a, b) it is true that $g(x) \neq 0$, we can require that the deleted neighborhood D of the limit application be such that $(x + \Delta x)$ is in (a, b). Thus we avoid division by zero. Taking the limits we have

$$D_x\left[\frac{f}{g}\right] = \frac{g(x)f'(x) - f(x)g'(x)}{[g(x)]^2}$$

as required. Again we remind you that a more careful analysis of limits will be given later. □

It is worth observing that this result can be explained geometrically in a manner not greatly different from that used to suggest the reasonableness of the formula for the derivative of the product. Consider $f(x)$ as the area of a rectangle with one side being of length $g(x)$ and the other side being of length $f(x)/g(x)$. We now wish to see the effect on this quotient if we change

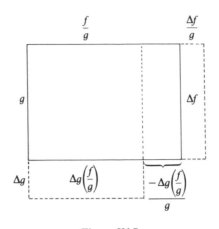

Figure IV.7

$f(x)$ and if we change $g(x)$. If we change $f(x)$ and let $g(x)$ remain fixed, we would have, as shown in Figure IV.7, an increase in area of $f(x)$ and consequently the side $f(x)/g(x)$ would be increased by $\Delta f/g(x)$, where Δf is the increment in f. On the other hand, if we were to increase $g(x)$ and let $f(x)$ remain unchanged, we would have an increase in area in the amount $\Delta g[f(x)/g(x)]$ to be diminished by reducing the length of $f(x)/g(x)$. Thus, the change in $f(x)/g(x)$ is $[-\Delta g(f(x)/g(x))]/g = -((\Delta g)f(x))/[g(x)]^2$. Putting these two changes together, we would have

$$\Delta\left[\frac{f(x)}{g(x)}\right] = \frac{\Delta f}{g(x)} - \frac{(\Delta g)f(x)}{[g(x)]^2} = \frac{(\Delta f)g(x) - (\Delta g)f(x)}{[g(x)]^2}.$$

We note that this shows the approximate change in the quotient. By thinking not of change but of rate of change this approximate rate of change brings us back to the formula for the derivative of a quotient. It should be emphasized that this is not a proof, but it does show that a geometric argument can be used to demonstrate the reasonableness of this result.

In the derivative of the product we have the *sum* of two terms, and since addition is commutative it makes no difference which one is written first. On the other hand, the numerator of the expression for the derivative of the quotient involves the *difference*, and since subtraction is *not* commutative it makes a great deal of difference whether one reverses the order of the terms. Be certain to keep the terms in correct order, for otherwise you will have an incorrect sign for all derivatives of quotients.

You will observe that we have omitted the operation of subtraction. We will now take care of that, by making use of theorems already derived. It will be easier to consider this in two steps. The first will be one that can save some work later on if you are alert to labor saving techniques. The second of which will provide us with the subtraction result.

Corollary 4.1. *If $f'(x)$ exists, then $D_x[cf(x)] = cf'(x)$ provided c is a constant.*

PROOF. Since $cf(x)$ is a product, we can apply Theorem 4.2 and obtain $D_x[cf(x)] = cD_x f(x) + f(x)D_x c$, but $D_x c = 0$ by the results of Section IV.3, since c is a constant. Hence, we have the desired result.   □

Corollary 4.2. *If $f'(x)$ and $g'(x)$ exist, then $D_x[f(x) - g(x)] = D_x f(x) - D_x g(x)$.*

PROOF. We can write $f(x) - g(x) = f(x) + (-1)g(x)$, and then apply Theorem 4.1 and Corollary 4.1 to obtain the desired result. Hence,

$$D_x[f(x) + (-1)g(x)] = D_x f(x) + D_x[(-1)g(x)]$$
$$= D_x f(x) + (-1)D_x g(x) = D_x f(x) - D_x g(x). \quad □$$

We are now able to differentiate the result of addition, subtraction, multiplication, or division provided we can differentiate the constituent parts of the function. You will find these results sufficiently useful that you should be able to repeat them without reference to notes. Their usefulness is illustrated in the following examples.

EXAMPLE 4.1. Find $D_x x^4$.

Solution. Since $x^4 = (x^2)(x^2)$ and since we know the derivative of x^2, we can use the product formula and write

$$D_x(x^4) = D_x[(x^2)(x^2)] = x^2 D_x(x^2) + [D_x(x^2)]x^2 = x^2(2x) + (2x)x^2 = 4x^3.$$

Notice that we did not have to use the definition of the derivative, and hence we didn't have to actually take any limits. Of course, the limits were taken in obtaining the derivative of (x^2), but we didn't have to repeat the matter of taking limits here.

EXAMPLE 4.2. Evaluate $D_x[(x^2 + 3)/(x^2 - \sqrt{x})]$.

Solution. Here we have a quotient, and we see that $f(x)$ of our formula is given by $f(x) = x^2 + 3$ while $g(x) = x^2 - \sqrt{x}$. Therefore

$$D_x\left[\frac{x^2 + 3}{x^2 - \sqrt{x}}\right] = \frac{(x^2 - \sqrt{x})D_x(x^2 + 3) - (x^2 + 3)D_x(x^2 - \sqrt{x})}{(x^2 - \sqrt{x})^2}$$

$$= \frac{(x^2 - \sqrt{x})(2x) - (x^2 + 3)\left(2x - \frac{1}{2\sqrt{x}}\right)}{(x^2 - \sqrt{x})^2}.$$

We have now done the differentiation, and only algebra remains. We will not write out the algebra here. It is very helpful to write out the respective terms carefully using either parentheses or brackets to indicate the various parts of the result as you go along.

With this much practice we will return to the illustration with which we started this section.

EXAMPLE 4.3. Find $f'(x)$ if $f(x) = 4x^2 - 3x + 7(x - 2)^2/x + \sqrt{24x}$.

Solution.

$D_x[4x^2 - 3x + 7(x - 2)^2/x + \sqrt{24x}]$

$\qquad = D_x(4x^2) - D_x(3x) + D_x[7(x - 2)^2/x] + D_x(\sqrt{24x})$

$\qquad = 4D_x(x^2) - 3D_x x + 7D_x\left[\dfrac{(x - 2)^2}{x}\right] + \sqrt{24}D_x\sqrt{x}$

$\qquad = 4(2x) - 3(1) + 7\left[\dfrac{xD_x(x^2 - 4x + 4) - (x - 2)^2 D_x x}{x^2}\right] + \sqrt{24}\left(\dfrac{1}{2\sqrt{x}}\right)$

$\qquad = 8x - 3 + 7\left[\dfrac{x(D_x(x^2) - 4D_x x + D_x 4) - (x - 2)^2(1)}{x^2}\right] + \dfrac{\sqrt{6}}{\sqrt{x}}$

$\qquad = 8x - 3 + 7\left[\dfrac{x[(2x) - 4(1) + (0)] - (x^2 - 4x + 4)}{x^2}\right] + \dfrac{\sqrt{6}}{\sqrt{x}}$

$\qquad = 8x - 3 + 7\left[\dfrac{2x^2 - 4x - x^2 + 4x - 4}{x^2}\right] + \dfrac{\sqrt{6}}{\sqrt{x}}$

$\qquad = 8x - 3 + 7\left[1 - \dfrac{4}{x^2}\right] + \dfrac{\sqrt{6}}{\sqrt{x}}$

$\qquad = 8x + 4 - \dfrac{28}{x^2} + \dfrac{\sqrt{6}}{\sqrt{x}}.$

It would have been possible to simplify the expression algebraically before differentiation and obtain the following:

$D_x[4x^2 - 3x + 7(x - 4 + 4/x) + 2\sqrt{6}\,\sqrt{x}]$

$\qquad\qquad\qquad = 8x + 4 + 28[x(0) - 1(1)]/x^2 + 2\sqrt{6}(1/2\sqrt{x}).$

$\qquad\qquad\qquad = 8x + 4 + 28[x(0) - 1(1)]/x^2 + 2\sqrt{6}(1/(2\sqrt{x})).$

The parentheses enclose results of differentiation where appropriate. This last result reduces to that obtained above.

EXERCISES

1. Differentiate each of the following:

(a) $4x^2 - 7x + 3\sqrt{x} - \sqrt{17}$
(b) $(4x^2 - 5)/(2x - 1/x)$
(c) x^3

(d) $x^{-3/2}$
(e) $14(x - 3)^2 - 5$
(f) x^{-1}
(g) x^{-3}
(h) $(x + 2)^3/(x - 3)^2$
(i) $4(x^2 - 2)^2 - 4(x^2 + 2)^2 + (4x)^2 + \sqrt{8}$
(j) $3x^3 + 4e^4 + 5\pi^5$

2. Differentiate each of the following:

(a) x^5
(b) x^{-5}
(c) $(x^4 - x^{-4})/(\sqrt{x} - 2)$
(d) $32/(4 - x)$
(e) $(3 - x)/(5 - x)$
(f) $(2x - 3)^2 e^7$

3. (a) When is the function $f(x) = x^4 - 2x^2 + 3$ increasing?
   (b) When is this function decreasing?
   (c) When is this function neither increasing nor decreasing? Interpret the performance of the function at these points.

4. Find the equation of the line which is tangent to $f(x) = 2(3x - 4)^2 + 7x - 3$ at the point where $x = 2$.

5. (a) Find the equation of the line which is tangent to the curve $f(x) - x^2$ at the point where $x = x_0$.
   (b) Show that the x-intercept of this line is $(x_0/2)$ and the y-intercept is $-f(x_0)$.

6. Let c be a non-zero constant. Sketch the curve whose equation is $xy = c$.

   (a) Find the equation of the line which is tangent to this curve at the point where $x = x_0$.
   (b) Find the x and y intercepts of the tangent line of part (a).
   (c) Show that the product of these two intercepts is $4c$.

7. (a) Find the equation of the line which is tangent to the curve whose equation is $y = x^3$ at the point where $x = 10$.
   (b) Find the y-coordinate of the point on the line for which $x = 9.8$.
   (c) Compare the value found in part (b) with $(9.8)^3$. Does this give a reasonable approximation for $(9.8)^3$?

8. State and prove a Theorem which would give the derivative of $1/g(x)$. Be sure to include any conditions which should be imposed upon $g(x)$.

9. (a) Find the rate of change of the area of a circle with respect to the radius of the circle.
   (b) Explain geometrically why this result is reasonable.
   (c) If we think of x as the *radius* of a square then the side (or *diameter*) of the square would be of length $2x$. This causes x to represent something analogous to the radius of part (a). Find the rate of change of area of this square with respect to a change in x.
   (d) Explain geometrically why this result is reasonable.

10. (a) Find the rate of change of the volume of a sphere with respect to the change in radius.
    (b) Explain geometrically why this result is reasonable.
    (c) Let x be the *radius* of a cube and hence the edge of the cube (which corresponds to a *diameter*) will be $2x$. Find the rate of change of the volume of this cube with respect to the value of x (the equivalent of the radius in part (a)).
    (d) Explain geometrically why this result is reasonable.

11. If $f(x)g(x)$ is a constant, show that $(f'(x)/f(x)) + (g'(x)/g(x)) = 0$.

P12. The force of attraction between two objects is given by the relation $F = GMm/r^2$ where G is a gravitational constant, M and m are the masses of the two objects, and r is the distance between the two objects.

   (a) Find the rate at which F is changing with respect to a change in r. [Your answer will involve the constants G, M, and m.]
   (b) Explain the sign of your result in terms of physics.

S13. In a certain situation the utility, U, is given by $U = 45Q - 6Q^2$ where Q represents the quantity of goods consumed.

   (a) Find the marginal utility with respect to quantity of goods.
   (b) Is the marginal utility ever negative?
   (c) How would you interpret the marginal utility in case it were negative?

S14. In a given situation $Q = 10,000 - 1.5p^2$ where Q is the quantity of goods demanded and p is the unit price of these goods. Find the elasticity of demand in this situation.

B15. Some biochemical research has shown that the reaction of a body to a drug is given by a relation of the form $R = D^2(a - bD)$ where R is the reaction of the individual to the drug, D is the dosage given, and a and b are constants. The constant a will ordinarily have a value which is half of the maximum dose that can be tolerated. The constant b is ordinarily $1/3$.

   (a) If the maximum dose of a certain drug that an individual can tolerate is 30 milligrams, find the rate at which the reaction changes with respect to changes in D.
   (b) For what values of D will the rate of reaction be increasing?
   (c) For what values of D will the rate of reaction be decreasing?
   (d) Show that for a dose which is half of the maximum dose that can be tolerated the rate of reaction appears to be increasing more rapidly than for other doses that might have been given.

IV.5   More About Derivatives

We have obtained several formulas concerning derivatives and we have also discovered the derivatives of a limited number of specific functions, such as the square function. Putting these together, it is possible to obtain many more formulas. For instance, we could treat $x^{19.5}$ as the product of 9 square functions, the identity function, and the square root function, and then with our formulas find the derivative. It would be more convenient to have a formula which would permit us to write down the derivative immediately. We will make progress in this direction with the following Theorem.

Theorem 5.1. *If $f(x) = x^n$ and n is either a non-zero integer or n is an integer plus one half, then $f'(x) = nx^{n-1}$.*

PROOF. We shall consider three cases.

Case 1. Let n be a non-negative integer. We already know that $D_x x = D_x x^1 = (1)x^0$. Using mathematical induction, we will let k be an integer for which the theorem is true, and obviously k could be one. We will assume that k is an integer such that $D_x x^k = kx^{k-1}$. Now we follow the induction technique and consider $D_x x^{k+1}$ as follows

$$D_x x^{k+1} = D_x[(x)(x^k)] = x(D_x x^k) + x^k(D_x x) = x(kx^{k-1}) + x^k(1)$$
$$= kx^k + x^k = (k+1)x^{(k+1)-1}.$$

This proves the result in this case, since the theorem is correct for the inductive set $S(1)$.

Case 2. This time we start with

$$D_x \sqrt{x} = D_x x^{1/2} = \frac{1}{2\sqrt{x}} = (1/2)x^{-1/2}$$
$$= (1/2)x^{(1/2)-1}.$$

For $n = 1/2$ the result holds. If we assume that k is a number for which the theorem holds, then $1/2$ is a suitable choice. Hence by the induction in Case 1 we see the formula is true for values of n which are $1/2$ increased by any positive integer. Hence, the theorem is true for the inductive set $S(1/2)$. As a result of Cases 1 and 2 we have the theorem for $1/2, 1, 3/2, 2, 5/2, 3, 7/2, \ldots$ and for all positive terms covered by the statement of the Theorem.

Case 3. Let n be a positive number for which the theorem is true. Consider the function $1/x^n = x^{-n}$. We have

$$D_x x^{-n} = D_x\left(\frac{1}{x^n}\right) = \frac{x^n D_x(1) - (1)D_x x^n}{[x^n]^2} = \frac{x^n(0) - nx^{n-1}}{x^{2n}}$$
$$= -n(x^{n-1-2n}) = (-n)x^{(-n-1)}.$$

$\square$

This completes the proof for all cases listed in the theorem. It is worth observing that the machinery we have developed is not all in vain, for it does simplify results a great deal. We are now able to find the derivative of any integral power of either x or its square root without resorting to the definition. One might suspect that this result would even apply to all possible exponents. This is indeed the case, but we do not yet have enough machinery to prove this easily. Note that one could apply the definition to obtain $D_x x^{1/3}$ and $D_x x^{2/3}$ and upon obtaining these results, use induction to obtain the corresponding results for all powers of cube roots. We have already completed the induction part of the proof.

Returning to the problem of differentiating $x^{19.5}$, we note that 19.5 is an exponent covered by our theorem. Hence, we have

$$D_x(x^{19.5}) = (19.5)x^{19.5-1} = 19.5x^{18.5}.$$

This is much less complex than evaluating

$$\lim_{\Delta x \to 0} \frac{(x + \Delta x)^{19.5} - x^{19.5}}{\Delta x}.$$

We now turn our attention to a different type of problem. We have found that the derivative of a constant is always zero. It will be helpful to know for later work that if the derivative of a function is zero for each value in the domain, then the function must be a constant. While this is intuitively clear (?), it requires a bit of proving to be certain that there can be no exceptions to this rule. In order to make the proof somewhat easier to follow, we will first prove two Lemmas. (A *lemma* is a small theorem which is used to prove another lemma or a theorem. Since we will have no reason to refer to this result later, we will not dignify it by calling it a theorem.)

Lemma 5.1. *If* $|DQf(a, b)| = c > 0$, *and if* $a < m < b$, *then either* $|DQf(a, m)| \geq c$ *or* $|DQf(m, b)| \geq c$.

PROOF. The equation $[f(b) - f(m)] + [f(m) - f(a)] = [f(b) - f(a)]$ is identically true. Therefore,

$$(b - m)\frac{f(b) - f(m)}{b - m} + (m - a)\frac{f(m) - f(a)}{m - a} = (b - a)\frac{f(b) - f(a)}{b - a}.$$

But this is just

$$(b - m)DQf(m, b) + (m - a)DQf(a, m) = (b - a)DQf(a, b)$$

or

$$\frac{b - m}{b - a} DQf(m, b) + \frac{m - a}{b - a} DQf(a, m) = DQf(a, b).$$

Now

$$\frac{b - m}{b - a} + \frac{m - a}{b - a} = 1,$$

and both of these fractions are positive since $a < m < b$. For convenience we can rename these fractions by the relations

$$\frac{b - m}{b - a} = k_1 \quad \text{and} \quad \frac{m - a}{b - a} = k_2,$$

giving

$$|k_1 DQf(m, b) + k_2 DQf(a, m)| = |DQf(a, b)| = c$$

with $k_1 + k_2 = 1$. If we assume that $|DQf(m, b)| < c$ and $|DQf(a, m)| < c$, then we have

$$|k_1 DQf(m, b) + k_2 DQf(a, m)| < k_1 c + k_2 c = (k_1 + k_2)c = c.$$

But this is a contradiction. Hence, it is not possible to have both $|DQf(m, b)|$ and $|DQf(a, m)|$ smaller than c, and therefore either $|DQf(m, b)|$ or $|DQf(a, m)|$ is at least as large as $|DQf(a, b)|$.   □

This lemma can be made intuitively clear if we consider a geometric interpretation. Suppose we let $DQf(a, b) = c > 0$. Then the slope of the line segment joining $(a, f(a))$ and $(b, f(b))$ is positive as indicated in Figure IV.8. Let m be some value between a and b. Then $(m, f(m))$ is on the segment joining $(a, f(a))$ and $(b, f(b))$ or above this segment or below this segment. If it is on the segment then $DQf(a, m) = DQf(m, b) = c$. If $(m, f(m))$ is above the segment, $DQf(a, m) > c$, and if the point is below the segment we have $DQf(m, b) > c$.

Lemma 5.2. *If $|DQf(a, b)| = c > 0$, and if the length of the interval (a, b) is denoted by $|(a, b)|$, then for each positive integer k there is an interval $(a_k, b_k) \subseteq (a, b)$ such that $|(a_k, b_k)| = 2^{-k}|(a, b)|$ and $|DQf(a_k, b_k)| \geq c$.*

PROOF. Let m_1 be the midpoint of (a, b). Then either $|DQf(a, m_1)|$ or $|DQf(m_1, b)|$ is greater than or equal to c by Lemma 5.1. Select the one that is greater than or equal to c (if both of them qualify, pick either one), and denote the left end of this subinterval by a_1 and the right end by b_1. Now $|DQf(a_1, b_1)| \geq c$, and $|(a_1, b_1)| = 2^{-1}|(a, b)|$. This gives us the first step in a proof by mathematical induction. Now, if we assume that the lemma is correct for some value k, that is $|DQf(a_k, b_k)| \geq c$ and $|(a_k, b_k)| = 2^{-k}|(a, b)|$, then let m_{k+1} be the midpoint of (a_k, b_k). By reasoning similar to that above, we have either $|DQf(a_k, m_{k+1})|$ or $|DQf(m_{k+1}, b_k)|$ greater than or equal to

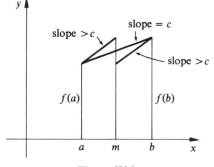

Figure IV.8

c, and each of the two subintervals is one half as long as (a_k, b_k). We will select the differential quotient which satisfies the inequality and label the end points of the interval a_{k+1} and b_{k+1} respectively. We now have the fact that $|DQf(a_{k+1}, b_{k+1})| \geq c$ and also that

$$|(a_{k+1}, b_{k+1})| = (1/2)(2^{-k})|(a, b)| = 2^{-k-1}|(a, b)| = 2^{-(k+1)}|(a, b)|.$$

Therefore, this relation is true for the inductive set $S(1)$, and hence the lemma is correct.   □

Theorem 5.2. *If* $f(x)$ *has derivative at every point of an interval* $[a, b]$ *and* $f'(x) = 0$ *for each value of* x *in* (a, b) *then* $f(x)$ *is a constant function.*

PROOF. If $f(x)$ were not a constant function, then there would be two values of x, say $x = c$ and $x = d$, such that $f(c) \neq f(d)$. Therefore, $DQf(c, d) \neq 0$. Let $|DQf(c, d)| = m > 0$. Then by Lemma 5.2 we have a sequence of intervals $I_k = (c_k, d_k)$ such that $|DQf(c_k, d_k)| \geq m$ and the length of I_k is the length of $I_o = |(c, d)|$ multiplied by 2^{-k}. Now we have a nest of intervals as shown in Figure IV.9 with each new interval occupying just one half of the preceding interval in the sequence. Since the intervals are getting shorter and shorter, this infinite sequence will ultimately distinguish a single point. To see that this is true, assume that we can have two distinct points in such a sequence. Assume further that they are e units apart. At some point in the sequence of intervals we will find an interval I_k of length less than e. Thus, it would be impossible to have both of our assumed points in I_k, and consequently the assumption that we could have two points must be false.

We will now designate this unique point X. It follows that $f'(X) \neq 0$. It is easy to see that we must have one of two cases. Either X is some value of a_k or b_k or else it is not. If X is an end point of some interval I_k, it is the endpoint of all subsequent intervals in the sequence. In each of these intervals we have the absolute value of the differential quotient at least m. Therefore, if the interval E of the definition of limit is any subset of $(-m, m)$ which includes 0 as an interior point, it would be impossible to find a deleted neighborhood D of X such that for all points in D the differential quotient is in E. Therefore we could not have $f'(X) = 0$. On the other hand, if we take the second case, that in which X is not an end point of any of the intervals I_k, we know that X must be an interior point of each I_k. Now we have either $DQf(a_k, X)$ or $DQf(X, b_k)$ is greater than m in absolute value for each value of k. With this information we can use the same argument we used in the

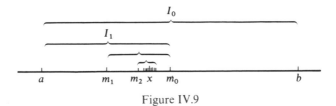

Figure IV.9

first case to show that we could not have $f'(X) = 0$. We can summarize our discussion by noting that if we assume that $f(x)$ is not a constant function on the interval $[a, b]$, there must be some point in this interval for which the derivative is not zero. Hence if the derivative is zero at every point of this interval, $f(x)$ must be a constant function throughout the interval. □

As we can see from the proof of Theorem 5.2, some simple results do not come easily. However, now that we have this result we can prove another Theorem very readily.

Theorem 5.3. *If both $f(x)$ and $g(x)$ have derivatives, and $f'(x) = g'(x)$ for all values of x in $[a, b]$, then $f(x)$ and $g(x)$ differ by at most an additive constant on (a, b).*

PROOF. By hypothesis $D_x[f(x) - g(x)] = f'(x) - g'(x) = 0$. Therefore $f(x) - g(x)$ is a constant. If we denote this constant by C, then we have $f(x) - g(x) = C$ or $f(x) = g(x) + C$. We can now state that a function has a zero derivative if and only if it is a constant function, and two functions have the same function for their derivative if and only if they differ by a constant function. □

EXERCISES

1. Differentiate each of the following functions:

   (a) $4x^7 - 3x^5 + 5/x^{10}$
   (b) $(7x^2/\sqrt{x}) - x^5(\sqrt{x}) + 3\sqrt{\pi}$
   (c) $(x^2 - 7x + 3x^{-2})/\sqrt{5x}$
   (d) $\sum_{k=1}^{20} (x^k/k)$
   (e) $\sum_{k=0}^{20} kx^k$
   (f) $(3 - x)^3$
   (g) $(4/3)\pi x^3$
   (h) $x^2/(1 + x^3)$
   (i) $(2x - 3)^2(4x^2 + 3)$.

2. Find the derivative of each of the following:

   (a) $4x^{7.5} - 3x^2 + 5/x^9$
   (b) $(x^2 - x^{-1})^3/x^4$
   (c) $1/(x^2 - x)$
   (d) $x^3(1 + x^4)$
   (e) $3x^5\sqrt{2x} - (x^2 + 1)^2$
   (f) $(x^{2.5} - 1)(1 - x^{-1})^2$
   (g) $x^3 - 4x^{-3} + 4^{-6}$
   (h) $x^3/(1 + x^4)$

3. (a) Show that $D_x(2x - 3)^k = 2k(2x - 3)^{k-1}$ for $k = 0$, 1, and 2.
   (b) Use mathematical induction to show that $D_x(2x - 3)^n = 2n(2x - 3)^{n-1}$ for all positive integers n.

4. If a and b are any real constants and n is a non-negative integer, use mathematical induction to show that $D_x(ax + b)^n = an(ax + b)^{n-1}$.

5. Let n be a non-negative integer or one-half plus a non-negative integer.

   (a) Find the equation of the line which is tangent to $y = x^{n+1}$ at the point (a, a^{n+1}).
   (b) Show that the point where this line intersects the x-axis divides the interval $[0, a]$ of the x-axis into two segments whose lengths have the ratio n to 1.
   (c) Show that the y-intercept is $(-n)$ times the y-coordinate of the point of tangency.

6. (a) Differentiate x^n if n is a positive integer.
   (b) Differentiate the result of part (a).
   (c) Differentiate the result of part (b), and continue successive differentiation until you have performed a total of n differentiations. Show that the result will be $n!$.
   (d) Show that if you were to differentiate at least $(n + 1)$ times altogether the derivative would be zero.
   (e) Show that if n had been one half plus an integer the derivative would never be zero for positive values of x.

7. In Exercise III.4.9 we showed that e^{-x} is approximated by

$$e^{-x} \doteq \sum_{k=0}^{n} \frac{(-x)^k}{k!}$$

and that the approximation is a rather good one if n is large.

   (a) Differentiate this approximation.
   (b) Examine your result and see if you can determine a probable derivative of e^{-x}.

8. In Exercise III.5.10 we showed that $\sin x$ and $\cos x$ can be approximated by the expressions

$$\sin x \doteq \sum_{k=0}^{n} \frac{(-1)^k x^{2k+1}}{(2k + 1)!} \quad \text{and} \quad \cos x \doteq \sum_{k=0}^{n} \frac{(-x^2)^k}{(2k)!}$$

with the approximations being good ones for large values of n.

   (a) Differentiate the expression for $\sin x$.
   (b) Examine your result and see if this is related to either $\sin x$ or $\cos x$. Hence make a guess as to the probable derivative of $\sin x$.
   (c) Differentiate the expression for $\cos x$.
   (d) Examine your result and see if this is related to either $\sin x$ or $\cos x$. Hence make a guess as to the probable derivative of $\cos x$.

9. Find all of the values of t for which $D_t(2t^6 - 3t^4 + 195782)$ is zero. What would be the significance of these points on a graph of $y = 2t^6 - 3t^4 + 195782$?

10. (a) Evaluate $D_x[(x + 2)/(x + 1)]$
    (b) Evaluate $D_x(x + 1)^{-1}$
    (c) Compare the two derivatives and determine from the derivatives any relation that must exist between $(x + 2)/(x + 1)$ and $(x + 1)^{-1}$.

M11. If the points $(1, 2)$ and $(2, 3)$ are on the graph of a continuous function, use the method of proof given for Theorem 5.2 to show that there must be some point on the graph between $(1, 2)$ and $(2, 3)$ at which the derivative is at least one.

M12. (a) Show that the derivative of $f(x) = [5x]$ is zero at all points except for multiples of $(1/5)$.
    (b) Explain why Theorem 5.2 does not apply in this case despite the fact that $f'(x) = 0$ for almost every value of x.

13. It will be shown later that $D_x \tan^2 x = D_x \sec^2 x$. What does this say about the values of $\tan^2 x$ and $\sec^2 x$? Does this agree with any trigonometric identity that you remember?

14. (a) Write out the derivative of $[f(x)g(x)h(x)]$ in terms of these three functions and their individual derivatives.
    (b) Show that $(f(x)/f'(x)) + (g(x)/g'(x)) + (h(x)/h'(x)) = 0$ if the product $[f(x)g(x)h(x)]$ is a constant.
    (c) Show that a similar result would hold for a product of five functions if the product had a constant value.

P15. An object is thrown upward and its height at any time is given by $h = 80 + 256t - 16t^2$.

    (a) What is its velocity at $t = 3$?
    (b) What is its acceleration at $t = 3$?
    (c) How fast does the acceleration change at $t = 3$?
    (d) How fast does the velocity change at $t = 3$?
    (e) When does the object hit the ground?
    (f) How high was the building from which the object was launched?
    (g) What is the domain which makes sense in this problem, assuming that the stop watch was started when the object was launched, and that the height is determined by this rule until the object hits the ground?

P16. If air resistance is ignored, and an object is thrown upward from the top of a building 1024 feet high with a velocity of 160 feet per second, its distance above the ground will be given by $h = 1024 + 160t - 16t^2$. The quantity t is measured in seconds and h is measured in feet.

    (a) How fast is the object going when it hits the ground?
    (b) How fast is the acceleration increasing?
    (c) How fast is the object going when it passes the top of the building on the way down?

P17. According to Boyle's law the product of the pressure and the volume of a gas held at constant temperature is constant. A cylinder having a movable piston initially has a volume of 100 in^3. If the initial pressure is 25 lbs/in^2 find the rate of change of pressure if the volume is increasing at the rate of 3 in^3/sec. Be sure to give the units involved.

P18. Assume that the parabola $y^2 = 4ax$ is a cross section of an automobile headlight which includes the axis of the headlight. Show that if the filament of the light is located at the focus $(a, 0)$ and is considered to be a single point, then all of the light rays which hit the reflector are reflected parallel to the axis of the headlight. [*Hint*: Sketch $y^2 = 4ax$. The angle that the initial beam makes with the tangent to the curve is equal to the angle that the reflected beam makes with the tangent to the curve.]

P19. A whispering gallery is formed by building a ceiling with an elliptical shape, such that all planes through the *axis* of the room are in the form of an ellipse. If the ellipse in each such plane has the equation $b^2x^2 + a^2y^2 = a^2b^2$, $(a > b > 0)$, show that anyone standing such that his mouth is at one focus can whisper, the sound waves will spread in a spherical pattern, bounce off of the reflecting ceiling, and then will all re-unite at the other focus with the same elapsed time for each such path. In this way a person whose ear is at the other focus would hear that which was whispered with the same intensity and no cancellation of waves, provided the ceiling had perfect reflectivity. [*Hint*: See the hint to Exercise 18. The foci are located at $(\sqrt{a^2 - b^2}, 0)$ and $(-\sqrt{a^2 - b^2}, 0)$.]

S20. The cost function for a firm is given by $C(n) = 100 + 1.2n - n^2 + 0.1n^3$.

   (a) Find the marginal cost.
   (b) Find the average cost per item [that is $C(n)/n$]. When is the average cost increasing?
   (c) When is the average cost decreasing?
   (d) Sketch the graph of the average cost using your knowledge of the intervals in which it is increasing and in which it is decreasing.
   (e) Where does the minimum value of the average cost appear to be?
   (f) For the value of n which makes the average cost a minimum, calculate both the average cost and the marginal cost.

B21. As mentioned in an earlier Exercise one theory of the excitation of the retina of the eye due to incident light is summarized in the relation $C(r) = r^{-k}$ where r is the distance from the center of the eye.

   (a) If $k = 1.5$, find the rate of change of the excitation with respect to the distance from the center of the eye.
   (b) What is the rate of change at $r = 0.1$ millimeters?
   (c) What is the rate of change at 0.01 millimeters?
   (d) What is the rate of change at 2 millimeters?

CHAPTER V

The Interrelation of Integration and Differentiation

V.1 The Fundamental Theorem of the Calculus

In the last chapter we left the subject of integration and embarked on the development of differentiation. Actually these two operations are very closely related. It will be our purpose in this section to develop this interrelationship. We will see that these two operations are almost inverse operations, and will develop the reason for the use of the word "almost."

Before trying to prove the Fundamental Theorem, we will need to remind ourselves of a few bits of information which are implicit in our earlier work. However, to make this information very clear here, we will proceed with the following lemmas.

Lemma 1.1. *If $f(x)$ is a continuous function on the interval $[a, b]$ and if c is any point in (a, b), then $\lim_{x \to c} f(x) = f(c)$.*

PROOF. This is effectively a restatement of our condition for continuity combined with the definition of limit. Since $f(x)$ is continuous on $[a, b]$, it is defined for each value on this interval, and hence $f(x)$ exists. Let E be the interval $(f(c) - e, f(c) + e)$ for some positive number e. By the definition of continuous founction there must be a positive number d such that if x is in the interval $D = (c - d, c + d)$, then $f(x)$ is in E. If we remove from D the single point $x = c$, we have a deleted neighborhood of c such that if x is in this deleted neighborhood then $f(x)$ is in E. This satisfies the Definition of the limit, and therefore proves the lemma. □

Lemma 1.2. *If $f(x)$ is a continuous function on the interval $[a, b]$ and if $f(x)$ does not change sign an infinite number of times as x increases from a to b,*

*then for any value $x = c$ in (a, b) there exist values $c_1 < c$ and $c_2 > c$ in (a, b)
such that $f(x)$ does not change sign in $[c_1, c]$ or in $[c, c_2]$.*

PROOF. Since $f(x)$ does not change sign an infinite number of times on
$[a, b]$ there must be a finite set of points at which it changes sign. Let
$x_1, x_2, \ldots, x_n$ be these points in (a, b) at which $f(x)$ does change sign as x
increases from a to b. If c is any value in (a, b), it must either be one of the n
points x_k or it must be in some interval (x_k, x_{k+1}). If $c = x_k$ for some value
of k, let c_1 be chosen in the interval (x_{k-1}, x_k) and c_2 in the interval (x_k, x_{k+1}).
Since $f(x)$ does not change sign in either of the two subintervals, the lemma
is satisfied. If c is not a point x_k then c must be in some subinterval (x_k, x_{k+1}).
In this case select c_1 in (x_k, c) and c_2 in (c, x_{k+1}). Since both c_1 and c_2 are in
the subinterval (x_k, x_{k+1}) and since $f(x)$ does not change sign in this interval,
it follows that this choice for c_1 and c_2 satisfies the lemma. Consequently
in any case it is possible to find c_1 and c_2 such that $f(x)$ takes on only one
sign in each of the two intervals (c_1, c) and (c, c_2). It should be noted that we
have not required that the sign of $f(x)$ be the same in both of these two
intervals. If c is a point x_k the signs will not be the same.   □

Before proceeding further we should comment on the fact that the variable
of integration used in the integral has no effect on the value of the integral.
Thus, we have

$$\int_1^2 x^3 \, dx^2 = \int_1^2 t^3 \, dt^2 = \int_1^2 y^3 \, dy^2 = \frac{2(2^5 - 1^5)}{5} = \frac{62}{5}.$$

More generally we have

$$\int_a^b f(x)dg(x) = \int_a^b f(t)dg(t) = \int_a^b f(y)dg(y).$$

Since we will wish to take the derivative of an integral in order to show
the relationship between differentiation and integration, it will be necessary
for us to establish the integral as a function which is not necessarily constant.
Otherwise the derivative would be zero in every case. We will do this by
noting that the integral $\int_a^b f(t)dg(t)$ is the integral over the interval $[a, b]$, and
if we were to let the value of b vary, we would expect to have different values
for the resulting integral. In order to emphasize this, we will write $F(x) =
\int_a^x f(t)dg(t)$ and note that $F(x)$, which is the integral over the interval $[a, x]$,
will be a function of the right hand endpoint of the interval. Thus, we can
expect $F(x)$ to be a function which in general will not be a constant. Since
$F(x)$ is an integral, we can now proceed with the differentiation of an integral
by attempting to find the derivative of $F(x)$.

Theorem 1.1 (The Fundamental Theorem of the calculus). *Let $f(t)$ be a
continuous function over the interval $[a, b]$ and let x be an interior point of this
interval. Let $g(t)$ be a function for which the derivative exists at every point*

of $[a, b]$ *and such that the derivative does not change sign an infinite number of times on* $[a, b]$. *Then*

$$D_x \int_a^x f(t)dg(t) = f(x)g'(x). \qquad (V.1.1)$$

PROOF. Since we have no previous experience with the derivative of an integral, we will have to revert to the definition of the derivative. In order to facilitate our discussion, let us denote the integral by $F(x)$, and thus we will start with

$$F(x) = \int_a^x f(t)dg(t),$$

and we will find $F'(x)$. Starting with the differential quotient, we will have

$$DQF(x, x + \Delta x) = \frac{1}{\Delta x}\left[\int_a^{x+\Delta x} f(t)dg(t) - \int_a^x f(t)dg(t)\right]$$

$$= \frac{1}{\Delta x}\int_x^{x+\Delta x} f(t)dg(t).$$

Since $g'(t)$ exists for all points in $[a, b]$ and since $g'(t)$ does not change signs an infinite number of times in $[a, b]$, we know that for any value of x in $[a, b]$ there is an interval to the left of x and one to the right of x in which $g'(t)$ does not change sign. Therefore, we know that $g(t)$ is monotonic in each of these two intervals. Therefore, for any value of Δx which is in either of these two intervals, $g(t)$ will be monotonic over the interval $\lfloor x, x + \Delta x \rfloor$ regardless of the sign of Δx. As a result, we can invoke the integral theorem of the mean and know that there must be some value z in the interval $(x, x + \Delta x)$ such that

$$\int_x^{x+\Delta x} f(t)dg(t) = f(z)\int_x^{x+\Delta x} dg(t) = f(z)[g(x + \Delta x) - g(x)].$$

Consequently we have

$$DQF(x, x + \Delta x) = \frac{1}{\Delta x}\int_x^{x+\Delta x} f(t)dg(t) = f(z)\frac{g(x + \Delta x) - g(x)}{\Delta x}$$

$$= f(z)DQg(x, x + \Delta x).$$

It should be emphasized that we have used the fact that Δx must be so chosen that the interval with end points x and $(x + \Delta x)$ will be an interval for which $g'(x)$ does not change sign, and consequently an interval for which $g(x)$ is monotonic. Furthermore, the value z will be some value within this interval. Since we will be taking the limit as Δx approaches zero, this causes no problem, for there is no harm in starting with a small value of Δx.

We are now ready to take the limit of the differential quotient in order to obtain $F'(x)$. We have a product here, and we have already observed in developing the derivative that the limit of a product is the product of the respective limits, provided the individual limits exist. We now have

$$F'(x) = \lim_{\Delta x \to 0} DQF(x, x + \Delta x) = \lim_{\Delta x \to 0} [f(z)DQg(x, x + \Delta x)]$$

$$= \left[\lim_{\Delta x \to 0} f(z) \right]\left[\lim_{\Delta x \to 0} DQg(x, x + \Delta x) \right]$$

$$= f(x)g'(x).$$

We obtain this final step by observing that z is in the interval $(x, x + \Delta x)$ and consequently as Δx approaches zero, z must approach x. The fact that z is closer to x then is $(x + \Delta x)$ permits us to state

$$\lim_{x + \Delta x \to x} f(z) = \lim_{z \to x} f(z) = f(x)$$

since Lemma 1.1 applies for the continuous function $f(x)$. On the other hand $\lim_{\Delta x \to 0} DQg(x, x + \Delta x)$ yields the derivative by definition provided the limit exists. We have assumed in the conditions for the theorem that $g'(x)$ does exist. Consequently the final question concerning this proof is resolved. $\square$

EXAMPLE 1.1. Differentiate $\int_a^x t^n \, dt^m$ where m and n are integers or halves of integers.

Solution. By the Fundamental Theorem of the calculus we have

$$D_x \int_a^x t^n \, dt^m = x^n D_x x^m = x^n(mx^{m-1}) = mx^{m+n-1}.$$

Alternatively we might have integrated and obtained

$$\int_a^x t^n \, dt^m = \frac{m(x^{m+n} - a^{m+n})}{(m+n)} = \frac{mx^{m+n}}{(m+n)} - \frac{ma^{m+n}}{(m+n)}.$$

If we differentiate this last result, and observe that the second term in this last expression is a constant, we see that the derivative is

$$\frac{m(m+n)x^{m+n-1}}{(m+n)} = mx^{m+n-1},$$

and this is the same result we had before. Hence we have verified the correctness of the Fundamental Theorem in this instance. We restricted the values of m and n in the original statement of the example in order that we could differentiate the result.

At first glance one might feel that this is an interesting result, but might also wonder why we have made it appear that this is a very important result. In fact, this theorem, relating differentiation and integration as it does, permits us to use our integral results to obtain corresponding derivative results and also permits us to use derivative results to obtain corresponding integral results. In fact we can use a variation of Example 1.1 to extend the formula for the derivative of a power function to include all rational exponents.

EXAMPLE 1.2. Prove $D_x x^n = nx^{n-1}$ for any rational number n.

Solution. If n is any rational number, we have the fact that

$$\int_0^x nt^{n-1}\, dt = \frac{nx^n}{n} = x^n.$$

By taking derivatives of each side of this result, we have

$$D_x \int_0^x nt^{n-1}\, dt = nx^{n-1}(1) = nx^{n-1} = D_x x^n$$

and this must be correct for any positive rational number n. If n is a negative rational number, then $(-n)$ is a positive rational number and we have

$$D_x x^n = D_x \frac{1}{x^{-n}} = -\frac{-nx^{-n-1}}{x^{-2n}} = nx^{n-1}. \tag{V.1.2}$$

We no longer have to be quite so restrictive in our use of this derivative formula. Note that we still have the restriction that n must be rational, but we will remove this at a later point. In this example we see the usefulness of being able to relate the derivative and the integral, for we were able to remove a major restriction with very little effort.

EXAMPLE 1.3. Prove

$$\int_a^x x^n\, dx^m = \frac{m(x^{m+n} - a^{m+n})}{(m+n)}$$

for any rational numbers m and n provided $m + n \neq 0$ and provided both x^m and x^n exist and are real at all points in the interval of integration.

Solution. The requirement $m + n \neq 0$ is needed in order to avoid a zero denominator. The second requirement is necessary if the result is to be real. From the Fundamental Theorem we know

$$D_x \int_a^x x^n\, dx^m = x^n(mx^{m-1}) = mx^{m+n-1}.$$

We also know that

$$D_x \left[\frac{m(x^{m+n} - a^{m+n})}{(m+n)} \right] = \frac{m(m+n)x^{m+n-1}}{(m+n)} = mx^{m+n-1}.$$

Since the two expressions have identical derivatives, the two expressions must differ by at most a constant. Hence we must be able to write

$$\int_a^x x^n \, dx^m = \frac{m(x^{m+n} - a^{m+n})}{m + n} + C$$

where C is some constant whose value we would like to know. However, since C is a constant, if we can determine it for one value of x we know it for all values of x. Let $x = a$. In this case we have

$$\int_a^a x^n \, dx^m = \frac{m(a^{m+n} - a^{m+n})}{m + n} + C$$

or $0 = 0 + C$. Therefore we know that $C = 0$ and

$$\int_a^x x^n \, dx^m = \frac{m(x^{m+n} - a^{m+n})}{m + n}. \tag{V.1.3}$$

We have now extended our previous results for both the integral and the derivative. We continue to have the restriction that $m + n \neq 0$. Thus we cannot evaluate

$$\int_a^b \frac{1}{x} \, dx = \int_a^b x^{-1} \, dx$$

at the present time. We will come back to this case later on.

   In Example 1.2 we used the fact that if two derivatives are equal, the functions of which they are derivatives can differ by at most a constant. This gives rise to the following theorem.

Theorem 1.2 (Alternate form of the Fundamental Theorem of the calculus). *If $F'(x) = f(x)g'(x)$, then*

$$\int_a^x f(t)dg(t) = F(x) - F(a)$$

and

$$\int_a^b f(x)dg(x) = F(b) - F(a). \tag{V.1.4}$$

PROOF. Since the derivative of the integral is given by the relation

$$D_x \int_a^x f(t)dg(t) = f(x)g'(x)$$

and since by hypothesis $F'(x)$ has the same derivative, then $F(x)$ must differ from the integral by at most an additive constant. Therefore, we have

$$\int_a^x f(t)dg(t) = F(x) + C. \tag{V.1.5}$$

This relation holds for any value of x for which the integral exists, and therefore it holds for $x = a$. But then the integral has a zero value and hence $F(a) + C = 0$. From this it follows that $C = -F(a)$, and (V.1.5) should read

$$\int_a^x f(x)dg(x) = F(x) - F(a).$$

If x takes on the value b, we have

$$\int_a^b f(x)dg(x) = F(b) - F(a). \qquad \square$$

We will find Theorem 1.2 to be very useful. For instance, we can evaluate

$$\int_2^5 3x^2\, dx$$

noting that $D_x x^3 = 3x^2$, and hence

$$\int_2^5 3x^2\, dx = 5^3 - 2^3 = 117.$$

In this case the derivative of x^3 is also the derivative of

$$\int_2^x 3t^2\, dt,$$

and the value of the integral is obtained at once by substitution. It is frequently convenient to use the following notation.

$$\int_2^5 3x^2\, dx = x^3\,\Big|_2^5 = 5^3 - 2^3 = 117.$$

In this case we have indicated the function corresponding to $F(x)$ in the theorem and followed with a *vertical line*. Note that this line follows the function. Since it is a line and not an integral sign, and since it follows the function, it should be clear that no further integration is required. We only need to use the limits of integration. In order to remind ourselves of these limits, we have placed them on the vertical line. The remaining step is one of evaluating the function $F(x)$ at the upper limit of integration and then subtracting from this value the value of $F(x)$ at the lower limit of integration. In general this result can be written

$$\int_a^b f(x)dg(x) = F(x)\,\Big|_a^b = F(b) - F(a).$$

This function $F(x)$ is often called the *anti-derivative of* $f(x)g'(x)$. This should indicate that instead of reading the statement $D_x F(x) = f(x)g'(x)$ from left to right we are to read it from right to left. In other words, we are looking for a function whose derivative has the value $f(x)g'(x)$. We might raise the question concerning which anti-derivative we should use. Actually it makes no difference, for if we were to use $F(x) + K$ for some constant K, we would then have

$$\int_a^b f(x)dg(x) = [F(x) + K]\Big|_a^b = [F(b) + K] - [F(a) + K]$$

$$= F(b) - F(a).$$

This is a perfectly general result, since any anti-derivative of $f(x)g'(x)$ can differ from any other anti-derivative only by a constant, such as the constant K we used in our illustration.

In order to further illustrate the role of the constant we have had to refer frequently to the fact that two anti-derivatives of the same function can differ only by a constant. Because of the role that the anti-derivative plays in integration, this constant is often referred to as the "constant of integration." The role of this constant can be further illustrated with a geometric illustration.

EXAMPLE 1.4. Find the equation of the curve $y = f(x)$ if the slope at any point $(x, f(x))$ is given by the function $3x^2 - 1$. In particular find the equation of the curve which passes through $(-2, 3)$.

Solution. Since the slope is obtained by evaluating the derivative, we know that $f'(x) = 3x^2 - 1$. Therefore, $f(x)$ is the anti-derivative of $3x^2 - 1$. From this we have $y = f(x) = x^3 - x + C$ where C is some constant. We have shown the curves corresponding to this equation for $C = 2, 1, 0, -1$, and -2 in Figure V.1. It is clear that there are many more curves each corresponding to some value of C.

We would not be able to determine which value of C is desired if we did not have some additional item of information. In this case we are told that we desire the curve which passes through the point $(-2, 3)$. If we substitute 3 for y and (-2) for x we have $3 = (-2)^3 - (-2) + C = (-8) + 2 + C$. From this it follows that $C = 9$. Therefore the particular equation we were asked to find is $y = x^3 - x + 9$. You should check to make certain that $(-2, 3)$ is a point on this curve.

It would be possible to obtain this specific equation using a somewhat different notation, but precisely the same logic. If we were to write

$$f(x) - f(-2) = \int_{-2}^x (3t^2 - 1)dt$$

this would lead us to the equation

$$y - 3 = (t^3 - t)\Big|_{-2}^x = (x^3 - x) - [(-2)^3 - (-2)]$$

$$= x^3 - x + 6.$$

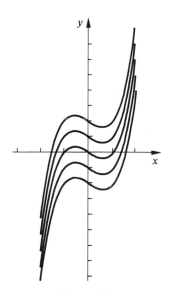

Figure V.1

From this our earlier result follows. That this is correct follows from Theorem 1.2, for $f(x)$ is the anti-derivative of $(3x^2 - 1)(1)$, $f(x) = y$, and $f(-2) = 3$ since the curve is to pass through the point $(-2, 3)$.

So far the specific examples we have considered in this section have used only the identity function for $g(x)$, despite the fact that we have been more general in our Theorems and discussions. Consider the following.

EXAMPLE 1.5. Show that

$$\int_0^x t^2 \, dt^3 = \int_0^x t^2(3t^2)dt = \int_0^x 3t^4 \, dt.$$

Solution. The respective derivatives of these expressions are $x^2(3x^2)$, $x^2(3x^2)(1)$, and $3x^4(1)$ where the derivative of the $g(x)$ is shown in parentheses in each instance. It is clear that these three derivatives are equal, and hence in each case the anti-derivative is given by $3x^5/5 + C$ for some constant C. However, since each of these integrals has the value 0 if $x = 0$, we know that the constant C must have the value zero.

It would have been possible, of course, to use our earlier formulas for the integral, but the machinery of the Fundamental Theorem can also be very helpful in this instance.

Example 1.5 suggests another theorem.

Theorem 1.3. *If the derivatives and integrals involved exist, then*

$$\int_a^x f(t)dg(t) = \int_a^x f(t)g'(t)dt.$$

PROOF. Observe that if we let $F(x)$ be defined by $F(x) = \int_a^x f(t)dg(t)$ we have $F'(x) = f(t)g'(t)$ by the Fundamental Theorem. Also, we have

$$D_x \int_a^x f(t)g'(t)dt = f(x)g'(x).$$

Since the derivatives have a common value, we now have

$$\int_a^x f(t)dg(t) = \int_a^x f(t)g'(t)dt + C.$$

If $x = a$ this becomes

$$\int_a^a f(t)dg(t) = \int_a^a f(t)g'(t)dt + C$$

or $0 = 0 + C$. Therefore $C = 0$, and the theorem is correct.   □

This last result suggests that many RS integrals can be written as Riemann integrals. You will remember that a Riemann integral is an RS integral for which $g(t) = t$ is the identity function. Therefore a table of Riemann integrals may be adequate for evaluating many integrals. This is certainly true. However, there are some functions $g(x)$ such that for at least some values in the interval of integration the derivative fails to exist. This result would obviously fail under those circumstances. We have also seen some problems in which it is somewhat easier to set up the integral using a function $g(x)$ other than the identity function. *If* one can reduce the integral to a Riemann integral, and *if* the Riemann integral is easier to evaluate, this Theorem is very helpful. It will prove to be helpful in other ways as well, as we will see later on when we try to expand our ability to integrate a variety of functions.

In this section we have frequently considered an integral over the interval $[a, x]$ of variable length in contrast to our former requirement that the interval be a fixed interval. This has led us to the relation

$$\int_a^x f(x)dg(x) = F(x) - F(a)$$

provided $F(x)$ is the anti-derivative of $f(x)g'(x)$. In this case the integral is itself a function. Here $F(a)$ is a constant determined by the value of a and the particular function $F(x)$. There are many instances in which we will not be certain of the value a, and consequently all we will know about the value of this integral is that the value is $F(x)$ increased or decreased by some constant. Given this situation, we could just as well write

$$\int f(x)dg(x) = F(x) + C.$$

· Again, it is presumed that $F(x)$ is the anti-derivative of $f(x)g'(x)$. Observe that we have omitted the limits of integration entirely. If we are not certain of the value of a, we might just as well replace $[-F(a)]$ by some constant C, and not even attempt to indicate limits of integration. An integral of this type is called an *indefinite integral* whereas the integral with specific limits is called the *definite integral*. The reason for this nomenclature should be clear.

EXAMPLE 1.6. Integrate $\int 2x^4 \, dx^5$.

Solution. Since $2x^4 D_x(x^5) = 2x^4(5x^4) = 10x^8$, and since $D_x[10x^9/9] = 10x^8$, we know that an anti-derivative of $f(x)g'(x)$ in this case is $[10x^9/9]$. Hence we have

$$\int 2x^4 \, dx^5 = \frac{10x^9}{9} + C.$$

This is an indefinite integral. We would be able to find a value for C only if we were supplied with some additional information giving the value of the integral for some particular value of x.

EXERCISES

1. Find the derivative of each of the following.

   (a) $\int_1^x e^{-3t} \, dt^{4/7}$
   (b) $\int_1^x \tan 3t \, dt^{2/3}$
   (c) $\int_{-\pi/8}^x (\sec 2t - \csc 2t)^2 dt^2$
   (d) $\int_1^x \csc(3\pi t/8) dt$
   (e) $\int_{-\pi}^x \sin 4t \, dt^2$
   (f) $\int_1^x (\cos 2t - \ln t) dt$
   (g) $\int_0^x (3 \sin 2t - 2)^3 \, dt^2$
   (h) $\int_{-1}^x e^{-2t} \sin 3t \, dt$

2. Find the derivative of each of the following.

   (a) $\int_2^x (e^t + \sin t) d(t^2 - t + 2)$
   (b) $\int_4^x \tan t \, d(\sqrt{t})$
   (c) $\int_1^{1.3} x^\pi \, d(\tan^3 x)$
   (d) $\int_2^x t^{-6} \, d(t^7)$
   (e) $\int_a^x e^{ct} \, dt$
   (f) $\int_1^x x^{-4} \, dx$
   (g) $\int_1^x x^{-2/3} \, dx$
   (h) $\int_{\pi/6}^{\pi/2} (\sin 2x)^{-3} \, d(\sin 2x)$
   (i) $\int_{-1}^x (x + 3)^{-5} \, dx$
   (j) $\int_1^x (2/\sqrt{x}) dx$

3. In each of the following show that the derivative of the integral is the same as the derivative of the function $F(x)$. From this information obtain the value of the integral.

(a) $\int_1^x t^3 \, dt^2; \quad F(x) = (2/5)x^5$

(b) $\int_1^x t^{2/3} \, dt; \quad F(x) = (3/5)x^{5/3}$

(c) $\int_1^x 3t^4 \, dt^2; \quad F(x) = x^6$

4. Evaluate each of the following indefinite integrals:

(a) $\int x^{3/4} \, dx^{2/3}$

(b) $\int e^{x/2} \, dx$

(c) $\int \sin x^2 \, dx^2$

(d) $\int (x - 2)^3 \, dx^2$

(e) $\int \sin(x/2) dx$

(f) $\int (x - 1)^4 \, dx$

(g) $\int \cos x^3 \, dx^3$

(h) $\int e^{\sin x} \, d(\sin x)$

5. (a) Show that $\int_1^x xe^{x^2} \, dx = (1/2) \int_1^x e^{x^2} \, dx^2$. Evaluate $\int_1^3 xe^{x^2} \, dx$.

(b) Show that $\int_0^x x^2 \sin x^3 \, dx = (1/3) \int_0^x \sin x^3 \, dx^3$. Evaluate $\int_0^1 x^2 \sin x^3 \, dx$.

(c) Show that $\int_0^x xe^{-3x^2} \, dx = (1/2) \int_0^x e^{-3x^2} \, dx^2$. Evaluate $\int_0^2 xe^{-3x^2} \, dx$.

6. The slope of a curve is given by the formula $(x^3 - 5x + 3)$ at a point $(x, f(x))$. Find the equation of this curve if the curve passes through the point $(2, 4)$.

7. Find the equation of the curve through the point $(3, 4)$ if the slope for any value x is given by $(x^{3/2} + \sin x - x^2)$.

8. Find the equation of the line which is tangent to the curve $y = x^3 + 31$ at the point $(-3, 4)$.

9. For each of the following functions find the intervals when the function is increasing, and when it is decreasing.

(a) $F(x) = \int_{-2}^x (t^2 - 4) dt^2$

(b) $F(x) = \int_0^x [t/(t^2 + 1)] dt$

(c) $F(x) = \int_{-3}^x e^{-x} \, dx^2$

10. Find an integral whose derivative is given by $(\sin x - \cos x)(2x)$. What restrictions, if any, need be placed on the lower limit of this integral?

11. Find an integral whose derivative is given by $(\tan x)(3x^2)$. What restrictions, if any, need be placed on the interval over which this integral is valid? [Hint: Is there any value for which the integrand might be expected to give any difficulty?]

12. (a) Show that $\int_a^x f'(t) dt = f(x) + C$ for some constant C.

(b) Show that $\int_a^x f'(g(t))g'(t) dt = \int_a^x f'(g(t)) dg(t) = f(g(x)) + C$ for some constant C.

13. Use the mean value theorem of the integral to show that there exists a value $x = c$ in the interval $[a, b]$ such that each of the following is true under the assumption

that the appropriate functions are continuous or monotonic. Indicate precisely these conditions in each case.

(a) $\int_a^b f(x)dg(x) = f(c)[g(b) - g(a)]$

(b) $\int_a^b f'(x)dg(x) = f'(c)[g(b) - g(a)]$

(c) $\int_a^b f(x)dg(x) = \int_a^b f(x)g'(x)dx = f(c)g'(c)[b - a]$

(d) $\int_a^b dg(x) = g'(c)[b - a]$

14. (a) Find the area under the curve $y = x^{3/2}$ from $x = 1$ to an arbitrary point x.

 (b) Differentiate your result of part (a) and show that the rate of change of the area as x moves toward the right is given by the value of y corresponding to the given value of x.

 (c) From the information of part (b) find the value of x at which the area is changing at a rate of 8 square units per unit change in x.

15. (a) Find a formula for the area inside $r = f(\theta)$ between $\theta = 0$ and an arbitrary value of the angle θ using polar coordinates.

 (b) Show that the rate at which the area of part (a) is changing with respect to a change in θ is given by $\frac{1}{2} fr^2$.

 (c) Illustrate the results of parts (a) and (b) with the curve having the equation $r = \theta$.

P16. If the location of a particle at time t is given by $s(t)$, the velocity by $v(t)$ and the acceleration by $a(t)$, use the fundamental theorem (differentiating each side of the equation) to show that the integral of the acceleration is the velocity and the integral of the velocity is the location.

P17. If an object moves with an acceleration (toward the center of the earth) of 32 ft/sec², find the velocity and location at time t if the initial velocity was upward at 8 ft/sec and the initial location was 256 feet above the earth.

P18. A spacecraft is traveling from the earth to the moon. It is acted upon by two gravitational fields in (nearly) opposite directions. The force pulling the space-craft toward the earth is given by $GM_e M_s/r^2$ where G is the universal gravitational constant, M_e is the mass of the earth, M_s is the mass of the spacecraft, and r is the distance from the center of the earth. The force pulling the spacecraft toward the moon is given by $GM_m M_s/(R - r)^2$ where in addition to the above symbols we have M_m as the mass of the moon and R as the distance between the center of the moon and the center of the earth. [$(R - r)$ is effectively the distance from the spacecraft to the moon.] Therefore, the force directed toward the earth at a distance, r, from the center of the earth is given by the equation

$$F = \frac{GM_e M_s}{r^2} - \frac{GM_m M_s}{(R - r)^2}.$$

This force is the net force pulling the spacecraft toward the earth. On this particular mission we are planning to crashland a (hopefully unmanned) spacecraft on the moon, and we are concerned with conservation of energy. We are anxious to get the spacecraft to the point at which the greater force is in the direction of the moon, but we are content to have an almost zero velocity at this point. If $M_e = 5.983 \times 10^{24}$ kilograms, $M_m = 7.237 \times 10^{22}$ kg., $G = 6.670 \times 10^{-11}$ (newton)(meters)²/(kgs)², and $R = 384393$ kilometers, find the minimal work required for a spacecraft of 200 kg. (The radius of the earth is approximately 6400 km and the radius of the moon 1700 km.) [*Hint*: Find the point at which the force changes from an earth directed force to one directed toward the moon. This is the point you must be able to reach. Also, remember that a physicist defines work as the force times the distance traveled, and since the force varies with the distance from the earth, you can only consider the distance to be a near constant for a very short interval. Hence, RS sums, etc., are indicated.]

S19. If the marginal cost of producing an item when the present production rate is n items per day is given by $M(n) = 50 + n + 0.0004n^2$, what is the cost of increasing production from 200 to 250 items per day?

S20. The cost of production of n items per day is given by the relation $C(n) = 5000 + 10n + 0.5n^2 - 0.001n^3$ dollars per day.

   (a) Find the marginal cost of production and interpret this function.
   (b) Using your result of part (a) find the cost of production. Note the need for a value for the constant of integration in this case. What information would ordinarily be supplied to obtain this value?

B21. The rate of population growth of a certain community is given by $R(t) = ae^{0.02t}$ where a is a constant. If it is known that the population of the community was 40,000 and growing at a rate of 2% in 1960, find a formula for the population, $P(t)$, of the community in the year, t. Use this to evaluate the population expected in 1980.

B22. In a colony of mice used for experimental purposes it is known that the number born on the t-th day is given by the relation $b(t) = 0.03p(t) - 0.00001[p(t)]^2$ where $p(t)$ is the population of the colony on the t-th day. It is also known that the number of deaths in the colony on the t-th day is given by $d(t) = 0.015p(t) - 0.00001[p(t)]^2$. It is also known that $p(0) = 2000$.

   (a) Find the rate of change of the mouse population per day.
   (b) Set up an integral which would give the total mouse population as a function of time. This integral will involve $p(t)$. You are not asked to evaluate this integral.
   (c) If you have a computer available, find the mouse population on the 10-th day.

M23. If $D_x \int_a^x f(t)dg(t) = D_x \int_b^x f(t)dg(t)$, then show that $\int_a^b f(t)dg(t)$ must be a constant.

V.2 Some More Derivatives and Integrals

In this section we will make a great deal of use of the Fundamental Theorem developed in the last section. We will start with some integrals developed in Chapter III and then take anti-integrals to obtain derivatives. We will use these derivatives together with some of our theorems of Section IV.4 to obtain more derivatives and then take anti-derivatives to obtain integrals. Perhaps you will begin to see why Theorem V.1.1 has been called the "Fundamental Theorem of the calculus."

EXAMPLE 2.1. From an earlier result we have $\int_0^x e^{ct}\,dt = (e^{cx}/c) - (1/c)$ where c is a constant other than zero. We now have

$$D_x \int_0^x e^{ct}\,dt = e^{cx}(1) = e^{cx} = D_x\left[\frac{e^{cx}}{c} - \frac{1}{c}\right] = \frac{1}{c}D_x e^{cx}$$

where the first two results come from the Fundamental Theorem and the last two are merely expressing the fact that we have the derivative of the result of integration and the simplication of this result. Using the third and the last of the equal results in the equation above and multiplying both sides by the constant c we have

$$D_x e^{cx} = ce^{cx}. \qquad (V.2.1)$$

This is a result we want to record, for it permits us to obtain the derivative, or rate of change, of the exponential function. If $c - 1$ we have the special case

$$D_x e^x = e^x. \qquad (V.2.1a)$$

It is interesting to find a function which is its own derivative, or which changes at a rate which is exactly equal to the value that it takes on at the point in question. You will remember that we carefully picked the value e as the one value such that the function would have a slope of one at the point $(0, 1)$.

After this success with the exponential function, perhaps we should try the trigonometric functions. We will follow the same pattern we followed above, but this time we will not put in quite as many words of explanation. You can merely copy the words above if you find that you need additional help.

EXAMPLE 2.2.

$$D_x \int_0^x \sin ct\,dt = \sin cx = D_x\left[\left(\frac{1}{c}\right)(1 - \cos cx)\right] = \left(-\frac{1}{c}\right)D_x \cos cx,$$

and hence

$$D_x \cos cx = -c \sin cx. \qquad (V.2.2)$$

Also

$$D_x \int_0^x \cos ct \, dt = \cos cx = D_x\left[\left(\frac{1}{c}\right)\sin cx\right] = \left(\frac{1}{c}\right)D_x \sin cx,$$

whence

$$D_x \sin cx = c \cos cx. \tag{V.2.3}$$

Note that with very little effort we have now obtained the derivatives of the exponential function and of the sine and cosine functions. Of course, we had to expend some work getting to the Fundamental Theorem but at this point it may almost seem worth it when you consider the alternative of finding the derivatives of these functions using the Definition. Now that we have these results, we can use the theorems of Section IV.4 to obtain the derivatives of the other four trigonometric functions.

EXAMPLE 2.3.

$$D_x \tan cx = D_x\left[\frac{\sin cx}{\cos cx}\right] = \frac{(\cos cx)(D_x \sin cx) - (\sin cx)(D_x \cos cx)}{\cos^2 cx}$$

$$= \frac{(\cos cx)(c \cos cx) - (\sin cx)(-c \sin cx)}{\cos^2 cx}$$

$$= \frac{c[\cos^2 cx + \sin^2 cx]}{\cos^2 cx} = \frac{c(1)}{\cos^2 cx}$$

$$= c \sec^2 cx. \tag{V.2.4}$$

Note that an anti-derivative of $\sec^2 cx = (1/c)\tan cx$, and consequently we can apply the Fundamental Theorem to obtain

$$\int_a^x \sec^2 ct \, dt = \left(\frac{1}{c}\right)[\tan cx - \tan ca] \tag{V.2.5}$$

or

$$\int_a^b \sec^2 cx \, dx = \left(\frac{1}{c}\right)[\tan cb - \tan ca]. \tag{V.2.6}$$

It is necessary to be careful in regard to the interval of integration for we cannot tolerate the situation in which the integrand might fail to exist. Hence, the interval $[a, x]$ in the first integral and the interval $[a, b]$ in the second integral must be picked rather carefully. For instance, the value $\pi/2$ should not be included in the interval. Similar observations should apply to all integrals. Later on we will find that it is possible under certain conditions to consider some integrals in which this condition is partially relaxed, but this is something that must be considered separately for each integral which fails to meet the requirements stated here.

EXAMPLE 2.4.

$$D_x \cot cx = D_x \left[\frac{\cos cx}{\sin cx} \right] = \frac{(\sin cx)(-c \sin cx) - (\cos cx)(c \cos cx)}{\sin^2 cx}$$

$$= \frac{-c(\sin^2 cx + \cos^2 cx)}{\sin^2 cx} = -c \csc^2 cx. \qquad (V.2.7)$$

Here we have done the same thing we did in the previous example, but we have not written out the steps in as much detail. In a manner similar to that above, we can also write

$$\int_a^x \csc^2 ct \, dt = \left(-\frac{1}{c} \right) \cot cx + \left(\frac{1}{c} \right) \cot ca \qquad (V.2.8)$$

and

$$\int_a^b \csc^2 cx \, dx = \left(\frac{1}{c} \right) [\cot ca - \cot cb]. \qquad (V.2.9)$$

You will note the reversal of order in the latter expression as a result of the negative sign which appeared in both the derivative and the basic integral.

EXAMPLE 2.5.

$$D_x \sec cx = D_x \left[\frac{1}{\cos cx} \right] = \frac{(\cos cx)(0) - (1)(-c \sin cx)}{\cos^2 cx}$$

$$= c \, \frac{\sin cx}{\cos cx} \frac{1}{\cos cx}$$

$$= c \tan cx \sec cx. \qquad (V.2.10)$$

We should observe that all of the algebraic simplification is not required to obtain the derivative, but it is usually done to obtain the results in a form that is easier to remember. We can now evaluate the integrals,

$$\int_a^x \sec ct \tan ct \, dt = \left(\frac{1}{c} \right) \sec cx - \left(\frac{1}{c} \right) \sec ca \qquad (V.2.11)$$

and

$$\int_a^b \sec cx \tan cx \, dx = \left(\frac{1}{c} \right) [\sec cb - \sec ca]. \qquad (V.2.12)$$

EXAMPLE 2.6.

$$D_x \csc cx = D_x \left[\frac{1}{\sin cx} \right] = \frac{(\sin cx)(0) - (1)(c \cos cx)}{\sin^2 cx}$$

$$= -c \, \frac{1}{\sin cx} \frac{\cos cx}{\sin cx}$$

$$= -c \csc cx \cot cx. \qquad (V.2.13)$$

This gives rise to the integral formulas

$$\int_a^x \csc ct \cot ct \, dt = \left(-\frac{1}{c}\right)\csc cx + \left(\frac{1}{c}\right)\csc ca \qquad (V.2.14)$$

and

$$\int_a^b \csc cx \cot cx \, dx = \left(\frac{1}{c}\right)[\csc ca - \csc cb]. \qquad (V.2.15)$$

We have not yet found any integrals for the tangent, cotangent, secant, or cosecant except as they appear in these strange products, but be patient. These require a bit more machinery. We shall obtain them in the next Chapter.

Note that we have been able to get a substantial number of formulas for derivatives and integrals based only on results already obtained. The fact that we can do this with so little effort should make you feel that you have learned quite a bit of calculus. You should make a list of all of the derivatives and the integrals that you have seen to date, for this list will be very handy in evaluating the integrals and the derivatives which may occur in any applications.

One final admonition. Until this time we have generally divided the work so that in Chapters II and III you knew that if you used calculus you must use integration, for you had not yet gotten to any material in this book on the derivative, and in Chapter IV it was natural to expect to use the derivative. Now, however, you may find that either of these two operations is used, and in fact both may be used in the same exercise. Hence you must always be on the lookout to determine which operation is appropriate. Remember we obtained the integral by considering sums and we discovered the derivative through the idea of rate of change. If you are applying the Fundamental Theorem, you should also be aware of whether you are starting with the derivative and trying to obtain the integral or starting with the integral and trying to obtain the derivative. Confusing? Not if you are careful to think through each problem before starting to work. Remember the old proverb "when pencil touches paper, mind goes out of gear." Be sure that you have your plan of attack in mind before you start solving any problem.

EXERCISES

1. (a) Make a list of all of the formulas we have derived for the derivative. This list should include the power function, the exponential function, and the six trigonometric functions.
   (b) Make a list of all of the integral formulas we have derived up to this point. It would be helpful to make this list match the formulas of part (a) in that the formula for the integral of the cosine would relate to the formula for the derivative of the sine.
   (c) Illustrate the relationship of each formula of part (a) with a formula of part (b) using the fundamental theorem.

2. Find the derivative of each of the following:

(a) $e^{3x} - \sin 4x + x^{12.5}$

(b) $e^{-x} \sin 2x + 0.5\sqrt[3]{x}$

(c) $3^{-2x} \cos 3x$

(d) $x^2 e^{3x}$

(e) e^{2x}/x^3

(f) $x^2 \cos x$

(g) $\sqrt{x} \sin(x/2)$

(h) $x^2 e^{2x}(\cos 3x - \sin 3x)$

(i) $\tan x - \sec 2x$

(j) $x^2 \cot x$

(k) $e^{-2x} \cos x$

(l) $\csc 2x \cot 2x$

(m) $\sec 2x \tan 2x$

3. Find the derivative of each of the following:

(a) $\sqrt[4]{x^7} - \sqrt[5]{x}/(x^3 - 3x + 2)$

(b) $\tan 3x + \cot 3x$

(c) $\sin(x + \pi/2) + \cos x$

(d) $\sqrt[3]{x} - 2$

(e) $x^{0.3} e^{0.4x} \sin 0.5x \cos 0.6x$

(f) $x^a e^{bx} \tan cx$ where a, b, and c are constants and a is rational.

4. Find the indefinite integral of each of the following:

(a) $\int \sin 3x \, dx$

(b) $\int \cos 4x \, dx$

(c) $\int \sec^2 5x \, dx$

(d) $\int \csc^2 2x \, dx$

(e) $\int \sec 2x \tan 2x \, dx$

(f) $\int \csc 3x \cot 3x \, dx$

(g) $\int e^{2x} \, dx$

(h) $\int x^{2/7} \, dx^{5/7}$

5. Find the indefinite integral of each of the following:

(a) $\int e^{\sin x} \, d(\sin x)$

(b) $\int e^{\sin x} \cos x \, dx$

(c) $\int \cos x \, d(\cos x)$

(d) $\int \sin x \, d(\sin x)$

(e) $\int \sin x \cos x \, dx$

(f) $\int \sin^3 x \cos x \, dx$

(g) $\int \tan x \, d(\tan x)$

(h) $\int \tan x \sec^2 x \, dx$

6. Find the indefinite integral of each of the following:

(a) $\int (e^{3x} - \sin 4x + \pi^2/x^4)dx$

(b) $\int (\sec^2 3x - \csc^2 4x)dx$

(c) $\int (\sec 2x \tan 2x - \cos 2x \tan 2x)dx$
   [*Hint*: Simplify the second term before trying to integrate.]

(d) $\int \sin 3x/(\cos^2 3x)dx$
   [*Hint*: Can you convert this to any other functions for which you know the integral?]

7. Evaluate each of the following:

(a) $\int_0^{\pi/4} \tan x \sec^2 x \, dx$

(b) $\int_0^4 e^{\sqrt{x}} \, d(\sqrt{x})$

(c) $\int_{1/6}^{1/3} \cot^2 \pi x \, dx$
   [*Hint*: $\csc^2 y - \cot^2 y = 1$]

(d) $\int_{-1}^8 \sqrt[3]{x}(x - 1)^2 dx$

(e) $\int_1^8 \sqrt[3]{x}(x - 1/x)^2 dx$
   Why could we not use the interval $[-1, 8]$ instead of $[1, 8]$ here?

8. Use the Fundamental Theorem to check each of the following to determine whether the indefinite integration is correct. If you find an error, correct either the integrand or the result of the integration in order to make a correct statement.

(a) $\int \sin x \, dx^2 = 2 \sin x - 2x \cos x + C$

(b) $\int \sin x \, dx = \cos x + C$

(c) $\int 5e^x \, d(\sin 2x) = 2e^x(\cos 2x + 2 \sin 2x) + C$

(d) $\int xe^x \, dx = xe^x - e^x + C$

(e) $\int (x^2 + 3)^2 dx = (1/3)(x^2 + 3)^3 + C$

9. (a) Find the equation of the line tangent to $f(x) = \cos x$ at $x = \pi/6$.

(b) Find the equation of the line tangent to $f(x) = \cos x$ at $x = 0$.

(c) When is $\cos x$ increasing?

(d) When is $\cos x$ decreasing?

10. (a) Draw the graph of each of the six trigonometric functions.

(b) Examine the graph of $\cos x$, $\cot x$, and $\csc x$ for values of x between 0 and $\pi/2$ (in the first quadrant) and indicate why these values indicate that the derivative of these functions should be negative.

11. (a) Show by differentiation that $\tan x$ and $\cot x$ are monotonic functions wherever they exist.

(b) Note the graph of these two functions and verify graphically that the information obtained by differentiation is correct.

12. (a) Find the derivative of $f(x) = xe^{-x}$.

(b) Show that $f(x)$ is positive when x is positive and negative when x is negative.

(c) Note the intervals in which $f(x)$ is increasing and when it is decreasing.

(d) Use the information of parts (b) and (c) to sketch the graph of $y = f(x)$.

(e) Show that $D_x(-xe^{-x} - e^{-x}) = xe^{-x}$.

(f) Using the result of part (e) find the area bounded by $y = f(x)$, $y = 0$ and $x = 10$.

(g) Show that there would be very little increase in the area of part (f) if "$x = 10$" were replaced by "$x = 100$."

13. Repeat Exercise 12 if $f(x) = x^2 e^{-x}$. In part (e) you should show that

$$D_x x^2 e^{-x} = -(x^2 + 2x + 2)e^{-x}.$$

14. Differentiate each of the following. Explain your results in each case.

(a) $\sin^2 x + \cos^2 x$
(b) $\sec^2 x - \tan^2 x$
(c) $\csc^2 x - \cot^2 x$

15. Sketch the graph of $y = \sec x$. Find the volume of revolution obtained by revolving the area bounded by $x = 0$, $x = \pi/4$, $y = \sec x$ and $y = 0$ about the x-axis.

16. Find the area bounded by the curve $r = \csc \theta$ and the vectors $\theta = \pi/6$ and $\theta = \pi/2$.

SB17. (a) The population of a given community is given by $P(t) = P(0)e^{0.025t}$ where $P(t)$ is the population of the community t years after its founding. Find the rate at which this community is growing.

(b) How long will it take this community to double its population at this rate of growth?

(c) What is the percentage rate of annual growth of this community?

(d) What would the formula have been for the population if the rate of annual growth had been 3%?

SB18. In 1920 R. Pearl and L. J. Reed published the formula

$$y = \frac{197273000}{1 + e^{-0.0313395t}}$$

in an effort to predict the population of the United States. In this equation t represents the number of years since April 1, 1914 and y represents the population of the United States. During the period 1790–1920 this formula was never off by more than 3.3% and it has continued to do a good job of predicting, being off by no more than 3.8% through 1950.

(a) Find the annual rate of growth of the population in 1790, 1860, and 1940.

(b) Using this formula what would be the maximum population the United States would ever attain?

(c) Find the rate of growth in 1980 using this formula as a predictor.

(d) What would be the predicted population of the U.S. in 1980 using this formula as a predictor?

SP19. A bit of information is a single item such as a dot or a dash in Morse code or such as a one or a zero in the binary system frequently used with computers and with data transmission. It is postulated that the marginal rate of information transmitted between a source and a destination with respect to the number of bits per second is given by $K(x) = Rt^{-x}$ where R and t are positive constants and x represents the complexity of the system in terms of the number of bits transmitted per second. Since $K(x)$ is a marginal rate, it represents the additional information that can be transmitted if the transmission rate is increased from x bits per second to $(x + 1)$ bits per second. If $f(x)$ represents the total amount of information that is transmitted per second when the transmission rate is x bits per second, find an expression for $f(x)$ as a function of x. In order to do this, you might observe that $f(0) = 0$ and then calculate the increase in total transmission with each additional increase in transmission rate. It should be pointed out that increasing the speed of transmission creates problems in that it is not unlikely that the rate of erroneous transmission would also increase. Hence $f(x)$ should not be expected to increase without limit.

P20. A pebble is dropped in a smooth pond of water. At a certain point in the pond the height of the water is observed to follow the function $h = 5e^{-0.05t}(\sin(\pi t/60) + 2\cos(\pi t/60))$ where t is measured in seconds and h is measured in inches from the equilibrium level of the pond.

(a) How fast is the height of the water changing when $t = 10$ seconds, and is the surface rising or falling?

(b) During what time intervals is the water rising at this point?

V.3   The Mean Value Theorem and the Differential

In the last chapter we started with the differential quotient and then went to the derivative by taking a limit. Thus, at least in some cases, there is certainly a connection between the differential quotient and the derivative. We wish to expand upon that relationship in this Section. However, we must be careful to observe that certain conditions are required before we can establish a useable relation. In Figure V.2 we have indicated the point P with coordinates $(a, f(a))$ or $(a, g(a))$ and the point Q with coordinates $(b, f(b))$ or $(b, g(b))$. We have also indicated two curves $y = f(x)$ and $y = g(x)$. We note that the slope of the segment $\overline{PQ}$ is given by $DQf(a, b) = DQg(a, b)$. We also note that there is a point $(c, f(c))$ at which the tangent line appears to be parallel to the line $\overline{PQ}$. In this case we would then have $f'(c) = DQf(a, b)$.

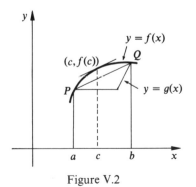

Figure V.2

On the other hand, we observe that there appears to be no similar point for the curve $y = g(x)$. Therefore, there seems to be a basic difference between the two functions $f(x)$ and $g(x)$. This difference results from the fact that $f'(x)$ is continuous over the interval $[a, b]$ while $g'(x)$ is not. This result is given explicitly in the following Theorem.

Theorem 3.1 (The mean value theorem). *If $f(x)$ is a function having a continuous derivative in the interval $[a, b]$, then there is some value, $x = c$, in the interval (a, b) such that $f(b) - f(a) = f'(c)(b - a)$.*

PROOF. We start with the integral $\int_a^b f'(x)dx$ and note that since $f'(x)$ is continuous in $[a, b]$ and x is monotonic, there is a value c in (a, b) such that

$$\int_a^b f'(x)dx = f'(c) \int_a^b dx = f'(c)(b - a).$$

We also have

$$\int_a^b f'(x)dx = f(b) - f(a)$$

since $f(x)$ is a function that has for its derivative $f'(x)$. Hence, we have

$$f(b) - f(a) = f'(c)(b - a). \tag{V.3.1} \quad \square$$

This Theorem will turn up in a surprising number of places and in various forms. We shall indicate a few such forms in this section. However, before we go further we should point out that the Theorem merely guarantees that at least one number, c, exists such that $f(b) - f(a) = f'(c)(b - a)$, and the theorem is only counting those numbers which exist as interior points of the interval (a, b). We have given no method for finding any such values. In fact, there is no easy way of finding these values. If the computation is feasible, we can find a value or values for c by calculating the values $f(a)$ and $f(b)$ and then solving the equation

$$f'(c) = \frac{f(b) - f(a)}{b - a} = DQf(a, b) \tag{V.3.2}$$

for the value or values of c which are in the interval (a, b). Unless $f'(c)$ is a rather simple function, there is little chance that a formula exists for the solution of (V.3.2) and often it is not necessary to obtain a solution.

EXAMPLE 3.1. Find the value of c which satisfies the mean value theorem if $f(x) = x^3 + x - 4$ and the interval involved is $[-1, 2]$.

Solution. Here we wish to find a value of c such that $f(2) - f(-1) = f'(c)[2 - (-1)]$. This is equivalent to the equation $6 - (-6) = (3c^2 + 1)(3)$, or $12 = 9c^2 + 3$. Since $c^2 = 1$ in this case, we must have $c = +1$ or $c = -1$.

Only the value $+1$ is satisfactory here, for the theorem requires that c be a point of the open interval $(-1, 2)$. The value $c = -1$ is a boundary point of the interval and not an interior point.

We should note in (V.3.2), that the average rate of change of $f(x)$ over the interval $[a, b]$, that is $DQf(a, b)$, is equal to the value of the derivative for some undetermined point within the interval (a, b). If the diameter (or length) of this interval is small, then the value of c is known without a large error. On the other hand, if the diameter of (a, b) is large, we may know very little about the location of c.

The mean value theorem is often expressed in the form

$$f(b) = f(a) + f'(c)(b - a). \tag{V.3.3}$$

This expresses the value of $f(b)$ in terms of the values of $f(a)$ and $f'(c)$. If we have even an approximate value for $f'(c)$, we may be able to obtain a suitable approximation for $f(b)$.

EXAMPLE 3.2. Find an approximate value of $e^{0.1}$.

Solution. In this case it is apparent that we should let $f(x) = e^x$ and let $b = 0.1$. Since we must know the value of $f(a) = e^a$, we should select a with some degree of care. We would like to have a fairly close to b and the rather obvious choice is $a = 0$, since we know that $e^0 = 1$. We can now fill in most of (V.3.3) and write

$$e^{0.1} = e^0 + f'(c)(0.1 - 0) = 1 + e^c(0.1).$$

We do not know the value of c, but we do know that it must be a value between 0 and 0.1. Since e^c does not vary by much in this interval, and since we know the value of e^0, it is appealing to use $c = 0$, even though we are sure that it is not the correct value. Thus, we would write $e^{0.1} \doteq 1 + e^0(0.1) = 1 + 0.1 = 1.1$. We have indicated the approximate value here with the dot over the equal sign. Since e^x is an increasing function and we have chosen to use a value for c which is smaller than any possibility for the correct value, we can be certain that our approximation is too small. Usually we are more than satisfied to have a value that is a fair approximation and would stop here. However, if you wished to improve this, you might try again, this time replacing c with 0.1 and using the approximation $e^{0.1} \doteq 1.1$. We would then have as a second approximation $e^{0.1} \doteq 1 + e^{0.1}(0.1) \doteq 1 + (1.1)(0.1) = 1.11$. We would expect this to be too large, for now we have used in place of c a value that is too large. If we were to take the average of 1.1 and 1.11, we would have 1.105. This would seem to be more reasonable. In fact, $e^{0.1} = 1.1051709$, and we have achieved a very good approximation.

Usually we will stop with the first value as an approximation, and this will be sufficiently accurate. It is rather clear that unless $f'(x)$ is changing rapidly in (a, b) and the value of $|b - a|$ is large, the choice of c will not be critical for a useable approximation.

A special case of the mean value theorem is sometimes given particular attention. It will be referred to from time to time, and consequently we will present it at this point.

Corollary 3.1 (Rolle's theorem). *If $f(a) = f(b) = 0$ and $f'(x)$ is continuous throughout the interval $[a, b]$ then there is a value $x = c$ in (a, b) such that $f'(c) = 0$.*

PROOF. Using the mean value theorem, substitute $f(a) = f(b) = 0$, and the result follows at once.   □

By a change of variable, we can write the mean value theorem in a slightly different manner. Thus $f(x + \Delta x) - f(x) = f'(c)[(x + \Delta x) - x] = f'(c)\Delta x$. Since the symbol Δ is used to denote a difference in the values of the quantity in question, we could write Δf to indicate the difference $f(x + \Delta x) - f(x)$. This equation would then reduce to

$$\Delta f = f'(c)\Delta x. \tag{V.3.4}$$

Since Δf represents the change in the value of the function over the interval $[x, x + \Delta x]$ and since Δx represents the change in x over the same interval, we note that the average rate of change of the function with respect to x is then $\Delta f / \Delta x = f'(c)$. This is just another way of writing $DQf(x, x + \Delta x)$. The main difficulty with all of these results is that we do not know the value of c other than the fact that it is in $(x, x + \Delta x)$. It is tempting to assign to c a known value which would give at least a reasonable approximation to the correct result in these situations. The rather obvious choice is to replace c with the value x, for while Δx can change as we consider intervals of different lengths, the point x remains fixed for any one discussion. If Δx is small, this should not cause a large error, or so it would seem. (The actual error created by replacing c by x would depend upon $f(x)$, the value of x, and the value of Δx.) This choice is the same one we made in Example 3.2.

We now return to (V.3.4) and note that replacing c by x gives us an approximation $\Delta f \doteq f'(x)\Delta x$. The right hand side would probably not give us the correct value, but hopefully it would be close. Note that the right hand side is a function of two variables, x and Δx. The function $f'(x)\Delta x$ is called the *differential of $f(x)$ with respect to x*, and its value will depend both on x and upon the length of the interval involved. Since this is not the value Δf, we will use another notation for the differential and write $df(x, \Delta x)$, $df(x)$, or simply df. The notation $df(x, \Delta x)$ would be logically correct, but we tend to abbreviate if there is no ambiguity. If we let $f(x)$ be the identity function, $f(x) = x$, we would have $df = dx = (1)\Delta x$, and it is clear that $\Delta x = dx$. For purposes of symmetry, we usually write $df = f'(x)dx$ instead of $df = f'(x)\Delta x$. This is summarized in the following definition.

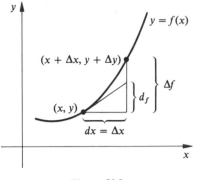

Figure V.3

Definition 3.1. If $f(x)$ is a function for which $f'(x)$ exists and is continuous on a given interval, the *differential of f with respect to x* on the interval is defined by the relation $df(x, dx) = f'(x)dx$. Where there is no ambiguity, the differential is abbreviated $df(x)$ or df.

We see by comparing Definition 3.1 with (V.3.4) that the differential is an approximation to Δf (as we had expected it to be). We can picture this, as shown in Figure V.3. If we take the interval of length $dx = \Delta x$ on the x-axis, we produce an actual change in $f(x)$ in the amount Δf. On the other hand, if we think of the rate of change as being a constant and equal to $f'(x)$, we would have the change indicated by the tangent to $f(x)$ at the point x. The values Δf and df are shown on the graph, and you should inspect them carefully to be certain that you understand their relationship to each other. Note that smaller values of $dx = \Delta x$ are apt to produce smaller variations between df and Δf. This, too, would be anticipated from our earlier discussion.

We can summarize much of our latest discussion by writing

$$f(a + \Delta x) = f(a) + f'(c)[(a + \Delta x) - a] \doteq f(a) + df(a), \quad (V.3.5)$$

where we have replaced x by a to emphasize that we are talking about a fixed point, and we have used $df(a)$ to denote $df(a, \Delta x)$. This result is frequently used for approximations.

EXAMPLE 3.3. Find an approximate value of $\sqrt{9.12}$.

Solution. Let $f(x) = \sqrt{x}$. Then (V.3.5) states that $f(9 + 0.12) \doteq f(9) + f'(9)(0.12)$ where we have used $a = 9$ and $\Delta x = 0.12$. Here $f'(9)(0.12)$ is the differential indicated by $df(a)$ in (V.3.5). Since $f'(x) = (1/2)x^{-1/2}$, $f'(9) = (1/2)(1/3) = 1/6$. Hence, we have $f(9.12) \doteq f(9) + (1/6)(0.12) = 3 + 0.02 = 3.02$. We can check this by noting that $3.02^2 = 9.1204$. This seems to be a very good approximation. (More accurately $\sqrt{9.12} = 3.01993$ and therefore our approximation is excellent in this case.)

At this time it is well to pause long enough to recall that we have seen this same notation before, and there was an indication that an explanation would follow in due course. Due course has now arrived for this particular topic. The notation for the integral includes $dg(x)$ or dx. In the RS sums we had $[g(x_k) - g(x_{k-1})]$, but if we let $\Delta x_k = x_k - x_{k-1}$, we note by the mean value theorem that $[g(x_k) - g(x_{k-1})] = g'(c)\Delta x_k$. In what might be termed the *limiting situation* that produces the lub or the glb, as the case may be, we would then be able to approximate this difference rather accurately with the differential, since the interval would be small. The error in using $g'(x_k)$ instead of $g'(c)$ would be small. Thus, we can replace $dg(x)$ with $g'(x)dx = dg(x, dx)$. Note that this agrees with the result we obtained in the fundamental Theorem when we obtained the derivative of the integral. There we had the fact that the differential, given by the product of the derivative and dx, was $g'(x)dx$. This is exactly $dg(x, dx)$.

We observe another result of the notation that we have developed. It is possible to write not only $\Delta f/\Delta x = f'(c)$, but it is also possible to write $df/dx = f'(x)$, since, by definition, we know that $df(x) = f'(x)dx$. This means that the derivative can be thought of as the quotient of two differentials. Note that it would not have been possible to obtain the differential of $f(x)$ without first obtaining the derivative, and consequently this does not provide an alternate means of obtaining derivatives. It does provide an alternate notation. There are many times when such a notation is useful. We will find this particularly true in Chapter XI. In case of doubt, we need only replace df by its value, and we will have $df/dx = f'(x)dx/dx = f'(x)$.

You should make use of this result where it is helpful, but be very cautious concerning the use thereof. While it happens that the first derivative (that is the result of a single differentiation of $f(x)$) is the quotient of differentials, if we were to differentiate the derivative we would find no simple counterpart for this result which would relate to the differentials of $f(x)$. Since we have occasion to differentiate derivatives in the next Section, this is something to keep in mind.

Before leaving the subject of the mean value theorem, we should state and prove the extended mean value theorem.

Theorem 3.2 (The extended mean value theorem). *If $f(x)$ and $g(x)$ are functions possessing continuous derivatives, and if $g'(x)$ is not zero in the interval $[a, b]$, then there exists a number c in (a, b) such that $(f(b) - f(a))/(g(b) - g(a)) = f'(c)/g'(c)$.*

PROOF. Since $g'(x)$ is continuous over the interval $[a, b]$ and since $g'(x) = 0$ for no value in $[a, b]$, then $g(b) - g(a) \neq 0$ in $[a, b]$. This follows, since there must be some number, c_1, in (a, b) such that $g(b) - g(a) = g'(c_1)(b - a)$, but the right hand side of this equation is not zero. Furthermore, since $g'(x)$ is continuous and is never zero, it follows that $g(x)$ must be monotonic throughout the interval. Therefore, it is appropriate to use the mean value theorem for integrals and obtain

$$\int_a^b \frac{f'(x)}{g'(x)}\, dg(x) = \frac{f'(c)}{g'(c)}\, [g(b) - g(a)]$$

for some value of c in (a, b). But since $dg(x) = g'(x)dx$ we have

$$\int_a^b \frac{f'(x)}{g'(x)} dg(x) = \int_a^b \frac{f'(x)}{g'(x)} g'(x)dx = \int_a^b f'(x)dx = f(b) - f(a).$$

Now equating these two values of the integral with which we started, we see that $[f'(c)/g'(c)][g(b) - g(a)] = f(b) - f(a)$ or $(f(b) - f(a))/(g(b) - g(a)) = f'(c)/g'(c)$ as required. Note that we have no difficulty with zero denominators by the hypothesis that $g'(x) \neq 0$ throughout the interval.   □

You will note in this extended mean value theorem if $g(x)$ is the identity function, we not only fulfill all of the hypotheses but we also have precisely the Theorem of the mean. This is the reason for the introduction of the adjective *extended* in this case.

We should include one application of the differential which is very useful in problems requiring a knowledge of the length of a curve or the area of a surface of revolution.

EXAMPLE 3.4. Find the length of the curve $y = x^{3/2}$ from $x = 1$ to $x = 9$.

Solution. We have drawn a sketch of the curve in Figure V.4 and have noted an interval of the curve with an approximation of the length of the portion of the curve. This portion is enlarged in Figure V.5. Note that the length of this portion is approximately given by the length of the line segment joining the points (x_{k-1}, y_{k-1}) and (x_k, y_k). If we denote the x and y increments by Δx and Δy respectively, we see that this line segment is of length $[(\Delta x)^2 + (\Delta y)^2]^{1/2}$. Using the right triangle in Figure V.5 we see that $\sqrt{(\Delta x)^2 + (\Delta y)^2} = \Delta x \sec \theta$ where θ is the angle formed by the segment with

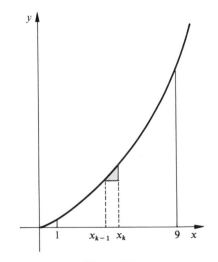

Figure V.4

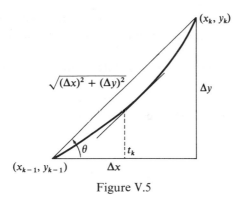

Figure V.5

the horizontal as indicated. We do not have a convenient method for obtaining the secant of this angle, but we can obtain the tangent, since the tangent is the slope, and we can relate the secant and tangent with an identity. Therefore, we would have $\sqrt{(\Delta x)^2 + (\Delta y)^2} = \Delta x \sec \theta = \Delta x(\tan^2 \theta + 1)^{1/2}$. The tangent is the slope of the line segment, but in view of the fact that $f(x) = x^{3/2}$ has a continuous derivative over the interval $[1, 9]$ we also know that there is some value of x between x_{k-1} and x_k for which $f'(x) = \tan \theta$. This follows, of course, from the mean value theorem. Therefore, we can write $\sqrt{(\Delta x)^2 + (\Delta y)^2} = \Delta x[(f'(t_k))^2 + 1]^{1/2}$ using t_k as an evaluation point. We are now ready to go back and add up all of the short segments. We thus obtain the RS sum

$$\sum_{k=1}^{n} [(f'(t_k))^2 + 1]^{1/2} \Delta x_k.$$

From this sum we obtain the integral $\int_1^9 [(f'(x))^2 + 1]^{1/2} \, dx$. In our case we have $f(x) = x^{3/2}$ and hence $f'(x) = 3x^{1/2}/2$. Therefore, the integral becomes

$$\int_1^9 \left(\frac{9x}{4} + 1\right)^{1/2} dx = \frac{3}{2} \int_1^9 \left(x + \frac{4}{9}\right)^{1/2} dx = \frac{3}{2} \int_{\sqrt{13/9}}^{\sqrt{85/9}} y^{1/2} \, dy$$

$$= \frac{3}{2} \left(\frac{2}{3}\right) y^{3/2} \Big|_{\sqrt{13/9}}^{\sqrt{85/9}} = \left(\frac{85}{9}\right)^{3/2} - \left(\frac{13}{9}\right)^{3/2}$$

$$= \frac{85\sqrt{85} - 13\sqrt{13}}{27} = 27.2885.$$

This example uses a great deal of the theory we have developed, and it is well worth taking a close look at all of the work that is involved, including the use of the mean value theorem, the development of the RS sum, the development of the integral, and the method by which the integral was evaluated. It is customary to represent the length of arc by the letter s and we can then write $\Delta s = ((\Delta x)^2 + (\Delta y)^2)^{1/2}$ or in differentials we can express this as $ds = (dx^2 + dy^2)^{1/2}$ with very little error. Any error in this last statement would disappear in effect as we proceed from the RS sum to the integral.

We will now examine the problem of determining the surface area created by revolving a curve about the axis.

EXAMPLE 3.5. Find the surface area created by revolving that portion of the curve $y = x^3$ between $x = 1$ and $x = 3$ about the x-axis.

Solution. This curve and the surface created are sketched in Figure V.6. Again we have partitioned the x-axis and shown the result of revolving that portion of the curve between (x_{k-1}, y_{k-1}) and (x_k, y_k) about the x-axis. We can use our work of Example 3.4 concerning arc length here, for we see that we essentially have the length of arc Δs revolved to form the surface of a truncated cone. The radius of the conical surface varies from y_{k-1} to y_k. If we use some radius r_k selected from this range, we have an approximate surface area $2\pi r_k(\Delta s)$ for the surface of this particular section of the total

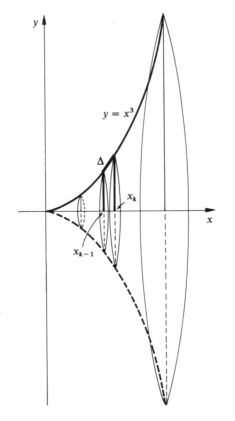

Figure V.6

surface we seek. We can add these portions up in an RS sum and obtain

$$\text{Approximate surface} = \sum_{k=1}^{n} 2\pi r_k(\Delta s) \doteq 2\pi \sum_{k=1}^{n} r_k[(f'(t_k))^2 + 1]^{1/2}\Delta x.$$

With continued refinement this yields the integral

$$S = 2\pi \int_1^3 f(x)[(f'(x))^2 + 1]^{1/2}\, dx,$$

where we have replaced r_k, actually some value of $y = f(x)$ in the interval $[x_{k-1}, x_k]$, by the value $f(x)$ which it must approach as the subintervals of the partition become smaller and smaller. In our case we have $f(x) = x^3$ and hence $f'(x) = 3x^2$. Thus, we wish to evaluate

$$S = 2\pi \int_1^3 x^3(9x^4 + 1)^{1/2}\, dx = 6\pi \int_1^3 \left(x^4 + \frac{1}{9}\right)^{1/2} x^3\, dx$$

$$= \frac{6\pi}{4} \int_1^3 \left(x^4 + \frac{1}{9}\right)^{1/2}\, dx^4$$

$$= \frac{3\pi}{2} \int_{10/9}^{730/9} y^{1/2}\, dy$$

$$= \frac{3\pi}{2} \left(\frac{2}{3}\right) y^{3/2} \Big|_{10/9}^{730/9}$$

$$= \frac{730\sqrt{730}\,\pi - 10\sqrt{10}\,\pi}{27}$$

$$= 729.33\pi.$$

This example also involves a lot of careful thought. Note the value of thinking of RS sums, for that allows us to think of the small sections, and thus determine results on a relatively simple scale. If we add these we get the RS sum. Successive refinements of the RS sum will bring us to the integral.

EXAMPLE 3.6. Find the length of arc of $r = 1 + \cos\theta$.

Solution. We have sketched this curve in Figure V.7. However, we can handle part of this problem without the use of the figure. It is always good practice to have such a sketch, for it enables one to verify that the result obtained is reasonably close to the correct result.

In this instance we know that the arc length of a curve can be obtained by evaluating the integral

$$\int_a^b ds$$

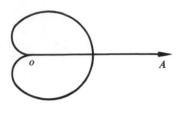

Figure V.7

with appropriate limits of integration. We also know that ds is given by the relation $ds = (dx^2 + dy^2)^{1/2}$. We have made use in the past of the relations $x = r \cos \theta$ and $y = r \sin \theta$ as the relations between the polar coordinates and the rectangular coordinates. Consequently, it would seem only necessary that we use these relations to obtain dx and dy. Since r is a function of θ, it is clear that for points on the cardioid of this problem x and y are also functions of θ, and the differentiation necessary to obtain the differentials dx and dy should then be carried out with respect to θ. Also, we note that both x and y are products, and it will be necessary to use the product formula for differentiation. Consequently, we will have

$$dx = [D_\theta(r \cos \theta)]d\theta = [r' \cos \theta - r \sin \theta]d\theta \quad \text{and}$$
$$dy = [D_\theta(r \sin \theta)]d\theta = [r' \sin \theta + r \cos \theta]d\theta.$$

Upon substituting these expressions in the expression for ds, we obtain

$$ds = [(r')^2(\cos^2 \theta + \sin^2 \theta) + r^2(\sin^2 \theta + \cos^2 \theta)]^{1/2} \, d\theta$$
$$= [(r')^2 + r^2]^{1/2} \, d\theta.$$

You should verify the algebra, for the *cross product* terms of dx^2 and dy^2 will negate each other. Upon using this expression for ds in the integral for arc length, we obtain

$$s = \int_\alpha^\beta [(r')^2 + r^2]^{1/2} \, d\theta.$$

In the particular case $r = 1 + \cos \theta$, we have $r' = (-\sin \theta)$ and therefore we have

$$s = \int_{-\pi}^{\pi} [(-\sin \theta)^2 + (1 + \cos \theta)^2]^{1/2} \, d\theta$$

$$= \int_{-\pi}^{\pi} (\sin^2 \theta + \cos^2 \theta + 1 + 2 \cos \theta)^{1/2} \, d\theta$$

$$= \int_{-\pi}^{\pi} (2 + 2 \cos \theta)^{1/2} \, d\theta.$$

If we remember that $\cos(\theta/2) = \sqrt{(1 + \cos \theta)/2}$, we realize that we can use this relation in the integral and thus obtain

$$s = \int_{-\pi}^{\pi} 2 \cos \frac{\theta}{2} \, d\theta = 4 \sin \frac{\theta}{2}\bigg|_{-\pi}^{\pi} = 4 \sin \frac{\pi}{2} - 4 \sin \frac{-\pi}{2}$$

$$= 4 - (-4) = 8.$$

We have used the interval $[-\pi, \pi]$ here in order that we would have that interval for which $\cos(\theta/2)$ is positive throughout the interval. Otherwise we would have had one portion of the integrand canceling another portion. This is something that must be watched carefully in the case of polar coordinates.

You should observe the manner in which differentials have been helpful in the development of Example 3.6. We were careful to ascertain that the expressions involved were actually functions of the single variable θ, but from this point on we could use the differentials and thus eliminate the necessity for a separate development in polar coordinates.

EXERCISES

1. Find the value of c which satisfies the requirements of the mean value theorem in each of the following cases:

   (a) $f(x) = x^2 - 4x + 3$ for the interval $[2, 5]$.
   (b) $f(x) = 4x - 5$ for the interval $[-2, 3]$.
   (c) $f(x) = x^3 - 3x^2 + 4x - 6$ for the interval $[-1, 5]$.
   (d) $f(x) = x^3 - 6x^2 + 11x - 6$ for the interval $[-2, 0]$.
   (e) $f(x) = x^3 - 6x^2 + 11x - 6$ for the interval $[0, 6]$.

2. Find the differential of each of the following functions:

   (a) $e^{3x} - \sin 2x$
   (b) $\tan 2x - x^2 e^{-x}$
   (c) $e^{-x/2}(\sin 2x - \cos 2x)$
   (d) x^{-1}
   (e) $(x^{3/2} - \sin x)$
   (f) $x^{3/2} \sin x$
   (g) $\sec 4x - \csc 2x$
   (h) $5x^{-3} - 2x^{-3/4}$

3. Use differentials to find an approximation for each of the following:

   (a) $\sqrt{15.95}$
   (b) $1/0.98$
   (c) $\sin 31°$ [Remember that you are in the habit of using differentiation formulas which are valid for radian measure only.]
   (d) $e^{-0.15}$
   (e) $\cos 32°$

(f) $(7.93)^{2/3}$

(g) The volume of a sphere of radius 9.87.

(h) tan 47°

(i) cot 47°

4. It would seem reasonable that a value of c which is the average of a and b would be close to the correct value of c to satisfy the mean value theorem. Test this hypothesis in the following cases.

(a) $f(x) = x^2$, $a = n$, $b = n + 1$ for $n = 2, 5, 10$, and 50
(b) $f(x) = x^2$, $a = n$, $b = n + 1$ for a general value of n
(c) $f(x) = x^3$, $a = 4$, $b = 6$
(d) $f(x) = x^3$, $a = n$, $b = n + 1$

5. (a) Find the differential of the area of a square whose sides are of length x.
   (b) Sketch the square and indicate an increment Δx in the length of each side.
   (c) Compare the value of the differential with the actual increase in area of the square geometrically.

6. (a) Find the differential of the area of a circle of radius r.
   (b) Sketch the circle and another circle with radius $r + \Delta r$.
   (c) Compare geometrically the differential with the increase in area shown in your sketch.

7. (a) Find the differential of the volume of a sphere of radius r.
   (b) Sketch the sphere of radius r and indicate the sphere obtained by increasing the radius by an amount Δr.
   (c) Compare the differential of part (a) with the actual increase in volume indicated in the sketch of part (b).

8. (a) Find the differential of the volume of a cube whose side is of length x.
   (b) Sketch the cube and indicate a cube whose side is of length $x + \Delta x$.
   (c) Compare the differential and the increment in the volume of the cube.

9. In the extended mean value theorem let $f(x) = \sin x$ and $g(x) = x$. If $a = 0$ and we let b approach zero, will this give insight into the limit $\lim_{b \to 0} (\sin b/b)$? If so, what insight will it give? You should note that we should not use this reasoning to obtain a value for this limit, since we used this limit in obtaining the derivative of $\sin x$. However, this might be a convenient way of reminding us of the value of this limit.

10. Find the length of the curve $y = 2x^{3/2}$ from $x = 1$ to $x = 4$.

11. Use the method of Example 3.4 to find the length of the curve $y = 2x - 3$ from $x = -2$ to $x = 3$.

12. Find the length of the catenary $y = \cosh x$ from $x = 0$ to $x = 3$.

13. (a) Sketch the curve for which all points have coordinates $(3 \cos \theta, 3 \sin \theta)$ with θ taking on all values from $\theta = 0$ to $\theta = 2\pi$.
   (b) Given that $x = 3 \cos \theta$ and $y = 3 \sin \theta$, show that $dx = -3 \sin \theta \, d\theta$ and $dy = 3 \cos \theta \, d\theta$.
   (c) Using the results of part (b), determine ds, the differential of arc length.

(d) Set up the integral for the arc length of the curve sketched in part (a) using the results of part (c).

(e) Find the length of this curve and check it against the result you would expect from geometry.

14. Find the surface area generated by revolving that portion of $y = 2x^3$ from $x = 0$ $x = 2$ about the x-axis.

15. Find the surface area generated by revolving the curve of Exercise 13 about the x-axis.

16. (a) If A is the area of a circle and C is the circumference of the circle, show that $dA = C\,dr$ where r is the radius of the circle.

(b) If V is the volume of a sphere and S is the surface of the sphere, show that $dV = S\,dr$ where r is the radius of the sphere.

(c) Give a geometric explanation of the results of parts (a) and (b).

17. Find the length of the curve $r = 2 \sin \theta$ in polar coordinates using the method of Example 3.6.

18. Find the length of the curve $r = 2 - 2 \cos \theta$.

19. Find the length of that portion of the curve $r = \sec \theta$ which is between $\theta = 0$ and $\theta = \pi/4$.

S20. Show that the definition of *marginal* given in Chapter II and in many books on economics is equivalent to the differential with $dn = 1$.

B21. Given that the excitation of the retina of the eye at a point r centimeters from the center of the retina is given by the formula $C(r) = r^{-k}$, find the approximate decrease in excitation which would be observed in going from a point 1 mm from the center of the retina to a point 1.1 mm away from the center of the retina. Assume that k is rational and between 1 and 2.

P22. Specifications called for a flywheel with a uniform mass of 2 pounds per square foot of cross section of the wheel and a radius of 4 feet. The wheel that was delivered had the correct density, but had a radius of 4.1 feet. Find the approximate difference in moment of inertia from the specification. If a variance of 1 % is permissible would you accept the wheel that was delivered? Why?

V.4   Extreme Values

It is customary, or so it seems, for people to want to find extreme values. For instance, we would like to find the conditions for making the greatest profit, for obtaining the greatest amount of pleasure, or perhaps we are interested in finding a method of manufacturing that will create the least amount of pollution. While we will not address ourselves to the larger problems mentioned in the preceding sentence, we will concern ourselves in this section with the problem of finding extreme values of a certain smaller class of functions, namely those functions which have continuous derivatives over the domain in which we are interested. As a consequence of our basic results, we will find that we can assist in the determination of extreme points by considering the possibility of differentiating derivatives.

Before we start our discussion we should give a definition of what we mean by extreme values.

Definition 4.1. A value $f(X)$ is said to be a *relative maximum value of $f(x)$ on the interval* $[a, b]$ provided there is some neighborhood D of X such that $f(X) \geq f(x)$ for all x in the intersection of D and $[a, b]$. The value $f(X)$ is said to be an *absolute maximum value of $f(x)$ on* $[a, b]$ if $f(X) \geq f(x)$ for all x in $[a, b]$. A *maximum value* is any value which is either a relative maximum value or an absolute maximum value. A similar definition holds for *relative minimum value*, *absolute minimum value*, and *minimum value*.

Definition 4.2. A value $f(X)$ is said to be an *extreme value of $f(x)$ on the interval* $[a, b]$ if it is either a maximum value or a minimum value.

These definitions make specific reference to a closed interval, but they would apply equally well to any domain of a function $f(x)$ whether it be closed, open, or perhaps the entire number line.

In attempting to find the extreme values of a function on a closed interval, we must consider three cases. If we examine the left hand end point $x = a$ we have no values of x to consider which are less than a. If we examine the right hand end point $x = b$ we have no values of x to consider which are greater than b. Finally, if we consider any other point of the closed interval we have points to the right and to the left of the point under consideration. We shall take these cases in turn.

Case 1. In order to determine whether $x = a$ is an extreme point we would do well to first consider $f'_+(x)$ in the vicinity of $x = a$. Note that we have taken the right hand derivative, and hence have used no points outside $[a, b]$ in obtaining the derivative. If $f'_+(a)$ exists and is positive, we know that the differential quotient $DQf(a, x) > 0$ for all x in some interval $[a, x)$ with $x > a$. Since $x > a$, we know that $(x - a) > 0$, and consequently $DQf(a, x) = (f(x) - f(a))/(x - a) \geq 0$ implies $f(x) \geq f(a)$ for all values of x in this small interval. Therefore, $f(a)$ must be a minimum value on the interval $[a, b]$ by definition. This can be seen easily by examining the sketch in Figure V.8. In view of the fact that we are ignoring all points to the left of $x = a$ we have the point $(a, f(a))$ as a minimum point provided the slope at this point is positive. Frequently a sketch of this simple variety will be of assistance in remembering the interpretation of the right hand derivatives as it relates to the possibility of an extreme value at the left hand end point of a closed interval.

In similar fashion we could show that if $f'_+(a) < 0$ we would have a maximum value $f(a)$. Since the denominator of the differential quotient

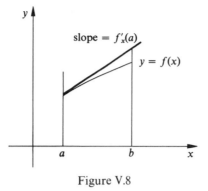

Figure V.8

must be positive the numerator must be negative. Therefore we have $f(x) \le f(a)$ and we have satisfied the definition for a relative maximum value. If $f'_+(a)$ fails to exist or if it is zero the discussion above does not hold. In this instance it would be necessary to make an individual examination for each case.

Case 2. We can consider the point $(b, f(b))$ at the right hand end of the interval in a manner similar to that used in Case 1. In this case we would need to use the left hand derivative, and in the differential quotient we have $(x - b) < 0$ for all permissible values of x. Therefore, if $f'_-(b)$ is positive we have a maximum point and if $f'_-(b)$ is negative we have a minimum point. It is again possible to draw a sketch similar to that of Figure V.8 to demonstrate that these results are reasonable. In a manner similar to that of Case 1 we observe that we will have to investigate each case individually if the left hand derivative is zero or fails to exist at $x = b$.

Case 3. We now consider the interior points of the interval $[a, b]$. Let $x = c$ be an interior point of this interval. If $f'(x)$ is continuous in the vicinity of $x = c$ and if $f'(c) > 0$, then there is a neighborhood of c such that $f'(x) > 0$ for all x in this neighborhood. If x is in this neighborhood, we then have $f(x) - f(c) = f'(d)(x - c)$ for some value d between x and c, and hence in the neighborhood. Therefore $f(x) < f(c)$ if $x < c$ and $f(x) > f(c)$ if $x > c$. Since we have values of x which are smaller than c and some which are larger than c in this neighborhood, we see that there are values of x in the neighborhood for which $f(x) < f(c)$ and values for which $f(x) > f(c)$. As a result $f(c)$ cannot be an extreme value. A similar argument would show that $f(c)$ is not an extreme value if $f'(c) < 0$ and $f'(x)$ is continuous in a neighborhood of $x = c$. Therefore the only possible interior extreme points occur when $f'(x)$ is not continuous (which would include places where it does not exist) or when $f'(x) = 0$. We can treat these two cases in similar fashion. If $(c, f(c))$ is a point for which either $f'(c) = 0$ or else $f'(c)$ fails to exist but $f(c)$ exists, consider a neighborhood D of c in which $f'(x)$ exists and is continuous except perhaps at $x = c$, and such that $f'(x) \ne 0$ if x

is in D but $x \neq c$. We have four possibilities. We will assume that x is in D throughout each of the following.

(i) $f'(x) > 0$ if $x < c$ and $f'(x) < 0$ if $x > c$. Using the mean value theorem we have $f(x) - f(c) = f'(d)(x - c)$ for a value d between x and c. If $x < c$ then $f'(d)$ is positive and the right side of the expression for the mean value theorem is negative and we have $f(x) < f(c)$. On the other hand if $x > c$ then $f'(d)$ is negative and the right side of this expression is still negative. Hence, we have shown that $f(c) > f(x)$ for any x in D, and consequently $(c, f(c))$ is a maximum point.

(ii) $f'(x) > 0$ if $x < c$ and $f'(x) > 0$ if $x > c$. In this case we observe that $f'(d)$ in our expression of the mean value theorem is always positive, and therefore we have a case similar to the case in which $f'(x)$ is positive throughout D. As we saw before, this does not yield an extreme value.

(iii) $f'(x) < 0$ if $x < c$ and $f'(x) < 0$ if $x > c$. This is similar to (ii) with the signs reversed and again fails to yield an extreme value.

(iv) $f'(x) < 0$ if $x < c$ and x is in D and $f'(x) > 0$ if $x > c$ and x is in D. This case resembles (i) with the signs reversed, and we can use an argument modeled after that of (i). This argument will show us that in this case we have a minimum value.

You will probably want to read the development of these three cases including the subcases of Case 3 more than once. It would help at this point to give an example.

EXAMPLE 4.1. Find the extreme points of $f(x) = x^3 - 3x^2 - 9x + 7$ in the interval $[-2, 8]$.

Solution. It is apparent that we already have two candidates for extreme points, namely the points for which $x = -2$ and $x = 8$, since we always have to consider the end points of a closed interval if a closed interval is specified. We can look for further candidates for the honor of being extreme points by examining the derivative $f'(x) = 3x^2 - 6x - 9$. The derivative is a polynomial and we see that there are no points of discontinuity of the derivative in the interval $[-2, 8]$. Hence we obtain no additional candidates (usually called *critical points*) through having points of discontinuity. Any additional critical points must then appear as zeros of $f'(x)$. We obtain these by considering

$$3x^2 - 6x - 9 = 3(x^2 - 2x - 3) = 3(x + 1)(x - 3) = 0.$$

It is apparent that the two additional critical points occur when $x = -1$ and $x = +3$. (If we had not been able to factor this quadratic, we could have used the quadratic formula or completing the square. Do not count on having all derivatives factor this nicely.) We now have a list of four critical points, and we can handle them as indicated.

$x = -2, f'(-2) = +15$. By Case 1 this is a minimum point $(-2, +5)$.
$x = -1$, the derivative is positive to the left of -1 and negative to the right. Therefore this is a maximum point $(-1, +12)$ by (i).
$x = +3$, the derivative is negative to the left of this point and positive to the right. Therefore this is a minimum point $(3, -20)$ by (iv).
$x = 8, f'(8) = 135$. By Case 2 this is a maximum point $(8, 255)$.

You will note that we are able to handle each situation with the detailed analysis that we carried through. However, it is difficult to keep every case in mind, and therefore an easier way of remembering all of these details is recommended. We can assist with a sketch as shown in Figure V.9. Note that we have indicated the critical values, -2, -1, $+3$, and $+8$ in the same order in which we would find them on the number line. We have next indicated the sign of the derivative in the intervals between successive critical points. Using these signs we have sketched in lines with the appropriate slopes. The status of the interior points is obvious. The status of the end points of the interval is also easily ascertained by remembering that we are not going to consider any values of x to the left of $x = -2$ nor any values of x to the right of $x = 8$. If you check carefully, you will see that we have considered the various cases of our earlier discussion, and each one is in agreement with a corresponding portion of a sketch similar to that shown in Figure V.9. In determining the signs of the derivative in the various intervals one can observe the fact that the factor $(x + 1)$ of $f'(x)$ is positive to the right of $x = -1$ and negative to the left and $(x - 3)$ is positive to the right of $x = 3$ and negative to the left. Therefore if we are left of $x = -1$ both factors are negative and the product must be positive. If we are between $x = -1$ and $x = 3$ one factor is positive and the other negative, thus giving a negative derivative. If we are to the right of $x = 3$, both factors are positive and hence the derivative is positive. On the other hand since $f'(x)$ is continuous in this case we can note that a continuous function changes sign only at the points where the function is zero. Hence between the zeros of $f'(x)$, or between the critical points here, the function must be of one sign and we only need to try such values as $x = -2$, $x = 0$, and $x = 4$ to determine the signs in the intervals in which we are interested.

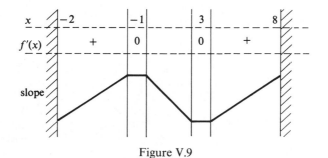

Figure V.9

An examination of the values that we have obtained for minima and maxima shows that the absolute minimum of this interval is -20 when $x = 3$, and hence $(-2, 5)$ is a relative minimum. In similar manner we have $(8, 255)$ as the absolute maximum point and $(-1, 12)$ is a relative maximum. If the interval for consideration had been $[-2, 3]$ the point $(-1, 12)$ would have been an absolute maximum. Also note that if we had considered the interval $[-2, 3]$ the derivative at the right hand end point would be zero, and it would be necessary to examine the interval immediately to the left of $x = 3$ in order to determine that $(3, -2)$ is a minimum point. It is necessary to check each critical point to determine its nature. This is clearly demonstrated in the following Example.

EXAMPLE 4.2. Find the extreme points of $f(x) = \tan x - x$.

Solution. In this case we have no end points, and we only need to consider the situation which holds in Case 3. We note that $f'(x) = \sec^2 x - 1$, and this is continuous except at points where x is an odd multiple of $(\pi/2)$ where neither $\tan x$ nor $f'(x)$ are defined. Since $f(x)$ is not defined we could hardly consider these values (which do not exist) as extreme values. Since these points are to be excluded, we need only consider the points where $f'(x) = 0$, and these points occur at all of the integral multiples of π. The question which now needs an answer involves the nature of these points which can be written as $(n\pi)$ with the understanding that n can be any integer. However, a little thought will tell us that $\sec^2 x$ is never smaller than one, and consequently $f'(x) = \sec^2 x - 1$ is never negative. Thus, if $f'(x)$ is not zero, it must be positive. We can sketch this in a manner similar to that used in Example 4.1, and we will have Figure V.10. We have indicated the breaks that occur at the odd multiples of $(\pi/2)$. It is apparent that this function has no extreme points, but rather points where the function stops momentarily in its otherwise always increasing movement.

EXAMPLE 4.3. Find the extreme points of $f(x) = x^3 - 12x$ in the interval $[-1, 2]$.

Solution. Here we again have end points, and we know at once that $x = -1$ and $x = 2$ will give critical points. If we differentiate we have $f'(x) = 3x^2 - 12 = 3(x + 2)(x - 2)$. This is certainly continuous and it is zero when $x = -2$ and when $x = +2$. We note that $x = -2$ is not in the interval

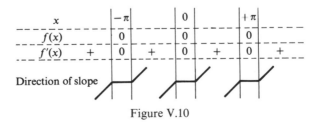

Figure V.10

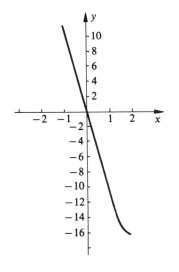

Figure V.11

in which we are interested here, and $x = +2$ is already a critical point. If we examine $f'(x)$, we see that $(x + 2)$ is positive throughout this interval, and $(x - 2)$ is always negative in this interval except for the point where $x = 2$. In this latter case $(x - 2) = 0$. Therefore in this interval we have $f'(x)$ is either negative or zero. The actual graph is shown in Figure V.11. From this we see that the point $(-1, 11)$ is a maximum point, as we could have foretold using the discussion of Case 1. We also see that $(2, -16)$ is a minimum point despite the fact that we have $f'(x) = 0$ at an end point of the interval. The fact that $f'(x)$ is negative in the vicinity of the end point is sufficient, for the function can only decrease as it moves toward the right or increase as it moves toward the left in the interval immediately to the left of $x = 2$.

It should be apparent by this time that we must consider each request for extreme points as an individual problem with an analysis for the particular problem at hand. It is true that we have been able to set up some general rules, but they cannot be applied without some care. It is also true that a little thought in the form of a sketch similar to those of Figures V.9 and V.10 can be helpful.

Our consideration of the character of the interior critical points can be somewhat simplified under certain conditions. If $x = c$ gives a critical point $(c, f(c))$ which is interior to the interval under consideration, we observe that this is a maximum point if and only if $f'(x)$ is positive immediately to the left of $x = c$ and negative immediately to the right. Thus, we have a maximum point only if $f'(x)$ is decreasing from a positive value through zero to a negative value as x increases through the value $x = c$. Therefore, the rate of change of $f'(x)$ must be negative if we are to have a maximum point. This suggests looking at the derivative of $f'(x)$. If $D_x f'(x)$ exists in the neighbor-hood of c, it must be negative if we have maximum point at $x = c$. In similar manner if $D_x f'(x)$ is positive in a neighborhood of $x = c$ we see that $f'(x)$ is an increasing function. Consequently if $f'(c) = 0$ we must have $f'(x)$ negative to the left of $x = c$ and positive to the right. These are precisely the conditions required for a minimum point. This appears to give us a very simple test for extreme values, for if we have an interior point and the deriva-tive of the derivative is negative when the derivative is zero, we have a maxi-mum value, and if the derivative of the derivative is positive when the deriva-tive is zero, we have a minimum point. We can test this by observing that $D_x f'(x)$ in Example 4.1 is given by $6x - 6$. For $x = -1$ this is negative and we have a maximum value while for $x = 3$ this is positive and we have a minimum value. It works! We could not use it on the end points of the interval, but those didn't require as much work anyway. In Example 4.2 we observe that $D_x f'(x) = 2 \sec^2 x \tan x$ (making use of the product rule for differentiation), and this is zero for all integral multiples of π. This is neither positive nor negative, but then we had no extreme points, so this does not appear to be a contradiction. In point of fact we should do further checking if this derivative of a derivative is zero, for this apparent lack of information is just that—lack of information. In this instance we would go back to the consideration we had earlier and look at the sign of the derivative on each side of the critical point. In Example 4.3 we had no interior critical points so this discussion would not apply.

This derivative of a derivative is called a *second derivative* and is usually denoted by $D_x^2 f(x)$ or $f''(x)$. The second derivative counterpart for

$$\frac{df}{dx} = f'(x) \quad \text{would be} \quad f''(x) = \frac{d\left(\frac{df}{dx}\right)}{dx} = \frac{d^2 f}{dx^2}.$$

Note that the last expression is not capable of simple interpretation in terms of dx and $df(x, dx)$. It is not difficult to see how one could obtain still more derivatives, and so indicate by larger *exponents* of D or by more primes on

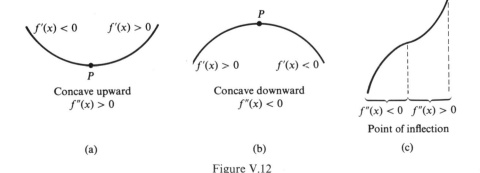

Figure V.12

the f. In the case of the n-th *derivatives*, the notation would be specifically

$$D_x^n f(x), \qquad f^{(n)}(x) \quad \text{or} \quad \frac{d^n f}{dx^n}.$$

EXAMPLE 4.4. Find the first five derivatives of $f(x) = x^7 + 3x^3 - e^{-2x}$.

Solution.

$$D_x f(x) = f'(x) = \frac{df}{dx} = 7x^6 + 9x^2 + 2e^{-2x}.$$

$$D_x^2 f(x) = f''(x) = \frac{d^2 f}{dx^2} = 42x^5 + 18x - 4e^{-2x}.$$

$$D_x^3 f(x) = f'''(x) = \frac{d^3 f}{dx^3} = 210x^4 + 18 + 8e^{-2x}.$$

$$D_x^4 f(x) = f^{iv}(x) = \frac{d^4 f}{dx^4} = 840x^3 - 16e^{-2x}.$$

$$D_x^5 f(x) = f^v(x) = \frac{d^5 f}{dx^5} = 2520x^2 + 32e^{-2x}.$$

Note the use of various notations for the derivative. In the notation $f^{(n)}(x)$ either the Roman numerals are used, usually not enclosed in parentheses, or Arabic numerals are used enclosed in parentheses to denote the fact that these are not powers of a function.

We will now take another look at the second derivative to obtain a geometric interpretation. If $f''(x) > 0$, this means that the first derivative is increasing, and we have the geometric situation in which the slope is increasing, as illustrated in Figure V.12(a). In this situation we say that the curve is *concave upward*. On the other hand if $f''(x) < 0$, then the slope is decreasing, and we have the situation shown in Figure V.12(b), in which case the curve is said to be *concave downward*. The concavity indicates the way the slope is changing. Note that a critical point will be a minimum if the curve is concave upward at the critical point, while it will be a maximum if the curve is concave downward at the critical point. If $f''(x) = 0$, the curve might be concave upward or downward, provided the second derivative is either positive on both sides of the point at which $f''(x) = 0$, or is negative on both sides respectively. On the other hand, it is perfectly possible that the second derivative changes from positive to negative or from negative to positive at the point where $f''(x) = 0$. Such a point is called a *point of inflection*, and geometrically it is a point at which the function changes its direction of concavity as in Figure V.12(c). One can test a value which makes $f''(x) = 0$ by noting whether the second derivative changes sign as x passes through the point, is positive on both sides, or is negative on both sides. A consideration similar to that used in showing that the second derivative will check the performance of the first derivative would show that the third derivative could be used in many cases to check the performance of the second derivative. There is no common geometric interpretation of the third derivative so we will stop the discussion at this point.

EXAMPLE 4.5. Find the extreme points, points of inflection, and intervals in which the curve is concave upward and the intervals of downward concavity for $f(x) = 3x^5 + 10x^3 - 45x + 7$. Sketch the graph.

Solution. $f'(x) = 15x^4 + 30x^2 - 45 = 15(x^2 + 3)(x^2 - 1)$. This is zero at $x = +1$ and $x = -1$ and at no other real points. We have $f''(x) = 60x^3 + 60x = 60x(x^2 + 1)$. This is positive when $x > 0$ and negative when $x < 0$. In particular $f''(+1) = 120 > 0$ and $f''(-1) = -120 < 0$. Hence the point $(1, -25)$ is a relative minimum of $f(x)$ and the point $(-1, +39)$ is a relative maximum. Also $f''(x) = 0$ only if $x = 0$, and the second derivative changes sign at $x = 0$. Hence $(0, 7)$ is a point of inflection. The graph is concave upward when $x > 0$ since $f''(x)$ is positive for $x > 0$. Similarly the curve is concave downward when $x < 0$. The slope is negative in the interval $(-1, 1)$ and positive elsewhere. All of this information is indicated in the graph of Figure V.13. Now that we have the information concerning slope, points of inflection, and extreme values it is relatively easy to sketch the graph. It would be well to utilize such information in sketching curves in the future.

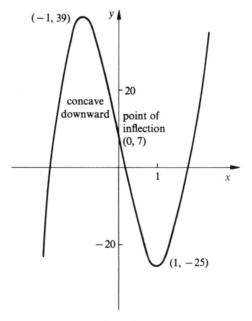

Figure V.13

We can emphasize many of the things we have discussed in this section with one more example.

EXAMPLE 4.6. Find the extreme points of the functions x^3, x^4, and x^5.

Solution. If $f(x) = x^3$, then $f'(x) = 3x^2$ and $f''(x) = 6x$. The only critical point occurs when $x = 0$, and here we have $f''(0) = 0$. Consequently we obtain no information concerning the nature of the critical point $(0, 0)$ from the second derivative. However, we observe that $f'(x)$ can never be negative, and consequently the slope is positive both to the right and left of the critical point. Hence, we do not have an extreme value. On the other hand $f''(x)$ does change sign as x goes from negative to positive, and therefore we change the direction of concavity. Thus $(0, 0)$ is a point of inflection. Since $f'''(x) = 6$, we have $f'''(0) = 6$ and therefore the second derivative has to be increasing at zero. Since $f''(0) = 0$, the increasing nature of this derivative must mean that it started out as negative and increased to become positive, again indicating a point of inflection.

If $f(x) = x^4$, then $f'(x) = 4x^3$ and $f''(x) = 12x^2$. Again we have $(0, 0)$ as the only critical point, and again we have no information coming from the second derivative test since $f''(0) = 0$. However, we can check $f'(x)$ and observe that it changes sign as x passes through 0 while $f''(x)$ can never be negative. A more careful examination reveals that $(0, 0)$ is a minimum point, but it is not a point of inflection. Here we would have $f'''(x) = 24x$, and since $f'''(0) = 0$ we obtain no aid from the third derivative in finding out whether $(0, 0)$ is a point of inflection.

If $f(x) = x^5$, we have $f'(x) = 5x^4$, $f''(x) = 20x^3$, and $f'''(x) = 60x^2$. Again $(0, 0)$ is the only critical point, and here each of the first three derivatives is zero at the origin. Again we have no information concerning the nature of this point unless we go back and check what happens on each side of the origin. In this case the origin is again a point of inflection. From a comparison of these cases you should observe that the zero values yield no information. For this reason you should remember the first method we had in order to have that to fall back on when needed.

EXERCISES

1. Find the extreme values of each of the following functions:

   (a) $x^2 - x$
   (b) $x^3 - 3x$
   (c) $x^4 - 2x^2$
   (d) $3x^5 - 5x^3$
   (e) $x^5 - 5x$
   (f) $3x^5 - 25x^3 + 60x - 4$
   (g) $3x^5 + 25x^3 + 60x + 4$
   (h) $x^3 + x + 1$

2. Find the extreme values of each of the following functions in the interval indicated.

   (a) $\sin x$ in the interval $[-2\pi, 2\pi]$.
   (b) $x + \cos x$ in the interval $[-4, 7]$.
   (c) xe^{-x} in the interval $[-3, 3]$.
   (d) $4x^6 - 6x^4 + 2$ in the interval $[-100, 100]$.
   (e) $3x^5 - 5x^3 - 17$ in the interval $[-6, 7]$.

3. Find the extreme values, the points of inflection, and the intervals in which the curve is concave upward and concave downward. Sketch the graph.

   (a) $f(x) = x^3 - 6x^2 + 11x - 6$ in the interval $[-2, 4]$.
   (b) $f(x) = x + (1/x)$ in the interval $[-4, 4]$.
   (c) $f(x) = x^2 e^{-x}$ in the interval $[-3, 3]$.
   (d) $f(x) = e^x \sin x$ in the interval $[-2\pi, 2\pi]$.
   (e) $f(x) = x + e^{-x}$ in the interval $[-3, 3]$.
   (f) $f(x) = x^3$ in the interval $[-2, 2]$.
   (g) $f(x) = \sqrt{x}$ in the interval $[0, 7]$.

4. Find all of the information you can about each of the following functions and sketch the graph of each.

   (a) $y = 2x/(1 + x^2)$
   (b) $y = \sin x + \cos x$
   (c) $y = (x - 3)^4 - 2(x - 3)^2$
   (d) $y = (x + 2)/(x + 1)$
   (e) $y = (8x)^{1/3}$
   (f) $y = x + (1/x^2)$

5. Find the second derivative of each of the following functions:

   (a) $\tan x$
   (b) $x^3 e^{2x}$
   (c) x^n
   (d) $\sec x$
   (e) $1/x$
   (f) $\cot x$
   (g) $\csc x$
   (h) $\sin 3x$

6. Find the first 5 derivatives of each of the following functions. See whether you can guess a formula for the n-th derivative. If you can guess a formula for the n-th derivative, prove that it is the correct formula using mathematical induction.

   (a) $f(x) = e^{-x}$
   (b) $f(x) = \sin 2x$
   (c) $f(x) = x^3$
   (d) $f(x) = (x + 2)^3/x^3$
   (e) $f(x) = 1/x$
   (f) $f(x) = x^{1/2}$
   (g) $f(x) = x^{10}$
   (h) $f(x) = xe^x$

7. (a) Find the 2nd and 3rd derivatives of x^2.
   (b) Find the 3rd and 4th derivatives of x^3.
   (c) Find the 4th and 5th derivatives of x^4.
   (d) Find the n-th and $(n + 1)$-st derivatives of x^n.

8. (a) Find an extreme point of $f(x) = x^{2/3}$.
   (b) Sketch the graph of $f(x) = x^{2/3}$.
   (c) Show that only the odd derivatives (that is the 1-st, 3-rd, 5-th, etc.) of $f(x)$ $= x^{2/3}$ are negative in some places and positive in others.

9. Find the extreme points of $f(x) = |x|$. Explain carefully why some of the methods described in this section fail to work for this function.

10. Let $f(x) = x^n$ where n is a positive integer.

   (a) Show that $f(x)$ has a minimum point if and only if n is even.
   (b) Show that $f(x)$ has a point of inflection if and only if n is odd.
   (c) Show that $f(x)$ never has more than one minimum point and never more than one point of inflection.
   (d) Show that $f(x)$ never has a maximum value unless the interval is restricted.

P11. If $s(t)$ is the location of a particle t seconds after a stopwatch is started, show that the velocity of the particle is given by $s'(t)$ and the acceleration is given by $s''(t)$.

P12. If a particle moves in such a manner that the distance from a fixed point is given by the relation $s(t) = 10 \sin t$, find the velocity of the particle and the acceleration of the particle. Find the location, velocity and acceleration when $t = 0, 1, \pi$, and $3\pi/2$.

P13. The specific weight of water at $t°$ centigrade is given with relatively little error by the equation $w = 1 + (5.3 \times 10^{-5})t - (6.53 \times 10^{-6})t^2 + (1.4 \times 10^{-8})t^3$. Find the temperature at which water has the greatest specific weight.

P14. The rate of a certain type of auto-catalytic reaction can be shown to obey the rule $v = kx(10 - x)$ where k is a constant dependent upon the substances involved and x is the amount of product formed from an original source of 10 grams. For what value of x is the rate, v, a maximum? [Note that the rate of reaction depends on the product of the amount of x formed and the amount of the original substance remaining.]

B15. It has been verified by use of X-rays that the diameter of the windpipe shrinks when one coughs. If we let r_0 be the radius of the windpipe when there is no differential pressure and let r be the radius when there is a differential pressure P, it has been shown that $(r_0 - r) = aP$ is a very good approximation to the actual observed performance provided P is not greater than $(r_0/2a)$. The constant a is a constant of proportionality. The resistance to flow for an ideal fluid is given by the relation $R = k/r^4$ where k is another constant of proportionality. Finally, the rate of air flow through the windpipe is given by the pressure divided by the resistance to flow.

(a) Show that the rate of air flow through the windpipe is given by the velocity of air flow multiplied by the cross sectional area of the windpipe.

(b) Find an expression for the velocity of flow as a function of the radius. This expression will contain the constants r_0, a, k, and π.

(c) Find the radius which produces the maximum velocity of flow through the windpipe. (It is worth noting that the increase in velocity due to coughing has the beneficial function of providing a force which can dislodge unwanted items which have found their way into the windpipe.)

S16. Assume that the cost per day of making n units is given by the equation $C(n) = 2000 + 10n - 0.01n^2 + 0.000001n^3$ dollars.

(a) When is the marginal cost of production an extreme value?

(b) Is the extreme marginal cost a maximum or a minimum?

(c) Is there an extreme value for the cost of production?

(d) What factors other than those stated above would you wish to know if you were planning the production rate for this firm?

(e) What conclusions concerning optimal production rate could you draw from the information given here?

S17. A plant is required to produce 50,000 gizmos per year. It is customary to make a batch of N gizmos and then keep them in stock until they run out at which time another batch of N gizmos is to be made. It costs $1,000 to get the plant ready for gizmo production for each batch that is made. The cost of production is $8.95 per gizmo, and there is no particular cost advantage in unit cost in making large or small batches. The cost of keeping gizmos in inventory is $0.75 per gizmo per year. The usage rate throughout the year is constant.

(a) Find the cost of manufacturing a batch of N gizmos.

(b) Find the storage cost connected with a batch of N gizmos assuming that these will be consumed in the $(N/50000)$-th part of a year, and on the average each gizmo is on the shelf for only one half of this period. (Why is this latter assumption realistic?)

(c) Find the total cost of a year's production of gizmos including inventory costs.

(d) Find the value of N which minimizes the annual cost of producing and stocking gizmos. If the value of N is such that $(50000/N)$ is not an integer, find the value of N which makes this an integer and which would give the minimum cost assuming that $(50000/N)$ must be an integer.

18. A tin can is to be made in the shape of a right circular cylinder.

(a) Find the ratio of the radius of the base and the height of the can for which the total amount of metal is minimum for a given volume. Assume that the ends and the sides are made of metal of the same thickness.

(b) Repeat part (a) if the bottom and top are to be made of metal which is twice as thick as that used for the sides.

(c) Repeat part (a) if the bottom is to be made of metal twice as thick as the metal used for the top and sides.

V.5 Newton's Method

We frequently need the values of x for which a function $f(x)$ is zero. This was the case in the last section when it was necessary to find the zeros of $f'(x)$ in order to find critical points. If the function is a polynomial we may be able to find linear factors of the polynomial, and these will lead us to at least some zeros. If $f(x)$ is a quadratic function we even have a formula for its solution. In some cases we can use other information as we did with the trigonometric functions in Example (V.4.2). None of these comments suggest a method which can be applied in the more general cases. The fact that no such general method exists was proven by E. *Galois* (1811–1832) when he showed that there can never be a formula for solving polynomial equations of degree greater than four by exact methods. It will be the purpose of this section to develop a method for obtaining very good approximations to the zeros. The method to be investigated here, that is *Newton's method*, will seldom give exact values but will usually give results with more than sufficient accuracy for any practical requirements.

We will start toward our solution by making a *guess* concerning what the correct value might be. Hopefully this will be a fairly good guess, but nevertheless it is a guess. It may sound better to call it an approximation, and this we will do, but we make no pretense that our first approximation is *the* answer. We can make this first estimate by drawing a graph of the function $y = f(x)$, for we observe that if $y = 0$, then $f(x) = 0$. However $y = 0$ for points on the x-axis, and consequently we only need locate with reasonable accuracy the values of x for which the curve crosses the x-axis to have good first approximations. As a matter of convenience we will designate this starting approximation by x_0. We would like to be able to find the exact value for which $f(x) = 0$, and while we will probably never have this exactly we can at least talk about it. We will designate this value by X. Consequently $f(X) = 0$. By the mean value theorem we have $f(X) - f(x_0) = f'(c)(X - x_0)$ for some value c between x_0 and X. If $f'(x) \neq 0$ for values of x in the neighborhood of X we can write

$$X = x_0 - \frac{f(x_0)}{f'(c)} \quad \text{since} \quad f(X) = 0. \tag{V.5.1}$$

Just as in the case of approximation using the differential, we do not know the value of c. If x_0 and X are fairly close together and if $f'(x)$ is not changing rapidly in this interval, it would seem reasonable to replace c by x_0, realizing that this will give us a value other than X. We can call the value that we get this way x_1, and hope that it is a better approximation than x_0. Thus, we would have the relation

$$x_1 = x_0 - \frac{f(x_0)}{f'(x_0)}. \tag{V.5.2}$$

In fact this is closely related to the method we used for obtaining approximations using differentials. Just as we found those approximations to be fairly good provided we started with a reasonable choice, so the usual case will have x_1 substantially closer to the desired result, X, than was x_0.

If this method will give us a value x_1 which is an improvement upon our original estimate x_0, it would seem reasonable that we should then use x_1 as a good estimate and determine a value x_2 by similar reasoning. Of course this method can be continued and we could write

$$x_{k+1} = x_k - \frac{f(x_k)}{f'(x_k)} \qquad (V.5.3)$$

for the $(k + 1)$-th approximation based on the k-th approximation. Using this relation we should be able to start with x_0 and obtain a sequence of values $\{x_0, x_1, x_2, \ldots, x_n\}$, each value being closer to the desired value X than its predecessors. In order to be certain that our wishful thinking (in replacing c by x_0) is correct, we must investigate further, but before doing so we give an example to illustrate the method we have proposed. It is this method which is known as *Newton's method*, and the fact that Newton's name is used indicates that this is a method that has a long history.

EXAMPLE 5.1. Use Newton's method to find the real root of $x^3 + x - 1 = 0$ correct to three decimal places.

Solution. Since $f(x) = x^3 + x - 1$, we have $f'(x) = 3x^2 + 1$. Our first task is that of locating a first approximation. In order to do this we observe that $f'(x)$ is always at least as large as one, and therefore the curve always has a positive slope. Furthermore, we observe that $f(0) = -1$ and $f(1) = +1$. We have shown these values on Figure V.14 to emphasize the point that this assures us of a zero between $x = 0$ and $x = 1$. (In view of the fact that the slope is always positive we are assured that there is no other real root, a fact that verifies the statement of the problem to the effect that we want *the* real root.)

We can make one of two choices here. We can either let $x_0 = 0$ or let $x_0 = 1$. Either one would serve us in this case. In order to demonstrate this, we will consider each possibility.

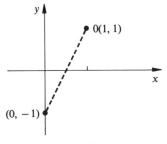

Figure V.14

Case 1. We have $x_0 = 0$. Then by (V.5.3) we have $x_1 = 0 - (f(0)/f'(0)) = 1/1 = 1$. Now we have $x_2 = 1 - (f(1)/f'(1)) = 1 - 1/4 = 0.75$. We can go further and obtain the next step $x_3 = 0.75 - (0.171875/2.6875) = 0.686047$ and then $x_4 = 0.682340$, $x_5 = 0.682328$, $x_6 = 0.682328$. At this point we see that the values appear to be repeating, at least to six decimal places. This is far more than sufficient to assure us that we have an accuracy to three decimal places. In fact, the change from x_4 to x_5 would have so little effect on the third decimal place that we could have stopped one step sooner. The answer in this case is 0.682.

Case 2. We have $x_0 = 1$, and we can proceed as above to obtain $x_1 = 0.75$, $x_2 = 0.686047$, $x_3 = 0.682340$. We are already at a point where we have the final value for the first three decimal places, namely 0.682.

It is clear that the second case was faster in this instance, but also that either one could be used. Since no one is fond of putting in unnecessary steps, it would be interesting to learn whether we could have told in advance that the second case would require less work. This gives us an additional incentive for taking another look at the development of Newton's method.

We now return to our development of Newton's method. In Equation (V.5.3) we had obtained an equation which gives us a value of x_{k+1} provided we have a value for x_k. Of course this was based on the fact that $f'(x) \neq 0$ in a neighborhood of X which includes all of the values of x_k, for otherwise we would have the problem of zero denominators. We have not resolved the general question of whether the successive values x_k will converge toward X and a limit. In order to do this we would like to know whether $|X - x_k|$ approaches zero as k increases. We will start by considering whether $|X - x_1|$ is smaller than $|X - x_0|$. From equations (V.5.2) and (V.5.1) we have

$$(x_1 - x_0) = -\frac{f(x_0)}{f'(x_0)} \quad \text{and} \quad (X - x_0) = -\frac{f(x_0)}{f'(c)}.$$

Upon dividing the first of these by the second, we obtain with the aid of some algebra

$$\frac{X - x_1}{X - x_0} = \frac{(X - x_0) - (x_1 - x_0)}{X - x_0} = \frac{f'(x_0) - f'(c)}{f'(x_0)}.$$

If $f'(x)$ has a continuous derivative in the neighborhood of X, the neighborhood under consideration, we can apply the mean value theorem and have $f'(x_0) - f'(c) = f''(x^*)(x_0 - c)$ for some value x^* between x_0 and c. With this result we can write

$$\frac{X - x_1}{X - x_0} = \frac{f''(x^*)}{f'(x_0)}(x_0 - c). \tag{V.5.4}$$

Note that c was between x_0 and X, and consequently x^* is between x_0 and X. This is a reminder of the reason we should be concerned about the behavior of $f(x)$ and its derivatives in the neighborhood of X. Since $f'(x_0) \neq 0$, we can certainly find some value x_0 as a starting value such that the right side of (V.5.4), and consequently the left side, is less than one in absolute value. However, this is equivalent to saying that $|X - x_1| < |X - x_0|$. Therefore, we have x_1 closer to the correct value than was x_0. If we now consider x_1 to be a new x_0 then x_2 is closer to X than was x_1 and (V.5.4) must continue to hold, thus making the next term in our sequence even nearer the value X. By continuing this argument we observe that we are assured of convergence provided we start with a value x_0 sufficiently close to X.

In order to consider whether we should start with a value x_0 which is larger than X or a value which is smaller than X, we can observe that in the neighborhood of X we have asked that $f'(x)$ not be zero. Since $f'(x)$ must be continuous, $f'(x)$ will not change sign in this neighborhood. If $f''(x)$ does not change sign, then the sign of the right hand side of (V.5.4) will depend upon the sign of $(x_0 - c)$. However, $(x_0 - c)$ has the same sign as $(x_0 - X)$. If the right hand side of (V.5.4) is positive and less than one, we have the fact that x_1 is between x_0 and X. Furthermore, by our assumptions concerning the signs of the first and second derivatives, all successive iterations will produce results which are on the same side of X and successively closer to X. On the other hand, if the right hand side of (V.5.4) is negative, x_1 will be on the opposite side of X from x_0. After the first iteration, that is after obtaining the value x_1, we can expect all remaining iterates to be on the same side of X.

We should clear up one final point. If $f''(x)$ is zero in the vicinity of X, it is possible that we have a point of inflection. Should the point of inflection be the point $(X, f(X))$, we observe that the right hand side of (V.5.4) will have the same sign on either side of X. If this sign is positive, we can select an appropriate starting point and be assured of convergence. However, if this sign is negative we see that the values x_k will alternate with one being on one side of X and the next one being on the opposite side. In this instance it is perfectly possible that one cannot find a starting value that will give a converging sequence of iterates. We will illustrate one such instance in Example 5.2. Before going on to this Example, however, it would be well to indicate the possible problems that would be produced if we had $f'(x) = 0$ in the neighborhood of X. If $f'(x) = 0$ at some point other than X and if this point were some x_k, we would have $f(x_k) \neq 0$ and then attempt to divide a non-zero value by zero. On the other hand if $f'(X) = 0$ it is possible that the method might still work, since we are not apt to ever reach the value X, and presumably the sign of $f'(x)$ would not be changing in the interval encompassed by the x_k's. While the situations in which you seek a zero of $f(x)$ and either the first or second derivative is zero in the immediate vicinity of the correct value are very rare, it is well to be alert to the possibility that such situations can exist.

EXAMPLE 5.2. Use Newton's method to find the solution of $x^{1/3} = 0$.

Solution. In this instance the solution is obvious, namely $x = 0$, but we are instructed to use Newton's method. Although we know the value we seek, let us start with something nearby. We might start with $x = 1$. On taking the required derivatives we have

$$x_{k+1} = x_k - \frac{x_k^{1/3}}{(\frac{1}{3})x_k^{-2/3}}.$$

If we let $x_0 = 1$, we have $x_1 = 1 - 1/(1/3) = 1 - 3 = -2$. This is a very sad state of affairs, for we have x_1 on the opposite side of the value we know to be correct, and we also find that it is twice as far from the correct value as the value we started with. We might have started with $x_0 = 0.001$ to see whether this would help. In this case we would have $x_1 = 0.001 - 0.1/(1/3)(100) = 0.001 - 0.003 = -0.002$. Again we have results as we had them before. Something is clearly amiss in this situation. If we simplify our expression for x_{k+1}, we see that we have

$$x_{k+1} = x_k - (x_k^{1/3})(3x_k^{+2/3}) = x_k - 3x_k = -2x_k$$

and we could have predicted these results from the start. This is certainly a case in which Newton's method will not work. Here we have

$$\frac{f''(x^*)}{f'(x_0)} = \frac{(-\frac{2}{9})(x^*)^{-5/3}}{(\frac{1}{3})(x_0)^{-2/3}} = \frac{-2x_0^{2/3}}{3(x^*)^{5/3}} = -\frac{2}{3}\left(\frac{x_0}{x^*}\right)^{2/3}\left(\frac{1}{x^*}\right).$$

Since x^* is nearer 0 then x_0, we see that the denominator becomes very small and hence the fraction becomes very large. This will be accentuated as we get closer to the value zero. Therefore, it should not be surprising that this method fails in this case. The negative sign is responsible for the fact that we are getting successive values on alternate sides of zero. It is disappointing, but we see that this method is not a panacea, for it does not work in every case. However, it is sufficiently efficient when it does work that it is one of the most frequently used methods for solving equations. It is also true that we seldom come across a case in which this method fails in dealing with applications.

It would be helpful to indicate a geometric development of Newton's method before we leave it. Let us consider the graph of $y = f(x)$ given in Figure V.15. Note that $x = X$ is the point at which the graph crosses the x-axis. We select a point x_0 which is not too far removed, (it could have been guessed from a sketch of the graph). Since $f'(x_0)$ is the slope of the line which is tangent to $y = f(x)$ at the point where $x = x_0$, we have from the triangle $P_0 P_1 Q_0$ the relation

$$f'(x_0) = \frac{f(x_0)}{x_0 - x_1}.$$

Consequently $x_0 - x_1 = f(x_0)/f'(x_0)$ or

$$x_1 = x_0 - \frac{f(x_0)}{f'(x_0)}.$$

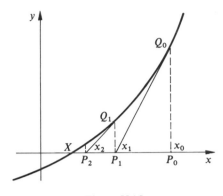

Figure V.15

This is just relation (V.5.2). Since we could continue this line of reasoning using triangle $P_1P_2Q_1$, we could again obtain steps which would lead us to (V.5.3). In Figure V.15 we observe that the graph of $f(x)$ in the vicinity of X has a positive slope and is concave upward. Therefore both the first and second derivatives are positive in this vicinity and it was proper to start with a value x_0 which is larger than X. If we had started with a value x_0 to the left of X in this case the tangent line would have crossed the x-axis at a point to the right of X. This gives a geometric explanation of the results we obtained earlier concerning the more efficient starting point. A sketch of the curve in question showing the direction of slope and the direction of concavity will often aid in determining whether the more efficient starting point should be to the right of X or to the left.

It is true that there are some cases in which Newton's method would fail to converge, but these are only the cases in which $f'(x)$ is not continuous or is zero near the root, in which our first approximation was not close enough, or in which we had a point of inflection at the root. In the latter case we can solve the equation $f''(x) = 0$ to obtain the root. Among the principal advantages of Newton's methods are its applicability to equations which are not polynomial equations and its relative efficiency with regard to the amount of computation required.

EXAMPLE 5.3. Find the extreme values of

$$f(x) = e^x + x^2.$$

Solution. Since we wish the extreme values of $f(x)$, and since this is not over a closed interval, we have only to look at the first derivative and find the zeros of $f'(x)$. Since $f'(x) = e^x + 2x$, our problem becomes that of finding the roots of $e^x + 2x = 0$. For convenience we will call this $g(x)$, in view of the fact that we are not asking for the zeros of $f(x)$ and we might easily forget at some point which function we are working with. Since e^x is never negative, we see that both $g'(x) = e^x + 2$ and $g''(x) = e^x$ are always positive.

From this we deduce the fact that $g(x)$ has at most one real root, and we should start with a guess to the right of the actual root. Since $g(0)$ is positive, we can proceed toward the left and observe that $g(-1) = e^{-1} - 2$ is surely negative, for the exponential portion of this expression is less than one. Therefore, we have a zero in the interval $(-1, 0)$, and since we should start at the right we could well start with $x_0 = 0$. We now have

$$x_1 = x_0 - \frac{g(x_0)}{g'(x_0)} = -\frac{1}{3} = -0.333333.$$

We can now proceed to the following sequence of steps

$$x_2 = -0.333333 - \frac{g(-0.333333)}{g'(-0.333333)} = -0.333333 - \frac{0.049865}{2.716531} = -0.351689$$

$$x_3 = -0.351689 - \frac{g(-0.351689)}{g'(-0.351689)} = -0.351689 - \frac{0.000120}{2.703499} = -0.351734.$$

If we continue we find that x_4 agrees with x_3 in the first six decimal places, and consequently we have apparently obtained an accuracy to six decimal places. Therefore, the extreme value of $f(x)$ must occur when $x = -0.351734$, and we have $f(-0.351734) = 0.827184$ as a minimum value of $f(x)$. The fact that it is minimum is a consequence of the positive second derivative of $f(x)$ or first derivative of $g(x)$.

While it is certainly true that the computation in this instance is not something you want to do with mental arithmetic, the method does give us a solution and does so with a very small number of steps. The actual computation could be carried out using results such as those in the tables of Appendix C, with a computer, or with a hand held calculator. Many of the hand held calculators have the ability to compute the exponential function, and that is all that was required for the example here.

EXERCISES

In some of the exercises you may wish to use a computer if one is available.

1. Find the real root of each of the following with an error no greater than 0.001:
   (a) $f(x) = x^3 + x^2 + 1 = 0$
   (b) $f(x) = x^3 + 3x - 7 = 0$
   (c) $f(x) = x^3 + 3x^2 + 1 = 0$
   (d) $f(x) = \cos x - x = 0$
   (e) $f(x) = e^x + x = 0$

2. Find the extreme values of each of the following accurate to three decimal places:
   (a) $x^4 + x^2 - 5x - 3$
   (b) $x^4 - x^3 + 10x$
   (c) $x^2 + 2 \sin x - 2$
   (d) $x^4 - 5x^2 + 4x - 7$

3. (a) Use Newton's method to find the fifth root of 75 to four decimal places. [*Hint*: this is a solution of the equation $x^5 = 75$.]
   (b) Use Newton's method to find the sixth root of 45 to four decimal places.
   (c) Find the extreme values of $f(x) = (x^2 + 1)e^{x^2-x}$ with an error no greater than 0.001.

4. The curve $y = \tanh x$ has a shape in the vicinity of the origin which is somewhat like that of $y = x^{1/3}$.

   (a) Would the problems of Example 5.2 occur in attempting to solve the equation $\tanh x = 0$ by Newton's method, starting with a non-zero value?
   (b) Does your response to part (a) depend upon the starting value selected?

5. Find the point in the first quadrant at which the curve $y = \cos x$ intersects the curve $y = \tan x$.

6. Show that if $f(x_k) = 0$ for some x_k in the sequence of successive iterates using Newton's method then $x_n = x_k$ for each value of n larger than k.

7. Find the smallest positive zero of $f(x) = x - 2 \sin x$ to three decimal places.

8. Find both real roots of $e^x - x - 2 = 0$ to five decimal places if you have a computer available and to three places otherwise.

9. Find three positive roots of $\tan x = x$ to three decimal places.

P10. A pendulum swings according to the law $s = e^{-at} \sin bt$ where t is time measured in seconds, s is the length of the arc swept out by the pendulum measured from the point at which the pendulum would remain at rest, and a and b are positive constants.

   (a) Find the time at which the pendulum is farthest from the point of equilibrium. Since it is assumed that the pendulum starts when $t = 0$, your answer should not be negative.
   (b) Find the time at which the velocity of the pendulum is maximum.
   (c) Find the time at which the acceleration is maximum.

S11. If the equation of a supply curve is $y = 80 + 2x + x^2$ and the equation of the corresponding demand curve is given by $y = 150 - 35e^{-0.5x}$, what is the equilibrium price under pure competition to the nearest penny?

V.6 The Chain Rule and Related Rates

Up to this point we have been somewhat restricted in the functions we could differentiate by use of available formulas. For instance, we could differentiate sin x, but we could not differentiate sin x^2. In similar fashion, we could differentiate $(4 - x^2)$, but not $(4 - x^2)^{1/2}$. Each of these could be differentiated if we had some means of differentiating the composite function $f(g(x))$. In the cases we have cited, we would have first $f(x) = \sin x$ and $g(x) = x^2$, and secondly $f(x) = x^{1/2}$ and $g(x) = (4 - x^2)$. With these examples before us it is not difficult to see that an endless list of similar examples could be written down. As always, when we mention something like this, it indicates that we are about to fill the void we have just indicated. This situation is no exception, and the next Theorem will take care of this deficiency.

Theorem 6.1 (The chain rule). *If $f(x)$ and $g(x)$ both possess derivatives, and if the range of $g(x)$ is included in the domain of $f(x)$, then*

$$D_x f(g(x)) = f'(g(x))g'(x),\qquad\qquad (V.6.1)$$

where by $f'(g(x))$ we mean that we are evaluating $f'(y)$ at the point where the value of y is $g(x)$.

PROOF. From Theorem III.3.2 we have

$$\int_a^x f'(g(t))dg(t) = \int_{g(a)}^{g(x)} f'(y)dy = f(g(x)) - f(g(a))$$

Differentiating the left side we have

$$D_x \int_a^x f'(g(t))dg(t) = f'(g(x))g'(x).$$

Differentiating the last term of our equation above, we have

$$D_x[f(g(x)) - f(g(a))] = D_x f(g(x))$$

since $f(g(a))$ is a constant, and its derivative is zero. Since the expressions we differentiated are equal, the derivatives must be equal, and we have the desired result. $\qquad\square$

We now go back to the first example we mentioned earlier. If $f(x) = \sin x$ and $g(x) = x^2$, we have $f'(x) = \cos x$ and $g'(x) = 2x$, but from this we have the fact that $f'(g(x)) = \cos g(x) = \cos x^2$ and hence $D_x \sin x^2 = (\cos x^2)(2x) = 2x \cos x^2$. It is suggested that you go through this computation more than once, not because it is not possible to follow the first time, but rather to insure that you see just how we are applying the result of the chain rule theorem. The flow chart of Figure V.16 may help in making the procedure clear.

Before we leave this example, we ought to consider the effect of this particular differentiation on a related integral. Since we know that $D_x \sin x^2 = 2x \cos x^2$, it follows that

$$\int_a^x \cos x^2(2x\,dx) = \sin x^2 - \sin a^2.$$

You will note that we have placed the $2x$ in parentheses with the differential, dx, for this makes more obvious the similarity to the general expression $f'(g(x))g'(x)dx$. That is, $f'(y) = \cos y$ and $g(x)$ is so picked that in the first

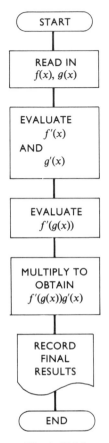

Figure V.16

part we have $f'(g(x))$, and in the remaining factor the particular $g'(x)$ associated with $g(x)$. Since $g'(x)dx = dg(x)$, it is apparent that this is merely a further application of the RS integral. In the illustration of this paragraph, for instance, we could have written

$$\int_a^x \cos x^2(2x\,dx) = \int_a^x \cos x^2\,d(x^2)$$

and then used Theorem III.3.2 directly. This would suggest that if we wish to use the fundamental Theorem to obtain the value of integrals, it may well be necessary to use careful observation in order to select the appropriate functions.

In order to further illustrate the application of the chain rule, we refer to the second example given in the opening paragraph of this section. If we wish $D_x(4 - x^2)^{1/2}$ we can let $f(x) = x^{1/2}$ whence $f'(x) = (1/2)x^{-1/2}$, and then let $g(x) = (4 - x^2)$ whence $g'(x) = (-2x)$. We have at once $D_x(4 - x^2)^{1/2} = (1/2)(4 - x^2)^{-1/2}(-2x)$. We might pause long enough to note that the differential can be obtained from the derivative just as before, namely, $d(4 - x^2)^{1/2} = (1/2)(4 - x^2)^{-1/2}(-2x)dx$. These results can be simplified but this form illustrates the application of the chain rule. Again, we could write

$$\int \frac{x\, dx}{\sqrt{4 - x^2}} = -\int \left(\frac{1}{2}\right)(4 - x^2)^{-1/2}(-2x)dx$$

$$= -\int d(4 - x^2)^{1/2} = -(4 - x^2)^{1/2} + C.$$

This illustrates the way by which the chain rule makes possible the evaluation of derivatives and integrals which were heretofore beyond our reach. We will use these techniques more in the remainder of this chapter and in Chapter VI.

In order to avoid confusion, we would do well at this point to compare the chain rule with our earlier work in differentiation. We had developed formulas which would yield $D_x(x^7) = 7x^6$. The question arises whether this requires the chain rule. If the chain rule were used we would have $D_x(x^7) = 7x^6 D_x x = 7x^6(1) = 7x^6$, and we would obtain the same result. This, in fact, is a correct way of viewing a derivative such as the one we have used here. In other words, the chain rule still applies, but in this case the $g'(x)$ turns out to have the value one, and multiplying by one does not alter the remaining portion of the derivative. Therefore, the answer to our question is "yes" with the pleasant proposition that since $g'(x) = 1$ we do not need to explicitly write out the $g'(x)$ term.

We will now discuss two uses of the chain rule that occur rather frequently in applications. The first one is rather subtle, and therefore deserves careful attention. If x changes with the passage of time, then x is actually a function of time, and we would more properly write $x = x(t)$. In this case an accurate representation of $f(x)$ would be $f(x(t))$. We see at once that we have a composite function. Since it is time that will control the value of x and hence $f(x)$, the appropriate derivative is then $D_t(f(t)) = f'(x(t))x'(t)$. It is understood that when we write $f'(x)$, we mean that we have the derivative of $f(x)$ with respect to x. To avoid ambiguity, we could write $D_t f(x(t)) = D_{x(t)} f(x(t))D_t x(t)$. With the understanding that this is what is intended, we shall continue, in accordance with normal practice, to use the form $D_t f(x) = f'(x)x'(t)$ with the explicit interpretation that x is, as discussed above, a function of t.

EXAMPLE 6.1. We have a ladder 25 feet long propped up against the house, and we wish to consider the effect as we push the bottom of the ladder in toward the house at a rate of 2 feet per minute with the top being able to slide up the side of the house as required by the geometry of the situation. Implicitly we will also assume that the ground is perfectly level, and hence that the side of the house makes a right angle with the ground. We wish to know

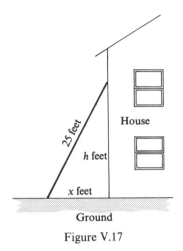

Figure V.17

what is happening to the top of the ladder when the bottom is still seven feet from the house.

Solution. The first thing to do in any problem of this type is to draw a picture, and this we have done in Figure V.17. It is not necessary that the picture be worth an *A* in a course in art, but it is necessary that it convey the essential properties associated with the problem, in this case the relation of the side of the house to the ground, and to the ladder. We are interested in the distance between the ground and the point at which the ladder contacts the house, a distance we have labeled *h* since it represents height. Since we are concerned with the height, we should obtain the height as a function of the other quantities which appear. By the Theorem of Pythagoras concerning right triangles, we know that $h^2 + x^2 = 25^2$, or that $h^2 = 625 - x^2$. We observe that if we are to obtain *h* as a function of *x*, we must take the square root, and in view of the fact that *h* is hardly likely to be a negative quantity, unless of course we are in the mining business, we can write $h = (625 - x^2)^{1/2}$. But now we must remember that *x* is changing with time, and therefore *x* is really a function of *t*. This, of course, implies that *h* is also a function of *t*, although somewhat indirectly. (This should not be a surprise considering the point that we are trying to illustrate here.) Since we are really interested in $D_t h$, in order to get the rate of change of height with respect to time, we might have replaced *x* by *x(t)* to remind us of this dependency on time. By the chain rule, we have $D_t h = (1/2)(625 - x^2)^{-1/2}(-2x)D_t x$. Before we go further, perhaps we should explain where these various terms came from, for there seem to be a great number of items multiplied together. As we go through this explanation, observe carefully, for it is a very good illustration of the lengths to which we sometimes have to go in applying the chain rule. If we let $f(x) = (625 - x^2)^{1/2}$, we can first consider this as being a composite function of the form $f_1(x) = x^{1/2}$ and $f_2(x) = 625 - x^2$. The subscripts here are merely to distinguish the different functions involved. Now $f(x) = f_1(f_2(x))$, and thus we can apply the rule for composite functions, obtaining $f'(x) = f'_1(f_2(x))f'_2(x)$, but $f'_1(x) = (1/2)x^{1/2}$, and $f'_2(x) = (-2x)$. Finally, since $D_t h = f'(x(t))D_t x$, we must multiply $f'(x)$ by $D_t x$ in order to have the desired $D_t h$.

After all of this discussion concerning the differentiation, we are now in a position to go back and attempt to solve the problem at hand. In our case we are interested in the rate of change of h when $x = 7$, (if you wonder where the 7, actually standing for seven feet, came from, go back and re-read the problem as stated), and when $D_t x = (-2)$ where the units here would be in feet per minute. Note that this derivative is negative, for the distance x is getting smaller as stated in the problem. Hence we have $D_t h = (1/2)(625 - 7^2)^{-1/2}(-2)(7)(-2)$ by merely substituting the various quantities in the result already obtained. A bit of arithmetic (or perhaps a lot) will reduce this to $D_t h = +7/12$. Thus we have found that the top of the ladder will be moving upward (since $D_t h$ is positive) at the rate of 7/12 feet per minute, or at the rate of 7 inches per minute at this particular point in time. Note that this gives us the *instantaneous rate* under these specific conditions.

It is clear from this example that we must keep track of the particular variable, in this case time, with respect to which differentiation is to be performed. When there are more than two variables in the problem, it is easy to forget which variable governs the rate of change, unless you are careful to write instructions to yourself in sufficient detail. It should also be noted that the expression for the derivative could have been simplified algebraically and we could have written

$$D_t h = \frac{-x}{(625 - x^2)^{1/2}} D_t x.$$

This would not have changed the result. The algebraic simplification is very helpful if one is going to have to obtain several arithmetic results, but the majority of people find it easier to simplify arithmetic than algebra. The point at which you perform the simplification is entirely up to you. Another method for obtaining $D_t h$ in Example 6.1. makes direct use of the relation $h^2 = 625 - x^2$. If we think of this equation as representing the relation between two functions of t, that is $h(t)$ and $x(t)$, we can write $[h(t)]^2 = 625 - [x(t)]^2$. Applying the chain rule here yields $2h(t)h'(t) = -2x(t)x'(t)$. Therefore

$$D_t h = h'(t) = -\frac{x(t)x'(t)}{h(t)} = -\frac{xD_t x}{\sqrt{625 - x^2}}.$$

Each of the methods for obtaining the derivative relies upon the chain rule.

We mentioned the fact that we would discuss two uses of the chain rule which occur frequently in applications. The first one had to do with rates of change dependent upon some underlying variable, such as time. It is often the case that such a variable occurs almost as an implicit variable instead of being explicitly written out as $x(t)$. The second of these two uses is somewhat similar. In this case we would have two variables, perhaps x and y, linked through their mutual dependence upon a third variable, usually called a *parameter*. The parameter is often denoted by t, but there is no reason for this other than a certain traditional notation. We will follow the tradition, however. We then have two equations, one giving an expression for x and the other for y. We can write these expressions as

$$\begin{cases} x = x(t), \\ y = y(t). \end{cases} \tag{V.6.2}$$

It would be convenient to solve one of these two equations for t in terms of either x or y and then substitute in the other equation, for this would give us a single familiar looking equation involving only x and y. However, this is not always possible, and besides there are very good reasons on occasion for retaining the relationship involving the parameter. Therefore, if we wish the derivative $D_x y$, we must find some way to work from the information given in (V.6.2). Since we wish the derivative with respect to x, the rather obvious thing to do is to differentiate each of the equations of (V.6.2) with respect to x and thus we would obtain through use of the chain rule

$$\begin{aligned} 1 &= x'(t)D_x t \\ D_x y &= y'(t)D_x t. \end{aligned} \tag{V.6.3}$$

If $D_x t = 0$, t must be a constant with respect to x and therefore our original relation is suspect. It would also be impossible to multiply $D_x t$ by $x'(t)$ and obtain one if the derivative were zero. Consequently we know that $D_x t \neq 0$ and we can divide the second of the equations in (V.6.3) by the first, thus obtaining

$$\frac{D_x y}{1} = \frac{y'(t)D_x t}{x'(t)D_x t}$$

or

$$D_x y = \frac{y'(t)}{x'(t)}. \tag{V.6.4}$$

Note that this derivative is a function of t, as were the original values of x and y. If the parameter represents time, we have the rate of change of y with respect to x represented as a function of time. This is frequently essential in applications.

EXAMPLE 6.2. Given the system of parametric equations

$$\begin{cases} x = x(t) = t^3 + t^2 \\ y = y(t) = e^{t^2} - t^3 \end{cases}$$

find $D_x y$.

Solution. Upon differentiating with respect to x in both equations and making liberal use of the chain rule in the case of the first term of y, we obtain

$$1 = 3t^2 D_x t + 2t D_x t = (3t^2 + 2t) D_x t$$

and

$$D_x y = e^{t^2}(2t)D_x t - 3t^2 D_x t = [2te^{t^2} - 3t^2]D_x t.$$

If $t \neq 0$ and $t \neq (-2/3)$, we can divide as indicated above and obtain

$$D_x y = \frac{D_x y}{1} = \frac{[2te^{t^2} - 3t^2]D_x t}{(3t^2 + 2t)D_x t} = \frac{2e^{t^2} - 3t}{3t + 2}.$$

It would be practically impossible to solve $x = t^3 + t^2$ for t as a function of x and then substitute this value of t in the equation for y, and it would be equally difficult to solve for t as a function of y in the second equation and substitute in the first. Therefore, this would seem to be the most expeditious method available for obtaining $D_x y$.

Parametric equations can be very useful for many purposes. One of these is illustrated in Example 6.3 when we attempt to obtain the equation of a curve formed by certain mechanical motions. In this case we will find that the parameter represents an angle. If we were to need the rate at which x and y are changing as the angle changes, we would have these rates in $D_t x$ and $D_t y$. On the other hand, if we wish the slope of the curve, we require $D_x y$. Again, it is necessary to keep in mind the particular rate needed for a given purpose. As we proceed, we find that we are able to do more things, but we also find that we must be ever more alert to know which of the many things are desired in a given situation.

EXAMPLE 6.3. A wheel of radius a is rolled around inside a circle of radius $4a$, the circle having its center at the origin. A spot on the rim of the wheel is painted white. If this spot is in contact with the circle at the point $(4a, 0)$, and if the wheel is then rotated around the inside of the circle with the rim of the wheel maintaining a friction contact with the circle, what is the equation of the spot? (See Figure V.18.) Find the slope of the curve when the wheel has gone 1/8 of the way around.

Solution. We assume that the spot starts at the point $P:(4a, 0)$ as directed. In order to obtain the equation, we must know the coordinates of the successive locations of the point as the wheel is rotated. Consider that the wheel has rotated, as indicated in Figure V.18, so that the center of the wheel is on the radius of the circle making an angle, t, with respect to the x-axis. The white spot will now be located at P'. The arc PQ must be equal to the arc QP' since there is no slippage. Since the radius of the circle is four times the radius of the wheel, the central angle involved on the wheel must be four times the central angle involved in the circle. Thus, the angle QRP' must be $4t$. Since angle QRS has t radians, SRP' must have $3t$ radians. Since the radius

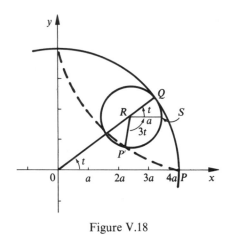

Figure V.18

RQ is of length a, the distance OR must be of length $3a$ and therefore the coordinates of R must be $(3a \cos t, 3a \sin t)$. The length of RP' is a, and therefore the coordinates of P' must be $(3a \cos t + a \cos 3t, 3a \sin t - a \sin 3t)$. This is obtained by considering the displacement from R to P'. We thus have the parametric equations

$$x(t) = 3a \cos t + a \cos 3t,$$
$$y(t) = 3a \sin t - a \sin 3t.$$

These can be simplified if we use the identities

$$\cos 3t = \cos^3 t - 3 \cos t \sin^2 t \quad \text{and} \quad \sin 3t = 3 \cos^2 t \sin t - \sin^3 t,$$

for we obtain

$$
\begin{aligned}
x(t) &= 3a \cos t + a \cos^3 t - 3a \cos t \sin^2 t \\
&= 3a(\cos t)(1 - \sin^2 t) + a \cos^3 t = 3a(\cos t)(\cos^2 t) + a \cos^3 t \\
&= 4a \cos^3 t
\end{aligned}
$$

and

$$
\begin{aligned}
y(t) &= 3a \sin t - 3a \cos^2 t \sin t + a \sin^3 t \\
&= 3a(\sin t)(1 - \cos^2 t) + a \sin^3 t = 3a(\sin t)(\sin^2 t) + a \sin^3 t \\
&= 4a \sin^3 t.
\end{aligned}
$$

These results when written without the intervening steps are

$$
\begin{cases}
x(t) = 4a \cos^3 t, \\
y(t) = 4a \sin^3 t.
\end{cases}
$$

To obtain the desired slope, we must evaluate $D_x y$. We can do this using the methods we have discussed above. Thus,

$$
\begin{cases}
1 = 12a \cos^2 t(-\sin t)D_x t \\
D_x y = 12a \sin^2 t(\cos t)D_x t.
\end{cases}
$$

Consequently,

$$D_x y = -\frac{\cos t}{\sin t} = -\cot t.$$

After 1/8 of a revolution, $t = (1/8)(2\pi) = \pi/4$, and hence $D_x y = -\cot(\pi/4) = -1$. Note the extent to which the parameter, t, helped in obtaining these equations. It is possible in this case to eliminate the parameter between the two equations, for we observe that $x^{2/3} + y^{2/3} = (4a)^{2/3}(\sin^2 t + \cos^2 t) = (4a)^{2/3}$. However the differentiation with the parametric equations is probably at least as easy as handling the fractional exponents.

As a last statement in this section, it is worth noting that the chain rule has been used in nearly every example. It permits us to differentiate many functions that we could not previously have handled. However, in the application of the chain rule, particularly in the case in which we have functions of functions, we must be careful to keep the various functions straight. It is suggested that you write out as many steps as necessary. If a little more writing insures a more accurate result, the additional writing is worthwhile. However, if you can handle several steps without writing them down (and this usually comes with practice), then it is obviously less work to do as much as you can do mentally with assurance that you are not making errors.

EXERCISES

1. Find the derivative of each of the following:

  (a) $e^{\sin x}$
  (b) $\sin e^x$
  (c) $(x^2 - 4x)^3$
  (d) $(x^3 - 4x + 2)^{2/3}$
  (e) $\tan^3 x$
  (f) $\tan^2 e^x$
  (g) $\sin^3 (e^{x^2})$
  (h) $\sec^2 x - \tan^2 x$ (Interpret your result.)

2. Find the differential of each of the following:

  (a) $[x^2 \sin(e^x + 2)]^{1/2}$
  (b) $\csc(e^{x^2})\sec(3 - x^2)$
  (c) $e^{\tan x + \sec x}$
  (d) $(x + 1/x)^{10}$
  (e) $\cot(x^2 + 3)^2$
  (f) $(3x - 1)/(x\sqrt{x^2 - 9})$
  (g) $2^{\sin x}$
  (h) $3^{x - \tan x}$

3. Evaluate:

(a) $D_x^4(e^{x^2})$

(b) $D_x^3 \csc x^3$

(c) $d(x^3 - x^{-3})^{-2}$

(d) $D_x^3 \tan x^2$

(e) $D_x^4(1/(1 - x^2))$

(f) $D_x^2(5^{x^5})$

4. Find $D_x y$ in each of the following:

(a) $\begin{cases} x = 4t^3 + 7t + 8 \\ y = t^4 - 7t^2 - 2 \end{cases}$

(b) $\begin{cases} x = 4 \tan t \\ y = 4 \sec t \end{cases}$

(c) $\begin{cases} x = \sec^3 t \\ y = \tan^3 t \end{cases}$

(d) $\begin{cases} x = e^{\sin t} \\ y = e^{\cos t} \end{cases}$

(e) $\begin{cases} x = \cos e^{3t} \\ y = t - \sin e^{2t} \end{cases}$

(f) $\begin{cases} x = t^2 - t \sin t \\ y = t^3 + t \cos t \end{cases}$

5. Find the approximate value of each of the following:

(a) $e^{(0.97)^2}$

(b) $(0.03^2 + 1)e^{0.03}$

(c) $e^{\sin 0.05}$

(d) $(25 - 4.05^2)^{1/2}$

(e) $\sec(0.08)^2$

(f) $\tan(0.9)^2$, $[(0.9)^2$ is near $\pi/4$.]

6. Find the equation of the line tangent to each of the following curves at the indicated point:

(a) $y = \sec(x - 2)^2$ at $(2, 1)$

(b) $y = e^{\tan x}$ at $(\pi/4, e)$

(c) $y = \cot(\pi e^x/6)$ at $(0, \sqrt{3})$

(d) $y = \cosh x^2$ at $(0, 1)$

(e) $y = x \sinh x$ at $(1, (e^2 - 1)/2e)$

(f) $y = (x^2 - 3x)^{1/2}$ at $(4, 2)$

7. Obtain an indefinite integral for each of the following: [Find a function $g(x)$ such that the integrand can be expressed as $f(x)g'(x)dx$.]

(a) $\int x^2 \sin x^3 \, dx$

(b) $\int \tan x \sec^2 x \, dx$

(c) $\int xe^{-x^2} \, dx$

(d) $\int (x^2 - 4)^9 x \, dx$

8. Integrate each of the following.

   (a) $\int \sin^2 x \cos x \, dx$
   (b) $\int \cos^3 x \sin x \, dx$
   (c) $\int e^{\sin x} \cos x \, dx$
   (d) $\int \tan^2 x \sec^2 x \, dx$
   (e) $\int x\sqrt{x^2 - 4} \, dx$
   (f) $\int x\sqrt{4 - x^2} \, dx$
   (g) $\int e^x \sin e^x \, dx$
   (h) $\int xe^{3x^2} \, dx$

9. The coordinates of a moving point are given by $(2^t t, t^2 - t)$ where t is the time since movement started.

   (a) Find the rate of movement in the horizontal direction.
   (b) Find the rate of movement in the vertical direction.
   (c) Find the direction of movement at a given time t_0.
   (d) Find the equation of the line tangent to this curve at the point determined by $t = 4$.

10. In a certain town the main point of interest is an intersection of a road that runs east and west and a road that runs north and south. One man leaves this intersection at noon heading east at 30 miles per hour. Thirty minutes later a second man leaves heading north at 40 miles per hour. How fast are the two increasing the distance between them at 2:00 pm?

11. Ship A is 12 miles due north of ship B at noon. Ship A is sailing due west at 16 miles per hour and ship B is sailing due north at 12 miles per hour.

    (a) Find the distance between the two ships as a function of the elapsed time since noon.
    (b) Find the time at which the two ships are nearest each other.

12. Find the surface of the sphere of radius a obtained by rotating $y = (a^2 - x^2)^{1/2}$ about the x-axis.

13. A balloon is losing air at the rate of 10 cubic inches per minute. If the balloon is 12 inches in diameter:

    (a) How fast is the radius decreasing?
    (b) How fast is the surface area decreasing?
    (c) Is the rate of change of the radius increasing or decreasing? By how much?

14. The surface of a balloon is increasing at the rate of 20 square inches per minute when the radius is 8 inches. How fast is the volume changing at this time?

15. A wheel of radius 15 inches rolls along the x-axis, starting at the origin. A spot on the outside edge of the tire is marked when it touches the origin. See Figure V.19.

    (a) If t represents the number of radians that the wheel has turned, find the x-coordinate of the spot and the y-coordinate of the spot after the wheel has turned t radians.
    (b) Find the time when the spot is moving most rapidly in the vertical direction.

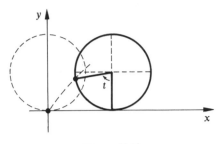

Figure V.19

(c) Find the time when the spot is moving most rapidly in the horizontal direction.
(d) Find the direction in which the spot is moving (the slope of movement) when $t = \pi/2$.

16. An object is fired at an angle α above the horizontal and with an initial speed of v_0 feet per second. Assume that there is no air resistance to slow the object down and that the acceleration due to the force of gravity is -32 feet per second per second, the negative sign indicating that the force is toward the earth.

(a) Find the parametric equations which give the location of the object at any time, t.
(b) Solve these simultaneously to find the equation of the trajectory in rectangular coordinates.
(c) Find how high the object goes and find its horizontal speed at the moment when the object is at its highest point.
(d) Find the horizontal and vertical components of speed with which it hits the ground.
(e) Show that without air resistance the maximum distance before hitting the ground is obtained when $\alpha = \pi/4$.

P17. A conical tank of water is 15 feet high and 20 feet in diameter at the top. Water is leaking out of this tank at the rate of 4 cubic feet per minute.

(a) How fast is the water level dropping when the maximum depth is 10 feet?
(b) If $t = 0$ when the tank was full, how fast is the water level dropping 10 minutes after the leak started?

P18. A certain water tower has a spherical shape with a radius of 12 feet. The water pressure on the bottom of the sphere is obtained by multiplying the depth of the water by 62.4 pounds per cubic foot. There are 231 cubic inches in a U.S. gallon.

(a) How fast is the pressure on the bottom changing when the depth is 18 feet if water is flowing in at the rate of 20 gallons per minute?
(b) If water is flowing out at the rate of 100 gallons per minute, find the depth at which the rate of change of bottom pressure is a minimum. Check this result against your intuitive answer to this question.

S19. If for a given situation the demand function is given by $y = 30 - x^{2/3}$, and the supply function is given by $y = 6 + x^{1/2}$, find the consumers surplus and the producers surplus if the market is operating under pure competition.

S20. If the population at time t (years) is given by $P(t) = P(C)e^{0.03t}$ and the food production is given by $F(t) = A + Be^{0.025t}$ where A, B and C are constants,

(a) Find the rate at which the amount of food is changing with respect to population.

(b) What is happening to our ability to feed people if these equations are approximately correct? What would you recommend to improve the situation?

V.7 Implicit Differentiation and Inverse Functions

There are many times when it is difficult to write the equation defining a function in the simple form $y = f(x)$. If we were given the equation $x^2y - x \sin y + e^{xy} = 197$, it is clear that we would have y depending on x. For given a value of x we have certainly determined a finite set of values from which y must be selected. If we limit the choice (by decree if necessary) such that there is only one value of y for each value of x, we have a function. However, it would be difficult to solve this equation for y to obtain the function in the form $y = f(x)$. If we find a situation such as this, the question then arises whether we are going to be denied the possibility of obtaining a derivative $D_x y$. As before, if we raise the question you do not expect us to answer it in the negative. Also, as before, you know that your expectations are fairly certain to be correct. Since in this equation we have two expressions that are equal for any admissible pair of values (x, y), the two expressions should change at the same rate. Therefore, if we differentiate both sides, the derivatives should be equal. As we look at the left side of this equation we see that we have to contend with both x and y, but then we realize that y is a function of x. If we need the derivative we can always designate this by $D_x y$ or perhaps more simply by y' where the latter notation is an obvious extension of our earlier use of $f'(x)$. With this in mind, and with the use of results from Section IV.4, we can proceed just as though we knew confidently what to expect.

In our illustration we have

$$[x^2(y') + y(2x)] - [x(\cos y)(y') + \sin y(1)] + e^{xy}[x(y') + y(1)] = 0$$

where we have included in parentheses the portion actually differentiated at each stage. Note that we have used liberally the results for a product, the chain rule, and many other results obtained earlier. Now that we have this result, the next question concerns what to do with it. It certainly looks far worse than anything we had before. But let us keep in mind the goal, namely finding the value of y', the derivative. This would imply that we should solve for y'. This is simple (?) algebra.

$$y'[x^2 - x \cos y + xe^{xy}] = -[2xy - \sin y + ye^{xy}]$$

or

$$y' = \frac{\sin y - 2xy - ye^{xy}}{x^2 - x \cos y + xe^{xy}}.$$

We note immediately that this method will not work if the coefficient of y' should be zero. However this would imply

$$x^2 - x \cos y + xe^{xy} = 0$$

and we have a simpler equation than the original with which to work. As in many cases, there are some additional fine points we should explore, but basically this example illustrates a method for obtaining derivatives even in cases in which we cannot solve for the function explicitly. When we can write $y = f(x)$, we say that we have an *explicit* expression. By contrast the method that we have just illustrated is called *implicit differentiation*. It involves differentiating both sides of an equation and then solving for the derivative of the function. Note that this would only be valid over such domains as may be appropriate for the function. We had done this same thing before, but when $y = f(x)$, we had $y' = f'(x)$ and there was no algebraic problem in solving for y'. If we were to consider the function $y = f(x)$ defined by means of the relation $x^2 + y^2 = 9$, we would, by following the procedure above, have $2x + 2yy' = 0$ or $y' = -x/y$. But you will note that the given expression has no meaning over a domain other than $[-3, 3]$. Also we only have a function in this illustration if we restrict our attention to the case in which $y \geq 0$ or the case in which $y \leq 0$. This corresponds to taking one or the other of the two square roots involved. Furthermore, the derivative fails to exist at points where $y = 0$.

You will note that we used implicit differentiation in handling parametric equations in the last section. We also find implicit differentiation of immeasurable help in obtaining the derivatives of inverse functions. Consider the function $f(x) = \sin x$. The graph is given in Figure V.20. The *inverse function* can be obtained from $y = \sin x$ by interchanging the x and the y and then solving for y. Thus, we would have $x = \sin y$. We would then write $y = \arcsin x$ where the notation arc sin x is one notation for the inverse function of sin x. (The notation $\sin^{-1} x$ is also in current use, but this has the disadvantage that one could misinterpret it to mean $1/(\sin x)$. For this reason we shall use the arc sin x notation here.) The graph of $y = \arcsin x$

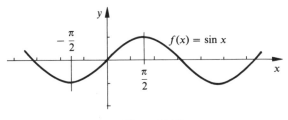

Figure V.20

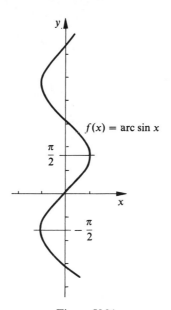

Figure V.21

is given in Figure V.21. Note that this is not a function, for there are many values of y for each value of x in the interval $[-1, 1]$. If we had limited the domain of $y = \sin x$ to $[-\pi/2, \pi/2]$, we would have considered only the portion indicated by the vertical lines in Figure V.20. The inverse function would then have used only the portion indicated by the horizontal lines in Figure V.21, and this would have given us a function. For this reason, we limit the range of the function $y = \text{arc sin } x$ to the interval $[-\pi/2, \pi/2]$.

Pursuant to the above discussion, we have

$$\sin(\text{arc sin } x) = x.$$

Using the chain rule, we have

$$\cos (\text{arc sin } x)D_x(\text{arc sin } x) = 1$$

or

$$D_x(\text{arc sin } x) = \frac{1}{\cos(\text{arc sin } x)} .$$

It is not very appealing to leave the right hand side in this shape. We observe that if we denote (arc sin x) by y for simplicity, we have cos y, but this could be written as

$$\cos y = (\cos^2 y)^{1/2} = (1 - \sin^2 y)^{1/2}.$$

Since $y = \text{arc sin } x$, then

$$\sin y = \sin(\text{arc sin } x) = x,$$

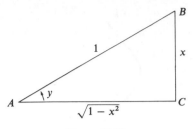

Figure V.22

and hence $\cos y = \sqrt{1 - x^2}$. We can illustrate this trigonometric sleight of hand with the use of Figure V.22. In this case we have labeled the angle A as y, for $y = \text{arc sin } x$ can be liberally translated as "y is the angle whose sine has the value x." In order to produce the correct sine, we have indicated that the side BC is of length x and the hypothenuse AB is of length 1. You will note that this does yield $\sin y = x$. From the Pythagorean theorem we now have the length of AC as $\sqrt{1 - x^2}$. A glance at Figure V.22 now tells us that $\cos y = \sqrt{1 - x^2}$. This result agrees with that we obtained using the trigonometric identities. We can now write our result as

$$D_x(\text{arc sin } x) = \frac{1}{\sqrt{1 - x^2}}.$$

We should observe here that since the cosine in this interval is never negative, we have no problem in determining which sign we should use in taking the square root. If we multiply both sides by dx, we obtain the corresponding differential of arc sin x. If we use the fundamental theorem, we can obtain the integral which yields arc sin x as a result. Thus, we have

$$D_x(\text{arc sin } x) = \frac{1}{\sqrt{1 - x^2}}$$

$$d(\text{arc sin } x) = \frac{dx}{\sqrt{1 - x^2}}$$

and

$$\int \frac{dx}{\sqrt{1 - x^2}} = \text{arc sin } x + C.$$

Although the arc sine function is defined over the closed interval, we find it necessary to delete the end points when differentiating in order to avoid meaningless derivatives. More will be said on this point in Chapter VII but for the moment we will not be concerned with either integration or differentiation over the interval $[-1, 1]$, despite the fact that the definition of the inverse function of the sine is valid over this interval.

We can handle each of the remaining five trigonometric functions in a similar manner. The trigonometric identities required to put each of the results in a more or less standard form are all derived from the familiar set

$$\sin^2 x + \cos^2 x = 1,$$

$$\tan^2 x + 1 = \sec^2 x,$$

and

$$\cot^2 x + 1 = \csc^2 x.$$

It is also possible, and probably easier, to use a triangle similar to the one in Figure V.22 in each case. It would be necessary to label the sides differently for each case, however. The results can be placed in a table as follows:

$D_x(\text{arc } \sin x) = \dfrac{1}{\sqrt{1 - x^2}}$	$d(\text{arc } \sin x) = \dfrac{dx}{\sqrt{1 - x^2}}$	$\displaystyle\int \dfrac{dx}{\sqrt{1 - x^2}} = \text{arc } \sin x + C$
$D_x(\text{arc } \cos x) = \dfrac{-1}{\sqrt{1 - x^2}}$	$d(\text{arc } \cos x) = \dfrac{-dx}{\sqrt{1 - x^2}}$	$\displaystyle\int \dfrac{dx}{\sqrt{1 - x^2}} = -\text{arc } \cos x + C$
$D_x(\text{arc } \tan x) = \dfrac{1}{1 + x^2}$	$d(\text{arc } \tan x) = \dfrac{dx}{1 + x^2}$	$\displaystyle\int \dfrac{dx}{1 + x^2} = \text{arc } \tan x + C$
$D_x(\text{arc } \cot x) = \dfrac{-1}{1 + x^2}$	$d(\text{arc } \cot x) = \dfrac{-dx}{1 + x^2}$	$\displaystyle\int \dfrac{dx}{1 + x^2} = -\text{arc } \cot x + C$
$D_x(\text{arc } \sec x) = \dfrac{1}{x\sqrt{x^2 - 1}}$	$d(\text{arc } \sec x) = \dfrac{dx}{x\sqrt{x^2 - 1}}$	$\displaystyle\int \dfrac{dx}{x\sqrt{x^2 - 1}} = \text{arc } \sec x + C$
$D_x(\text{arc } \csc x) = \dfrac{-1}{x\sqrt{x^2 - 1}}$	$d(\text{arc } \csc x) = \dfrac{-dx}{x\sqrt{x^2 - 1}}$	$\displaystyle\int \dfrac{dx}{x\sqrt{x^2 - 1}} = -\text{arc } \csc x + C$

Note that in the case of the integrals, we have only three distinct integrals, but each one has two possible results. Based upon previous work this would indicate that (arc sin x) and (−arc cos x) must differ by a constant. Can you find the constant in this case? Would the same constant occur in the case of (arc tan x) and (−arc cot x), and in (arc sec x) and (−arc csc x)? It is interesting that we would apparently have two different results, but in fact the results are not different. The fact that the constant, C, is to be determined gives us quite a bit of leeway. This will frequently lead us to results that appear to be different but which are in fact the same. Hence, from this point on it may be more difficult for you to compare answers with someone else and determine whether you are right or wrong. You may both be right, but the answers may just have a different appearance.

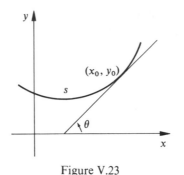

Figure V.23

EXAMPLE 7.1. Find $D_x y$ if $y = (\text{arc tan } e^x)^2$.

Solution. This is certainly a composite function, and hence we must use the chain rule as well as our newly developed formulas. Here we would have

$$D_x(\text{arc tan } e^x)^2 = 2(\text{arc tan } e^x)D_x(\text{arc tan } e^x)$$

$$= 2(\text{arc tan } e^x)\frac{1}{1 + e^{2x}} D_x e^x$$

$$= \frac{2e^x(\text{arc tan } e^x)}{1 + e^{2x}}.$$

We see in this example that we need all of the results so far. With an increasing number of results available, it also becomes increasingly important to make certain that you do not skip steps, for each step in any solution is important.

Now that we have developed formulas for handling the inverse trigonometric functions, we can consider the question of the rate at which the angle which a curve makes with the horizontal changes as we move along the curve. This rate is called the *curvature* of the curve at the point in question and is frequently denoted by K. We must first clarify precisely what it is that we are seeking. Let $y = f(x)$ be the equation of the curve. We will pick a particular point on the curve (x_0, y_0), as shown in Figure V.23. Denote by θ the angle which the line tangent to the curve at (x_0, y_0) makes with the positive x-axis. Let s be the length of the curve from some given point, such as the point at which the curve crosses the y-axis. The origin for measuring s is immaterial, as you will see. We are now concerned with finding $D_s\theta$ or the rate at which the angle θ changes with respect to the length of arc traversed. (We will assume that all derivatives in this discussion are to be evaluated at (x_0, y_0).) Since $\tan \theta = D_x y$, we know that $\theta = \text{arc tan } D_x y$. Hence we have

$$D_x\theta = \frac{D_x(D_x y)}{1 + (D_x y)^2} = \frac{D_x^2 y}{1 + (D_x y)^2}.$$

In Section V.3 we found that $ds = \sqrt{dx^2 + dy^2}$. Since $ds = D_x s\, dx$ by definition, and similarly $dy = D_x y\, dx$, we have $ds = D_x s\, dx = \sqrt{dx^2 + [D_x y\, dx]^2}$ $= \sqrt{1 + [D_x y]^2}\, dx$, and consequently $D_x s = \sqrt{1 + (D_x y)^2}$. If we consider θ to be a function of s and s a function of x, the chain rule states that $D_x \theta = D_s \theta D_x s$. Therefore we have

$$K = D_s \theta = \frac{D_x \theta}{D_x s} = \frac{\dfrac{D_x^2 y}{1 + (D_x y)^2}}{[1 + (D_x y)^2]^{1/2}}$$

$$= \frac{D_x^2 y}{[1 + (D_x y)^2]^{3/2}} = \frac{y''}{[1 + (y')^2]^{3/2}}. \qquad (V.7.1)$$

The last expression is obtained by a change of notation. The reciprocal of curvature is called *the radius of curvature*.

$$R = \frac{1}{K} = \frac{[1 + (y')^2]^{3/2}}{y''}. \qquad (V.7.2)$$

EXAMPLE 7.2. Find the curvature and radius of curvature of $x^2 + y^2 = 25$.

Solution. Since no specific point is given, we will assume that we are to find the curvature and radius of curvature at a general point. We need both y' and y''. Differentiating we have $2x + 2yy' = 0$ or $y' = -x/y$. Note that this is not valid if $y = 0$. Also, we should note whether we have the upper semicircle or the lower semicircle to obtain the correct sign for y'. To obtain y'' we will differentiate y'. Thus we have

$$y'' = -\frac{y - xy'}{y^2} = -\frac{y - x\left(-\dfrac{x}{y}\right)}{y^2} = -\frac{[y^2 + x^2]}{y^3} = -\frac{25}{y^3}.$$

Upon substitution in (V.7.1) we have

$$K = \frac{-\dfrac{25}{y^3}}{\left[1 + \left(-\dfrac{x}{y}\right)^2\right]^{3/2}} = \frac{-\dfrac{25}{y^3}}{\left[\dfrac{(y^2 + x^2)}{y^2}\right]^{3/2}} = -\frac{25}{y^3} \cdot \frac{y^3}{125} = -\frac{1}{5}.$$

This shows that in the case of the circle the curvature is independent of the point being considered. The negative sign arises from the fact that the curve is concave downward when y is positive and upward when y is negative. The radius of curvature is the reciprocal of the curvature, and consequently we see that the radius of curvature is five in absolute value. Furthermore it, too, is a constant. This arises from the fact that we are dealing with a circle and the radius of the circle is five. If we had taken a different curve, such as parabola, the radius of curvature would not have been a constant, for the (circle) most closely approximating the curve would vary in size from point to point along the curve.

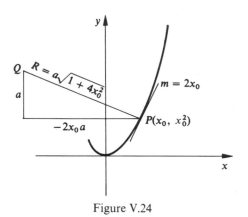

Figure V.24

In Example 7.2 we see that this does indeed give us the radius of the circle. In general this gives the radius of a circle called the *osculating circle* (kissing circle). The osculating circle is the one that most closely fits the curve at the point in question in that it is tangent to the curve and the direction of the tangent line is changing at the same speed on both the circle and the curve at that point.

EXAMPLE 7.3. Find the equation of the curve formed by the centers of the osculating circle to the curve $y = x^2$.

Solution. Since the osculating circle at a point (x_0, y_0) shares a common tangent with the curve and has a radius equal to the radius of curvature, we can locate the center of the circle corresponding to this point P in Figure V.24 by proceeding from P in the direction perpendicular to the tangent at P and toward the *inside* of the curve a distance equal to the radius of curvature. This will serve as an outline for our solution of the problem.

We see first that the slope of the parabola at (x_0, y_0) is given by $y' = 2x_0$, and hence the slope of the radius will be $(-1/2x_0)$ since the slope of perpendicular lines are negative reciprocals of each other. Since $y'' = 2$, we have the radius of curvature, that is the reciprocal of the curvature, given by $R = \frac{1}{2}(1 + 4x_0^2)^{3/2}$. If the point Q in Figure V.24 represents the center of curvature, we know that the slope of PQ is $(-1/2x_0)$ and the length of PQ is R, the radius of curvature. Let the vertical distance from P to Q be given by the unknown constant a. Then the horizontal distance must be $-2x_0 a$ as shown in order to obtain the desired slope. The distance PQ is then obtained by the Pythagorean theorem. Putting all of this information together, we have

$$a\sqrt{1 + 4x_0^2} = \frac{(1 + 4x_0^2)\sqrt{1 + 4x_0^2}}{2}$$

or

$$a = \tfrac{1}{2} + 2x_0^2.$$

Consequently we have for the coordinates (x, y) of Q the relations

$$\begin{cases} x = x_0 - 2x_0 a = x_0 - x_0(1 + 4x_0^2) = -4x_0^3 \\ y = x_0^2 + a = x_0^2 + \frac{1}{2} + 2x_0^2 = \frac{1}{2} + 3x_0^2. \end{cases}$$

Notice that this is a pair of parametric equations with parameter x_0, the x-coordinate of the point of tangency of $y = x^2$. This curve is called the *evolute* of the curve $y = x^2$.

EXERCISES

1. Find y' in each of the following:

   (a) $x^3 y - y^2 + 3x = 4y^3$
   (b) $e^{xy} = \sin x - \cos y$
   (c) $x^3 + x^2 y + y^3 = 45$
   (d) $ye^x + xe^y = 15$
   (e) $\tan(x^2 y) = x^3 + y^3$
   (f) $x^3 + y^3 = 3xy$

2. Differentiate each of the following:

   (a) arc sin e^x
   (b) arc tan x^2
   (c) arc csc (e^{x^2})
   (d) arc cot $(4 - x)$
   (e) arc cos(sin x)
   (f) arc sec($x^2 + 3$)
   (g) $e^{\text{arc tan } x}$
   (h) (arc sec $x)^3$

3. Evaluate:

   (a) $\int_0^1 (1 + x^2)^{-1} dx$
   (b) $\int ((x + 2)^2 + 1)^{-1} dx$ [*Hint*: Use Corollary 2 of Theorem III.3.3.]
   (c) $\int ((x + 3)\sqrt{x^2 + 6x + 8})^{-1} dx$ [*Hint*: What integral does this resemble?]
   (d) $\int_0^{1/2} ((\text{arc sin } x)/\sqrt{1 - x^2}) dx$
   (e) $\int (e^x/\sqrt{1 - e^{2x}}) dx$ [*Hint*: Let $e^x = y$.]

4. We know in general that a function of its inverse function is the identity function. Therefore the derivative should be one. Show that this is the case in each of the following by computing the derivatives using the chain rule.

   (a) sin(arc sin x)
   (b) arc sin(sin x)
   (c) $\sqrt{x^2}$
   (d) $(\sqrt{x})^2$
   (e) sec(arc sec x)

(f) $f(g(x))$ where $f(x) = x/(x - 3)$ and $g(x)$ is the inverse of $f(x)$ if it exists.

(g) cot(arc cot x)

(h) csc(arc csc x)

(i) arc tan(tan x)

5. Find the equation of the tangent at $(3, 2)$ to the curve given by the relation $2x^2 - 3xy - y^2 + 2x - 3y + 4 = 0$.

6. Find the length of the curve $y^2 - 4y + 4 - x^3 = 0$ from the point $(4, 10)$ to $(0, 2)$.

7. Find the curvature and radius of curvature of $y = x^3$ at the point $(2, 8)$.

8. Find the equation of the evolute of $y = x^3$.

9. Find the curvature and radius of curvature of $4x^2 + 9y^2 = 36$ at the point (x_0, y_0).

10. Show that the curvature of the line $ax + by + c = 0$ is zero.

11. (a) Find the curvature of the curve whose equation is $xy = 1$ at the point with coordinates (x_0, y_0).

    (b) Find the point (x_0, y_0) on $xy = 1$ at which the curvature is maximum.

    (c) Show that the point found in part (b) bisects the segment joining the origin and the center of curvature corresponding to this point.

    (d) Show that the radius of curvature for this curve has no maximum value.

12. Find y' if $y^n = x^m$ for positive integers n and m. Show that this gives an alternate derivation of the differentiation formula for the power function.

S13. If P is the price per unit for a certain commodity and Q is the quantity sold, the revenue is PQ.

    (a) Find the marginal revenue in terms of P and Q.

    (b) Show that the marginal revenue can be expressed as $PQ'(1 - 1/E)$ where E is the elasticity of demand.

    (c) Show that the marginal revenue is negative if and only if $E < 1$.

    (d) From this information, indicate the significance of a negative marginal revenue.

S14. Find the total revenue, the marginal revenue, and the elasticity of demand if the demand function is given by $Q = 25/(P + 6)$.

P15. A particle of mass m moves along the x-axis with velocity $v = D_t x$. If it started with initial velocity of v_0 and initial position of x_0, and if it moves such that the relation between its velocity and position is given by the equation $m(v^2 - v_0^2) = k(x_0^2 - x^2)$, show that the effective force on the particle, $mD_t v$ is given by $(-kx)$ where k is a constant of proportionality.

V.8 Exponentials and Logarithms

We made one glaring omission in the last section, for we omitted the inverse of the function e^x. That this was not unintentional may have occurred to you. This inverse brings in a function which has many uses, and it merits a section of its own. Since e^x is monotonic strictly increasing for all values of x, then there must be a unique inverse function without any restriction on the domain of the exponential function. We will denote this inverse function by $L(x)$, and hence

$$e^{L(x)} = L(e^x) = x.$$

Using implicit differentiation and the chain rule,

$$D_x e^{L(x)} = e^{L(x)} L'(x) = 1,$$

and from this

$$xL'(x) = 1,$$

or

$$L'(x) = \frac{1}{x}. \tag{V.8.1}$$

We pause to investigate the domain of $L(x)$. Since $e^x > 0$ for all values of x, and since the range of e^x is the domain of $L(x)$, then the domain of $L(x)$ must consist of just the positive real numbers. This is fortunate, for this insures that there will be no division by zero in evaluating $L'(x)$. It also indicates that $L(x)$ is monotonic strictly increasing since $L'(x) > 0$ for all values of x in the domain of $L(x)$.

We can expect $L(x)$ to behave like a logarithm, and in fact to be a logarithm, for one definition of the logarithm is based on exponents. We can derive some of these same results, however, from the information we have at hand. For instance, we know that $e^{L(1)} = 1 = e^0$, and hence $L(1) = 0$. We also have

$$\int_1^a \frac{1}{x}\,dx = \int_1^a L'(x)dx = L(a) - L(1) = L(a).$$

Therefore $L(x)$ is given by the relation

$$L(x) = \int_1^x \frac{dt}{t}. \tag{V.8.2}$$

Since we have methods for finding rather accurate values of integrals, whether we can formally integrate them or not, we can use the computational schemes suggested in Section 6 of Chapter II to obtain values for the logarithm.

We can readily derive the laws of logarithms from the information we have before us. If we consider $L(ax)$, then by the chain rule we have

$$D_x L(ax) = \frac{1}{ax} D_x(ax) = \frac{1}{ax} a = \frac{1}{x}$$

but this is the same as $L'(x)$, and hence

$$L(ax) = L(x) + C$$

where C is some constant. We observe that if $x = 1$ we have

$$L(a) = L(1) + C = 0 + C = C,$$

or

$$L(ax) = L(x) + L(a).$$

This is merely the statement that the logarithm of the product is the sum of the logarithms of the factors. In similar fashion, if we consider

$$D_x L(x^n) = \left(\frac{1}{x^n}\right) D_x(x^n) = \left(\frac{1}{x^n}\right)(nx^{n-1}) = n\left(\frac{1}{x}\right) = nD_x L(x) = D_x[nL(x)],$$

we have

$$L(x^n) = nL(x) + C.$$

Let x take on the value one. We then have $C = 0$, and consequently the rule $L(x^n) = nL(x)$.

In this last case we were forced to restrict our attention to rational values of n, for we have only derived the derivative of x^n for rational values. However, we can use our present results to remove this restriction. Observe that the integral of e^x was obtained with no restrictions on the domain, and hence by the Fundamental Theorem we had no restrictions on the derivative of e^x. Since e^x is a continuous function, there are no restrictions on the domain of $L(x)$ other than $x > 0$. Therefore, if $x > 0$ we can define x^n for any real value of n by the relation $x^n = e^{nL(x)}$, and in this case we have

$$D_x(x^n) = D_x(e^{nL(x)}) = e^{nL(x)}\left[n\left(\frac{1}{x}\right)\right] = x^n\left[\frac{n}{x}\right] = nx^{n-1}.$$

If n is negative, $(-n)$ is positive. Then

$$D_x x^n = D_x\left(\frac{1}{x^{-n}}\right) = -\frac{(-n)x^{-n-1}}{x^{-2n}} = nx^{n-1}$$

by the quotient rule and the fact that the differentiation formula holds for all real positive values of the exponent. Thus we have finally removed all of the restrictions on the exponent n in x^n, and we find that the derivative formula obtained in Section IV.5 for a restricted set of values, n, is a correct formula for all real values of n.

We have used the notation $L(x)$ in order to emphasize the functional nature of the logarithm, although we have not hesitated to call it the logarithm. We should now point out that this particular function is the logarithm to the base e, and this is referred to as the *natural logarithm*. The logarithm to the base 10, usually called the *common logarithm*, can be obtained from the natural logarithm by multiplying by a constant value, namely the logarithm of e to the base 10, or the common logarithm of e. Since both types of logarithms are used frequently, it is convenient to use different notations in order to keep them distinct in our minds. To this end we will denote the *common logarithm of* x by log x and the *natural logarithm of* x by ln x. Thus, the abbreviation log will denote that we are using base 10, and the abbreviation ln will indicate base e. This formalizes a similar discussion in Chapter III. For the purpose of the calculus, we will prefer the base e, just as we preferred radian measure for the trigonometric functions. This is due primarily to the fact that we have simpler formulas and do not have to carry along some very unappetizing multiplicative factors for both the integration and differentiation of the functions.

We should also note at this point that we have filled in another hole. We had developed the integral formula

$$\int_a^b x^n \, dx = \frac{1}{n+1} [b^{n+1} - a^{n+1}] \tag{V.8.3}$$

and at the time that we developed this we had required that n be restricted to rational numbers and only to those for which this expression would be meaningful. With the work of this section we have now expanded our results so that we see this formula to be correct for all real values of n except for the glaring case of $n = -1$ in which case we would have a zero denominator. Observe, however, that if $n = -1$, we then have the integral

$$\int_a^b x^{-1} \, dx = \int_a^b \frac{1}{x} \, dx = \ln b - \ln a = \ln \frac{b}{a}, \tag{V.8.4}$$

and the only restriction here is that both a and b be positive. If they are both negative, it is possible to substitute $(-y)$ for x, so that $x = -y$, then the limits for y will be positive. Not only is $x = -y$, but also $dx = (-1)dy = -dy$, and hence the integrand appears to be unchanged. Therefore if the limits of integration are both negative, say $-c$ and $-d$, we can handle the situation by substituting $x = -y$ to obtain

$$\int_{-c}^{-d} \frac{1}{x} \, dx = \int_c^d \frac{1}{(-y)} \, d(-y) = \int_c^d \frac{1}{y} \, dy = \ln d - \ln c = \ln \frac{d}{c}. \tag{V.8.5}$$

If the limits of integration are of opposite sign, then $x = 0$ somewhere within the interval of integration, and the integrand would fail to exist at $x = 0$. The result would therefore not be useful.

If we consider the indefinite integral, we have no way in advance of knowing whether we wished to consider this for positive values of x or negative values. Consequently, we can handle both at once by indicating that we will use the absolute value of the expression in question. Hence, we write

$$\int \frac{1}{x} \, dx = \ln |x| + C. \tag{V.8.6}$$

This will be used in the future in the case of both the definite and the indefinite integral. We must still be careful to insure that we do not have the situation in which the integrand may fail to exist.

EXAMPLE 8.1. Evaluate $D_x \log x$.

Solution. Since $x = e^{\ln x} = 10^{\log x} = (e^{\ln 10})^{\log x}$ we can equate the exponents on e and obtain $\ln x = (\ln 10)(\log x)$. From this we have $\log x = \ln x/\ln 10$, and consequently $D_x \log x = D_x(\ln x/\ln 10) = (1/\ln 10)(1/x) = 0.43429(1/x)$. The inconvenient coefficient which appears in this case is a prime reason for the general preference for natural logarithms over common logarithms when the calculus is involved.

With these results, we have completed the consideration of the differentiation of the functions which are frequently referred to as *elementary*. We have also considered the integrals for many new cases in the various sections of this chapter. It is apparent that there are many functions which we would still find somewhat difficult to handle, and we will devote the next chapter to some additional techniques that may ease the problem of finding derivatives and integrals. With all of the formulas we have to date—and all of these will be used in the pages ahead and in the various applications of the calculus—it is almost obvious that one must stop and think to insure using the correct one. As we have admonished in the past, stop before embarking on any problem and think through that problem carefully. In other words, analyze the situation at hand and plan your method of attack rather than plowing ahead without thinking. If you plow a deep rut, it may be very difficult to get out, even though you may be aware that the rut leads in the wrong direction. Human minds seem to work in strange ways, and a mental path once started seems to present a magnet which draws all future attempts into the same path, whether it be correct or not.

EXERCISES

1. Differentiate each of the following:

   (a) $\ln(\sin x)$
   (b) $\ln(\sec x + \tan x)$
   (c) $\ln(x^3 - 4x + 2)$
   (d) $\ln[x^3(4x - 5)^2]$
   (e) $\ln(\sec x)$
   (f) $\ln(\cos x)$
   (g) $\ln(x + 5)^4$
   (h) $x \ln x$

2. Differentiate each of the following:

   (a) $e^{\ln x} - x^2$
   (b) $\ln(\csc x^2)$

   (c) $x^5 \ln x$
   (d) $x \ln x^5$
   (e) $\ln(e^x - 2)$
   (f) $x \ln x - x$
   (g) $\ln \cot 4x$
   (h) $\ln(\cot x \csc x)$

3. Integrate each of the following:

   (a) $\int_1^3 [(4 + x)/x] dx$
   (b) $\int ((3 - x)^3/x^2) dx$
   (c) $\int (\sum_{k=-5}^{5} x^k) dx$
   (d) $\int_1^2 (x + 3/x)^3 dx$
   (e) $\int ((x^2 + x)^2/x) dx$
   (f) $\int (\cos x/\sin x) dx = \int (1/\sin x) d(\sin x)$
   (g) $\int_0^{\pi/4} \tan x \, dx = \int_0^{\pi/4} (\sin x/\cos x) dx$
   (h) $\int [\sin 3x/(\cos 3x - 4)] dx$

4. Evaluate each of the following:

   (a) $D_x^5 \ln(1 - x)$
   (b) $D_x \sin(\ln x)$
   (c) $\int_1^e ((\ln x)/x) dx = \int_1^e \ln x \, d(\ln x)$
   (d) $\int ((\sin(\ln x))/x) dx$
   (e) $D_x^4 \log(1 - x)$
   (f) $D_x \ln(\ln(\ln x))$
   (g) $\int_1^4 (2x/(x^2 + 1)) dx$
   (h) $D_x^2(\ln 3x)^3$

5. Use differentials to approximate each of the following:

   (a) $\ln 0.95$
   (b) $\ln 1.1$
   (c) $\ln \sec 0.04$
   (d) $\ln(\tan 0.75)$ [This is near $\pi/4$.]
   (e) $\ln(\sin 27°)$
   (f) $\ln[1 - (0.08)^2]$

6. Find the root of each of the following equations accurate to three decimal places:

   (a) $x + \ln x = 0$
   (b) $x^2 + \ln x = 0$
   (c) $x^3 + \ln x = 0$
   (d) $e^x + \ln x = 0$

7. Set up the integral for the length of the curve $y = \ln \sec x$ from $(0, 0)$ to $(\pi/3, \ln 2)$. Use a numerical method to find an approximate value for this length.

8. (a) Show that $\ln 2 < 1 < \ln 3$ by approximating the value of the integral in (V.8.2).

(b) Show that the logarithm is a strictly increasing function.

(c) Using part (b) show that the logarithm must have an inverse function.

(d) Show that $DQ \ln(a, a + 1)$ is a monotonic function of x for any positive constant a. Is this increasing or decreasing?

9. Let $g(x)$ be the greatest integer function $g(x) = [x]$. Evaluate each of the following. [You may have to go back to RS sums.]

(a) $\int_{0.5}^{10} \ln x \, dg(x)$

(b) $\int_{0.5}^{10} \ln 3x \, dg(x)$

(c) $\int_{0.5}^{10} \ln x \, dg(3x)$

(d) $\int_{0.5}^{10} \ln 3x \, dg(3x)$

P10. Work is defined as the product of a force and the distance through which the force operates.

(a) If we have a gas in a cylinder pressing against a piston, show that the force on the piston is the pressure of the gas multiplied by the area of the piston against which the gas is pressing.

(b) Show that the work done in moving the piston so as to compress the gas is approximated by the pressure multiplied by the change in volume for small changes in volume.

(c) Find the work done in compressing an ideal gas in a cylinder from 30 cubic inches to 20 cubic inches if the gas was initially at a pressure of 20 lbs per square inch, and the temperature remains constant. [An ideal gas obeys the law $PV = nRT$ where P is pressure, V is volume, n is the number of moles of gas, R is the universal gas constant, and T is temperature measured in degrees Kelvin.]

SB11. (a) The rate of decay of a radioisotope is proportional to the amount present. This can be stated as $D_t A = -kA$ where A is the amount present at time, t, and k is a constant. The negative sign appears due to the fact that A is decreasing. Write this equation in terms of differentials, and then obtain an integral by which you can evaluate A in terms of t.

(b) The half life of carbon (C^{14}) is 5570 years (that is after 5570 years there will be half as much radioactive C^{14} present as there was initially). If a fossil is found in which there is only two percent as much C^{14} as one would normally expect, what is the age of the fossil. (This is the method upon which carbon dating is based.)

SB12. The rate of growth of population is frequently postulated to be proportional to the present population. If the annual rate of increase in population is two per cent of the current population and the world population is currently assumed to be about 4,000,000,000, when will the world population be thirty billion? This assumes that there is no change in the rate of increase.

SB13. There are four frequently postulated growth curves. These are

(i) $P(t) = P(0)e^{Rt}$,

(ii) $P(t) = K - [K - P(0)]e^{-Rt}$,

(iii) $P(t) = KP(0)/(P(0) + [K - P(0)]e^{-KRt})$,
   and
(iv) $P(t) = ae^{(-be^{-Rt})}$,

and are known as *exponential growth, exponential growth to constant value, logistic growth*, and *Gompertz's growth curve* respectively. In each case K and R are constants and $P(t)$ is the population at time, t.

(a) Find the rate of growth in each case.
(b) Determine which of these apparently approach a limiting population size, and determine what it appears to be. [In (ii) and (iii) you should consider both the case in which $K < P(0)$ and $K > P(0)$.]

B14. (a) If you have a population of cells in an ideal situation where they reproduce by simple division, and where the division occurs at specified time intervals without variation, show that the population of the r-th generation is $P(r)$ = $P(0)2^r$.
   (b) If the time for one generation is T, and the variable time is denoted by t, then we can rewrite this equation as $P(t) = P(0)2^{t/T}$. Find the rate of increase of the population as a function of time.
   (c) For what value of T will the growth rate be 2% per year? Compare this result with that obtained in Exercise 12.

M15. (a) Show that $(\log e)(\ln 10) = 1$
   (b) Show that $\int a^x \, dx = a^x(\log e/\log a) + C$
   (c) Show that $D_x a^x - a^x(\log a/\log e) + C$

CHAPTER VI

Techniques of Differentiation and Integration

VI.1 A Look Back

In the last three chapters we have accumulated a rather large number of formulas for both differentiation and integration. It is the purpose of this chapter to bring these results together, and to present ways by which these results can be extended. It would therefore be helpful to place the results obtained so far in tabular form for easy reference. We list here the results for both derivatives and integrals, pairing them off where this is appropriate. The differential can be easily obtained from the derivative, and for that reason will not be listed in this table.

(1a) $D_x[cf(x)] = cf'(x)$

$$\int cf(x)dg(x) = c \int f(x)dg(x)$$

(1b) $\qquad\qquad \int f(x)d[cg(x)] = c \int f(x)dg(x)$

(2a) $D_x[f(x) \pm g(x)] = f'(x) \pm g'(x)$

$$\int [f_1(x) \pm f_2(x)]dg(x) = \int f_1(x)dg(x) \pm \int f_2(x)dg(x)$$

(2b) $\qquad\qquad \int f(x)d[g_1(x) \pm g_2(x)] = \int f(x)dg_1(x) \pm \int f(x)dg_2(x)$

(3) $D_x[f(x)g(x)] = f(x)g'(x) + g(x)f'(x)$

(4) $D_x[f(x)/g(x)] = (g(x)f'(x) - f(x)g'(x))/[g(x)]^2$

(5a) $D_x f(g(x)) = f'(g(x))g'(x)$

$$\int f(x)dg(x) = \int f(x)g'(x)dx$$

(5b) $\qquad\qquad \int_a^b f(h(x))dg(h(x)) = \int_{h(a)}^{h(b)} f(y)dg(y)$

(6a) $D_x(c) = 0 \qquad \int 0 dg(x) = C$

(6b) $D_x(x) = 1 \qquad \int dg(x) = g(x) + C$

(6c) $D_x(x^n) = nx^{n-1} \quad \int x^n dx = x^{n+1}/(n+1) + C, n \neq -1$

(7) $D_x(e^x) = e^x$ $\quad\quad \int e^x dx = e^x + C$

(8) $D_x(\ln x) = 1/x$ $\quad\quad \int (1/x)dx = |\ln x| + C$

(9) $D_x(\sin x) = \cos x$

$$\int \cos x \, dx = \sin x + C$$

(10) $D_x(\cos x) = -\sin x$

$$\int \sin x \, dx = -\cos x + C$$

(11) $D_x(\tan x) = \sec^2 x$

$$\int \sec^2 x \, dx = \tan x + C$$

(12) $D_x(\cot x) = -\csc^2 x$

$$\int \csc^2 x \, dx = -\cot x + C$$

(13) $D_x(\sec x) = \sec x \tan x$

$$\int \sec x \tan x \, dx = \sec x + C$$

(14) $D_x(\csc x) = -\csc x \cot x$

$$\int \csc x \cot x \, dx = -\csc x + C$$

(15) $D_x(\arcsin x) = 1/\sqrt{1 - x^2}$

$$\int (1/\sqrt{1 - x^2})dx = \arcsin x + C, |x| < 1$$

(16) $D_x(\arccos x) = -1/\sqrt{1 - x^2}$

$$\int (1/\sqrt{1 - x^2})dx = -\arccos x + C, |x| < 1$$

(17) $D_x(\arctan x) = 1/(x^2 + 1)$

$$\int (1/(x^2 + 1))dx = \arctan x + C$$

(18) $D_x(\text{arc cot } x) = -1/(x^2 + 1)$

$$\int (1/(x^2 + 1))dx = -\text{arc cot } x + C$$

(19) $D_x(\text{arc sec } x) = 1/x\sqrt{x^2 - 1}$

$$\int (1/x\sqrt{x^2 - 1})dx = \text{arc sec } x + C, |x| > 1$$

(20) $D_x(\text{arc csc } x) = -1/x\sqrt{x^2 - 1}$

$$\int (1/x\sqrt{x^2 - 1})dx = -\text{arc csc } x + C, |x| > 1$$

As you will agree, this is a rather impressive list. You should keep this list at hand, or better yet be able to derive each result. However, this list is deficient in some rather obvious places as well as in some places that are not so obvious. We do not have an integral of the tangent, cotangent, secant, or cosecant, for instance, and there are many other functions that you might think of for which we do not have integrals, such as $(1 - x^2)^{1/2}$. In the next section we will introduce the use of logarithms for the purpose of simplifying some differentiations and making possible some others. In the remaining sections of this chapter we will provide methods for filling in some of the gaps in integration, such as those referred to above. The fact that some of the results may be a bit less than obvious should not surprise you, however, for there are many integrals for which no one can find a formula. (There are also some continuous functions for which no derivative exists.) We will not dwell further on the negative, but will amplify our arsenal of differentiation and integration techniques so that we can handle a wider spectrum of problems in the future.

One point should be made here. It seems that we have been dealing primarily with the Riemann integral rather than with the Riemann–Stieltjes integral. However, you will remember that

$$\int_a^b f(x)dg(x) = \int_a^b f(x)g'(x)dx.$$

If we consider $f(x)g'(x)$ as the integrand, we have the Riemann integral. While this is dependent upon the existence of $g'(x)$, and while the RS integral can exist without requiring that $g'(x)$ exist, the majority of the *nice* cases occurring in applications are covered by this use of the Riemann integral to evaluate the RS integral. Hence the results we have obtained are much more general than might appear at first glance.

We will close this section with some examples of ways in which the use of the listed results can be extended. Some of these review methods used in Chapter V.

EXAMPLE 1.1. Evaluate

$$\int ((x + 2)/(x^2 + 4x + 9))dx$$

Solution. This apparently does not fit any of the results listed earlier. However, if we would consider the use of the integral (5b) above, we might let $g(x) = x^2 + 4x + 9$, and then replace $g(x)$ by y. This requires that $dg(x) = g'(x)dx = (2x + 4)dx = 2(x + 2)dx$, and we note that we could rewrite the integral as

$$\int \frac{(x + 2)dx}{x^2 + 4x + 9} = \int \frac{2(x + 2)dx}{2(x^2 + 4x + 9)} = \frac{1}{2}\int \frac{dg(x)}{g(x)} = \frac{1}{2}\int \frac{dy}{y} = \frac{1}{2}\ln|y| + C$$

$$= \frac{1}{2}\ln|x^2 + 4x + 9| + C$$

$$= \ln\sqrt{x^2 + 4x + 9} + C.$$

Observe that the integration became very easy, simply by a substitution and the use of a standard result. Note, also, that we obtained a logarithm for the result, even though the integrand gave no apparent hint that such would be the case.

While it is frequently true that an expression in the denominator leads to a logarithm in the result, it is not always true, and you should observe very carefully the considerations which distinguish the various cases that can arise.

EXAMPLE 1.2. Evaluate

$$\int \frac{(x + 2)dx}{(x^2 + 4x + 9)^{5/2}}$$

Solution. This integral appears not too dissimilar from the integral of Example 1.1. This suggests that we might try the same technique. However, we find that we have a slight alteration in the resulting integral after substitution. This requires that we use a different one of the results from our list.

$$\int \frac{(x + 2)dx}{(x^2 + 4x + 9)^{5/2}} = \frac{1}{2} \int [g(x)]^{-5/2} \, dg(x) = \frac{1}{2} \int y^{-5/2} \, dy$$

$$= \frac{1}{2} \frac{y^{(-5/2)+1}}{(-5/2) + 1} + C$$

$$= \frac{1}{2} \frac{[g(x)]^{(-5/2)+1}}{(-5/2) + 1} + C$$

$$= -\frac{1}{3} (x^2 + 4x + 9)^{-3/2} + C.$$

Thus if the exponent of the term in the denominator is one, we have the unique case for which the result (6c) fails to work. It is the single case in which the logarithm appears as a result of integration. Failure to note the distinction between these two cases is one of the more frequent causes of error by the unobservant.

The same principle of substitution that we have seen in the last two examples can be seen in a slightly different way in the following example.

EXAMPLE 1.3. Evaluate $\int x^2 \sin x^3 \, dx$.

Solution. It is apparent that we are dealing with a composite function, $\sin (x^3)$. We should use $f(x) = \sin x$ and let $g(x) = x^3$. In order to use the integral result (5a), we will need $g'(x)$, but we see that $g'(x) = 3x^2$, and we already have the x^2 in the integrand. Hence we write

$$\int x^2 \sin x^3 \, dx = \int \frac{1}{3} \sin x^3 (3x^2) dx = \frac{1}{3} \int \sin x^3 \, d(x^3) = -\frac{1}{3} \cos x^3 + C.$$

Observe that if we had not had the x^2 already present this method would have failed, for while we can let a constant factor filter through the integral sign, it is not correct to do the same thing for a non-constant factor. In this case we would have had to attempt something else. We will be pursuing other techniques in the remainder of this chapter. It should be noted that integrals exist for which no method known to man will work. In this case we could resort to numerical integration, provided of course the integral is a definite integral.

EXAMPLE 1.4. Evaluate $\int xe^{-x^2}\, dx$.

Solution. Here we have an integrand that has something in common with integral formula (7), but which is not precisely the same as that result. We could handle this, however, if we were to let $g(x) = (-x^2)$, in which case we would have $g'(x) = (-2x)$. Because the additional factor, x, is present we can manage this case. It is possible to provide the factor (-2) since (-2) is a constant. Note that if the x were not present we would be in trouble. Our results now appear as

$$\int xe^{-x^2}\, dx = \int\left(-\frac{1}{2}\right)e^{-x^2}(-2x)dx = -\frac{1}{2}\int e^{-x^2}\, d(-x^2) = -\frac{1}{2}e^{-x^2} + C.$$

The moral to these examples is that there are many ways in which we can expand the use of our table, but it will require keeping your eyes open and your mind alert. We will be dealing with other, less obvious, substitutions and techniques in the remaining sections of this chapter. There is no set rule by which one can determine what method to use at any given time other than *experience*, but keep trying, for this is the only way that one can gain experience.

EXERCISES

1. Differentiate:

    (a) $x^2 \ln 3x + e^{-x^2/2}$
    (b) arc tan(sin x).
    (c) 4^x
    (d) $(x^2 - 3\cos^3 x^3)/\sqrt{x^2 + 2}$
    (e) x^3 arc sec$(x^2 - 3x)$
    (f) $e^{\sin x}$ arc cos x^2
    (g) ln(csc $4x$)
    (h) x ln(arc csc x^3)

2. Integrate each of the following:

    (a) $\int x \sec^2 x^2\, dx$
    (b) $\int ((x - 1)/(x^2 - 2x + 2))dx$
    (c) $\int e^{(1-x)}\, dx$
    (d) $\int x^2 \cos(x^3 + \pi)dx$
    (e) $\int (1/\sqrt{1 - x^2})dx$
    (f) $\int (x/\sqrt{1 - x^2})dx$
    (g) $\int (x/(1 - x^2))dx$
    (h) $\int (1/\cos^2 2x)dx$

3. Evaluate each of the following:

    (a) $\int_0^1 (x^3/(4 - x^4))dx$
    (b) $\int_0^{\sqrt{2}/2} (x/\sqrt{1 - x^4})dx$

(c) $\int_{\pi/6}^{\pi/4} ((\sin x)/(\cos^2 x))dx$

(d) $\int_0^{\pi/4} e^{\tan x} \sec^2 x \, dx$

(e) $\int_1^e (\sin(\ln x)/x)dx$

(f) $\int_{-2}^3 \tan(\arc \tan x)dx$

(g) $\int_{\pi/4}^{\pi/3} \tan^3 x \sec^2 x \, dx$

(h) $\int_2^3 x(x^2 - 2)^3 dx$

4. Find the area bounded by the curve $xy = 4$, $x = 1$, and $x = n$

   (a) For $n = 5$
   (b) For $n = 10$.
   (c) For $n = 100$.
   (d) For $n = 1000$.
   (e) How large might this area become if n is allowed to increase indefinitely?

5. The area bounded by $xy = 4$, $x = 1$ and $x = n$ is revolved about the x-axis to form a solid of revolution.

   (a) Find the volume of this solid if $n = 5$.
   (b) Find the volume of this solid if $n = 10$.
   (c) Find the volume of this solid if $n = 100$.
   (d) Find the volume of this solid if $n = 1000$.
   (e) How large might this volume become if n is allowed to increase indefinitely?

6. The area bounded by $y = \sec x$, $x = \pi/4$, $x = \pi/3$, and $y = 0$ is revolved about the x-axis. Find the volume of revolution.

7. Set up the integral for the length of the curve $y = \ln(4 \csc x)$ from the point $(\pi/6, \ln 8)$ to the point $(\pi/2, \ln 4)$. Approximate this length numerically.

8. Find the length of the curve $y = x^{3/2} + 2$ from $(1, 3)$ to $(4, 10)$.

9. Find the area bounded by the curve $r = \csc \theta$, $\theta = \pi/6$, and $\theta = \pi/2$.

10. Find approximate values of

    (a) $\sec 47°$.
    (b) $\arc \cos(0.52)$
    (c) $\tan(7/13)$
    (d) $\arc \sec 1.93$

11. Using the fact that $\sec^2 x \geq 1$ show that $\tan x \geq x$ if x is in the interval $(0, \pi/2)$ and $\tan x \leq x$ if x is in the interval $(-\pi/2, 0)$.

12. A ship is sailing at 20 miles per hour due north and staying just 4 miles from a north–south shore line. A man is standing on the shore watching the ship. How fast must his eyes be turning to watch the ship when the ship is 4 miles farther north than he is?

13. A picture four feet high is hanging on a wall with the bottom six feet above the floor. How far away from the wall should a person stand in order that the vertical image of the picture would cover as wide an angle on the retina of his eyes as possible?

VI.2 Logarithmic Differentiation

We need to fill one gap in our differentiation formulas. We can differentiate a function of the type x^n where we have a variable raised to a constant power. We can also differentiate a function of the type a^x where we have a constant raised to a variable power. We might raise the question concerning both the constant raised to a constant power and the variable raised to the variable power. The first of these two cases is one of those relaxing kind that you can dash off without any pencil and paper, and thus look like a genius, for you know at once that a constant raised to a constant power is a constant, and therefore the derivative must be zero. *Quod Erat Demonstrandum* (the deed is done!). The other problem gives pause for thought, however, and perhaps you were thinking of it while we were doing the easy one. In this case we are trying to find the derivative of a function such as

$$f(x) \doteq [g(x)]^{h(x)}.$$

This is a somewhat frightening function however you view it. But we are not vanquished yet, for we remember the last Section of Chapter V, and decide to use logarithms. Therefore we write

$$\ln[f(x)] = \ln([g(x)]^{h(x)}) = h(x)\ln[g(x)].$$

The last term is a product and we know how to differentiate a product. Hence, we can use implicit differentiation and obtain

$$\frac{1}{f(x)}f'(x) = h(x)\left[\frac{1}{g(x)}\right] + h'(x)\ln[g(x)].$$

We multiply both sides by $f(x)$ and have the expression for $f'(x)$. We already know $f(x)$ as a function of $g(x)$ and $h(x)$. Consequently we can express $f'(x)$ completely as a function of $g(x)$, $h(x)$, $g'(x)$, and $h'(x)$ by replacing $f(x)$ by $[g(x)]^{h(x)}$. The final result becomes

$$f'(x) = h(x)[g(x)]^{h(x)-1}g'(x) + h'(x)[g(x)]^{h(x)}\ln[g(x)]. \qquad \text{(VI.2.1)}$$

It is not suggested that this be memorized. The method for obtaining this is one that is worth noting. By taking the logarithm of each side, we have transformed that which did not conform to any of our formulas into a more familiar function. We had the machinery needed after all. We might observe, of course, that if $g(x) = c$, then $g'(x) = 0$ and (VI.2.1) reduces to $f'(x) = (\ln c)c^{h(x)}h'(x)$. This is exactly what we would have had using the methods of Chapter V and the chain rule. Note that if $c = e$, then $\ln e = 1$ and this is the chain rule amplification of formula (7) of Section VI.1. On the other hand, if $h(x) = c$, then $h'(x) = 0$ and we have $f'(x) = c[g(x)]^{c-1}g'(x)$ which is the corresponding amplification of formula (6).

While it is not profitable to remember the derivative of a variable raised to a variable power (for the simple reason that this does not occur very frequently in applications) the method can be useful. We illustrate this in the following example.

EXAMPLE 2.1. Evaluate $D_x[x^3 \sin^2 x(4 - x^2)^{1/2}]/[(x - 2)^5\sqrt{\cos x}]$.

Solution. This is a quotient, so we could use formula (4) and follow this with a liberal use of the formulas for the product and the chain rule. The resulting algebra required to simplify the result would probably keep you busy for some time. Let us try logarithms, however, for with the title of this section we would be surprised if they did not enter in. We would then differentiate

$$\ln f(x) = 3 \ln x + 2 \ln(\sin x)$$
$$+ (1/2)\ln(4 - x^2) - 5 \ln(x - 2) - (1/2)\ln(\cos x).$$

This becomes, upon differentiation,

$$\frac{1}{f(x)}f'(x) = \frac{3}{x} + \frac{2 \cos x}{\sin x} + \frac{-2x}{2(4 - x^2)} - \frac{5}{x - 2} - \frac{-\sin x}{2 \cos x}$$

$$= \frac{3}{x} + 2 \cot x - \frac{x}{4 - x^2} - \frac{5}{x - 2} + \frac{\tan x}{2}.$$

Therefore

$$f'(x) = \left[\frac{3}{x} + 2 \cot x - \frac{x}{4 - x^2} - \frac{5}{x - 2} + \frac{\tan x}{2}\right]\left[\frac{x^3 \sin^2 x\sqrt{4 - x^2}}{(x - 2)^5\sqrt{\cos x}}\right].$$

We have not had to use results for products or quotients.

The use of the logarithm can frequently be helpful in making the differentiation of complicated functions easier. Therefore, the results are less likely to have errors. Note, however, that the logarithm is helpful primarily in those cases in which we have products and quotients, for it is *NOT* true that we can apply logarithms easily to sums and differences. This is another of our labor saving devices, but only in those places in which the logarithm is appropriate.

With these results and some ingenuity on your part, you should now be able to differentiate any function you are likely to see in applications. If all else fails, of course, you can go back to the definition of the derivative, but the formulas and relations we have developed can very probably make life easier for you if they are used correctly. Remember to keep your eyes open for the easiest, most direct way in each case. The fewer the steps, the fewer the chances for making mistakes in copying, writing something down, or in just plain thinking.

EXERCISES

1. Differentiate:

   (a) $\sqrt{x^3 - 3} \sin^2 2x/7e^{4x}$
   (b) $e^{3x} \sin 4x + 17$
   (c) $e^{3x} \sin(4x + 17)$
   (d) $x^{\sin^2 x}$

(e) $(2x/\sqrt{x^2 - 3}) + (4 - x)e^x$

(f) $(x + 2)/(x - 1)$

(g) $(x^3 + 3x^2 - 7x)/(6 + e^x)$

(h) $4^{\sin x} + x^\pi - \sqrt{\pi}$

2. Differentiate:

(a) x^x

(b) $x^{(x^x)}$

(c) $x^{(x^{(x^x)})}$

(d) $(\tan x)^{\sin x} + (\sin x)^{\tan x}$

(e) $(\sin x)^{\cos x}$

(f) $\ln(x^{\sin x})$

(g) $(\ln x)^{\sin x}$

(h) $x^{\arctan x}$

3. Evaluate each of the following:

(a) $D_x \ln(\ln(\ln x))$

(b) $\int dx/(x(\ln x)(\ln(\ln x))(\ln(\ln(\ln x))))$

(c) $D_x \ln(\ln(\ln(\ln x)))$

(d) $\int dx/(x(\ln x)(\ln(\ln x))(\ln(\ln(\ln x)))(\ln(\ln(\ln(\ln x)))))$

4. Find $D_x^2(x^x)$, $D_x^3(x^x)$, and $D_x^4(x^x)$.

5. (a) Show that a solution of $x^x = e$ is also a solution of $x \ln x = 1$.

   (b) Find a solution of $x^x = e$ accurate to three decimal places.

6. Find the extreme points of $f(x) = xe^{-x^2}$.

7. Find the maximum value of $x^{\sin x}$ in the interval, $(0, \pi]$. Obtain this result with an error no greater than 0.01.

8. If $f(x) = e^{-x} \sin x$, use differentials to approximate each of the following:

(a) $f(0.1)$

(b) $f(-0.1)$

9. Find $D_x y$ if $x(t) = (\sin t)^{\tan t}$ and $y(t) = (\cos t)^{\cos t - t^3}$

10. Let $f(x) = (x - a)^k g(x)$ with $g(a) \neq 0$. Show that $(x - a)^{k-1}$ is a divisor of $f'(x)$, but $(x - a)^k$ is not a divisor of $f'(x)$.

11. Let $f(x) = g_1(x)g_2(x)g_3(x) \cdots g_k(x)$. Show that

$$\frac{f'(x)}{f(x)} = \sum_{j=1}^{k} \frac{g_j'(x)}{g_j(x)}.$$

12. (a) Let $f(x) = (x - a_1)(x - a_2)(x - a_3) \cdots (x - a_n)$. Use the results of Exercise 11 to show that

$$f'(x) = f(x) \sum_{j=1}^{n} \frac{1}{x - a_j} = \sum_{j=1}^{n} \frac{f(x)}{x - a_j}.$$

(b) Use the result of part (a) to find the derivative of

$$f(x) = x(x - 1)(x - 2)(x - 3)(x - 4)(x - 5)(x - 6)$$

(c) Without substituting any values in $f'(x)$ show that $f'(x) \neq 0$ if $x = 0, 1, 2, 3,$ 4, 5, or 6.

(d) If $a = 0, 1, 2, 3, 4, 5,$ or 6 show that $f'(a)$ is the product of

$$(a - 0)(a - 1)(a \quad 2)(a - 3)(a - 4)(a - 5)(a - 6)$$

from which the single factor $(a - a)$ has been removed.

VI.3 Integration of Trigonometric Functions

The integration of trigonometric functions can be very useful in a surprisingly large number of integration problems, as we shall see in a later section. In order to prepare for the onslaught of trigonometric integrals, we will take time in this section to obtain additional formulas and to illustrate the use of trigonometric identities. At the outset it might be worth noting that if you have ever wondered in the past concerning the fact that you were exposed to trigonometric identities, that wonderment should be removed by the time you finish this section. Remember that an identity is an equation which is correct for every element of the domain, and hence either of the two identical expressions can be employed without altering the problem. We will start with an easy one.

EXAMPLE 3.1. Evaluate $\int \tan x \, dx$.

Solution. We note that

$$\int \tan x \, dx = \int \frac{\sin x \, dx}{\cos x}$$

$$= \int \frac{-d(\cos x)}{\cos x}$$

$$= -\ln|\cos x| + C = \ln \left| \frac{1}{\cos x} \right| + C$$

$$= \ln|\sec x| + C. \tag{VI.3.1}$$

We can check this result very easily by differentiating $\ln \sec x + C$ and seeing that we then have

$$\frac{\sec x \tan x}{\sec x} = \tan x.$$

Therefore our solution is correct. There would have been no harm in retaining $[-\ln|\cos x|] + C$, but the result $\ln|\sec x| + C$ is more frequently seen.

EXAMPLE 3.2. Evaluate $\int \cot x \, dx$

Solution. Here

$$\int \cot x \, dx = \int \frac{\cos x \, dx}{\sin x} = \int \frac{d(\sin x)}{\sin x} = \ln|\sin x| + C. \quad \text{(VI.3.2)}$$

Since the result is already positive, there is no reason for modifying it. If you happen to prefer $-\ln|\csc x| + C$, this is all right too.

The two results above indicate a rather strange state of affairs, for we seem to find that we have both logarithms and trigonometric functions appearing in the same solution. If you haven't yet arrived at the point where nothing is surprising, you should be getting to that state soon, for even stranger results may occur. These, in fact, were rather straight forward, and we had no major difficulties in getting to them. We had only to note that we had precisely the case in which the integral is a logarithm. We now proceed to the secant and cosecant. These are not quite so obviously tied to the logarithm. We will take the cosecant first, because oddly enough that is the easier one to handle. We will make use of the formulas for double angles. We will consider x as a double angle and then $x/2$ will be the primary angle. The double angle formulas, or the half angle formulas as the case may be, are very useful relations in many integration problems involving trigonometric functions. Let us see how they work.

EXAMPLE 3.3. $\int \csc x \, dx$.

Solution. We can write

$$\int \csc x \, dx = \int \frac{1}{\sin x} \, dx.$$

Now by the double angle formulas, we have the fact that

$$\sin x = \sin 2\left(\frac{x}{2}\right) = 2 \sin\left(\frac{x}{2}\right)\cos\left(\frac{x}{2}\right)$$

and therefore

$$\int \csc x \, dx = \int \frac{1}{2\left(\sin \dfrac{x}{2}\right)\left(\cos \dfrac{x}{2}\right)} \, dx$$

$$= \int \frac{\left(\sec^2 \dfrac{x}{2}\right)\dfrac{1}{2} \, dx}{\tan \dfrac{x}{2}}.$$

We multiplied both numerator and denominator of the integrand by $\sec^2(x/2)$ to obtain in the denominator (without reference to the "2" which is taken into the numerator to form the $1/2$)

$$\left(\sin\frac{x}{2}\right)\left(\cos\frac{x}{2}\right)\left(\sec^2\frac{x}{2}\right).$$

Since the secant is the reciprocal of the cosine we obtain, upon simplification,

$$\frac{\sin\dfrac{x}{2}}{\cos\dfrac{x}{2}} = \tan\frac{x}{2}.$$

Now we have in the numerator of this integrand just $d(\tan(x/2))$, and therefore this becomes

$$\int \csc x\, dx = \int \frac{d\left(\tan\dfrac{x}{2}\right)}{\tan\dfrac{x}{2}} = \ln\left|\tan\frac{x}{2}\right| + C.$$

This last result is perfectly correct, but it is more common to use some additional identities and eliminate the reference to $x/2$. Since

$$\tan\frac{x}{2} = \frac{1 - \cos x}{\sin x} = \csc x - \cot x,.$$

we have the more common form for this integral

$$\int \csc x\, dx = \ln|\csc x - \cot x| + C. \tag{VI.3.3}$$

We realize that this dose of trigonometric identities is a big one, but you will find these all derived in Appendix A. It might be well to take a look at the relations listed there, for we will be using them a great deal in this chapter. Have you ever experienced the feeling that you learn one course when you take the next one? Perhaps this is the time to become familiar with the more common identities from trigonometry. They aren't hard, but it does take some getting used to them in order to be able to toss them around like old friends. We will see what can be done toward making them just that—old friends. Having engaged in this bit of legerdemain, let us now try the secant.

EXAMPLE 3.4. Evaluate $\int \sec x\, dx$.

Solution. We can use the relation between functions of complementary angles and write $\sec x = \csc(\pi/2 - x)$. Our integral becomes $\int \csc(\pi/2 - x)dx$.

We now let $f(x) = \csc x$ and $g(x) = \pi/2 - x$. We have the composite function again, and since $g'(x) = -1$, we can write

$$\int \sec x \, dx = \int (-1)\csc\left(\frac{\pi}{2} - x\right)d\left(\frac{\pi}{2} - x\right)$$

$$= -\ln\left|\csc\left(\frac{\pi}{2} - x\right) - \cot\left(\frac{\pi}{2} - x\right)\right| + C$$

$$= -\ln|\sec x - \tan x| + C$$

$$= \ln\left|\frac{1}{\sec x - \tan x}\right| + C$$

$$= \ln\left|\frac{\sec x + \tan x}{\sec^2 x - \tan^2 x}\right| + C$$

$$= \ln|\sec x + \tan x| + C. \tag{VI.3.4}$$

The last four steps are for simplification only. The final step results from the fact that $\sec^2 x = \tan^2 x + 1$ is an identity.

Note that each of the results along the way is correct, although not as neat as we might like. In case you are wondering, there are other methods for deriving the integrals of the secant and cosecant, but they involve the multiplication of the numerator and denominator of the integrand by $(\sec x + \tan x)$ and $(\csc x - \cot x)$ respectively. If we had had a crystal ball with the answer showing, we would have seen these at once. Without the insight to produce such weird *integrating factors* however, the methods we have used produce the same results, and also give some insight into the use of formulas which we will find helpful from time to time.

The next set of trigonometric results will include such integrands as $\sin^2 x$ and $\cos^2 x$. Again we use some of those identities that you have always enjoyed. Specifically we will use

$$\sin^2 x = (1 - \cos 2x)/2 \quad \text{and}$$
$$\cos^2 x = (1 + \cos 2x)/2. \tag{VI.3.5}$$

You might note that these check to the extent that $\sin^2 x + \cos^2 x = 1$. If you are hazy concerning their derivation, consult Appendix A again. You will observe that these identities make integrable some things that did not appear to be integrable.

EXAMPLE 3.5. Evaluate $\int \sin^2 x \, dx$ and $\int \cos^2 x \, dx$.

Solution. If we try to use the power function formula, we would try to handle an integral of the type $\int y^2 \, dy = \int (\sin x)^2 \, d(\sin x)$. Note that we do not have anything that could possibly provide $d(\sin x)$ for us. Consequently,

this leads to a dead end street. You will not be in the least surprised, however, in view of the suggestion above to find that if we write

$$\int \sin^2 x \, dx = \int (1/2)(1 - \cos 2x) dx = (1/2) \left[\int dx - \int \cos 2x \, dx \right]$$

we can evaluate the last expression and obtain

$$\int \sin^2 x \, dx = \frac{x}{2} - \frac{\sin 2x}{4} + C.$$

We have used, of course, the fact that $\cos 2x \, dx = (1/2)\cos 2x \, d(2x)$, for
without this we could not have handled the last integral properly. It is also
possible to replace $\sin 2x$ by its equivalent expression, $2 \sin x \cos x$, but
this is not necessary. We find that a similar method works with the integral
of $\cos^2 x$, and we have

$$\int \cos^2 x \, dx = \frac{x}{2} + \frac{\sin 2x}{4} + C.$$

Make certain that you know how to evaluate this, for this will provide a
test of your understanding of the derivation we wrote out in detail.

The results so far obtained permit us to integrate many of the functions
that appear in applications. Although we would prefer to be able to handle
anything that comes to hand, it should be pointed out that this last wish is
one that cannot be granted. No matter how long you may pursue the study
of mathematics, there will be integrals that you cannot evaluate with formal
techniques. This is one fact that makes it slightly more pallatable to go to
greater lengths when we have to, for if we can possibly find a formal result
it does save going back to the RS sums. Through the years mathematicians
have found many methods that in some instances would appear to be tricks
for performing various integrations. We will not attempt to give all of them,
but we will enlarge our arena of operations with regard to trigonometric
integrals here. In case you are wondering why we are spending so much
time first on the trigonometric integrals before starting on what would seem
to be a better starting point, namely the algebraic functions, you will find
out in Section VI.5 that one of the most powerful methods for handling
many algebraic integrands is through the medium of the trigonometric
integrals. Surprising? Perhaps, but then many of the results which we obtain
might appear surprising at first glance.

With this brief pause for catching our breath, we forge ahead. Our next
consideration will be integrals of the type

$$\int \sin^m x \cos^n x \, dx$$

where m and n are non-negative integers. Later on we will find methods by
which we can reduce such integrals to expressions involving integrals with

smaller exponents, provided m and n are large enough to make this profitable. For the moment, however, we will face the problem directly.

EXAMPLE 3.6. Evaluate $\int \sin^m x \cos^n x \, dx$ where m and n are non-negative integers.

Solution. We will consider essentially two cases in our solution. In the first case we will assume that either m is odd, n is odd, or both m and n are odd. It is not difficult to determine that the second case will involve the situation in which both m and n are even.

Case 1. Since either m or n is odd, let us assume m to be odd, and then $m - 1$ is even. (The same argument would hold if we had selected n to be odd.) Now we write

$$\int \sin^m x \cos^n x \, dx = \int \sin^{m-1} x \cos^n x (\sin x \, dx)$$

$$= \int (\sin^2 x)^{(m-1)/2} \cos^n x [-d(\cos x)]$$

$$= -\int (1 - \cos^2 x)^{(m-1)/2} \cos^n x \, d(\cos x).$$

But if we let $y = \cos x$ this last result is equivalent to

$$-\int (1 - y^2)^{(m-1)/2} y^n \, dy,$$

and we have the integral of a polynomial. The power function formula can now be used. Then each y can be replaced with $\cos x$. Note that our prime requirement for using the substitution $\sin^2 x = 1 - \cos^2 x$ with a chance of a tractable result is merely that the power of $\sin^2 x$ be an integer. This will happen provided m is odd.

In similar fashion if n were odd, we would use $d(\sin x) = \cos x \, dx$. This would require one of the factors of the form $\cos x$, and since n is odd, we would have an even power of $\cos x$ remaining.

Case 2. If both m and n are even integers, then the method employed in Case 1 would not help, for we would find ourselves faced with $\sqrt{1 - \sin^2 x}$ or $\sqrt{1 - \cos^2 x}$ in the integral. If we stop long enough to consider the identities we have already used we would realize at once that if we were to replace $\sin^2 x$ and $\cos^2 x$ by $(1 - \cos 2x)/2$ and $(1 + \cos 2x)/2$ respectively, we have a great deal more algebra to contend with but we have cut in half the size of the exponents involved. Furthermore, since both m and n are even, we have not introduced any obnoxious square roots. We can now apply Case 1 or reapply Case 2, noting that if the latter situation occurs, we are always lowering the powers involved. At some point the exponents must come down to the case of a sine or cosine to the first power or to the second power, and we know how to handle each of these cases.

EXAMPLE 3.7. Evaluate $\int \sin^2 x \cos^4 x\, dx$.

Solution. In this case we have both exponents even. We therefore have an illustration of Case 2. Hence

$$\int \sin^2 x \cos^4 x\, dx = \frac{1}{8} \int (1 - \cos 2x)(1 + \cos 2x)^2\, dx$$

$$= \frac{1}{8} \int (1 - \cos^2 2x)(1 + \cos 2x)dx$$

$$= \frac{1}{8} \int (1 + \cos 2x - \cos^2 2x - \cos^3 2x)dx.$$

(Note that in the second step we made use of the fact that the product of the sum and the difference of two variables is the difference of their squares. This eases the algebra somewhat. It is always well to look for such labor saving items.) We now have four integrals, the first presenting no problem, the second being a standard result, the third being an application of an earlier example in this section and the fourth being an illustration of Case 1 of the last example given.

Upon using these separate techniques, we obtain

$$\int \sin^2 x \cos^4 x\, dx = \frac{1}{8}\left[x + \left(\frac{\sin 2x}{2}\right) - \left(\frac{x}{2} + \frac{\sin 4x}{8}\right)\right.$$

$$\left. - \left(\frac{\sin 2x}{2} - \frac{\sin^3 2x}{6}\right)\right] + C.$$

This can be simplified, but only to the extent of combining like terms. You should, of course, check each of the results listed in parentheses above to make certain that you know how they are obtained.

The next question arises concerning the possibility that other trigonometric functions appear in the integrand. The identities $\tan^2 x + 1 = \sec^2 x$ and $\cot^2 x + 1 = \csc^2 x$ would apply if we had only secants and tangents, or only cosecants and cotangents. If there is any other mixture, it is probably preferable to first put the integrand entirely in terms of sines and cosines using the relations

$$\tan x = \frac{\sin x}{\cos x}, \quad \cot x = \frac{\cos x}{\sin x}, \quad \sec x = \frac{1}{\cos x}, \quad \csc x = \frac{1}{\sin x}.$$

The next attempt should be to determine whether it is possible to use a sine or a cosine in combination with dx to obtain $d(\sin x)$ or $d(\cos x)$. If this can be done and if the remainder of the integrand can then be put in terms of the function for which we have the differential, we have reduced the problem to one that can be handled as an algebraic function by the substitution $y = \sin x$ or $y = \cos x$. If this latter case is not possible, it may be necessary to use integration by parts (which will be discussed later in this chapter), to use half angle or double angle formulas, as we did on the integral of the cosecant, or to use any other device that may suggest itself. It is indeed possible to find some integrands for which there is no technique by which we can find a formal result, but you should not use this as an excuse for giving up at an early point in your endeavors. You will find that it is very seldom (if ever) that a problem in the book or one given by your instructor fails to have some technique that will apply. The situation is not quite so nice in applications, for there it is possible to find integrals for which we will have to use numerical methods. It is usually worth a reasonable amount of effort, however, to try to find *some* method for formal integration before giving up.

EXAMPLE 3.8. Evaluate $\int \tan^4 2x \, dx$.

Solution. We have no formula which involves the fourth power of the tangent. However, we do have an identity that relates the tangent and the secant. Using this, we obtain

$$\int \tan^4 2x \, dx = \frac{1}{2} \int \tan^2 2x (\sec^2 2x - 1) d(2x)$$

$$= \frac{1}{2} \left[\int \tan^2 2x \, d(\tan 2x) - \int (\sec^2 2x - 1) d(2x) \right]$$

$$= \frac{1}{2} \left[\frac{\tan^3 2x}{3} - \tan 2x + 2x \right] + C$$

$$= \frac{1}{6} \tan^3 2x - \frac{1}{2} \tan 2x + x + C.$$

Before we end this section, we have one other case that demands our attention. We have discussed the possibility of integrating products of sines and cosines provided the angles involved are the same for all trigonometric functions in the integrand. The possibility of handling something like $\int \sin 2x \cos 3x \, dx$ is one that we have not considered. There is always the possibility, of course, of replacing $\sin 2x$ by $2 \sin x \cos x$ and of replacing $\cos 3x$ by $\cos^3 x - 3 \sin^2 x \cos x$, but this would seem to be a lot of work, and you may have forgotten the formulas for the functions of triple angles. (You could go back to complex numbers, of course, and derive these results using the expansion of binomials, but that doesn't seem to be a very appetizing method either.) By this time you already suspect that we have something in mind or this would not have been brought up. We should allay your suspicions at once with the following Example.

EXAMPLE 3.9. Evaluate $\int \sin 4x \cos 7x \, dx$.

Solution. This is the same type we mentioned above, but the prospects of substituting expressions in $\sin x$ and $\cos x$ for $\sin 4x$ and $\cos 7x$ seem even less appetizing than the substitutions we mentioned earlier. In thinking back over the various identities that we know, we look for one that involves the product of a sine and a cosine of different angles, and two come to mind. We have

$$\sin(A + B) = \sin A \cos B + \cos A \sin B$$

and   (VI.3.6)

$$\sin(A - B) = \sin A \cos B - \cos A \sin B.$$

The only problem is that we have more terms on the right side than we would like, but let's move on and see what we can do with the results. We can write

$$\sin(4x + 7x) = \sin 4x \cos 7x + \cos 4x \sin 7x$$

and

$$\sin(4x - 7x) = \sin 4x \cos 7x - \cos 4x \sin 7x.$$

We wish to keep the product $\sin 4x \cos 7x$, but not the product $\cos 4x \sin 7x$. We can accomplish this by the simple act of adding the two expressions. We obtain $\sin 11x + \sin(-3x) = 2 \sin 4x \cos 7x$. Thus, we have,

$$\sin 4x \cos 7x = \frac{\sin 11x - \sin 3x}{2}$$

and as a result we can complete the integration by writing

$$\int \sin 4x \cos 7x \, dx = \frac{1}{2} \int [\sin 11x - \sin 3x] dx$$

$$= \frac{1}{2} \left[\frac{-\cos 11x}{11} - \frac{-\cos 3x}{3} \right] + C$$

$$= \frac{1}{6} \cos 3x - \frac{1}{22} \cos 11x + C.$$

We managed to avoid a requirement for the reduction of functions of multiple angles. If we have a product of a sine and a cosine, we can use the identities (VI.3.6). If we have the product of two sines involving different angles or of two cosines involving different angles, we could use in the same manner

$$\cos(A + B) = \cos A \cos B - \sin A \sin B$$

and   (VI.3.7)

$$\cos(A - B) = \cos A \cos B + \sin A \sin B.$$

In this case we would add the two expressions to retain only the product of cosines and we would subtract the first from the second in order to eliminate the product of cosines and retain the product of sines. It is strongly suggested that you keep this technique in mind. Do not try to memorize the results, for this case does not appear often enough in practice to warrant remembering the additional formulas. These results do occur from time to time, however, and it is necessary to be able to handle them when they do occur.

It is worth reminding ourselves that each integration can be checked by differentiation. It is very much worthwhile to make a practice of checking in this way.

EXERCISES

1. Evaluate:

   (a) $\int \sin 2x \cos x \, dx$

   (b) $\int_0^\pi \sin^4 3x \, dx$

   (c) $\int \tan^6 2x \, dx$

   (d) $\int \tan^5 4x \sec 4x \, dx$

   (e) $\int x \sin^3 x^2 \, dx$

   (f) $\int e^{-x} \, dx$

   (g) $\int_0^{\pi/2} \cos^6 x \, dx$

   (h) $\int \cos^3 2x \sin^4 2x \, dx$

2. Evaluate:

   (a) $\int \tan x \sin x \, dx$

   (b) $\int \cos 4x \cos 5x \, dx$

   (c) $\int \sec^4 2x \, dx$

   (d) $\int \sin x \sec^3 x \, dx$

   (e) $\int \tan^2 x \cos^3 x \, dx$

   (f) $\int_0^{\pi/3} \sin 7x \sin(x/2) dx$

   (g) $\int_0^{\pi/2} \cos 3x \sin 3x \, dx$

   [There are three ways of handling part (g).]

3. Evaluate:

   (a) $\int \sin 5x \cos 2x \, dx$

   (b) $\int \sin 3x \sin 7x \, dx$

   (c) $\int \sin 2x \tan 2x \sec 2x \, dx$

   (d) $\int \sin^4 3x \cos^2 3x \, dx$

   (e) $\int \cos 4x \cos 7x \, dx$

   (f) $\int \cos 3x \sin 4x \, dx$

   (g) $\int \sin^3 4x \cos^4 4x \, dx$

   (h) $\int \sin x \cos^8 x \, dx$

4. Evaluate:

(a) $\int (1/\cos 4x)dx$

(b) $\int (\cos 2x + \sec 2x)^2 dx$

(c) $\int (\cos 3x - \cos 4x)^2 dx$

(d) $\int (\sin x)^{-4} \cos^3 x \, dx$

(e) $\int (\tan x + 3)^3 dx$

(f) $\int (\csc x - 2)^2 dx$

(g) $\int (\cot x - \csc x)^2 dx$

(h) $\int e^{\cot 3x} \csc^2 3x \, dx$

5. The lemniscate has the equation $r^2 = \cos 2\theta$. Find the area in one loop of the lemniscate.

6. Find the area inside the cardioid $r = 2 - 2 \sin \theta$.

7. Find the area inside the smaller loop of the limacon $r = 4 - 8 \sin \theta$.

8. Find the area which is inside the outer loop of the limacon $r = 2 + 4 \cos \theta$, but which is outside the inner loop of this curve.

9. The area bounded by $y = \sin x$, $x = 0$, $x = \pi$, and the x-axis is revolved about the x-axis. Find the volume of revolution.

10. Find the length of the curve $y = \ln \sec x$ from $(0, 0)$ to $(\pi/3, \ln 2)$.

11. Find the length of the closed curve with the parametric equations $x = 5 \cos t$, $y = 5 \sin t$.

12. (a) Find the area bounded by $y = \csc 2x$, $x = \pi/8$, $x = \pi/4$, and the x-axis.

(b) Find the first moment of the area of part (a) about the x-axis.

(c) Divide the result of part (b) by the result of part (a) to find the y-coordinate of the centroid of the area.

(d) Find the volume of revolution formed by revolving the area of part (a) about the x-axis.

(e) Find the product of the area of part (a) and the circumference of a circle whose radius is the y-coordinate of the centroid.

(f) Compare the results in parts (d) and (e). The fact that these results are equal was known to the Greeks and was stated as a *Theorem of Pappus*.

13. If m and n are non-negative integers,

(a) Evaluate $\int_c^{c+2\pi} \cos mx \sin nx \, dx$

(b) Evaluate $\int_c^{c+2\pi} \cos mx \cos nx \, dx$ if $m \neq n$.

(c) Evaluate $\int_c^{c+2\pi} \sin mx \sin nx \, dx$ if $m \neq n$.

(d) Evaluate $\int_c^{c+2\pi} \cos mx \, dx$ if $m \neq 0$

(e) Evaluate $\int_c^{c+2\pi} \sin mx \, dx$.

VI.4 Integration by Algebraic Substitution

We concentrated on the integration of trigonometric functions in the last section. We will turn our attention to the use of substitution in this section and the next. We found that substitution was helpful starting with the substitution theorem in Chapter III, so this is nothing new. We will pay particular attention to the problem of substitution here, however, for it frequently helps to state a problem in a slightly different way (and this is essentially what we do in substitution).

Before proceeding, however, it is well to mention a few of the difficulties which you may find. This is not to make the problem difficult, but simply to let you know that you will not be the first one to meet these difficulties. The first question is whether a given integrand can be helped by substitution. The answer is that it is not always possible to be certain in advance, but "experience will help". The second question concerns *which* substitution is appropriate. The only all-inclusive answer is that the one that works is appropriate. Unfortunately there is no formal way of stating what will work in every case. Therefore, the sum and substance of all of this discussion can be capsulized by suggesting that you try what you think might work. If it works that is fine. If it doesn't work, see whether you can gain some insight from the failure of your first attempt which may suggest an alternative that will work. If after several attempts none of your efforts are successful, then go back to RS sums or to numerical methods. The one comforting thing about a textbook or a classroom situation is that the majority of the problems presented therein have some technique which permits solution, and therefore you can start with some small assurance that success is possible in the problems found here. The experience you gain handling these problems will stand you in good stead when it comes to dealing with integrations which may arise in applications. The more standard applications, in fact, seldom require anything that cannot be handled by the methods of this chapter. In part that may be due to the fact that the persons involved in developing various applications were guided to a great extent toward those integrals they could evaluate, and they failed to consider any others. With the advent of the computer we can use numerical methods for definite integrals which are otherwise intractable, and consequently our applications need not be limited. In the future you may have the job of breaking with tradition and opening up new areas of application in the field of your choice.

We now turn toward the actual business of integrating algebraic integrands. We will again do this by giving examples.

EXAMPLE 4.1. Evaluate $\int ((2x^3 + 3)/\sqrt{x - 1})dx$

Solution. This integrand looks bad with the square root in the denominator. Since most of us are not particularly intrigued by having an excess number of square roots around, it would be fine to be able to use some device, however strange, to get rid of the square root. We remember that we have a result concerning integration which states $\int f(g(x))dg(x) = \int f(y)dy$. With this clue, we proceed by letting $y = \sqrt{x - 1}$. Consequently we have an indication of the type of substitution that we will attempt. (Remember we do not know whether it will succeed until we have tried it.) If we start with this substitution, the easier method to follow would be to evaluate x in terms of y, and then to make the substitutions as they are required in the problem as given. Hence, $y = \sqrt{x - 1}$, $x = y^2 + 1$, and $dx = 2y\,dy$. (Remember how to find differentials?) We are now ready to make our substitutions. We obtain the integral

$$\int \frac{2(y^2 + 1)^3 + 3}{y} \, 2y \, dy = 2 \int [2y^6 + 6y^4 + 6y^2 + 5]dy$$

$$= 2\left[\frac{2y^7}{7} + \frac{6y^5}{5} + \frac{6y^3}{3} + 5y\right] + C.$$

Since the problem was posed in terms of the variable x instead of y, we have one of two choices. If the integral had been a definite integral with limits of integration, we would have varied the limits of integration by modifying the interval $[a, b]$ to the interval $[\sqrt{a - 1}, \sqrt{b - 1}]$. Since the integral is not a definite integral, we will proceed to replace each y by $\sqrt{x - 1}$. It is admitted that the latter is not as appealing, for a number of strange exponents (such as 7/2, 5/2, 3/2, 1/2) occur, but we can help somewhat by making use of a little very straightforward algebra. Note that this result could be written in terms of y as

$$\frac{2y}{35}[10y^6 + 42y^4 + 70y^2 + 175] + C.$$

Within the brackets we have only even powers of y. Therefore we can replace y^2 by $(x - 1)$, and we do not have to worry about radicals in this simplification. We then have

$$\frac{2\sqrt{x - 1}}{35}[10(x - 1)^3 + 42(x - 1)^2 + 70(x - 1) + 175] + C.$$

The resulting bracket can be simplified rather easily to obtain

$$\frac{2\sqrt{x - 1}}{35}[10x^3 + 12x^2 + 16x + 137] + C.$$

As before, we notice that the simplification is not essential to obtain the correct answer, although the replacement of y is necessary in order to provide an answer in terms of the original variable. You can check that this is a solution by differentiating the result. Our trick of watching the algebra will frequently save a great deal of time and frustration in the job of simplification, and we commend it to you. Remember the old adage "using head saves paper and pencil". We might even have added time and energy. It is always wise to check the result by differentiation.

In many cases in which we have a square root a substitution of the type indicated is helpful. In fact, this type of substitution is often helpful with cube roots, etc. There are some criteria which could be mentioned indicating when this is more likely to work and when it is not, but in general you will find that it takes less time to try it and see whether it works than to apply the criteria. We now try this same example using an alternate substitution, just to show you some of the various methods available to us.

Alternate Solution. In the example above, let us try the substitution $y = x - 1$. This avoids the situation where y represents a square root, but still leaves the square root of y as a part of the integrand. If we follow through, however, we see that we have the following set of relations.

$$y = x - 1,$$
$$x = y + 1,$$
$$dx = dy.$$

Then, we have, since $x^3 = (y + 1)^3$,

$$\int \frac{2(y + 1)^3 + 3}{\sqrt{y}}\, dy = \int [2y^3 + 6y^2 + 6y + 5]y^{-1/2}\, dy$$

$$= \int [2y^{5/2} + 6y^{3/2} + 6y^{1/2} + 5y^{-1/2}]dy$$

$$= \frac{4y^{7/2}}{7} + \frac{12y^{5/2}}{5} + \frac{12y^{3/2}}{3} + \frac{10y^{1/2}}{1} + C$$

$$= \frac{2y^{1/2}}{35}[10y^3 + 42y^2 + 70y + 175] + C.$$

It can be seen upon substitution of $y = x - 1$ that we have the same result we had before. As for the preferred method, we leave it to you. Since both work, either one of them is permissible. It is merely a matter of personal choice. We suggest, however, that you become familiar with both methods, for then you will have additional techniques at your disposal. By using both from time to time, you will soon find out which one you prefer.

If there are radicals involved it is often possible to remove them by suitable substitution. One thing you might observe in general is that if you can get rid of what seems to be the worst part of the integrand, then perhaps the remainder will fall in line. You can use substitution to replace any *one* expression in the given variable by another variable. However, once you have used that one choice, you are then forced to use a combination of algebra and calculus (in connection with replacing the differential) for all other substitutions in the expression. You have no more leeway with which to work. Since you are in the position of having only one choice to make, it is up to you to make that choice count. The aim of all our substitutions is to take integrands that do not correspond to any formulas we have seen so far and to try to put them in a form that is recognizable.

There are many cases in which we can simplify results by substitutions which are not distinguished by the presence of radical signs. Consider the following case.

EXAMPLE 4.2. Evaluate $\int ((x^3 + 2x - 7)/(x - 1)^2)dx$.

Solution. The problem here is that we have a fraction, and there appears to be little that we can do about it. It would not be quite so bad if the denominator were merely $(x - 1)$, for then we could divide the numerator by the denominator (using long division or synthetic division), and we could investigate the resulting polynomial and fraction. However, since the denominator is the square of $(x - 1)$, this presents additional complications. It is helpful to remove such difficulties. Perhaps the easiest way of doing this is with the mathematical version of sweeping the difficulties under the rug. This version is again, as you might suspect since it is mentioned in this Section, the method of substitution. Let us replace $(x - 1)$ by y. We will have

$$y = x - 1,$$
$$x = y + 1,$$
$$dx = dy.$$

The integral becomes

$$\int \frac{x^3 + 2x - 7}{(x - 1)^2} dx = \int \frac{(y + 1)^3 + 2(y + 1) - 7}{y^2} dy$$

$$= \int \frac{y^3 + 3y^2 + 5y - 4}{y^2} dy$$

$$= \int [(y + 3) + \frac{5}{y} - (4y^{-2})]dy$$

$$= \frac{y^2}{2} + 3y + 5 \ln y + 4y^{-1} + C$$

$$= \frac{1}{2}(x - 1)^2 + 3(x - 1) + 5 \ln(x - 1) + \frac{4}{x - 1} + C.$$

You will note that we removed the difficulties rather easily this way. It is always easier to have powers of a single variable in the denominator although the expressions are equivalent throughout. Thus, if we have powers of an expression, the integration may frequently be aided by an appropriate substitution.

We will consider one more example.

EXAMPLE 4.3. Evaluate $\int ((x^5 - 4x^2)/\sqrt[5]{1 - x^3})dx$.

Solution. In this instance we see a fifth root, something that no one is anxious to face. However, we do have a single term in the denominator, and this gives us some hope. We will try a substitution.

$$y = 1 - x^3$$
$$x^3 = 1 - y$$
$$3x^2 \, dx = -dy.$$

It is true that we could have solved for x, but this would have involved cube roots. You might try doing that. However, we note that the numerator can be written $x^2(x^3 - 4)$, and hence we can think of the integral as

$$\int \frac{(x^3 - 4)3x^3 \, dx}{3\sqrt[5]{1 - x^3}} = \int \frac{(1 - y) - 4}{3y^{1/5}} \, dy$$

$$= -\frac{1}{3} \int (y^{4/5} + 3y^{-1/5})dy$$

$$= -\frac{1}{3} \left[\frac{5}{9} y^{9/5} + \frac{15}{4} y^{4/5} \right] + C$$

$$= -\frac{5}{27} (1 - x^3)^{9/5} - \frac{5}{4} (1 - x^3)^{4/5} + C.$$

The results may not be pretty but they work. That is what counts. It is true that the substitution worked in this instance only because we had the extra "x^2" in the numerator, but the ability to integrate often rests on just such fortuitous circumstances. You would find that letting y equal the fifth root would have worked in much the same manner, and success would have required the same fortuitous circumstance in having the x^2 when it was needed.

You should not be under the illusion that we always invoke substitutions when we have an algebraic expression in the integrand. It is always possible that the integrand may be one of the formulas we listed in Section VI.1. It is also possible that simple algebraic manipulations will suffice. Consider the following Example.

EXAMPLE 4.4. Evaluate $\int ((x^5 - 3x^4 + 7x^2 - 4x + 2)/(x^2 + 1))dx$.

Solution. In this instance the denominator of the integrand is reminiscent of the formula which results in arc tan x. However, the numerator is far from being the simple "1" which we would desire. In this case we can go back to that old algebraic operation called division. If we divide the numerator by the denominator, we obtain a quotient of $(x^3 - 3x^2 - x + 10)$

with a remainder of $(-3x - 8)$. Therefore we can write

$$\int \frac{x^5 - 3x^4 + 7x^2 - 4x + 2}{x^2 + 1}\, dx$$

$$= \int \left(x^3 - 3x^2 - x + 10 - \frac{3x}{x^2 + 1} - \frac{8}{x^2 + 1} \right) dx$$

$$= \frac{x^4}{4} - x^3 - \frac{x^2}{2} + 10x - \frac{3}{2} \ln(x^2 + 1) - 8 \arctan x + C.$$

Needless to say we have used our earlier results liberally in integrating the result of the division, but we see that the division did make it possible to evaluate the given integral. Never overlook the possibility that some algebraic operation, such as division, may aid in getting to the point where you can apply one of the available formulas.

There is no single path that we can recommend in attempting to evaluate an integral. You will need to use everything at your command. Practice is the only route to success, for if you practice sufficiently you will develop a feeling concerning what is likely to work. It is true that there are tables of integrals. No matter how lengthy they may be, they are never complete, however. If you use such a table you have two hurdles to master. The first of these is to find the correct formula in the table. The second will usually be that of making some substitution or algebraic manipulation which will bring the integral you have at hand into conformity with the integral in the table. Either the formula fits *exactly* or it does not apply at all. In this case *close* does not count.

EXERCISES

1. Evaluate. Check each problem by differentiation:
   (a) $\int_{-1}^{1} ((x^3 - 4x^2 + 7x - 8)/(2x + 3)^3)\, dx$
   (b) $\int_0^{\pi/3} (\sin x \cos^3 x/\sqrt{\cos x})\, dx$
   (c) $\int x^2 (1 - x)^{-1/3}\, dx$
   (d) $\int (x/\sqrt{2 + 3x})\, dx$
   (e) $\int ((x^2 + 4x)/\sqrt{2 - 3x})\, dx$
   (f) $\int_{-1}^{0} x^8 (1 - x^3)^{-13}\, dx$
   (g) $\int (x/\sqrt{1 + x^2})\, dx$
   (h) $\int ((x - 2)/(x - 1)^{2/3})\, dx$

2. Evaluate:
   (a) $\int x\sqrt{1 + x^2}\, dx$
   (b) $\int (x^3/\sqrt{1 - x})\, dx$
   (c) $\int (x^5/(x^3 + 1))\, dx$
   (d) $\int ((x^4 + 7x^2 - 3x)/(x^2 + 1))\, dx$

(e) $\int ((\sqrt{x+1}+2)/(\sqrt{x+1}-2))dx$

(f) $\int (x^7/(4-x^2)^2)dx$

(g) $\int ((7x+4)/\sqrt{3x-5})dx$

(h) $\int ((4x-3)/\sqrt{1-x^2})dx$

3. Integrate each of the following:

(a) $\int ((x+1)/\sqrt{1-x^2})dx$

(b) $\int ((-4x+3)/\sqrt{1-x^2})dx$

(c) $\int ((16x^4-24x^3+6x-5)/(2x-1))dx$

(d) $\int (x^2+3)\sqrt{2-x}\,dx$

(e) $\int ((x^2-3x+4)/(3-x))dx$

(f) $\int ((\arctan x)/(x^2+1))dx$

(g) $\int ((e^{3x}-e^x)/(e^x+2))dx$

(h) $\int ((\sin\sqrt{x})/\sqrt{x})dx$

4. Integrate each of the following:

(a) $\int ((x^2+4x)/(2x-1)^{3/2})dx$

(b) $\int ((x^3+3x^2+4x-7)/(x-2))dx$

(c) $\int ((x^3+3x^2-4x-7)/(x^2+1))dx$

(d) $\int x^3\sqrt{2x+1}\,dx$

(e) $\int (x^3-3x+2)(2-3x)^{3/2}\,dx$

(f) $\int (x^3+3x)(2-x^2)^{2/3}dx$

(g) $\int x(x-3)^{3/2}dx$

(h) $\int \sqrt{x}(2x+1)^2dx$

5. Evaluate each of the following:

(a) $\int_0^4 (x^3/\sqrt{25-x^2})dx$

(b) $\int_3^6 x\sqrt{7-x}\,dx$

(c) $\int_0^1 ((16x^3-16x^2+4x+4)/(2x-3))dx$

(d) $\int_1^4 (x^2/(5-x))dx$

(e) $\int_1^8 x\sqrt[3]{9-x}\,dx$

(f) $\int_0^3 (x^2+x+1)\sqrt{x+1}\,dx$

6. Find the area bounded by $y = \sqrt{x-1}/x$, $y = 0$, $x = 5$, and $x = 10$.

7. (a) Evaluate $\int ((x^3-4x^2+2x-3)/(x-1)^3)dx$.

   (b) Show that you can use the methods we have obtained so far to integrate any integral of the type $\int (f(x)/(ax-b)^n)dx$ where $f(x)$ is a polynomial, where a and b are constants and n is a positive integer.

8. Evaluate $\int ((x^3-4x^2-4x+1)/(4x^2-12x+10))dx$ using the substitution $y = 2x - 3$. Notice that this substitution is essentially equivalent to completing the square in the denominator.

9. Evaluate.

(a) $\int (1 + e^x)^{-1} dx$ using the substitution $y = e^{-x}$. Note that this substitution will work whereas $y = e^x$ would not help.

(b) $\int_0^{(1/2)\ln 3} (e^x + e^{-x})^{-1} dx$ using $y = e^x$. Also try using $y = e^{-x}$.

S10. For a given firm the demand curve is given by $y = 80 - (x^2/\sqrt{x^2 + 225})$ and the supply curve is given by $y = 4 + ((x^2 + 200)/0.5x)$.

(a) Show that the equilibrium quantity supplied under pure competition is twenty.

(b) Find the equilibrium price under pure competition.

(c) Set up the integral for the consumer's surplus.

(d) Find the producer's surplus.

VI.5 Integration by Trigonometric Substitution

We have made reference to the use of trigonometry as an aid to integration. We will now attempt to produce results which vindicate the advanced billing. We will be principally concerned with expressions of the type $(a^2 + x^2)$, $(a^2 - x^2)$, and $(x^2 - a^2)$. Note that the last two might imply that $a \geq x$ and $x \geq a$ respectively, for it is not unnatural to assume that the expressions are positive in each instance. Unless a square root, or some root of even order, is involved, however, one can use either of the two substitutions we will suggest.

Experience in mathematics at times brings about rather curious chains of thought, and for many when there is the sum or difference of two squared quantities there is the vision of a right triangle with a corresponding use of the Pythagorean theorem. (It goes without saying, of course, that this connection would not be mentioned here if we were not going to exploit it.) As you will note in Figure VI.1, we have the relations $a^2 + b^2 = c^2$, $a^2 = c^2 - b^2$, and $b^2 = c^2 - a^2$. We can select values for two of the three sides of the triangle, and the third side is then determined. In assigning values to two sides we must keep in mind the fact that the hypotenuse cannot be shorter than either of the other two sides. These expressions are not at all unlike the ones mentioned in the preceding paragraph. This explains why these expressions conjure up visions of right triangles. Let us now proceed to illustrate some of the ways in which we can use such visions.

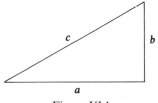

Figure VI.1

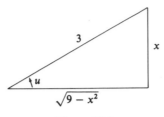

Figure VI.2

EXAMPLE 5.1. Evaluate $\int \sqrt{9 - x^2} \, dx$.

Solution. We note that this is of the type $a^2 - x^2$ with $a = 3$. Therefore, we will draw a triangle and label it just for this problem, as in Figure VI.2. It is apparent that the hypotenuse must be of length 3, for the radicand cannot be negative and therefore 3 must be the length of the longest side. It is not clear whether the length x should be marked on the vertical or horizontal leg of the triangle. Actually, it makes no difference in obtaining a result. To illustrate this we will solve this particular problem both ways. We will use the labeling of Figure VI.2 for the first solution and consider that the vertical side is of length x, whence the horizontal side is of length $\sqrt{9 - x^2}$. We then have the following equations.

$$\sin u = \frac{x}{3} \quad \text{or} \quad x = 3 \sin u,$$

$$\sqrt{9 - x^2} = 3 \cos u,$$

and

$$dx = 3 \cos u \, du.$$

(This last expression is obtained by differentiating $x = 3 \sin u$.) The integral becomes

$$\int (3 \cos u)(3 \cos u \, du) = 9 \int \cos^2 u \, du.$$

This is one of the integrals we learned how to handle in Section VI.3. Therefore we can write

$$9 \int \cos^2 u \, du = \frac{9}{2} \int (1 + \cos 2u) du = \frac{9}{2} \left[u + \frac{\sin 2u}{2} \right] + C.$$

This is not quite what we would like, for if the problem is given to us in terms of x, we should leave the result in terms of x (unless of course we had a definite integral in which case we would use the limits and have a numeric answer). Now $\sin 2u = 2(\sin u)(\cos u)$, and returning to Figure VI.2, we

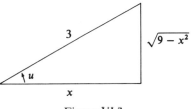

Figure VI.3

see that we can revert to an expression in terms of x as follows:

$$\int \sqrt{9 - x^2}\, dx = \frac{9}{2}\left[\text{arc sin}\,\frac{x}{3} + \frac{x}{3}\frac{\sqrt{9 - x^2}}{3}\right] + C$$

$$= \frac{9}{2}\,\text{arc sin}\,\frac{x}{3} + \frac{1}{2}x\sqrt{9 - x^2} + C.$$

This follows from the fact that $\sin u = x/3$ and $\cos u = (\sqrt{9 - x^2})/3$. Although it is weird looking, we do have a result. You should check by differentiation to see whether this does produce the integrand.

For the other possible labeling of the triangle, we would reverse the position of x and $\sqrt{9 - x^2}$ as shown in Figure VI.3 and then our substitutions would have been

$$\cos u = \frac{x}{3},$$

$$x = 3\cos u,$$

$$\sqrt{9 - x^2} = 3\sin u,$$

and

$$dx = -3\sin u\, du.$$

Our integral would have become

$$-9\int \sin^2 u\, du.$$

The integration would have been handled in a similar way, and the same final result would be obtained. We leave it to you to carry on from here, since at this point it becomes routine. (You will have to remember, of course, the relation between arc $\sin(x/3)$ and arc $\cos(x/3)$). As a further note on this example, we should indicate that in the case of the definite integral $\int_{3/2}^{3} \sqrt{9 - x^2}\, dx$ we can make the substitution $x = 3\sin u$ or $u = \text{arc sin}(x/3)$ and obtain the following result.

$$\int_{3/2}^{3} \sqrt{9 - x^2}\, dx = 9 \int_{\arc \sin[(3/2)/3]}^{\arc \sin(3/3)} \cos^2 u\, du = 9 \int_{\arc \sin 1/2}^{\arc \sin 1} \cos^2 u\, du$$

$$= 9 \int_{\pi/6}^{\pi/2} \cos^2 u\, du = \frac{9}{2}\left(u + \frac{1}{2}\sin 2u\right)\Big|_{\pi/6}^{\pi/2}$$

$$= \frac{9}{2}\left[\left(\frac{\pi}{2} + \frac{1}{2}\sin\frac{2\pi}{2}\right) - \left(\frac{\pi}{6} + \frac{1}{2}\sin\frac{2\pi}{6}\right)\right]$$

$$= \frac{9}{2}\left[\frac{\pi}{3} - \frac{1}{4}\sqrt{3}\right].$$

If we had first rewritten the result in terms of x, we would have obtained

$$\left[\frac{9}{2}\left(\arc \sin\frac{3}{3} + \frac{3}{3}\frac{\sqrt{9-9}}{3}\right) - \frac{9}{2}\left(\arc \sin\frac{3/2}{3} + \frac{\sqrt{9-(9/4)}}{3}\right)\right]$$

$$= \frac{9}{2}\left[\frac{\pi}{2} - \frac{\pi}{6} - \frac{1}{6}\sqrt{\frac{27}{4}}\right] = \frac{9}{2}\left[\frac{\pi}{3} - \frac{\sqrt{3}}{4}\right].$$

Note that these last two results are identical, and we have a verification of our earlier statements. You might also observe that if we were to evaluate $\int_0^3 \sqrt{9 - x^2}\, dx$ we would obtain $9\pi/4 = (1/4)\pi(3)^2$, which is precisely what we would have expected using the formula for the area of a quadrant of a circle of radius 3: (Check to see that this is the integral that would have arisen in trying to obtain the area of the quarter circle.)

Note that we could label the right triangle in either of the two ways that would make sense in this situation, that we were able to do the trigonometric integration and that we needed to make the substitution back to the original variable only if we started with an indefinite integral. We had to put the length "3" on the hypotenuse, but we could choose whether we would label the vertical side "x" or label the horizontal side "x". When we represented the horizontal side by "x", we found that we were dealing with the *co-functions* for the purpose of taking differentials. We therefore had negative signs in the differentials. The results were exactly the same in the end, and if you do not mind dealing with negatives, the choice of side for length "x" is of no importance. If you wish to avoid the differentials of the co-functions, you will find that you can do so by the following rule:

(a) If the hypotenuse is "x", then label the horizontal side as the constant. The only choice that is forced on you is the labeling of the hypotenuse, for that is the longest side, and must therefore have the length which is the largest of the three quantities x, a, and $\sqrt{x^2 - a^2}$.

(b) If the hypotenuse is not labeled x, then label the vertical side as "x".

We now proceed to consider another case:

EXAMPLE 5.2. Evaluate $\int (x^2 - 16)^{-1/2}\, dx$.

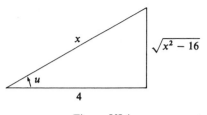

Figure VI.4

Solution. In this case we have the square root of a positive quantity only if $|x| > 4$, and hence we should place x on the hypotenuse as in Figure VI.4. In order to avoid the differential of a negative quantity we will follow our rule and label the triangle with 4 on the horizontal side as shown in this figure. We now have the following substitutions:

$$x = 4 \sec u,$$

$$\sqrt{x^2 - 16} = 4 \tan u,$$

and $dx = 4 \sec u \tan u \, du$. Upon substitution the integral becomes

$$\int \frac{4 \sec u \tan u \, du}{4 \tan u} = \int \sec u \, du$$

$$= \ln|\sec u + \tan u| + C.$$

Using the relationships which are indicated by the triangle, we have

$$\int \frac{dx}{\sqrt{x^2 - 16}} = \ln\left|\frac{x}{4} + \frac{\sqrt{x^2 - 16}}{4}\right| + C$$

$$= \ln|x + \sqrt{x^2 - 16}| + [C - \ln 4]$$

$$= \ln|x + \sqrt{x^2 - 16}| + K$$

where K is also a constant, namely the first constant of integration diminished by the natural logarithm of four. It is just as easy to evaluate the second constant as the first, and there is no reason whatsoever for unduly complicating the appearance of the constant.

EXAMPLE 5.3. Evaluate $\int (x^2 + 25)^{-1/2} \, dx$.

Solution. This appears to be almost the twin of the last example, but the presence of the sum rather than the difference in the radicand makes a great deal of difference in the labeling of the triangle. Here we would not be able to place either "x" or "5" on the hypotenuse, for neither is as large as $\sqrt{x^2 + 25}$. Since "x" and "5" would be the labels of the two legs, we

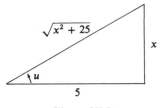

Figure VI.5

would, if we follow the rule, place "x" on the vertical leg as shown in Figure VI.5. With the substitutions indicated we have the relations

$$x = 5 \tan u,$$

$$dx = 5 \sec^2 u \, du,$$

and

$$\sqrt{x^2 + 25} = 5 \sec u.$$

This yields the integral

$$\int \frac{5 \sec^2 u \, du}{5 \sec u} = \int \sec u \, du$$

$$= \ln|\sec u + \tan u| + C.$$

Upon substitution we have

$$\int \frac{dx}{\sqrt{x^2 + 25}} = \ln\left| \frac{\sqrt{x^2 + 25}}{5} + \frac{x}{5} \right| + C$$

$$= \ln|\sqrt{x^2 + 25} + x| + [C - \ln 5]$$

$$= \ln|\sqrt{x^2 + 25} + x| + K.$$

The reasoning concerning the constant parallels that of the preceding example.

In order to illustrate the three possible cases, we have stayed with the more straightforward examples up to this point. However, as this remark may indicate, there are other less obvious cases in which trigonometric substitution may be the easiest method to use. Consider the following illustration.

EXAMPLE 5.4. $\int (x/\sqrt{x^2 - 2x})dx$.

Solution. We note that the denominator in this case apparently does not have the square root of the sum or difference of two squares, but it does have a rather messy radicand. We can aid this by a technique that is extremely helpful in any problem having a quadratic expression not written as the sum

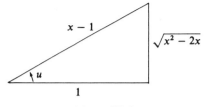

Figure VI.6

or difference of two squares. We will *complete the square* in the radicand. We can write $x^2 - 2x = (x - 1)^2 - 1 = (x^2 - 2x + 1) - 1$ where we have been guided by the fact that we can obtain $x^2 - 2x$ as the first two terms of the square of $(x - 1)$, since the square of a binomial includes the square of the first term, in this case the square of x, and twice the cross product—that is twice the produce of the first and second term of the binomial. Here we have $(-2x)$ which is twice $x(-1)$, our choice being dictated by the fact that the first term of the binomial has already been taken to be "x". Finally we have the square of the second term, $(-1)^2 = +1$. We add and then subtract the same constant in order not to change the value of the original expression. Hence, our integral could be written $\int (x/\sqrt{(x - 1)^2 - 1})dx$. This would indicate that we should draw a right triangle as shown in Figure VI.6. We now have the substitutions

$$x - 1 = \sec u$$

$$\text{or} \quad x = 1 + \sec u,$$

$$dx = \sec u \tan u \, du,$$

$$\text{and} \quad \sqrt{x^2 - 2x} = \tan u.$$

This gives rise to the integral

$$\int \frac{(1 + \sec u)\sec u \tan u \, du}{\tan u} = \int (\sec u + \sec^2 u)du$$

$$= \ln|\sec u + \tan u| + \tan u + C.$$

We substitute terms in x for those in u and obtain

$$\int \frac{x \, dx}{\sqrt{x^2 - 2x}} = \ln|x - 1 + \sqrt{x^2 - 2x}| + \sqrt{x^2 - 2x} + C.$$

Again, you are invited to differentiate this expression and simplify it in order to show that we have the original integrand as the resulting derivative. We have again followed our rules concerning the labeling of the legs of the right triangle. If we had reversed the labeling of the two legs we would have been dealing with cotangents and cosecants, and the resulting differential would have a negative sign. However, the result after substituting back in terms of "x" would have been precisely the same. As mentioned before,

if we had started with a definite integral we could have changed the limits
and not had to make the back substitution to terms involving "x".

In general, if you find that you have an expression that involves both the
square of a variable and a term involving the first power of the same variable,
the process of completing the square will yield either the sum or difference of
two squares, and the result can be related to the triangle. You should also
observe that our earlier work on the integration of trigonometric functions
is of great value here. The problem of returning to the original variable was
much simplified by looking at the simple drawing of a triangle that we made
and carefully labeled. This matter of drawing pictures, where appropriate,
cannot be stressed too strongly, for it helps keep the substitutions clear.

Completing the square has many applications in mathematics not the
least of which is its use in integration. In order to make use of tables of
integrals you will frequently have to complete the square before you find a
formula which is applicable. The coefficients are not always integers, either,
nor is it true that the exponents must always be one and two. For instance, an
expression such as $(3x^4 - 7x^2)$ can be written

$$3[x^4 - (7/3)x^2] = 3[(x^2 - 7/6)^2 - 49/36].$$

We have thus found an expression in which we have the difference of two
squares, thus permitting the use of the right triangle with the hypotenuse
being labeled $(x^2 - 7/6)$.

We will finish this section with one more example, one not having a
square root involved, but also illustrating the use of the right triangle type
of substitution.

EXAMPLE 5.5. Evaluate $\int ((x^5 - 7x^4 + 3x^3 + 2x^2 - 6x + 1)/(x^2 + 1)^2)dx$.

Solution. While there are no radicals here, we do note that we have an
expression which is the sum of two squares, namely $x^2 + 1$, and it is located
in the denominator. Were it in the numerator there would be little problem
but having it in the denominator presents problems that we would prefer
to ignore. With this in mind, let us again try the right triangle. (You are
aided here by the fact that this section refers to trigonometric substitutions,
and therefore you are looking for some way to bring in the trigonometric
functions. Would you recognize this if it were to appear at the end of some
subsequent section?) We will use the triangle shown in Figure VI.7, and
have the substitutions

$$x = \tan u,$$

$$dx = \sec^2 u \, du,$$

$$\sqrt{x^2 + 1} = \sec u, \quad \text{and}$$

$$(x^2 + 1) = \sec^2 u.$$

Using these relations, our integral becomes

$$\int \frac{(\tan^5 u - 7 \tan^4 u + 3 \tan^3 u + 2 \tan^2 u - 6 \tan u + 1)\sec^2 u \, du}{\sec^4 u}$$

$$= \int \left(\frac{\sin^5 u}{\cos^3 u} - 7 \frac{\sin^4 u}{\cos^2 u} + 3 \frac{\sin^3 u}{\cos u} + 2 \sin^2 u - 6 \sin u \cos u + \cos^2 u\right) du.$$

We have the necessary means for integrating each term in this integrand. We can simplify our work a great deal by integrating the sum of the first, third and fifth terms and then integrating the sum of the other three terms.

$$\int \left(\frac{\sin^5 u}{\cos^3 u} + 3 \frac{\sin^3 u}{\cos u} - 6 \sin u \cos u\right) du$$

$$= -\int \left[\frac{(1 - \cos^2 u)^2}{\cos^3 u} + 3 \frac{(1 - \cos^2 u)}{\cos u} - 6 \cos u\right] d(\cos u)$$

$$= -\int [(\cos u)^{-3} + (\cos u)^{-1} - 8 \cos u] d(\cos u)$$

$$= -\left[\frac{(\cos u)^{-2}}{-2} + \ln \cos u - 8 \frac{\cos^2 u}{2}\right] + C$$

$$= \frac{1}{2} \sec^2 u - \ln \cos u + 4 \cos^2 u + C$$

and

$$\int \left[-7 \frac{\sin^4 u}{\cos^2 u} + 2 \sin^2 u + \cos^2 u\right] du$$

$$= \int \left[-7 \frac{(1 - \cos^2 u)}{\cos^2 u} \sin^2 u + \sin^2 u + (\sin^2 u + \cos^2 u)\right] du$$

$$= \int [-7 \tan^2 u + 8 \sin^2 u + 1] du$$

$$= \int [-7(\sec^2 u - 1) + 8 \sin^2 u + 1] du$$

$$= -7 \tan u + \frac{8}{2}\left(u - \frac{1}{2}\sin 2u\right) + 8u + C$$

$$= -7 \tan u + 4(u - \sin u \cos u) + 8u + C$$

$$= -7 \tan u + 12u - 4 \sin u \cos u + C.$$

Since the original problem was the sum of the two smaller problems, we will add the two smaller results to obtain the result of integration, noting that only one additive constant is necessary in the result. It is not difficult to use the triangle to find the appropriate substitutions in terms of "x". Be

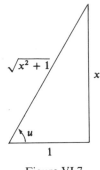

Figure VI.7

careful to note that our method of doing some algebra prior to integration is solely in the interests of saving work. Such work-reducing organization is noticed only if you stay alert. Of course, each of the terms in the original expression could have been integrated using other techniques, but the result would have been the same. It would also have been possible to use long division in the original integrand obtaining a linear expression and a fraction having a cubic numerator and a quartic denominator. It is suggested that you try different techniques on this integral to find the method which seems most reasonable to you. We have left you the task of replacing the variable u with the corresponding functions in terms of x.

There has been no intention here of trying to indicate that trigonometric solutions are the answer for all difficult integrations. They are frequently helpful. Consider all available techniques and then select the one that gives evidence of obtaining the desired result in the most direct manner.

EXERCISES

1. Integrate:

   (a) $\int (x^2 + 9)^{-1/2} dx$
   (b) $\int (x^2 + 9)^{-3/2} dx$
   (c) $\int (9 - x^2)^{-1} dx$
   (d) $\int (9 - x^2)^{-1/2} dx$
   (e) $\int (x/\sqrt{x^2 + 9}) dx$
   (f) $\int (x/(9 - x^2)) dx$
   (g) $\int (\sqrt{9 - x^2})^5 dx$
   (h) $\int (9 - x^2)^{3/2} dx$

2. Evaluate:

   (a) $\int_3^4 ((x^3 + 3x - 2)/\sqrt{x^2 - 4}) dx$
   (b) $\int_0^1 ((x^3 - 3)/\sqrt{x^2 + 1}) dx$
   (c) $\int_1^2 ((x + 3)/\sqrt{x^2 + 2x}) dx$

(d) $\int_5^6 [1/((x-2)\sqrt{x^2-4x})]dx$

(e) $\int_0^4 \sqrt{8x-x^2}\,dx$

(f) $\int_{1/4}^1 (\sqrt{x^2+2x}/x)dx$

(g) $\int_0^1 x\sqrt{8x-x^2}\,dx$

(h) $\int_{-4}^0 (\sqrt{x^2-8x}/(x-4))dx$

3. Evaluate:

(a) $\int_1^4 [1/(x^2\sqrt{x^2+9})]dx$

(b) $\int_0^4 (x^5/\sqrt{x^2+9})dx$

(c) $\int_{\sqrt{3}}^4 (1/x\sqrt{x^2+9})dx$

(d) $\int_0^4 (x^3/\sqrt{x^2+9})dx$

(e) $\int_0^4 (x^2+9)^{-1}dx$

(f) $\int_0^4 (x/(x^2+9))dx$

(g) $\int_0^4 (x^2/(x^2+9))dx$

(h) $\int_{\sqrt{3}}^4 (1/x(x^2+9))dx$

4. Evaluate:

(a) $\int_1^2 (1/\sqrt{10x-x^2})dx$

(b) $\int_1^2 (x/\sqrt{10x-x^2})dx$

(c) $\int_1^2 (x^2/\sqrt{10x-x^2})dx$

(d) $\int_1^2 (1/(10x-x^2))dx$

(e) $\int ((1+x)/(10x-x^2))dx$

(f) $\int (1/((x-5)(10x-x^2)))dx$

5. Evaluate each of the following by trigonometric substitution and show that the results agree with the formulas we developed earlier.

(a) $\int (1-x^2)^{-1/2}dx$

(b) $\int (x^2+1)^{-1}dx$

(c) $\int (1/x\sqrt{x^2-1})dx$

6. (a) Integrate: $\int ((4-x)/\sqrt{4x-x^2})dx$

(b) Evaluate: $\int_{-2}^1 ((7x^3-3x+4)/(x^2+x+1))dx$

(c) Use trigonometric substitutions to integrate

$$\int ((x^3-4x^2-4x+1)/(4x^2-12x+10))dx.$$

Compare your results with those of Exercise VI.4.8.

7. (a) $\int (e^x/(e^{2x}+1))dx$

(b) $\int (a^2-x^2)^{-1}dx$

(c) $\int (x^2-a^2)^{-1}dx$

(d) $\int ((x-1)/\sqrt{1-x^{2/3}})dx$

(e) $\int (12x-9x^2)^{-1/2}dx$

(f) $(x^2+a^2)^{-1}dx$

8. Verify that the instructions for avoiding negative signs in trigonometric substitutions work as stated in this section.

9. (a) Find the area of that portion of the circle $x^2 + y^2 = 100$ which is between the lines $x = 0$ and $x = 5$.
   (b) Sketch the circle and the two lines of part (a). Find the area of this portion of the circle geometrically by combining a sector of the circle and a triangle. Do the two results agree?

10. Find the length of the curve $y = \ln x$ between $x = 2$ and $x = 4$.

11. (a) Find the area completely enclosed by the curve $xy = 4$ and the line $x + y = 5$.
    (b) Find the volume swept out if this area is revolved about the x-axis.

12. (a) Find the first moment of a circle of radius 5 about a diameter.
    (b) Find the second moment of this circle about a diameter.
    (c) Find the first moment of this circle about a line tangent to the circle.
    (d) Find the second moment of this circle about a line tangent to the circle.

13. Find the centroid of the first quadrant of $x^2 + y^2 = a^2$.

14. Find the area of the region enclosed by one loop of the curve $x^2 = y^4 - y^6$.

15. (a) Find the area in the first quadrant bounded by $x^4y^2 = x^2 - 9$ and the line $x = 5$. [*Hint*: $x^2 - 9$ can not be negative in this case.]
    (b) Find the volume of the solid generated by revolving this area about the y-axis.
    (c) Find the centroid of the area of part (a).
    (d) Find the centroid of the volume of part (b).

16. (a) Find the area of the ellipse $4x^2 + 9y^2 = 36$.
    (b) Find the volume of revolution formed by revolving this about the x-axis.
    (c) Find the volume of revolution formed by revolving the area of part (a) about the y-axis.

VI.6 Integration by Parts

We have very carefully avoided any mention of integrating products up to this point. It is true that in obtaining the fundamental theorem we did have two functions involved in the integrand, but this is of less than far reaching assistance since one of the two functions must be in the form of a differential, that is $dg(x)$, or else in the somewhat expanded form $g'(x)dx$. We did, on the other hand, have a result for the differentiation of a product, and perhaps, in view of the very close link between integration and differentiation, we could bring this into play. Since, by the fundamental theorem, we can use the antidifferentials to aid in the evaluation of integrals, we will examine the differential of a product. We remember that

$$d[f(x)g(x)] = f(x)dg(x) + g(x)df(x).$$

Upon integration, we have

$$f(x)g(x) = \int f(x)dg(x) + \int g(x)df(x).$$

We can only indicate the integration to be performed on the right hand side of the equation, for we do not know what the results should be in the absence of knowledge of the specific functions involved. We omitted the additive constant, but since there are integrals still to be evaluated, we can defer writing the " $+ C$" until we have evaluated the final integral. In the meantime we require the integrals which remain unintegrated to carry the burden of the constant. Since we now have two integrals instead of one, the question immediately arises whether we are any better off than before. The answer must be a resounding "maybe". In point of fact, we use this relationship in a slightly different way from that indicated by the equation above. Let us write

$$\int f(x)dg(x) = f(x)g(x) - \int g(x)df(x) \qquad \text{(VI.6.1)}$$

and then think of starting with the integral on the left side. We will break up the original integrand in such a way that we can obtain an $f(x)$ and a $g(x)$ which produce the original integrand, $f(x)dg(x)$, and such that the integral on the right hand side is more amenable to treatment—either in that it can be handled directly or that it is simpler in some sense of the word. We will illustrate this with Examples. Before going to the Examples, however, we note that since $dg(x) = g'(x)dx$, we are really looking for some way of determining an $f(x)$ and a $g'(x)$ such that their product gives the original integrand and the product $g(x)f'(x)$ will give us an integrand which is easier to handle. The main part of this particular game or technique, then, is to select the $f(x)$ and $g(x)$ in such a way that this relationship will be helpful to us. Fortunately, there are a few general guidelines for some of the more common integrals. We will indicate these in the Examples.

EXAMPLE 6.1. Evaluate $\int (\ln x)dx$.

Solution. In this case we have a rather easy choice to make. Since we have no formula at all for finding the anti-derivative of $(\ln x)$, and this would be necessary if we were to take $g'(x) = \ln x$, it would seem rather clear that the better choice would be $f(x) = \ln x$. This leaves $dg(x) = dx$, or $g(x) = x$. We can then write by relation (VI.6.1)

$$\int (\ln x)dx = x(\ln x) - \int (x)d(\ln x) = x \ln x - \int (x)\left(\frac{1}{x}dx\right)$$

$$= x \ln x - \int dx = x \ln x - x + C.$$

Note that with this choice of $f(x)$ and $g(x)$ we were fortunate enough to obtain a second integral which we could easily integrate. In the case of integration of functions such as those involving ln x, arc sin x, arc cos x, arc tan x, arc cot x, arc sec x, and arc csc x, the choice of $f(x)$ and $g(x)$ is rather clear. We cannot select one of these functions as $g'(x)$, since we do not know the anti-derivative of these functions. While these functions do occur from time to time and we do need to know how to handle them, they do not occur frequently enough that it would pay us to memorize their anti-derivatives. Consequently, this method, known as *integration by parts* since we do part of it at a time, is the one you will be most likely to use in such cases. This comes as close as anything that we have to providing us with a method for integrating a product.

In Example 6.1 we found $g'(x) = 1$. It is not always true that $g'(x)$ is so simple.

EXAMPLE 6.2. Integrate $\int x^3$ arc sin $x\ dx$.

Solution. In this case we must let $f(x) = $ arc sin x for the same reasons cited above. Therefore we must have $g'(x) = x^3$. From this we have $g(x) = x^4/4$, and hence

$$\int x^3 \text{ arc sin } x\ dx = \frac{x^4}{4} \text{ arc sin } x - \frac{1}{4}\int \frac{x^4\ dx}{\sqrt{1 - x^2}}.$$

The integral on the right side of this equation is one that we can handle by trigonometric substitution, letting $x = \sin u$. We leave this one to you to handle. Our purpose here is to illustrate the way in which integration by parts can replace an apparently intractable problem with one for which we have adequate techniques available.

If we consider the choice for $f(x)$ and $g(x)$ in the integral $\int x^2 \sin x\ dx$ it is apparent that we can let $f(x) = \sin x$ and $g'(x) = x^2$ or alternatively we can let $f(x) = x^2$ and $g'(x) = \sin x$. Let us consider the consequences of each choice. We note that we will ultimately have to deal with $f'(x)$ and $g(x)$. Whether we differentiate or find the anti-derivative of $\sin x$ makes little difference, for we will have either $\cos x$ or $(-\cos x)$, and these are equally easy (or difficult) to deal with. On the other hand, if we consider x^2, one option would yield $2x$ and the other would yield $x^3/3$. It would seem that in this case we should make the choice that would have us differentiate x^2. The result would not necessarily be something simple, but we would probably rather deal with $\int 2x \cos x\ dx$ whether it be positive or negative, than with $(1/3)\int x^3 \cos x\ dx$, regardless of sign. Note that we continue to harp on the fact that a bit of thinking along the way may save some work before we are through. If you don't believe it, try both methods here and see what happens.

We now go to a different situation, one in which the resulting integral seems to be as bad as the one we started with, regardless of which way we choose our functions. We should be happy if the resulting integral is not worse, but the fact that it is not better is still cause for some cogitation.

EXAMPLE 6.3. $\int e^x \sin x \, dx$.

Solution. In this case we can hardly gain either way, for if we let $f(x) = e^x$, then $f'(x) = e^x$. We don't change much if we let $g'(x) = e^x$. Similarly, starting with the sine we will obtain the cosine, either way we go. Since we can't simplify things by thinking, let us just get down to work and have it done with. We will arbitrarily pick $f(x) = e^x$ and $g'(x) = \sin x$ because that is the order in which they happen to be written. No harm would have been done had you made the other choice. Now we have $f'(x) = e^x$ and $g(x) = (-\cos x)$. We could insert the additive constant here, and it might be instructive to do it, but you would find that the results would not be altered by so doing. We now have

$$\int e^x \sin x \, dx = e^x(-\cos x) - \int (-\cos x)e^x \, dx.$$

We seem to have a result that is just about the same type as the one we started with. Having gone this far, however, we shouldn't quit here. We are now faced with the same choice we had before. We can either reverse the choice this time or retain the choice that got us this far. We will continue with the same relation. We would suggest you try the other one, but don't be surprised if you come out with the fact that the original integral is equal to itself—a not surprising result. We will let $f(x) = e^x$ and $g'(x) = \cos x$, whence we have $f'(x) = e^x$ and $g(x) = \sin x$. This will produce

$$\int e^x \sin x \, dx = (-e^x \cos x) + \left[e^x \sin x - \int \sin x(e^x)dx \right]$$

$$= (-e^x \cos x) + e^x \sin x - \int e^x \sin x \, dx.$$

We have now come back to the same integral we had in the beginning. However, on the right hand side it appears with a negative sign, and by adding this integral to each side we obtain (with the additive constant added this time, for now we have no remaining integrals on the right side)

$$2 \int e^x \sin x \, dx = e^x[\sin x - \cos x] + 2C.$$

We introduced $2C$ since we looked ahead and suspected we might wish to take half of each side. We finally have the result we sought in the first place, namely

$$\int e^x \sin x \, dx = (1/2)e^x(\sin x - \cos x) + C.$$

In this case we were able to obtain the result we wanted despite the fact that we seemed to return to the starting point. Had we made the opposite choice for the second integration by parts, then everything would have cancelled out and we would have found we had little to show for our efforts.

EXAMPLE 6.4. Evaluate $\int \sec^3 x \, dx$.

Solution. While this does not appear to be amenable to anything, we note that if $g'(x) = \sec^2 x$, then $g(x) = \tan x$, and we are left with $f(x) = \sec x$ and $g'(x) = \sec x \tan x$. We therefore have

$$\int \sec^3 x \, dx = (\sec x)(\tan x) - \int (\tan x)(\sec x \tan x)dx$$

$$= \sec x \tan x - \int \sec x \tan^2 x \, dx$$

$$= \sec x \tan x - \int \sec x \, (\sec^2 x - 1)dx$$

$$= \sec x \tan x - \left[\int \sec^3 x \, dx - \int \sec x \, dx \right].$$

Now however, we are right back where we began, for we again have the integral of $\sec^3 x$ on the right side. It is true that we also have the integral of $\sec x$, but we know how to handle that, so that doesn't bother us. We can do the same type of thing we did in Example 6.3, and obtain

$$2 \int \sec^3 x \, dx = \sec x \tan x + \ln|\sec x + \tan x| + 2C,$$

or

$$\int \sec^3 x \, dx = (1/2)[\sec x \tan x + \ln|\sec x + \tan x|] + C.$$

We do not by any means wish to indicate that you should perform every integration by this method, but it helps in many cases. It is the only method applicable in a large number of applications. It pays to plan your attack before starting, though, for you should stop to consider whether there is any a priori reason for making one choice as opposed to another one for the $f(x)$ and $g(x)$. This is as close as we will come to integrating products.

EXERCISES

1. Integrate:

   (a) $\int \csc^3 2x \, dx$
   (b) $\int \tan^3 3x \, dx$
   (c) $\int e^{-x} x^3 \, dx$
   (d) $\int_1^2 e^{2x} \cos 3x \, dx$

(e) $\int_1^2 x^{10} \ln x \, dx$

(f) $\int_0^1 x^2 e^x \, dx$

(g) $\int_0^{\pi/3} x \sin x \, dx$

(h) $\int_0^{\pi/3} x \cos x \, dx$

2. Integrate and check your results by differentiation:

(a) $\int x^3 \arcsin x \, dx$

(b) $\int x^2 \sin x \, dx$

(c) $\int x \arctan x \, dx$

(d) $\int x \operatorname{arc} \csc x \, dx$

3. Integrate:

(a) $\int \csc^4 x \, dx$

(b) $\int x^{100} \ln x \, dx$

(c) $\int \arcsin x \, dx$

(d) $\int \arccos x \, dx$

(e) $\int \arctan x \, dx$

(f) $\int \operatorname{arc} \cot x \, dx$

(g) $\int \operatorname{arc} \sec x \, dx$

(h) $\int \operatorname{arc} \csc x \, dx$

4. Use integration by parts to obtain results similar to those of Section VI.3 in each of the following cases.

(a) $\int \cos^2 x \, dx$ [*Hint*: Let $f(x) = g'(x) = \cos x$.]

(b) $\int \cos 4x \sin 7x \, dx$

(c) $\int \sin 3x \sin 5x \, dx$

5. Evaluate

(a) $\int_{\pi/6}^{\pi/3} x \sin^2 x \, dx$

(b) $\int_{\pi/6}^{\pi/4} x \cos^2 x \, dx$

(c) $\int_{\pi/6}^{\pi/4} x \tan^2 x \, dx$

(d) $\int_{\pi/6}^{\pi/4} x \cot^2 x \, dx$

(e) $\int_{\pi/6}^{\pi/4} x \sec^2 x \, dx$

(f) $\int_{\pi/6}^{\pi/4} x \csc^2 x \, dx$

6. Evaluate:

(a) $\int_0^{1/2} x^2 \arcsin x \, dx$

(b) $\int_0^{1/2} x \arccos x \, dx$

(c) $\int_0^1 x \operatorname{arc} \cot x \, dx$

(d) $\int_1^2 x \operatorname{arc} \sec x \, dx$

7. (a) Prove that

$$\int x^m \sin x \, dx = -x^m \cos x + m x^{m-1} \sin x - m(m-1) \int x^{m-2} \sin x \, dx.$$

(b) Use the result of part (a) to evaluate $\int_0^{\pi/4} x^2 \sin x \, dx$.

(c) Use the result of part (a) to evaluate $\int_0^{\pi/6} x^6 \sin x \, dx$.

(d) For which exponents would the result of part (a) be helpful?

(e) Find a formula for $\int x^m \cos x \, dx$ similar to that of part (a).

8. (a) Evaluate $\int x^n e^x \, dx$ for $n = 0, 1, 2, 3$, and 4.

(b) Can you guess what the result would be for $n = 5$?

(c) Evaluate part (a) for $n = 5$ and check your guess.

9. (a) Find a formula for $\int e^{ax} \cos bx \, dx$ where a and b are real constants.

(b) Find a formula for $\int e^{ax} \sin bx \, dx$ where a and b are real constants.

10. (a) Evaluate $\int (1/x)dx$ by parts using $f(x) = 1/x$ and $g'(x) = 1$.

(b) Does this prove that $1 = 0$? State your reasons for your answer.

(c) Evaluate $\int_1^e (1/x)dx$ by parts and then answer the question in part (b) with regard to this result.

11. (a) Use integration by parts to show that $\int_0^x e^t \, dt = xe^x - \int_0^x te^t \, dt$.

(b) Using the fact that $\int_0^x e^t \, dt = e^x - 1$, show that

$$-1 = -e^x + xe^x - \int_0^x te^t \, dt.$$

(c) Continue the *reverse* integration by parts and show that

$$-1 = -e^x + xe^x - \frac{x^2}{2!}e^x + \frac{x^3}{3!}e^x - \int_0^x \frac{t^3}{3!}e^t \, dt.$$

(d) Use mathematical induction to show

$$-1 = -e^x + xe^x - \frac{x^2}{2!}e^x + \frac{x^3}{3!}e^x - \cdots - (-1)^n \frac{x^n}{n!}e^x + (-1)^n \int_0^x \frac{t^n}{n!}e^t \, dt.$$

(e) Multiply by e^{-x} and show that if $0 \le x \le 1$ then the integral of part (d) can be made arbitrarily small by making n large in order to obtain the result

$$e^{-x} = 1 - x + \frac{x^2}{2!} - \frac{x^3}{3!} + \frac{x^4}{4!} - \cdots.$$

In what way does it help to have $x \le 1$?

12. (a) Find the length of the portion of $f(x) = x^2$ from $x = 2$ to $x = 4$.

(b) Find the surface of revolution formed by revolving the curve of part (a) about the x-axis.

13. (a) Find the area bounded by $y = \arcsin x$, $y = 0$, and $x = 0.5$.

(b) Find the area bounded by $y = \arctan x$, $y = 0$, and $x = 1$.

(c) Find the area bounded by $y = \text{arc sec } x$, $y = 0$, and $x = 2$.

(d) Find the centroid of part (a).

(e) Find the x-coordinate of the centroid of part (b).

(f) Find the x-coordinate of the centroid of part (c).

14. (a) Find the area bounded by $y = (\arcsin x)^2$, $y = 0$, and $x = 0.5$.

(b) Find the y-coordinate of the centroid of the area of part (a).

VI.7 Integration by Partial Fractions

If you have been keeping track of the various cases we have considered, you are aware of the fact that our techniques of integration apply rather well if we do not have fractions, or if the fractions have a denominator with at worst a power of a single expression. A denominator such as $(ax + b)^k$ is not bad, nor is one like $(ax^2 + bx + c)^k$. (In the latter case, we would probably complete the square if $b \neq 0$, but we can handle that.) With this introduction you can anticipate that we intend to consider integrals in which the denominator is a product of such factors. This is in fact the case. Our method of attack will involve expressing the integrand as the sum of two or more fractions, each of which has a denominator of the type we have just mentioned. In other words, we will consider here the problem not of adding fractions, but rather the problem of expressing a fraction as the sum of two or more new fractions, each of which is easier to integrate.

Before proceeding further, let us go back to a bit of ancient history. Using a result attributed to Euclid, it can be shown that if we have two integers, such as 15 and 77, integers which have no prime factor in common, we can find two other integers, p and q, such that $15p + 77q = 1$. In this case we could use $p = 36$ and $q = -7$ to obtain this result. We are not concerned at the moment with the method of finding p and q, although this is an interesting problem and not a hard one. You can find this result in any book on *number theory*, a topic which is really concerned with the theory of integers and not all numbers. However, the important fact here is that for any two integers that have no prime factor in common we can always find integer coefficients p and q as we did in this case, such that the linear combination of the two given integers has the value one. Now, polynomials have many more things in common with integers than might meet the eye, as you can observe by seeing which of the field properties are satisfied by each of these two systems. It can also be shown (although we will not do it here) that for any two polynomials, $f(x)$ and $g(x)$, which have no polynomial of degree at least one as a common factor, it is possible to find other polynomials $p(x)$ and $q(x)$ such that $p(x)f(x) + q(x)g(x) = 1$. The proof of this fact is almost identical with the corresponding proof for integers. This stems from a result which you will find under the title *The Euclidean Algorithm* if you try to look it up.

We are now ready to consider the *method of partial fractions*. We will use this result ascribed to Euclid in the development of the method. Let us consider that we have an integrand of the form $h(x)/(f(x)g(x))$ where $f(x)$, $g(x)$, and $h(x)$ are polynomials, such that no two of them have a factor in common of degree as great as one. (This is the type of integrand we will be considering in this section.) With these assumptions, we see that we can find polynomials $p(x)$ and $q(x)$ such that

$$p(x)f(x) + q(x)g(x) = 1.$$

From this result we have at once that

$$\frac{1}{f(x)g(x)} = \frac{q(x)}{f(x)} + \frac{p(x)}{g(x)}.$$

If we multiply both sides by $h(x)$ we obtain

$$\frac{h(x)}{f(x)g(x)} = \frac{h(x)q(x)}{f(x)} + \frac{h(x)q(x)}{g(x)}.$$

We can show (but won't here) that if $h(x)$ is of lower degree than the product $[f(x)g(x)]$, that is if the fraction on the left side of the equation is *proper*, it is possible to find polynomials $p(x)$ and $q(x)$ such that each of the two fractions on the right side of the equation is proper. This shows us that if we have a rational expression and if the denominator is the product of two *relatively prime* factors, then it is possible to express our original expression as the sum of two rational functions, each with one of the original denominator factors as a denominator. (*Relatively prime* merely indicates that the two factors have no factor in common of degree as great as one.) Furthermore, if the original expression is proper, we can express this as the sum of two proper fractions. In fact we could do the same thing if there were more than two relatively prime factors in the denominator. It would appear that we could break large fractions (in the sense of complicated denominators) up into the sum of two or more smaller fractions, the latter having more chance of being integrable. This we will do in the Examples that follow.

Before going to the Examples, we should continue this discussion just a little further. It can be shown that any polynomial with real coefficients can be factored into polynomials of either first or second degree with real coefficients. The factors which are not relatively prime will be powers of first or second degree polynomials. Consequently we can factor the denominator into products of powers (perhaps the first power in some cases) of linear and quadratic factors. We can then break the integrand up into the sum of fractions, each of which has for its denominator a power of a linear or a quadratic factor. We have already found out how to handle such integrals by algebraic or trigonometric substitutions, and consequently we have essentially solved the problem when we know that we can break the fraction up in this way. This evades the question, of course, of factoring the denominator polynomials, but if necessary you can go back to Newton's method and find the real roots and use this with the factor theorem (which states that if r is a root, then $x - r$ is a factor) to find the linear factors. We will not try to handle the general problem of factorization here, and in many applications the factoring may present no problem.

EXAMPLE 7.1. Integrate $\int (x^3 - 8)^{-1} \, dx$.

Solution. Since we have a cubic term in the denominator, this does not suggest a right triangle, and there seems to be no other ready solution.

However, we can factor $x^3 - 8$ and obtain

$$(x - 2)(x^2 + 2x + 4).$$

Since the integrand is a proper fraction, we expect to express this integral as the sum of two proper fractions. Since one of them will have a first degree denominator, the numerator must be of lower degree, and hence a constant. Since the other denominator is of degree two, the numerator will be of degree no greater than one. Thus, taking the most general possible case and using capital letters to represent the coefficients which are as yet undetermined we have

$$\frac{1}{x^3 - 8} = \frac{1}{(x - 2)(x^2 + 2x + 4)} = \frac{A}{x - 2} + \frac{Bx + C}{x^2 + 2x + 4}.$$

From this we get

$$1 = A(x^2 + 2x + 4) + (Bx + C)(x - 2) \qquad \text{(VI.7.1)}$$

by multiplying through by $(x - 2)(x^2 + 2x + 4)$. We observe that this is equivalent to

$$1 = (A + B)x^2 + (2A - 2B + C)x + (4A - 2C).$$

This will be true for all values of x only if

$$A + B = 0$$
$$2A - 2B + C = 0 \qquad \text{(VI.7.2)}$$
$$4A - 2C = 1.$$

This is a system of three equations in three unknowns, and we solve this to obtain $A = 1/12$, $B = -1/12$, and $C = -1/3$. Therefore, we are called upon to evaluate

$$\int \frac{dx}{x^3 - 8} = \int \left(\frac{1/12}{x - 2} + \frac{(-1/12)x - (1/3)}{x^2 + 2x + 4} \right) dx$$

$$= \frac{1}{12} \int \frac{dx}{x - 2} - \frac{1}{12} \int \frac{x + 4}{x^2 + 2x + 4} \, dx.$$

These last two integrals are similar to those we handled earlier. The first one is a standard type. The second one can be broken up algebraically by writing the denominator as $[(x + 1)^2 + 3]$. This gives

$$\int \frac{dx}{x^3 - 8} = \frac{1}{12} \ln|x - 2| - \frac{1}{12} \int \frac{(x + 1) + 3}{x^2 + 2x + 4} \, dx$$

$$= \frac{1}{12} \ln|x - 2| - \frac{1}{24} \int \frac{2x + 2}{x^2 + 2x + 4} \, dx - \frac{1}{4} \int \frac{dx}{(x + 1)^2 + (\sqrt{3})^2}$$

$$= \frac{1}{12} \ln|x - 2| - \frac{1}{24} \ln|x^2 + 2x + 4| - \frac{1}{4\sqrt{3}} \arctan \frac{x + 1}{\sqrt{3}} + C.$$

The important point to notice here is that we are able to take the more complicated integral and break it up into integrals which can be handled by methods we discussed earlier. (Note that trigonometric substitutions could have been used on the second integral.)

We should also observe that in equation (VI.7.1) we could use a slightly different technique to find some of the unknown quantities if not all of them. Since this relation is to hold for every value of x, it must hold in particular for the value $x = 2$. However, $x = 2$ is chosen precisely because this makes the factor $(x - 2) = 0$, and hence (VI.7.1) reduces to $1 = A(2^2 + (2)(2) + 4) = 12A$. This immediately yields the value $A = 1/12$. If we then use this in equations (VI.7.2), the solution becomes almost immediate using only the first and third equations. A particular choice of x will frequently help determine at least one of the unknown numerator coefficients used in this method. The particular value of x to be used is selected such that as many as possible of the coefficients will be zero. We will observe this again in the next Example.

EXAMPLE 7. 2. Integrate $\int (x^3/(x^2 - x - 2))dx$.

Solution. In this case the integrand is not a proper fraction, and therefore our previous discussion does not immediately apply. However, we can divide x^3 by $(x^2 - x - 2)$, and obtain

$$\frac{x^3}{x^2 - x - 2} = x + 1 + \frac{3x + 2}{x^2 - x - 2}.$$

Now we have the problem of evaluating

$$\int \frac{x^3\, dx}{x^2 - x - 2} = \int \left(x + 1 + \frac{3x + 2}{x^2 + x + 2} \right) dx$$

$$= \int (x + 1)dx + \int \frac{(3x + 2)}{x^2 - x - 2} dx$$

and we see that only the fraction causes any difficulty. The fraction we now have is a proper fraction, however, and we are able to handle this by the method of partial fractions. Since $(x^2 - x - 2) = (x - 2)(x + 1)$, and since these two factors have no factors of degree as great as one in common, we will start with the equation

$$\frac{3x + 2}{x^2 - x - 2} = \frac{A}{x - 2} + \frac{B}{x + 1}.$$

The A and B represent constant numerators. The fractions added must be proper and this requires that the numerators be of lower degrees than the first degree. This yields $3x + 2 = A(x + 1) + B(x - 2)$. From this we can deduce, as before, that $A + B = 3$ and $A - 2B = 2$. We could also note that when $x = 2$ we have $8 = 3A$ and when $x = -1$ we have $-1 = -3B$. These expressions must give the value of A and B to be used, for A and B are constants. Therefore, we have $A = 8/3$ and $B = 1/3$. Note that we have the solution of the two equations in A and B, but we obtained the solution without solving the system of simultaneous linear equations. There are instances as we saw in Example 7.1 in which this method will not give sufficient information to obtain the complete solution. Where it will work, it provides another labor saving device, and if it will not permit getting the values of all of the unknowns, it will usually obtain some. These will make the solution of the system of equations less arduous. Any help is worth something in solving larger systems of equations. With these values for A and B, we now have

$$\int \frac{x^3 \, dx}{x^2 - x - 2} = \int \left(x + 1 + \frac{8}{3} \frac{1}{x - 2} + \frac{1}{3} \frac{1}{x + 1} \right) dx$$

$$= \frac{1}{2} x^2 + x + \frac{8}{3} \ln|x - 2| + \frac{1}{3} \ln|x + 1| + C$$

$$= \frac{x^2}{2} + x + \frac{1}{3} \ln|(x - 2)^8 (x + 1)| + C.$$

We might observe that this could also have been done by trigonometric substitution. We consider one more example in order to illustrate the technique for determining the nature of the numerators of our "partial fractions" and for evaluating the constants in these numerators.

EXAMPLE 7.3. Integrate $\int ((x^2 + 3)/(x^3(x - 1)^2(x + 1)))dx$.

Solution. The relatively prime factors of the denominator are x^3, $(x - 1)^2$, and $(x + 1)$. Note that the denominator is of degree six and the numerator of degree two, and hence we have a proper fraction. Therefore we do not need to divide as we did in the last example. Furthermore, since the degree of the denominator is six, we can expect to provide for a possible degree of five in the numerator, and this implies that we would have six coefficients to determine. The mere thought of this probably sends a chill down the spine, but let us proceed. We must consider the equation

$$\frac{x^2 + 3}{x^3(x - 1)^2(x + 1)} = \frac{Ax^2 + Bx + C}{x^3} + \frac{Dx + E}{(x - 1)^2} + \frac{F}{x + 1},$$

where we have written the three numerators to indicate in each case the largest possible degree that the particular numerator could have and still yield a proper fraction. Upon multiplication we have

$$x^2 + 3 = (Ax^2 + Bx + C)(x - 1)^2(x + 1)$$
$$+ (Dx + E)x^3(x + 1) + Fx^3(x - 1)^2$$
$$= (Ax^2 + Bx + C)(x^3 - x^2 - x + 1)$$
$$+ (Dx + E)(x^4 + x^3) + F(x^5 - 2x^4 + x^3)$$
$$= (A + D + F)x^5 + (-A + B + D + E - 2F)x^4$$
$$+ (-A - B + C + E + F)x^3$$
$$+ (A - B - C)x^2 + (B - C)x + C.$$

From this we obtain the system of equations

$$\begin{cases} A + & D + & F = 0 \\ -A + B + & D + E - 2F = 0 \\ -A - B + C + & E + F = 0 \\ A - B - C & = 1 \\ B - C & = 0 \\ C & = 3. \end{cases}$$

We note that we can start with the last equation and work up to obtain $C = 3$, $B = 3$, and $A = 7$. In the first step of our work the substitution $x = 1$ would have given us $4 = (D + E)(2)$ or $D + E = 2$, and $x = -1$ would give $4 = -4F$ or $F = -1$. Note that this combination of events permits us to start with the first equation to obtain $D = -6$, and then use the second to obtain $E = 8$. These values check in each of the equations. The combination of methods has eased greatly the problem of solving the system of equations. We now use these values to rewrite our problem

$$\int \frac{x^2 + 3}{x^3(x - 1)^2(x + 1)} dx = \int \frac{7x^2 + 3x + 3}{x^3} dx$$
$$+ \int \frac{-6x + 8}{(x - 1)^2} dx + \int \frac{-1}{x + 1} dx.$$

This can be handled rather easily for we note that the first integral on the right hand side is merely the integral

$$\int (7x^{-1} + 3x^{-2} + 3x^{-3}) dx = 7 \ln|x| - 3x^{-1} - (3/2)x^{-2} + C_1$$

where C_1 is an additive constant as before. The second integral can be handled by the substitution $(x - 1) = u$, or alternatively we could write

$$\int \frac{-6x + 8}{(x - 1)^2} dx = \int \frac{-6(x - 1) + 2}{(x - 1)^2} dx$$
$$= \int [-6(x - 1)^{-1} + 2(x - 1)^{-2}] d(x - 1)$$
$$= -6 \ln|x - 1| - 2(x - 1)^{-1} + C_2.$$

Also

$$\int \frac{-1}{x+1} \, dx = -\ln|x+1| + C_3.$$

Thus we have

$$\int \frac{x^2 + 3}{x^3(x-1)^2(x+1)} \, dx = 7 \ln|x| - \frac{3}{x} - \frac{3}{2x^2} - 6 \ln|x-1|$$

$$- \frac{2}{x-1} - \ln|x+1| + C$$

$$= \ln \left| \frac{x^7}{(x-1)^6(x+1)} \right| - \frac{3}{x} - \frac{3}{2x^2} - \frac{2}{x-1} + C$$

where we have replaced the sum of the three constants $(C_1 + C_2 + C_3)$ by C to signify that we have an additive constant which is as yet undetermined. Note that the C we used here is not the same C we used earlier in the problem to represent an undetermined coefficient. In general it is not a wise thing to use the same symbol for two different values in the same problem, but *if* you keep it straight no harm is done. We mention this matter just to remind you that you should be careful not to go back and make a substitution for a constant which has been confused with another constant.

We find it necessary to use partial fractions when the denominator is a polynomial that does not yield to such substitutions as the trigonometric substitutions, and when the numerator is also a polynomial. Note that we have shown how to obtain the simpler fractions such that the denominators are powers of single expressions. Since we can factor the denominator polynomial into factors which are of first or second degree, we can use either algebraic or trigonometric substitutions to deal with the simpler fractions thus obtained. For rather obvious reasons, we usually try other methods first, but the method of partial fractions is a helpful method in case other methods fail. This is not to say that this method, nor for that matter all of the knowledge of integration that is available, is adequate for every possible integral, but each method increases the likelihood that one can successfully evaluate the integral at hand without having to resort to numerical techniques. Sometimes we can even use this method to avoid integration by parts, although we leave it to you to determine which you prefer to avoid. This is illustrated in the next example.

EXAMPLE 7.4. $\int \sec^3 x \, dx$.

Solution. It is frequently helpful to change trigonometric functions to expressions involving only sines and cosines. In this case we have $\int (1/\cos^3 x) dx$. We observe that we do not have the required differential of $\cos x$. We might try multiplying numerator and denominator by $\cos x$, just to see what would happen, and then note that we obtain $\cos^4 x$ in the denominator. This permits us to use an identity and introduce sines instead of cosines in the denominator. We now have $\int (1 - \sin^2 x)^{-2} d(\sin x)$. This can be handled by the method of partial fractions, as one might expect from the fact that it is introduced here. This will be easier to see if we let $u = \sin x$. We have $\int (1 - u^2)^{-2} du$. This leads us to the relation

$$\frac{1}{(1 - u^2)^2} = \frac{1}{(1 - u)^2(1 + u)^2} = \frac{Au + B}{(1 - u)^2} + \frac{Cu + D}{(1 + u)^2}$$

or

$$1 = (Au + B)(1 + u)^2 + (Cu + D)(1 - u)^2.$$

Using the values $u = -1$, $u = 0$, and $u = 1$, we obtain respectively

$$1 = (-C + D)4,$$
$$1 = B + D,$$
$$1 = (A + B)4.$$

Noting that the coefficient of u^3 is $(A + C)$, and this must be zero, we have a fourth equation $A + C = 0$, without doing much multiplying of algebraic expressions. The solution here yields $A = -1/4$, $C = 1/4$, and $B = D = 1/2$. Thus, our integral becomes

$$\int \frac{du}{(1 - u^2)^2} = \int \frac{-1/4u + 1/2}{(1 - u)^2} du + \int \frac{1/4u + 1/2}{(1 + u)^2} du$$

$$= \frac{1}{4} \int \frac{(-u + 1) + 1}{(1 - u)^2} du + \frac{1}{4} \int \frac{(u + 1) + 1}{(1 + u)^2} du$$

$$= \frac{1}{4} \int \frac{1}{1 - u} du + \frac{1}{4} \int \frac{du}{(1 - u)^2} + \frac{1}{4} \int \frac{1}{1 + u} du + \frac{1}{4} \int \frac{du}{(1 + u)^2}$$

$$= -\frac{1}{4} \ln|1 - u| + \frac{1}{4(1 - u)} + \frac{1}{4} \ln|1 + u| - \frac{1}{4(1 + u)} + C$$

$$= \frac{1}{4} \ln\left|\frac{1 + u}{1 - u}\right| + \frac{1}{4}\left[\frac{1}{1 - u} - \frac{1}{1 + u}\right] + C$$

$$= \frac{1}{4} \ln\left|\frac{1 + u}{1 - u}\right| + \frac{1}{4}\left[\frac{2u}{1 - u^2}\right] + C.$$

Upon replacing u by sin x, we have

$$\int \sec^3 x \, dx = \frac{1}{4} \ln \left| \frac{1 + \sin x}{1 - \sin x} \right| + \frac{\sin x}{2(1 - \sin^2 x)} + C.$$

Upon "simplifying", this becomes

$$\int \sec^3 x \, dx = \frac{1}{4} \ln \left| \frac{(1 + \sin x)^2}{1 - \sin^2 x} \right| + \frac{\sin x}{2 \cos^2 x} + C$$

$$= \frac{1}{2} \ln \left| \frac{1 + \sin x}{\cos x} \right| + \frac{1}{2} \frac{1}{\cos x} \frac{\sin x}{\cos x} + C$$

$$= \frac{1}{2} \ln |\sec x + \tan x| + \frac{1}{2} \sec x \tan x + C.$$

This is precisely the same result we obtained in the last section. Note that it is frequently possible to use different methods to arrive at the same result. The choice you make should be based upon your own preference, but you can hardly exercise any preference unless you have reasonable familiarity with each of the methods which could apply.

EXERCISES

1. Integrate using partial fractions:

(a) $\int (x^3 + 8)^{-1} dx$

(b) $\int (x^5/(x^2 - x - 6)) dx$

(c) $\int ((x^7 + x^2)/(x^4 - 16)) dx$

(d) $\int (x^2 - a^2)^{-1} dx$

(e) $\int (a^2 - x^2)^{-1} dx$

(f) $\int (x^4 - 1)^{-1} dx$

2. Evaluate:

(a) $\int_1^2 ((x - 2)/x^4(x + 1)) dx$

(b) $\int_1^2 ((x^2 - 2x)/(x^3 - 7x - 6)) dx$

   [Hint: To factor the denominator, try finding some of the roots of $x^3 - 7x - 6 = 0$.]

(c) $\int_0^1 ((x^2 + 1)/(x^4 + x^2 + 1)) dx$

   [Hint: Note that $x^4 + x^2 + 1 = (x^4 + 2x^2 + 1) - x^2$.]

(d) $\int_0^2 (x^2 - 2x - 3)^{-1} dx$

   by at least two methods.

(e) $\int_0^1 (x + 1)^{-3}(x - 2)^{-2} dx$

3. Integrate by the method of partial fractions:

(a) $\int ((x^2 - 4x + 1)/(x^2 - 1)(x - 2))dx$
(b) $\int (4x/(x^2 - 2x + 5)(x^2 + 2x + 5))dx$
(c) $\int ((x^3 - 2x^2 + 5x)/(x - 1)^3(x + 1)^2)dx$
(d) $\int ((x^2 + x + 3)/(x^3 - 3x - 3x - 2))dx$
(e) $\int ((4x + 1)/(x^2 + 2x + 2)(x - 1)^2)dx$
(f) $\int ((x^2 + 1)/(x^3 - x))dx$

4. Integrate by the method of partial fractions:

(a) $\int ((17 - 2x)/(2x + 3)(4x^2 + 4x + 17))dx$
(b) $\int (16/(x^4 - 16x^2))dx$
(c) $\int (16/(x^4 - 4x^3))dx$
(d) $\int (16/(x^4 - 8x))dx$
(e) $\int (6x^2/(x^6 - 1))dx$
(f) $\int ((60x^2 - 48)/(x^8 - 5x^6 + 4x^4))dx$

5. Integrate by the method of partial fractions:

(a) $\int (x^5/(x^2 + 2x - 8))dx$
(b) $\int ((24x^3 + 120)/(x^5 - 5x^3 + 4x))dx$
(c) $\int ((x^3 + 2x^2 + 3x + 4)/(x^2 - 4)(x^2 + 4))dx$
(d) $\int_0^1 ((x^4 + 3x)/(x + 1)^3)dx$
(e) $\int_0^2 ((x^5 + 3x^2)/(x^2 + 1)^2)dx$
(f) $\int_0^1 (x^2/(x + 1)^3(x^2 + 1)^2)dx$

6. Use the method illustrated in Example 7.4 to integrate $\int \csc^3 x\, dx$.

7. Find the area under the curve $x^3y - xy = 4$ which is above the x-axis and between $x = 2$ and $x = 4$.

8. Find the volume formed by revolving the area of Exercise 7 about the x-axis.

9. If $f(x)$ and $g(x)$ are real polynomials, and if we assume the truth of the Theorem which states that every real polynomial can be factored into factors with real coefficients with each factor being of degree one or degree two, show that we can carry out the integration of every integral of the form $\int (f(x)/g(x))dx$. You should consider both the case of proper fractions and improper fractions.

10. Integrate $\int \sec^5 x\, dx$ using the method of Example 7.4. Show that this method would work for $\int \sec^n x\, dx$ if n is an odd positive integer.

BP11. In a paper by S. Hecht in 1934, it was postulated that the photochemical process by which a retinal substance decomposes when exposed to light is given by the equation $D_t x = kI(A - x) - px^2$ where x is the present amount of the substance, t is the time measured in seconds, I is the intensity of light involved, A is the concentration of the substance prior to the incidence of the light, and k and p are constants. Find an expression for x as a function of t. [*Hint*: Convert to differentials and solve for dt before integration.]

S12. One model for population growth, based on the fact that there is a limit which the population can approach, but which the population cannot exceed, is expressed by $P'(t) = kP(t)[L - P(t)]$ where k is a positive constant and L is the limit to which the population will tend.

(a) If we assume that the arable land in the world will support 20,000,000,000 people at a subsistence level and that this is an upper limit to the population the world can support (a number that is not well-established), and if we use the fact that the population in 1850 was about 1 billion, whereas in 1970 it was 3.5 billion, what would the predicted population be in the year 2000?

(b) When will the population reach 10 billion? [*Hint*: You can use whatever interval of time you wish as being equivalent to one unit. Thus, it would be possible to consider one unit as a decade, a century, or any other unit of time that might simplify the computation. Convert to differentials and solve for dt before integrating.]

B13. When ether is administered as an anaesthetic, it is known that the effect is directly dependent upon the concentration of ether in the brain. If c_i, c_a, and c_v are the concentrations of ether in the brain, the arterial blood, and the venous blood respectively and if c_a is constant, Q_i is the amount of ether taken up by the brain, and F_i is the arterial flow in liters per minute, it has been postulated that the rate at which Q_i changes is given by the equation $D_t Q_i = F_i c_a - F_i c_v$. It should be apparent that $Q_i = c_i v_i$ where v_i is the volume of blood in the brain.

(a) Find c_i as a function of time and show that c_i approaches c_a as time increases. [In the case of ether, which has great solubility in the blood, the implicit assumption that the arterial concentration is constant is approximately correct.] [*Hint*: Convert to differentials.]

(b) Show that it is reasonable to assume $c_i = c_v$ in this problem.

VI.8 Recursion Relations and Related Results

In many problems it is possible to find relations which replace the initial problem with a partial result and a somewhat simpler residue problem. For instance, if one wished to evaluate the integral $\int \sin^9 x \cos^8 x \, dx$, the task would be long and cumbersome by the methods we discussed in the section on trigonometric methods. It would be far easier if we had some scheme by which we could express this result in terms of integrals with successively smaller exponents, until finally we came to a familiar case, such as one involving exponents of one or two, or better yet, an exponent of zero. More generally, we will investigate the case where the exponents are m and n, with the understanding that m and n are non-negative integers, and specifically we will consider the evaluation of $\int \sin^m x \cos^n x \, dx$. Using integration by parts, we have

$$\int \sin^m x \cos^n x \, dx = \int (\cos^{n-1} x \sin^m x) d(\sin x)$$

$$= \frac{1}{m+1} \sin^{m+1} x \cos^{n-1} x$$

$$- \int \frac{\sin^{m+1} x}{m+1} (n-1)\cos^{n-2} x \, d(\cos x)$$

$$= \frac{1}{m+1} \sin^{m+1} x \cos^{n-1} x$$

$$+ \frac{n-1}{m+1} \int \sin^{m+2} x \cos^{n-2} x \, dx$$

$$= \frac{1}{m+1} \sin^{m+1} x \cos^{n-1} x$$

$$+ \frac{n-1}{m+1} \left[\int \sin^m x \cos^{n-2} x \, dx - \int \sin^m x \cos^n x \, dx \right].$$

In the last step of this extended development we replaced $\sin^{m+2} x$ with $\sin^m x(1 - \cos^2 x)$, and hence

$$\sin^{m+2} x \cos^{n-2} x = \sin^m x \cos^{n-2} x (1 - \cos^2 x)$$

$$= \sin^m x \cos^{n-2} x - \sin^m x \cos^n x.$$

Since $\int \sin^m x \cos^n x \, dx$ appears on each side of the result above, we can solve for this integral and obtain

$$\left[1 + \frac{n-1}{m+1} \right] \int \sin^m x \cos^n x \, dx = \frac{1}{m+1} \sin^{m+1} x \cos^{n-1} x$$

$$+ \frac{n-1}{m+1} \int \sin^m x \cos^{n-2} x \, dx$$

or

$$\int \sin^m x \cos^n x \, dx = \frac{1}{m+n} \sin^{m+1} x \cos^{n-1} x$$

$$+ \frac{n-1}{m+n} \int \sin^m x \cos^{n-2} x \, dx. \quad \text{(VI.8.1)}$$

Similarly

$$\int \sin^m x \cos^n x \, dx = \frac{-1}{m+n} \sin^{m-1} x \cos^{n+1} x$$

$$+ \frac{m-1}{m+n} \int \sin^{m-2} x \cos^n x \, dx. \quad \text{(VI.8.2)}$$

Returning to our earlier problem we can write

$$\int \sin^9 x \cos^8 x \, dx = \frac{1}{17} \sin^{10} x \cos^7 x + \frac{7}{17} \int \sin^9 x \cos^6 x \, dx$$

$$= \frac{1}{17} \sin^{10} x \cos^7 x + \frac{7}{17} \left[\frac{1}{15} \sin^{10} x \cos^5 x \right.$$

$$\left. + \frac{5}{15} \int \sin^9 x \cos^4 x \, dx \right].$$

This can be continued until we finally have

$$\int \sin x \, dx$$

as the integral to be evaluated. A relation which permits us to continue in this fashion, not finishing the problem in a single step, but always coming out with a result which resembles the original problem and which is simpler than the one we started with, is called a *recursion relation*. We have developed two such relations for this particular integral. One of these reduces the power of the sine, and the other reduces the power of the cosine. One may wish to reduce the lower of the two original powers first, or alternatively one may wish to reduce the odd power first (if just one of them is odd). The last suggestion would insure the requisite differential for the final integration without recourse to the use of identities.

The relations developed above are very useful for the evaluation of certain definite integrals. If we consider $\int_0^{\pi/2} \sin^m x \cos^n x \, dx$, and note that sin $0 = 0$ and cos $\pi/2 = 0$, we observe that the two relations above become

$$\int_0^{\pi/2} \sin^m x \cos^n x \, dx = \frac{n-1}{m+n} \int_0^{\pi/2} \sin^m x \cos^{n-2} x \, dx \quad (\text{if } n \geq 2)$$

$$= \frac{m-1}{m+n} \int_0^{\pi/2} \sin^{m-2} x \cos^n x \, dx \quad (\text{if } m \geq 2).$$

Note the special case

$$\int_0^{\pi/2} \sin^2 x \, dx = \int_0^{\pi/2} \sin^2 x \cos^0 x \, dx = \frac{2-1}{2+0} \int_0^{\pi/2} \sin^0 x \cos^0 x \, dx$$

$$= \frac{1}{2} \int_0^{\pi/2} dx = \frac{1}{2} \left(\frac{\pi}{2} \right) = \frac{\pi}{4}.$$

Similarly

$$\int_0^{\pi/2} \cos^3 x \, dx = \frac{3-1}{3+0} \int_0^{\pi/2} \cos x \, dx = \frac{2}{3} \left[\sin \frac{\pi}{2} - \sin 0 \right] = \frac{2}{3}.$$

These results are helpful since many applications require the use of the interval $[0, \pi/2]$, and the evaluation can be obtained very easily by means of (VI.8.1) or (VI.8.2). In fact, we have a theorem concerning such integrals. This theorem has been known for a long time, being credited to *John Wallis* (1616–1700). In the expressions (a) and (b) of the theorem the factors involving m will all be even or odd depending on whether $(m - 1)$ is even or odd, the factors involving n will all be even or odd depending on whether $(n - 1)$ is even or odd, and the factors involving $(m + n)$ will be even or odd depending on whether $(m + n)$ is even or odd.

Theorem 8.1 (Wallis' theorem). *If m and n are integers, each greater than one, then*

$$(a) \int_0^{\pi/2} \sin^m x \, dx = \int_0^{\pi/2} \cos^m x \, dx = \frac{(m - 1)(m - 3) \cdots 2 \text{ or } 1}{m(m - 2) \cdots 1 \text{ or } 2} \alpha$$

and

$$(b) \int_0^{\pi/2} \sin^m x \cos^n x \, dx$$

$$= \frac{[(m - 1)(m - 3) \cdots 2 \text{ or } 1][(n - 1)(n - 3) \cdots 2 \text{ or } 1]}{(m + n)(m + n - 2)(m + n - 4) \cdots 2 \text{ or } 1} \alpha$$

where $\alpha = \pi/2$ if m is even in (a) or if both m and n are even in (b), and otherwise $\alpha = 1$.

OUTLINE OF PROOF. In case (a) show the theorem is true if $m = 1$ or 2 by straight calculation. Assume the theorem is true for some value k, then consider the case for $m = k + 2$ using the recursion relation and the fact that $\sin 0 = \cos \pi/2 = 0$. In case (b) show the theorem is true if $m = n = 1$, $m = n = 2$, $m = 1$ and $n = 2$, and for $m = 2$ and $n = 1$. Then apply the same type of mathematical induction on m and on n separately.   □

This result can be very useful. For instance, we note that

$$\int_0^{\pi/2} \sin^9 x \cos^8 x \, dx = \frac{[(8)(6)(4)(2)][(7)(5)(3)(1)]}{(17)(15)(13)(11)(9)(7)(5)(3)(1)} (1) = \frac{128}{109395}.$$

This is far easier than trying to go through any process involving integration, whether with recursion relations or without.

Theorem 8.2. *If m and n are non-negative integers, then*

$$\int_0^{\pi/2} \sin^m x \cos^n x \, dx = (-1)^n \int_{\pi/2}^{\pi} \sin^m x \cos^n x \, dx$$

$$= (-1)^{m+n} \int_{\pi}^{3\pi/2} \sin^m x \cos^n x \, dx$$

$$= (-1)^m \int_{3\pi/2}^{2\pi} \sin^m x \cos^n x \, dx.$$

OUTLINE OF PROOF. Use the periodicity of the sine and cosine and note the signs in the quadrants involved. ∎

Since integrals which involve more than one quadrant can be decomposed into the sum of integrals over single quadrants, Theorem 8.2 greatly expands the scope of Wallis' theorem.

Substitution may also aid in extending the above results.

EXAMPLE 8.1. Evaluate $\int_{\pi/2}^{3\pi/4} \sin^{12} 2x \, dx$.

Solution. If we use the substitution $y = 2x$, we have $dx = \frac{1}{2} dy$ and our integral becomes

$$\frac{1}{2} \int_{\pi}^{3\pi/2} \sin^{12} y \, dy.$$

This is a direct application of Theorem 8.2, for we note that $m = 12$ and hence

$$\frac{1}{2} \int_{\pi}^{3\pi/2} \sin^{12} y \, dy = \frac{(-1)^{12}}{2} \int_{0}^{\pi/2} \sin^{12} y \, dy = \frac{1}{2} \frac{(11)(9)(7)(5)(3)(1)}{(12)(10)(8)(6)(4)(2)} \frac{\pi}{2}$$

$$= \frac{231\pi}{4096}.$$

Another instance in which a recursion relation is very likely to be helpful occurs when the integrand includes an exponential function as a factor. This is dealt with in the following theorem. Note that in the outline of the proof we have given only general directions, for we realize that by this time you are a past master at using integration by parts.

Theorem 8.3. *Let $f(x)$ be a function such that the first n derivatives exist. Then*

$$\int e^{cx} f(x) dx = e^{cx} \left[\frac{1}{c} f(x) - \frac{1}{c^2} f'(x) + \frac{1}{c^3} f''(x) - \cdots + \frac{(-1)^{n-1}}{c^n} f^{(n-1)}(x) \right]$$

$$+ \frac{(-1)^n}{c^n} \int e^{cx} f^{(n)}(x) dx. \tag{VI.8.3}$$

OUTLINE OF PROOF. Use integration by parts and mathematical induction. ∎

As a corollary of this theorem, we have for the special case of the polynomial of degree n an even nicer result.

Corollary 8.1. *Let $f(x)$ be a polynomial of degree n, then*

$$\int e^{cx} f(x)dx = e^{cx} \sum_{k=0}^{n} \frac{(-1)^k}{c^{k+1}} f^{(k)}(x) + C \qquad \text{(VI.8.4)}$$

where $f^{(0)}(x) = f(x)$.

Note that in all cases in which we have indicated derivatives, we have placed the order of the derivatives (that is in the "n" of the n-th derivative) in parentheses to aid in distinguishing the order from an exponent. This is not true, of course, where we have used the primes as we have done for the lower ordered derivatives. It is also customary to denote by the zero-th order derivative the original function, for this frequently simplifies the writing of summations, as it did above. Theorem 8.3 does not obviate the necessity of ultimately evaluating an integral except in the case in which some derivative has a constant value zero, as in the case of the polynomial. However, it will frequently simplify the problem of integration.

It is possible to use Theorem 8.3 in a manner reminiscent of some earlier problems when we obtained an integral on the right side of the equality which resembled the one on the left.

EXAMPLE 8.2. Evaluate $\int e^{3x} \cos 2x\, dx$.

Solution. We have $c = 3$ and $f(x) = \cos 2x$. Consequently

$$\int e^{3x} \cos 2x\, dx = e^{3x} \left[\frac{1}{3} \cos 2x - \frac{1}{9}(-2 \sin 2x) \right]$$

$$+ \frac{(-1)^2}{9} \int e^{3x}(-4 \cos 2x)dx.$$

We note that $\int e^{3x} \cos 2x\, dx$ appear on both sides of this equation, and hence we can solve for the value of this integral. This gives us

$$\frac{13}{9} \int e^{3x} \cos 2x\, dx = \frac{e^{3x}}{9} [3 \cos 2x + 2 \sin 2x] + \frac{13C}{9}.$$

We have added the strange looking constant for we no longer have an integral on the right side of this equation and consequently we need a constant of integration. Furthermore we can look ahead and notice that we will want to divide by $(13/9)$ to obtain the desired result, and the coefficient $(13/9)$ for C at this stage will permit us to avoid a fractional coefficient at the next step. Finally we have

$$\int e^{3x} \cos 2x\, dx = \frac{e^{3x}}{13} [3 \cos 2x + 2 \sin 2x] + C.$$

One final result that will be of assistance in evaluating definite integrals is special to the extent that it only applies when the interval of integration has zero as a midpoint. While the theorem itself is not used often, the corollary can be very helpful.

Theorem 8.4. *If the integrals exist,* $\int_{-a}^{a} f(x)dx = \int_0^a [f(x) + f(-x)]dx$

PROOF.

$$\int_{-a}^{0} f(x)dx = \int_a^0 f(-x)d(-x) = -\int_a^0 f(-x)dx = \int_0^a f(-x)dx.$$

Hence

$$\int_{-a}^{a} f(x)dx = \int_0^a f(x)dx + \int_{-a}^0 f(x)dx = \int_0^a f(x)dx + \int_0^a f(-x)dx$$

$$= \int_0^a [f(x) + f(-x)]dx. \qquad \square$$

There are many places in which this is very helpful. For instance, since $\sin(-x) = -\sin x$, we would have at once

$$\int_{-a}^{a} \sin x\, dx = \int_0^a [\sin x + \sin(-x)]dx = \int_0^a [\sin x - \sin x]dx = 0$$

without further ado. Many functions have properties similar to those of $\sin x$, and we take note of these in the following definition.

Definition. A function is an *even function* if $f(-x) = f(x)$ for every value of x in the domain of $f(x)$. A function is an *odd function* if $f(-x) = -f(x)$ for every value of x in the domain of $f(x)$.

Note that $\sin x$ is an odd function. Any polynomial consisting only of even powers is an even function, and a polynomial consisting only of odd powers is an odd function. We have the following corollary.

Corollary 8.2. *If* $f(x)$ *is an even function, then*

$$\int_{-a}^{a} f(x)dx = 2\int_0^a f(x)dx.$$

If $f(x)$ *is an odd function, then*

$$\int_{-a}^{a} f(x)dx = 0.$$

This corollary will frequently ease the problem of evaluating integrals over intervals having zero as the midpoint. This case occurs with some frequency in certain types of applications.

EXAMPLE 8.3. Evaluate $\int_{-2}^{2}(x^4 + x^3 - x \sin x + e^x)dx$.

Solution. We note that x^4 is an even function and x^3 is an odd function. Since the interval $(-2, 2)$ is centered about the origin we can use Corollary 8.2. We might also note that if $f(x) = x \sin x$, then

$$f(-x) = (-x)\sin(-x) = (-x)(-\sin x) = x \sin x = f(x).$$

Hence, $f(x) = x \sin x$ is an even function. On the other hand the value of e^{-x} is not equal to e^x even in absolute value. Hence e^x will not be aided by this new result. Using this information we have

$$\int_{-2}^{2} (x^4 + x^3 - x \sin x + e^x)dx$$

$$= 2 \int_{0}^{2} x^4 \, dx + 0 - 2 \int_{0}^{2} x \sin x \, dx + \int_{-2}^{2} e^x \, dx$$

$$= \frac{2(2)^5}{5} - 2[-x \cos x + \sin x]\big|_0^2 + e^2 - e^{-2}$$

$$= \frac{64}{5} + 4 \cos 2 - 2 \sin 2 + e^2 - e^{-2}$$

$$= \frac{64}{5} + 4(-0.416) - 2(0.909) + 7.389 - 0.135$$

$$= 16.572.$$

EXERCISES

1. Evaluate:

   (a) $\int \sin^5 x \cos^6 x \, dx$
   (b) $\int_0^{\pi/2} \sin^5 x \cos^6 x \, dx$
   (c) $\int \sin^4 x \cos^5 x \, dx$
   (d) $\int \sin^4 x \cos^6 x \, dx$
   (e) $\int_0^{\pi/2} \sin^7 x \cos^4 x \, dx$
   (f) $\int_0^{\pi/2} \sin^8 x \, dx$
   (g) $\int_0^{\pi/2} \cos^7 x \, dx$
   (h) $\int_0^{\pi/2} \sin^4 x \cos^4 x \, dx$

2. Evaluate:

   (a) $\int_0^{\pi/4} \sin^7 2x \, dx$
   (b) $\int_0^{\pi/6} \cos^4 3x \, dx$
   (c) $\int_{-\pi/2}^{\pi/2} \cos^5 2x \sin^8 2x \, dx$
   (d) $\int_0^{\pi/2} \cos^7 x \tan^4 x \, dx$
   (e) $\int e^{3x}(x^4 - 3x^2 + 2x + 7)dx$

(f) $\int e^{2x} \sin 3x \, dx$

(g) $\int e^{-x} \cos x \, dx$

(h) $\int e^{x^2}(x^4 + x^2 + 1)x \, dx$

3. (a) Classify each of the six trigonometric functions as even or odd.

(b) If n is an integer, show that $(x^{2n} \sin x)$ is odd and $(x^{2n} \cos x)$ is even.

(c) Evaluate $\int_{-1}^{1} (\sin 3x - x^2 + \tan x + x \sec x) dx$.

(d) Show that the products $(\sin x \sinh x)$ and $(\cos x \cosh x)$ are both even.

(e) Show that the fact that $(f(x)g(x))$ is even proves that if $f(x)$ is odd then $g(x)$ is odd, and if $f(x)$ is even, then $g(x)$ is even provided $f(x) \neq 0$.

(f) Show that it is possible to find a function $f(x)$ and a function $g(x)$ such that neither is even or odd but their product is even.

4. Find the area inside the curve $r = 1 - \cos^3 \theta$ between $\theta = 0$ and $\theta = \pi/2$.

5. (a) Evaluate $\int_0^x e^t t^5 \, dt$.

(b) Evaluate $\int_0^x e^t t^n \, dt$ if n is a positive integer.

(c) Find a formula for $\int_0^x e^{-t} t^n \, dt$ if n is a positive integer.

6. Evaluate each of the following:

(a) $\int_{-2}^{2} (\sum_{k=0}^{8} x^k) dx$

(b) $\int_{-1}^{1} (\sum_{k=0}^{3} x^k \sin^k x) dx$

(c) $\int_{-1}^{1} (\sum_{k=0}^{4} x^k \cos^k x) dx$

(d) $\int_{-\pi/4}^{\pi/4} (\sum_{k=0}^{10} \sin^k 2x) dx$

(e) $\int_{-\pi/2}^{\pi/2} (\sum_{k=0}^{10} \cos^k x) dx$

(f) $\int_{-\pi/4}^{\pi/4} (\sum_{k=0}^{3} \tan^k x) dx$

7. Evaluate:

(a) $\int_{-4}^{4} (16 - x^2)^{5/2} dx$

(b) $\int_{-1}^{1} x^6(1 - x^2)^3 dx$

(c) $\int_0^4 x^3(4x - x^2)^{1/2} dx$

(d) $\int_0^8 (8x - x^2)^{7/2} dx$

(e) $\int_{-2}^{2} x(4 - x^2)^{3/2} dx$

(f) $\int_{-5}^{5} x^4(25 - x^2)^{7/2} dx$

(g) $\int_0^6 x^3(6x - x^2)^{3/2} dx$

(h) $\int_0^{10} (10x - 4x^2)^6 dx$

8. (a) Sketch $f(x) = x^2 e^x$.

(b) Find the values of x for which $x^2 e^x = \int_0^x t^2 e^t \, dt$.

(c) Give a geometric explanation of the values obtained in part (b).

9. (a) Prove that $\int_0^{2\pi} \sin^m x \cos^n x \, dx = 0$ unless both m and n are even.

(b) For what values of m and n is it true that $\int_0^\pi \sin^m x \cos^n x \, dx = 0$?

(c) For what values of m and n is it true that $\int_{-\pi/2}^{\pi/2} \sin^m x \cos^n x \, dx = 0$?

10. (a) Find the area bounded by $y = \sin x$, $y = 0$, and $x = \pi$.

(b) Find the volume of revolution formed by revolving the area of part (a) about the x-axis.

11. Derive the recursion relation by which the exponent of $\sin x$ is reduced in the integral $\int \sin^m x \cos^n x \, dx$.

12. Prove Theorem 8.1.

13. Prove Theorem 8.2.

14. Prove Theorem 8.3.

VI.9 A Soliloquy on Techniques

In this chapter we have tried to indicate some of the more common techniques whereby the work of evaluating derivatives and integrals can be made possible and less difficult. Since not all functions have derivatives, and not all functions possess integrals which can be evaluated in terms of elementary formulas, it should not be surprising that we have not covered all possible cases. There are many special techniques which apply in a small number of cases, cases which arise less frequently in applications. We have made no effort to include these, but rather have tried to stay with methods which have wide application. There are special substitutions, for instance, which are very helpful when needed, but which are needed with insufficient frequency that you are apt to forget them between uses. We will cover one such substitution in the next paragraph. There is also the possibility of using tables of derivatives and integrals. These can be very helpful, but a relatively small table of integrals may well have more than 1000 formulas. It becomes difficult, then, to find the correct formula in the table. Frequently it is necessary to make substitutions, complete the square, or do some other bit of manipulation before the integral you have will match one in the table. This is not to discourage the use of tables, but rather to be honest in indicating that tables do not remove a requirement for knowing methods whereby one integral can be shown to be equivalent to another in what might be called a standard form. We should also keep in mind that for those integrals where it is not possible to find an appropriate entry in the tables and which cannot be integrated using any or all of the techniques mentioned here, we may use the numerical methods discussed in Chapter II. Since integrals requiring numerical methods occur with reasonable frequency these days, we will discuss numerical methods in Chapter X with particular reference to methods which are more efficient than those in Chapter II.

In order to illustrate the variety of substitutions that may be invoked in the process of integration, let us consider

$$\int \frac{1 + \cos x}{2 - \sin x + \cos x} \, dx.$$

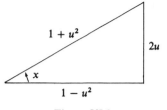

Figure VI.8

Some astute individual observed that in cases such as this we can invoke the half angle results, much as we did in the integration of the cosecant. In this particular case we use the substitution

$$u = \tan(x/2).$$

By using the double angle formula for tangents we obtain

$$\tan x = \frac{2 \tan(x/2)}{1 - \tan^2(x/2)} = \frac{2u}{1 - u^2}.$$

with this information, we can draw a triangle (much as we did before) as shown in Figure VI.8. Note that by the Pythagorean theorem we have the square of the hypotenuse represented by

$$(2u)^2 + (1 - u^2)^2 = 4u^2 + 1 - 2u^2 + u^4$$
$$= 1 + 2u^2 + u^4$$
$$= (1 + u^2)^2.$$

We can now obtain values for sin x and for cos x. We have $\sin x = 2u/(1 + u^2)$ and $\cos x = (1 - u^2)/(1 + u^2)$. Furthermore, since $x/2 = \arctan u$, we have $x = 2 \arctan u$, and consequently $dx = (2/(1 + u^2))du$. Our original in-integral becomes, upon substitution

$$\int \frac{1 + \dfrac{1 - u^2}{1 + u^2}}{2 - \dfrac{2u}{1 + u^2} + \dfrac{1 - u^2}{1 + u^2}} \frac{2\,du}{1 + u^2} = \int \frac{4\,du}{(1 + u^2)(3 - 2u + u^2)}$$

where the last result is obtained by algebraic simplification of the integrand. We recognize that the last result is now in a form in which we can use partial fractions. This becomes

$$\int \frac{1 + u}{1 + u^2}\,du + \int \frac{1 - u}{3 - 2u + u^2}\,du = \frac{1}{2} \ln \left| \frac{u^2 + 1}{u^2 - 2u + 3} \right| + \arctan u + C.$$

We have used techniques discussed in this chapter for the integration, and we have also done some simplification by combining the logarithmic terms.

Now we replace the terms involving u by the appropriate terms involving x, and have

$$\frac{1}{2} \ln \left| \frac{\tan^2 \frac{x}{2} + 1}{\tan^2 \frac{x}{2} - 2 \tan \frac{x}{2} + 3} \right| + \frac{x}{2} + C$$

$$= \frac{1}{2} \ln \left| \frac{1}{\sin^2 \frac{x}{2} - 2 \sin \frac{x}{2} \cos \frac{x}{2} + 3 \cos^2 \frac{x}{2}} \right| + \frac{x}{2} + C$$

$$= \frac{x}{2} - \frac{1}{2} \ln |2 - \sin x + \cos x| + C.$$

You will observe that we have used many of the methods discussed in this chapter and many identities in a single example. It is suggested that you check this by differentiation.

One final bit of advice on differentiation and integration is in order. Be very certain that you take a careful look at the problem before you, think of the effects of using each of the various tools you now have available, and then make the selection of the one you wish to use. If one fails, try another of the possible alternatives. If all of them fail, then resort to numerical methods. The principal thing to keep in mind is that the more you use these methods, the easier their use becomes. If it seems that we have been fortunate in quoting only the correct (or perhaps a correct) method in each case, just stop to think of the time that was probably spent in practice by the author. The same thing would apply to an instructor in class who always seems to know just what to do, for he (or she) has probably had a great deal of practice, too. Unfortunately, there is only one way to obtain this experience, and that is by doing problems. Try to learn from each one. A post-mortem on each one to determine why one method worked, and another one failed is not a bad idea.

EXERCISES

1. Integrate:

(a) $\int ((2 - \sin x)/(1 + \cos x))dx$
(b) $\int_{\pi/6}^{\pi/2} (3/(1 + \cos x))dx$
(c) $\int (2 \sin x/(1 + \sin x))dx$

2. Differentiate:

(a) $f(x) = x^3(\text{arc sin } x)/\sqrt[3]{4 - x^2}$
(b) $f(x) = 4x(\text{arc sec } x)\sqrt[3]{8x} + x^2$
(c) $f(x) = (\text{arc tan } e^x)/(4x - x^3)\csc^4 x$

3. Integrate:

   (a) $\int \sec^2 2x \tan 2x \, dx$
   (b) $\int \sec^4 x \sin x \tan^2 x \, dx$
   (c) $\int \cos^8 x \sin^8 x \, dx$

4. Evaluate:

   (a) $\int_0^{\pi/4} \sin^5 x \cos^6 x \, dx$
   (b) $\int_{\pi/6}^{\pi/2} \cos^7 x \, dx$
   (c) $\int_{\pi/6}^{\pi/3} \cos^8 x \sin^4 x \, dx$
   (d) $\int_{\pi/4}^{\pi/2} \sin^5 x \cos^7 x \, dx$
   (e) $\int_0^{\pi/6} \sin^5 3x \cos^4 3x \, dx$

5. Integrate:

   (a) $\int e^{3x}(\cos 5x + 2 \sin 5x)dx$
   (b) $\int e^{-2x}(x^4 - 7x^3 + 2x^2 + 18x - 20)dx$
   (c) $\int e^{-3x}\sqrt{3 - x} \, dx$ [Obtain four terms plus (or minus) another integral.]
   (d) $\int e^{x/2} \ln x \, dx$ [Follow the instructions for part (c).]

6. Evaluate:

   (a) $\int_{-\pi/2}^{\pi} \sin^5 x \cos^4 x \, dx$
   (b) $\int_{-\pi/4}^{\pi/4} \sin^5(x/2)\cos^8(x/2)dx$
   (c) $\int_0^{\pi/2} \cos^{16} x \sin^{14} x \, dx$
   (d) $\int_0^{\pi} \sin^9(x/2)\cos^{16}(x/2)dx$

7. Evaluate each of the following:

   (a) $\int \left(\sum_{k=0}^5 \tan^k x \right)dx$
   (b) $\int \left(\sum_{k=0}^5 \cot^k x \right)dx$
   (c) $\int \left(\sum_{k=0}^5 \sec^k x \, dx \right)dx$
   (d) $\int \left(\sum_{k=0}^5 \csc^k x \right)dx$

8. Evaluate each of the following:

   (a) $\int \sin^5 x(\cos x)^{-6}dx$
   (b) $\int \sin^3 2x \cos^4 x \, dx$
   (c) $\int \sin 3x \sin 5x \, dx$
   (d) $\int \sin^3 x(\cos x)^{-5}dx$
   (e) $\int \sin 5x \cos 8x \, dx$
   (f) $\int \cos 4x \cos 9x \, dx$

9. Evaluate each of the following:

   (a) $\int ((x^4 + 1)/(x^3 - x))dx$
   (b) $\int ((13x + 6)/(x^3 - x^2 - 6x))dx$
   (c) $\int ((x^7 + 2)/(x^3 + x))dx$
   (d) $\int ((x^2 - 3x + 12)/(x^3 - x^2 + 4x - 4))dx$

10. Find the area inside one loop of the curve $r = \cos^2 \theta$.

11. Using the relation involving dx and dy in terms of dr and $d\theta$, find an integral for the length of arc in polar coordinates. Find the entire length of the curve $r = 1 + \cos \theta$.

12. Find the second moment of the circle $x^2 + y^2 = 16$ about the vertical line $x = 6$. [Set up the RS sum, obtain the integral, and integrate.]

13. Two roommates, each taking calculus are involved in an argument. They have integrated

$$\int \csc^3 3x \cot 3x \, dx$$

and one answer reads $[-1/6 \csc^2 3x + C]$ whereas the other reads

$$[-1/6 \cot^2 3x + C].$$

How did each one obtain his answer? Who was right? How would you settle this argument?

P14. A rod has a mass at each point of e^{-x^2} grams per centimeter where x is the distance in centimeters from one end of the rod. Find the first moment of this rod about the end where $x = 0$ if the rod is 3 centimeters long.

S15. A demand curve for a certain commodity has the equation

$$y = \frac{157 + x}{x^3 + x^2 + x + 1}.$$

Find the consumers surplus if the equilibrium price is $4.

B16. Many of the applications published in biological research since 1908 have been based upon a growth curve developed by T. B. Robertson. This curve is determined by the relation $D_t P^{(t)} = (1 + b)bkP(0)e^{-kt}/(1 + be^{-kt})^2$ where b and k are positive constants, $P(t)$ is the population at any time, t, and $P(0)$ is the population at $t = 0$. Find an equation for $P(t)$ as a function of t.

CHAPTER VII

Limits and Related Topics

VII.1 Reflections—and a Look Ahead

We have frequently commented that we were considering proofs or derivations in a fairly complete fashion, but that we would wait until Chapter VII
to take a more careful look. Now we have arrived. You may have feelings of
anticipation, dread, foreboding, or just wonderment at the whole thing. If you
have made a careful inventory of those places in which we have made reference to this chapter, you will note that they all have to do with continuity, limits
or limiting processes. We do not propose here to go back and consider each
specific reference individually, but rather to develop the concepts which are
necessary to remove any doubts about the validity of the earlier arguments.
You will note that the methods used here will follow very closely those used
before. Consequently only slight changes in wording are required to put the
earlier proofs in the more traditional language associated with limits.

As we have indicated in one manner or another throughout discussions
involving limits, the one central theme we must consider when we say that
the limit of $f(x)$ as x approaches a is the number L is that we must be assured
that as x gets closer and closer to a (although we have no reason for requiring
or permitting that x ever take on the value a) we must have $f(x)$ getting
closer and closer to L. The main problem in setting up a mathematical
definition of limit, then, is the problem of making our concept of *closer* such
that there is no ambiguity. In our every day speech we may say that we were
close to Podunk if we drove within fifty miles—or perhaps within two
hundred miles if the driving were rugged. On the other hand, we do not
feel that we have parked sufficiently close to the curb in parallel parking
unless we are within 18 inches (or preferably within 6 inches) of the curb.
Therefore, *close* can have many meanings. This lack of precision is not

appropriate for a definition of limit. Being within even 6 inches is certainly not approaching the curb as a limiting position. We probably wouldn't wish to approach the curb as a limiting position anyway if we had white walled tires. We solved this dilemma in Definition IV.2.1 by saying that for *any* open set E which included L as an interior point we wanted to insure that $f(x)$ not only could be found within E, but would also remain in E provided x were in a corresponding *deleted* neighborhood of a. Note that we have again mentioned the fact that we do not require, and might not desire, that x be permitted to take on the value a. For instance, zero was rather obviously an undesirable value for Δx in the case of the limits of the differential quotients in differentiation. Our earlier definition is correct and we will not alter it in substance. For purposes of proofs, however, there is a slight variation on this definition which will be of assistance. In this rewording of the definition we will consider that the sets E and D are symmetrically located about L and a respectively, and we will denote the radius of the two sets (that is half of the diameter of the intervals) by ε and δ respectively. It is a matter of sophistication, of course, to use the Greek letters here, but don't let that worry you. These letters, the lower case *epsilon* and *delta* respectively, merely represent numbers. Since these two letters are used for this purpose in almost every publication involving limits you have the advantage that once you have become familiar with this usage you will not have to modify it in the future, even to the extent of changing the notation used. (You might raise the question whether we selected the letters E and D for the sets of our earlier definition with the ε and δ in mind. We will let you speculate on this matter.) Now, since ε and δ are the radii of two intervals, and hence represent lengths of half intervals, there is no reasonable interpretation for these quantities unless they are positive. (We certainly would not have any sets to work with if they were zero, and the negative choice does not present any decent interpretation.) However, we will usually state that they are positive as a reminder.

With these comments in mind, we suggest that you think in terms of the sets we used earlier as well as in terms of the ε's and δ's, for a combination of the two approaches may well be helpful. Note that the main difference that we are bringing about is that of requiring that the sets E and D be symmetric about the points of principal interest. We could take precisely the portion of our earlier sets, E and D, for which we can go equally far in both directions from L and a respectively, and therefore consider our new sets as subsets of the earlier ones.

EXERCISES

1. Let E be an interval such that L is an interior point of E.

   (a) Show that it is possible to find a value ε such that the interval $(L - \varepsilon, L + \varepsilon)$ is a subinterval of E.

(b) Show that the subinterval of part (a) is equivalent to the interval determined by $|L - x| < \varepsilon$.

(c) Illustrate parts a and b if E is the interval $[3, 6]$ and $L = 4$. Indicate the interval of choices you have available for ε.

2. Let D be a deleted neighborhood of a point $x = a$.

(a) Show that it is possible to find a value δ such that $0 < |x - a| < \delta$ is a subset of D.

(b) If D is the interval $(0.5, 1.2)$ from which the point $x = 0.95$ has been deleted, find the value of δ in part (a) such that the deleted neighborhood of part (a) would be a subset of D.

3. (a) In Section (III.5) we proved that $\lim_{x \to 0} (\sin x)/x = 2\pi/U$. Write out a proof of this result using the language of ε and δ.

(b) In Section (III.5) we proved that $\lim_{x \to 0} (1 - \cos x)/x = 0$. Write out a proof of this result using the language of ε and δ.

4. (a) In Section (IV.4) we developed a formula for the derivative of a product the proof of which required a limit. Write out the proof that the limit exists using the language of ε and δ.

(b) In Section (IV.4) we developed a formula for the derivative of a quotient the proof of which required a limit. Write out the proof that the limit exists using the language of ε and δ.

5. (a) If a person takes a single dose of medicine the concentration of the medicine is built up in the person's bloodstream and then gradually diminishes to the point where it would not be possible to determine that the person had taken the medicine. If we consider that the concentration is approaching a value of zero, state this in the language of limits. Note that in this case D would not be the usual deleted neighborhood, but would be an open interval extending from some fixed time on through all future time. How could you word this?

(b) If a capacitor is connected to a battery, a charge builds up on the plates of the capacitor, and approaches some maximum charge. Show that the charge is approaching a limit in the sense in which we have used the term *limit*, and express this in a manner similar to that requested in part (a).

VII.2   Limits

We indicated in Section VII.1 that we will now use the Greek letters, ε and δ, to indicate closeness. Therefore, Definition IV.2.1 becomes

Definition 2.1. The limit of the function $f(x)$ as x approaches the value a [denoted by $\lim_{x \to a} f(x)$] has the value L if and only if for *any* $\varepsilon > 0$ there is a value $\delta > 0$ such that if $0 < |x - a| < \delta$, then $|f(x) - L| < \varepsilon$.

You will observe that we have indicated the set E of the earlier definition as the interval $(L - \varepsilon, L + \varepsilon)$ and the deleted neighborhood D as the interval

$(a - \delta, a + \delta)$ where we have excluded $x = a$ by the simple device of not permitting $|x - a|$ to have the value zero. The sequence of events is still the same. For *any* $\varepsilon > 0$ (which means that ε could be as large as you might wish or as small as you may care to make it), we have a limit *only* if we can be assured that there is *some* $\delta > 0$ such that when x is the interval $(a - \delta, a + \delta)$ *and* $x \neq a$ we know that $f(x)$ must then (for all values of x meeting these conditions) be in the interval $(L - \varepsilon, L + \varepsilon)$. This definition can be used in two ways. The more obvious way is that of showing that a function does have a limit. We will frequently consider, however, a function that does have a limit. We will then know that these conditions must hold, and therefore for any ε there must be a δ satisfying the relations stated in the definition. You will see both uses in this and subsequent sections.

We now prove a theorem which is of very great assistance in differentiation using the definition of a derivative.

Theorem 2.1. *If $f(x) = g(x)$ for all values of x in (a, b) with the exception of the single point $x = c$ in (a, b), and if $\lim_{x \to c} f(x)$ exists, then $\lim_{x \to c} g(x)$ also exists and the two limits are equal.*

PROOF. Since $\lim_{x \to c} f(x)$ exists, then for any value of $\varepsilon > 0$ there is a corresponding value $\delta > 0$ such that when $0 < |x - c| < \delta$ it follows that $|f(x) - L| < \varepsilon$. L is the value such that $L = \lim_{x \to c} f(x)$. L exists by hypothesis. Since $x = c$ is excluded from consideration, and since for any value of x in the interval other than $x = c$ we have $g(x) = f(x)$, we can then write for this same choice of ε and δ that when $0 < |x - c| < \delta$ we have $|g(x) - L| < \varepsilon$. Therefore, by the definition, we have $\lim_{x \to c} g(x) = L = \lim_{x \to c} f(x)$. $\square$

Note that we have used the definition in both directions in this one proof, for we have used the fact that $f(x)$ has a limit to assure the existence of precisely the conditions that in turn assure the existence of the limit of $g(x)$.

We have used the results of this theorem many times. When we took the limit in obtaining the derivative of a function such as $f(x) = x^2$ we were faced with the limit

$$\lim_{\Delta x \to 0} \frac{(x + \Delta x)^2 - x^2}{\Delta x} = \lim_{\Delta x \to 0} \frac{2x\Delta x + (\Delta x)^2}{\Delta x}.$$

We now observe that

$$\frac{2x\Delta x + (\Delta x)^2}{\Delta x} = 2x + \Delta x$$

for all values of Δx *except* for $\Delta x = 0$. Since we wish the limit as Δx approaches zero, we can use this theorem to note that $[2x\Delta x + (\Delta x)^2]/\Delta x$ and $(2x + \Delta x)$ have the same limit. Since it is not difficult to determine the limit of $2x + \Delta x$, we have simplified our problem a great deal. It would be worth your time to review all of the theorems concerning derivatives to note the number of times that we made use of this result.

Up to this point we have only considered limits as x approaches a point in some open interval, or in other words as x approaches a finite value. There are many instances in which we desire to investigate what happens as x increases without any stopping point (or as $x \to \infty$, to use generally accepted notation). Note that ∞ is *not* a number, and our use of this notation merely means that we will keep on going forever, or that however far we may have gone is not far enough. At first glance an investigation of this type would seem to be non-productive, for it is not unreasonable to expect that most functions would just disappear if we permit x to increase without any bounds whatsoever. That this is not true, however, can be seen by considering the function $f(x) = e^{-x} = 1/e^x$, for as x increases without limit, then e^x increases, but this would require that $f(x)$, which is the reciprocal of e^x, would decrease toward the value zero. This function occurs very frequently in application. In the decay of radioactive materials or in the amount of a medicine remaining in the system after a given time, this is the function which is descriptive of the results. This would indicate that in time the amount of radioactive material or the amount of medicine in the body would drop below any discernable amount. In order to include the situation where the variable increases without bound we state the following Definition.

Definition 2.2. Let $f(x)$ be a function defined for all values $x \geq a$ where a is some finite number. Then $\lim_{x \to \infty} f(x)$ exists and has the value L (written $\lim_{x \to \infty} f(x) = L$) provided for any number $\varepsilon > 0$ we can find a number $X \geq a$ such that if $x > X$, then $|f(x) - L| < \varepsilon$. A similar definition holds in the case of $\lim_{x \to -\infty} f(x)$ with a reversal of the inequalities involving X.

In the evaluation of such limits, we frequently have a case similar to the one described above, namely that of having a denominator which increases without limit. For this reason it would be convenient to have an established result that will permit us to consider such cases more easily.

Theorem 2.2. *If $f(x)$ is a function and if for any positive number, M, there is a number X such that when $x > X$, it is true that $|f(x)| > M$, then $\lim_{x \to \infty} [1/f(x)] = 0$.*

PROOF. Consider any $\varepsilon > 0$. Since ε is non-zero, then $1/\varepsilon$ exists. Let $M = 1/\varepsilon$. by hypothesis there is a value X such that for all $x > X$ it is true that $|f(x)| > M$. If we divide both sides of this inequality by the positive quantity $M|f(x)|$, we have $\varepsilon = 1/M > 1/|f(x)| = |(1/f(x)) - 0|$ for all values $x > X$. This proves the theorem. $\qquad\qquad\square$

We illustrate the use of this theorem with an example.

EXAMPLE 2.1. Evaluate $\lim_{x \to \infty} [x^2/(x^3 + x + 2)]$.

Solution. Let a be any positive number. When $x > a$, $x^2/(x^3 + x + 2) = 1/(x + x^{-1} + 2x^{-2})$. Therefore

$$\lim_{x \to \infty} \left[\frac{x^2}{(x^3 + x + 2)} \right] = \lim_{x \to \infty} \left[\frac{1}{\left(x + \dfrac{1}{x} + \dfrac{2}{x^2} \right)} \right].$$

For any $M > 0$ if $x \geq M$, then $x + 1/x + 2/x^2 > M$. Therefore, $\lim_{x \to \infty} [x^2/(x^3 + x + 2)] = 0$ by Theorem 2.2.

At this point we should make some general observations concerning the application of the definition. Since the entire problem of defining limits hinged on the matter of stating in a precise manner that we were sufficiently *close* to the value which was a candidate for the limiting value, the use of the number ε is in a sense a semantic device. There would have been no harm in using 2ε if that had seemed more convenient. The principal consideration is in the fact that we have the ability to require that this quantity, whatever its name, can be as small as we demand in accordance with our need for closeness in the particular case at hand. In some proofs we have the option of adjusting the size of the interval in advance, perhaps by using $\varepsilon/2$, and thus obtaining the final degree of closeness as ε, or of starting with ε and then obtaining a final degree of closeness of 2ε. Unless one has a crystal ball, or has gone through the proof before, the latter case is probably easier to obtain and is perfectly correct. Another fact that we should consider is that we do not demand in any sense that we have the largest value of δ that will fulfill the conditions of the Definition. All that is required is that we show that at least one positive value of δ exists for which the conditions are satisfied. In fact, any smaller value would also work, for we would have a deleted neighborhood which is a subset of one for which everything works well. If everything is well behaved in the larger neighborhood, it certainly must in the subset. We illustrate this in the following example.

EXAMPLE 2.2. Use the definition to find the value of $\lim_{x \to 2} (x^2 - 4x + 5)$.

Solution. In this case it would seem very likely that the limit is 1 just by noting what would appear to be the limit of each term in the function $x^2 - 4x + 5$. Using this bit of *intuition*, we will try to determine a value of δ corresponding to a given value of ε such that the definition will apply. We need ultimately to be able to make a statement of the type "whenever x is such that $0 < |x - 2| < \delta$, then it *must* follow that $|(x^2 - 4x + 5) - 1| < \varepsilon$, or equivalently that $|x^2 - 4x + 4| < \varepsilon$." It is clear that this is correct provided $|x - 2| < \sqrt{\varepsilon}$. Since $\varepsilon > 0$, $\sqrt{\varepsilon}$ is a real number. This gives us a clue as to a value we might use for δ. We can let $\delta = \sqrt{\varepsilon}$ being careful to select the positive square root. It follows that if $0 < |x - 2| < \delta$, then we are certain that $|(x^2 - 4x + 5) - 1| < \varepsilon$. We have shown both that the required limit exists and that the limit has the value one. If we had chosen any δ that is smaller than $\sqrt{\varepsilon}$ we could have made the same final statement. In other words, it is not at all necessary that we choose the largest possible δ. Since we have to be able to make the selection of a δ for any ε, we are then certain that we could have chosen ε to be 0.01, 0.001, 0.00317, or any other positive quantity we might wish and then could have found a suitable value for δ. If we had selected ε be 0.01, we could use in this case any choice we would like for δ just as long as $0 < \delta \le 0.1$.

We observe that in a situation in which we must prove that a limit exists, it is probably easier if we find a candidate for the limit (hopefully the successful candidate). We should then be able to find conditions governing the selection of a value for δ such that we fulfill all of the requirements for the application of the definitions. In the application of the definition we must be able to use an *arbitrary* value for ε, one for which there are no pre-conditions other than $\varepsilon > 0$.

In some instances we may know a limit exists but may not be able to find the value of the limit. We might wish to find an approximation for this value. We might specify an acceptable error tolerance and then find an approximation with an error no greater than this tolerance. Here we usually specify that we wish to find the value *to within an epsilon of* the prescribed amount. While this is a slight variation in the use of terminology, you should be aware that this usage does occur.

By implication we have indicated that we are only concerned about the case of a limit as x approaches a finite number a such that a is an interior point of some interval in which $f(x)$ is defined. It is perfectly possible that we might be interested in $\lim_{x \to a} f(x)$ where x is only defined over the interval (a, b). You will remember we mentioned this problem in Section IV.2. We do not have to include a in the interval, for it is not necessary that $f(a)$ be defined in order to obtain $\lim_{x \to a} f(x)$. Since in this case we do not have any value of x for which $0 < |x - a| < \delta$ and also $x - a < 0$, we are only able to include those points for which x exceeds a. Here we can talk about a limit, but we have something that is a little less complete than is required by Definition VII.2.1. As before we will denote this result by writing $\lim_{x \to a^+} f(x)$. In similar fashion, if $f(x)$ is defined over the interval (a, b), and we wish the limit as x approaches b, we could write $\lim_{x \to b^-} f(x)$ with the same considerations but for the fact that this time x is to the left of b. Such *one-sided limits* are of value in several instances, and the results obtained for the more general limits apply to the one-sided limits. In one sense the special case $\lim_{x \to \infty} f(x)$ is a one-sided limit, although in this case the value of x is unbounded rather than headed for a finite value. This also makes plausible the one-sided derivatives in which we took either the left handed or the right handed limits of the differential quotient. These were used, you will remember, in our discussion of the extreme values at the end points of a closed interval.

EXERCISES

1. Evaluate each of the following, citing each Theorem used:

   (a) $\lim_{x \to \infty}((n + 1)/n)$
   (b) $\lim_{n \to 5}(2/n)$
   (c) $\lim_{x \to 1}((n^2 + 1)/n)$
   (d) $\lim_{n \to \infty}((4 - 5n^2)/n^3)$

2. Show that each of the following limits exist by finding a δ as a function of ε:

   (a) $\lim_{x \to 2}(x^2 - 3x - 1)$
   (b) $\lim_{x \to -3}(4 - x^2)$
   (c) $\lim_{x \to (-1/3)}(9x^2 - 6x + 1)$
   (d) $\lim_{x \to 0}(x/\cot x)$

3. State carefully the definition of $\lim_{x \to -\infty} f(x) = L$.

4. Does $\lim_{x \to 0}(1/x)$ exist? Does $\lim_{x \to -\infty}(1/x)$ exist? Give reasons for each answer. Find the value of the limits which exist.

5. (a) Determine whether $\lim_{x \to \infty}((\sin x)/x)$ exists. Find the limit if it exists.
   (b) Determine whether $\lim_{x \to 0} x \sin(1/x)$ exists. Find the limit if it exists.

6. Use the definition of the derivative to find $f'(x)$ if $f(x) = x^2 - x + (1/x)$. State all of the limit theorems you use in the evaluation of the associated limit.

7. Prove that $\lim_{x \to a}(x^2 - 3x + 1)$ exists for each finite value of a.

8. (a) If $f(x) = \sqrt{x - 2}$, the domain would ordinarily consist only of real numbers not less than 2. Give a complete proof that $\lim_{x \to 2+} f(x)$ exists, and show where this differs from the case in which the domain of $f(x)$ would include an interval of which 2 is an interior point.
   (b) Using instructions similar to those for part (a) of this exercise, prove that $\lim_{x \to 3-} \sqrt{9 - x^2}$ exists.

9. (a) Determine whether $\lim_{x \to 0} (|x|/x)$ exists. If it does not, do the two one-sided limits exist? Explain fully.
   (b) Do the same for $\lim_{x \to 0}(\sqrt{x^2}/x)$.

10. A person walks half way to his destination, and then walks half of the remaining distance, and then half of the remaining distance, etc. Show whether he approaches his destination as a limit.

11. (a) Evaluate $\lim_{x \to 0} |x|$. Prove that your result is correct.
    (b) Is this a continuous function in the vicinity of $x = 0$? [Remember the definition of a continuous function in Chapter I.]

B12. Let T_k be the number of individuals possessing a certain trait in a given culture in the k-th generation. Let N_k be the number in the k-th generation that do not possess this trait. For this particular mysterious trait it has been conjectured that the corresponding numbers for the $(k + 1)$-th generation are given by

$$T_{k+1} = 0.8T_k + 0.3N_k,$$
$$N_{k+1} = 0.2T_k + 0.7N_k.$$

(a) Show that $(T_{k+1} + N_{k+1}) = (T_k + N_k)$ for each permissible value of k.
(b) If $T_0 = N_0 = 250$, find T_1, N_1, T_2, and N_2.
(c) Find the values of T and N such that if $T_k = T$ and $N_k = N$ then $T_{k+1} = T$ and $N_{k+1} = N$.
(d) If T and N are the values obtained in part (c), show that if $T_k = T - \varepsilon$ then $T_{k+1} = T - 0.5\varepsilon$ and if $N_k = N + \varepsilon$ then $N_{k+1} = N + 0.5\varepsilon$.
(e) Prove that $\lim_{k \to \infty} T_k = T$ and $\lim_{k \to \infty} N_k = N$.

VII.3 Theorems Concerning Limits

You have undoubtedly noticed that with almost every concept introduced we have stopped and obtained a few theorems before we have gone very far. This has proven to be helpful, for the theorems have simplified later considerations. It also means that we have to derive a given result only one time instead of re-deriving it on each of the several occasions. This is a pattern in the method of operation of mathematics, for if a technique seems to work in one case, the same technique will probably work in many others. Hence, it would be well to consider some *appropriate* theorems. You should not be overly surprised either at their inclusion, or with the results.

Theorem 3.1. *Let $f(x)$ and $g(x)$ be functions such that $\lim_{x \to a} f(x) = F$ and $\lim_{x \to a} g(x) = G$ where F and G are finite, and where a is either finite or infinite. Then*

 (i) $\lim_{x \to a} [f(x) + g(x)] = F + G$,
 (ii) $\lim_{x \to a} cf(x) = cF$, *(where c is a constant)*,
 (iii) $\lim_{x \to a} [f(x)g(x)] = FG$,
 (iv) $\lim_{x \to a} [1/f(x)] = 1/F$ *(if $F \neq 0$)*.

PROOF. We will first consider the case in which a is finite. Given the value $\varepsilon > 0$ we can find $\delta_f > 0$ and $\delta_g > 0$ such that if $0 < |x - a| < \delta_f$ we have $|f(x) - F| < \varepsilon$ and if $0 < |x - a| < \delta_g$ we have $|g(x) - G| < \varepsilon$. If we let δ be the smaller of δ_f and δ_g, then for $0 < |x - a| < \delta$ we have both $|f(x) - F| < \varepsilon$ and $|g(x) - G| < \varepsilon$ at the same time. Now if $0 < |x - a| < \delta$ we have

$$|[f(x) + g(x)] - [F + G]| = |[f(x) - F] + [g(x) - G]|$$
$$\leq |f(x) - F| + |g(x) - G| < \varepsilon + \varepsilon = 2\varepsilon.$$

(Note that we have made use of the triangle inequality for absolute values.) Had we felt it essential to obtain this final bound as precisely ε, we could just as well have started with $\varepsilon/2$ in the first sentence of the proof. Therefore, we have a proof of part (i) of the theorem. This situation was one of those referred to in Section VII.2. For (ii) we note that $|cF(x) - cF| = |c| \cdot |f(x) - F|$, and hence when $0 < |x - a| < \delta$ we have $|cf(x) - cF| < |c|\varepsilon$. As before we have the desired proof, for if $c \neq 0$ we could have started with $\varepsilon/|c|$ if we had felt it necessary to obtain precisely ε. If $c = 0$, then we have $|cf(x) - cF| = 0$, and this is certainly less than any positive quantity.

In the proof of (iii), we will let δ be chosen in a manner similar to that used in the proof of (i) with the exception that we shall also require that δ be small enough that when $0 < |x - a| < \delta$, we can be certain that $|f(x) - F| < \varepsilon$, $|g(x) - G| < \varepsilon$, and also that $|g(x)| < |G| + 1$. Using the technique we have used in the past of inserting a convenient term and then subtracting the same term, we have

$$|f(x)g(x) - FG| = |f(x)g(x) - Fg(x) + Fg(x) - FG|$$
$$= |g(x)[f(x) - F] + F[g(x) - G]|$$
$$\leq |g(x)| \cdot |f(x) - F| + |F| \cdot |g(x) - G|.$$

Now when $0 < |x - a| < \delta$ we have

$$|g(x)| \cdot |f(x) - F| < (|G| + 1) \cdot \varepsilon$$

and

$$|F| \cdot |g(x) - G| < |F| \cdot \varepsilon.$$

Hence

$$|f(x)g(x) - FG| < (|G| + 1)\varepsilon + |F|\varepsilon = \varepsilon[|G| + |F| + 1].$$

If we let $M = |G| + |F| + 1$, we have for $0 < |x - a| < \delta$ the fact that $|f(x)g(x) - FG| < M\varepsilon$, and the result is proven.

The proof of (iv) is slightly more tricky in that we do not know how close to zero the various denominators involved may go. Since $F \neq 0$, we know that if we take $\varepsilon < |F/2|$, then for the corresponding δ we are certain that within the interval $0 < |x - a| < \delta$, $f(x)$ is in the interval $(F - (F/2), F + (F/2))$ or $(F/2, 3F/2)$. We have selected $F/2$ in this case to insure that $f(x)$ not only is not permitted to be zero, but also that it is some assured distance from being zero. When $0 < |x - a| < \delta$ we are certain that $|f(x)| > |F/2|$, and therefore $|f(x)F| > F^2/2$. Consequently within this interval $|(1/f(x)) - (1/F)| = |(F - f(x))/Ff(x)| < |f(x) - F|/(F^2/2)$. We have replaced the denominator by a positive number smaller than the smallest possible absolute value that could occur in the original expression. This leads us to $|(1/f(x)) - (1/F)| < 2\varepsilon/F^2$ and since $2/F^2$ is a finite constant we have again proven our result. It is true that the result might have looked a bit nicer in the end had we required that $|f(x) - F|$ be smaller than $(\varepsilon F^2/2)$, but this would have required more foresight than can be expected of either the author or the reader. It is one of those bits of wisdom that one obtains by constructing the proof. Then if one rewrites it, it looks so good. We will not try to obfuscate things here by writing out the refined proof in lieu of the one above, for you would probably have had the feeling that someone somewhere had a crystal ball that you had not been given access to. The proof of this theorem in the case in which a is infinite follows the same general pattern. $\square$

With Theorem 3.1 we have many of the basic properties we need to develop some additional results.

Corollary 3.1. If $\lim_{x \to a} f(x) = F$ and $\lim_{x \to a} g(x) = G$, then

$$\lim_{x \to a} [f(x) - g(x)] = F - G \quad \text{and} \quad \lim_{x \to a} \left[\frac{f(x)}{g(x)} \right] = \frac{F}{G}, \quad (\text{if } G \neq 0).$$

PROOF. Consider the functions $[f(x) + (-1)g(x)]$ and $[f(x) \cdot (1/g(x))]$, and use the results of Theorem 3.1. □

With the use of Theorem 3.1 and Corollary 3.1 we are now in a position to go back to the development of the derivatives and demonstrate anew that all of the limits taken are correct. We would now construct proofs using epsilons and deltas instead of intervals and deleted neighborhoods, but this changes neither the logic nor the validity of the proofs. There are two additional relations we should consider before we conclude this discussion. The first of these is taken care of by the following theorem.

Theorem 3.2. If $f(x) \leq g(x)$ for all values of x in the interval (a, b), if c is in (a, b), and if $\lim_{x \to c} f(x) = F$ and $\lim_{x \to c} g(x) = G$ then $F \leq G$. This theorem is also correct if c is replaced by ∞.

PROOF. Since $f(x) \leq g(x)$ for all points in the interval, then $f(x) - g(x) \leq 0$ for all points in the interval. However $\lim_{x \to c} [f(x) - g(x)] = F - G$ by the corollary. Now if $F - G > 0$, let $\varepsilon = (F - G)/2$ and we know there exists a value $\delta > 0$ such that when $0 < |x - c| < \delta$ then

$$|[f(x) - g(x)] - [F - G]| < \varepsilon < |F - G|.$$

This cannot happen if $[f(x) - g(x)]$ is negative. Therefore, it is not true that $F > G$. Consequently by Theorem I.3.1 we have $F \leq G$. The proof when c is replaced by ∞ is similar. □

Corollary 3.2. If $f(x) \leq g(x) \leq h(x)$ for all values of x in the interval (a, b), if c is in (a, b), and if $\lim_{x \to c} f(x) = \lim_{x \to c} h(x) = L$, then $\lim_{x \to c} g(x) = L$. The result also holds if c is replaced by ∞.

PROOF. Since $\lim_{x \to a} f(x) = \lim_{x \to a} h(x) = L$, we know that for any $\varepsilon > 0$ there is a value of $\delta > 0$ such that when $0 < |x - a| < \delta$ we have both $|f(x) - L| < \varepsilon$ and $|h(x) - L| < \varepsilon$. Since $g(x)$ is between $f(x)$ and $h(x)$, then $g(x)$ must also be in the interval $(L - \varepsilon, L + \varepsilon)$, and consequently we not only know that $g(x)$ has a limit, but we also know that $\lim_{x \to a} g(x) = L$.

While we have already proved the corollary, it would be profitable for other purposes to show that the limit must be L in a slightly different way. By Theorem 3.2 we know that if $\lim_{x \to a} g(x) = K$, then $K \leq L$ since $g(x) \leq h(x)$ in the interval. We also know that $K \geq L$ since in this interval

$g(x) \geq f(x)$. If $K \geq L$ and also $K \leq L$, then the only possible simultaneous result is that $K = L$. Again we have the proof of the corollary. The proof for the limit as $x \to \infty$ is similar. $\square$

This corollary is used fairly often. You may remember earlier uses although we approached it then from a more intuitive point of view. Since it is so important, it should probably have a name. It has been variously called the *squeeze theorem*, the *sandwich theorem*, and undoubtedly many other things. This is the same result we used in determining the fact that when the upper and lower integrals are equal, we can start with any RS sum to obtain the integral. In this case the upper RS sums would be represented by $h(x)$ and the lower RS sums would be represented by $f(x)$, whereas the general RS sums would be represented by $g(x)$. You will remember that in obtaining $\lim_{x \to 0} ((\sin x)/x)$ we used the squeeze theorem. It is also worth noting that if we can prove that two quantities have the relationship that each is not larger than the other, the two must be equal.

So far we have been concerned with what happens with the arithmetic operations of addition, subtraction, multiplication, and division, and with certain other related items. We now turn to the limit of the composition of functions.

Theorem 3.3. *If $f(x)$ and $g(x)$ are functions for which $\lim_{x \to a} f(x) = b$ and $\lim_{x \to b} g(x) = c$ exist, and if $g(b) = c$, then $\lim_{x \to a} g(f(x)) = c$.*

PROOF. We start with $g(x)$, since this is the function which appears as the principal function in the final result. If we are given $\varepsilon > 0$, then we know that there is a number $\varepsilon_1 > 0$ such that when $0 < |x - b| < \varepsilon_1$, we are certain of having $|g(x) - c| < \varepsilon$. On the other hand, given $\varepsilon_1 > 0$, we know that we have a value $\delta > 0$ such that when $0 < |x - a| < \delta$ we have $|f(x) - b| < \varepsilon_1$. (This does not preclude the possibility that $f(x) = b$ for all values of x in the deleted neighborhood $0 < |x - b| < \varepsilon_1$.) All we need to do now is to string the results together, for we have all of the information before us. In other words, if $0 < |x - a| < \delta$ then $|f(x) - b| < \varepsilon_1$, but then since $|f(x) - b| < \varepsilon_1$, we have $|g(f(x)) - c| < \varepsilon$. Note that we are covered in the case $f(x) = b$, for we have $g(b) = c$ by hypothesis. This can be shortened to the statement that when $0 < |x - a| < \delta$ then $|g(f(x)) - c| < \varepsilon$. This is precisely the statement we need to prove the theorem. $\square$

The last result is very helpful in such cases as those in which we wish to evaluate $\lim_{x \to 1} \sin(\ln x)$, for here we can see that $\sin x$ is the counterpart of $g(x)$ just as $\ln x$ is the counterpart of $f(x)$. If we can obtain the limits of the more elementary functions we can obtain the limit of the composite function.

EXAMPLE 3.1. Find the value of $\lim_{x \to 1}((x^3 - 1)/(\sqrt{x} - 1))$.

Solution. We first wish to prove the apparently obvious fact that $\lim_{x \to 1} x = 1$. In order to do this we only need to use the definition of the limit and let $\delta = \varepsilon$. For any positive value of ε we will then know that if $0 < |x - 1| < \delta$ then $|x - 1| < \varepsilon$ and we have established the limit $\lim_{x \to 1} x = 1$. We now look a bit further ahead and see trouble on the horizon for both the numerator and the denominator appear to be heading toward zero in this case. With this in mind we will multiply both numerator and denominator by $(\sqrt{x} + 1)$, and we will have

$$\lim_{x \to 1} \frac{x^3 - 1}{\sqrt{x} - 1} = \lim_{x \to 1} \frac{(x - 1)(x^2 + x + 1)(\sqrt{x} + 1)}{(x - 1)}$$

$$= \lim_{x \to 1} (x^2 + x + 1)(\sqrt{x} + 1).$$

Here we have factored $(x^3 - 1)$, and then divided both numerator and denominator by the factor $(x - 1)$. The latter is possible since we have specifically required that $x \neq 1$, and hence $(x - 1) \neq 0$. The limits are equal by Theorem VII.2.1. We see that $\sqrt{x}$ is a composite function, being the square root function of the identity function. Hence the limit of the square root is the square root of the limiting value of the identity function or $\lim_{x \to 1} \sqrt{x} = \sqrt{1} = 1$. Furthermore $\lim_{x \to 1} x^2 = 1^2 = 1$ by Theorem 3.1, and the limit of the sum $(x^2 + x + 1)$ is the sum of the limits by Theorem 3.1. Therefore, we have

$$\lim_{x \to 1} \frac{x^3 - 1}{\sqrt{x} - 1} = \lim_{x \to 1} (x^2 + x + 1)(\sqrt{x} + 1) = (1^2 + 1 + 1)(1 + 1) = 6.$$

We have gone into a great deal of detail in this example, but it illustrates that we do have the information to back up the statements we made earlier. We do not often go to such lengths in writing out the details when we need a limit, but *this* is the type of reasoning which is involved in taking a limit. You should be aware of this in order that you can check carefully any limit which might otherwise be suspect.

EXAMPLE 3.2. Evaluate $\lim_{x \to \infty}((x^3 - 27)/(x^3 + 3x))$.

Solution. It is apparent that this situation is embarassing in that both numerator and denominator increase without limit as x increases without limit. However, we can divide the numerator and denominator by x^3, the largest power of x appearing in this expression. If we do this, we will have

$$\lim_{x \to \infty} \frac{x^3 - 27}{x^3 + 3x} = \lim_{x \to \infty} \frac{1 - (27/x^3)}{1 + (3/x^2)}$$

since all of the values of x en route to the limit are finite. We have no worries about dividing by zero, for we need only consider those values of x which are very large in connection with this limit. We have theorems which show that all of the terms with x's in the denominator approach zero, and hence we have the fact that

$$\lim_{x \to \infty} \frac{x^3 - 27}{x^3 + 3x} = \lim_{x \to \infty} \frac{1 - (27/x^3)}{1 + (3/x^2)} = \frac{(1 - 0)}{(1 - 0)} = 1.$$

Observe that the division by the highest power of x present insures that each of the terms in the resulting fraction will either be constant or will approach zero. In this case we have no difficulty with the matter of *infinity*.

EXERCISES

1. Evaluate each of the following limits. In each case show that your result is correct and indicate why each step you make is permissible.

   (a) $\lim_{x \to 4}((x - 4)/(\sqrt{x} - 2))$
   (b) $\lim_{x \to 2}((x^2 - 4)/(x^2 + 4))$
   (c) $\lim_{x \to 0}((x^2 - 4)/(x^2 + 4))$
   (d) $\lim_{x \to 3}((x^2 - 4x + 3)/(x - 3))$
   (e) $\lim_{x \to 3}((x^2 - 4x + 3)/(x^3 - 27))$
   (f) $\lim_{x \to 0} \ln(\cos x)$

2. Evaluate each of the following limits. In each case show that your result is correct and indicate why each step you make is permissible.

   (a) $\lim_{x \to \infty}((2x - 3)/(x + 2))$
   (b) $\lim_{x \to \infty} e^{-2x}$
   (c) $\lim_{x \to \infty}((x^2 + 2)/x)$
   (d) $\lim_{x \to \infty}((\sin x)/x)$
   (e) $\lim_{x \to \infty}((x^2 - 3x + 2)/(x^3 + 4))$
   (f) $\lim_{x \to \infty} e^{2x}$

3. (a) Prove $\lim_{n \to \infty}(1/n!) = 0$.
   (b) Prove $\lim_{n \to \infty}(2^n/n!) = 0$.
   (c) Prove $\lim_{n \to \infty}(17^n/n!) = 0$.
   (d) Prove $\lim_{n \to \infty}(x^n/n!) = 0$ for any constant value x.

4. (a) Prove $5^n < 3^n + 5^n < (2)(5^n)$ for any positive integer n.
   (b) Prove $\lim_{n \to \infty} 2^{1/n} = 1$.
   (c) Use the results of parts (a) and (b) to prove $\lim_{n \to \infty} \sqrt[n]{3^n + 5^n} = 5$.
   (d) Use reasoning similar to that of parts (a), (b), and (c) to evaluate

   $$\lim_{n \to \infty} \sqrt[n]{2^n + 3^n + 5^n + 19^n}.$$

5. Evaluate each of the following limits.

   (a) $\lim_{x \to 1}((x^2 - 1)/(x - 1))$
   (b) $\lim_{x \to 1}((x - 1)^2/(x - 1)^3)$
   (c) $\lim_{x \to 1}((x^4 - 1)/(x^2 - 1))$
   (d) $\lim_{x \to 1}((x^3 - 1)/(x^2 - 1))$
   (e) $\lim_{x \to 1}((x - 1)^2/(x^2 - 1))$
   (f) $\lim_{x \to 1}((x - 1)^4/(x - 1)^2)$

6. Let $f(x)$ be a function such that $f(x) = x$ if x is a rational number and $f(x) = 0$ if x is an irrational number.

   (a) Show that $\lim_{x \to 0} f(x) = 0$.
   (b) Show that $\lim_{x \to a} f(x)$ fails to exist if $a \neq 0$.

7. (a) If $\lim_{x \to a} f(x)$ exists and is positive and if $\lim_{x \to a} g(x) = 0$ show that $\lim_{x \to a}(f(x)/g(x)) = \infty$. (This is equivalent to showing that as x approaches the value a then $f(x)/g(x)$ will become larger than *any* positive value M that one might select.)
   (b) Show that $\lim_{x \to (\pi/2)^-} \tan x = \lim_{x \to (\pi/2)^-}((\sin x)/(\cos x)) = \infty$.
   (c) Show that $\lim_{x \to (\pi/2)^+} \tan x = -\infty$.
   (d) Show why it was necessary to consider one-sided limits in parts (b) and (c).

8. Some tables of trigonometric functions give the value ∞ for $\tan 90°$, $\cot 0°$, $\sec 90°$, and $\csc 0°$.

   (a) Show that each of these can be interpreted correctly if one considers this value to be the limit as the angle approaches the limiting value from within the first quadrant.
   (b) Show that these values would be incorrect if one were to consider the two-sided limit.

9. If $f(x)$ is the greatest integer function, $f(x) = [x]$, show that $\lim_{x \to a} f(x)$ exists for any value a which is not an integer, and that $\lim_{x \to a^-} f(x)$ exists but $\lim_{x \to a} f(x)$ does not exist if a is an integer.

10. (a) Show that $\lim_{x \to 0} f(x)$ exists if $f(x) = |x|$. ,
    (b) Show that the derivative of $f(x) = |x|$ exists for any value of x other than $x = 0$.
    (c) Show that $\lim_{x \to 0^+} f'(x)$ and $\lim_{x \to 0^-} f'(x)$ exist, but they are not equal.

VII.4 Continuity

Our pursuit of the subject of limits may appear to concentrate on the obvious or on the other hand it may seem to be making hard work out of what might have been easy. The purpose of all this, however, is to remove the question marks that have appeared from time to time in our development of the integral and the derivative. It should be clear that the theorems we have considered to date will cover most of the problems we glossed over in connection with limits. However, we also made a great deal of use of continuity and of the intermediate value property. We will now try to give additional backing for the statements we made concerning these concepts.

Definition 4.1. A function, $f(x)$, is said to be *continuous at $x = a$* provided

   (i) $f(a)$ exists,
   (ii) $\lim_{x \to a} f(x)$ exists,
   (iii) $\lim_{x \to a} f(x) = f(a)$.

It is true that the third requirement of the definition implies the first two, but it would be difficult to overemphasize the requirement that the function must exist at the point of continuity and the limit of the function must also exist. Definition 4.1 defines continuity at a single point but we have often used the fact that a function is continuous over an interval. For this we have the following definition.

Definition 4.2. A function $f(x)$ is said to be *continuous over an interval* (a, b) provided it is continuous at every point in (a, b).

The functions we have been dealing with have in general been continuous functions, although there have been a few that were not continuous. Since continuous functions are used in many places, as indicated by the number of theorems requiring that a function be continuous, it would be helpful to develop some methods which would aid in identifying continuous functions.

Theorem 4.1. *If $f(x)$ and $g(x)$ are continuous in the interval (a, b) then $[f + g](x)$ and $[fg](x)$ are continuous in (a, b) and $[f/g](x)$ is continuous provided $g(x)$ is not zero in the interval (a, b).*

OUTLINE OF PROOF. Use the algebra of functions and Theorem VII.3.1.   □

Theorem 4.2. *If $f(x)$ is continuous on the interval $[a, b]$, and if $f(a) \neq f(b)$, then for any value, d, between $f(a)$ and $f(b)$ there is a point $x = c$ in (a, b) such that $f(c) = d$.*

PROOF. Since $f(a) \neq f(b)$ we know that either $f(a) < f(b)$ or $f(a) > f(b)$. The proof would be essentially the same in either case, but in order to make the development somewhat more concrete we will assume $f(a) < f(b)$. Since the value d is between $f(a)$ and $f(b)$, we then have $f(a) < d < f(b)$. If we pick $\varepsilon < d - f(a)$, we are assured by the continuity of $f(x)$ in $[a, b]$ that there exists a δ such that for values of x in $[a, a + \delta]$ we have $f(x)$ in $(f(a) - \varepsilon, f(a) + \varepsilon)$. However $f(a) + \varepsilon < d$. Therefore, we are certain that there are intervals $[a, x_k)$ such that $f(x) < d$ for each value of x in the interval. Let I be the set of all intervals $[a, x_k)$ such that $f(x) < d$ for all values of x in the interval. Since $x_k < b$ for each x_k, we have a set of values $\{x_k\}$ with an upper bound, and therefore by the least upper bound axiom $\{x_k\}$ must have a least upper bound. We will denote this least upper bound by c.

    Since $f(x)$ is continuous in $[a, b]$, $f(x)$ is continuous at $x = c$. If $f(c) < d$, there must be an interval of which c is an interior point such that $f(x) < d$ for any value of x in the interval. This is easily seen if we consider $\lim_{x \to c} f(x) < d$ and let ε be any value such that $\varepsilon < d - f(c)$. In this case c is not the least upper bound of the set of x_k, for $f(x) < d$ for all values of x in the interval $[a, c + \delta)$. On the other hand if $f(c) > d$ we could use a similar argument to show that there are values of x smaller than $x = c$ such that $f(x) > d$.

Since neither $f(c) < d$ nor $f(c) > d$ can possibly hold, it must follow that $f(c) = d$. This proves the theorem.   □

Theorem 4.3. *If $f(x)$ is continuous on the closed interval $[a, b]$, $f(x)$ is bounded on $[a, b]$.*

PROOF. Suppose $f(x)$ does not have an upper bound on $[a, b]$. If we divide $[a, b]$ into two half intervals, $f(x)$ is unbounded on the left hand half interval, the right hand half interval, or on both half intervals. We will select a half interval on which $f(x)$ is unbounded, and if the function is unbounded on both sides, we will take the left hand half interval. The half interval selected will be denoted by $[a_1, b_1]$. We will repeat this process and obtain an interval $[a_2, b_2]$ which is either the left or right half of $[a_1, b_1]$ and such that $f(x)$ is unbounded on $[a_2, b_2]$. If we continue this, we obtain a sequence of points $a, a_1, a_2, a_3, \ldots$ which is monotonic increasing, and which is bounded by b. This sequence must therefore have a lub. Let the lub be denoted by c. Now $f(c)$ exists by the hypothesis that $f(x)$ is continuous in $[a, b]$. If we take any ε, say $\varepsilon = 1$, then we know that there is a value of $\delta > 0$ such that in the interval $(c - \delta, c + \delta)$ it is true that $f(x)$ is in the interval $(f(c) - 1, f(c) + 1)$, or $f(x) < f(c) + 1$. But since the intervals $[a_k, b_k]$ are of length 2^{-k} times the length of $[a, b]$, it follows that for some value of k the entire interval $[a_k, b_k]$ must be in the interval $(c - \delta, c + \delta)$. This gives us a contradiction, for in the interval $[a_k, b_k]$ the function is unbounded, but on the other hand it is bounded by $f(c) + 1$. Therefore, the function must have an upper bound. A similar proof shows that $f(x)$ has a lower bound over $[a, b]$. (This proof is similar to the proof of Theorem IV.5.2).   □

You may remember we mentioned that a function continuous over a closed interval actually attains its extreme points in that interval.

Theorem 4.4. *If $f(x)$ is continuous on the closed interval $[a, b]$, there is some value c in $[a, b]$ such that $f(c)$ is the least upper bound of all values of $f(x)$ on $[a, b]$. A similar result holds for the greatest lower bound.*

PROOF. By Theorem 4.3 we know that $f(x)$ is bounded on $[a, b]$ and consequently there must be a lub. Let M be the lub of $f(x)$ on $[a, b]$, and assume that there is no value of x in $[a, b]$ for which $f(x) = M$. Consequently $M - f(x) > 0$ for all points on $[a, b]$. Since M is a constant function and therefore continuous, $M - f(x)$ is continuous. Since $M - f(x) > 0$ on $[a, b]$, then $1/[M - f(x)]$ is continuous on $[a, b]$, and must have an upper bound by Theorem 4.3. Let this bound be U. This means that $1/[M - f(x)] \leq U$ for all values of x in $[a, b]$, and a bit of algebra then establishes that $f(x) \leq M - 1/U$ for all points in $[a, b]$. Consequently, $(M - 1/U)$ is an upper bound for $f(x)$ on this interval and M could not have been a lub for $f(x)$ on the interval. This contradicts the assumption that $f(x)$ did not attain its lub. A similar result would hold for the greatest lower bound.   □

Corollary 4.1 (Intermediate value theorem). *If $f(x)$ is continuous on $[a, b]$ and if m is a glb for $f(x)$ on this interval while M is a lub for $f(x)$ on $[a, b]$, then for any value d in the interval (m, M) there is a value c in $[a, b]$ such that $f(c) = d$.*

OUTLINE OF PROOF. Use Theorems 4.2 and 4.4.   □

We now have sufficient information that we can go back and reassure our-selves that the proofs leading to the integrals and the derivatives developed in the last several chapters are all rigorously correct. While we did not anticipate any difficulties, it is reassuring to be able to assure ourselves that there are no difficulties in the limits we have taken, nor in our conclusions concerning the continuous functions assumed in many of the theorems. All of this work is now helpful in establishing the continuity of a large number of functions.

Theorem 4.5. *If a function $f(x)$ has a derivative at the point $x = c$, then $f(x)$ is continuous at $x = c$.*

PROOF. Since $f'(c)$ exists, we know that $\lim_{\Delta x \to 0}((f(c + \Delta x) - f(c))/\Delta x) = \lim_{x \to c}((f(x) - f(c))/(x - c))$ exists. We want to show $\lim_{x \to c} f(x) = f(c)$. This is equivalent to showing $\lim_{x \to c}[f(x) - f(c)] = 0$. We can obtain this by considering

$$\lim_{x \to c} [f(x) - f(c)] = \lim_{x \to c} \left[\frac{f(x) - f(c)}{(x - c)} (x - c) \right]$$

$$= \lim_{x \to c} \left[\frac{f(x) - f(c)}{(x - c)} \right] \lim_{x \to c} (x - c)$$

$$= f'(c) \cdot 0 = 0. \qquad \qquad □$$

This theorem will help a great deal. We have shown that the derivatives of a large number of functions exist, either by using the definition of the derivative, using theorems concerning derivatives, or using the Fundamental Theorem. Since each function possessing a derivative at a point must be continuous at that point, we have now established the continuity of a very large number of functions. Continuous functions include the power function with the single exception of the point $x = 0$ in case the exponent is negative, the exponential function, the trigonometric functions where they exist, the logarithm for positive arguments, etc. Furthermore, by the theorems established in this section we know that algebraic combinations of continuous functions are continuous and the composition of continuous functions gives a continuous function, provided the composition is meaningful.

If a function is not continuous at some point, there is no reason to even begin to look for a derivative at that point. As a matter of interest we point out that the converse of this theorem is not true, for there are examples of continuous functions for which no derivative exists. One of the first of these was discovered by the German mathematician *K. W. T. Weierstrass* in the last century. It would not be surprising if we were to suggest that such a function has rather peculiar properties. If you are interested in pursuing this topic, you might consult a book in the history of mathematics or a book in more advanced mathematical analysis. There are no common applications of such functions, and we will not try to explore their properties here. You should bear in mind, however, that functions not continuous at a point do not have a derivative at that point, and conversely that if a function does have a derivative at every point throughout an interval, then the function is continuous throughout the interval. We have *not* said anything about whether the derivative is continuous.

Some consideration is due with regard to the integrability of functions. Since there are many functions for which no formulas exist, and since numerical methods of obtaining approximations for integrals are time consuming to say the least, there is no point in pursuing the matter of evaluating an integral if we can determine in advance that it is not likely to exist. The result that we will develop is a one-sided result, for it gives a basis for assuring the existance of an integral, but it is not to be construed as giving a basis for showing that an integral may not exist. It provides the type of condition which the mathematicians would call a *sufficient* condition, but not a *necessary* condition.

Theorem 4.6. *If $f(x)$ is continuous and has a finite number of extreme values in $[a, b]$, and if $g(x)$ is monotonic throughout $[a, b]$, then $\int_a^b f(x)dg(x)$ exists.*

PROOF. Our problem here is merely that of showing that the upper and lower integrals are equal, for we know that we have a lub for the lower RS sums and a glb for the upper RS sums. Hence we have upper and lower integrals. In order to show that these two integrals are equal, it will suffice to show that there is some partition $P[a, b]$ such that $U(P, f, g) - L(P, f, g)$ is less than an arbitrary positive number ε or some constant multiple thereof. (Note that by taking a suitable fraction of ε for a starting value we could always make this proof neat and end up by making the difference less than ε.) We will assume $g(x)$ is increasing, although a similar proof would be easily constructed if $g(x)$ were decreasing. Being given $\varepsilon > 0$, we will now divide the interval $[m, M]$, where m is the absolute minimum of $f(x)$ in $[a, b]$ and M is the absolute maximum of $f(x)$ in this interval. We will let $y_0 = m$, $y_1 = m + \varepsilon$, $y_2 = m + 2\varepsilon$, and continue in this fashion until we finally have

$$y_{n-1} = m + (n - 1)\varepsilon \quad \text{where} \quad y_{n-1} < M \quad \text{and} \quad y_{n-1} + \varepsilon \geq M.$$

We will let $y_n = M$. Note that this partition is not concerned, at least at the moment, with the interval $[a, b]$, but we will come to that. We do note that the successive points in this partition are just ε units apart with the possible exception of the last two points which are no farther apart than ε. Since $f(x)$ is continuous, there are a finite number of points which fulfill the requirement that $f(x) = y_j$ for each value of j unless $f(x) = y_j$ on some closed interval. If in the case of a coincident segment we use only the left endpoint and otherwise we take all of the values of x such that $f(x) = y_j$ for some j, we have a finite number of points. Let these points, in ascending order, be labeled $x_1, x_2, \ldots, x_{n-1}$ with $x_0 = a$ and $x_n = b$. It is this partition of $[a, b]$ which we wish to use in our proof. Note that in the interval $[x_{k-1}, x_k]$ the smallest value of $f(x)$ and the largest value of $f(x)$ can differ by no more than ε. We know that

$$U(P, f, g) - L(P, f, g) \leq \sum_{k=1}^{n} (M_k - m_k)[g(x_k) - g(x_{k-1})]$$

where M_k and m_k are respectively the maximum value and the minimum value of $f(x)$ in $[x_{k-1}, x_k]$. Since $0 \leq (M_k - m_k) \leq \varepsilon$ we have

$$U(P, f, g) - L(P, f, g) \leq \varepsilon \sum_{k=1}^{n} [g(x_k) - g(x_{k-1})] = \varepsilon[g(b) - g(a)].$$

But $g(b) - g(a)$ is a constant, and hence we have the upper and lower sums differing by less than a constant multiple of ε. Since the upper and lower integrals differ by no more than this amount, we have the machinery for using our limiting procedures and we see that the difference of the upper and lower integrals must be zero. Thus we have the proof.   □

Corollary 4.2. *If $f(x)$ is bounded and is continuous except at perhaps a finite number of points in the interval $[a, b]$, and if furthermore $f(x)$ has only a finite number of extreme values, while $g(x)$ has at most a finite number of points at which it changes from monotone increasing to monotone decreasing and vice versa, then $\int_a^b f(x)dg(x)$ exists.*

PROOF. Consider a partition of $[a, b]$ including all of the points at which $f(x)$ is not continuous and all of the points at which $g(x)$ changes its direction of monotonicity. This partition must include only a finite number of points by our hypothesis. We can obtain the integral in each interval of the partition, and then using Theorem III.2.4 we note that the resulting corollary is true.

□

It is true that there are integrals of continuous functions that do not have a finite number of extreme points, and it is possible to obtain a more general theorem. However, this result will suffice for the applications that we will be using, and it does give a certain assurance that many integrals exist and their evaluation is worth the effort. Since we stated that it was possible to use the Riemann integral to evaluate many of the RS integrals, you may wonder why we chose this statement of the theorem and corollary. Note that in our statement we have not required that $g(x)$ be a function possessing a derivative, nor have we even required continuity for $g(x)$. In the relation which permits us to evaluate an RS integral by using a corresponding Riemann integral, we would have had to be able to differentiate $g(x)$. There are many areas, including statistics, where applications involve functions for which the RS integral is essential.

EXERCISES

1. Show by the definition that the identity function is continuous.

2. Using the theorem concerning the product of continuous functions and using mathematical induction show that any positive integral power of the identity function is continuous.

3. Show that a strictly monotone continuous function has a continuous inverse function.

4. (a) Show that $f(x) = |x|$ is continuous at $x = 0$.
   (b) Show that $f'(x)$ is not continuous at $x = 0$.
   (c) If we were to define a function $g(x) = f'(x)$ when $x \neq 0$ and $g(0) = a$ for some constant a, show that no value of a could make $g(x)$ continuous at $x = 0$.

5. (a) Using integration and the fact that $|\cos x| \leq 1$, show that $|\sin x| < |x|$.
   (b) Use the definition of continuous function to show that $\sin x$ is continuous at $x = 0$.
   (c) Use the result of part (a) and integration to show that $|1 - \cos x| < x^2/2$.
   (d) Show that $\cos x$ is continuous at $x = 0$.
   (e) Show that $\sin x - \sin a = 2 \sin((x - a)/2)\cos((x + a)/2)$ and $\cos a - \cos x = 2 \sin((x - a)/2)\sin((x + a)/2)$.
   (f) Use the results of parts (b), (d), and (e) to show that $\sin x$ and $\cos x$ are continuous at $x = a$.

6. (a) Show that $f(x) = 1/x$ is continuous in the open interval (a, b) if $0 < a < b$.
   (b) Show that $f(x)$ has no upper bound in the interval $(0, b)$.
   (c) Explain the result of part (b) and compare it with the result of Theorem 4.3.
   (d) Show that $f(x)$ is continuous in the interval $(0, b)$ and yet $f(x)$ is unbounded in this interval.

7. (a) Show that $f(x) = \tan x$ and $g(x) = \sec x$ are continuous in the interval $(-\pi/2, \pi/2)$.
   (b) Show that $f(x)$ and $g(x)$ are unbounded in this interval.
   (c) Show that $f'(x)$ and $g'(x)$ are continuous but unbounded in this interval.

8. A function $f(x)$ is defined such that $f(x) = 0$ if x is a rational number and $f(x) = x$ if x is an irrational number.

   (a) Show that this function is continuous at $x = 0$.
   (b) Determine whether this function is continuous at any point other than $x = 0$.
   (c) Does this function have a derivative at any point?

C9. (a) Write a computer program to evaluate $DQf(4, 4 + h)$ if $f(x) = \sqrt{x}$ and $h = (2)^{-k}$ for $k = 0, 1, 2, 3, \ldots, 50$.
   (b) Using your program of part (a) can you determine $f'(4)$?
   (c) In what way does the number system used in the computer cause problems in evaluating the limit?
   (d) Is the computer's version of $f(x)$ a continuous function? Explain.

B10. One can find a quotation similar to "the bacterial population of a certain culture is given by the equation $p(t) = K(1 - e^{-0.1t})$ where t is measured in hours and K is a positive constant such as $K = 10^7$."

  (a) Is $p(t)$ a continuous function?
  (b) If population is considered to be the number of whole organisms in the culture, show that population is a step function.
  (c) Discuss the way in which a continuous function can be used to represent a discontinuous function.

S11. (a) Show that since most functions in economics represent numbers of dollars or quantity of goods, these functions must be step functions in the literal sense.
  (b) Show that it is theoretically possible to find a continuous function in which the error at any time would be no greater than one half of the largest step in the step function.
  (c) If the number of dollars or units is large, show that the relative error in using the continuous function to represent the step function is likely to be small.

P12. (a) One theory concerning the excitation of the atom concludes that the energy of the atom is always an integral number of minute units each one of which is called a *quantum*. Show that a function giving the energy of an atom as a function of time under this theory must be a step function.
  (b) If the atom is in an excited state, that is the atom has a large number of quanta of energy, show that the relative error in approximating the energy by a continuous function would be small.
  (c) Show that the derivative of the theoretical energy function fails to exist at those times when a change in energy occurs.
  (d) Since the energy is a step function, show that it is possible to evaluate the number of quantum seconds of time-energy of the atom using the RS integral, but it is not possible to do this using the Riemann integral.

VII.5 Inverse Functions

If a function has a derivative at some point, it has been shown that the function must be continuous at that point. This information has proven helpful in determining that a large class of functions is continuous. In our previous development we have been concerned not only with functions, such as the exponential function, the trigonometric functions, and the algebraic functions, but we have also been concerned with their inverses. We used implicit differentiation to find the derivatives of the inverse functions where they existed, and it seems reasonable that the implicit differentiation would work, but we did not give a rigorous proof that such would be the case. It is our purpose in this section to demonstrate that the derivatives of inverse functions we have derived are correct in a more rigorous sense.

Theorem 5.1. *Let $f(x)$ be a function with a non-zero, continuous derivative for each value of x in the interval $[a, b]$. The inverse function, $f^{-1}(x)$ exists on $[a, b]$ and has a continuous derivative in this interval.*

PROOF. Since $f'(x)$ is non-zero and continuous in $[a, b]$, it is either positive throughout the interval or negative throughout the interval. In either case, $f(x)$ is monotonic and $f^{-1}(x)$ must exist. Let $a \leq x_1 < x_2 \leq b$. By the mean value theorem we have $f(x_2) - f(x_1) = (x_2 - x_1)f'(x^*)$ where $x_1 < x^* < x_2$. Now we have $(f(x_2) - f(x_1))/(x_2 - x_1) = f'(x^*) \neq 0$. We can write

$$\frac{x_2 - x_1}{f(x_2) - f(x_1)} = \frac{f^{-1}(f(x_2)) - f^{-1}(f(x_1))}{f(x_2) - f(x_1)}$$

$$= \frac{f^{-1}(y_2) - f^{-1}(y_1)}{y_2 - y_1} = \frac{1}{d'(f^{-1}(y^*))}$$

where $y_1 = f(x_1)$, $y_2 = f(x_2)$, and $y^* = f(x^*)$. Also y^* is between y_1 and y_2 since $f(x)$ is monotonic. Such values of y exist by the intermediate value Theorem. If we now take the limit as x_2 approaches x_1, x^* must approach x_1. By the continuity of $f(x)$, y_2 must approach y_1 and y^* must approach y_1. Therefore,

$$D_y f^{-1}(y) = \lim_{y_2 \to y_1} \frac{f^{-1}(y_2) - f^{-1}(y_1)}{y_2 - y_1} = \frac{1}{f'(f^{-1}(y_1))}. \qquad \square$$

This theorem states that the derivative of the inverse function exists under the stated conditions. However, we have been careful to observe those conditions. You will note that in the case of the exponential function the derivative exists, is continuous (for it is still the exponential function) and is always positive. In the case of the inverse trigonometric functions we have been careful to avoid the trouble spots. The sine function would have a zero derivative at $x = -\pi/2$ and at $x = \pi/2$, but we have avoided these points in establishing the inverse function. We have treated the other five inverse trigonometric functions in similar manner. Therefore, by Theorem 5.1 the derivatives exist, and by Theorem VII.4.5 these functions are continuous. Now we are in a position to use our Theorems concerning the sum, difference, product, quotient, and composition of continuous functions to show that those functions which we claimed to be continuous were in fact continuous. Note that the only problem here concerns division. Points at which the denominator function of a quotient are zero cannot be points of continuity.

Now that we have done so much work with so little apparent effort (don't forget the effort in getting to this point) you can appreciate the value of some of the mathematical machinery we have developed. As a consequence of the fact that these functions are continuous the intermediate value Theorem applies. We also have a very helpful device for evaluating many limits. Since $\lim_{x \to a} f(x) = f(a)$ if $f(x)$ is continuous at $x = a$, we can evaluate the limits by evaluating the functions. It should be observed that this method of evaluating limits is only good in the case of continuous functions, however. Be certain to check whether the function involved is continuous before you try to take a limit by mere substitution in the function.

We should mention the fact that a continuous function might only be defined over a domain which does not include all real numbers. For instance it is necessary to restrict the domain to the interval $(0, \infty)$ in the case of the logarithm function. Other restrictions are required for the inverse trigonometric functions and for power functions which involve square roots, fourth roots, etc. The statement of the domain is an integral part of the definition of the function, and it is necessary to keep this in mind. If one wishes to deal with a continuous function further restrictions may be necessary. A quotient of continuous functions, for instance, is continuous only if we exclude all points of the domain at which the denominator is zero.

EXERCISES

1. (a) Let $f(x) = x^2$ for non-negative values of x. Find the inverse function in this case and show that there is no ambiguity in obtaining this function.
   (b) Evaluate $f'(x)$ and then $f'(f^{-1}(x))$.

2. (a) Let $f(x) = \sin x$ and determine a domain over which $f(x)$ has an inverse function.
   (b) Use the result in the proof of Theorem 5.1 to find the derivative of $f^{-1}(x)$ and show that this agrees with the results of Chapter V.
   (c) Do the same thing for $f(x) = \cos x$.
   (d) Do the same thing for $f(x) = \tan x$.
   (e) Do the same thing for $f(x) = \cot x$.
   (f) Do the same thing for $f(x) = \sec x$.
   (g) Do the same thing for $f(x) = \csc x$.

3. (a) Let $y = f(x)$ be defined by the relation $x^2 + y^2 = 25$. Show that $f(x)$ is a function if we restrict our attention to either the upper semicircle or the lower semi-circle.
   (b) Show that $f(x)$ has an inverse if we restrict our attention to that portion of the circle in any one of the four quadrants.
   (c) Show that the function $f(x)$ determined in the first quadrant is continuous and has a continuous inverse.
   (d) Show that the derivatives of both $f(x)$ and $f^{-1}(x)$ exist for the portion of the circle in the first quadrant but not on either the x-axis or the y-axis. Show that one or the other of the derivatives fails to exist on the axes.

4. (a) Let $g(x) = f^{-1}(x)$. Use the results of Theorem 5.1 to show that $g'(f(x))f'(x) = 1$.
   (b) Demonstrate the result of part (a) with $f(x) = e^x$.
   (c) Demonstrate the result of part (a) with $f(x) = \sin x$.

5. (a) Let $f(x) = (\sin x)/x$. Show that $f(x)$ is undefined when $x = 0$, but is continuous at every other point.
   (b) If $g(x) = f(x)$ when $x \neq 0$ and $g(0) = 1$, show that $g(x)$ is continuous for all real values of x under the assumption that the unit of angular measurement is the radian.
   (c) Does the derivative of $g(x)$ exist at $x = 0$?

6. Find all points of discontinuity for each of the following functions.

   (a) $(\sin x + \tan x)/(x^3 + 9x)$
   (b) $\sqrt{x^2 + 4}$
   (c) $\sqrt{x^2 - 4}$
   (d) $\tan x - \cot x$
   (e) $x \sin(1/x)$
   (f) $\ln(x + 2)$
   (g) $\ln(x^2 + 1)$
   (h) $(2x + 3)/(x^4 - 5x^2 + 4)$

7. Let $f(x)$ be an *almost identity function* in that $f(x)$ is the value of x rounded to 5 decimal places. Thus, $f(0.534) = 0.534$ but $f(2.1234567) = 2.12346$.

   (a) Is $f(x)$ continuous?
   (b) Does $f'(x)$ exist everywhere? Anywhere?
   (c) Does the intermediate value theorem hold for $f(x)$?
   (d) Is $f(x)$ bounded over any closed interval?
   (e) Would the answers to the above questions be altered if the 5 decimal places were replaced by n decimal places for any positive integer n?

8. (a) If $f(x) = |x|$, find $f''(x)$ and indicate where it exists.
   (b) Over what domain is $f''(x)$ continuous?
   (c) Over what domain is $f'''(x)$ continuous?

9. Let $f(x) = x$ and $g(x) = [x]$ be the largest integer which does not exceed x.

   (a) Is $f(x)$ continuous?
   (b) Is $g(x)$ continuous?
   (c) What is the value of $f'(x)$ where it exists and where does it exist?
   (d) What is the value of $g'(x)$ where it exists and where does it exist?
   (e) Evaluate $\int_0^5 f(x)dg(x)$ if it exists.
   (f) Evaluate $\int_0^5 g(x)df(x)$ if it exists.

10. Let $f(x) = x^3$ over the domain $[-2, 3]$.

   (a) Determine whether all of the hypotheses of Theorem 5.1 are satisfied.
   (b) Determine whether the inverse of $f(x)$ exists.
   (c) Determine whether the inverse of $f(x)$ has a derivative at every point where it exists.
   (d) Is the derivative of $f^{-1}(x)$ continuous?
   (e) Do the results you have obtained in this exercise disprove Theorem 5.1? Explain.

11. Answer the five parts of Exercise 10 if $f(x) = x^2$.

SB12. One often sees a graph of data in which the data points are plotted and these points are then connected with straight line segments.

   (a) Is the function represented by such a graph continuous?
   (b) Does such a function have a derivative at every point? At any point?
   (c) Is the derivative (if it exists) continuous?

VII.6 Methods for Handling Some Obnoxious Limits

In Section VII.3 we discussed some theorems concerning limits and in Section VII.4 we indicated that the matter of evaluating limits is very easy if we have continuous functions. However, there are cases which occur all too frequently which do not yield easily to any of the techniques developed so far. The particular cases we have in mind here will be rational functions with differentiable numerators and denominators, such that the numerator and denominator simultaneously approach zero as a limit, or simultaneously fail to exist. Of course, as you might expect, we are interested in the value of the rational function at precisely the point where this catastrophe happens.

The first of the results we give here was discovered by *Jean Bernoulli* (1667–1748) very shortly after the development of the calculus by Newton and Leibniz. It first received wide dissemination in the first book published on the calculus, one written by *Guilame Francois Marquis de l'Hôpital* (1661–1704) in 1696, and it has henceforth been known as *l'Hôpital's rule*. That this problem in limits arose early in the development of the calculus is not surprising, for in our evaluation of the limit of the differential quotient we faced the fact that we were dealing with a fraction in which both numerator and denominator approach zero at precisely the wrong point, namely the one at which we wanted to evaluate the limit. We were able to prove a theorem that would get us out of this dilemma in many cases, but it was not always possible to find a direct method for evaluating the limit. In the evaluation of $\lim_{x \to 0}((\sin x)/x)$, for instance, we had to resort to geometric chicanery. We will consider the general l'Hôpital's rule in separate theorems, for this will ease the development.

Theorem 6.1. *If $f(x)$ and $g(x)$ have continuous derivatives in the neighborhood of $x = a$, if $\lim_{x \to a} f(x) = \lim_{x \to a} g(x) = 0$, and if $\lim_{x \to a}[f'(x)/g'(x)] = L$, then*

$$\lim_{x \to a} \left[\frac{f(x)}{g(x)} \right] = \lim_{x \to a} \left[\frac{f'(x)}{g'(x)} \right] = L. \tag{VII.6.1}$$

PROOF. Before we proceed we should note that we have the quotient of the derivatives and *not* the derivative of the quotient in the right hand of the two limits in the conclusion. The quotient $f(x)/g(x)$ does not represent a continuous function at $x = a$, for the denominator approaches zero, and, in fact, we have no assurance that either $f(a)$ or $g(a)$ exists. We will take care of this latter possibility by introducing two new functions, $F(x)$ and $G(x)$, such that $F(x) = f(x)$ if $x \neq a$, and $F(a) = 0$, and $G(x) = g(x)$ if $x \neq a$, and $G(a) = 0$. Therefore, by Theorem VII.2.1 we know that

$$\lim_{x \to a} \left[\frac{f(x)}{g(x)} \right] = \lim_{x \to a} \left[\frac{F(x)}{G(x)} \right]$$

provided the second limit exists. Furthermore, we know that $F(x)$ and $G(x)$ have continuous derivatives in a neighborhood of $x = a$. Therefore, the extended mean value theorem (Theorem V.3.2) can be applied and we have

$$\frac{F(x) - F(a)}{G(x) - G(a)} = \frac{F'(c)}{G'(c)}$$

with c between x and a. However, since $F(a) = G(a) = 0$, we can write, for values of x other than $x = a$,

$$\frac{F(x)}{G(x)} = \frac{F'(c)}{G'(c)} = \frac{f'(c)}{g'(c)}.$$

The last expression results from the fact that $F(x)$ and $f(x)$ have equal derivatives. A similar statement applies to $G(x)$ and $g(x)$. Given any $\varepsilon > 0$, there exists a number $\delta > 0$ such that when $0 < |x - a| < \delta$ then $|(f'(x)/g'(x)) - L| < \varepsilon$. But we now have all that we require to complete the proof, for if $0 < |x - a| < \delta$ then certainly $|c - a| < \delta$ since c is between x and a, and hence $|c - a| < |x - a|$. Therefore

$$\left| \frac{F(x)}{G(x)} - L \right| = \left| \frac{f'(c)}{g'(c)} - L \right| < \varepsilon$$

and

$$\lim_{x \to a} \left[\frac{F(x)}{G(x)} \right] = L.$$

This is equivalent to the statement

$$\lim_{x \to a} \left[\frac{f(x)}{g(x)} \right] = \lim_{x \to a} \left[\frac{f'(x)}{g'(x)} \right] = L,$$

and the desired conclusion has been obtained. $\square$

EXAMPLE 6.1. Find the value of $\lim_{x \to 0}((\sin x)/x)$ using l'Hôpital's rule.

Solution. We already know this value, but we now have a means of obtaining the result in a different manner. We observe here that $\lim_{x \to 0} \sin x = 0$ and $\lim_{x \to 0} x = 0$. Therefore we have satisfied the hypotheses of our theorem. Since $D_x(\sin x) = \cos x$ and $D_x(x) = 1$, we have by l'Hôpital's rule $\lim_{x \to 0}((\sin x)/x) = \lim_{x \to 0}((\cos x)/1)$. Since $\lim_{x \to 0} \cos x = 1$, we now have $\lim_{x \to 0}((\sin x)/x) = 1/1 = 1$.

This is a much easier method of evaluating this limit than the method used in Chapter III, but we should observe that we had to use the derivative of $\sin x$ and this derivative could not have been obtained without first knowing the value of this limit. Therefore, this method could not have come first. However, it may prove useful to you in remembering the value of this limit for future use.

EXAMPLE 6.2. Evaluate $\lim_{x\to 1}((e^x - e)/(x^2 - 1))$.

Solution. Since $\lim_{x\to 1}(e^x - e) = 0$ and $\lim_{x\to 1}(x^2 - 1) = 0$, we can again use l'Hôpital's rule. In this case we will use the fact that $D_x(e^x - e) = e^x$ and $D_x(x^2 - 1) = 2x$. Furthermore $\lim_{x\to 1} e^x = e$ and $\lim_{x\to 1} 2x = 2$. Therefore we have

$$\lim_{x\to 1} \frac{e^x - e}{x^2 - 1} = \lim_{x\to 1} \frac{e^x}{2x} = \frac{e}{2}.$$

This result would have been much more difficult to obtain by other methods.

Theorem 6.1 covers the case in which both $f(x)$ and $g(x)$ approach zero as limits. There are also instances in which both are unbounded as x nears the value a.

Theorem 6.2. *If $f(x)$ and $g(x)$ have continuous derivatives in a neighborhood of $x = a$, if $\lim_{x\to a} f(x) = \lim_{x\to a} g(x) = \infty$, and if $\lim_{x\to a}[f'(x)/g'(x)] = L$, then*

$$\lim_{x\to a} \left[\frac{f(x)}{g(x)}\right] = \lim_{x\to a} \left[\frac{f'(x)}{g'(x)}\right] = L. \qquad (VII.6.2)$$

PROOF. Since we are dealing with functions as their values increase without limit, one can expect the proof to be a bit more intricate. We will handle this by considering the right hand limit and the left hand limit separately. If each of the one-sided limits obeys the conclusion of the theorem, then the two-sided limit must also obey this conclusion. The two one-sided limits can be proven in similar fashion, and therefore we will consider in detail only the proof of one of these two one-sided cases. Arbitrarily we will pick the left-hand limit, and assume initially that we have $\lim_{x\to a}[f'(x)/g'(x)] = L$. Then for any $\varepsilon > 0$, there must be a value of $\delta > 0$ such that for any value of x in $(a - \delta, a)$ it is true that $|(f'(x)/g'(x)) - L| < \varepsilon$. Let $x^* = a - \delta$, and then let x be any point in the interval (x^*, a) as indicated in Figure VII.1. By the extended mean value theorem $(f(x) - f(x^*))/(g(x) - g(x^*)) = (f'(c))/(g'(c))$ where c is the interval (x^*, x). Therefore

$$\left| \frac{f(x) - f(x^*)}{g(x) - g(x^*)} - L \right| < \varepsilon. \qquad (VII.6.3)$$

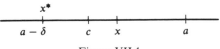

Figure VII.1

Define the function $h(x)$ by the equation

$$h(x) = \frac{1 - \dfrac{f(x^*)}{f(x)}}{1 - \dfrac{g(x^*)}{g(x)}}. \tag{VII.6.4}$$

We can rewrite (VII.6.3) as $|(f(x)/g(x))h(x) - L| < \varepsilon$ and this is valid for all values of x in the interval (x^*, a). Since $f(x^*)$ and $g(x^*)$ are finite numbers and $\lim_{x \to a^-} f(x) = \lim_{x \to a^-} g(x) = \infty$ it follows from (VII.6.4) that $\lim_{x \to a} h(x) = 1$. Since $h(x)$ is a continuous function we can select a value $\delta_1 > 0$ such that when $|x - a| < \delta_1$ we are certain that $|h(x) - 1| < \varepsilon$ and $h(x) > 1/2$. Therefore

$$\left| \left[\frac{f(x)}{g(x)} - L \right] h(x) \right| = \left| \frac{f(x)}{g(x)} h(x) - Lh(x) \right| \le \left| \frac{f(x)}{g(x)} h(x) - L \right|$$
$$+ |L[1 - h(x)]| < \varepsilon + |L|\varepsilon.$$

From this we obtain

$$\left| \frac{f(x)}{g(x)} - L \right| < \frac{(1 + |L|)\varepsilon}{h(x)} < 2(1 + |L|)\varepsilon.$$

This proves the result for the left-hand limit. A similar proof for the right-hand limit will complete the proof of the theorem. $\quad\square$

Theorems 6.1 and 6.2 do not cover all the limits we would like to consider. In particular we note that these theorems only consider the case in which x approaches a finite limit. There are instances in which x increases without bound (or decreases without bound). Since we find $f(\infty)$ so distasteful that we aren't even willing to discuss it, that immediately obviates the possibility of using Theorems 6.1 or 6.2 in a situation in which x is increasing or decreasing without bound. This discussion should indicate that we are ready for another theorem.

Theorem 6.3. *If $f(x)$ and $g(x)$ have continuous derivatives for all values of $x > X$ for some number X, if $\lim_{x \to \infty} f(x) = \lim_{x \to \infty} g(x) = b$, where $b = 0$ or $= \infty$ and if $\lim_{x \to \infty}[f'(x)/g'(x)] = L$, then $\lim_{x \to \infty}[f(x)/g(x)] = \lim_{x \to \infty}[f'(x)/g'(x)] = L$.*

PROOF. The method used before of defining new functions and invoking the extended mean value theorem does not seem to apply here, but since we have previous results we will try to use them. We will replace x by $1/t$, for when x gets very large then t gets very small. Furthermore,

$$\lim_{x \to \infty} f(x) = \lim_{t \to 0^+} f\left(\frac{1}{t}\right) = \lim_{x \to \infty} g(x) = \lim_{t \to 0^+} g\left(\frac{1}{t}\right) = b.$$

We have the case we considered in the preceding theorem, for t approaches zero, which is certainly a finite limit. We invoke the earlier result and obtain

$$\lim_{x \to \infty} \left[\frac{f(x)}{g(x)} \right] = \lim_{t \to 0^+} \left[\frac{f(1/t)}{g(1/t)} \right] = \lim_{t \to 0^+} \left[\frac{f'(1/t)(-1/t^2)}{g'(1/t)(-1/t^2)} \right] = \lim_{x \to \infty} \left[\frac{f'(x)}{g'(x)} \right] = L.$$

Note that we have indicated the limits as t approaches zero from the positive side. This is a point you should consider. A similar proof can be constructed for the case in which x is decreasing without bound, and you should construct this one just to be certain you are in complete control of the method we have employed. Have you noted how frequently mathematicians, just as jurists, call on precedents?   □

These three theorems cover all of the cases we need consider here. In effect they state that if the numerator and the denominator simultaneously approach zero or increase without bound as we approach a value $x = a$, regardless of whether a is finite or infinite, the existence of a limit of the quotient of the derivatives establishes the fact that the original quotient had a limit, and both have the same limit. We must be careful to observe the condition that both numerator and denominator approach zero, or that both are unbounded simultaneously. L'Hôpital's rule is not applicable in any other case. These two situations are often referred to as *indeterminate forms* and are listed as 0/0 or ∞/∞. These latter terms are meaningless if taken literally, but they describe the situation at hand. Other indeterminate forms include $\infty - \infty$, 0^0, 0^∞, ∞^0, and 1^∞. These are all loose descriptions, indicating the limits of the individual functions composing the indeterminate form. While these latter forms are not expressed in such a way that l'Hôpital's rule is applicable, we can frequently use algebraic manipulations to arrive at an expression for which this rule is suitable.

EXAMPLE 6.3. Evaluate $\lim_{n \to \infty}(1 + (x/n))^n$.

Solution. This does not resemble the function to which l'Hôpital's rule applies. However, if we take the limit of $\ln(1 + (x/n))^n = n \ln(1 + (x/n))$ we can apply this result to the problem at hand for we have already discussed the limits of composite functions. This last form gives us the product of two limits. One of these factors increases without bound while the other one approaches zero. There is a big question concerning whether the increasing value or the decreasing value will dominate. Let us see how this might be handled, remembering that we have only the three Theorems with which to work. We can consider either the expression

$$\frac{\ln\left(1 + \dfrac{x}{n}\right)}{\dfrac{1}{n}} \quad \text{or the expression} \quad \frac{n}{1/\ln\left(1 + \dfrac{x}{n}\right)}$$

It would appear that the first of these two options is preferable, for while it is bad to have to differentiate $(1/n)$, it seems even worse to have to differentiate the reciprocal of the logarithm. (We should pause to note here that since it is n that is increasing without stopping, it is with respect to n that we must differentiate). We can proceed as follows:

$$\lim_{n\to\infty}\left[\ln\left(1+\frac{x}{n}\right)^n\right] = \lim_{n\to\infty}\left[n\ln\left(1+\frac{x}{n}\right)\right]$$

$$= \lim_{n\to\infty}\frac{\ln\left(1+\frac{x}{n}\right)}{n^{-1}} = \lim_{n\to\infty}\frac{\left[\left(1+\frac{x}{n}\right)^{-1}(-xn^{-2})\right]}{-n^{-2}}.$$

We have applied L'Hôpital's rule to the third expression in this sequence, since in this instance both numerator and denominator approach zero. Again we remind you that the differentiation was with respect to n, for n is controlling the limit in this instance. Upon performing the indicated simplication

$$\lim_{n\to\infty}\left[\ln\left(1+\frac{x}{n}\right)^n\right] = \lim_{n\to\infty}\frac{x}{1+\frac{x}{n}} = x.$$

Hence,

$$\lim_{n\to\infty}\left(1+\frac{x}{n}\right)^n = \lim_{n\to\infty}\exp\left(\ln\left(1+\frac{x}{n}\right)^n\right) = \exp\left(\lim_{n\to\infty}\left[\ln\left(1+\frac{x}{n}\right)^n\right]\right) = e^x.$$

This is a rather surprising result, for the original question was basically one that appeared to consider taking a number approaching one and raising it to an ever increasing power. It shows that one cannot always trust intuition. In fact there are some books that use this definition to derive the value of e^x in the first place. This is a perfectly valid approach.

The example which we have just considered illustrates the fact that there are many cases in which it may not appear that L'Hôpital's rule will work, but where, in fact, it is just this rule that will enable us to handle the situation. You will note that here we have two such indeterminate forms in a single problem—that of 1^∞ and that of $0 \cdot \infty$, where we have used the symbols 1, 0 and ∞ as we indicated earlier. The main thing required is a careful analysis of what we are faced with, what we are able to handle, and an investigation of the methods by which we can go from the given expression to the desired result. Consider as another instance the case in which we appear to have the problem of obtaining the difference of two functions, each of which is approaching an infinitely large value (whatever that is).

EXAMPLE 6.4. Evaluate $\lim_{x\to 0}(\csc x - \cot x)$.

Solution. Since both csc 0 and cot 0 fail to exist, this would appear to be impossible. However, the question has been posed, and we would find it most embarrassing to try to duck it. Can we possibly put this in one of the forms covered by a theorem in this section? This would require that we rephrase it by asking for the limit of a quotient. But then we observe that

$$\lim_{x \to 0} (\csc x - \cot x) = \lim_{x \to 0} \left(\frac{1}{\sin x} - \frac{\cos x}{\sin x} \right) = \lim_{x \to 0} \left(\frac{1 - \cos x}{\sin x} \right),$$

and this fits Theorem 6.1 nicely. Consequently we can proceed

$$\lim_{x \to 0} (\csc x - \cot x) = \lim_{x \to 0} \left(\frac{1 - \cos x}{\sin x} \right) = \lim_{x \to 0} \frac{\sin x}{\cos x} = \lim_{x \to 0} \tan x = 0,$$

and we have the required result.

There are certain problems which occur frequently, and to which we should give some thought. If we are faced with the necessity of taking the limit of an expression having an exponential as a factor (and these seem to occur with greater regularity than we could wish), we frequently find that L'Hôpital's rule provides the most direct method.

EXAMPLE 6.5. Evaluate $\lim_{x \to \infty} e^{-x} x^{3/2}$.

Solution. We observe that e^{-x} is approaching zero, but that $x^{3/2}$ is getting larger without bound. Again we have the question of where (if at all) this might settle down. Since our methods here only apply when we have a quotient, and both the numerator and the denominator are approaching the same reprehensible limit, be it zero or infinity, we should see whether we can transform our problem into one having these properties. Probably the most obvious approach is to consider

$$\lim_{x \to \infty} e^{-x} x^{3/2} = \lim_{x \to \infty} \frac{x^{3/2}}{e^x} = \lim_{x \to \infty} \frac{(\frac{3}{2}) x^{1/2}}{e^x},$$

but we note here that this last expression, obtained by applying the rule, is also indeterminate, for both numerator and denominator increase without stopping. This last expression looks better, though, for the exponent in the denominator is no worse and the one in the numerator is smaller (if that is a virtue). We should apparently try again and obtain

$$\lim_{x \to \infty} e^{-x} x^{3/2} = \lim_{x \to \infty} \frac{(\frac{3}{2})(\frac{1}{2}) x^{-1/2}}{e^x}.$$

This is the time to stop, for while the form may not appear greatly different, the substance is very different. In this case the numerator is no longer increasing without bound. L'Hôpital's rule will not apply here, for the proper conditions are not present. You should be acutely aware of checking at each stage to make certain whether you can continue to apply the rule (although there are many times when we will need to continue for more stages than we found necessary here). Using this last result, we see that we can write

$$\lim_{x \to \infty} e^{-x} x^{3/2} = \lim_{x \to \infty} \frac{\frac{3}{4}}{e^x x^{1/2}} = 0,$$

since the denominator contains two factors, each of which increases without limit. If we had been given an exponent for x which was very large (such as 219), we would have had to continue the application of this rule for a very large number of stages before finally coming to a limit which was more amenable to evaluation.

EXAMPLE 6.6. Evaluate $\lim_{x \to 0} x^{\tan x}$.

Solution. This is a limit of the 0^0 variety. We can best handle this by considering

$$\lim_{x \to 0} \ln x^{\tan x} = \lim_{x \to 0} (\tan x)(\ln x).$$

This is of the $0 \cdot \infty$ type, and therefore we must attempt to convert this to a quotient suitable for treatment by l'Hôpital's rule. Following the line of reasoning used in Example 6.3, it would seem preferable to take the reciprocal of tan x, that is cot x. Therefore we will have

$$\lim_{x \to 0} \ln x^{\tan x} = \lim_{x \to 0} (\tan x)(\ln x) = \lim_{x \to 0} \frac{\ln x}{\cot x} = \lim_{x \to 0} \frac{x^{-1}}{-\csc^2 x}.$$

This last expression does not seem to help, but with some algebra and the use of trigonometric relations we can write

$$\lim_{x \to 0} \left(\frac{-x^{-1}}{\csc^2 x} \right) = \lim_{x \to 0} \left(\frac{-\sin^2 x}{x} \right) = \lim_{x \to 0} \frac{-2 \sin x \cos x}{1} = \frac{0}{1} = 0.$$

We used l'Hôpital's rule in going from the third to the fourth step in this sequence of equalities. Therefore we are now ready to finish our assignment, for we can write

$$\lim_{x \to 0} x^{\tan x} = \lim_{x \to 0} e^{\ln(x^{\tan x})} = e^0 = 1.$$

It is not true that L'Hôpital's rule is the answer to *all* problems involving the limits of quotients, but it is helpful in a surprisingly large number of cases. We admonish you, however, that you should size up the particular situation which faces you and then act accordingly. Do not, for instance, assume that just because an exercise comes at the end of this section it must necessarily use L'Hôpital's rule, nor even if it can be so handled that this is the best or shortest method.

EXERCISES

1. Find each of the following limits (if they exist):

   (a) $\lim_{x \to 0}(\sin x)^x$
   (b) $\lim_{x \to 0} x^{\sin x}$
   (c) $\lim_{x \to 0}((x - \sin x)/x^3)$
   (d) $\lim_{x \to \pi}((\pi - x)/\sin x)$
   (e) $\lim_{x \to 2}((x^2 - 4)/(x^2 - 2x))$
   (f) $\lim_{x \to 0}((1 - \cos x)/x^2)$
   (g) $\lim_{x \to \pi/2}(\sec x - \tan x)$
   (h) $\lim_{x \to \pi/2}(\sec^2 x - \tan^2 x)$

2. Find each of the following limits (if they exist):

   (a) $\lim_{x \to \infty} xe^{-x}$
   (b) $\lim_{x \to \infty} x^2 e^{-x}$
   (c) $\lim_{x \to \infty} x^4 e^{-x}$
   (d) $\lim_{x \to \infty} \sqrt{xe^{-x}}$
   (e) $\lim_{x \to \infty} x^{5/3} e^{-x}$
   (f) $\lim_{x \to \infty} x^n e^{-x}$ for any real number n.
   (g) $\lim_{x \to \infty} x^3 e^{-x^2}$
   (h) $\lim_{x \to \infty} x^n e^{-x^2}$ for any real number n.

3. Find each of the following limits (if they exist):

   (a) $\lim_{x \to 0} x^{\cot x}$
   (b) $\lim_{x \to 0}(\cot x)^x$
   (c) $\lim_{x \to 4}((x^3 - 64)/\sqrt{x} - 2)$
   (d) $\lim_{x \to \infty}((2x - 4)/3x^2)$
   (e) $\lim_{x \to 0}((1/x) - \csc x)$
   (f) $\lim_{x \to 0}((1/x) - \cot x)$
   (g) $\lim_{x \to 0} x^2 \sin(1/x)$
   (h) $\lim_{x \to \infty}((x^3 + x^2 + x + 1)/(2x^3 - 3x^2 - \pi x))$

4. Evaluate each of the following:

   (a) $\lim_{x \to 0} x^{\ln x}$
   (b) $\lim_{x \to \infty} x^{\ln x}$
   (c) $\lim_{x \to 0}(\ln x)^x$
   (d) $\lim_{x \to \infty}(\ln x)^x$
   (e) $\lim_{n \to 0}((\ln n)/n)$
   (f) $\lim_{n \to \infty}((\ln n)/n)$
   (g) $\lim_{n \to 0} n(\ln n)$
   (h) $\lim_{n \to \infty} n(\ln n)$

5. Evaluate $\lim_{x \to \infty} f(x)$ for each of the following functions:

   (a) $f(x) = \arctan x$
   (b) $f(x) = \text{arc cot } x$
   (c) $f(x) = \text{arc sec } x$
   (d) $f(x) = \text{arc csc } x$

(e) $f(x) = \sinh x$
(f) $f(x) = \cosh x$
(g) $f(x) = \tanh x$
(h) $f(x) = \coth x$
(i) $f(x) = \operatorname{sech} x$
(j) $f(x) = \operatorname{csch} x$

6. (a) Can l'Hôpital's rule be used to evaluate $\lim_{n \to \infty}(e^n/n!)$?
   (b) If $f(n) = e^n/n!$, show that $f(n + 1) = (e/(n + 1))f(n)$.
   (c) Use the result of part (b) to evaluate $\lim_{n \to \infty} f(n)$.

7. Let $f(n) = (1 + (1/n))^n$.

   (a) Evaluate $f(2)$
   (b) Evaluate $f(3)$
   (c) Evaluate $f(5)$
   (d) Show that the results of parts (a), (b), and (c) are approaching the value of e. Compare $f(n)$ with the function of Example 6.3 and explain this result.
   (e) Expand $f(n)$ by the binomial Theorem and obtain the first six terms.
   (f) Show that the limit of the six terms obtained in part (e) taken as n increases without limit will be $1 + 1 + (1/2) + (1/3!) + (1/4!) + (1/5!)$.
   (g) How would you expect the expression of part (f) to continue if you took additional terms? Find the sum of the first eleven terms carrying your arithmetic to seven decimal places and compare your result with the value of e.

8. (a) If $f(x)$ is a polynomial, a is finite, and $\lim_{x \to a} f(x) = 0$, show that $(x - a)$ is a factor of $f(x)$.
   (b) If $f(x)$ and $g(x)$ are polynomials and if $\lim_{x \to a} f(x) = \lim_{x \to a} g(x) = 0$, show that $f(x)$ and $g(x)$ have common factors $p(x)$ such that either

$$\lim_{x \to a} \frac{f(x)}{p(x)} \quad \text{or} \quad \lim_{x \to a} \frac{g(x)}{p(x)} \quad \text{is not zero.}$$

   (c) Show that in the case in which $f(x)$ and $g(x)$ are polynomials and $\lim_{x \to a}(f(x)/g(x))$ is indeterminate, the limit can be found without recourse to l'Hôpital's rule provided a is finite.

9. (a) Show that $\lim_{x \to 0}((e^{3x} - 1)/(e^x - 1)) = 3$.
   (b) Show that $f(x) = ((e^{3x} - 1)/(e^x - 1))$ is a continuous function at every point except the point where $x = 0$.
   (c) Show that the function $g(x)$ defined such that $g(x) = f(x)$ if $x \neq 0$ and $g(0) = 3$ is a continuous function for all real values of x.

S10. Population is sometimes assumed to follow the relation $P(t) = P(1 - e^{-rt})$ where P and r are positive constants and $P(t)$ is the population t years after a starting time.

   (a) Show that as time goes on the population approaches a limiting value.
   (b) Find the limiting value.
   (c) Show that the ratio of the population in the $(t + 1)$-st year to the population in the t-th year approaches a limit as t increases without bound. Would you have anticipated this result?

VII.7 Improper Integrals

There are many applications in which integrals arise which cause difficulties for reasons other than the problem of finding an anti-derivative of the integrand. In general these problems will occur with definite integrals, and will be of the type in which either the interval over which the integration is to be performed is unbounded, or else the integrand fails to exist at some point in the closed interval which is the interval of integration. This latter situation may well come to pass at one of the end points of the closed interval, that is to say at one of the limits of integration. We will classify such integrals with the following definition.

Definition. A definite integral $\int_a^b f(x)dg(x)$ is said to be an *improper integral* provided either

(i) one of the limits of integration, either a or b or both, is infinite, or
(ii) the integrand is *unbounded* at some point in the interval $[a, b]$.

We shall handle these two cases in essentially the same way, and the method to be used will be illustrated in the examples that follow. We should point out, however, that we will be taking the coward's way out in this instance (sometimes wisdom is better than bravery) by considering the integral almost to the point of difficulty, and then sneaking up on the troublesome point using the limits we have been discussing in the earlier sections of this chapter. Having been honest enough to note just what we have in mind, the next thing is to watch this process at work. Observe how close we can come to a troublesome point while still managing to avoid it.

EXAMPLE 7.1. Evaluate

$$\int_0^\infty \frac{1}{x^2 + 1} \, dx.$$

Solution. This is a case in which one of the limits of integration is infinite, and hence comes under our heading of an improper integral. Following the general technique outlined above, we will handle this by integrating over the interval $[0, M]$ where M is some rather large number. Since M is finite it represents a case we can presumably handle without difficulty. After evaluating the integral

$$\int_0^M \frac{1}{x^2 + 1} \, dx = \arctan M - \arctan 0 = \arctan M,$$

we will see what happens if M is allowed to increase without bound. This causes the upper limit of the integral to come toward an upper limit of ∞. We will consider

$$\lim_{M \to \infty} \int_0^M \frac{1}{x^2 + 1} \, dx = \lim_{M \to \infty} \arctan M.$$

As the tangent of the angle increases, the angle must approach a right angle. Hence we have

$$\lim_{M \to \infty} \int_0^M \frac{1}{x^2 + 1}\, dx = \lim_{M \to \infty} \text{arc tan } M = \frac{\pi}{2}.$$

In this case there are several things we should investigate, things we deferred until we had had an opportunity to illustrate the technique in general. First of all, the integrand is not discontinuous at any finite point, for the denominator is a polynomial and it cannot be zero in the interval. Therefore we had no problems of the second type indicated in our definition. The only problem could occur at the one limit, namely the upper limit. We could not reasonably expect to reach this limit with the most complex RS sums. Further, we should note that x starts from the point $x = 0$ and increases without limit, thus indicating that we consider arc tan M only for finite positive values which can be reached by a continuous motion in the positive direction, starting with zero. Thus, the angle represented by arc tan M would be an increasing angle, as M increases, and would start as an acute angle. The first point at which M can become infinite must be the point (or angle) such that we have just $(\pi/2)$ radians. This rules out some of the other angles for which the tangent is also infinite. This type of reasoning is frequently necessary, particularly in cases involving the inverse trigonometric functions. Whenever the limiting value can be assumed more than once there must be a decision as to which of the various possible values should be selected.

We now turn our attention to another type of improper integral. Consider the following example.

EXAMPLE 7.2. Evaluate $\int_2^5 (3 - x)^{-2}\, dx$.

Solution. We note here that the value $x = 3$ causes difficulty, for at this point the integrand fails to exist. Moreover, we notice that $x = 3$ is within the interval $[2, 5]$. Hence, we must do something to avoid it. We will do this by considering two positive numbers, δ_1 and δ_2, which are ordinarily taken to be rather small, since they represent the amount by which we will be missing the point $x = 3$. Furthermore, we will be letting each of these get smaller, and thus if we start them at small values no damage is done. You will note that we have carefully labeled δ_1 and δ_2 as separate entities, for there is no reason why they should be the same size. In no case do we want to consider that they can "cancel each other out." We now consider the *almost* integral (which includes almost all of that which we were asked to evaluate)

$$\int_2^{3-\delta_1} \frac{1}{(3-x)^2}\, dx + \int_{3+\delta_2}^5 \frac{1}{(3-x)^2}\, dx.$$

Observe that we have omitted only the interval $(3 - \delta_1, 3 + \delta_2)$ from the interval of integration which we were given. We are now able to evaluate this almost integral and obtain

$$\int_2^{3-\delta_1} \frac{1}{(3-x)^2}\,dx + \int_{3+\delta_2}^5 \frac{1}{(3-x)^2}\,dx = \left[\frac{1}{3-(3-\delta_1)} - \frac{1}{3-2} \right]$$
$$+ \left[\frac{1}{3-5} - \frac{1}{3-(3+\delta_2)} \right]$$
$$= \left[\frac{1}{\delta_1} - 1 \right] + \left[-\frac{1}{2} + \frac{1}{\delta_2} \right].$$

Since we wish to know what happens when this narrow portion of the interval $[2, 5]$ which we have omitted is squeezed as hard as possible, we will take the limits as δ_1 and δ_2 approach zero. In this way the amount that is cut out of the interval will approach zero. Thus, we have

$$\lim_{\delta_1 \to 0} \int_2^{3-\delta_1} \frac{1}{(3-x)^2}\,dx + \lim_{\delta_2 \to 0} \int_{3+\delta_2}^5 \frac{1}{(3-x)^2}\,dx$$
$$= \lim_{\delta_1 \to 0} \left[\frac{1}{\delta_1} - 1 \right] + \lim_{\delta_2 \to 0} \left[-\frac{1}{2} + \frac{1}{\delta_2} \right].$$

These last limits fail to exist, for the expressions are unbounded. Therefore, we simply say that the given integral *does not exist*. In effect we indicate that we are now aware that we have been sent on the mathematical version of a wild goose chase.

The methods used in these two examples are the same in principle, though slightly different in implementation. The whole point is to cut out the offending portion and then let it creep back in via the limit route. If the limits exist, we have a value which we can proudly display as the value of the integral. If the limits fail to exist, we can point a finger at the person who proposed the problem and think such thoughts as are appropriate to the occasion. We might note in Example 7.2 that if we had let the offending point $x = 3$ go unnoticed and then evaluate the integral, we would have ended up with a value of $-\frac{3}{2}$. This in itself should have startled us, because we would have integrated a positive integrand in the positive direction and obtained a negative result. That the integrand is positive follows from the fact that it is a square of a non-zero number (except at $x = 3$ of course). From this example we see that it is perfectly possible to obtain incorrect results, provided we let little things like the non-existence of the integrand at some point escape our attention. We would repeat the admonition to be alert at all times. It would be bad if such situations were to give answers which are accepted as correct, but which produce bridges, economies, societies, or health remedies which are defective.

If the integrand fails to exist at any point it frequently does so at one of the end points, and occasionally at both of them. The same principle will apply, for we only need to exclude a very small interval involving the point in question, and then take the limit as the length of the excluded section approaches zero. This has some rather interesting facets in the case of integrals which involve integration by parts and which involve recursion relationships. Consider the following example.

EXAMPLE 7.3. Evaluate $\int_0^\infty e^{-x} x^n \, dx$, where n is positive number.

Solution. We note trouble here, for we could write $e^{-x} x^n = x^n/e^x$, and as x increases we have both the numerator and denominator increasing without bound. Therefore, we have double trouble at the upper limit, for in the first place the upper limit is much larger than we like, and in the second place this fact induces trouble. There is the further fact that we will probably either have to use integration by parts or use the recursion relation developed in Section VI.8. That was developed by integration by parts as you remember. Using integration by parts we proceed as follows.

$$
\begin{aligned}
\int_0^\infty e^{-x} x^n \, dx &= \lim_{M \to \infty} \int_0^M e^{-x} x^n \, dx \\
&= \lim_{M \to \infty} \left[e^{-M} \left(\frac{1}{-1} M^n \right) - e^0 \left(\frac{1}{-1} 0^n \right) \right] \\
&\quad - \lim_{M \to \infty} \left[\frac{1}{-1} \int_0^M e^{-x} n x^{n-1} \, dx \right] \\
&= \lim_{M \to \infty} \left[-\frac{M^n}{e^M} \right] + n \int_0^\infty e^{-x} x^{n-1} \, dx.
\end{aligned}
$$

We have made use of the limits of integration, re-interpreting the last integral by going from the limit of a proper integral back to the improper integral of the form with which we started, but with the exponent for x reduced by one. Our problem first is to evaluate the limit that still remains, and we note that M^n/e^M is of a form that is susceptible to L'Hôpital's rule. One of our problems here is that we do not know the value of n, for n was given merely as a positive number. However, we can consider the string of limits

$$
\lim_{M \to \infty} \frac{M^n}{e^M} = \lim_{M \to \infty} \frac{n M^{n-1}}{e^M} = \lim_{M \to \infty} \frac{n(n-1) M^{n-2}}{e^M} = \cdots
$$

and observe that with successive differentiation the denominators remain the same and the numerators are possessed of increasingly complicated coefficients. However the power to which M is raised is being eroded. Since n is a finite number, after less than $(n + 1)$ steps we will either have a numerator with a zero-th power, or one with a negative power of M. In this case the result is no longer indeterminate, and with a denominator increasing and a numerator that can do no more than remain constant, we have a limit of zero. Note that we have a type of mathematical induction in reverse, for the size of the number involved is decreasing, and we did not have to know that n was an integer. This particular integral has many uses in which n is not an integer, as we shall note presently. We are now in the position where we can summarize the results of this solution with the statement

$$\int_0^\infty e^{-x}x^n \, dx = n \int_0^\infty e^{-x}x^{n-1} \, dx.$$

This particular function is known as the *gamma function*, and is defined, using the Greek capital gamma as

$$\Gamma(n + 1) = \int_0^\infty e^{-x}x^n \, dx.$$

From the derivation above, we note that $\Gamma(n + 1) = n\Gamma(n)$. It is worth noting that

$$\Gamma(1) = \int_0^\infty e^{-x}x^{1-1} \, dx = \int_0^\infty e^{-x} \, dx = \lim_{M \to \infty} \int_0^M e^{-x} \, dx$$

$$= \lim_{M \to \infty} [-e^{-M} - (-e^{-0})] = \lim_{M \to \infty} \left(\frac{-1}{e^M}\right) + 1 = 1.$$

We have $\Gamma(2) = 1\Gamma(1) = 1$, $\Gamma(3) = 2\Gamma(2) = 2$, $\Gamma(4) = 3\Gamma(3) = 6$. We can continue in this fashion to obtain values of the gamma function for all positive integral arguments. In fact, you can use mathematical induction to show that we have here a function, which has values for at least some non-integral values of n and which for positive integers give us the factorials. Thus, we have $\Gamma(n + 1) = n!$. Our relation which connects $\Gamma(n)$ and $\Gamma(n + 1)$ can also be taken backward by solving for $\Gamma(n)$ in terms of $\Gamma(n + 1)$, but this would break down if we tried to find values of the gamma function for non-positive integers. This function has sufficient importance in many applications, such as statistics, that it has been tabulated for values of n in the interval $[1, 2]$. A short table is given in Appendix C. With the recursion relation which we derived in this example we are able to obtain all other values of this function, except for non-positive integers. (This function does not exist for non-positive integers.)

EXAMPLE 7.4. Find the value of $\Gamma(4.2)$ and of $\Gamma(-2.8)$.

Solution. $\Gamma(4.2) = (3.2)\Gamma(3.2) = (3.2)(2.2)\Gamma(2.2) = (3.2)(2.2)(1.2)\Gamma(1.2)$. But we can find the value of $\Gamma(1.2)$ in Table 3 of Appendix C. The table gives us the value $\Gamma(1.2) = 0.9182$. Therefore, we have

$$\Gamma(4.2) = (3.2)(2.2)(1.2)(0.9182) = 7.7570.$$

This is rounded to four decimal places, since we do not have more than four place accuracy with which to start. For $\Gamma(-2.8)$ we have to go in reverse noting that $(-2.8)\Gamma(-2.8) = \Gamma(-1.8)$. Carrying on with this line of thought, we have

$$(-2.8)(-1.8)(-0.8)(+0.2)\Gamma(-2.8) = \Gamma(1.2) = 0.9182.$$

By division, we have $\Gamma(-2.8) = -1.1386$. Note the way in which we have used our recursion formula.

If you have ever wondered about a function that behaved like a factorial, but permitted fractional arguments, you have now become acquainted with one. That this function has a long history is attested to by the fact that it was dealt with at length by *Leonhard Euler* (1707–1783); and it is understood to have been known by Wallis at an earlier date. This also gives an illustration of a very important function in mathematics with rather common applications which is not a power function, a trigonometric function, an exponential function, or a logarithm. This function is not one of those called an *elementary function*, but that does not mean it is unimportant or that it is one with which we cannot work without elaborate machinery.

It would be possible to conjure up many more illustrations of improper integrals. Note, however, that they all are handled by a single technique, and they exist or fail to exist depending on the behavior of the limits that are involved as we try to sneak up on the offending point. It is important to keep your eyes open lest you let an improper integral slip by you. If you do let one slip, you may well be able to find an answer, but the answer could well be wrong.

Exercises

1. Evaluate each of the following integrals (if they exist):

    (a) $\int_0^1 (1 - x)^{-1/2} dx$

    (b) $\int_0^1 (1 - x^2)^{-1} dx$

    (c) $\int_5^\infty (1/x\sqrt{x^2 - 9}) dx$

    (d) $\int_3^5 (1/x\sqrt{x^2 - 9}) dx$

    (e) $\int_3^\infty (1/x\sqrt{x^2 - 9}) dx$

    (f) $\int_2^4 (x - 1)^{-1} dx$

    (g) $\int_0^\infty e^{-x}(\sin x + \cos x) dx$

    (h) $\int_{-2}^2 |x| dx$

2. Evaluate each of the following integrals (if they exist):

    (a) $\int_0^2 (x - 1)^{-1} dx$

    (b) $\int_1^2 (x - 1)^{-3/2} dx$

    (c) $\int_0^{\pi/2} \tan x \, dx$

    (d) $\int_0^\pi \sec x \, dx$

(e) $\int_2^\infty (x - 1)^{-1} dx$

(f) $\int_2^\infty (x - 1)^{-2} dx$

(g) $\int_2^\infty (x - 1)^{-1/2} dx$

(h) $\int_2^\infty (x - 1)^{-1/3} dx$

3. Evaluate each of the following integrals (if they exist):

(a) $\int_0^\infty (x/(x^2 + 1)) dx$

(b) $\int_{-1}^2 x^{-2/3}\, dx$

(c) $\int_0^4 (x^2 - x)^{-1} dx$

(d) $\int_{-2}^2 (x^2 - 1)^{-1} dx$

(e) $\int_{-\infty}^0 e^x x^4\, dx$

(f) $\int_{-\infty}^\infty x e^{-x^2}\, dx$

(g) $\int_0^\infty (x^{3/2}/(x^2 - 1)) dx$

(h) $\int_0^\infty e^{-x/2}\, dx$

4. Evaluate each of the following integrals (if they exist):

(a) $\int_0^4 (4 - x)^{-2} dx$

(b) $\int_0^{\pi/2} (1 - \cos x)^{-1} dx$

(c) $\int_0^{\pi/2} (1 - \cos x)^{-2} dx$

(d) $\int_1^\infty ((\ln x)/x^2)) dx$

(e) $\int_0^{\pi/2} \sec^2 x\, dx$

(f) $\int_1^\infty x^2 \ln x\, dx$

(g) $\int_0^{\pi/2} ((\sin^2 x)/(1 - \cos x)^2) dx$

(h) $\int_0^\infty (x^4 - 16)^{-1} dx$

5. (a) Let $\int_0^{10} x^n\, dx = A(n)$. Show that $A(n)$ exists for all positive values of n.
   (b) Find the negative values of n for which $A(n)$ exists.
   (c) Find a formula for $A(n)$ for those values of n for which $A(n)$ exists.
   (d) Show that there is no smallest value of n for which $A(n)$ exists, but there is a largest value for which it does not exist. What is this value?

6. (a) Let $\int_{10}^\infty x^n\, dx = B(n)$. Show that $B(n)$ exists for $n = -3$.
   (b) Find the values of n for which $B(n)$ exists.
   (c) Show that there is no largest value of n for which $B(n)$ exists, but there is a smallest value for which it does not exist. What is this value?
   (d) Show that there is just one value of n for which $A(n)$ of Exercise 5 and $B(n)$ of this exercise fail to exist.
   (e) Show that there is no value of n for which both $A(n)$ of Exercise 5 and $B(n)$ of this exercise exist.

7. Prove that $\Gamma(n) = (n - 1)!$ for any positive integer n.

8. It can be shown that $\int_0^\infty e^{-u^2}\, du = \sqrt{\pi}/2 = 0.8862269255$.

   (a) Use the substitution $u^2 = x$ and show that this integral is equivalent to $1/2\ \Gamma(0.5)$.
   (b) Use the information of part (a) to evaluate $\Gamma(2.5)$ to nine decimal places.

    (c) Evaluate $\Gamma(-2.5)$ to nine decimal places.

    (d) Evaluate $\Gamma(1.5)$ and check your results with Table 3 of Appendix C.

9. (a) Show that $\Gamma(n) > 0$ for all values of $n > 0$.

    (b) Show that $\Gamma(n) < 0$ for values of n such that $-1 < n < 0$.

    (c) Show that $\Gamma(n) > 0$ for values of n such that $-2 < n < -1$.

    (d) Show that $\Gamma(n)$ fails to exist for integral values of n which are not positive.

    (e) Show that $\Gamma(n)$ is positive if n is in the interval $(a, a + 1)$ where a is a negative even integer, and $\Gamma(n)$ is negative in such an interval if a is a negative odd integer.

    (f) Use the information given in parts (a–e) to sketch the graph of $y = \Gamma(x)$.

10. (a) Sketch the curve $y = 1/x$.

    (b) Find the area (if it exists) under the curve $y = 1/x$ to the right of the line $x = 1$.

    (c) Find the volume of revolution formed by revolving the area of part (b) about the x-axis (provided the volume exists).

11. (a) Sketch the curve $y = 1/(x^2 + 1)$.

    (b) Find the area (if it exists) between this curve and the x-axis.

    (c) Find the volume of revolution formed by revolving the area of part (b) about the x-axis (if the volume exists).

12. (a) Sketch the curve $y = x/(x^2 + 1)$.

    (b) Find the area in the first quadrant under the curve whose equation is given in part (a) (if it exists.)

    (c) Find the volume of revolution formed by revolving the area of part (b) about the x-axis (if the volume exists).

13. Let $f(x)$ be a function with the property that there exists positive real numbers M and n such that $|f(x)| < Mx^n$ for all values of x in the interval $[a, \infty)$. Show that

$$\int_a^\infty e^{-x} f(x) dx$$

exists.

P14. (a) If the maximum amount of salt that can be dissolved in a certain glass is 20 grams, the time required to dissolve G grams $(G < 20)$ is given by the integral

$$\int_0^G \frac{k \, dg}{(20 - g)}$$

    where k is a constant. Find the time required to dissolve 10 grams of salt in the water.

    (b) Find the time required to dissolve 19.9 grams of salt in the glass.

    (c) Find the time required to dissolve the last 0.1 grams of salt in the water.

    (d) Find the time required to dissolve the first gram, the second gram, the third gram, and so forth through the twentieth gram in the water.

SB15. In many cases involving population it is assumed that there is a maximum population, M, that can be supported by the environment. As the population approaches M the support mechanisms available are so strained that fewer individuals survive the increased competition for food or other environmental support. If the rate of growth would ordinarily be r, one of the laws of growth states that the time required to increase from a population P_1 to a population P_2 is given by

$$t = \frac{1}{r} \int_{P_1}^{P_2} \frac{M\, dP}{P(M - P)}.$$

(a) If M is 20 billion and $r = 0.03$, find the time required to increase from a population of 4 billion to a population of 8 billion.
(b) How long would it take to increase from 4 billion to 19 billion?
(c) Compare the time required to increase from one billion to two billion with the time required to increase from four billion to five billion.
(d) Find the time required to increase from four billion to the maximum population that can be supported.

P16. The power emitted in the discharge of an electronic device is given by $P = e^{-0.25t}t^3$ where P is measured in watts (joules per second) and t is measured in seconds.

(a) Find the work (measured in watt-seconds) done in the first five seconds.
(b) Find the work done in the first ten seconds.
(c) Find the work done in the first 100 seconds.
(d) How much work is done starting at $t = 0$ and continuing without stopping?
(e) How long would it take to perform 99% of the work determined in part (d)?

CHAPTER VIII

Converting Data to Functions

VIII.1 An Illustration

In our discussions up to this point we appear to have assumed that functions would always be available when we need them, and we merely need to know what to do with them. Upon reflection it is apparent that this is not correct. In any application which you may face you will be required to furnish the functions. Much of the time you may have data, perhaps obtained in a laboratory or from a questionnaire, and you need a function that will describe the data. It will be our purpose in this chapter to consider this matter and to find some techniques that will be helpful. You should be aware, however, that one needs to know the discipline from which the data is obtained in order to obtain the best possible function in any given case.

We will start with an illustration. We will attempt to find a function, $f(x)$, which will describe the data

x	0	1	2	3
$f(x)$	1	0	-1	0

The data is simple enough that we should be able to find a function rather easily. If we were fortunate, we might try $f(x) = (x^3/3) - x^2 - (x/3) + 1$ and we would find that this would work. However, we might have tried $f(x) = \cos(\pi x/2)$ and this works also. In fact, we could find a very large number of functions that would satisfy our demands. This rather strange situation may seem a bit more clear if we examine Figure VIII.1. It is clear that any graph that will pass through the four given points will meet the requirements we set for ourselves. In practice we will need to know something about the type of function that would best fit the problem at hand.

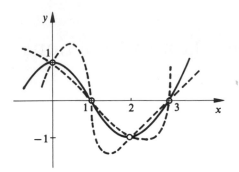

Figure VIII.1

In presenting you with two solutions for our problem you will suspect that we have used a bit of sleight of hand. However, the development of these solutions follows in a very straightforward manner. If we have four points, we have sufficient information to give four equations. Four equations are sufficient to determine the values of four parameters. We should then word the problem so that we have four parameters to determine. We can start by supposing that the function would look like

$$f(x) = a_3 x^3 + a_2 x^2 + a_1 x + a_0.$$

Since

$$f(0) = 1, f(1) = 0, f(2) = -1, \quad \text{and} \quad f(3) = 0,$$

we now have the four equations

$$a_3 0^3 + a_2 0^2 + a_1 0 + a_0 = f(0) = \quad 1$$
$$a_3 1^3 + a_2 1^2 + a_1 1 + a_0 = f(1) = \quad 0$$
$$a_3 2^3 + a_2 2^2 + a_1 2 + a_0 = f(2) = -1$$
$$a_3 3^3 + a_2 3^2 + a_1 3 + a_0 = f(3) = \quad 0.$$

These can be written

$$a_3 \qquad\qquad\qquad = \quad 1$$
$$a_3 + \ a_2 \ + a_1 + a_0 = \quad 0$$
$$8a_3 + 4a_2 + 2a_1 + a_0 = -1$$
$$27a_3 + 9a_2 + 3a_1 + a_0 = \quad 0.$$

If we solve these we will obtain $a_3 = 1/3, a_2 = -1, a_1 = -1/3$, and $a_0 = 1$. Upon inserting these values in the function $f(x)$, we obtain the first choice indicated as a potential solution to our problem.

The core of the discussion of the preceding paragraph centered around the fact that we had four points and therefore could determine four coefficients (or parameters). In fact, we could have used almost any four functions $f_1(x), f_2(x), f_3(x), f_4(x)$ and then tried to obtain the four coefficients in the relation

$$f(x) = a_1 f_1(x) + a_2 f_2(x) + a_3 f_3(x) + a_4 f_4(x).$$

This will give rise to the four equations

$$
\begin{aligned}
a_1 f_1(0) + a_2 f_2(0) + a_3 f_3(0) + a_4 f_4(0) &= f(0) = +1 \\
a_1 f_1(1) + a_2 f_2(1) + a_3 f_3(1) + a_4 f_4(1) &= f(1) = \quad 0 \\
a_1 f_1(2) + a_2 f_2(2) + a_3 f_3(2) + a_4 f_4(2) &= f(2) = -1 \\
a_1 f_1(3) + a_2 f_2(3) + a_3 f_3(3) + a_4 f_4(3) &= f(3) = \quad 0.
\end{aligned}
\tag{VIII.1.1}
$$

It is possible that you might select functions such that (VIII.1.1) would not have a unique solution, but such cases are rare. In the paragraph above we used powers of x for the functions $f_i(x)$. If we had let $f_i(x) = \cos((i - 1)\pi x/4)$, we would have had $f_1(x) = \cos 0 = 1, f_2(x) = \cos((2 - 1)\pi x/4) = \cos(\pi x/4)$, $f_3(x) = \cos(2\pi x/4)$, and $f_4(x) = \cos(3\pi x/4)$. In this case our four equations would have had the solution $a_1 = a_2 = a_4 = 0$ and $a_3 = 1$, giving us the second function we proposed as a solution to our problem. Other possible solutions, depending on our choice of the functions $f_i(x)$ would be

$$f(x) = -\frac{1}{6} + \frac{2}{3}\cos\frac{\pi x}{3} + \frac{2}{3}\cos\frac{2\pi x}{3} - \frac{1}{6}\cos \pi x$$

and

$$f(x) = x^4 - \frac{17}{3}x^3 + 10x^2 - \frac{19}{3}x + 1.$$

This last solution has more than four terms, and this opens the possibility of having an even greater number of solutions than might have appeared before. We now find we are not restricted to a single choice even in the case where we decide to use nothing but polynomials.

Before proceeding further we should stop to take stock. Given any set of four functions we can with very few exceptions find a linear combination of the four functions which will include the four given data points in the solution set of the linear combination. It is possible, of course, that we could not find coefficients for some particular set of four functions, but we are not apt to run across such a set. In fact, there is nothing magic about the number four. If we were given ten points, we would merely need to find a linear combination of ten functions. The algebra would be somewhat more terrifying, but the solution could be found. Since we have such a variety of solutions available to use through our selection of a set of functions, we should give some consideration to the selection of the particular set we would need for a particular case. Here we must consider the nature of the problem itself. If the problem suggests a periodic solution, it would be reasonable to bring in sines and cosines. If the problem is one involving radioactive decay or growth of the type anticipated in dealing with population, we would expect to use exponential functions. When we have no prior knowledge of what to expect we ordinarily use power functions of the form x^n for these tend to simplify the algebra. It cannot be emphasized too strongly, however, that the selection of the set of functions can be almost arbitrary from the standpoint of the mathematics, but it takes a knowledge of the source from which the problem came to make the best choice for this function set.

The method we have used in this section is straightforward and will work in case no other method is available. We will investigate an alternative method for handling this problem in the next section, but the alternate method will handle only the case of polynomials. Therefore, you have the method for the general case. It is worth noting that the number of coefficients must equal the number of data points, and this method would be somewhat difficult to carry through if we had a very large number of data points. It is also true that with the large number of data points the resulting function would probably be rather complicated, and we might wish an alternative to such a function. We will find such an alternative in Section VIII.3.

Exercises

1. Find a polynomial that passes through the points

x	-2	1	4
$f(x)$	-13	5	41

2. Find a polynomial that passes through the points

x	-2	-1	1	2
$f(x)$	-3	0	1	-2

3. Find a polynomial that passes through the points

x	-2	0	1	3
$f(x)$	0	2	0	20

4. The function $f(x) = a + b \cos x + c \sin x$ passes through the points

x	0	$\pi/4$	$\pi/2$
$f(x)$	3	2	1

Find the function.

5. Find the polynomial that passes through the points

x	-3	-2	-1	1	2	3
$f(x)$	4	1	0	0	1	4

Note that this appears to be an even function. Can you use this information to ease your computation?

6. Find a polynomial that passes through the points

x	-3	-2	-1	1	2	3
$f(x)$	-5	0	1	-1	0	5

Note that this function appears to be an odd function. Can you use this to ease your computation?

7. (a) If we desire a function of the form $f(x) = ae^{bx}$ show that we can express this relationship by the equation $\ln f(x) = bx + \ln a$. Note that this expression is linear in x.
   (b) Using your result of part (a) find the values of a and b such that $f(1.7) = 3$ and $f(2.5) = 2$.
   (c) Find $f(x)$ of the form given in part (a) which satisfies the requirements of part (b).

8. A certain store has sales of one item which average $s(t)$ dollars per day at time t. This particular item is seasonal in nature, and therefore we could expect that $s(t)$ would be of the form $s(t) = at + b \sin(\pi t/6) + c \cos(\pi t/6)$ where t is the number of months from the first of the year. If the average sales figures in thousands of dollars per day are given by the table

Time	January 1	April 1	July 1
Average sales	9	12.3	16

Find the function $s(t)$.

9. Measurements of the amount of a certain medication remaining in the blood stream t hours after ingestion give the following data.

t	0	1	2
Amount (mg)	18	12.6	9

Since this is expected to be a decaying quantity one might propose that the amount should be given by an equation of the form

$$A(t) = a(2^{-t/2}) + b(2^{-t}) + c(2^{-3t/2}).$$

Find the values of a, b, and c which would cause $A(t)$ to fit this data. Using your results what should be the amount in the blood stream 6 hours after ingestion?

10. (a) Find the cubic polynomial that passes through the data

x	1	2	4	5
y	4	3	1	0

   (b) Show that the polynomial $f(x) = x^4 - 12x^3 + 49x^2 - 79x + 45$ also passes through this data.
   (c) Subtract your result of part (a) from $f(x)$ of part (b) and determine the zeros of this difference.
   (d) Show how you could get other polynomials passing through the four points of part (a).
   (e) Show why you would expect only one polynomial of degree less than four to pass through these four points.

VIII.2   The Method of Lagrange

There are many techniques by which one can obtain a polynomial which passes through a given set of data points. Those which require that the x-values be located at equal intervals, or equivalently that the x-values be the points of a regular partition of the interval covered by the data, are generally simpler. However, since it is not our purpose here to engage in a full scale course in numerical analysis, we will settle for a single method which may be slightly longer in its application, but which is more inclusive in the number of cases which it will serve. This method is attributed to the French mathematician *J. L. Lagrange* (1736–1813) who worked in many areas of mathematics during the latter half of the 18th and the first years of the 19th century. It is a rather complete reversal of the method of Section VIII.1. There we had assumed the functions $f_i(x)$ were given and we determined the coefficients. In the method of Lagrange we assume that the coefficients are given and we try to find the functions $f_i(x)$. Does this sound strange? It is not quite as bad as it might be, for we will restrict our attention to polynomials. Furthermore, we are going to impose the condition that we wish the simplest possible result—namely the polynomial of lowest possible degree that will pass through the points in question.

In order to illustrate this technique, we again use the four points of the last section, but note carefully that with four data points we should expect a polynomial of at most degree three (one less than four). That this is true can be seen from our earlier work, for we note that there are four coefficients to be determined by the four pieces of information we have if we have a polynomial of degree one less than four. The coefficients of the first three powers of x together with the constant term give the four coefficients which we need to determine. Now we can express $f(x)$ as follows

$$f(x) = (1)f_1(x) + (0)f_2(x) + (-1)f_3(x) + (0)f_4(x).$$

We note that it is the four polynomials $f_i(x)$ that are now to be determined. The coefficients are the values of $f(x)$ which were given us initially. That is fine, provided we can manage to find $f_1(x)$ in such a way that $f_1(x)$ is zero at each of the data points except the first one, and likewise find $f_2(x)$ such that $f_2(x)$ is zero at each of the data points except the second one, etc. Thus, we are asking for a polynomial function such that $f_i(x)$ is zero at all data points except for the i-th data point, and that $f_i(x_i) = 1$ where x_i is the x-value of the i-th data point. At first glance this sounds impossible, but having gone this far, we shouldn't give up yet. To be specific, we note that we wish the polynomial $f_1(x)$ to obey the four stipulations $f_1(1) = f_1(2) = f_1(3) = 0$, and $f_1(0) = 1$. We know nothing at all about any points other than the data points. If $g(a) = 0$ and $g(x)$ is a polynomial, then $(x - a)$ is a factor of $g(x)$. That this is true should be clear when we consider that upon dividing $g(x)$ by $(x - a)$ we obtain a quotient $q(x)$ and a remainder, r. The remainder will be constant since we are dividing by a first degree divisor. We have $g(x) = (x - a)q(x) + r$. Since $g(a) = 0$ we have $g(a) = (a - a)q(a) + r$, and this can only be true if $r = 0$, whence $g(x) = (x - a)q(x)$, or $(x - a)$ is a factor of $g(x)$. This short diversion was to indicate that we now know that $(x - 1)(x - 2)(x - 3)$ must be a factor of $f_1(x)$. But now we already have a cubic polynomial and three is the degree we are hoping not to exceed. Therefore, we would satisfy the first three of the four requirements on $f_1(x)$ if we were to be able to write $f_1(x) = c(x - 1)(x - 2)(x - 3)$ where c is a constant still to be determined. But remember that we still have one more condition to satisfy, namely $f_1(0) = 1$. This leads to

$$f_1(0) = 1 = c(0 - 1)(0 - 2)(0 - 3) \quad \text{or} \quad 1 = -6c.$$

Therefore, we know that

$$f_1(x) = \frac{(x - 1)(x - 2)(x - 3)}{-6}.$$

In similar fashion we could observe that

$$f_2(x) = \frac{(x - 0)(x - 2)(x - 3)}{(1 - 0)(1 - 2)(1 - 3)} = \frac{x(x - 2)(x - 3)}{+2},$$

$$f_3(x) = \frac{(x - 0)(x - 1)(x - 3)}{(2 - 0)(2 - 1)(2 - 3)} = \frac{x(x - 1)(x - 3)}{-2},$$

and

$$f_4(x) = \frac{(x - 0)(x - 1)(x - 2)}{(3 - 0)(3 - 1)(3 - 2)} = \frac{x(x - 1)(x - 2)}{+6}.$$

Note the form for each of these polynomials. The numerator is a polynomial which becomes zero at each data point other than the i-th point, and the denominator is equal to the value the numerator assumes at the i-th point. This is rather ingenious when you stop to think about it. It means that for the i-th point we can be certain that the value of $f(x)$ will be precisely the value we wish, for $y_i f_i(x)$ is the only term in the expansion that is not extinguished. The polynomial coefficient at that point takes on the value one. If we complete our calculations in this case we obtain

$$f(x) = (1)\frac{x^3 - 6x^2 + 11x - 6}{-6} + (0)\frac{x^3 - 5x^2 + 6x}{+2}$$

$$+ (-1)\frac{x^3 - 4x^2 + 3x}{-2} + (0)\frac{x^3 - 3x^2 + 2x}{+6}$$

$$= x^3[(-1/6) + (1/2)] + x^2[(1) + (-2)] + x[(-11/6) + (3/2)] + [(1) + (0)]$$

$$= x^3(1/3) - x^2 - x/3 + 1.$$

This is one of the functions we obtained in Section VIII.1. Here we did not have to solve a large system of equations, although we did have to multiply a few linear factors. We have a procedure for obtaining the *undetermined polynomials*. We had to solve for the *undetermined coefficients* in the earlier approach.

It would be well now to go back and summarize what we have done using a slight generalization of notation. Assume that we have been given a set of data of the form (x_i, y_i) where the subscript i take on successively the values 1, 2, 3, ... as far as we may need to go. If the total number of them is n, then we should have a polynomial of degree $(n - 1)$. Let us define the *Lagrangian coefficient polynomial* by the relation

$$f_i(x) = \frac{\text{the product of the } (x - x_j) \text{ for each } j \text{ other than } i}{\text{the product of the } (x_i - x_j) \text{ for each } j \text{ other than } i}.$$

We have described the numerator and denominator rather than trying to write each of them out in some form of notation. It is clear that $f_i(x_j)$ will have the value zero if $i \neq j$, for one of the numerator factors will be zero. On the other hand if $i = j$, then the value of $f_i(x_j)$ will be one, for the numerator and denominator will be identical, and neither will be zero. The latter follows since the very fact that we can look for a function implies that no two of the x-values are equal. Now that we have defined the Lagrangian coefficient polynomial, we are in a position to obtain $f(x)$, the polynomial we were seeking in the first place, for we have

$$f(x) = \sum_{i=1}^{n} y_i f_i(x). \tag{VIII.2.1}$$

Note that if $x = x_j$ is any one of the data points, then all of the summands but one are zero, and that one has the value y_j. Consequently $f(x_j)$ has the value y_j and we have a polynomial that matches the data perfectly.

This is a fine method, and you notice that we have one with universal (?) application. It will work in theory for any set of data as long as no two of the x's are equal. As before, however, it is a moot point whether one would desire a polynomial of extremely high degree. For this reason, this method is seldom used when there are a large number of data points. It is useful if there are relatively few data points, and in instances in which we want a very good fit for the data in a small portion of the domain. We will be returning to this result in Chapter X as a basis for other results. All in all, while this method appeared to be rather strange when we were asked to find functions, it has turned out to be a rather nice one.

You may have observed in the problem we did at the beginning of the section that we did not have to evaluate $f_2(x)$ and $f_4(x)$ since they had zero coefficients. If so, you were alert. For purposes of illustration we thought it best not to take such short cuts the first time around. There are many times in mathematics when you can shorten the amount of work involved by noting that some value will have a zero coefficient.

If we are given only isolated data, we find that we can obtain a polynomial that will describe that data without error. There is the gnawing
question, however, whether the polynomial is the function we really wanted.
You will remember our earlier discussion to the effect that there are an
infinite number of possibilities for functions which fit the data, and the
selection of the polynomial of lowest degree is only one possibility. What
might the error be if we had made the wrong selection? Let us assume that
$F(x)$ is the correct function and that $f(x)$ is the polynomial that we obtained
by the method of Lagrange. If we could find the error, $E(x)$, such that $F(x) =
f(x) + E(x)$, we would know how far off we were in using $f(x)$. Unfortuna-
tely, it is not always an easy matter to find $E(x)$. (If we could find the error
we would have the exact function, thus eliminating the need for the
Lagrangian interpolation.) We do know that if our data is correct, then for
the data points we have $E(x_j) = F(x_j) - f(x_j) = 0$ since the correctness of
our result at the data points implies that $f(x)$ has the correct value at these
points. Hence, if we have data points $(x_1, f(x_1))$, $(x_2, f(x_2))$, $(x_3, f(x_3))$,
and $(x_4, f(x_4))$, we would have $E(x_1) = E(x_2) = E(x_3) = E(x_4) = 0$. Consequently $E(x)$ must be of the form

$$E(x) = a(x - x_1)(x - x_2)(x - x_3)(x - x_4)$$

where a may well depend upon the value of x. While we do not have sufficient
information to permit us to evaluate a in general, we can obtain some
insight into its value for a specific value of x. We start by noting that we can
now write $F(x)$ as

$$F(x) = f(x) + a(x - x_1)(x - x_2)(x - x_3)(x - x_4). \quad \text{(VIII.2.2)}$$

Since we will be trying to find out something about the error for a particular
value of x, say x_0, we are then trying to ascertain something about the
value of a which will make (VIII.2.2) correct when $x = x_0$. If a is the particular value such that

$$F(x_0) = f(x_0) + a(x_0 - x_1)(x_0 - x_2)(x_0 - x_3)(x_0 - x_4)$$

we know that the function $g(x)$ defined by

$$g(x) = F(x) - f(x) - a(x - x_1)(x - x_2)(x - x_3)(x - x_4) \quad \text{(VIII.2.3)}$$

is zero when $x = x_0, x_1, x_2, x_3$, or x_4. There is no point in considering the
case in which x_0 is one of the data points, and consequently we can assume
that $g(x)$ is zero at five distinct points. These five can be considered as a
partition having four subintervals, none of them being of zero length. By
Rolle's theorem we know that $g'(x)$ must be zero in the interior of each of
the four subintervals, and hence $g'(x)$ must be zero at four distinct points
in the interval determined by the five points x_0, x_1, x_2, x_3, and x_4. Therefore
by another application of Rolle's theorem the derivative of $g'(x)$, that is
$g''(x)$, must be zero at some point in the interior of each of three subintervals.
Following this line of argument $g'''(x)$ must be zero at two distinct points

and $g^{iv}(x)$ must be zero at one point in this interval. Let us call the point at which the fourth derivative is zero $x = c$. (This is not to say that the fourth derivative may not be zero at more than one point, but rather we are certain it is zero at one point. We can choose any one of the points at which $g^{iv}(x) = 0$ to be c.)

After this lengthy discussion we are ready to return to our examination of (VIII.2.3). Here we have

$$g^{iv}(c) = 0 = F^{iv}(c) - f^{iv}(c) - aD_x^4[(x - x_1)(x - x_2)(x - x_3)(x - x_4)].$$

Since $f(x)$ is cubic, the fourth derivative is zero. In the product of four linear factors we see that we have the fourth derivative of x^4 followed by some linear combination of the fourth derivatives of powers lower than the fourth power. $D_x^4(x^4) = 4! = 24$, and the other derivatives in this expression are zero. Therefore our fourth derivative reduces to

$$g^{iv}(c) = 0 = F^{iv}(c) - 24a. \qquad \text{(VIII.2.4)}$$

From this we have

$$a = \frac{1}{4!} F^{iv}(c).$$

Consequently the error term is given by

$$E(x_0) = \frac{1}{4!} F^{iv}(c)(x_0 - x_1)(x_0 - x_2)(x_0 - x_3)(x_0 - x_4). \quad \text{(VIII.2.5)}$$

Note that the value of c depends upon the value of x_0. Since we have no way of determining the particular value of c which is correct, we usually have to content ourselves with using the maximum value which $F^{iv}(x)$ assumes in the interval spanned by the data points and the point x_0, and then be satisfied that we have an upper bound for the error.

We could follow the same line of reasoning if we had n points instead of 4 points. In this case we would have the relation.

$$E(x) = \frac{1}{n!} F^{(n)}(c)\left[\prod_{j=1}^{n} (x - x_j)\right].$$

The capital letter *pi*, the Greek letter corresponding to *p*, the first letter of *product*, indicates the product in a manner similar to that in which sigma indicates sums.

As a final note, it is frequently necessary to evaluate $f(x)$ for a value of x other than one of the data points. We are continuing to use the notation $f(x)$ to represent the interpolated polynomial. In this case it may be easier to replace x in (VIII.2.1) by the specific value for which we want $f(x)$. If we are given the data poins (1, 3), (2, 4), and (3, 4) and asked to find the value of the function when $x = 1.5$, we could well write

$$f(1.5) = \frac{(1.5 - 2)(1.5 - 3)}{(1 - 2)(1 - 3)}(3) + \frac{(1.5 - 1)(1.5 - 3)}{(2 - 1)(2 - 3)}(4)$$

$$+ \frac{(1.5 - 1)(1.5 - 2)}{(3 - 1)(3 - 2)}(4)$$

$$= \frac{(-0.5)(-1.5)}{(-1)(-2)}(3) + \frac{(0.5)(-1.5)}{(1)(-1)}(4) + \frac{(0.5)(-0.5)}{(2)(1)}(4)$$

$$= 1.125 + 3.00 - 0.50 = 3.625.$$

Since we are using three points, this is equivalent to a quadratic expression, and hence we have used *quadratic interpolation*. If we had used two points, we would have been using linear interpolation, the kind that you have used in connection with tables of logarithms and trigonometric functions in the past. Interpolation may well be more accurate when it is of higher degree, but this will depend upon the data. Note that in this last case the error would be equal to

$$\frac{1}{3!}F'''(c)(1.5 - 1)(1.5 - 2)(1.5 - 3) = \frac{0.375}{6}F'''(c) = 0.0625F'''(c).$$

If we have reason to suspect that the third derivative of the function that we should be using is very small, then we can be rather happy about our result. Without some knowledge of the source of the problem, however, we have little knowledge of $F(x)$, and the error term may be helpful, but will not be as informative as we would like.

EXERCISES

1. (a) Find the cubic polynomial which passes through the data points

   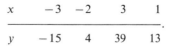

x	-3	4	-1	1
y	2	3	4	5

   (b) Show that you would have the same result if the first two data points were interchanged so that (4, 3) came before $(-3, 2)$.
   (c) Find an expression for the error in evaluating y when $x = 2$.

2. (a) Find the cubic polynomial which passes through the data points

x	-3	-2	3	1
y	-15	4	39	13

   (b) Check each of the data points to be certain that the polynomial obtained in part (a) does pass through each of the points.

3. (a) Find the quartic polynomial that passes through each of the following data points

x	3	1	-2	0	-1
y	3.25	-0.75	-3	1	-0.75

(b) Show that the sum of the Lagrangian coefficient polynomials is one in this case.
(c) If you had known that the sum would be one in part (b), show how you could have avoided the computation of one of the coefficient polynomials in part (a).
(d) Using the fact that the sum of the coefficient polynomials is one in this case, show how this could be used as a check on the accuracy of your computation.

4. (a) Find the quadratic polynomial which passes through the data points $(\pi/3, \cos(\pi/3))$, $(-\pi/3, \cos(-\pi/3))$, and $(0, \cos 0)$.
(b) Find $f(\pi/6)$ using the polynomial of part (a).
(c) Find the maximum error you could expect in evaluating $f(\pi/6)$ using your polynomial,
(d) Show that the actual error is less than the calculated maximum error in this case.
(e) The cosine has been shown to be an even function, and with the data points taken symmetrically about $x = 0$ we have even data. What affect does this have on the resulting polynomial?
(f) Earlier we had shown that $\cos x$ is approximated by the relation

$$\cos x = 1 - \frac{x^2}{2} + \frac{x^4}{24} - \cdots$$

Show that your polynomial of part (a) is greater than the first two terms of this expression but less than the first three terms.

5. According to the U.S. Census, the population of the United States during the period 1940–1970 is given by the following table:

year	1940	1950	1960	1970
population	131,669,275	150,697,361	179,323,175	203,235,298

(a) It is customary to use data of this type to the nearest million. Doing this for simplicity find a cubic polynomial which will give the population of the United States during this period. You may wish to further simplify the arithmetic by counting decades and letting 1940 be -1, 1950 be 0, 1960 be 1, and 1970 be 2. (You could start with 1940 equivalent to 0 or 1, but this would require larger numbers in the computation).
(b) Using your result of part (a) find the expected population in 1980.
(c) Find an expression for maximum error of calculation for 1980.
(d) Since the correct function is the actual population, can you make any estimates concerning the size of the fourth derivative in this time interval? (If you have used time in terms of decades you must remember to think of the fourth derivative in terms of this time unit.)

6. According to the Federal Power Commission the amount of electric energy produced
   in the United States during the period 1940–1970 is given by the following table:

year	1940	1950	1960	1970
energy (megawatts)	141,837,010	329,141,343	753,350,271	1,531,608,921

   (a) Find a polynomial which will give the amount of electric energy used as a func-
       tion of time.
   (b) How much of your work for Exercise 5 can be applied to this exercise?
   (c) Use your result of part (a) to find the anticipated power usage in 1980.
   (d) Find an expression for maximum error in your computation of part (c).

7. (a) Use the values of $\cos(\pi/6)$ and $\cos(\pi/4)$ with linear interpolation to find an
       approximate value of $\cos(2\pi/9)$.
   (b) Show that it makes no difference whether you use degrees or radians for this
       computation as long as the values of the cosine used are correct in each case.
   (c) Find $\cos(2\pi/9)$ by quadratic interpolation using the values of part (a) and also
       $\cos(\pi/3)$.
   (d) Indicate why it would be better to use $\cos(\pi/3)$ as a third value in the computa-
       tion of part (c) than to use cos 0.
   (e) Find the error in your computation using the method of this section and check
       it against the value of cos 40° found in the table.

8. (a) Using $x_1 = 0$, $x_2 = \pi/6$, and $x_3 = \pi/2$, find the quadratic polynomial which
       best approximates $f(x) = \sin x$.
   (b) Use the result of part (a) to approximate $f(5\pi/18)$.
   (c) Find the maximum error in the computation of part (b).
   (d) Use the result of part (a) to approximate $f(3\pi/5)$.
   (e) Find the maximum error in the computation of part (d).
   (f) Explain why the errors of parts (c) and (e) are not equal. Why would you expect
       one to be larger, and which one would you have expected to be larger?

9. (a) Find the quadratic polynomial which passes through the following points.

x	$-h$	0	h
y	y_1	y_2	y_3

   (b) Integrate this function over the interval $[-h, h]$. Observe that the value of the
       integral depends only on the values y_1, y_2, and y_3.
   (c) Let $y = x^2 + 3x - 4$ and $h = 1$. Use your values of y for $x = -h$, 0, and h as
       the values y_1, y_2, and y_3 of part (b). Substitute these values in the results of
       part (b) to approximate $\int_{-h}^{h} (x^2 + 3x - 4)dx$.
   (d) Integrate $\int_{-h}^{h} (x^2 + 3x - 4)dx$ and check the accuracy of your approximation of
       part (c).

10. (a) Find the quadratic polynomial which passes through the points

x	$a - h$	a	$a + h$
y	y_1	y_2	y_3

   (b) Integrate this function over the interval $[a - h, a + h]$.

(c) Show why the results of part (b) should equal the results of part (b) in Exercise 9.
(d) Use your results of part (b) to evaluate $\int_{a-h}^{a+h} (x^2 + 3x - 4)dx$ and check the accuracy of your approximation by integrating the actual integral.

11. (a) Use linear interpolation to obtain $\Gamma(1.09)$ with the data from Appendix C.
    (b) Use quadratic interpolation to obtain $\Gamma(1.09)$ using the data from Appendix C.. Which points should you use here?
    (c) Use cubic interpolation to obtain $\Gamma(1.09)$.
    (d) Compute the error in each of the three cases and determine which of the three results, parts (a), (b), and (c), should be most accurate.

12. (a) Let $f(x)$ be the constant function $f(1) = 1$. Then each value of $f(x)$ is one. Use Lagrangian interpolation to find a fourth degree polynomial which approximates $f(x)$ using $x_j = j$ for $j = 1, 2, 3, 4, 5$.
    (b) Show that the sum of the Lagrangian coefficient polynomials in this case is one.
    (c) Show that the sum of the Lagrangian coefficient polynomials for this $f(x)$ must be one regardless of the data points selected.
    (d) Show that the Lagrangian coefficient polynomials do not depend upon the values of $f(x)$.
    (e) Show that the sum of the Lagrangian coefficient polynomials must be one for each Lagrangian interpolation.
    (f) Show how the result of part (e) can be used as a check on the accuracy of your computation of the Lagrangian coefficient polynomials.

13. (a) If the data points are in order so that $x_i < x_j$ if $i < j$, show that the error term will be smaller if we seek $f(x_0)$ for x_0 near the middle of the interval $[x_1, x_n]$ than it would be if x_0 were near either end of the interval.
    (b) Show that the error of interpolation is apt to be less if the number of points, n, is large and the length of the interval $|x_n - x_1|$ is small.
    (c) Show that linear interpolation is best when the curvature is smallest, and therefore the curve is nearly a straight line.
    (d) Show that quadratic interpolation has no error if the function is a linear function or if its graph is a parabola with a vertical axis.

VIII.3 The *Almost* Function

We have hinted at difficulties in obtaining functions from data. We first indicated that you must make a choice of the type of function you wish to use. Then we mentioned the horrible possibility that you might have so many data points that the polynomial (if that is what you used) would be of very high degree, and the amount of work involved would be more than you would care to contemplate. In such a case it might be worth thinking of a function that did not go through each of the points (x_i, y_i), but rather went close to each one. This would sacrifice some accuracy but would reduce the complexity of the function a great deal. Although we would be more likely to apply such techniques in cases in which we have a great deal of data, we will use as an illustration a situation in which we have a small amount of data. This suggests that we use the data of Section VIII.1 again. In view of the fact that we can obtain a perfect approximation with a cubic function, perhaps we ought to practice our *almost* technique with a quadratic function, for it is apparent from Figure VIII.1 that it is not likely that a quadratic function would fit these four points exactly. We assume then that we wish to have $f(x)$ of the form $f(x) = ax^2 + bx + c$. Hence we would like to have

$$f(0) = \quad 1 = a(0)^2 + b(0) + c = \qquad\qquad c,$$
$$f(1) = \quad 0 = a(1)^2 + b(1) + c = \quad a + \quad b + c,$$
$$f(2) = -1 = a(2)^2 + b(2) + c = 4a + 2b + c,$$
$$f(3) = \quad 0 = a(3)^2 + b(3) + c = 9a + 3b + c.$$

A little work will show that there are no three numbers a, b, and c such that all four of these equations are true for the same set of three values. Following the hints given above, we could then suggest that we want the equalities to be *almost* correct. In other words, we might wish to have some combination of

$$a(0)^2 + b(0) + c - f(0) = \qquad\qquad c - f(0),$$
$$a(1)^2 + b(1) + c - f(1) = \quad a + \quad b + c - f(1),$$
$$a(2)^2 + b(2) + c - f(2) = 4a + 2b + c - f(2),$$

and

$$a(3)^2 + b(3) + c - f(3) = 9a + 3b + c - f(3),$$

be as small as possible. The question then is what combination should we use. The sum of the four expressions is not appealing, for we could make the sum zero (or even negative) but still have the individual summands rather large. This follows since some could be large in the positive direction, thus negating others which could be large in the negative direction. Perhaps we should use the sum of the absolute values. But then arises the question of how we should make the sum of the absolute values small. Remember that we were able to get minimum values by finding derivatives (a method which might possibly work here). The derivative of the absolute value is not bad most of the time, but at the point where the absolute value is zero we have a point for which no derivative exists. Perhaps this method wouldn't work well after all. At least the sum of the absolute values had the virtue that each contribution was positive and therefore no one value could *cancel* another value. If we pursue this further, we think of the squares of these expressions, for the squares of real numbers cannot be negative (hence no cancelling out) and we know how to differentiate the squares. We will try making the sum of the squares of the four expressions a minimum.

Now that we have decided to use the squares we will continue and see whether other problems develop. To do things in the obvious way, although not necessarily in the easiest way, we wish to find the values of a, b, and c such that

$$[c - f(0)]^2 + [a + b + c - f(1)]^2 + [4a + 2b + c - f(2)]^2$$
$$+ [9a + 3b + c - f(3)]^2$$

has the smallest possible value. Upon multiplication and combination of terms we then wish to minimize

$$H = 98a^2 + 14b^2 + 4c^2 + 72ab + 28ac + 12bc + 8a + 4b + 2,$$

(VIII.3.1)

where we have replaced $f(0), f(1), f(2)$, and $f(3)$ by $1, 0, -1$, and 0 respectively. (Later on we will find an easier way out which will help in case we have more data points. For the moment this was at least feasible, if not pleasant.) We now face the question of making this a minimum. The difficulty arises in that we have three variables here, and our previous work concerned[*] functions that depend on only one variable. We wish to emphasize at this point that we are not attempting to be completely rigorous in developing a method for obtaining a minimum here, but the methods employed are correct, and the reasoning needs only a little bolstering to answer the queries of the doubting. Let us suppose that we have determined values for b and for c that we wish to stick with, and it is only a that can vary. "Voilá" as they would say in Paris! We now have a function in which only one variable is actually a variable and we have for the moment fixed the other variables (now only so-called variables for they are constant by our assumption that we have determined values for them). Hence we can apply our previous work relating to finding minima. We have only to differentiate (VIII.3.1) with respect to a, and determine the value of a that would make this a minimum. Since for this differentiation we would have b and c as constants, they will be so treated in obtaining the result. Hence we have

$$\frac{\partial H}{\partial a} = 196a + 72b + 28c + 8.$$

Notice the funny things that we have used in the symbol for this derivative. They only faintly resemble the dH/da we would use if we were to utilize the notation which indicates that the derivative is the quotient of the differentials. These strange looking things, called *rounds*, are used to indicate that we have played tricks on our original function and have decided that although the function, H, is dependent upon three variables we will hold two of them fixed, and only allow one of them, namely a, to be a genuine variable for this particular operation. Such a derivative is called a *partial derivative*. It does not indicate the total rate of change in the function, but indicates the change that is possible as a single variable, in this case a, is permitted to vary.

So far we seem to have overcome in some way most of the difficulties, but there are some things that are still nagging us. For instance, we said that we would consider that we had determined values for b and c. But had we? If so, what values were they to assume in order that H would have a minimum value? Also, we obtain a minimum by setting the derivative equal to zero and finding the value of the variable, in this case a, which makes the derivative zero. This is, of course, not necessarily going to yield a minimum point, but it will certainly indicate critical points. We have no worries about some of the other problems that might have occurred, since H is a polynomial as far as a is concerned, and polynomials have nice continuous derivatives. The best we can do here is obtain a in terms of b and c. But perhaps we could then find partial derivatives of H with respect to b and with respect to c and determine b in terms of a and c, and determine c in terms of a and b. This would at least give us three relations in three unknowns, and this is better than nothing. We now have

$$\frac{\partial H}{\partial a} = 196a + 72b + 28c + 8,$$

$$\frac{\partial H}{\partial b} = 28b + 72a + 12c + 4,$$

and

$$\frac{\partial H}{\partial c} = 8c + 28a + 12b.$$

Our reasoning thus far would indicate that each of these ought to be zero to insure the best possible choice for a in terms of b and c, of b in terms of a and c, and of c in terms of a and b. Thus, we would solve the system

$$196a + 72b + 28c = -8$$
$$72a + 28b + 12c = -4$$
$$28a + 12b + 8c = 0.$$

Solving this system we obtain $a = 0.5$, $b = -1.9$, and $c = 1.1$. Hence we would have the function

$$f(x) = 0.5x^2 - 1.9x + 1.1.$$

We have not investigated to determine whether these points actually do yield a minimum for H. This is a much more complex task when there are several variables. However, we note that we have only one candidate, and that the sum of squares can grow and grow, whence it is not likely to have a maximum. Since it cannot be negative it is bounded below and therefore must have a minimum. Here we have a case in which we resort to the origin of the mathematical problem to find a rationale for our decision that we have a minimum.

Now that we have this function, we ought to investigate whether it provides a satisfactory approximation to our data. We can perform straightforward calculations and observe that $f(0) = 1.1$ instead of 1, that $f(1) = -0.3$ instead of 0, that $f(2) = -0.7$ instead of -1, and that $f(3) = -0.1$ instead of 0. When we consider that we are trying to place a parabola through four points, perhaps this is not so bad. We do note that the variations are respectively $+0.1$, -0.3, $+0.3$ and -0.1 from the values we desired. The sum of these errors is zero, which is a good sign. The sum of the squares of these errors is only 0.2, and that could be worse. In fact, this is the best fit that we can get for this data in the sense that we have made the sum of the squares a minimum as shown in Exercise 1. It is known as the *least squares fit*. The function $f(x)$ in this case is known also as the *regression function*. This does not indicate that the function has regressive attributes in the usual meaning of the word regressive.

The regression curve is often preferred to the result obtained by interpolation in cases in which a large amount of data is involved. Since there are frequently slight errors in data, the regression curve tends to smooth out these discrepancies and give a curve which may well bear much more resemblance to the correct curve than the interpolated curve. If, for instance, we had 50 data points and desired a polynomial through the data points, the polynomial would probably be of degree 49 if we were to use interpolation. This is rather obviously far too complicated for most applications. A quadratic or cubic polynomial might well remove some of the minor errors that frequently creep into data at the time such data is collected. Therefore, in practice you can expect to use the method of regression the vast majority of the time.

We have mentioned many things in this section, and have even indicated that there are methods for doing some of the steps involved in obtaining the regression function in an easier manner. These items merit more discussion, and consequently we will devote the next sections to elaboration of the work we have begun here.

EXERCISES

1. (a) Show that H in (VIII.3.1) can be written $(7a + 3b + 2c)^2 + 5(3a + b + 0.4)^2 + (2a - 1)^2 + 0.2$. [Note: The decomposition of H into a sum of squares has one less variable in each succeeding expression.]
   (b) Show that this is always at least as large as 0.2.
   (c) Also, show that this has a minimum if and only if the equations

$$7a + 3b + 2c \quad\quad = 0$$
$$3a + b \quad\quad + 0.4 = 0$$
$$2a \quad\quad\quad - 1.0 = 0$$

   are satisfied.
   (d) Find the solution of this system of equations and check with the solution found in the text.

2. (a) Using the data

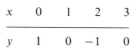

x	0	1	2	3
y	1	0	−1	0

find the linear function $f(x) = ax + b$ which best fits this data in the least squares sense.

(b) Find the sum of the deviations of the given function and the computed values for the four data points.

(c) Find the sum of squares of the deviations of the given function from the computed values for the four data points.

(d) Compare this sum of squares with 0.2, the sum of squares obtained using the quadratic function. Explain why one is larger. Explain which one should be larger.

(e) Sketch the data and also the line whose equation is obtained in part (a).

3. (a) Using the data of Exercise 2 find the best cubic approximation for this data.

(b) Compare your result with the results found in Section VIII.1. Explain the results of your comparison.

4. (a) Find the line of regression for the data

x	1	2	4	5
y	7	5	3	3

(b) Sketch the line of part (a) and plot each of the data points.

(c) Mark the vertical segments joining the points and the line in your sketch.

(d) Show that the lengths of the vertical segments marked off in part (c) are the deviations between the calculated values for each value of x and the given value.

(e) Find the sum of the deviations of part (d).

5. (a) Find the quadratic curve of regression for the following data

x	−3	4	−2	0
y	16	5	13	2

(b) Find the sum of the deviations between the given data points and the values calculated using the results of part (a).

(c) Plot the points and sketch the curve on the same axes. Does this sketch look reasonable as an approximation to the data?

6. (a) Using the data

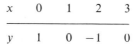

x	0	1	2	3
y	1	0	−1	0

find the best possible approximation for the data if $f(x)$ is to have the form $f(x) = a + bx + (c/(x + 1))$.

(b) Find the sum of the deviations between the actual and the calculated data for the four given data points of part (a).

(c) Is the sum of the squares of the deviations smaller than that obtained in Exercise 2 for the same data?

7. (a) Find the quadratic regression function which approximates $f(x) = \sin x$ using the data points $x = 0$, $x = \pi/6$, $x = \pi/4$, and $x = \pi/3$.
   (b) Using the results of part (a) find the value of $\sin(\pi/12)$ and compare your results with the value in the tables of Appendix C.
   (c) If you replace x in the quadratic equation of part (a) by $\pi y/180$ show that you have the corresponding quadratic expression with y measured in degrees.

8. (a) Show that the equation $f(x) = ae^{bx}$ is equivalent to the equation $\ln f(x) = \ln a + b^x$. (This was also considered in Exercise VIII.1.7).
   (b) Using the result of part (a) show that we can obtain the exponential function by linear regression if we let the desired function be $(\ln f(x))$, the constant term $(\ln a)$, and the coefficient of the first degree term (b).
   (c) Using this approach, find the exponential function which best approximates the data

x	1	2	4
y	3	8	80

   in the least squares sense.
   (d) Using the exponential relation obtained in part (c) calculate the deviations between the calculated and given data and then determine their sum.
   (e) Explain why the sum of the deviations is likely to be other than zero in this case.

9. (a) A more general form of the exponential equation is given by the relation $f(x) = ae^{bx+c}$. Show that this is equivalent to the relation of Exercise 8 where $\ln a$ is replaced by $(c + \ln a)$.
   (b) Show that the use of the equation of part (a) is equivalent to the use of the equation of Exercise 8 with a different choice for the coefficient a.
   (c) Use regression to find the best exponential equation to explain the data

x	0	1	3	4
y	20	7.4	1	0.4

   (d) Sketch the exponential curve and plot the given data points. Also find the sum of the squares of the deviations between the given values for y and the calculated values for y. Indicate whether you think this curve is a good fit for the data.

10. (a) Given n data points, show that it is possible to find a polynomial of degree $(n - 1)$ by regression for which the deviations at the data points will all be zero.
    (b) Show that if you are given n data points and want a polynomial of degree n there will be an infinite number of choices for which the deviations will all be zero.
    (c) If you have n data points and try to find a polynomial of degree n by regression, show why you can expect to have a system of $(n + 1)$ equations in $(n + 1)$ unknowns for which no unique solution exists.
    (d) Illustrate the results of parts (b) and (c) by trying to find a quadratic regression curve for the data

x	2	5
y	7	6

11. Find the partial derivatives $\partial z/\partial x$ and $\partial z/\partial y$ in each of the following cases:
    (a) $z = x^2 + y^2$
    (b) $z = (x + y)^2$
    (c) $z = x^3 + 3x^2y + y + 3xy^2 - y^3$
    (d) $z = (x^2 + xy + y^2)^2$
    (e) $z = \sin(xy)$
    (f) $z = e^{xy}$
    (g) $z = \sin(x + y)$
    (h) $z = \tan^3(2x - 3y)$

12. (a) Find $x(\partial z/\partial x) + y(\partial z/\partial y)$ if $z = x^2 - 4xy - 8y^2$.
    (b) Show that your result in part (a) is $2z$.
    (c) Note that each term of z is of degree two since xy is the product of two first degree terms.
    (d) Show that $x(\partial z/\partial x) + y(\partial z/\partial y) = 3z$ if $z = x^3 + 3xy^2 - 4y^3$.

13. Find the partial derivatives $\partial f/\partial x$, $\partial f/\partial y$, and $\partial f/\partial z$ for each of the following:
    (a) $f(x) = xy + xz + yz$
    (b) $f(x) = x^2 + y^2 + z^2 - xyz$
    (c) $f(x) = \sin(xyz) + x(e^y)(\ln z)$
    (d) $f(x) = (xy + xz + yz)^3$

VIII.4 Regression Functions

We will now develop a more efficient organization of the work required to obtain the regression function. While there are still some things that need to be said about partial derivatives, we will take what we need on faith in this section. There will be more discussion of partial derivatives in the next section. In this way we will not interrupt our train of thought.

In Section VIII.3 we were able to find the quadratic regression polynomial for the four points with which we started, but we were more than thankful that we did not have more than four points and that the polynomial was of no higher degree. We were doing just what was indicated—computing the sum of the squares of the differences between the calculated values and the given values, and then minimizing this sum of squares. Now let us take a look at this development from a more sophisticated angle and see whether we can't find an easier method. We have found that it sometimes helps to have a little more theory on which to proceed, for then results seem to come more easily. In fact, it might help if we considered the case for something more general than a polynomial. We can think of the polynomial situation as being a special case.

We illustrate the development by considering that we wish to express the function (as accurately as possible) in the form

$$f(x) = a_1 f_1(x) + a_2 f_2(x) + a_3 f_3(x) = \sum_{k=1}^{3} a_k f_k(x).$$

Such a form is a *linear combination* of the three functions $f_1(x)$, $f_2(x)$ and $f_3(x)$. (It would not be difficult to see how we would proceed if we were to desire more than three such functions, or perhaps less than three.) If we are given n points (x_j, y_j), $(j = 1, 2, 3, \dots, n)$ and we assume that $n > 3$, we will need to minimize the sum of the squares of the n terms knowing that each term is of the form

$$[a_1 f_1(x_j) + a_2 f_2(x_j) + a_3 f_3(x_j) - y_j] = \left[\sum_{k=1}^{3} a_k f_k(x_j) - y_j\right].$$

(While it is not difficult to write out the expression with three functions the sigma form would make it easier to follow if more than three such functions were present.) This means that we wish to minimize the function

$$H = \sum_{j=1}^{n} [a_1 f_1(x_j) + a_2 f_2(x_j) + a_3 f_3(x_j) - y_j]^2 = \sum_{j=1}^{n} \left[\sum_{k=1}^{3} a_k f_k(x_j) - y_j\right]^2.$$

This is the same type of expression we had before. So far there is no difference between what we did in Section VIII.3 and what we are doing here. Remember that we multiplied this all out in Section VIII.3, and the multiplication was the step that required much of the work. If partial derivatives behave as we would like to have them behave, we know that the derivative of a sum is the sum of the derivatives, and we can differentiate each of the n summands separately and add the results. This may not seem to be helpful, but let us go ahead. If we differentiate with respect to a_1 and for this differentiation think of a_2 and a_3 as being constants, we will have

$$\frac{\partial H}{\partial a_1} = \sum_{j=1}^{n} 2[a_1 f_1(x_j) + a_2 f_2(x_j) + a_3 f_3(x_j)] \frac{\partial}{\partial a_1} [a_1 f_1(x_j)$$
$$+ a_2 f_2(x_j) + a_3 f_3(x_j)]$$

$$= \sum_{j=1}^{n} 2[a_1 f_1(x_j) + a_2 f_2(x_j) + a_3 f_3(x_j) - y_j] f_1(x_j)$$

$$= 2 \sum_{j=1}^{n} [a_1 (f_1(x_j))^2 + a_2 f_1(x_j) f_2(x_j) + a_3 f_1(x_j) f_3(x_j) - y_j f_1(x_j)].$$

The summation is with respect to the points and in view of the fact that we have a finite number of terms, we can write

$$\frac{\partial H}{\partial a_1} = 2\left[a_1 \sum_{j=1}^{n} (f_1(x_j))^2 + a_2 \sum_{j=1}^{n} f_1(x_j) f_2(x_j)\right.$$
$$\left. + a_3 \sum_{j=1}^{n} f_1(x_j) f_3(x_j) - \sum_{j=1}^{n} y_j f_1(x_j)\right].$$

The first of our three equations to be used for solving for a_1, a_2, and a_3 is

$$a_1 \sum_{j=1}^{n} (f_1(x_j))^2 + a_2 \sum_{j=1}^{n} f_1(x_j) f_2(x_j) + a_3 \sum_{j=1}^{n} f_1(x_j) f_3(x_j) = \sum_{j=1}^{n} y_j f_1(x_j).$$

We obtain this equation, of course, by using the same logic we used before, namely we cannot have a minimum value for H with respect to the choice of a_1 unless $\partial H/\partial a_1 = 0$. In the process of arriving at the equation, we have divided by two and we have moved the term which does not involve any of the a's to the right side of the equal sign.

Before we go further, let us examine what we have for any trends that may be developing. We note that we obtained this equation by differentiating with respect to a_1, and that we had a factor $f_1(x_j)$ occuring in each of the sums in addition to the factor $f_i(x_j)$ that was already there as a multiplier of a_i. (We are aware of the awkward fact that we have a factor *in addition to*, but such is the way the English language seems to work, and we are breaking new ground in mathematics but not in English in this book.) This thought concerning the formation of the equation we have so far obtained leads us to suspect that the second and third equations might appear as follows:

$$a_1 \sum_{j=1}^{n} f_2(x_j)f_1(x_j) + a_2 \sum_{j=1}^{n} (f_2(x_j))^2 + a_3 \sum_{j=1}^{n} f_2(x_j)f_3(x_j) = \sum_{j=1}^{n} y_j f_2(x_j)$$

and

$$a_1 \sum_{j=1}^{n} f_3(x_j)f_1(x_j) + a_2 \sum_{j=1}^{n} f_3(x_j)f_2(x_j) + a_3 \sum_{j=1}^{n} (f_3(x_j))^2 = \sum_{j=1}^{n} y_j f_3(x_j).$$

If you carry through the necessary manipulations, you will find that these are correct. This does not solve the system of equations, of course, but it certainly helps in obtaining it. We might even make a table to help keep the arithmetic straight. If we go back to the data we were using in the last section we could set up the following table. (Remember there we used $f(x) = a_1 x^2 + a_2 x + a_3$ and consequently $f_1(x) = x^2, f_2(x) = x$, and $f_3(x) = 1$.)

x	y	f_1	f_2	f_3	f_1^2	$f_1 f_2$	$f_1 f_3$	f_2^2	$f_2 f_3$	f_3^2	$y f_1$	$y f_2$	$y f_3$
0	1	0	0	1	0	0	0	0	0	1	0	0	1
1	0	1	1	1	1	1	1	1	1	1	0	0	0
2	−1	4	2	1	16	8	4	4	2	1	−4	−2	−1
3	0	9	3	1	81	27	9	9	3	1	0	0	0
					98	36	14	14	6	4	−4	−2	0

Now our system of equations can be written down easily. We have

$$98a_1 + 36a_2 + 14a_3 = -4$$
$$36a_1 + 14a_2 + 6a_3 = -2$$
$$14a_1 + 6a_2 + 4a_3 = 0.$$

Except for the fact that we have divided each equation by two, this is precisely the system of equations we obtained before. For the purpose of our table it helps to simply compute each of the functional values $f_k(x_j)$, for we need these anyway, and then we do not have to compute any one of them more than one time. You will note that several of the sums we obtained are used for more than one coefficient. Consequently we have found quite a labor saving device. There is no reason why the functions had to be polynomials in this case (although the majority of people prefer polynomials because they seem to be somewhat easier to manipulate). We could let

$$f_1(x) = \sin\frac{\pi x}{2}, \; f_2(x) = \cos\frac{\pi x}{2}, \;\; \text{and} \;\; f_3(x) = 1.$$

In this case we would have had the following.

x	y	f_1	f_2	f_3	f_1^2	f_1f_2	f_1f_3	f_2^2	f_2f_3	f_3^2	yf_1	yf_2	yf_3
0	1	0	1	1	0	0	0	1	1	1	0	1	1
1	0	1	0	1	1	0	1	0	0	1	0	0	0
2	-1	0	-1	1	0	0	0	1	-1	1	0	1	-1
3	0	-1	0	1	1	0	-1	0	0	1	0	0	0
					2	0	0	2	0	4	0	2	0

The corresponding system of equations becomes

$$2a_1 + 0a_2 + 0a_3 = 0,$$
$$0a_1 + 2a_2 + 0a_3 = 2,$$
$$0a_1 + 0a_2 + 4a_3 = 0.$$

Here the solution is $a_1 = 0$, $a_2 = 1$, and $a_3 = 0$. Some arithmetic shows that this particular choice, namely

$$f(x) = (0)\sin\frac{\pi x}{2} + (1)\cos\frac{\pi x}{2} + (0) = \cos\frac{\pi x}{2},$$

fits exactly, and is one of the choices we displayed in Section VIII.1. In this case since the fit is perfect the sum of the squares is zero. Since the sum of the squares is made a minimum and since it cannot be negative, if there is an exact fit available with the functions we have proposed, we will find it by this method.

There is one thing we should watch out for (isn't there always?). We might have selected $f_1(x) = \sin x$, $f_2(x) = \cos x$, and $f_3(x) = \sin(x + (\pi/4))$. If we did this, we would in reality have had only two functions instead of the three that we thought we had. Since

$$\sin\left(x + \frac{\pi}{4}\right) = \sqrt{(1/2)}\sin x + \sqrt{(1/2)}\cos x,$$

we observe that the inclusion of $f_3(x)$ did not expand our repertoire for $f_3(x)$ can be expressed as a linear combination of $f_1(x)$ and $f_2(x)$. In this case we would end up with an indeterminate system of equations, and there would be no unique solution available to us.

Now that we have a rather formal method for obtaining the regression function, a few comments are in order. First of all, we have very carefully not indicated what kind of functions the f's must be. Their selection would usually be dependent on the source of the data. We are not even differentiating the f's, and consequently they do not have to have the properties required of differentiable functions. They do, of course, have to have values for each of the data points in order that we can perform our calculations, and it is presumed that they should have other values as well or there would seem to be little point in all of this work. As a rule, the functions chosen will be differentiable, for then we can utilize the calculus. Although we developed the table for the case of three functions, it is not hard to see that this same development would work for any number of functions. You should remember, of course, that if the number of functions is as great as the number of data points, and if the functions selected are independent in that no expression of a linear nature can be found that will connect them, the resulting regression function will pass through all of the data points, and we no longer have just an approximation of the data. We could have used this method in place of the Lagrangian method in Section VIII.2.

Since the case in which $f(x) = a_1 x + a_2$ is used so frequently, we will give a specific solution for this linear case. Here $f_1(x) = x$ and $f_2(x) = 1$. Consequently the equations for obtaining the coefficients are

$$a_1 \sum_{j=1}^{n} x_j^2 + a_2 \sum_{j=1}^{n} x_j = \sum_{j=1}^{n} x_j y_j$$

$$a_1 \sum_{j=1}^{n} x_j + n a_2 = \sum_{j=1}^{n} y_j.$$

Since n terms are to be added in each case, and since the sum of n ones is n, the n appears in the second of these two equations. If we solve these two equations for a_1 and a_2, we get

$$a_1 = \frac{n\left(\sum_{j=1}^{n} x_j y_j\right) - \left(\sum_{j=1}^{n} x_j\right)\left(\sum_{j=1}^{n} y_j\right)}{n\left(\sum_{j=1}^{n} x_j^2\right) - \left(\sum_{j=1}^{n} x_j\right)^2}$$

$$\text{(VIII.4.1)}$$

$$a_2 = \frac{n\left(\sum_{j=1}^{n} x_j^2\right)\left(\sum_{j=1}^{n} y_j\right) - \left(\sum_{j=1}^{n} x_j\right)\left(\sum_{j=1}^{n} x_j y_j\right)}{n\left(\sum_{j=1}^{n} x_j^2\right) - \left(\sum_{j=1}^{n} x_j\right)^2}.$$

The line $f(x) = a_1 x + a_2$ obtained using these values is the *line of regression*. In view of the fact that this is the simplest case of regression that is likely to have any meaning, this is the case which is used most frequently. In fact, it is used so frequently that when people talk about *regression* or the *curve of regression* without being more specific, they usually mean the line of regression.

EXAMPLE 4.1. Find the line of regression for the data

x	1	2	4	5	6	8
$f(x)$	7	6	5	4	3	2

Solution. In this case $n = 6$ and we can calculate the values of the various summations required in (VIII.4.1). These are

$$\sum_{j=1}^{6} x_j = 1 + 2 + 4 + 5 + 6 + 8 = 26,$$

$$\sum_{j=1}^{6} x_j^2 = 1 + 4 + 16 + 25 + 36 + 64 = 146,$$

$$\sum_{j=1}^{6} y_j = 7 + 6 + 5 + 4 + 3 + 2 = 27,$$

$$\sum_{j=1}^{6} x_j y_j = 7 + 12 + 20 + 20 + 18 + 16 = 93.$$

Therefore

$$a_1 = \frac{(6)(93) - (26)(27)}{(6)(146) - (26)^2} = \frac{-144}{200} = -0.72.$$

Similarly

$$a_2 = \frac{(146)(27) - (26)(93)}{200} = \frac{1524}{200} = 7.62.$$

Note that we have used the fact that the two denominators are the same, and hence we did not have to recompute the denominator for the evaluation of a_2. We now have the line $y = (-0.72)x + 7.62$. This line is sketched in Figure VIII.2. This line would give the computed data

x	1	2	4	5	6	8
y	6.9	6.18	4.74	4.02	3.3	1.86

The discrepancies between the given data and the computed data are (-0.1), $(+0.18)$, (-0.26), $(+0.02)$, $(+0.3)$, and (-0.14) respectively. The sum of these discrepancies is zero, and therefore we can suspect that the computation was done without error. The fact that the sum of the discrepancies is zero does not guarantee accuracy, but it is one of the checks that you can use.

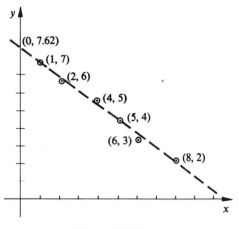

Figure VIII.2

The method we have given in this section is particularly well suited to use on computers, for the computer can easily provide the necessary summations, and then the solution of the resulting system of simultaneous linear equations. If you will be working with substantial amounts of data, you may very well wish to write a single program that could be useful on each of the sets of data that you may have in the future.

EXERCISES

1. (a) Given the following table find the line of regression.

x	-2	-1	0	3	2	5
y	-3	-2	-2	6	2	22

   (b) Using the data of part (a) find the quadratic regression curve.

2. Use the method of regression to fit a curve of the type $f(x) = a + b \sin x + c \cos x$ to the data

x	$-\pi/2$	0	$\pi/4$	$2\pi/3$
$f(x)$	1	2	1.3	3.9

3. (a) Find a line of regression for the data

x	-9	-7	-5	-1	0
y	125	70	30	-1	-1

   (b) Find a quadratic regression curve which best fits this data.

4. (a) Find the line of regression corresponding to the data

x	1	2.5	4	6	8	9.5
y	5	22	45	95	160	220

(b) Find the quadratic curve of regression corresponding to this data.

5. (a) Find the line of regression corresponding to the data

x	-2	-1.5	-1	-0.5	0
y	9	3	-1	-3	-3

(b) Find the quadratic curve of regression corresponding to this data.

6. (a) Find the line of regression corresponding to the data

x	-2	-1	0	2	3	4	6.5	10
y	-2	1	4	10	13	16	24	35

(b) Find the quadratic curve of regression corresponding to this data.

S7. The Bureau of Labor Statistics of the U.S. Department of Labor has given the following data on production workers in major industry.

Year	1960	1965	1970	1972
Workers	16,796,000	18,062,000	19,349,000	18,933,000

(a) Find the line of regression for this data.
(b) Find the quadratic curve of regression for this data.
(c) Find the best exponential curve of regression. (That is $f(x) = ae^{bt}$.)
(d) If you had no later data than that given in this exercise and you wished to run for re-election in 1980, which of these curves would you use to paint the best possible picture?
(e) If you had no later data than that given in this exercise and you wished to run against the incumbent which of these curves would you use?
(f) If you were either candidate, how would you explain the discrepancy between the figures you used and those used by your opponent?

C8. (a) Write a program for the computer that will obtain the curve of regression using as many as 100 points and as many as five independent functions. If function statements are used it should be relatively easy to modify the program from one run to the next to change the independent functions used, and it should also be possible to change the number of independent functions used.
(b) Use your program to obtain the fourth degree polynomial curve of regression using for data the values of sin x at one degree intervals from 0° to 90°.
(c) Use your program and the population of the United States taken in each decennial census since 1790 to find the exponential curve of regression for the population as a function of time.
(d) Use your result of part (c) to predict the population in 1980 and in 2000.
(e) Sketch your exponential curve and plot the actual population figures on the same set of axes. Use your knowledge of history to explain the major varia-

tions between the expected exponential growth and the actual population at specific times periods. Your explanation might involve the fact that the United States was in a recession in 1893 or other facts that could have an affect on population growth.

BC9. A culture of drosophila was started and an estimate was made of the number of individuals in the culture at two day intervals. The data for the first 38 days is given in the following table.

Age of culture	Number of Drosophila	Age of culture	Number of Drosophila
0	20	20	2599
2	33	22	3013
4	55	24	3268
6	94	26	3434
8	155	28	3514
10	259	30	3532
12	431	32	3545
14	720	34	3559
16	1202	36	3562
18	2009	38	3562

(a) The data for the first 18 days fit the exponential curve $P(t) = ae^{bt}$ fairly well. Find this curve as a curve of regression using the data from the first 18 days.

(b) The data for the last 12 days fit the curve $P(t) = c - fe^{gt}$ fairly well where c, f, and g are constants. Since c is the limiting population, it would appear from the data that c is about 3562 or very slightly larger. Find the values of f and g using regression techniques for $c = 3562, 3563, 3564,$ and 3565.

(c) Check the sum of squares of the deviations for each of the four curves you obtain in part (b) and determine which one appears to be best. (This type of estimation is not unusual in situations involving an upper population limit.)

C10. Using your program of Exercise 8 find the fourth degree curve of regression approximating $\cos x$ using not less than fifty points selected from the interval $[-\pi/2, \pi/2]$.

C11. (a) Find the quadratic curve of regression for the data

x	y	x	y	x	y
-9.5	-2	-2.5	-6	5	33
-9	-3.5	-2	-5	5.5	37
-8.5	-5	-1.5	-3.5	6	41.5
-7.5	-7	-1	-2	7	51
-6	-9	0	2	7.5	56
-5	-9	0.5	4	8	61
-4	-8.2	2	12	8.5	67
-3	-7	3.5	22	10	84

(b) Plot the points of part (a).

(c) On the same set of axes graph carefully the quadratic result of part (a).

(d) Indicate the *smoothing effect* of regression and show how it provides the essential nature of the curve without including all of the small fluctuations in the data.

S12. A company has obtained a great deal of data relating the price of a given commodity to the demand at that price and to the willingness of producers to supply items at that price. This data is given in the table below.

price	demand	supply
$10.00	2000	100
11.00	1900	250
12.00	1750	450
14.00	1000	1100
16.00	450	2100

(a) Plot the data and make a guess as to the type of curve which best fits the demand function and the type of curve which best fits the supply function.

(b) Use regression to obtain the best linear curve in the demand and supply situations.

(c) Plot the linear curves and indicate how well you think they fit the data. Would other curves have been significantly better?

(d) Using your results of part (b), find the price of equilibrium under pure competition.

(e) Find the consumer's surplus and the producer's surplus.

VIII.5 Partial Derivatives

We made no pretense of being rigorous in our introduction of partial derivatives earlier in this chapter, nor do we intend to go into the many details that would be involved in filling in all of the holes here. In order to simplify our discussion further we will assume all partial derivatives to be continuous in this section and the next. Partial derivatives have wide spread use, however, and it is well to stop and take stock of what they appear to be on the basis of what we have said so far. Geometrically we might think of them in the following way. Let $z = f(x, y)$ be a function of both x and y. If we try to draw a graph, we would have a three dimensional graph of the type shown in Figure VIII.3. Note that the domain is the horizontal plane, and the range is the vertical axis. For each point P_0 on the plane (a point in the domain) we have a number in the range which indicates the height of the corresponding point P above or below the horizontal coordinate plane. If we try to differentiate $f(x, y)$, we find a dilemma at once, for we are asking for the slope of the surface formed, and it is not obvious how one would obtain such a

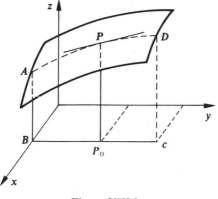

Figure VIII.3

slope at this point. We could, of course, think of placing a vertical plane through the point at which we wish the derivative, and this plane would cut the surface in a curve. We could then go back to our former work and talk about the slope of that curve. This is possible, but we will not do this in general. Rather we will consider just two specific vertical planes through the point. These will be planes which are parallel to the coordinate axes. In the one case we will take the plane which is parallel to the y and z axes as indicated by $ABCD$ in Figure VIII.3. Therefore, in the plane $ABCD$ we have z as a function of y alone, and we can take the derivative of z with respect to y using the same development we used in obtaining the derivative in Chapter IV. However, this is precisely what we denoted as the *partial derivative* earlier, for in this plane we have held x constant. Hence, this is the derivative that we denoted by $\partial z/\partial y$ in view of the fact that we are holding all of the independent variables constant with the exception of y. The result of this differentiation with respect to y can, of course, be a function of both x and y, and we could differentiate the resulting function either with respect to x or with respect to y. The fact remains, however, that the geometric interpretation of the partial derivative $\partial z/\partial y$ is the slope of the line which is both tangent to the surface at P and is also in a plane parallel to the y-z plane (the plane containing the y-axis and the z-axis).

We can phrase the discussion of the last paragraph a bit differently. Let (x_0, y_0, z_0) be the coordinates of the point P in Figure VIII.4. If we move in a plane parallel to the y-z plane, the x-coordinate remains fixed. Therefore, if we wish to determine the rate of change of z as y moves to the right or left, but as x remains fixed, we can consider a change in y, and this change will induce a corresponding change in z. Thus, we can move from the point (x_0, y_0, z_0) to the point $(x_0, y_0 + \Delta y, z_0 + \Delta z)$, and the rate of change will be given by the differential quotient

$$DQf_y(P, Q) = \frac{f(x_0, y_0 + \Delta y) - f(x_0, y_0)}{\Delta y}.$$

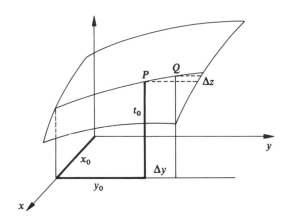

Figure VIII.4

If we take the limit of this differential quotient as Δy approaches zero, we have the partial derivative with respect to y. The discussion of this paragraph parallels the preceding work on derivatives with the exception that there is an additional variable, x, which is just going along for the ride. Since the partial derivative is now defined by

$$\frac{\partial f(x, y)}{\partial y}\bigg|_{\substack{x = x_0 \\ y = y_0}} = \lim_{\Delta y \to 0} DQf_y(P, Q) = \lim_{\Delta y \to 0} \frac{f(x_0, y_0 + \Delta y) - f(x_0, y_0)}{\Delta y}$$

we see that there is little to distinguish this from our earlier work with derivatives but for the fact that the x is considered as a constant. Therefore, we can expect all of our earlier results to continue to hold. This is a relief, for it means that we can build upon what we already know. As we did before, we can drop the notation (x_0, y_0) which was used to emphasize that we take the derivative at a point, and use the more general (x, y).

We could just as well have taken a plane parallel to the x-z plane. We would then have had a constant value for y and would have had x as the one independent variable. In this case it is possible to repeat all of our work of this section to date and obtain the slope of the curve formed by the intersection of the surface $z = f(x, y)$ and the plane parallel to the x-z plane. In like manner we could consider the differential quotient obtained by considering the two points (x_0, y_0, z_0) and $(x_0 + \Delta x, y_0, z_0 + \Delta z)$. The limit of this differential quotient as Δx approaches zero will give us the partial derivative of the function $f(x, y)$ with respect to x, $\partial z / \partial x$. This derivative, too, could be a function of both x and y, and therefore subject to further differentiation.

Since it is possible to simplify our geometric problem by taking the intersection of a plane, such as the x-z plane, with the curve $z = f(x, y)$,

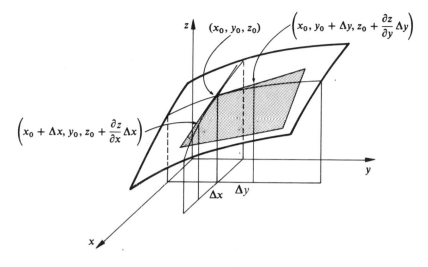

Figure VIII.5

we can frequently make use of our work in two dimensions to suggest methods for tackling three dimensional problems. If we seek the plane which is tangent to the surface at the point (x_0, y_0, z_0), we can obtain this plane by noting that it has to contain the line which is tangent to the surface at this point and in the plane parallel to the x-z plane and also the line which is tangent to the surface and in the plane parallel to the y-z plane. Since the slope of the line is given by the derivative, we know that a point (x, y, z) is on the line tangent to the curve and parallel to the x-z plane if and only if $(z - z_0)/(x - x_0) = \partial z/\partial x$ where the partial derivative is evaluated at the point (x_0, y_0, z_0). In other words if $x - x_0 = \Delta x$, and if $y = y_0$ (since y is not changed as we move in the x-z plane), then $z = z_0 + (\partial z/\partial x)\Delta x$. Therefore the point $(x_0 + \Delta x, y_0, z_0 + (\partial z/\partial x)\Delta x)$ is in the plane as is (x_0, y_0, z_0) as shown in Figure VIII.5. In similar fashion we can use the intersection of the surface and the y-z plane, and we have as the slope of the curve tangent to this cross section $(z - z_0)/(y - y_0) = \partial z/\partial y$. From this we can assure ourselves that $(x_0, y_0 + \Delta y, z_0 + (\partial z/\partial y)\Delta y)$ is also in the tangent plane.

From geometry we know that three points determine a plane, and consequently, since we now have three distinct points (if Δx and Δy are not zero) we have sufficient information to obtain the equation of the plane. Just as the line, a flat configuration, is determined by a first degree relation, so the plane, also a flat configuration, is determined by a first degree relation. Hence, the general equation of the plane must be $ax + by + cz = d$. Using methods similar to those of Section VIII.1, we can determine these coefficients by substituting in the values of x, y, and z and then solving for the coefficients. We will not worry yet about the fact that there are apparently four coefficients and only three relations. Using the three points we know to be on the plane we will have

$$ax_0 \qquad\qquad + by_0 \qquad\qquad\qquad + cz_0 \qquad\qquad\qquad = d$$

$$a(x_0 + \Delta x) + by_0 \qquad\qquad + c\left(z_0 + \frac{\partial z}{\partial x} \Delta x\right) = d$$

$$ax_0 \qquad\qquad + b(y_0 + \Delta y) \quad + c\left(z_0 + \frac{\partial z}{\partial y} \Delta y\right) = d.$$

Subtracting the first equation from the second and from the third gives the following two equations

$$a\Delta x + c\frac{\partial z}{\partial x} \Delta x = 0$$

$$b\Delta y + c\frac{\partial z}{\partial y} \Delta y = 0.$$

Since we have assumed that neither Δx nor Δy is zero, we can divide the first and second equations by Δx and by Δy respectively. This gives us values for a and b in terms of c and the partial derivatives as follows.

$$a = -c\frac{\partial z}{\partial x}, \qquad b = -c\frac{\partial z}{\partial y}.$$

We now have the values of a and b in terms of c and the partial derivatives. In order to find the value of d, we can note that $ax + by + cz = d$ for any point (x, y, z) on the plane. Furthermore we have $ax_0 + by_0 + cz_0 = d$. It follows that

$$ax + by + cz = ax_0 + by_0 + cz_0$$

or

$$a(x - x_0) + b(y - y_0) = -c(z - z_0).$$

If we now substitute the values of a and b in this last equation we have

$$-c\frac{\partial z}{\partial x}(x - x_0) - c\frac{\partial z}{\partial y}(y - y_0) = -c(z - z_0).$$

If c had been zero in the original equation for the plane, the plane would have been parallel to the z-axis, and the slopes involved would not have existed. Therefore, in the majority of cases we can anticipate that c will not be zero. This is certainly the case if the partial derivatives exist. Therefore, we have the equation of the *tangent plane*

$$z - z_0 = \frac{\partial z}{\partial x}(x - x_0) + \frac{\partial z}{\partial y}(y - y_0). \qquad\qquad (VIII.5.1)$$

EXAMPLE 5.1. Find the equation of the plane tangent to the curve $z = x^2 + xy - y^2$ at the point $(2, 3, 1)$.

Solution. In this case we have $\partial z/\partial x = 2x + y$ and $\partial z/\partial y = x - 2y$. At the point $(2, 3, 1)$ these become 7 and (-4) respectively. Consequently the tangent plane has the equation

$$7(x - 2) - 4(y - 3) = z - 1$$

or

$$z = 7x - 4y - 1.$$

If we examine (VIII.5.1) and think of $z - z_0$ as an increment Δz, and treat each of the other increments in similar fashion, we can rewrite (VIII.5.1) as

$$\Delta z = \frac{\partial z}{\partial x} \Delta x + \frac{\partial z}{\partial y} \Delta y.$$

This looks very much like the situation we dealt with in developing the differential, but this time we have two independent variables. It would seem reasonable in this case to define the differential of z to be

$$dz = \frac{\partial z}{\partial x} dx + \frac{\partial z}{\partial y} dy. \tag{VIII.5.2}$$

This is, in fact, the differential. We have not been rigorous in our development of this work, but it should appear intuitively clear that we are paralleling our earlier development of the differential with relatively minor modifications. The value dz in (VIII.5.2) is sometimes referred to as the *total differential* in view of the fact that this includes both the independent variables, x and y. Just as the differential in the case of a function of a single variable gave corrections for the dependent variable, in that case y, which was on the line tangent to the curve, so the total differential gives corrections which are on the plane tangent to the surface. In fact, we can use the differential to obtain approximations in a manner similar to that in which we previously used the differential for such purposes.

EXAMPLE 5.2. Approximate the value of $e^{0.03} \sin(32\pi/180)$.

Solution. Let $z = f(x, y) = e^x \sin y$. Since we know the value of e^0, let $x_0 = 0$, and since we know the value of $\sin (30\pi/180)$, let $y_0 = 30\pi/180 = \pi/6$. In this case $dx = 0.03 - 0 = 0.03$ and $dy = (32\pi/180) - (30\pi/180) = \pi/90$. Therefore

$$dz = \frac{\partial z}{\partial x} dx + \frac{\partial z}{\partial y} dy = \left(e^0 \sin \frac{\pi}{6} \right)(0.03) + \left(e^0 \cos \frac{\pi}{6} \right)\left(\frac{\pi}{90} \right)$$

$$= 0.015 + 0.0302 = 0.0452.$$

Since we now have the approximate change in z, and since $z_0 = e^0 \sin(\pi/6) = 0.5$, we know that the approximate value of $e^{0.03} \sin(32\pi/180) = 0.5 + 0.0452 = 0.5452$. The actual value should be 0.5461, and therefore the error is 0.0009 or 0.164%. This is a good approximation for a function that is not overly easy to evaluate without some aid such as a calculator or at the very least some tables.

At this point it might be well to introduce alternative notations. We have been using the standard notation for partial derivatives, but as a matter of convenience we frequently write $f_x(x, y)$ or $f_1(x, y)$ for the partial derivative of $f(x, y)$ with respect to x. The use of the subscript "1" is intended to indicate that the derivative is with respect to the first of the variables listed. In order to lessen the likelihood of misunderstanding, we shall restrict ourselves to the notations $\partial z/\partial x$ and $f_x(x, y)$ or perhaps f_x in the case of derivatives with respect to x and to the notation $\partial z/\partial y$ and $f_y(x, y)$ or f_y in the case of derivatives with respect to y. Thus, we can rewrite (VIII.5.2) as

$$df(x, y) = f_x(x, y)dx + f_y(x, y)dy.$$

You will remember that we proposed at an earlier point that the apparent variables in a problem, in our most recent cases this would be x and y, might actually be functions of some common parameter. We might designate this parameter t as we did in our earlier discussion of parametric equations. If both x and y are functions of t and if $z = f(x, y)$, then z is a function of t. Consequently we have x, y, and z as functions of a single variable t, and we should be able to get regular derivatives of a single variable for each of these functions. Of course, it is possible to substitute, for if $x = x(t)$ and $y = y(t)$ then it follows that $z = f(x, y) = f(x(t), y(t))$ and we can express z directly as a function of t. However, this is often inconvenient, and we may lose some information concerning the intermediate behavior of x and y by doing this. Therefore, it would be helpful to have some method for obtaining the derivative of z with respect to t, $D_t z$, without recourse to the method of substitution. In the following development we will use the notation dz/dt, dx/dt and dy/dt for their introduction will suggest the form of the result and may be useful in remembering this form.

By the definition of the derivative we have

$$\frac{dz}{dt} = \lim_{\Delta t \to 0} \frac{f(x(t + \Delta t), y(t + \Delta t)) - f(x(t), y(t))}{\Delta t}$$

$$= \lim_{\Delta t \to 0} \frac{f(x(t + \Delta t), y(t + \Delta t)) - f(x(t), y(t + \Delta t)) + f(x(t), y(t + \Delta t)) - f(x(t), y(t))}{\Delta t}$$

$$= \lim_{\Delta t \to 0} \frac{f(x(t + \Delta t), y(t + \Delta t)) - f(x(t), y(t + \Delta t))}{\Delta t}$$

$$+ \lim_{\Delta t \to 0} \frac{f(x(t), y(t + \Delta t) - f(x(t), y(t))}{\Delta t}.$$

In going from the first step above to the second step we have merely added and then subtracted $f(x(t), y(t + \Delta t))$, and hence have changed nothing. This is done in order that we can think of changing only one variable at a time, something that we have been doing before in our rationale for using partial derivatives. The last step in the above development is to break the preceding expression into two parts, the first of which has only the function x varying as a result of changes in t and the second of which has only y varying. Now, if $dx/dt \neq 0$ at the point for which we wish the derivative, then as t changes, so does x. Consequently $[x(t + \Delta t) - x(t)] \neq 0$ for sufficiently small values of Δt. As a result it is appropriate to write

$$\lim_{\Delta t \to 0} \frac{f(x(t + \Delta t), y(t + \Delta t)) - f(x(t), y(t + \Delta t))}{\Delta t}$$

$$= \lim_{\Delta t \to 0} \left[\frac{f(x(t + \Delta t), y(t + \Delta t)) - f(x(t), y(t + \Delta t))}{x(t + \Delta t) - x(t)} \frac{x(t + \Delta t) - x(t)}{\Delta t} \right]$$

$$= \left[\lim_{\Delta t \to 0} \frac{f(x(t + \Delta t), y(t + \Delta t)) - f(x(t), y(t + \Delta t))}{x(t + \Delta t) - x(t)} \right]$$

$$\times \left[\lim_{\Delta t \to 0} \frac{x(t + \Delta t) - x(t)}{\Delta t} \right].$$

In this last expression we note that the first limit is merely the derivative of $f(x(t), y(t))$ with respect to $x(t)$, or in earlier notation $f_{x(t)}(x(t), y(t))$. The second limit is our familiar dx/dt. Therefore, we can write the first limit as $(\partial f/\partial x)(dx/dt)$. Using similar reasoning we can write the second limit as $(\partial f/\partial y)(dy/dt)$. Therefore we have

$$\frac{dz}{dt} = \frac{\partial z}{\partial x} \frac{dx}{dt} + \frac{\partial z}{\partial y} \frac{dy}{dt}. \tag{VIII.5.3}$$

You will observe that the dx and ∂x give the appearance of cancelling in this case. While this is not quite right, it does give some help in remembering (VIII.5.3). The result given in (VIII.5.3) is often called the *total derivative of z with respect to t.*

We have, of course, assumed that all of the required limits exist. If they do not, we will fail to have a derivative. We also assumed that we did not get into trouble by taking only the case where $dx/dt \neq 0$. If we had considered $dx/dt = 0$ the numerator of the corresponding limit would also be zero. Consequently we would have an indeterminate form. This should not give us trouble, however, for if the derivative is zero, then x does not change, and consequently the first term of the total derivative would have to be zero. Thus, the expression (VIII.5.3) holds whenever the desired derivatives exist.

EXAMPLE 5.3. Find the total derivative of $f(x, y) = x^2y - 2x + 3y - 4$ using both the direct method of substitution and the method of the total derivative if $x(t) = t - 2$ and $y(t) = t^2$.

Solution. If we use the total derivative, we have

$$\frac{df}{dt} = \frac{\partial f}{\partial x}\frac{dx}{dt} + \frac{\partial f}{\partial y}\frac{dy}{dt},$$

and in this case $\partial f/\partial x = 2xy - 2, \partial f/\partial y = x^2 + 3, dx/dt = 1$ and $dy/dt = 2t$. Substituting these results in the expression for the total derivative we have

$$\frac{df}{dt} = (2xy - 2)(1) + (x^2 + 3)(2t).$$

Upon substitution in this expression, we obtain

$$\frac{df}{dt} = [2(t - 2)(t^2) - 2] + [(t - 2)^2 + 3][2t] = 4t^3 - 12t^2 + 14t - 2.$$

On the other hand, if we substitute directly in the expression for $f(x, y)$ we will have

$$f(x, y) = (t - 2)^2(t^2) - 2(t - 2) + 3t^2 - 4 = t^4 - 4t^3 + 7t^2 - 2t.$$

It is easy to see that the derivative of this last expression with respect to t gives us the same value we obtained by the method of the total derivative. It is also apparent that we had somewhat less algebra to perform using the total derivative. Frequently it is helpful to retain the form of the total derivative for the additional insight it may shed on the problem at hand.

EXERCISES

1. Find the indicated derivatives in each of the following cases:

   (a) f_x and f_y if $f(x, y) = x^2 \sin y - y^2 \cos x$
   (b) f_x and f_y if $f(x, y) = xe^y \sin xy + x^2y^3$
   (c) $f_x(2, 3)$ and $f_y(3, 2)$ if $f(x, y) = x^2y + xy^2 + (x^2/y)$
   (d) f_x, f_y, and f_z if $f(x, y, z) = x^2y + y^2z + z^2x$
   (e) f_r and f_θ if $f(r, \theta) = r^2 \sin 2\theta$
   (f) f_r and f_θ if $f(r, \theta) = re^\theta$.

2. Find the equation of the plane tangent to each of the following surfaces at the indicated point:

   (a) $z = x^2 + y^2$ at $(2, 3, 13)$
   (b) $z = x^2 - xy + y^2$ at $(-2, -3, 7)$
   (c) $z = x^2 + 4y^2 - 4x + 8y - 7$ at $(2, -1, -15)$
   (d) $z = 12/xy$ at $(2, 3, 2)$

3. Find the equation of the plane tangent to each of the following surfaces at the indicated point:

(a) $z = \sqrt{16 - (x - 1)^2 - (y + 2)^2}$ at $(1, -2, 4)$
(b) $z = \sqrt{49 - x^2 - (2y - 4)^2}$ at $(3, 1, 6)$
(c) $z = ye^x - y \arcsin(x - y)$ at $(1, 0.5, (e/2) - (\pi/12))$
(d) $z = \arctan x^2 y$ at $(2, 0.25, \pi/4)$

4. Find the differential in each of the following cases:

(a) $f(x, y) = \ln(x + 2y)^3$
(b) $f(x, y) = (2xy - x + y)^4$
(c) $f(x, y) = x^2 - 2y^3 + 3xy$
(d) $f(x, y) = (\ln xy)/(x - y)$
(e) $f(r, \theta) = r \cos 2\theta$
(f) $f(x, y, z) = x^3 y^2 z + xe^y \sin z$

5. Approximate the value of each of the following:

(a) $4.1e^{-0.06}$
(b) $(9.94)(14.1)$
(c) $\sqrt{101 - (4.05)^2 - (6.94)^2}$
(d) $(\arcsin 0.53)/3.97$
(e) $\sin 27° \tan 47°$
(f) $\sin 28° \cos 62° \tan 43°$

6. (a) If $z - f(x, y)$ has an extreme value at the point (x_0, y_0, z_0) and the partial derivatives of z are continuous in a region of which this point is an interior point show that the equation of the plane which is tangent to $z = f(x, y)$ at the extreme point must have the equation $z = z_0$.
(b) If (x_0, y_0, z_0) is an extreme point of $z = f(x, y)$ and if z has continuous partial derivatives in the vicinity of this extreme point, show that all approximations starting from the extreme point will give z_0 as the value of z.
(c) Find an extreme point of $z = x^2 + xy + y^2 - 4x - 5y + 2$ and show that the facts indicated in parts (a) and (b) are correct in this case.

7. In each of the following cases show that $x(\partial z/\partial x) + y(\partial z/\partial y) = nz$. Compare the value of n with the degree of $f(x, y)$ in the equation $z = f(x, y)$.

(a) $z = x^2 + xy - y^2$
(b) $z = 14x + 32y$
(c) $z = x^4 - y^4 + (x + 2y)^4$
(d) $z = \sin(x/y) + \ln(y/x)$
(e) $z = x^3 + 3x^2 y + 7xy^2 - 4y^3$
(f) $z = x^2 + x^{3/2}y^{1/2} - x^{-2/5}y^{12/5} - xy$
(g) $z = x^4 - (xy + y^2)^2$
(h) $z = (x - y)/(x^2 + y^2)$

8. Find the total derivative with respect to t in each of the following cases.

(a) $z = x^2 y + xy^2 - e^{xy}$ if $x = t^2 - 2t$ and $y = 2 - 3t$.
(b) $z = x^2 + 2xy + y^2$ if $x = 2 + \cos(\pi t/8)$ and $y = 3 + \sin(\pi t/8)$.
(c) $z = (2x + 3)/(3y - 2)$ if $x = e^t + t$ and $y = e^{-t} - t$.
(d) $z = (x + 2y)^3 - (x^2 - (1/y))^3$ if $x = (t + 2)^2$ and $y = (t + 3)^3$.

9. (a) Show that the total derivative can be extended to encompass functions with more than two independent variables.

   (b) Find $D_t f(x, y, z)$ if $f(x, y, z) = xy + yz + zx$ and $x = t$, $y = t^2$ and $z = t^3$.

   (c) If $f(x_1, x_2, x_3, x_4)$ is a function of four variables, and if $x_i = t$ for $i = 1, 2, 3$, or 4 but each of the other $x_j = 0$, show that the total derivative $df/dt = df/dx_i$.

10. In determining arc length we had the integrand $\sqrt{dx^2 + dy^2} = \sqrt{1 + (dy/dx)^2}\, dx$. We also had the relation between rectangular and polar coordinates $x(r, \theta) = r \cos \theta$ and $y(r, \theta) = r \sin \theta$.

   (a) Find dx and dy as total differentials of $x(r, \theta)$ and $y(r, \theta)$.

   (b) Substitute these total differentials in the expression for arc length and obtain the corresponding expression for arc length in polar coordinates.

   (c) Using this expression find the arc length of the curve $r = 3 \sin \theta$.

   (d) Find the length of the cardioid $r = 1 - \cos \theta$.

S11. The Cobb-Douglas production function in its traditional form is given by the relation $Q = aL^b C^{1-b}$ where Q is the quantity produced, L is the labor required to produce the output, and C is the capital required to produce the output. The values a and b are constants with a positive and b in the interval $(0, 1)$ in the usual case.

   (a) The units for Q, L, and C are units that are appropriate to the case at hand. For instance, Q may be measured in bushels, L in man-hours, and C in number of machines. Show that the units for a in this case should be bushels/(man-hours)^b(machines)$^{1-b}$ in order to make the units match in the given equation.

   (b) If we write $Q = Q(L, C)$, show that $Q(kL, kC) = kQ(L, C)$ for any constant k. A function which behaves in this way is said to be *homogeneous of degree one*.

   (c) Show that $Q = L(\partial Q/\partial L) + C(\partial Q/\partial C)$.

   (d) If $b = 0.75$, under what ratio of labor to capital would the production increase more rapidly with further increase in labor than with further increase in capital?

S12. A certain model of the economy is expressed in the equations

$$Y = C + I + G$$
$$C = a + b(Y - T)$$
$$T = c + dY$$

where Y is the national income, C is consumption of goods, I is the amount of investment, G is government expenditure, T is the amount of taxes collected, and $a, b, c,$ and d are constants with $0 < b < 1$ and $0 < d < 1$.

   (a) Show that consumption is a fixed amount augmented by a percentage of the amount of income left after taxes.

   (b) Show that the amount collected for taxation is a fixed amount plus a percentage of the national income.

   (c) Express the national income as a function of investment and government expenditure but no other variables.

   (d) Show that the assumptions made in this model assure that $\partial Y/\partial G$ will be positive.

(e) Show that if c is allowed to vary, then $\partial Y/\partial c$ must be negative.

(f) If the tax rate, d, is allowed to vary show that

$$\left(\frac{\partial Y}{\partial d}\right)(bc - I - G - a) > 0.$$

(g) Using the results of parts (d), (e), and (f) explain what variables should change and in what direction to increase the national income.

BP13. The heat Q (calories) liberated by mixing x (gmol) of sulphuric acid with y (gmol) of water is given by

$$Q - \frac{17860xy}{1.798x + y}.$$

If we have 6 gmol of acid and 4 gmol of water,

(a) Find the heat generated.
(b) Find the increase in heat generated if the amount of acid is increased by 2%.
(c) Find the increase in heat generated if the amount of water is increased by 1% using the original amount of acid.
(d) Find the increase in heat generated if we start with the given amounts and increase the acid by 2% and the water by 1% at the same time.

VIII.6   More About Partial Derivatives

In Sections VIII.3 and VIII.4 we used partial differentiation to find extreme values in order to obtain curves of regression. We attempted to parallel the methods used with functions of a single variable in setting the derivatives equal to zero and finding the critical points. However, we relied on a very intuitive argument to determine that we did have a minimum point. It is true that we had a second derivative test that would usually indicate whether we had a maximum or a minimum in the case of one variable. The situation is not quite as simple when more variables are involved. We shall take a look at this problem in the case of two variables in the course of this section.

As you might expect, there are many more possibilities with functions of several variables than with functions of a single variable. In order to point these out, we will consider some examples.

EXAMPLE 6.1. Sketch the surface $z = x^2 + y^2$ and find any extreme values that this function may have.

Solution. In sketching the surface we note that insofar as x and y are concerned we have the equation of a circle of radius $\sqrt{z}$. This is easily seen by examining the equation $(\sqrt{z})^2 = x^2 + y^2$. Thus, if we think of taking horizontal planes at a height z units above the x-y plane, we will have a circle in this plane with center on the z-axis. Furthermore, the surface does

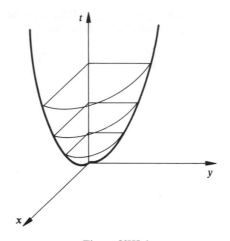

Figure VIII.6

not exist when z is negative. Therefore, we have the surface indicated in Figure VIII.6. We might also note that the cross sections in the x-z plane and in the y-z plane are parabolas. This helps in making our sketch.

From the sketch we can determine that the point $(0, 0, 0)$ is a minimum point, but we should proceed. If we take the partial derivatives, we have $\partial z/\partial x = 2x$, $\partial z/\partial y = 2y$. It is easy to see that these are both zero only when $x = y = 0$, then $z = 0$ and we are led to the origin as the only critical point. In order to assure ourselves that this is a minimum we might decide to take the second derivatives. In this case we would have

$$\frac{\partial}{\partial x}\left(\frac{\partial z}{\partial x}\right) = \frac{\partial^2 z}{\partial x^2} = 2 \quad \text{and} \quad \frac{\partial}{\partial y}\left(\frac{\partial z}{\partial y}\right) = \frac{\partial^2 z}{\partial y^2} = 2.$$

Observe that we have used a notation for the second derivative similar to that used earlier where we had written

$$\frac{d^2 f}{dx^2} \quad \text{to represent} \quad \frac{d}{dx}\left(\frac{df}{dx}\right).$$

In the case of partial derivatives we could also differentiate $\partial z/\partial x$ with respect to y, and in this case we would write

$$\frac{\partial}{\partial y}\left(\frac{\partial z}{\partial x}\right) = \frac{\partial^2 z}{\partial y \partial x}.$$

Note that we read from right to left for the order of differentiation. Fortunately the order of differentiation makes no difference with the vast majority of functions, and we do not have to be too careful in remembering whether we differentiated first with respect to x and then y or first with respect to y and then x. In the case of the second derivatives we can also use the notation $f_{xx}(x, y)$ to indicate the result of taking the derivative with respect to x, and then differentiating the result with respect to x. In other words $f_{xx}(x, y) = \partial^2 f/\partial x^2$. In similar fashion $f_{xy}(x, y) = \partial^2 f/\partial y \partial x$ and $f_{yy}(x, y) = \partial^2 f/\partial y^2$. Higher order partial derivatives can be expressed in an analogous manner. In this Example we notice that both $f_{xx}(0, 0)$ and $f_{yy}(0, 0)$ are positive, and therefore we would expect, based upon our earlier work, that we have a minimum value. Everything appears to check out. The fact that we have overlooked something is not apparent here. However, let us try another example.

EXAMPLE 6.2. Sketch the surface $z = x^2 + y^2 - 4xy$ and find any extreme values that this function may have.

Solution. In this case we will look for the extreme values first. If we take the partial derivatives, we have $\partial z/\partial x = 2x - 4y$ and $\partial z/\partial y = 2y - 4x$. A little algebra will show that these are both zero only when $x = y = 0$. At this point this result would appear to coincide with the result of Example 6.1. In fact, we might go further and try to check this result by examining the second derivatives. We would again have $\partial^2 z/\partial x^2 = 2$ and $\partial^2 z/\partial y^2 = 2$ as before. Thus, our first indication would again suggest a minimum value. However, if $x = y = 0$ then $z = 0$, but if $x = y = 1$ we have $z = -2$ and this is clearly smaller than $z = 0$. Therefore, something is amiss.

In order to find out what is wrong, it is time to sketch the surface. If $x = 0$ we have the equation $z = y^2$, and consequently the cross section in the y-z plane is a parabola opening upward. If $y = 0$ we have the equation $z = x^2$, and we have a similar parabola in the x-z plane. These are both shown in Figure VIII.7. On the other hand if $x = y$, the equation becomes $z = -2x^2$, and consequently if we take the cross section including the z-axis and the line $x = y$ (this is actually the plane $x = y$) we have a parabola opening downward with the projection on the x-z plane being $z = -2x^2$ and the projection on the y-z plane being $z = -2y^2$. This is also shown in Figure VIII.7. For these projections we have a negative second derivative, indicating that we have a maximum point at the origin on this cross section. In this case the origin is a critical point, but it is apparent that it is neither a maximum nor a minimum. Hence it is not an extreme point.

Example 6.2 indicates that we must check a critical point by examining the cross sections of the surface formed by planes including the vertical line through the critical point and proceeding in each possible direction from this vertical line. Thus, we need to be able to find the partial derivatives in each direction from a given point. This sounds like a big order, but in fact we can use information gathered in the last section. If the critical point is the point (x_0, y_0, z_0) then the plane we need must first be vertical. This

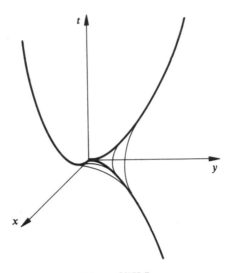

Figure VIII.7

will happen if the equation of the plane includes z with a zero coefficient, for in this case a change in z does not affect the equation, and vertical movement from a point on the plane continues to produce points on the plane. If we wish to take the plane which makes an angle θ with the x-z plane, we can describe it by the parametric equations

$$x = x_0 + t \cos \theta$$
$$y = y_0 + t \sin \theta.$$

Here the parameter t represents the distance measured horizontally from the point (x_0, y_0, z) for any value z. This is shown in Figure VIII.8. In this case the total derivative gives the slope of the curve made by the intersection of the surface and the plane through (x_0, y_0, z) which makes an angle θ with the x-z plane. The total derivative in this case is

$$\frac{dz}{dt} = \frac{\partial z}{\partial x} \cos \theta + \frac{\partial z}{\partial y} \sin \theta.$$

This is called the *directional derivative* for it is the derivative in the direction indicated by the angle θ. Since t is a measure of distance, the derivative retains the correct units to qualify as a slope.

This directional derivative gives us freedom to take any vertical plane cross section and investigate the behavior of the curve. Since our concern in Example 6.2 related to checking the second derivative in each direction, we will investigate the behavior of the directional derivative and obtain the following theorem.

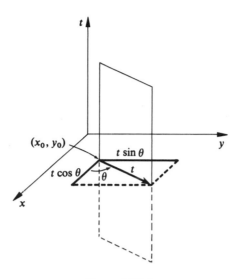

Figure VIII.8

Theorem 6.1. *If (x_0, y_0) is a point at which the first order partial derivatives of $f(x, y)$ are both zero and if the second partial derivatives of $f(x, y)$ exist and are continuous in the neighborhood of (x_0, y_0) let*

$$M = \left(\frac{\partial^2 f}{\partial x^2}\right)\left(\frac{\partial^2 f}{\partial y^2}\right) - \left(\frac{\partial^2 f}{\partial x \, \partial y}\right)^2.$$

If $M > 0$ the point (x_0, y_0) is a maximum point if $\partial^2 f/\partial x^2 < 0$ and is minimum if $\partial^2 f/\partial x^2 > 0$. If $M < 0$ then (x_0, y_0) is not an extreme point. If $M = 0$ there is no indication from this test whether the point is an extreme point or not.

PROOF. We are given the fact that $\partial f/\partial x = 0$ and $\partial f/\partial y = 0$ at the point (x_0, y_0), and therefore we have a critical point. Since the directional derivative is given by

$$\frac{df}{dt} = \frac{\partial f}{\partial y} \cos \theta + \frac{\partial f}{\partial y} \sin \theta,$$

we see at once that the directional derivative is also zero at this point. This reenforces the notion that this is a critical point. Using our earlier results we can obtain information which will assist us in determining whether this is an extreme point by examining the second derivative. However, Example 6.2 has shown us that we must examine this second derivative in each possible direction from the point (x_0, y_0). Therefore we must examine the behavior of the directional derivative. We must do this, however, in the direction in which the first derivative was taken. Thus we must consider

$$\frac{d}{dt}\left(\frac{df}{dt}\right) = \frac{d^2f}{dt^2} = \frac{\partial}{\partial x}\left(\frac{\partial f}{\partial x}\cos\theta + \frac{\partial f}{\partial y}\sin\theta\right)\cos\theta$$

$$+ \frac{\partial}{\partial y}\left(\frac{\partial f}{\partial x}\cos\theta + \frac{\partial f}{\partial y}\sin\theta\right)\sin\theta$$

$$= \frac{\partial^2 f}{\partial x^2}\cos^2\theta + \frac{\partial^2 f}{\partial x\,\partial y}\sin\theta\cos\theta$$

$$+ \frac{\partial^2 f}{\partial y\,\partial x}\cos\theta\sin\theta + \frac{\partial^2 f}{dy^2}\sin^2\theta$$

$$= \frac{\partial^2 f}{\partial x^2}\cos^2\theta + 2\frac{\partial^2 f}{\partial x\,\partial y}\cos\theta\sin\theta + \frac{\partial^2 f}{\partial y^2}\sin^2\theta.$$

In the last step, we used the fact that the second order partial derivative which involves one differentiation, with respect to each of the independent variables, gives the same result regardless of the order of differentiation. As we noted earlier this is correct for almost all of the functions used in applications. We will have no occasion to deal with any function for which this is not true.

Now we have the directional second derivative, but it is much more complicated than we would like. We must still determine whether this is always positive or always negative. Before considering the general question we first note that we certainly will have insufficient information to determine whether we have an extreme value if $\partial^2 f/\partial x^2 = 0$. Thus we need only consider the case where $\partial^2 f/\partial x^2 \neq 0$. Furthermore, we can only have affirmative information concerning the presence of an extreme point if the product $(\partial^2 f/\partial x^2)/(d^2 f/dt^2) > 0$, for otherwise these two second derivatives either have a zero value or have opposite signs. In the former case we have inconclusive information and in the latter case we know we do not have an extreme point. Therefore, we will consider this product, and will investigate whether this product must be positive for each value of θ. If this product is positive we have a maximum value if either (and consequently both) of the second derivatives is negative, and similarly we will have a minimum value if either of the second derivatives is positive. The product gives us

$$\frac{\partial^2 f}{\partial x^2}\frac{d^2f}{dt^2} = \left(\frac{\partial^2 f}{\partial x^2}\right)^2\cos^2\theta + 2\frac{\partial^2 f}{\partial x^2}\frac{\partial^2 f}{\partial x\,\partial y}\cos\theta\sin\theta + \frac{\partial^2 f}{\partial x^2}\frac{\partial^2 f}{\partial y^2}\sin^2\theta.$$

This does not look much more appealing, but we can use the technique of completing the square and obtain

$$\frac{\partial^2 f}{\partial x^2}\frac{d^2f}{dt^2} = \left[\frac{\partial^2 f}{\partial x^2}\cos\theta + \frac{\partial^2 f}{\partial x\,\partial y}\sin\theta\right]^2 + \left[\left(\frac{\partial^2 f}{\partial x^2}\right)\left(\frac{\partial^2 f}{\partial y^2}\right) - \left(\frac{\partial^2 f}{\partial x\,\partial y}\right)^2\right]\sin^2\theta.$$

$$\text{(VIII.6.1)}$$

The coefficient of $\sin^2 \theta$ is the expression denoted by M in the statement of the theorem. Thus, we have here a term which is a square, and which therefore cannot be negative and a term of the form $M \sin^2 \theta$. Since $\sin^2 \theta$ cannot be negative, we see that (VIII.6.1) cannot be negative if M is positive. We still are not certain that it would not be possible to have both of the terms in (VIII.6.1) zero at the same time, and consequently we are not quite done.

Since we have assumed that $\partial^2 f / \partial x^2 \neq 0$, the first term of (VIII.6.1) can be zero only if

$$\cot \theta = -\frac{\partial^2 f / \partial x\, \partial y}{\partial^2 f / \partial x^2}.$$

The cotangent here must have a finite value. Therefore θ is not a multiple of π, and consequently $\sin \theta \neq 0$. Therefore for this value of θ the expression (VIII.6.1) must have the same sign that M has. If M is positive, (VIII.6.1) is positive for all values of θ, and we have obtained the result we desired. On the other hand if M is negative we need only let θ have the value for which the first term of (VIII.6.1) is zero and the product of the two second derivatives will be negative, a situation which assures us that we do not have an extreme value.

To demonstrate that we have an indeterminate situation if $M = 0$, consider the two functions $f(x, y) = x^3 + y^3 + x^2 y$ and $g(x, y) = x^4 + y^4$. In each case the only critical point occurs at $(0, 0)$ and in each case the value of M is zero. We also note that $f(0, 0) = g(0, 0) = 0$. However, $f(x, -x) = -x^3$. If x is positive we have values of the function less than zero and if x is negative we have values larger than zero. Therefore the origin can be neither a maximum nor a minimum. Note that this is true regardless of how small x may be in absolute value, and consequently this is true for points arbitrarily close to $(0, 0)$. On the other hand, $g(x, y)$ is the sum of two fourth powers, and can never be negative. Therefore $(0, 0)$ must be a minimum value. It is possible then to have either an extreme value or no extreme value in cases involving $M = 0$. It is worth emphasizing here that M is evaluated at the critical point, in this case $(0, 0)$, and the expression represented by M might have non-zero values for other points. $\square$

While it would be possible to obtain theorems which are somewhat similar to Theorem 6.1 for more than two variables, we will not attempt to find such theorems. This theorem gives an indication of the techniques required in the analysis of functions of several variables. We will leave further results to a later course. Before closing this section it might be interesting to see how one might apply some of these techniques to a problem which could come up in manufacturing.

EXAMPLE 6.3. A certain company is making two products and it is necessary to make a decision as to how many of each product should be made in order to maximize the profits of the company. It is known that the first product can be sold for $7.00 and the second one for $5.00. It is proposed that we make x of the first product and y of the second. It is further known that the cost of manufacture in dollars is given by the expression

$$C = \frac{3x^2 + 2xy + 3y^2}{100}.$$

How many of each should be made each day to make the profit a maximum?

Solution. The income to be derived from the sale of x of the first product and y of the second is given by $7x + 5y$ dollars. The profit must therefore be given by

$$P(x, y) = 7x + 5y - \frac{3x^2 + 2xy + 3y^2}{100}.$$

In order to find the critical point, we need to differentiate. Here we have

$$P_x(x, y) = 7 - 0.06x - 0.02y \quad \text{and} \quad P_y(x, y) = 5 - 0.02x - 0.06y.$$

If we solve the two equations

$$0.06x + 0.02y = 7$$
$$0.02x + 0.06y = 5$$

we have $x = 100$ and $y = 50$. This is the only critical point, and therefore we might rely on intuition and be satisfied. However, it never hurts to check. In this case we have

$$P_{xx}(x, y) = -0.06, \qquad P_{yy}(x, y) = -0.06, \quad \text{and} \quad P_{xy}(x, y) = -0.02.$$

Therefore $M = (-0.06)(-0.06) - (-0.02)^2 = 0.0036 - 0.0004 = 0.0032$. This is positive and therefore we have an extreme value. Furthermore, $P_{xx}(100, 50)$ is negative and this extreme point must be a maximum. Consequently, we should manufacture 100 of the first product and 50 of the second each day.

Before leaving this solution it would be enlightening to check this result another way. We might see what the profit would be if we manufacture according to the results obtained, and then investigate the results if we vary this somewhat. We observe that $P(100, 50) = \$475.00$ by straight substitution in $P(x, y)$. On the other hand we have $P(99, 50) = P(101, 50) = P(100, 49) = P(100\ 51) = \$474.97$. A slight movement in any direction from our result would then appear to diminish the profit, and our theory seems to be vindicated.

EXERCISES

1. Find all second and third order partial derivatives for each of the following functions. In each case show that $f_{xy}(x, y) = f_{yx}(x, y)$.

(a) $f(x, y) = x^3 + x^2y + xy^2 + y^3$
(b) $f(x, y) = x^2 \sin(xy) + y^2 \cos(xy)$
(c) $f(x, y) = e^x \ln y - \arctan xy^2$
(d) $f(x, y) = x^2 \tan y - y^2 \cot x$
(e) $f(x, y) = x^3 \sec y + (3xy/(x - y))$

2. Find the directional derivative in each of the following cases. Also find the angle θ such that the directional derivative is an extreme for the point where $x = 2$ and $y = 3$. Determine whether the extreme value is a maximum or a minimum.

(a) $f(x, y) = x^3 - y^3$
(b) $f(x, y) = x^3 - x^2y + xy^2 - y^3$
(c) $f(x, y) = (x - y)/(x + y)$
(d) $f(x, y) = x^2 + y^2 - 4x + 6y - 2$

3. Find the extreme values (if any) of each of the following functions and determine whether the value is a maximum or a minimum by means of Theorem 6.1.

(a) $f(x, y) = x^2 + 2y^2 - 4x - 6y + 3$
(b) $f(x, y) = 3x^2 - 4xy + y^2 - 2x + 3y$
(c) $f(x, y) = 4x^2 - 3y^2 - 12xy + 3x$
(d) $f(x, y) = (x^2 - 2xy + 4y^2)/xy$

4. Find the extreme values of each of the following functions and determine whether they are maximum or minimum.

(a) $f(x, y) = x^3 + x^2y + xy^2 + y^3 - 6x - 6y$
(b) $f(x, y) = x^4 - 2x^2 + y^2 - 2y$
(c) $f(x, y) = x^4 + y^4 - 2x^2 - 2y^2$

5. (a) Show that there are six second partial derivatives and ten third order partial derivatives of a function of three independent variables.
   (b) Find all of the second and third order partial derivatives of $f(x, y, z) = x^2y + y^2z + z^2x + xyz$.
   (c) How many fourth order partial derivatives would you expect this function to have? How many of them would have the same value in this case? What would be the common value? Could you tell this without performing the differentiation?

P6. The general gas law is given by the relation $PV = nRT$ where P is pressure, V is volume, n and R are constants and T is absolute temperature.

   (a) Find the percentage change in P if V is increased by 2% and T is decreased by 3%.
   (b) Find the total derivative of P with respect to time in terms of the rates of change of V and T with respect to time.
   (c) Show that any changes in pressure and volume are adiabatic (that is T remains constant) if and only if any percentage change in P is matched by an equal percentage change in the opposite direction in V.

S7. Among the indicators that can be used in predicting crime rates is one which has the form $C = 5w^3 + 10p^3 - 17w - 18p - 19wp$. In this case w is the number of millions of dollars spent on welfare programs and p is the number of millions of dollars spent on the prison system. The coefficients have been modified for ease of computation.

(a) Find any extreme values of $C(p, w)$, and determine whether they are maxima or minima.

(b) What should be the expenditures on welfare and prisons if one wants a minimum crime rate?

S8. A certain firm manufactures calculators and digital watches. They can sell the calculators for $15.00 and the digital watches for $12.50. The cost of manufacturing c calculators per day and w watches per day is given by

$$C(c, w) = \frac{5c^2 + 4cw + 4w^2}{100}.$$

(a) How many calculators and how many watches should be made per day to maximize the profit?

(b) Show that if the cost of production is cut in half, one should double the number of watches made and the number of calculators made in order to maximize the profit with the new cost figures.

(c) If the selling price had to be reduced by one cent for each calculator and each watch made beyond 100 per day due to saturation of the market, what would be the optimal production to maximize profit?

S9. (a) As given in the last section, the Cobb–Doublas production function is given by the relation $Q = aC^{1-b}L^b$ where b is in the interval $(0, 1)$. Find the total differential of Q.

(b) If R is the ratio of labor to capital, express Q as a function of C and R.

(c) Find the total differential of Q in terms of C and R.

(d) In the Cobb–Douglas equation there are three terms that can be varied, C, L, and b. Show there is no triple of values for these three terms which yield a critical point, for the value of L that makes one partial derivative zero makes another one fail to exist.

(e) If it is assumed that $C + L = k$ where k is constant, find the value of b which will produce a critical point. Can you classify this point?

S10. If we consider only the number of units produced and the tax rate, the daily profit $\pi(x, t)$ of a firm might well be given by the equation $\pi(x, t) = xp(x) - C(x) - xt$ where $p(x)$ is the amount received for each unit when x units per day are available, $C(x)$ is the cost of producing x units, and t is the tax per unit.

(a) Find $p(x)$ if the units can be sold at a base price of $60 per unit, but the price is reduced by one cent per unit for each unit over 100 per day as a result of approaching saturation of the market.

(b) Find $C(x)$ if there is an overhead cost of $500 per day and there is a materials and labor cost of $30 per unit.

(c) Find the number of units that should be produced per day as a function of the tax rate if we desire maximum profit.

(d) Find the number that should be produced if the tax rate is 20% of the selling price.

(e) Consider $\pi_{xt}(x, t)$ and then determine whether production should increase or decrease with an increase in tax rate in order to continue to have maximum profit.

(f) Find the extreme value (if any) of $\pi(x, t)$ if both x and t are allowed to vary. Why is this solution of little value in practical terms in any real situation?

CHAPTER IX

Infinite Series

IX.1 The Problem

At one time or another nearly everyone has seen the problem which proposes that although we start toward some place, we never get there. The reasoning, of course, is that we first go half way, then we go half the remaining distance, namely one fourth of the way, and after that half the remaining distance, that is one eighth of the way, etc. We can get close enough for almost any intended purpose, but we cannot actually arrive. This is equivalent to considering the non-ending sum $1/2 + 1/4 + 1/8 + 1/16 + \cdots$. From our work in Chapter I we realize that this is a geometric progression, and therefore the sum of the first n term is

$$\frac{(1/2)^{n+1} - 1/2}{(1/2) - 1} = 1 - (1/2)^n.$$

We observe that the distance that separates us from our goal is $(1/2)^n$, and after a sufficient number of terms this will be small enough to ignore, whatever that may mean. We are tempted to say, then, that this sum of an infinite number of terms has the value one.

Two problems arise immediately concerning this brief description of an old problem. The first concerns the fact that we are jumping to some type of conclusion in saying that the sum is one, when we know full well we can never add enough terms to get to one. In fact, if we hadn't had such a nice set of numbers, there would have been a very great question concerning what the set of numbers might have for its sum, if indeed there were any number that could possibly fulfill the requirements of a *sum*. The second problem centers around whether we have gained anything by expressing "1" in such a complicated way, or alternatively whether we can find such expressions that could be useful in providing us with results we could not otherwise obtain (and certainly "1" is not in that class). Before we face up to these problems, a bit of terminology is in order. This will make the problem appear to be more sophisticated (whether it is or not). Such an unending sum is called an *infinite series*, and the term infinite series implies not only an infinite number of terms, but also that this infinite number of terms is to be added up. The fact that a sum (or anything that could be called a sum) exists is equivalent to saying *the infinite series converges to a limit, or to a sum*. Thus, we can rephrase our problem by saying that we are concerned whether we can find any convergent infinite series that provides us with information that we could not have obtained more easily by other means. That the answer is "yes" should be apparent from the mere fact that we are discussing the question.

We will start with the problem of convergence, and then proceed to the matter of finding infinite series of assistance to us. In that way we will have available techniques for testing series when we need them. Hence, the first question we will consider is that of determining whether a series will converge. In order to do this, we start with the idea of an infinite sequence. An infinite sequence is merely a set of numbers occuring in some given order such that for each number there is a sequel or following number. We might consider the sequence 1, 1/2, 1/3, 1/4, 1/5, . . . or the sequence 3, 5, 4, −9, 0, 0, 0, 0, ... In the latter case there appears to be no rhyme or reason for the sequence of numbers, but the mere fact that they follow in a definite order is sufficient. We also note that the second sequence appears to quit at some point, and all of the terms thereafter are zero. Had we not written the zeros in, we would have had a terminating sequence. In the first case, there seems to be much more method in our madness, and we note that if we are to follow the pattern that has been set we would never run out of terms. This is an infinite sequence.

It is possible to describe a sequence as a function. We can consider the first sequence to be merely the set of values of the function $\{f(n)\} = \{1/n\}$ where the domain is the set of positive integers. The terms are written in the order in which they would appear if we were to take ascending values of the domain element. Since we have a function, and since the argument of the function, in this case n, is increasing without limit, we are on familiar ground, for we know what is meant by

$$\lim_{n \to \infty} f(n) = \lim_{n \to \infty} \frac{1}{n}.$$

This has a limit of zero. Therefore, it would appear to be eminently reasonable to say that the limit of the infinite sequence we wrote down first is zero. The limit is not affected by the fact that the domain of $f(n)$ consists only of integers. We can still apply Definition VII.2.2. Since we are concerned with the terms that appear in the *ad infinitum* part of the sequence, we could establish that the limit of the second sequence is also zero. Thus, we have the fact that a sequence is a function having for its domain a set of integers which is bounded below, and the limit of the sequence is the limit as the domain element increases without bounds. More formally, we have

Definition 1.1. An *infinite sequence* is a set of ordered quantities $\{f(n)\}$ having a domain consisting of a set of integers which is bounded below. The limit of the sequence is $\lim_{n \to \infty} f(n)$.

The limit of the sequence exists if and only if $\lim_{n \to \infty} f(n)$ exists. The problem of determining whether a sequence has a limit has now been changed to the problem of determining whether a function has a limit as the argument increases without bound. We handled the latter problem in Chapter VII and so we already have this under control. Note that we referred to the terms in functional notation as $f(n)$. If we are given the fofmula for the n-th term, we have, in fact, been given the function. We might consider the sequence

$$\left\{ \frac{1}{1}, \frac{2}{1}, \frac{4}{2}, \frac{8}{6}, \frac{16}{24}, \dots, \frac{2^n}{n!}, \dots \right\}$$

Here we have actually been given the formula, for $f(n) = 2^n/n!$. In this case we started with $n = 0$, but that falls within the definition, for we only said that the domain had to be bounded below. If it is bounded below by zero, that is just as valid as having it bounded below by one. The choice of the lower bound for the domain is, in fact, a matter of convenience, for we can see what starting point would be most helpful to us in simplifying the formula for the n-th term. If the formula is not given, we may be called upon to do some educated guessing. This is usually done by taking a look at the terms which are given and then trying to determine what formula might have been used that would have given precisely those terms. It is frequently made somewhat easier if the numbers are written out without simplification. If the sequence above had been simplified by reducing all fraction to lowest terms, $1, 2, 2, 4/3, 2/3, \dots$ it would have been somewhat less likely that you would have found this particular formula for $f(n)$. On the other hand, as we found in Chapter VIII, there are many different formulas that will go through any given set of data points. We usually try to find the *simplest* one. While we would not be inclined to view the choice as *simple*, the function

$$g(n) = \frac{n - \left[\dfrac{n}{3}\right] + \left[\dfrac{n}{4}\right] - 3\left[\dfrac{n}{5}\right]}{\left[\dfrac{n+2}{3}\right] + \left[\dfrac{n}{4}\right]}$$

would describe the terms that are given numerically above. The brackets indicate the greatest integer function. If this function were used, it would be more difficult to determine the limit of the sequence. It is not suggested that you try anything this bizarre, but this does give an indication that different people might view a sequence indicated by a few numeric terms and come to different conclusions concerning the sequence.

We now go back to our definition of limit and say that the sequence has a limit, L, (or converges to L) if and only if there is some number N such that for all n greater than N it is true that the elements of the sequence are all closer to L than a previously prescribed amount $\varepsilon > 0$. In other words we can set forth the definition of convergence as follows.

Definition 1.2. A *sequence* $\{f(n)\}$ *converges to a limit* L if and only if for any $\varepsilon > 0$ there is an integer N such that for each n, $n > N$, it is true that $|f(n) - L| < \varepsilon$.

While this definition reaffirms Definition VII.2.2, the re-statement in this form may be helpful.

EXAMPLE 1.1. Determine whether the following sequence is convergent. If it is convergent, find the limit $\{2, 3/2, 4/3, 5/4, \ldots\}$.

Solution. Since the formula for the n-th term is not given, we can only guess at the form which this should take. However, the way in which this is written would certainly indicate that here we could have $f(n) = (n + 1)/n$. If this is the case, then we can also write $f(n) = 1 + (1/n)$, and it would seem intuitively clear that the limit should be 1. In order to be certain, we use the definition. Since the difference between 1 and $f(n)$ is $1/n$, we need to insure that there is some N such that for any $n > N$ we have $1/n < \varepsilon$. Since $n > N$ insures that $1/n < 1/N$, we know that making $1/N < \varepsilon$ will give the result we want. However $1/N < \varepsilon$ is equivalent to $N > 1/\varepsilon$. Thus if we make $N > 1/\varepsilon$ we can be sure that for $n > N$ we have $|f(n) - 1| = |(1 + (1/n)) - 1| = 1/n < 1/N < \varepsilon$. Since $\varepsilon > 0$ we have no problem with dividing by zero. Thus we have established that one is the limit we are seeking, and we have also shown that the sequence is convergent. Of course we could have used any value of N that we wished as long as it met the criteria we had established, and consequently we only have a lower bound on the choice of N, not an upper bound. In other words, there is no unique N that we must use, but there is usually a lower bound to the N which will satisfy the requirements of the definition in establishing the fact that the sequence actually does converge.

EXAMPLE 1.2. Determine whether the sequence $\{\ln 1, \ln 2, \ln 3, \ln 4, \ln 5, \ldots\}$ converges, and if it is convergent, find the limit.

Solution. It would appear here that our function is $f(n) = \ln n$. It would also appear that the terms are getting larger, and there seems to be no limit in sight. (Intuition is not always reliable, but it certainly presents a good starting point.) Let us see whether we can prove that this gets larger than any finite number. Suppose we consider any number L as a possible limit. If we write the inequality $f(n) > L$, we realize that this is merely stating that ln $n > L$, or $n > e^L$ since the logarithm is a strictly increasing function. But this is possible for any number L, for we can always find an integer larger than e^L for any finite value of L. Now if we consider $\ln(e^k n) > L + k$, we see that by going further in the sequence we can exceed L by as much as we may wish. Since this same argument could be used concerning any supposed finite limit L, we see that this could not have a finite limit. We say that this is unbounded, or if we wish we can give up and say that the *limit is* ∞ where the figure 8 became weary and lay down for a rest. We might as well say, "This has no limit", for it hardly has one that is useful. In such a case we say the sequence diverges.

EXAMPLE 1.3. Determine whether the sequence $\{1, -1, 1, -1, 1, -1, \ldots\}$ converges.

   Solution. This is another type of sequence, and a strange one at that. It is content with oscillating between $+1$ and -1. In fact, we could write either $f(n) = (-1)^n$, or we could write $f(n) = \cos n\pi$, and in either case count our first term as $f(0)$. We can only let n be an integer and hence these two definitions of the function are equivalent. Since we can take ε as small as we please, just as long as it is positive, we could let $\varepsilon = .5$ (this is not really very small). However, there is no number which is within one-half a unit of both $+1$ and -1. It follows that if we select any L such that $|1 - L| < \varepsilon$, we do not have $|-1 - L| < \varepsilon$. Similarly if we satisfy $|-1 - L| < \varepsilon$, we fail to satisfy $|1 - L| < \varepsilon$. Therefore, we cannot find any place in the sequence in which two successive terms will satisfy $|f(n) - L| < \varepsilon$. Consequently there can be no N such that for values of $n > N$ we have this relation holding for all values of n. This sequence does not converge and therefore we call this a divergent sequence.

   You note that we have two types of divergent sequence here, one going off into the unknown and uncharted areas of infinity, and the other oscillating and not willing to settle down.

   There are two theorems which will be helpful at this point. The first is one that we used in the development of the integral. You will remember that the sums $U(P, f, g)$ and $L(P, f, g)$ formed sequences which were bounded and monotonic, and which therefore had limits.

Theorem 1.1. *Any bounded, monotonic sequence converges.*

OUTLINE OF PROOF. Note that any increasing sequence which is bounded must have a lub. Prove that this is the required limit. A similar proof is possible for decreasing sequences. $\square$

Corollary. *Any bounded sequence for which a positive integer N exists such that all terms after the n-th form a monotonic sequence must be a convergent sequence.*

The next Theorem is due to the very prolific French mathematician *Augustin-Louis Cauchy* (1789–1857).

Theorem 1.2. *Let the n-th term of a sequence be denoted by a_n. This sequence converges if and only if for any $\varepsilon > 0$ there is an integer N such that for any n and m, both greater than N, it is true that $|a_m - a_n| < \varepsilon$.*

OUTLINE OF PROOF. If the sequence has a limit, L, then for any $\varepsilon > 0$ we can find N such that $|a_n - L| < \varepsilon$ and $|a_m - L| < \varepsilon$ for $n, m > N$. It follows that we have $|a_m - a_n| = |(a_m - L) - (a_n - L)| \leq |a_m - L| + |a_n - L| < 2\varepsilon$ for any m and n which satisfy the conditions of the theorem. On the other hand, for the N satisfying the conditions of the Theorem, we know that all terms beyond the N-th term are in the interval $(a_n - \varepsilon, a_n + \varepsilon)$ provided $n > N$. This requires that the sequence be bounded, both above and below. Therefore, there is a lub for the set of terms beyond the N-th term and also a glb. This lub and glb differ by not more than 2ε. This requires that the sequence converge.   □

We started to talk about series and ended up discussing sequences. This may seem a bit disconcerting, but we will get to series in the next section. You will be happy at that point that we took this diversion. With regard to determining limits, there are no lengths to which we wouldn't be willing to go to find them if they exist or to find out whether they do or do not exist. We have illustrated three sequences. We have not taken such problems as would require L'Hôpital's rule and some of the other sophisticated items of Chapter VII, but don't be afraid to call in such machinery if it is appropriate.

EXERCISES

1. In each of the following sequences write the first six terms. Determine whether the sequence converges. If the sequence does converge, find the limit.

   (a) $\{f(n)\} = \{(0.9)^n\}$ starting with $n = 0$.
   (b) $\{f(n)\} = \{(n + 1)^2/n(n - 1)\}$ starting with $n = 2$.
   (c) $\{f(n)\} = \{n^3 e^{-n}\}$ starting with $n = 1$.
   (d) $\{f(n)\} = \{n!/n\}$ starting with $n = 1$.
   (e) $\{f(n)\} = \{n!/15^n\}$ starting with $n = 0$.
   (f) $\{f(n)\} = \{(n + 1)(n + 3)/(n + 2)(n + 4)\}$ starting with $n = 0$.

2. In each of the following sequences write the first six terms. Determine whether the sequence converges. If the sequence converges, find the limit.

   (a) $\{\arctan n\}$ starting with $n = 0$.
   (b) $\{\text{arc sec } n\}$ starting with $n = 1$.
   (c) $\{e^{-n}\}$ starting with $n = 0$.
   (d) $\{\ln n\}$ starting with $n = 5$.
   (e) $\{e^{-n} \ln n\}$ starting with $n = 1$.
   (f) $\{(\sin n)/n\}$ starting with $n = 1$.

3. (a) If $f(n) \le g(n)$ for each value of n, show that the limit of $\{f(n)\}$ must be no greater than the limit of $\{g(n)\}$ provided both limits exist.
   (b) If $\{g(n)\}$ has a limit and if $f(n) \le g(n)$ is a monotonic increasing function, show that $\{f(n)\}$ must have a limit.
   (c) If $f(n) = (n-1)/(n+1)$ and $g(n) = n/(n+1)$ show that $f(n) < g(n)$ for each nonnegative value of n.
   (d) Using the functions of part (c) show that $\{f(n)\}$ and $\{g(n)\}$ both converge and they both have the same limit.

4. Let $f(n) = \sum_{k=1}^{n} (0.5)^k$.

   (a) Write down $f(0), f(1), f(2), f(3), f(4),$ and $f(5)$.
   (b) Does the sequence $\{f(n)\}$ converge?
   (c) If the sequence converges, what is the limit of the sequence?
   (d) Does the starting value of n have anything to do with whether this sequence converges?
   (e) Does the starting value of n have anything to do with the limit to which this sequence converges (if it does converge)?

5. Let $f(n) = \sum_{k=0}^{n} ((-1)^k/(2k+1))$.

   (a) Write down $f(0), f(1), f(2), f(3), f(4),$ and $f(5)$.
   (b) Does the sequence $\{f(n)\}$ converge?
   (c) Find an upper bound and a lower bound for all terms in the sequence.
   (d) Can you find the limit of this sequence if it converges?
   (e) Is it necessary to be able to write down the limit of a sequence in order to prove that the sequence converges?

6. Let $x_0 = 30$, and let the sequence $\{x_k\}$ be the sequence of values obtained by using Newton's method to find the zeros of $f(x) = x^2 - 30$.

   (a) Write out the first six terms of $\{x_n\}$.
   (b) Show that this sequence has a glb.
   (c) Show that this sequence converges.
   (d) To what limit does this sequence converge?

7. One wishes to find a zero of $f(x)$ by Newton's method starting with a given value x_0. Let (a, b) be an interval such that x_n is in (a, b) for each value of x_n obtained by Newton's method. Assume that $f'(x)$ and $f''(x)$ do not change sign in (a, b).

   (a) Show that the sequence $\{x_n\}$ starting with $n = 1$ is monotonic.
   (b) Show that the sequence $\{x_n\}$ is bounded if there is a zero of $f(x)$ in the interval (a, b).
   (c) Show that the sequence $\{x_n\}$ must converge under the conditions given if there is a zero of $f(x)$ in (a, b).

8. Let $f(x)$ be a function for which $f'(a)$ exists. Let $\{x_n\}$ be a sequence which converges to the limit a.

   (a) Show that $\{DQf(a, x_n)\}$ converges to $f'(a)$ provided we exclude the possibility that $x_n = a$.
   (b) What is the relation between the interval E we used in our earlier development of the derivative as the limit of a differential quotient and the number N we have used in our definition of limit in this section?

(c) Have we said anything in this exercise which relates to our earlier specification of a deleted neighborhood? If so, what?

(d) Is the concept of a deleted neighborhood essential if we use the approach suggested here to obtain the derivative?

9. (a) Show that the sequence $\{2.81, 2.8181, 2.818181, 2.81818181, \ldots\}$ converges.

   (b) Find the limit to which this sequence converges.

   (c) Show that any decimal number can be considered as the limit of a sequence such as $\{1, 1.4, 1.41, 1.414, 1.4142, 1.41421, \ldots\}$.

SB10. It has been suggested that populations grow generally according to the law $P(t) = A - (A - B)e^{-kt}$ where A and B are constants, and where k is a positive constant considerably smaller than one.

   (a) Write the first five terms of $\{P(t)\}$ starting with $t = 0$.

   (b) Does this sequence appear to be approaching a limit?

   (c) If this sequence converges, what is the apparent limit?

   (d) What is the result of increasing the value of k?

SB11. A table, such as a table of populations, diameter of a tree for successive years, or GNP is a set of values in a sequence dictated by the time at which the data was taken.

   (a) Does the data of such a table conform to our definition of sequence?

   (b) Would it be reasonable to ask whether such a table would converge?

   (c) Would the fact that as time goes on additional entries can be placed in such a table have any influence on your answer? If so, how?

P12. A damped simple harmonic motion, such as the linear movement of a moving weight at the bottom of a pendant spring, is given by the equation

$$x(t) = 10^{-0.05t} \sin \frac{\pi t}{3}.$$

In this case t is measured in seconds and $x(t)$ is measured in centimeters.

   (a) Write out the first 6 terms of $\{x(t)\}$ starting with $t = 0$.

   (b) Is this sequence monotonic?

   (c) Is this sequence converging?

   (d) If this sequence is converging, what is its limit?

   (e) Does the speed of oscillation change with time?

IX.2 Convergence of Infinite Series

In the last section we discussed infinite sequences. We now turn our attention to infinite series.

Definition 2.1. If $\{a_k\}$ is an infinite sequence, then

$$\sum_{k=1}^{\infty} a_k = a_1 + a_2 + a_3 + \cdots$$

is an *infinite series*.

Note carefully that we do not imply addition when we talk about an infinite sequence, but we are discussing the sum of an infinite number of terms when we talk about infinite series.

The fact that we started this chapter with a section on infinite sequences suggests we intend to use such sequences as a vehicle for handling infinite series. This is precisely what we plan to do. We will start by considering a sequence of *partial sums*.

Definition 2.2. If

$$\sum_{k=1}^{\infty} a_k$$

is an infinite series, the sequence $\{s_n\}$ where $s_n = \sum_{k=1}^{n} a_k$ is called a sequence of *partial sums of the infinite series*.

It is clear that s_n is the sum of the first n terms of the sequence $\{a_k\}$. As n increases, s_n comes closer to approximating the sum of the infinite series. Thus, we have apparently transformed the problem of determining whether an infinite series converges to that of determining whether an infinite sequence converges. Likewise we have transformed the problem of finding the sum of a converging infinite series to the problem of finding the limit of a converging infinite sequence. We will clarify this with another Definition.

Definition 2.3. The infinite series $\sum_{k=1}^{\infty} a_k$ is said to converge to a limit S if and only if the infinite sequence of partial sums $\{s_k\}$ converges to S.

Since we have already considered the problem of determining whether a sequence converges, it would seem that we have the matter of converging series rather well under control. However with the exception of a very few series, such as the geometric series, there is no nice formula for the n-th partial sum, and the problem of determining convergence is not quite as easy as it first appeared. On the other hand, if we cannot be assured that a given series will converge there would seem to be little reason for further investigation of that series. This is not quite true, but it holds in the vast majority of cases. We must therefore seek other information which will aid us in determining whether a series converges or diverges.

Our first result concerning the convergence of a series will be of a one-sided nature, as indicated in the following theorem.

Theorem 2.1. *If it is not true that*

$$\lim_{j \to \infty} a_j = 0, \quad then \quad \sum_{j=1}^{\infty} a_j$$

diverges (or in other words fails to converge).

OUTLINE OF PROOF. Using the notation above, consider the difference $s_j - s_{j-1} = a_j$, and then apply the Cauchy condition (Theorem IX.1.2) for convergence of a sequence for the special case in which $m - n = 1$. □

This theorem does not tell us when a series *will* converge, but rather gives us a condition for determining when it will *not* converge. However, this is frequently useful information, too. It might seem on the surface of things that if the n-th term approaches zero that would be sufficient to guarantee convergence, but unfortunately this is not the case. Consider the following example.

EXAMPLE 2.1. Determine whether the infinite series $S = \sum_{k=1}^{\infty} (1/k)$ converges.

Solution. This series is known as the *harmonic series*. If we write out a few terms we have $S = 1 + 1/2 + 1/3 + 1/4 + \cdots$. It would appear that the terms are getting smaller, and therefore it might be reasonable to assume that the series would converge. In fact we note that $\lim_{k \to \infty} (1/k) = 0$ and therefore Theorem 2.1 certainly does not give any indication that the series would not converge. However, we explicitly warned against making the assumption that it would converge and we must investigate further. We do this by comparing this with a second series which we will designate S_1. In order to facilitate the comparison we will write several terms in a manner that makes the comparison very straightforward.

$$S = 1 + \tfrac{1}{2} + \tfrac{1}{3} + \tfrac{1}{4} + \tfrac{1}{5} + \tfrac{1}{6} + \tfrac{1}{7} + \tfrac{1}{8} + \tfrac{1}{9} + \tfrac{1}{10} + \cdots \qquad \text{(IX.2.1)}$$
$$S_1 = 1 + \tfrac{1}{2} + \tfrac{1}{4} + \tfrac{1}{4} + \tfrac{1}{8} + \tfrac{1}{8} + \tfrac{1}{8} + \tfrac{1}{8} + \tfrac{1}{16} + \tfrac{1}{16} + \cdots$$
$$= 1 + \tfrac{1}{2} + (\tfrac{2}{4}) \quad + \quad (\tfrac{4}{8}) \quad + \quad (\tfrac{8}{16}) + \cdots$$

Since S_1 has an infinite number of terms, it is clear that we can continue to combine enough terms to add as many halves as we wish. Thus S_1 increases without limit. On the other hand we note that for each term in S_1 there is a corresponding term in S, and the corresponding term in S is at least as large as the term in S_1. Therefore the value of S must be at least as great as the value of S_1. The only conclusion possible is that (IX.2.1) diverges. The fact that the terms approach zero was not sufficient to insure convergence.

We can gain one positive result from Theorem 2.1, however, for there is one type of series in which convergence follows provided the terms approach zero as we proceed along the series. We should precede this theorem with a definition.

Definition 2.4. An infinite series is an *alternating series* provided the product of any two consecutive terms is negative.

Theorem 2.2. *If* $A = \sum_{j=0}^{\infty} (-1)^j a_j$ *is an alternating series and* $\lim_{j \to \infty} a_j = 0$, *and if there is some value of* N *such that* $\{a_j\}$ *is monotonic for all* $j > N$, *then the series is convergent.*

OUTLINE OF PROOF. Since the series is alternating, either $a_j > 0$ for every j or $a_j < 0$ for every j. This follows since the alternating signs are taken care of by the factor $(-1)^j$. We will consider the case in which $a_j > 0$. Since the terms are monotonic if $j > N$, we know that they must be monotonic decreasing, since they are positive and approaching zero as a limit. It will be convenient to let $M = N$ if N is even and let $M = N + 1$ if N is odd. We can now write the alternating series in either of two ways.

$$A = \sum_{j=0}^{M-1} (-1)^j a_j + (a_M - a_{M+1}) + (a_{M+2} - a_{M+3}) + \cdots \quad \text{(IX.2.2)}$$

or

$$A = \sum_{j=0}^{M} (-1)^j a_j - (a_{M+1} - a_{M+2}) - (a_{M+3} - a_{M+4}) - \cdots. \quad \text{(IX.2.3)}$$

Since the monotonicity assures us that the values in the parentheses are positive, we see that A has a value that is bounded above by

$$\sum_{j=0}^{M} (-1)^j a_j \quad \text{and is bounded below by} \quad \sum_{j=0}^{M-1} (-1)^j a_j.$$

These results assure us that the partial sums of (IX.2.2) form a monotonic increasing sequence with an upper bound. Therefore we have a lub. In similar fashion in (IX.2.3) we have partial sums forming a monotonic decreasing sequence which is bounded below, and which therefore has a glb. The fact that the a_j are approaching zero as the series progresses assures us that the lub and glb referred to must be equal. Therefore the series must converge to this common limiting bound. If we considered the case in which the a_j were negative a similar proof could be used.   □

The net result we have at the present time from two theorems and one example is a bit confusing. We know that a series will not converge unless the sequence of successive terms of the series converges to zero. On the other hand this information assures us of convergence only in the case of the alternating series. It would be helpful to develop some additional techniques for determining whether a series will converge. We can now go back to the method we used in Example 2.1 and set the method of comparing two series forth somewhat more explicitly. In fact we will state this method as a theorem.

Theorem 2.3. *If* $A = \sum_{j=0}^{\infty} a_j$ *is an infinite series of non-negative terms, and if* $B = \sum_{j=0}^{\infty} b_j$ *is also a series of non-negative terms, and if* $a_j \geq b_j$ *for each value of* j, *then if* B *is divergent so is* A, *and if* A *is convergent so is* B.

OUTLINE OF PROOF. Consider the two sequences of partial sums, and note that each sequence is an increasing sequence. If A has a limit, B is bounded and must have a limit. If B fails to have a limit, then A has no upper bound. $\square$

You should observe that knowing the series A to be divergent would give no information concerning the series B. If B is convergent, we have no information concerning A. The majority of errors that occur in attempting to apply this Theorem occur when one of these last two cases occurs. We have talked about a series of non-negative terms here. Such a series permits us to use the method of comparison in a straightforward fashion. The fact that comparison is not as easily applied to a series in which the terms can vary in sign should be apparent in view of the fact that the alternate series

$$\sum_{j=1}^{\infty} \frac{(-1)^{j+1}}{j}$$

converges by Theorem 2.3 while the harmonic series, the corresponding series with only positive terms,

$$\sum_{j=1}^{\infty} \frac{1}{j}$$

diverges as shown in Example 2.1. This indicates that the presence of negative signs may alter significantly the character of an infinite series. The fact that the convergence in the case in which all terms are non-negative is very decisive is indicated in the next theorem. However, before starting the theorem, we should define two more terms.

Definition 2.5. If $A = \sum_{j=0}^{\infty} a_j$ represents an infinite series, and if the series $\sum_{j=0}^{\infty} |a_j|$ is convergent, then A is said to be *absolutely convergent*. If A converges, but is not absolutely convergent, then A is said to be *conditionally convergent*.

The story now begins to unfold. The illustration in Section I.1 was conditionally convergent. Did this have anything to do with the strange result we were able to obtain by using it? Perhaps, but on with the next theorem.

Theorem 2.4. *If an infinite series is absolutely convergent, it is convergent.*

PROOF. Let $A = \sum_{j=1}^{\infty} a_j$ be the given series, let $r_k = \sum_{j=1}^{k} a_j$ and let $s_k = \sum_{j=1}^{k} |a_j|$. By hypothesis the sequence $\{s_k\}$ converges and hence for any $\varepsilon > 0$ there is an integer N such that if $n > m > N$ we have $|s_n - s_m| < \varepsilon$ by Theorem IX.1.2. But

$$s_n - s_m = \sum_{j=m+1}^{n} |a_j|.$$

Therefore we have

$$\varepsilon > |s_n - s_m| = \sum_{j=m+1}^{n} |a_j| \geq \left| \sum_{j=m+1}^{n} {}^* a_j \right| = |r_n - r_m|.$$

Therefore, with the aid of the triangle inequality in the above relation we can use Theorem IX.1.2 to show convergence of the given series. $\square$

It is now apparent that if we can show any series to be absolutely convergent, we have shown it to be convergent, regardless of the presence or the placement, of negative signs. Much of our work from this point on in attempting to establish convergence will therefore be devoted to series having only positive terms, for this is tantamount to considering absolute convergence. Before we leave this line of thought we should insert one additional result.

Corollary 2.1. *If A is a series of positive terms and if B is a series consisting of the same positive terms (perhaps in a different order) and if A converges to a limit L then B converges to L.*

OUTLINE OF PROOF. Let the j-th term of A be denoted by a_j and the j-th term of B by b_j. For any $\varepsilon > 0$ there is a value N such that

$$\sum_{j=N+1}^{\infty} a_j < \varepsilon \quad \text{and hence} \quad \sum_{j=1}^{N} a_j > L - \varepsilon.$$

Since B includes all of the terms of A there is some number $M \geq N$ such that each a_j, $1 \leq j \leq N$, is included among the b_k, $1 \leq k \leq M$. Therefore

$$\sum_{k=1}^{M} b_k \geq \sum_{k=1}^{N} a_j > L - \varepsilon.$$

Since all of the terms of A are in B and vice versa, we must have

$$\sum_{k=M+1}^{\infty} b_k \leq \sum_{j=N+1}^{\infty} a_j < \varepsilon.$$

This last result tells us that B converges, and we know that $\sum_{k=1}^{M} b_k \leq L$, since all of the terms in this sum are in A. From this it follows that B must converge to L. $\square$

Definition 2.6. If A and B are the sums of the convergent infinite series indicated by

$$A = \sum_{j=0}^{\infty} a_j \quad \text{and} \quad B = \sum_{j=0}^{\infty} b_j,$$

and if $s_j = a_j + b_j$ and $p_j = \sum_{k=0}^{j} a_k b_{j-k}$, then the series $S = \sum_{j=0}^{\infty} s_j$ is defined to be the *sum of the series A and B* and the series $P = \sum_{j=0}^{\infty} p_j$ is defined to be the *(Cauchy) product of A and B.*

It is possible to define the product of two series in more than one way, and this is the reason for establishing the specific definition of product we will be using. This definition will be particularly useful when we consider power series later in the chapter.

Theorem 2.5. *If $A = \sum_{j=0}^{\infty} a_j$ and $B = \sum_{j=0}^{\infty} b_j$ are two convergent series, their sum converges to $(A + B)$.*

OUTLINE OF PROOF. Show that for any $\varepsilon > 0$ there is a positive integer N_a such that if $n > N_a$, then $|\sum_{j=0}^{n} a_j - A| < \varepsilon$, and similarly there is an integer N_b such that for $n \geq N_b$ it follows that $|\sum_{j=0}^{n} b_j - B| < \varepsilon$. Let N be the larger of N_a and N_b, and then show that the series representing the sum is within the interval $(A + B - 2\varepsilon, A + B + 2\varepsilon)$. $\qquad\square$

We can use this result to prove a corollary which explains many things.

Corollary 2.2. *If*

$$A = \sum_{j=1}^{\infty} a_j$$

is a conditionally convergent series, the series obtained by deleting all positive terms without changing the order of occurrence of the remaining terms, and the series obtained in a similar manner by deleting the negative terms will each diverge.

OUTLINE OF PROOF. If the series of positive terms converges, it is absolutely convergent. If the series of negative terms converges, it is the negative of a series of positive terms, and therefore is absolutely convergent. If each of these series is absolutely convergent, the original series is the sum of two absolutely convergent series and is therefore absolutely convergent. If the series of negative terms converges to $(-N)$ but the series of positive terms diverges, the partial sums of the given series are bounded below by $p_k - N$ where p_k is a partial sum of the series of positive terms. Since N is presumed to be finite and the sequence $\{p_k\}$ has no upper bound, it follows that $\{p_k - N\}$ has no upper bound, and the given series will diverge, contrary to hypothesis. A similar argument can be used if the series of negative terms is assumed to diverge and the series of positive terms to converge. Consequently, it is not possible that either the series of positive terms or the series of negative terms will converge if the given series is conditionally convergent. $\qquad\square$

Now we have the full story. If a series is absolutely convergent we can place the terms in any order we wish as long as we include all of the terms, and the sum will not be altered. On the other hand, if the series is conditionally convergent any change in the order may well affect the sum of the series. In fact, if you give the matter some thought you could convince yourself that given a conditionally convergent series you can arrange the terms in some order such that you can obtain any desired number as a sum. It is now clear that the insertion of parentheses in (I.1.1) did no harm, but the change in the order of the terms due to the rearrangement between (I.1.4) and (I.1.5) was the fatal error.

Theorem 2.6. *If $\sum_{j=0}^{\infty} a_j$ converges to L, then $\sum_{j=0}^{\infty} ca_j$ converges to cL where c is any constant.*

OUTLINE OF PROOF. Use the sequence of partial sums and the generalized distributive law. Note that for the $\varepsilon > 0$ with which you start, the resulting inequality will involve $c\varepsilon$.   □

We will now consider a more general theorem concerning products.

Theorem 2.7. *The Cauchy product of two absolutely convergent infinite series is absolutely convergent, and converges to the product of the sums of the two series.*

OUTLINE OF PROOF. Let $A = \sum_{j=0}^{\infty} a_j$ and $B = \sum_{j=0}^{\infty} b_j$ be two convergent series of positive terms. The series

$$\sum_{j=0}^{\infty} Ba_j$$

will converge to AB. If we replace B in each term of this last series by the infinite series having B as a sum, we then have a series which includes as terms every product that can be formed using one factor from A and one from B. Therefore, since there is no duplication of terms in a Cauchy product, AB is an upper bound for the monotonic increasing sequence of partial sums of the Cauchy product. Therefore the product converges. We note that A and B are series of positive terms, and hence the partial sums form an increasing sequence in each case. Define N as in the proof of Theorem 2.5. Then we have

$$\sum_{j=0}^{N} a_j > A - \varepsilon \quad \text{and} \quad \sum_{j=0}^{N} b_j > B - \varepsilon.$$

However, $\sum_{j=0}^{2N} p_j$ includes all of the products that can be obtained from multiplying $\sum_{j=0}^{N} a_j$ and $\sum_{j=0}^{N} b_j$ and some additional products as well. (We are using p_j as defined in Definition 2.6.) Therefore, we are assured that

$$\sum_{j=0}^{2N} p_j > (A - \varepsilon)(B - \varepsilon) = AB - \varepsilon(A + B) + \varepsilon^2 > AB - \varepsilon(A + B).$$

Now we know that the product is bounded above by AB, but is within a finite multiple of ε of the value AB. Hence the sum of the product series must be AB.   □

The statement of Theorem 2.7 demands that both series be absolutely convergent. It can be shown, although we will not do it here, that the theorem would also be correct if one of the two series is absolutely convergent and the other one is conditionally convergent.

EXAMPLE 2.2. Find the sum and product of the series

$$\sum_{k=1}^{\infty} \frac{1}{k!} \quad \text{and} \quad \sum_{k=1}^{\infty} \frac{1}{k^2}.$$

Solution. Each of these series can be shown to be absolutely convergent, and hence the sum and product will be convergent. The sum will be

$$\left[\sum_{k=1}^{\infty} \frac{1}{k!}\right] + \left[\sum_{k=1}^{\infty} \frac{1}{k^2}\right] = \sum_{k=1}^{\infty} \left[\frac{1}{k!} + \frac{1}{k^2}\right]$$

$$= [1+1] + \left[\frac{1}{2} + \frac{1}{4}\right] + \left[\frac{1}{6} + \frac{1}{9}\right] + \cdots$$

Notice that the addition involves the sum of the two first terms, the sum of the second terms, etc. For the product we will have

$$\left[\sum_{k=1}^{\infty} \frac{1}{k!}\right]\left[\sum_{k=1}^{\infty} \frac{1}{k^2}\right] = [1 \cdot 1] + \left[\frac{1}{2} \cdot 1 + 1 \cdot \frac{1}{4}\right] + \left[\frac{1}{6} \cdot 1 + \frac{1}{2}\frac{1}{4} + 1 \cdot \frac{1}{9}\right] + \cdots$$

$$= 1 + \frac{3}{4} + \frac{29}{72} + \cdots$$

Here we have been careful to take the product of the first term, then the sum of the two products obtained by multiplying terms whose subscripts would add up to three, then the sum of products of those terms in which the subscripts would add up to four, etc. If you check, these products are precisely the ones we listed in our definition.

We now have theorems which will permit us to do some arithmetic with series, and we have other theorems that will assist us in determining whether a series converges or not. The comparison test would be particularly useful, but it would require knowledge of some series which are convergent and some which are divergent for purposes of testing. We have found that the harmonic series is divergent, and this is useful information, but this would not help us in establishing the fact that a series would converge. Another series that we might use is the geometric series. In this case each partial sum is the sum of a geometric progression, and since we have a formula for the sum of a geometric progression we have a good chance of being able to handle this case. We carry this out in Example 2.3.

EXAMPLE 2.3. Determine the character of convergence of the geometric series

$$\sum_{j=0}^{\infty} ar^j.$$

Solution. If $a = 0$, all partial sums are zero and the sum of the series is 0 regardless of the value of r. If $a \neq 0$ and $|r| = 1$, $|ar^j| = |a| \neq 0$ for each value of j and the series diverges by Theorem IX.2.1. If $a \neq 0$ and $|r| > 1$, $|ar^j| \geq |a| > 0$ for each value of j and the series diverges by Theorem IX.2.1. If $a \neq 0$ and $0 < |r| < 1$, let s_k be defined by

$$s_k = \sum_{j=0}^{k} ar^j = \frac{a}{1-r} - \frac{a}{1-r} \cdot r^{k+1}.$$

But

$$\lim_{k \to \infty} r^{k+1} = 0$$

and consequently $\lim_{k \to \infty} s_k = a/(1-r)$.

Therefore, the geometric series converges if and only if $|r| < 1$ and then it converges to the sum $a/(1-r)$.

That venture paid off well, for we found out not only some series which converge, but also some which diverge. We will let the geometric series and the harmonic series $1 + 1/2 + 1/3 + 1/4 \ldots$ suffice for this section. This latter series is called *the* harmonic series, although there are, in reality, several series of the harmonic type. They can all be obtained by using reciprocals of the terms in any arithmetic progression (provided none of the terms in the arithmetic progressions are zero), and they can all be shown to be divergent using a method somewhat similar to that employed in this section. They are called harmonic, for they have a relation to certain musical scales. We will find in the next section that the harmonic series is a special case of a larger class of series.

EXERCISES

1. Determine whether each of the following series converges, and if it does, determine the limit (if possible):

    (a) $\sum_{j=0}^{\infty} (1/3^j)$
    (b) $\sum_{j=0}^{\infty} (1/(-3)^j)$
    (c) $\sum_{j=0}^{\infty} ((\sin j)/2^j)$
    (d) $\sum_{j=0}^{\infty} \cos \pi j$

2. Determine whether each of the following series converges, and find the limit if it converges (if possible):

    (a) $(1/1) + (1/2) + (1/4) + (1/8) + (1/16) + \cdots.$
    (b) $(1/2) + (1/3) + (1/5) + (1/9) + (1/17) + \cdots.$
    (c) $(1/2) + (1/4) + (1/6) + (1/8) + (1/10) + (1/12) + \cdots.$

(d) $(1/1) + (1/3) + (1/5) + (1/7) + (1/9) + (1/11) + \cdots$.

(e) Show that your results in parts (c) and (d) and the convergence of

$$\sum_{k=1}^{\infty} \frac{(-1)^k}{k}$$

verify the corollary of Theorem 2.5.

3. Determine whether each of the following series converges:

(a) $\sum_{k=2}^{\infty} ((-1)^k/(\ln k))$

(b) $\sum_{n=1}^{\infty} n^{-0.5}$

(c) $\sum_{n=0}^{\infty} (n/(n+1))$

(d) $\sum_{n=0}^{\infty} e^{-0.02n}$

(e) $\sum_{n=0}^{\infty} ((\sin n)/2^n)$

(f) $\sum_{n=0}^{\infty} \arctan n$

4. The series $\sum_{k=0}^{\infty} (1/k!)$ converges to the value e.

(a) With this information use multiplication of series to find a series with limit e^2.

(b) Use addition of series to find a series with limit $e^2 + e$.

(c) Obtain a series with limit $(e + 1)$.

(d) Multiply the series with limit e by your result of part (c) and see whether you obtain the same series as that obtained in part (b) or whether you have a different series converging to $e^2 + e$.

5. (a) Express $2.6413131313\ldots$ as a constant plus an infinite series of the form $2.64 + 0.0013 \sum_{k=0}^{\infty} (1/10^{2k})$.

(b) Express $2.6413131313\ldots$ as the quotient of two integers using the sum of part (a).

(c) Using the technique expressed in parts (a) and (b) express $13.478478478\ldots$ as the quotient of two integers.

(d) Let N be a number which in decimal form consists of some sequence of digits followed by a repeating sequence of digits in the manner of the numbers in parts (a) and (c). Show that any such number can be expressed as the quotient of two integers, and that the number is rational by definition.

6. (a) Evaluate $\int_0^{\infty} e^{-x} d[x]$ where $[x]$ is the greatest integer function.

(b) For what values of k does $\int_0^{\infty} e^{kx} d[x]$ exist?

(c) For values of k for which the result in part (b) exists, find the value of the integral as a function of k.

7. A ball is dropped from a height of 8 feet. Each time it hits the floor it bounces to a height which is three-fourths of the height from which it last fell. Find the total distance the ball travels from the moment it is dropped until it comes to rest.

8. If $\sum_{k=0}^{\infty} a_k$ is an alternating series and if the absolute values of the successive terms form a monotonic decreasing sequence, show that the sum of the series is between any two successive partial sums. Use this result to show that any partial sum differs from the sum of the series by less than the first term omitted.

9. Prove Theorem 2.1.

10. Prove Theorem 2.2.

11. Prove Theorem 2.3.

12. Prove Theorem 2.5.

13. Prove the Corollary of Theorem 2.5.

14. Prove Theorem 2.6.

C15. (a) Using the computer find the sum of the first 100 terms of the harmonic series.
     (b) Find the sum of the first 1000 terms.
     (c) Find the sum of the first 5000 terms.
     (d) Do the results you have obtained give any information by themselves con-
         cerning convergence?
     (e) Explain why it is difficult to use numerical results to try to prove theorems
         concerning convergence.

C16. (a) Using the computer find N_n, the number of terms of the harmonic series such
         that the partial sum s_{N_n} is at least as large as n, for $n = 2, 3, 4, 5, 6, 7, 8, 9$, and
         10.
     (b) Find the ratio N_{n+1}/N_n for $n = 2, 3, \ldots, 9$ from your results in part (a).
     (c) Is the sequence of ratios found in part (b) approaching a limit? If so what limit
         is it approaching?
     (d) Determine whether there is any relation between $\sum_{j=1}^{N_n} (1/j)$ and $\int_1^n (1/x)dx$.
     (e) See whether the information found from your investigation of part (d) will
         shed any light on the results observed in part (c).

IX.3 The Integral Test and Ratio Test

A little thought will show that there is a very close link between an infinite
series and the integral. In obtaining the integral we used summations with an
increasing number of terms in each successive sum as we refined our intervals. It should not be surprising, then, if we were to relate integrals and series
here. In order to facilitate our discussion in this section we will consider all
terms of each series to be positive. This is a conservative assumption for if the
series converges with all positive terms, it is absolutely convergent. We have
already shown that this implies that the series must converge regardless of any
modifications made in the signs of various terms. Furthermore, we wish to
consider that there is some term beyond which all successive terms form a
monotonic decreasing sequence. This does not stretch our imaginations
either, for we know that if a series converges the sequence of successive terms
must approach zero as a limit. We also know that the first few terms of a
series do not affect the status of the series as far as convergence is concerned.
Therefore, we do not alter the effectiveness of the results we will be developing
by assuming that the terms of the series are non-negative and that beginning
with N-th term they form a monotonic decreasing sequence.

With these assumptions already made, consider the series,

$$A = \sum_{n=0}^{\infty} a_n.$$

It will be convenient to think of having a function $f(n) = a_n$. Such a function will always exist, although it is not always easy to write such a function down in explicit terms. We will ignore this latter problem for the moment and assume we have such a function. We could then rewrite our series as

$$A = \sum_{n=0}^{\infty} f(n).$$

With this introduction we are ready to prove our first theorem of the section.

Theorem 3.1. *If $A = \sum_{n=0}^{\infty} f(n)$ is an infinite series such that $f(n)$ is non-negative and monotonic decreasing for all values of n larger than some fixed integer N, then A converges or diverges depending on whether $\int_{N}^{\infty} f(n)dn$ exists or fails to exist.*

PROOF. Since $f(n)$ is monotonic decreasing provided $n \geq N$ we know that $f(n) \geq f(n + 1)$ for all such values of n. Define the function $M(x)$ such that $M(x) = f(n)$ for any integer n if $n \leq x < n + 1$, and define $m(x)$ such that $m(x) = f(n + 1)$ if $n < x \leq (n + 1)$ as shown in Figure IX.1. These functions are such that $M(x) \geq f(x) \geq m(x)$ for all values of $x > N$. Consequently,

$$\int_{N}^{\infty} M(x)dx \geq \int_{N}^{\infty} f(x)dx \geq \int_{N}^{\infty} m(x)dx.$$

By the manner in which $M(x)$ and $m(x)$ were formulated it follows that

$$\int_{N}^{\infty} M(x)dx = \sum_{n=N}^{\infty} \int_{n}^{n+1} M(x)dx = \sum_{n=N}^{\infty} f(n) \cdot 1 = \sum_{n=N}^{\infty} f(n)$$

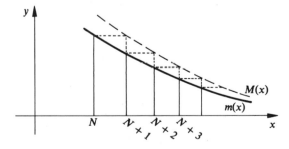

Figure IX.1

and

$$\int_N^\infty m(x)dx = \sum_{n=N}^\infty \int_n^{n+1} m(x)dx = \sum_{n=N}^\infty f(n+1) \cdot 1$$

$$= \sum_{n=N+1}^\infty f(n) \cdot 1 = \sum_{n=N+1}^\infty f(n).$$

Hence, we have the relation

$$\sum_{n=N}^\infty f(n) \geq \int_N^\infty f(x)dx \geq \sum_{n=N+1}^\infty f(n)$$

and the difference between the first and last term is precisely the value $f(N)$. The important item to notice is that if the improper integral

$$\int_N^\infty f(x)dx$$

exists, this is an upper bound for the portion of the infinite series from the $(N + 1)$-st term on, and therefore the series must converge. On the other hand, if the improper integral fails to exist, then since the sum of the terms of the infinite series from the N-th term on exceeds the integral the series must diverge. This proves the theorem.   □

On the face of things, this would seem to be the answer to all our problems, particularly as they relate to absolute convergence of infinite series. This is a test that gives positive results—either the series does converge, or it does not converge. The difficulty arises in obtaining a function $f(n)$ for which we can evaluate the improper integral. For instance, if we have a factorial in each of the terms of the series, and this is a rather frequent situation, we are in trouble since we have no basis for integrating the factorial. It is true that we found a function called a gamma function in Chapter VII which seems to have all of the properties of the factorial, but then we would have the problem of trying to integrate a function which has an improper integral in the integrand—and this is not too inviting. However, the integral test is a big step forward. Among other things it will provide us with series which will be useful in the comparison test.

EXAMPLE 3.1. Test for convergence the infinite series

$$\sum_{n=1}^\infty \frac{1}{n^p}$$

where p is a positive constant.

Solution. The function that will serve the purpose of making it possible to use the integral test is rather obviously $f(n) = 1/n^p$.

While we indicate that it is obvious, it should not go without checking, for we must observe first of all that this gives us the right value, and secondly that this is monotonic decreasing. Since this does satisfy our hypotheses, we will use $f(n) = n^p$, and check whether the integral $\int_1^\infty x^{-p} \, dx$ exists. We have two cases to consider. If $p \neq 1$, then we have

$$\int_1^\infty x^{-p} \, dx = \lim_{M \to \infty} \int_1^M x^{-p} \, dx = \lim_{M \to \infty} \left[\frac{1}{p-1} - \frac{1}{(p-1)M^{p-1}} \right].$$

Our denominators cause no trouble, since $p \neq 1$, and we observe that if $p > 1$, the last term approaches zero, for the exponent $(p - 1)$ is positive. However, if $p < 1$, we have the exponent in the denominator negative, or equivalently we would have a power of M in the numerator, and this would then increase beyond all bounds. In this case the integral would fail to exist. As a result of our deliberations so far, we know that the given series will converge if $p > 1$ and diverge if $p < 1$. We still have no information on $p = 1$, since we have excluded this from our considerations. If we go back to the integral with $p = 1$, we have

$$\int_1^\infty x^{-1} \, dx = \lim_{M \to \infty} \int_1^M \frac{1}{x} \, dx = \lim_{M \to \infty} [\ln M - \ln 1].$$

This limit also fails to exist. Hence we have the fact that this series is divergent for $p = 1$. (You might note that the case for which $p = 1$ is exactly the one we considered in the last section, and we came out with the same conclusion. At least we are consistent.) Therefore, we know that the initial series is convergent if $p > 1$ and divergent if $p \leq 1$. This series, frequently referred to as the *p-series*, is a good one to keep in mind for purposes of comparison.

EXAMPLE 3.2. Test the series

$$\sum_{n=0}^\infty \frac{1}{n^2 + 1}$$

for convergence.

Solution. Note that the n-th term of this series is smaller than the n-th term of the p-series for $p = 2$. Thus $1/(n^2 + 1) < 1/n^2$ since the denominator of the first fraction is larger than the denominator of the second. However, since the p-series is convergent for $p = 2$, we know that the given series is convergent by Theorem (IX.2.3).

It should be pointed out that we do not know what the sum (or limit) is in this case, but we do know the series converges. Usually we are interested in knowing whether the series will converge, for if it does we could use calculating or computing equipment to obtain an approximation to the sum of the series. It should be pointed out, if you have not already observed it, that it is possible to write a program which will come out with a nice result for the sum of a series, even though the series may *not* converge. A computer has no way of discerning divergence. Therefore, it is essential that you know whether a sum exists in order to know whether to attempt to find its value or even more important that you know whether to accept a purported sum as having a chance of being correct. In other words it would be rather futile to put in a lot of work trying to find even an approximation for the sum if the sum did not exist. Therefore, the information concerning convergence is worth working for.

You may well ask whether we might not capitalize on the fact that going from $p = 1$ to a value very slightly greater than one took us from the divergent to the convergent case. Would it be possible that this might give us some boundary series and we could then compare everything with this "last divergent series" or "first convergent series", which ever we had. The answer, unfortunately, is that this is not possible. Given any divergent series we can find a more slowly divergent series (using the intuitive meaning of the word "slowly" here), and given any convergent series we can find a more slowly converging series. Some of these will be indicated in the exercises at the end of this section. Despite the fact that we have not taken care of all cases, however, we do have an additional series for purposes of applying the comparison test. The geometric series and the p-series, are the most frequently used series for comparison purposes. Be certain that you know how to apply them.

As an example of the use of the comparison test with the geometric series, we develop one more frequently used convergence test. It should be made clear at this point that there are a great many convergence tests, and we will not discuss all of them. If we did, this entire book could easily be devoted to the problem of testing series for convergence. However, there are certain tests that seem to take care of a very large percentage of the problems that arise in practice, and as we have done in other areas where we selected only a small number of the vast possible number of techniques, we will mention only this small number of tests here.

Theorem 3.2. *Given the infinite series* $\sum_{n=1}^{\infty} a_n$, *define R by the relation* $R = \lim_{n \to \infty} a_{n+1}/a_n$ *if the limit exists. If* $R < 1$, *the series converges. If* $R > 1$ *the series diverges. If* $R = 1$ *this test gives no information.*

OUTLINE OF PROOF. If $R > 1$, there is some value, N, such that for all $n > N$ it is true that $|a_{n+1}/a_n| > 1$. This is equivalent to the statement that $|a_{n+1}| > |a_n|$. If this is true, then the n-th term cannot be approaching zero and the series must be divergent. On the other hand, if $R < 1$ consider an ε such that $\varepsilon < (1 - R)/2$. There is then a value, N, such that for all $n > N$, it is true that $|a_{n+1}/a_n| < R + \varepsilon$. Denote by S the value $R + \varepsilon$, and note that $S < 1$. From the N-th term on the given series can be compared with the geometric series $|a_N| + |a_N|S + |a_N|S^2 + |a_N|S^3 + \cdots$ and we note that each term of the geometric series is at least as large as the absolute value of the corresponding term of the given series. Since $S < 1$, the geometric series is convergent, and hence the given series must be convergent. To show that $R = 1$ gives no information, note that the p-series gives a limiting ratio of $R = 1$ for each value of p, and for some of these series we have convergence while for others we have divergence. Therefore, there is no information to be derived from this test if $R = 1$.   □

The test we have just introduced is called the *ratio test*. It is very much used, for many reasons, one of them being the fact that it is probably the easiest one to try since it is not necessary to dig into your repertoire of convergent and divergent series, nor is it necessary to evaluate improper integrals that at times are less than inviting. Consider, for example, the following illustration.

EXAMPLE 3.3. Test the series

$$\sum_{n=1}^{\infty} \frac{2^n}{n!}$$

for convergence.

Solution. Note that this would not have been handled easily by the earlier tests we derived, for we do not know how to integrate a factorial, and we do not have any ready series for purposes of comparison. Here we note that $a_{n+1} = (2^{n+1})/(n+1)!$ and $a_n = (2^n)/n!$. We can now determine R by

$$R = \lim_{n \to \infty} \frac{(2^{n+1})/(n+1)!}{2^n/n!} = \lim_{n \to \infty} \frac{2}{n+1} = 0.$$

In the algebraic simplification we have used the fact that $(n + 1)! = (n + 1)(n!)$, since $n!$ (except for the special definition of $0!$) is the product of the first n positive integers. Since $R < 1$, we know this series converges. (This type of manipulation with the factorials will serve you well on many occasions, so take a close look at it.)

There are many other tests, and you may become sufficiently interested to look up several of them and note the situations for which they ease the job of testing for convergence. The ones we have considered will be adequate for our purposes.

We should mention one very important type of series before terminating this particular section. Many series are not just series of constants, as we have considered up to the present, but are series of functions. Thus, the particular value of each term will depend on the value of some variable. For instance, we might consider the infinite series

$$\sum_{n=1}^{\infty} \frac{(-1)^{n+1} x^n}{n}.$$

This is, in fact, a series we will see later on, although we will derive it at that point and you do not need to remember it here. We do want to know, however, for which values of x this will converge. Note that each of the terms of the series is a function of x. Hence the value that each term will assume depends on the value that x assumes. It would be rather horrible to contemplate having to test the series for convergence for each value of x. Ordinarily the first test that one would apply in this case, partly because practice has shown this to be an efficient approach, is the ratio test. Now remember that x is presumed fixed for the particular series we are investigating. Hence, we have

$$R = \lim_{n \to \infty} \left| \frac{[x^{n+1}/(n+1)]}{x^n/n} \right| = \lim_{n \to \infty} |x| \left| \frac{n}{n+1} \right| = |x|.$$

(You might use L'Hôpital's rule on the limit of $n/(n+1)$, but there are easier ways of doing it.) Observe that x is a constant in the process of taking this limit. In other words, we have $R = |x|$ for our particular value of x. If $|x| < 1$, we know this series will converge and will in fact be absolutely convergent. If $|x| > 1$ it will diverge. The question remains concerning what happens when $|x| = 1$, or when $x = 1$ and when $x = -1$. We can check these by straight substitution, for if $x = 1$, we have the series $1 - 1/2 + 1/3 - 1/4 + 1/5 - \cdots$, and we know this converges. On the other hand if $x = -1$, we have the series $-1 - 1/2 - 1/3 - 1/4 - 1/5 \cdots$ and this is the negative of the harmonic series, and hence is known to diverge. Therefore, we know that the given series will converge if x is in the interval $(-1, 1]$, or $-1 < x \le 1$. This interval is called the *interval of convergence* of the given series. We were able to handle essentially all possible values of x in one operation this way, and this is not an inconsiderable advantage of the ratio test. In general, we can expect to use the ratio test to determine open intervals of convergence, and open intervals of divergence, but we will usually have to consider the end points of these intervals as special cases of infinite series of constant terms. This results from the fact that at these end points R is usually one.

EXERCISES

1. Test each of the following series for convergence:
    (a) $\sum_{n=1}^{\infty} (1/\sqrt{n^3 + 1})$
    (b) $\sum_{n=1}^{\infty} (n/\sqrt{n^3 + 1})$
    (c) $\sum_{n=2}^{\infty} ((n \cos n\pi)/(n^2 - 1))$
    (d) $\sum_{n=2}^{\infty} (1/n(\ln n))$
    (e) $\sum_{n=10}^{\infty} (1/n(\ln n)(\ln \ln n))$
    (f) $\sum_{n=0}^{\infty} (e^n/n!)$
    (g) $\sum_{n=0}^{\infty} (n^n/n!)$
    (h) $\sum_{n=1}^{\infty} (1/n(n + 1))$

2. Find the interval of convergence of each of the following series:

   (a) $\sum_{n=1}^{\infty} ((2x - 1)^n/n)$
   (b) $\sum_{n=1}^{\infty} (x^n/n!)$
   (c) $\sum_{n=1}^{\infty} (x^n/n^2)$
   (d) $\sum_{n=1}^{\infty} ((\cos n\pi x)/n^2)$
   (e) $\sum_{n=1}^{\infty} ((-1)^n/nx)$
   (f) $\sum_{n=0}^{\infty} (x^n)(n!)$

3. Find the interval of convergence:

   (a) $\sum_{n=0}^{\infty} ((-x^2)^n/(2n)!)$
   (b) $\sum_{n=0}^{\infty} (x(-x^2)^n/(2n + 1)!)$
   (c) $\sum_{n=1}^{\infty} ((\ln x)/n)$
   (d) $\sum_{n=2}^{\infty} (x^n/(n^2 - 1))$
   (e) $\sum_{n=1}^{\infty} x^n$
   (f) $\sum_{n=1}^{\infty} n^x$

4. Find the interval of convergence of each of the following series:

   (a) $\sum_{n=0}^{\infty} ((x - 2)^n/n!)$
   (b) $\sum_{n=1}^{\infty} ((x^2 - 4)^n/n)$
   (c) $\sum_{n=1}^{\infty} (x^n(\sqrt{n} + 1)/n^2)$
   (d) $\sum_{n=1}^{\infty} (x^n(\sqrt{n} + 1)/n)$
   (e) $\sum_{n=0}^{\infty} n(2x - 3)^n$
   (f) $\sum_{n=2}^{\infty} ((3x + 4)^n/(n^2 - 1))$
   (g) $\sum_{n=2}^{\infty} ((2 - x)^n/n(\ln n))$
   (h) $\sum_{n=1}^{\infty} ((-x)^n/n(n!))$

5. (a) Find the first five terms of the series for $f(x)g(x)$ if $f(x) = \sum_{n=1}^{\infty} (x^n/n)$ and $g(x) = \sum_{n=0}^{\infty} (x^n/n!)$.
   (b) Find the first five terms of $(\sum_{n=0}^{\infty} (x^n/n!))^2$.
   (c) Write out the first five terms of $A = \sum_{n=0}^{\infty} (x(-x^2)^n/(2n + 1)!)$.
   (d) Write out the first five terms of A^2.
   (e) Write the first six terms of $B = \sum_{n=0}^{\infty} ((-x^2)^n/(2n)!)$.
   (f) Write out the first six terms of B^2.
   (g) Obtain the first five terms of $A^2 + B^2$. Can you think of two other functions that behave this way?

6. (a) The series $\sum_{n=1}^{\infty} (x^n/n)$ converges in the interval $-1 \le x < 1$. Show that this series is absolutely convergent in the interval $-1 < x < 1$.
   (b) The p-series converges if $p = -1.01$ and diverges if $p = 1$. Use differentials to find the difference between $n^{-1.01}$ and n^{-1} if $n = 10{,}000$.
   (c) Find the percentage difference between $n^{-1.01}$ and n^{-1} if $n = 10{,}000$.

B7. A drug is administered to a patient in doses of c_o milligrams at intervals of T hours. An active residue in the amount of $c_o e^{-T}$ milligrams remains from the original dose after T hours. Hence at the time of the initial dose the active amount is c_o, after the second dose the amount is $c_o + c_o e^{-T}$, at the time of the third dose the amount is $c_o + (c_o + c_o e^{-T})e^{-T} = c_o + c_o e^{-T} + c_o e^{-2T}$, etc.

(a) Find the amount after the k-th dose.

(b) Does the active amount approach a limiting value?

B8. If one is given a dose of R units of radioactive iodine of atomic weight 131 at noon on one day, a strength of $11R/12$ units will remain at noon of the following day.

(a) If a person is given 0.05 units each day at noon, find the amount in the body immediately after the second dose.

(b) Find the amount in the body immediately after the fifth dose.

(c) If this dosage is continued, will the amount in the body at any time approach a limiting value? If so, what is it?

9. (a) Show that $\sum_{n=20}^{\infty} (1/n(\ln n))$ diverges.

(b) Show that $\sum_{n=20}^{\infty} (1/n(\ln n)(\ln \ln n))$ diverges, and use the comparison test to show that the terms in this series are smaller than the terms in the series of part (a).

(c) Show that $\sum_{n=20}^{\infty} (1/n(\ln n)(\ln \ln n)(\ln \ln \ln n))$ diverges and use the comparison test to compare this series with that of part (b).

(d) Use the results indicated by this Exercise to demonstrate that it is always possible to find a *more slowly diverging series* if it is to be of this general form.

IX.4 Power Series

Having spent some time learning how to determine whether an infinite series is convergent, we should now find out how to obtain such a series. Before starting on this venture it is well to state carefully the assumptions we will make and the notation we will use. We will assume we are starting with a function $f(x)$ which possesses a continuous n-th derivative. This implies by earlier theorems that $f(x)$ also has a continuous first derivative, a continuous second derivative, and so on through a continuous $(n-1)$-st derivative. We shall denote the n-th derivative by $f^{(n)}(x)$ where the order of the derivative is given by the superscript in parentheses. We will also make use of the integral theorem of the mean which states that if $f(x)$ is continuous and $g(x)$ is monotonic in (a, b) then there exists a value c in (a, b) such that $\int_a^b f(x)dg(x) = f(c)(g(b) - g(a))$ as well as the version of the Fundamental Theorem which states $\int_a^b f'(x)dx = f(b) - f(a)$. These results are all familiar, but it is well occasionally to restate such results in order that we know precisely what we are assuming.

With the assumptions indicated above we can now write

$$\int_a^x f^{(n)}(x)dx = f^{(n)}(c)(x - a)$$

with c between a and x and

$$\int_a^x f^{(n)}(x)dx = f^{(n-1)}(x) - f^{(n-1)}(a).$$

These two results combine to give us the relation

$$f^{(n)}(c)(x - a) = f^{(n-1)}(x) - f^{(n-1)}(a). \tag{IX.4.1}$$

Before going further it would be well to observe that since a is a constant, the value of c will depend upon the value of x. Thus, if x were to change, it is reasonable to expect c to change, and therefore c is no more a constant than is x. We can emphasize the likelihood that c is a variable by replacing c with z. This we shall do in the sequel.

Since (IX.4.1) is a relation involving functions, we can integrate again. Note that $f^{(n-1)}(a)$ is a constant, and that $f^{(n-1)}(x)$ can be treated just as we treated $f^{(n)}(x)$ in the preceding step. Consequently we have

$$\int_a^x \left[\int_a^x f^{(n)}(x)dx \right]dx = \int_a^x f^{(n)}(z)(x - a)dx$$

$$= \int_a^x f^{(n-1)}(x)dx - \int_a^x f^{(n-1)}(a)dx. \tag{IX.4.2}$$

Considering the separate parts of (IX.4.2) we have

$$\int_a^x f^{(n)}(z)(x - a)dx = f^{(n)}(z_1) \int_a^x (x - a)dx = f^{(n)}(z_1)\frac{(x - a)^2}{2}$$

with z_1 between a and x,

$$\int_a^x f^{(n-1)}(x)dx = f^{(n-2)}(x) - f^{(n-2)}(a),$$

and

$$\int_a^x f^{(n-1)}(a)dx = f^{(n-1)}(a)(x - a).$$

Therefore we have from (IX.4.2)

$$f^{(n)}(z_1)\frac{(x - a)^2}{2} = [f^{(n-2)}(x) - f^{(n-2)}(a)] - f^{(n-1)}(a)(x - a). \tag{IX.4.3}$$

Before we leave this step it is worth noting that in (IX.4.2) we indicated we were taking the integral of an integral. In other words, we integrated the interior integral (with the corresponding interior dx) first and this result became the integrand for the second integral. Such a result is called an iterated integral. In many cases iterated integrals have values which are the same as those for a similar result called a multiple integral. We will use the term iterated integral here, for that is descriptive of the process we are using.

The two sides of the equation (IX.4.3) represent two ways of writing the value of the iterated integral of (IX.4.2).

Our terminology would suggest that we intend to continue this process. In fact, it would seem reasonable to integrate n successive times, for two times permitted us to reach the $(n - 2)$-th derivative and it would seem reasonable that n times would get to the $(n - n)$-th derivative or the function $f(x)$ itself. We will write out one more step, however, before trying to generalize. It is always good practice to write out several steps to insure that you know just what is happening, and we are trying to practice what we preach.

$$\int_a^x \left[\int_a^x \left[\int_a^x f^{(n)}(x)dx \right]dx \right]dx = \int_a^x f^{(n)}(z_1) \frac{(x - a)^2}{2} \, dx$$

$$= \int_a^x f^{(n-2)}(x)dx - \int_a^x f^{(n-2)}(a)dx$$

$$- \int_a^x f^{(n-1)}(a)(x - a)dx \qquad \text{(IX.4.4)}$$

or

$$f^{(n)}(z_2) \frac{(x - a)^3}{3!} = f^{(n-3)}(x) - f^{(n-3)}(a) - f^{(n-2)}(a)(x - a)$$

$$- f^{(n-1)}(a) \frac{(x - a)^2}{2!}.$$

We have z_2 within the interval of integration or between a and x. The pattern is now beginning to emerge. If we continue this process and integrate n successive times we have on the left side

$$\int_a^x \cdots \left[\int_a^x \left[\int_a^x \left[\int_a^x f^{(n)}(x)dx \right]dx \right]dx \right] \cdots dx.$$

The resulting equation can be written

$$f^{(n)}(z_{n-1}) \frac{(x - a)^n}{n!} = f^{(0)}(x) - \sum_{k=1}^{n} f^{(n-k)}(a) \frac{(x - a)^{n-k}}{(n - k)!}. \qquad \text{(IX.4.5)}$$

We have written $f^{(0)}(x)$ to indicate the derivation, but since the result of integrating the n-th derivative n times must be the original function we could write in equivalent fashion $f(x)$. Following our earlier discussion it is also apparent that z_{n-1} is some value in the interval bounded by a and x. For convenience we will replace z_{n-1} by Z, and then with a slight rearrangement of (IX.4.5) we can write

$$f(x) = \sum_{k=0}^{n-1} f^{(k)}(a) \frac{(x - a)^k}{k!} + f^{(n)}(Z) \frac{(x - a)^n}{n!}. \qquad \text{(IX.4.6)}$$

We have taken the liberty of changing the index of summation, but you will note that we have described exactly the same terms we had in the earlier expression. While this is not yet an infinite series, it approaches that if n is very large. The one disturbing item in this result is the presence of Z, a value which we cannot accurately determine. However, in the majority of cases of interest the final term in this expression, that is the one involving the n-th derivative, will be sufficiently small that we will not have much concern over the value of Z. We shall investigate this in some of the examples to follow.

The result (IX.4.6) which we have so laboriously obtained is a very useful one. It was first published by *Brook Taylor* (1685–1731) in 1715. Despite his insight in obtaining this result (probably by a different method) he had not yet considered the problem of convergence. Taylor seems to have been unaware that the rudiments of this development were discovered earlier by the Scotsman *James Gregory* (1638–1675). Taylor gets the credit, however, for the result in (IX.4.6) is known as *Taylor's theorem*. Often we find that $\lim_{n \to \infty} f^{(n)}(Z)((x - a)^n/n!) = 0$, and in this case we can write (IX.4.6) as

$$f(x) = \sum_{k=0}^{\infty} f^{(k)}(a) \frac{(x - a)^k}{k!}. \qquad \text{(IX.4.7)}$$

We now have an infinite series, and this series is known as *Taylor's series*. The special case in which $a = 0$ is known as *Maclaurin's series* after *Colin Maclaurin* (1698–1746) who published it in 1742 although the same result was known 25 years earlier by *James Stirling* (1692–1770). The dates given indicate that this was developed very soon after the publication of the basic elements of the calculus by Newton and Leibniz.

EXAMPLE 4.1. Expand $f(x) = e^x$ using Taylor's theorem with $a = 0$.

Solution. This is a good one to start with, for we find that each derivative is equal to e^x. Hence, each $f^{(j)}(a) = e^0 = 1$, and we have

$$e^x = \sum_{j=0}^{n-1} \frac{(x - 0)^j}{j!} + e^Z \frac{(x - 0)^n}{n!} = \sum_{j=0}^{n-1} \frac{x^j}{j!} + e^Z \frac{x^n}{n!}.$$

If we examine this result, we find that since Z must be some constant for each value of x, and in fact must be between 0 and x, then e^Z is no larger than e^x if x is positive and no larger than $e^0 = 1$ if x is negative. Hence, if we try to evaluate e^x for any specific value of x, we know that e^Z is some constant, and in any case we have an upper bound for its possible value. Again, assuming that x is some fixed value we can show that $x^n/n!$ approaches zero as n gets larger without limit. In view of the fact that e^Z is a constant, we then know that $\lim_{n \to \infty} e^Z(x^n/n!) = 0$. Therefore, if we take a sufficient number of terms, we can ultimately reach a point where the *remainder term*, that is the term involving the n-th derivative, is as small as we please. We can thus evaluate e^x with any desired degree of accuracy for any value of x. If you look at Exercise

III.4.9, you will find that you were really obtaining the Maclaurin's expansion for the function e^{-x}. On the basis of the above discussion we can go from Taylor's theorem to Taylor's series and write

$$e^x = \sum_{n=0}^{\infty} \frac{x^n}{n!}.$$

From this example we see that for some functions, at least, it is possible to obtain a Taylor's expansion, and to use the remainder term to test for convergence. When we talk about convergence, of course, we are referring to the infinite series. We started with a finite number of terms. It was possible, however, to think of going from step to step with increasing values of n, and thus constructing an infinite series. We had the net gain here that we had a measure of possible error at each step.

In general we can obtain an expression using Taylor's theorem for any function with a sufficient number of continuous derivatives. If we think of that portion of (IX.4.6) which excluded $f^{(n)}(Z)(x - a)^n/n!$ we have a partial sum of a Taylor's series. If the *remainder term*, $f^{(n)}(Z)(x - a)^n/n!$, is small, the partial sum must be close to the correct value of the function for the particular values of x used. If we show that the remainder term approaches zero as n increases without limit, we have shown that it would be profitable to take more terms in a Taylor's theorem expression, and hence to take the Taylor's series. Using this line of reasoning we are able to obtain an expansion for each of a large variety of functions as an infinite power series. We can frequently test for convergence by an examination of the remainder term, although in some instances the remainder term will be more complicated than we would wish to deal with. While it is true that the infinite series for e^x would converge for any value of x, it is also true that the series expansion for many functions will converge only over some finite interval. You should always determine whether a series converges before using it.

At this point we must put in a disclaimer. What we have shown in the case of e^x is that we can find the value of e^x for any real value of x to any desired degree of accuracy in theory. In practice, it is possible to do this provided we carry the required accuracy for each term in all computations. Hence, if the arithmetic is done by means of decimals, we would carry an appropriate number of decimal places if the particular term is not a terminating decimal. This indicates that we must be certain in advance that the number of decimal places carried is sufficient to prevent contamination by any round-off error in the arithmetic. Exact computation would imply that we would use an infinite number of terms, each with an infinite number of decimal places. Is it any wonder that we settle for an approximation? For practical purposes we use only a finite number of decimal places in any given situation anyway. Consequently, the fact that we obtain approximations causes no trouble. Except in rare circumstances we end up in practice with some approximation.

Since we can make the remainder term as small as we please, we can continue until any error that is present results only from the round-off error.

Thus, to calculate the value of e, we would have $x = 1$ in e^x, and hence

$$e = \sum_{j=0}^{\infty} \frac{1}{j!} \doteq 1 + 1 + .5 + .1666667 + .0416667 + .0083333$$

$$+ .0013889 + .0001984 + .0000248 + 0000028$$

$$+ .0000003 + \cdots,$$

the remaining terms being zero to at least seven decimal places. In this situation, as in some others, it is possible to simplify the work with appropriate organization. This computation can be made somewhat easier to follow by placing the numbers in a column and dividing the numbers by the successive integers in the manner shown. The sum of the quotients will give us the desired value. Note that we could easily have carried additional digits. Observe that we have rounded the results.

1	1.0000000
2	1.0000000
3	0.5000000
4	0.1666667
5	0.0416667
6	0.0083333
7	0.0013889
8	0.0001984
9	0.0000248
10	0.0000028
	0.0000003
sum =	2.7182819

The resulting value of e is given by $e = 2.7182819$. The last digit is in error by one if we round off the value of e, but by slightly less than one in actuality. The individual round-off errors have in some instances been positive and in others negative, and the sum of the positive errors about equalled the sum of the negative errors. The fact that we run into this computational complication does not refute the value of the Taylor's expansion, but rather brings into focus some of the problems of reality in carrying out computations. We might check further in this instance and observe that in this instance the theoretical error would be $e^Z/11!$, and since Z is between zero and one, we know that $e^Z < 3$ even if we are most generous. Therefore, the value of the error in stopping at this point can under no circumstances be as great as 10^{-7} by direct computation. All of the error in writing out seven decimal places must result from the round-off effects.

One additional item should be mentioned in connection with accuracy in computation. If we were to evaluate e^{-8} using a digital computer, we could expect rather inaccurate results, for while the series would certainly converge, the computer would bring in difficulties in that the computer uses only a fixed number of significant digits for each number. When we evaluate the term $8^7/7!$ we have 2,097,152/5,040 which is approximately equal to 400. This would require that of the fixed number of digits available we use three for the integral part, and hence we would have three fewer digits in which to carry the decimal part of the number. Furthermore, since e^{-8} involves a negative exponent, we have the subtraction of two numbers of approximately equal size. This eliminates many significant digits. This does not indicate that we should not use the computer, but rather that we have to exercise care in the use of the computer. It is not sufficient to be content with the knowledge of the theoretical value of the error term since round-off error will make a large contribution in many instances. Therefore, we should pay attention to the method used in developing the computer program. In determining values of e^x it is probably better to compute values of e^x for x in the interval $[0, 1]$ and then multiply the result by such powers of e as are required to obtain the particular value we seek.

Our development so far indicates that we can expand any function which has the necessary continuous derivatives. We should, of course, determine whether the resulting series will converge. In some instances the differentiation may not be overly pleasant to contemplate. For example, if we started with $f(x) = \tan x$, then $f'(x) = \sec^2 x$, and $f''(x) = 2 \sec^2 x \tan x$. We already have a product rearing its ugly head. Although we have a function which can be expanded in a Taylor's series, it is sometimes difficult to carry out the necessary computation to obtain the series. In some cases there are easier ways. We will mention a few of these in the next section.

EXERCISES

1. Find the Maclaurin series for each of the following functions and determine the interval of convergence for the series.
   (a) $\sin x$
   (b) $\cos x$
   (c) $x \sin x$
   (d) $x^3 - 3x^2 + 4x - 2$
   (e) $\tan x$
   (f) $\sqrt{x + 1}$
   (g) e^{3x}
   (h) e^{x^2}

2. Find the Taylor's series for each of the following functions about the indicated value of a and determine the interval of convergence of the series.
   (a) $\sin x$ about $a = \pi/6$
   (b) $\ln x$ about $a = 1$
   (c) $\cos x$ about $a = \pi/2$
   (d) $\sqrt{x}$ about $a = 9$
   (e) x^4 about $a = -2$
   (f) $\csc x$ about $a = \pi/2$

3. (a) Expand $\ln(1 + x)$ in a Maclaurin's series and find the interval of convergence.
   (b) Expand $\ln(1 - x)$ in a Maclaurin's series and find the interval of convergence.
   (c) By subtracting your result in part (b) from your result in part (a) obtain $\ln((1 + x)/(1 - x))$. What is the interval of convergence?
   (d) For what value of x does $(1 + x)/(1 - x) = 5$.
   (e) Using your results in part (d) and (c) evaluate $\ln 5$ to four decimal places.
   (f) If n is any positive number show that it is possible to use the method of this exercise to find $\ln n$. Include an assurance that all numbers x involved are within the interval of convergence of the series in part (c).

4. (a) Obtain the first four terms and the remainder term of the Maclaurin's expansion for $\tan x$. (Some of these terms may be zero.)
   (b) Using the first four terms of the Maclaurin's expansion of part (a) find the value of $\tan(\pi/6)$.
   (c) Use the remainder term of part (a) to find a bound for the error in your approximation of part (b).

5. (a) Expand $(a + bx)^n$ in a Maclaurin's series were a, b, and n are real constants.
   (b) Find the interval of convergence of this series.
   (c) Show that your result in part (a) is the binomial series.
   (d) If $a = 4$, $b = -0.3$, $n = 0.5$, and $x = 1$ use the remainder term to find how many terms would be required if the error were to be less than 0.000001.
   (e) Using your results of parts (a) and (d) find the value of $\sqrt{3.97} = \sqrt{4 - 0.03}$ with an error no greater than one millionth.

6. (a) Find the Maclaurin's expansion for $\sinh x$.
   (b) Find the Maclaurin's expansions for e^x and e^{-x}.
   (c) Using the definition of $\sinh x$ and the results of part (b) obtain the series for $\sinh x$.
   (d) Compare your results in parts (a) and (c).
   (e) Show that $\sinh x$ is an odd function.
   (f) Note whether the exponents of the Maclaurin's expansion of $\sinh x$ are all odd, all even, or mixed.

7. (a) If $f(x)$ is a polynomial of degree n, show that the Taylor's expansion will have a finite number of non-zero terms.
   (b) If all of the non-zero terms of the expansion of $f(x)$ are used, show that the remainder term indicates an error of zero.
   (c) Show that it is possible using a Taylor's expansion to express any polynomial $f(x)$ as a polynomial $g(x - a)$ for any real value a.
   (d) If $f(x)$ has a zero between a and $a + h$, show that $g(x - a)$ would have a zero between 0 and h.
   (e) Show that if h is sufficiently small (perhaps less than 0.1) the terms in the polynomial $g(x - a)$ of degree 2 or more can be ignored and a reasonable approximation of the zero of $g(x - a)$ can be obtained.
   (f) Show that the result of part (e) can be used to find a reasonable approximation of the zero of $f(x)$.

8. (a) Obtain the Maclaurin's expansion for $\sin x$.
   (b) It is known that $\sin x$ is an odd function. What can you observe about the exponents in your expansion.

(c) Differentiate the series for sin x and compare your result with the Maclaurin series for cos x.

(d) Differentiate the series for cos x and compare your results with the series for sin x.

(e) Square the series for sin x and the series for cos x and add the results. Could you have anticipated the result of this computation? ·

9. (a) Find a Maclaurin's series for arc tan x.

(b) Find the interval of convergence of your series.

(c) Since arc tan $1 = \pi/4$, find an infinite series which converges to $\pi/4$.

(d) From your result of part (c) find an infinite series which converges to π.

(e) Use trigonometric identities to show that arc tan $\frac{1}{2}$ + arc tan $\frac{1}{3}$ = arc tan 1.

(f) Use your series to obtain values for arc tan $\frac{1}{2}$ and arc tan $\frac{1}{3}$.

(g) Use your results in parts (e) and (f) to obtain a value for π.

C10. (a) Find the error term in the Maclaurin's series for cosh x.

(b) Find the number of terms of the Maclaurin's series for cosh x which would be required if we are going to limit x to the interval $[0, 1]$ and we want an error no greater than 0.000001.

(c) Write a computer program to evaluate cosh x for $x = 0, 0.1, 0.2, \ldots, 1$ with an error no greater than 0.000001 in each value. Have your program print the number of the first term in which the value is less than 0.0000001.

(d) Compare the number of terms used for each computation in part (c) with the number of terms predicted in part (b). Note that there is a more stringent requirement in part (c) in order to take care of any possible accumulation of round-off error or the fact that subsequent terms may add up to a total greater than 0.000001.

(e) From your comparison in part (d) note that the number of terms computed in part (b) is a maximum number, and you may frequently not require that number of terms.

M11. Write out in detail the proof of (IX.4.6) in the case where $n = 4$.

M12. (a) Expand ln x in a Taylor's expansion about $x = e$.

(b) Find the interval of convergence of this series.

(c) Examine ln x and determine whether ln x is defined for all values within the interval of convergence.

(d) Examine ln x and determine whether ln x is defined for any values outside the interval of convergence.

(e) Note that the interval of convergence usually extends an equal distance on either side of the value about which the expansion takes place. Would your results of part (d) explain, at least in part, the limitations on the interval of convergence?

IX.5 More About Power Series

In the last section we found that for any function with a sufficient number of continuous derivatives we could find a Taylor's expansion. At points where the remainder term is sufficiently small we can use this expansion to evaluate the function. However, the process of successive differentiations is not always a pleasant one to contemplate. The question arises whether there is an easier way of obtaining the series. For instance, if we wished a series for arc tan x, it is not overly pleasant to think of using the functions arc tan x, $1/(1 + x^2)$, $[-2x/(1 + x^2)^2]$, etc. At this point we are already well into the differentiation of fractions. Would it help to go in the other direction and take anti-derivatives? We might start with $1/(1 + x^2)$ and integrate. That would certainly give us arc tan x, but what do we have when we get it? Let us see.

$$\frac{1}{1 + x^2} = \frac{1 - (-x^2)^n}{1 + x^2} + \frac{(-x^2)^n}{1 + x^2}$$

$$= [1 - x^2 + x^4 - x^6 + \cdots + (-x^2)^{n-1}] + \frac{(-1)^n x^{2n}}{1 + x^2}.$$

For any value of x it is true that $x^{2n}/(1 + x^2) < x^{2n}$. Since we have a finite number of terms, we can integrate, and know that the integral of the sum is the sum of the integrals. Therefore, we can integrate, and have

$$\text{arc tan } x = \int_0^x \frac{dt}{1 + t^2}$$

$$= \int_0^x [1 - t^2 + t^4 - t^6 + \cdots + (-t^2)^{n-1}] dt + (-1)^n \int_0^x \frac{t^{2n}}{1 + t^2} dt$$

$$= x - \frac{x^3}{3} + \frac{x^5}{5} - \frac{x^7}{7} + \cdots + (-1)^{n-1} \frac{x^{2n-1}}{2n - 1} + (-1)^n \int_0^x \frac{t^{2n}}{1 + t^2} dt$$

for any finite value of n. This is an exact relationship. However,

$$\left| \int_0^x \frac{t^{2n}}{1 + t^2} dt \right| < \left| \int_0^x t^{2n} dt \right| = \left| \frac{x^{2n+1}}{2n + 1} \right|$$

and if $|x| \leq 1$, this term can be made as small as we please by taking n sufficiently large. Therefore if $|x| \leq 1$ we can let n increase without limit and we have an infinite series which is a convergent series. Thus, we seem to have a series which will represent arc tan x without having to perform the differentiation we might have anticipated. In view of the fact that we integrated a finite number of terms, we have no question concerning the validity of the finite operations performed. We have also established that we can obtain the value of arc tan x to any desired accuracy if $|x| \leq 1$.

The result we have obtained for arc tan x involves a polynomial plus a remainder term. Since a polynomial has a finite number of terms, we are able to integrate term by term and be certain that that portion of the work is correct. It is possible to obtain an upper bound on the integral of the error term and then determine whether this would approach zero for certain values of x. In this way we bypassed the problem of trying to integrate an infinite number of terms. While we will not discuss here the problems of integrating or differentiating an infinite series, we will note that this is frequently risky business. If one is integrating power series and uses only values of x inside the interval of convergence of the series, a discussion similar to that for arc tan x would show that we would have a convergent series which would converge to the function we had in mind. If we do try to integrate term by term or to differentiate term by term, we should at least check the resulting series to determine whether it converges, and if it does converge to determine the interval of convergence.

We now appear to have two ways of obtaining a series for arc tan x. We have illustrated one in the preceding paragraphs. We can also use Taylor's expansion. Do these give two different series? It should be rather easy to see that if we have expanded one in powers of $(x - a)$ and the other in powers of $(x - b)$ where $a \neq b$, they must at least appear different. If, on the other hand, they are both in powers of $(x - a)$, we will have to do a bit more looking.

Theorem 5.1. *If the two series*

$$\sum_{j=0}^{\infty} a_j(x - c)^j \quad and \quad \sum_{j=0}^{\infty} b_j(x - c)^j$$

both converge absolutely to the same function $f(x)$, in some open interval which includes $x = c$ as an interior point, then $a_j = b_j$ for each value of j.

OUTLINE OF PROOF. Since the two series approach the same function as a limit throughout some neighborhood of $x = c$, we know that the two series should be equal when x is replaced by c. This leads us immediately to $a_0 = b_0$. Consider the function

$$f_1(x) = \frac{f(x) - a_0}{x - c}$$

which is defined at every point in the interval except $x = c$. Now

$$\lim_{x \to c} f_1(x)$$

must be unique if it exists. Since,

$$f_1(x) = \sum_{j=1}^{\infty} a_j(x - c)^{j-1} = \sum_{j=1}^{\infty} b_j(x - c)^{j-1},$$

we can show that

$$\lim_{x \to c} f_1(x) = a_1 \quad and \quad \lim_{x \to c} f_1(x) = b_1.$$

Therefore $a_1 = b_1$. Continue by mathematical induction. $\square$

With this theorem at our disposal, we know that we would have the same series for arc tan x regardless of whether we used the short method or the longer method as long as both series involve powers of $(x - 0) = x$. For many of the inverse trigonometric functions it is possible to use a method similar to that used for arc tan x. In the case of the inverse secant and cosecant it would be necessary to make a substitution in order to have a suitable interval of convergence since zero could not be included in the interval of convergence. The substitution $x = 1/y$ would work in this instance, but note that this would then cause a reduction to arc cos y and arc sin y respectively. It is also possible to expand the derivative of $\ln(1 + x)$ and of $\ln(1 - x)$ and then integrate to obtain the series for $\ln(1 + x)$ and $\ln(1 - x)$.

If we wish to obtain the expansion for tan x, it would seem inviting to use tan $x = (\sin x)/(\cos x)$. This again brings up the question of performing arithmetic operations with series. We have shown that addition (and subtraction) are possible if the series involved are convergent. We have also shown that multiplication is valid with absolutely convergent series. As might be anticipated, division is more likely to cause trouble. If the quotient is absolutely convergent, however, we can feel reasonably confident, for in that case we could check the division by multiplication of series much as we have checked other divisions in the past. In special situations it may help to note the manner in which the remainder terms enter the manipulation. A careful investigation will show in this case that the quotient of the Maclaurin series for the sine by the Maclaurin series for the cosine gives the Maclaurin series for the tangent in the interval in which the quotient series converges. In cases of doubt, return to the definition of the Taylor's expansion, but it is possible at times to find an easier way of deriving the power series expansion of a particular function. Be certain that you have a valid reason for anticipating that the result will be correct, however. Above all check to make certain that the series with which you are working is convergent for the interval in which you are interested.

At an early point in the development of the calculus Leonhard Euler, a very prolific mathematician and the one to whom we owe such symbols as $f(x), e, \Sigma$, and i, discovered a very interesting relation between the exponential function and the trigonometric functions. This relationship is so very useful in many areas of mathematics and mathematical applications that we will consider it before terminating this section. We know that the Maclaurin's series

$$e^{cx} = \sum_{k=0}^{\infty} \frac{(cx)^k}{k!}$$

converges for all values of x, and for any real constant c. It would take relatively little effort, using our definition of absolute value as distance in the Argand diagram and using the definition of limit we developed in Section VII.2, to show that this series converges in case c is a complex number. For our purposes we will let $c = i$. The series then becomes

$$e^{ix} = \sum_{k=0}^{\infty} \frac{(ix)^k}{k!} = \sum_{k=0}^{\infty} \frac{(-x^2)^k}{(2k)!} + i \sum_{k=0}^{\infty} \frac{x(-x^2)^k}{(2k + 1)!}. \qquad \text{(IX.5.1)}$$

For the latter expression we have made use of the fact that $i^{4n} = 1$ for any integer n and $i^{4n+1} = i$, $i^{4n+2} = -1$, and $i^{4n+3} = -i$. If your memory reaches back to some of the results contained in the Exercises of the last section, you will remember that

$$\cos x = \sum_{k=0}^{\infty} \frac{(-x^2)^k}{(2k)!} \quad \text{and} \quad \sin x = \sum_{k=0}^{\infty} \frac{x(-x^2)^k}{(2k+1)!}.$$

We can then take the results of (IX.5.1) and write

$$e^{ix} = \cos x + i \sin x. \tag{IX.5.2}$$

This is the *Euler relation*. We will be making use of it in Chapter XI, as well as in our discussion here. In particular, note that we now have $(\cos kx + i \sin kx) = e^{ikx} = (e^{ix})^k = (\cos x + i \sin x)^k$ for any real value, k, and hence we have a generalized version of DeMoivre's theorem. This explains the similarity of the developments of Sections III.4 and III.5. Euler was very much impressed with one particular result, namely $e^{i\pi} = \cos \pi + i \sin \pi = -1$, for he felt that with the inclusion of e, i, π, and -1 in a single relation he had something very fundamental combining some of the key units of mathematics.

If we observe that

$$\begin{cases} e^{ix} = \cos x + i \sin x \\ e^{-ix} = \cos x - i \sin x, \end{cases} \tag{IX.5.3}$$

it is easy to show

$$\sin x = \frac{e^{ix} - e^{-ix}}{2i} \quad \text{and} \quad \cos x = \frac{e^{ix} + e^{-ix}}{2}. \tag{IX.5.4}$$

These results permit us to derive many trigonometric identities without recourse to the right triangle or coordinate geometry. (Note the similarity to the definitions of sinh x and cosh x.) It is possible to define the trigonometric functions in terms of the exponential function and to derive all of trigonometry therefrom, although it would not appear as natural as the approach using right triangles.

One further item we can deduce from the Euler relation concerns logarithms. You remember that the logarithm is the inverse function of the exponential function. Thus, if we wish to obtain $\ln[r(\cos x + i \sin x)]$, we merely consider

$$\begin{aligned} \ln[r(\cos x + i \sin x)] &= \ln r + \ln(\cos x + i \sin x) \\ &= \ln r + \ln(e^{ix}) \\ &= \ln r + ix. \end{aligned} \tag{IX.5.5}$$

There is one difficulty here. Since an increase in x by any integral multiple of 2π would not alter the correctness of the given equations, the logarithm no longer has a unique value. We can handle this as we handled the case with the inverse trigonometric functions, however, for we can simply limit the domain of x in this relation to $[0, 2\pi)$. If one is willing to adjust the formula for the addition of two logarithms so that the value of x in the result (IX.5.5) is always in the interval $[0, 2\pi)$, it is perfectly possible to work with the logarithms of negative and complex numbers. (Remember that you can put any complex number into polar form, and hence carry out the computation indicated above.) You will observe that the logarithms of positive real numbers are real but all other logarithms have an imaginary component. (Could this be the reason that you were always restricted to the positive real numbers when using logarithms in the past?) We see that the logarithm of -1 is given by $\ln(-1) = \pi i$, and $\ln[(-1)^2] = 2 \ln(-1) = 2\pi i$. Since $2\pi i$ is outside the domain we would subtract $2\pi i$ and obtain the more frequently seen expression $\ln(+1) = 0$.

EXERCISES

1. (a) Consider the series for arc tan x obtained in this section. How many terms are required to obtain six decimal digit accuracy if $x = 1$?
   (b) Evaluate arc tan $\frac{1}{2}$ to six decimal places.
   (c) Evaluate arc tan $\frac{1}{3}$ to six decimal places.
   (d) Use the method of Exercise 9 of the last section to obtain the value of arc tan 1 to six decimal places using the results of parts (b) and (c).
   (e) Use the result of part (d) to obtain the value of π to at least five decimal places.

2. (a) Use division to expand $(1/(1 - x))$ in a series of ascending powers of x.
   (b) Integrate your result of part (a) to obtain a series for $\ln(1 - x)$.
   (c) Find the interval of convergence of the series of part (b).
   (d) Using $x = 0.5$, determine first six decimal places of $\ln(0.5)$.
   (e) Using the result of part (d) obtain ln 2 to six decimal places.

3. (a) Using the method of division, expand $(2/(1 - x^2))$ in an infinite series of ascending exponents.
   (b) Integrate your result of part (a) to obtain a series for $\ln((1 + x)/(1 - x))$.
   (c) Find the radius of convergence of your series of part (b).
   (d) Using $x = 0.5$ determine the first seven decimal places of ln 3.

4. (a) Expand $(1 - x^2)^{-1/2}$ in a binomial series.
   (b) Integrate the series of part (a) and obtain a series for arc sin x.
   (c) Determine the interval of convergence of the series of part (b).
   (d) Is there any logical reason in connection with the inverse sine function why the interval of convergence should be limited?
   (e) Evaluate arc sin 0.5 to six decimal places. Determine this value in both radians and degrees. (Use a conversion factor to obtain the result in degrees.)
   (f) Use your result of part e to evaluate π.

5. (a) Expand e^{-t^2} in a Maclaurin series.
   (b) Find the interval of convergence of the series of part (a).
   (c) Evaluate $\int_0^{0.5} e^{-t^2} \, dt$ to six decimal places. This is a very important integral for which there is no known method of direct evaluation.
   (d) Show that $\int_0^x e^{-t^2} \, dt$ converges for all values of x.

C6. Use the results of Exercise 5 to make a table of values of $\int_0^x e^{-t^2}\, dt$ for $x = 0, 0.1,$ $0.2, \ldots, 2.0$ correct to six decimal places.

7. (a) Obtain a Maclaurin's expansion for $\sin \sqrt{x}$.
   (b) Use your expansion of part (a) to find the first five decimal places of

$$\int_0^{0.5} \sin \sqrt{x}\, dx.$$

8. (a) Expand $e^{\sin x}$ in a Maclaurin's series in terms of $\sin x$.
   (b) Evaluate $\int_0^{\pi/2} e^{\sin x}\, dx$ correct to six decimal places. (Note the potential use of Wallis' theorem in this case.)

9. (a) Show that $\cos x = \cosh(ix)$.
   (b) Show that $\sin x = (-i)\sinh(ix)$.
   (c) Show that $\cosh x = \cos(ix)$.
   (d) Show that $\sinh x = (-i)\sin(ix)$.
   (e) On the assumption that the trigonometric identities of Appendix A are valid for complex arguments, find the corresponding identities for the hyperbolic functions.

10. (a) Find the quadratic equation $(x - e^{im})(x - e^{-im}) = 0$ satisfied by $x = e^{im}$ and $x = e^{-im}$.
    (b) Show that the quadratic expression of part (a) has no real zeroes, if $m \neq 0$.
    (c) Solve the quadratic equation of part (a) by completing the square or by means of the quadratic formula. Compare your results with the information given in part (a).

M11. Review the derivation of formulas for the integral of the exponential function in Section III.4 and the integral of the sine and cosine in Section III.5. Show that these two developments are essentially the same development if we use imaginary exponents in Section III.4 and use the Euler relation.

12. (a) Find the value of $\ln(1 + i)$.
    (b) Evaluate the cube roots of $(1 + i)$ by taking one-third of $\ln(1 + i)$ and then using the exponential function. Remember that there are three cube roots.
    (c) Find the value of $\ln(1 - i\sqrt{3})$ and $\ln(-2i)$.
    (d) Add the values obtained in part (c) and determine whether the result is actually $\ln(-2i - 2\sqrt{3})$.

13. (a) Express $2 - 2i$ as a number of the form re^{ix}, giving the value of r and x.
    (b) Show that any number of the form $(a + bi)$ can be expressed in the form re^{ix}.
    (c) Show that any quadratic polynomial with real coefficients has two real zeros (possibly equal zeros) or else has zeros of the form re^{ix} and re^{-ix}.
    (d) If the polynomial $ax^2 + bx + c$ has no real zero, show that it does have zeros re^{iy} and re^{-iy} where $ar^2 = c$ and $b = -2ar \cos y$.

P14. *The Bessel function of the first kind of index k is a very useful function in many areas of the physical sciences. It is defined as an infinite series, and is given by the relation*

$$J_k(x) = \sum_{j=0}^{\infty} \frac{(-1)^j x^{k+2j}}{2^{k+2j} j! (k + j)!}.$$

(a) Determine the interval of convergence of this series.

(b) Assuming that differentiation term by term is valid within the interval of convergence, show that this function satisfies the relation

$$x^2 D_x^2 J_k(x) + x D_x J_k(x) + (x^2 - k^2) J_k(x) = 0.$$

[*Hint*: Use the terms involving x^k, the terms involving x^{k+2}, the terms involving x^{k+4}, and then the general term.]

IX.6 Iterated Integrals

In Section 4 we derived the Taylor's expansion with the aid of an iterated integral. We do not propose to give an extended discussion of this very important part of mathematics, but it would be of assistance for many application to provide an intuitive background for simple applications of the iterated integral. As stated in Section IX.4 you will frequently hear the term *multiple integral*. In the majority of instances the multiple integral and the iterated integral will be synonymous in value if not in definition. Only those who persevere to the more esoteric applications of mathematics in their respective disciplines are apt to come across examples in which the distinction is one of importance. For the record, however, we are here discussing the iterated integral and not the multiple integral. This simplifies the problem considerably, for it means that we have a succession of integrals and each successive integral has the result of the preceding integration as its integrand. The successive integrations may all be with respect to the same variable, or they may be with respect to different variables. The limits of integration of the first integral evaluated may involve variables which are variables for an integration to be performed later. Does this sound confusing? If it does, go back to the RS sums for each of the successive integrands. The procedures we developed earlier for proceeding from the RS sums to the integral will continue to hold, but for the fact that we will have to obtain lubs and glbs more than once in a given problem.

EXAMPLE 6.1. Use iterated integrals to find the volume of the solid bounded by the plane $2x + 3y + 4z = 12$ and the three coordinate planes (the three planes each of which includes a pair of axes).

Solution. It is helpful to first draw a sketch of the volume involved. It is easy to see that if $x = 0$ and $y = 0$, then $z = 3$ and the plane must cross the z-axis at the point where $z = 3$. In similar fashion the plane will cross the x-axis where $x = 6$ and the y-axis where $y = 4$. These three points are sufficient to determine the plane. It is also possible to find the line in which the plane $2x + 3y + 4z = 12$ intersects the x-y plane, for in the x-y plane we have $z = 0$. When $z = 0$, the equation of intersection becomes $2x + 3y = 12$. All of this information is accumulated in Figure IX.2.

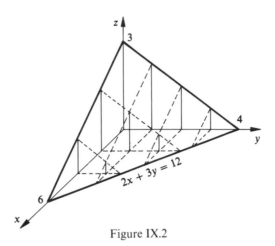

Figure IX.2

In order to proceed with the integration we will partition the portion of the x-axis involved, and also partition the interval of the y-axis that is involved. Through the partition points on the x-axis, we will draw planes that are parallel to the y-z plane, and through the partition points on the y-axis we will draw planes that are parallel to the x-z plane. We now have a grid in the x-y plane with columns over each of the rectangles that have been defined in the x-y plane. We will add the volumes of the columns, each of which has as its base a rectangle formed by an interval on the x-axis and an interval on the y-axis, and which has for its height the height of some point on the given plane $2x + 3y + 4z = 12$ which is above the base rectangle used for the columns. We have indicated the basis for RS sums although it is not yet apparent what method we can use for adding up the volumes in the large number of columns which result from the combination of the two partitions. In order to bring some order out of the chaos we will organize our work by first adding the columns in the y-direction for each interval indicated by the x-partition. This means we will first add all of the columns, going from left to right, which are between two successive planes, each plane being parallel to the y-z plane and determined by successive partition points in the partition of the x-axis. This will give a sum for each interval in the partition of the portion of the x-axis used. After these sums have been obtained, we shall add these sums up, thus obtaining as a final result the sum of the volumes in all the columns determined by the combination of the two partitions. Thus, in Figure IX.2 we will be adding the strips from left to right as the diagram is drawn, and each of these additions will produce a thin slab. We will then add the slabs obtained by multiplying a cross sectional area by the length of a subinterval in the x-direction, starting at the back with $x = 0$ and coming forward along the x-axis to the front vertex of the tetrahedron. Thus, we will have the RS sums

$$\sum_{j=1}^{n} \left[\sum_{i=1}^{m} f(r_j, s_i)(y_i - y_{i-1}) \right](x_j - x_{j-1}) \qquad \text{(IX.6.1)}$$

where the values s_i are the elements of the evaluation set associated with the partition in the y-direction (or the partition of the portion of the y-axis that is involved) and the values r_j are the elements of the evaluation set associated with the partition in the x-direction. It should be noted that n is the number of subintervals in the x-direction and m is the number of intervals in the y-direction. Also $f(x, y)$ is the height of the column if taken at the point (x, y) in the x-y plane. With this explanation, it is apparent that the summation inside the brackets provides directions for the computation necessary to obtain an approximation of the cross sectional area for any cross section in a plane parallel to the y-z plane. Having completed this summation we are now in a position to consider the outer summation, that is the one for which the summands include a summation. This will give us the sum of the terms which individually represent a cross sectional area multiplied by a thickness, hence the volume of a slab. The total result, then, is an approximation to the volume in question.

This result involves an RS sum for which the summands are also RS sums. It should not be surprising that we will first determine a glb for the upper sums of the summation within the bracket and then a lub for the lower sums of the same summation to establish the summation in the bracket through successive refinements of the y-interval involved. This determines an integral. For this purpose x is held constant, since this occurs in a single cross section parallel to the y-z plane. This will eliminate the difficulty we would have if both x and y were free to vary within the same integral. Now that we have re-written the summands for the outer RS sum, we can take the lub of the lower sums and the glb of the upper sums for the outer summation to determine whether the outer summation can be replaced by an integral. In this case there will be no y-variables appearing explicitly, for the summations with regard to y were taken care of in the first integration. Thus, we will have an iterated integral, and in this case it is clear that our decision to hold x constant for the inner integration was correct as was the decision to perform the inner integration first in order to have the integrand for the outer integration.

We have only sketched the reasoning involved in carrying this request for volume through to the iterated integral, but we have missed in this sketch one very important ingredient. If we look again at Figure IX.2, we see that in this case the length of the interval we are to use for the variable y is dependent upon the value of x being used. When x is small the interval is rather long but as x increases the interval becomes shorter, and finally becomes an interval of length zero. Thus, the interval of integration for the interior, or first integration, depends on the value of x, and the value of x is a part of the second integration. In order to determine the limits of integration we will start from the inside and work our way out. We see that the height, which we labeled $f(x, y)$ for purposes of setting up the RS sums, is given to us by the size of z for the given value of x and y. Hence, we would have the height given by $z = (1/4)(12 - 2x - 3y)$ for any point (x, y). We have indicated in (IX.6.1) that the interval in the y-direction is divided into m subintervals, or the partition has $m + 1$ points. The big question in writing down the integral is to determine what interval we have so partitioned. In this case it is not difficult to see that for any value of x we have the largest possible value of y at the time when $z = 0$, or when we are in the x-y plane. Thus, the relation between x and y which governs the largest value of y to be used for any given x is the relation $2x + 3y = 12$, where we have used $z = 0$ in the original equation. Hence, for any value of x in the partition we have y starting at the left with $y = 0$ and going to the right to, but not beyond, $y = (1/3)(12 - 2x)$. Since the first integration, that is the evaluation of the inside integral, is to be performed as though x were a constant, the interval of integration should be $[0, (1/3)(12 - 2x)]$, where these are values which y assumes and it is understood that x is constant for this integration. Finally, since the result of the first integration gives us the cross sectional area involved for a fixed value of x, the final integration gives us the volume, and must include all of the slabs starting at the back with $x = 0$ and proceeding to the front. Now the front point, that is the largest value which x can have in this solid, occurs when both y and z are zero, or when $2x = 12$, whence $x = 6$. Therefore, the interval for this integration is $[0, 6]$, and we now have the iterated integral,

$$\int_0^6 \left[\int_0^{(1/3)(12-2x)} \frac{1}{4}(12 - 2x - 3y)dy \right] dx. \qquad \text{(IX.6.2)}$$

We have included the brackets to show the relation between (IX.6.1) and (IX.6.2), but these would ordinarily be omitted in practice. As far as the evaluation of the iterated integral is concerned, the procedure would be just as though they were present. Thus, we would have

$$\frac{1}{4} \int_0^6 \left[\int_0^{4-(2x/3)} (12 - 2x - 3y)dy \right] dx$$

$$= \frac{1}{4} \int_0^6 \left[12\left(4 - \frac{2}{3}x\right) - 2x\left(4 - \frac{2}{3}x\right) - \frac{3}{2}\left(4 - \frac{2}{3}x\right)^2 \right] dx$$

$$= \frac{1}{4} \int_0^6 \left(24 - 8x + \frac{2}{3}x^2 \right) dx$$

$$= \frac{1}{4} \left[24(6) - 4(6)^2 + \frac{2}{9}(6)^3 \right] = 12.$$

Since the volume concerned is a pyramid, we can check our result using a formula from solid geometry. The base of the pyramid is a right triangle with legs of length 4 and 6. Hence the area of the base is $(1/2)(4)(6) = 12$ square units. The altitude of the pyramid is 3 and hence the volume should be $(1/3)(3)(12) = 12$ cubic units. This result agrees with that obtained by integration, and provides us with a check. While the geometric approach would be easier in this case the method of integration would be more useful in general. It is not to be expected that we will have the convenient geometric formula for every solid that we meet.

Before we leave this example it should be noted that we could just as well have added first with respect to x, holding y constant, and then added with respect to y. The important thing to observe is that we arrange our computation so that we add all columns once and only once when we consider the sums, and the integration should reflect the result of carrying this reasoning through the successive refinements to the integral.

EXAMPLE 6.2. Find the mean value of z on the surface $2x + 3y + 4z - 12$ in the first octant.

Solution. Following our earlier work concerning mean value, we know that the mean value of z multiplied by the base area should give the volume. In other words, we can obtain the mean value of z as

$$\frac{\int_0^6 \int_0^{(1/3)(12-2x)} \frac{1}{4}(12 - 2x - 3y)dy\, dx}{\int_0^6 \int_0^{(1/3)(12-2x)} dy\, dx}.$$

We know the numerator has the value 12, and we can determine that the denominator also has the value 12. Hence the mean value in this case is one. This is equivalent to saying that a prism having the same base and having an altitude of one would have the same volume as does the pyramid of our example.

Note that the mean value is defined in a manner completely analogous to the definition of mean value of Section III.3.

EXAMPLE 6.3. Find the mass of the ellipsoid $(x^2/a^2) + (y^2/b^2) + (z^2/c^2) = 1$ if the density at any point (x, y, z) in the ellipsoid is given by $|xyz|$ units of mass per cubic unit of volume.

Solution. Since the density is given by the absolute value, the mass is symmetric about each of the coordinate planes. Consequently we can evaluate the mass in the first octant and multiply the result by 8 in order to obtain the total mass. This will remove the requirement for keeping track of the absolute value throughout the problem. We will now sketch the ellipsoid in the first octant, and then partition the interval $[0, a]$ of the x-axis with n subintervals, the interval $[0, b]$ of the y-axis with m subintervals, and the interval $[0, c]$ of the z-axis with p subintervals. If we draw in the planes through the partition points, each plane being parallel to a coordinate plane, we will divide the solid into a large number of rectangular parallelopipeds (or boxes). A representative box is shown in Figure IX.3. The mass of this box is approximated by $r_j s_k t_l [x_j - x_{j-1}][y_k - y_{k-1}][z_l - z_{l-1}]$ units of mass. We can now proceed to add the approximations for all of the little boxes, and thus obtain an approximation for the mass of the ellipsoid. If we follow the procedures of Example 6.1, we will proceed to the triple iterated integral.

$$\int_0^c \int_0^{b(1-(z^2/c^2))^{1/2}} \int_0^{a(1-(y^2/b^2)-(z^2/c^2))^{1/2}} xyz\, dx\, dy\, dz. \qquad \text{(IX.6.3)}$$

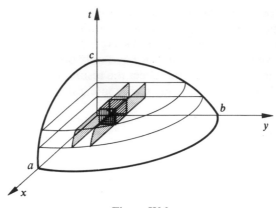

Figure IX.3

In this case we have not used the brackets to indicate the order of evaluation, but the order should be apparent by comparison with the work of Example 6.1.

We have not commented on the selection of limits of integration in (IX.6.3). Since the first addition is indicated as that in which we hold y and z constant, we must add the masses of the boxes from the y-z plane to the surface for each choice of subinterval along the y-axis and along the z-axis. The largest x to be used for each y and z will vary, and can be determined from the equation of the ellipsoid. Thus, we would have $x^2/a^2 = 1 - (y^2/b^2) - (z^2/c^2)$ or $x = a(1 - (y^2/b^2) - (z^2/c^2))^{1/2}$ since x cannot be negative. Once we have evaluated the integral (equivalent to taking the sum) with respect to x, we now have a set of columns starting at $x = 0$ and proceeding away from the paper as indicated in Figure IX.3. As we have written (IX.6.3) we should then add these columns horizontally, holding z constant. The columns will be shorter when y is larger, of course, but they come all the way to the right to the place in which the surface intersects the y-z plane. This occurs when $x = 0$. Consequently we are concerned with the y limits determined by the equation $(y^2/b^2) + (z^2/c^2) = 1$ or equivalently $y = b(1 - (z^2/c^2))^{1/2}$. Again we are using the fact that we are concerned only with the first octant. Finally, after having integrated twice, we have the *slabs* referred to in Example 6.1 and these start with the bottom when $z = 0$ and the top one occurs when $z = c$. The latter result should be clear from the fact that the high point occurs when both x and y are zero.

We now have the job of evaluating (IX.6.3). Holding y and z constant, we have for the first integration

$$\int_0^c \int_0^{b(1-(z^2/c^2))^{1/2}} \int_0^{a(1-(y^2/b^2)-(z^2/c^2))^{1/2}} xyz \, dx \, dy \, dz$$

$$= \int_0^c \int_0^{b(1-(z^2/c^2))^{1/2}} \frac{a^2 yz}{2} \left(1 - \frac{y^2}{b^2} - \frac{z^2}{c^2}\right) dy \, dz.$$

This can be carried one step further as follows

$$\frac{a^2}{2} \int_0^c \int_0^{b(1-(z^2/c^2))^{1/2}} z\left(1 - \frac{y^2}{b^2} - \frac{z^2}{c^2}\right)\left(-\frac{b^2}{2}\right)d\left(-\frac{y^2}{b^2}\right)dz$$

$$= \frac{a^2 b^2}{8} \int_0^c \left(1 - \frac{z^2}{c^2}\right)^2 z\, dz = \frac{a^2 b^2 c^2}{48}.$$

From this computation we see that the mass of this ellipsoid is $8(a^2 b^2 c^2/48)$ or $(a^2 b^2 c^2/6)$ units of mass.

Note that we have used the concept of RS sums as we did earlier in setting up the integrals. The matter of determining the limits of integration is a little more complicated in this instance, but it can be handled easily if you stop to think about what you are doing. It is particularly helpful to have at least a rough sketch before you when trying to determine these limits of integration, just as it is when determining the integrand itself.

We have used Examples 6.1 and 6.3 for purposes of illustrating the use of iterated integrals. They have involved the volume and the mass of three dimensional solids. In one case we used two independent variables, and in the other case we used three. It should be apparent that the iterated integral will apply in any case in which we desire a summation involving more than a single independent variable, and for which it is possible to refine the RS sums involved and obtain corresponding integrals. The development of the integrand is often relatively straightforward. Usually the more awkward portion of the problem is in the establishment of the limits of integration. In determining the limits of integration it is helpful to use sketches, and it is essential that one think carefully concerning which variables are to be kept constant for a given integration, the maximum extent of the variable which is not held constant for the particular integration, etc.

It would be well to consider again the matter of the order of summation (or equivalently the order of integration). In Example 6.3, for instance, it would have been possible to integrate first with respect to x and then z and then y, or with respect to y and then x and then z. In fact there are six possible orders of integration with three variables and 24 orders with four variables. It is frequently true that one order of integration may be easier to carry out than another. This is something that can only be checked in individual cases, however, and therefore no rules will be given here.

EXERCISES

1. Evaluate:

(a) $\int_{-1}^2 \int_1^{x+3} x^2 y\, dy\, dx$

(b) $\int_{-\pi/6}^{\pi/3} \int_0^{\cos x} e^{-y} \sin x\, dy\, dx$

(c) $\int_0^4 \int_0^{3z} \int_0^{y+z}(x + y + z)dx\, dy\, dz$

(d) $\int_0^x \int_0^x \int_0^x e^{-x}(dx)^3 = \int_0^x \{\int_0^x [\int_0^x e^{-x} dx] dx\} dx$

(e) $\int_0^{\pi/2} \int_0^{\sin y} xy \, dx \, dy$

(f) $\int_0^5 \int_0^x \sqrt{x^2 - y^2} \, dy \, dx$.

2. Find the volume in the first octant bounded by $4x^2 + 9y^2 + 16z^2 = 16$.

3. In Example 6.1 set up five other iterated integrals for obtaining this volume. [One is obtained by reversing the order of summation, and the others by considering columns parallel to the x and y axes]. Evaluate each of these.

4. (a) Sketch that portion of the elliptic cone $z^2 = x^2 + 4y^2$ for which $z \geq 0$.
   (b) Sketch the plane $z = 4$.
   (c) Find the volume contained in the solid bounded by the surfaces of parts (a) and (b).

5. (a) Find the volume contained within $9x^2 + 16y^2 + 36z^2 = 144$ using a double iterated integral.
   (b) Find the same volume using a triple iterated integral.

6. (a) Sketch the portion of the hyperbolic paraboloid $z = xy$ in the first octant.
   (b) Sketch the cylinder $x^2 + y^2 = 25$ in the first octant.
   (c) Find the volume bounded by the surfaces of parts (a) and (b) contained in the first octant.

7. (a) Find the volume of the ellipsoid $(x^2/a^2) + (y^2/b^2) + (z^2/c^2) = 1$ using iterated integration.
   (b) Show that the volume of a sphere can be obtained as a special case of the result of part (a).
   (c) Show that if $a + b + c$ is constant, the greatest volume occurs when the three semi-axes are of equal length, and the consequent solid is a sphere.
   (d) Using the result of Example 6.3 find the mean value of the density of the ellipsoid. This is obtained by dividing the mass by the volume.

C8. Write a program to approximate the mass requested in Example 6.3. Use at least 10 intervals in each direction. Obtain both upper and lower bounds, and show that the result obtained in Example 6.3 is between your bounds.

9. A square tile was cast while the form made an angle with the ground. As a result the density of the tile at each point is given by the relation $\rho(x, y) = 0.2 + 0.04(x + 2y)$ pounds per square inch of surface.

   (a) If the tile is 9 inches on a side and if it is assumed that the x and y axes are sides of the tile, find the weight of the tile.
   (b) Find the average density of the tile.

S10. The Cobb–Douglas production function states that $Q = aL^k C^{1-k}$ where Q is the output, L the quantity of labor, and C the quantity of capital, each of these quantities being measured in dollar equivalents. Both a and k are constants. If $a = 3$ and $k = 0.5$ find the mean value of output for all inputs of labor such that the total amount of labor and capital together does not exceed $10,000.

M11. Given the relation $(x_1/a_1) + (x_2/a_2) + (x_3/a_3) + (x_4/a_4) = 1$,

   (a) Show that the four "points" $(a_1, 0, 0, 0)$, $(0, a_2, 0, 0)$, $(0, 0, a_3, 0)$, and $(0, 0, 0, a_4)$ satisfy the equation.

(b) If these points are considered as though they were points on coordinate axes and the given equation is then the equation of something replacing a plane, set up an iterated integral (with either three or four integrations) which would obtain the equivalent of volume.

(c) Show that the "volume" is equal to the product of the four intercepts divided by four factorial.

(d) Generalize this problem to one having five variables and see whether the statement equivalent to that of part (c) is valid.

(e) Show that this result would apply for the length of a single segment, the area of a triangle in two dimensions, and the volume of a pyramid in three dimensions.

B12. The flow rate, F, in milliliters per second along a small tube with a pressure difference of P dynes per square centimeter between the tube ends is given by the equation $F = (\pi P r^4)/(8as)$ where r is the radius of the tube in centimeters, s is the length of the tube in centimeters, and a is the viscosity of the liquid measured in poise. (Because of the r^4 factor, the resistance is very important in determining flow rate.) Assuming that the cardiac output of a small dog is 40 milliliters per second and the viscosity of blood is 0.027 poise, find the average pressure drop in the aorta if the aorta may vary in length from 35 to 45 cm and if the diameter of the aorta can vary from 8 to 12 mm.

P13. The current in a simple electrical circuit is given by the equation $i = i_0 e^{(-Rt/L)} + (E/R)$ amperes where i_o is the initial current, the electrical resistance is R ohms, the electromotive force is E volts, and the inductance is given by L henries. The value of L remains constant, but it is known that the environmental conditions permit R to vary from 30 to 34 ohms whereas E can vary from 6 to 6.5 volts. If $i_0 = 0.5$ amperes and $L = 0.4$ henries, find the mean (or expected) value of i when $t = 1$ second.

IX.7 Change of Variable in Iterated Integrals

In the last section we developed the iterated integral. It is worth noting that we often had the pair of differentials $dx\,dy$ in the double integrals. It should be apparent that the product of an increment in the x-direction by an increment in the y-direction gives us an increment of area. In the uses we made of this integral, this is the interpretation which well fits our development. In some instances you may even find this product written as dA, emphasizing the area concept. Since this is the usage we had in mind, it would seem reasonable to wonder whether we might not be able to obtain an expression for dA in coordinate systems other than the Cartesian coordinates. In this section we will answer this question in the affirmative. We will develop in some detail but without complete rigor, the case for two variables. We will mention the method that would be used for more than two variables.

As has happened frequently in the past it will help to prove some preliminary results we will need later. This we do with the following lemma.

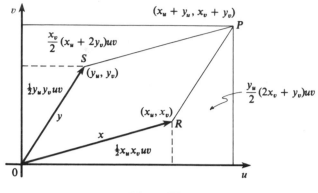

Figure IX.4

Lemma. *In a rectangular coordinate system with u and v axes let U be the length of a unit on the u-axis and V the length of a unit on the v-axis. If X and Y are two vectors in this coordinate system, the area of the parallelogram having X and Y as two adjacent sides is given by $|x_u y_v - x_v y_u||UV|$ where (x_u, x_v) and (y_u, y_v) are the coordinates of X and Y respectively.*

PROOF. Consider the picture shown in Figure IX.4. The point P has coordinates $(x_u + y_u, x_v + y_v)$, and the dimensions of the rectangle $ORPS$ are $(x_u + y_u)U$ and $(x_v + y_v)V$. Consequently the area of the rectangle is given by $(x_u x_v + x_u y_v + x_v y_u + y_u y_v)UV$. Now the area of the parallelogram can be obtained from the area of the rectangle if we subtract the two trapezoids and the two triangles shown in the figure. The triangles have areas of $(x_u x_v/2)UV$ and $(y_u y_v/2)UV$. The trapezoids have areas of $(x_v/2)(x_u + 2y_u)UV$ and $(y_u/2)(2x_v + y_v)UV$. Upon subtracting these areas we have $(x_u y_v - x_v y_u)UV$ as the remaining area, that is the area of the parallelogram we were seeking. □

We are now ready to consider the matter of changing variables in our iterated integrals. If we wish to replace x and y with suitable values of two other variables, say u and v, we will stipulate that when v is held constant and u is permitted to change, the direction of change must be perpendicular to that which would hold if u were held constant and v permitted to change. This will provide us with perpendicular directions of motion, and hence will permit us to use our lemma.

Theorem 7.1. *If $x = x(u, v)$ and $y = y(u, v)$ are differentiable functions and if the u-v coordinate system is such that at any point an increase in u is in a direction perpendicular to that determined by an increase in v, then $dx\,dy = |(\partial x/\partial u)(\partial y/\partial v) - (\partial x/\partial v)(\partial y/\partial u)|\,du\,dv$.*

PROOF. Since increases in u and v occur in perpendicular directions, du and dv can be considered as perpendicular vectors if they are sufficiently small. The projections of dx on the du and dv directions are $(\partial x/\partial u)du$ and $(\partial x/\partial v)dv$ respectively. Similarly the projections of dy are $(\partial y/\partial u)du$ and $(\partial y/\partial v)dv$. Consequently the lemma asserts that the area in the parallelogram with consecutive sides dx and dy must be $|(\partial x/\partial u)(\partial y/\partial v) - (\partial x/\partial v)(\partial y/\partial u)|\ du\ dv$. However, since dx and dy are perpendicular by convention, the area of the parallelogram is $dx\ dy$, for the parallelogram is in reality a rectangle. This proves our theoerem.   □

The expression $(\partial x/\partial u)(\partial y/\partial v) - (\partial x/\partial v)(\partial y/\partial u)$ is known as the *Jacobian* and is named for *Carl Gustav Jacob Jacobi* (1804–1851). This ignores the fact that Cauchy had used this expression prior to its use by Jacobi. This is often written as a determinant, and given the abbreviation $\partial(x, y)/\partial(u, v)$. Thus you will often find this written as

$$\frac{\partial(x, y)}{\partial(u, v)} = \begin{vmatrix} \dfrac{\partial x}{\partial u} & \dfrac{\partial x}{\partial v} \\[2ex] \dfrac{\partial y}{\partial u} & \dfrac{\partial y}{\partial v} \end{vmatrix}.$$

We are now ready to return to our problem of changing variables. Consider the following example.

EXAMPLE 7.1. Find the expression for dA in polar coordinates.

Solution. In the case of polar coordinates $x = r \cos \theta$ and $y = r \sin \theta$. A glance at Figure IX.5 will indicate that an increase in θ would move us further around a circle having the center at the pole. On the other hand an increase in r will move us away from the pole along a radius. Since the radius is perpendicular to the circle, we have met the requirements for use of the theorem. Of course a continued increase in θ would carry us along a circular arc instead of along a straight line, but if we confine our interests we have done very little injustice to the requirements for the theorem. Since we would want to use small increments in developing the integral, this will not affect our employment of the result in integration.

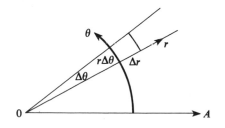

Figure IX.5

We could also note that if r is held constant $dx = -r \sin \theta$ and $dy = r \cos \theta$. Thus, $D_x y = -\cot \theta$. On the other hand if we were to let θ be constant we would have $dx = \cos \theta$ and $dy = \sin \theta$. Then we would have $D_x y = \tan \theta$. Since these two derivatives are negative reciprocals, the directions they represent are prependicular. This provides us with an alternate method for showing that we have fulfilled the requirements of our theorem.

In this case we have $\partial x / \partial r = \cos \theta$ and $\partial x / \partial \theta = -r \sin \theta$. Also we have $\partial y / \partial r = \sin \theta$ and $\partial y / \partial \theta = r \cos \theta$. Therefore,

$$dx \, dy = [(\cos \theta)(r \cos \theta) - (-r \sin \theta)(\sin \theta)]dr \, d\theta = r \, dr \, d\theta.$$

We can verify this by noting in Figure IX.5 that an increment $\Delta\theta$ produces a small arc of length $r\Delta\theta$. When this is multiplied by an increment Δr, we have $r \, \Delta r \, \Delta\theta$ as an approximation for the area of what appears to be almost a rectangle. If we were to think of having taken a partition and then refined it such that the values of Δr and $\Delta\theta$ are very small, it is not difficult to see that the result derived using the Jacobian agrees with what we should have anticipated.

EXAMPLE 7.2. Find the mass of a plate in the shape of a cardioid $r = 1 + \cos \theta$ if the density at those points r units from the pole is $6r$ grams per square centimeter.

Solution. Thinking of the RS sums using small units of area we see that we would like to add the increments of area, each multiplied by six times the distance r from the pole. The cardioid is shown in Figure IX.6 with a small increment of area to illustrate this concept. The RS sum will then lead us to the integral $\iint_A 6r \, dA$ where the integration is to take place over the entire cardioid. By Example 7.1 we know that in the case of polar coordinates we have $dA = r \, dr \, d\theta$. We also see from the cardioid that if we add in the r direction, holding θ constant, we can start at the pole and go out to the cardioid. If we do this we will obtain the mass of a small sector. We can add the sectors by integrating with respect to θ from $\theta = 0$ to $\theta = 2\pi$. Conse-

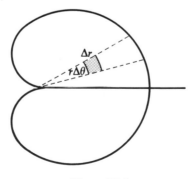

Figure IX.6

quently, we have the integral

$$\int_0^{2\pi} \int_0^{1+\cos\theta} 6r \, r \, dr \, d\theta = \int_0^{2\pi} \int_0^{1+\cos\theta} 6r^2 \, dr \, d\theta = 2\int_0^{2\pi} (1+\cos\theta)^3 \, d\theta$$

$$= 2\int_0^{2\pi} \left[2\cos^2\left(\frac{\theta}{2}\right)\right]^3 d\theta$$

$$= 16\int_0^{2\pi} \cos^6\left(\frac{\theta}{2}\right) d\theta.$$

If we use the substitution $\theta = 2u$, we will have

$$\text{Mass} = 16(2)\int_0^\pi \cos^6 u \, du.$$

Since $\cos^6 u$ is always positive, we can see that the integral over the second quadrant is the same as that over the first, and consequently we have

$$\text{Mass} = 32(2)\int_0^{\pi/2} \cos^6 u \, du = 64\,\frac{5\cdot 3\cdot 1}{6\cdot 4\cdot 2}\frac{\pi}{2} = 10\pi \text{ grams.}$$

Note the use of Wallis' formula in this case.

It is worth noting that this example illustrates the relative ease of computation in polar coordinates of a result that would be messy in rectangular coordinates. With the aid of the theorem we are able to make use of polar coordinates when it is appropriate. There are times when polar coordinates can help out in unexpected ways. Consider the following example.

EXAMPLE 7.3. Find the value of the integral

$$\int_0^\infty e^{-x^2} \, dx.$$

Solution. This certainly does not appear to come under the heading of this Section, for it is written in a single variable and is an improper integral. However, this is the one we have referred to from time to time as being essential in statistics and probability, but one for which we do not have a formula. We will proceed by giving it a name and calling it P. Note that we can write

$$P = \int_0^\infty e^{-x^2} \, dx = \int_0^\infty e^{-y^2} \, dy$$

since the variable used as the variable of integration does not affect the result. Now consider the square of this integral

$$P^2 = \int_0^\infty e^{-x^2} \, dx \int_0^\infty e^{-y^2} \, dy = \int_0^\infty \int_0^\infty e^{-(x^2+y^2)} \, dx \, dy.$$

The last expression is correct, for if you notice that holding y fixed for the first integration gives just P as the result of this integration. However P is some constant, and we are trying to find its value. Therefore in the second integration P would be factored outside the integral sign.

Using the fact that $x^2 + y^2$ in rectangular coordinates is r^2 in polar coordinates, and with the use of Example 7.1 we can write

$$P^2 = \int_0^{\pi/2} \int_0^\infty e^{-r^2} r \, dr \, d\theta.$$

The limits occur because we note that we were covering the area spanned by the combination of all positive x's and all positive y's, that is the entire first quadrant. This is equivalent to using all positive values of r and letting θ turn through the first quadrant. However, we can evaluate this iterated integral and obtain

$$P^2 = \int_0^{\pi/2} \int_0^\infty e^{-r^2} r \, dr \, d\theta = \int_0^{\pi/2} \lim_{M \to \infty} \frac{(-e^{-r^2})}{2} \bigg|_0^M d\theta = \frac{\pi}{4}.$$

Consequently we have the desired result $P = \sqrt{\pi/2}$.

Note that we could use this method if the integral P covered the interval $[0, \infty]$, but we could not have done this for any shorter interval. It is rather surprising that this integral can be evaluated over this one interval of infinite length. This also illustrates the lengths to which we must go at times to obtain values which are of very frequent use.

You will find in future courses that the theorem we have developed for change of variables is too restrictive. The requirement that the differentials operate in mutually perpendicular directions is not necessary. If the differentials do not represent perpendicular directions, the concept of area of a rectangle is no longer valid, but there are many applications of the iterated integral where such an interpretation is not appropriate. You will also find that our results apply with some generalization if we have more than two variables. While we will give no further proofs here, it will be of interest to indicate results you will see in some later course. If we have three functions instead of two, and if

$$x = x(u, v, w)$$

$$y = y(u, v, w)$$

$$z = z(u, v, w),$$

we would have $dx \, dy \, dz = (\partial(x, y, z)/\partial(u, v, w)) du \, dv \, dw$. The notation

$(\partial(x, y, z)/\partial(u, v, w))$ again refers to the Jacobian in determinant notation as

$$\frac{\partial(x, y, z)}{\partial(u, v, w)} = \begin{vmatrix} \dfrac{\partial x}{\partial u} & \dfrac{\partial x}{\partial v} & \dfrac{\partial x}{\partial w} \\[2mm] \dfrac{\partial y}{\partial u} & \dfrac{\partial y}{\partial v} & \dfrac{\partial y}{\partial w} \\[2mm] \dfrac{\partial z}{\partial u} & \dfrac{\partial z}{\partial y} & \dfrac{\partial z}{\partial w} \end{vmatrix}.$$

If you note the way in which the determinant is written the first row has derivatives of the first function, the second row of the second function, etc. The first column has derivatives with respect to the first variable, the second with respect to the second variable, etc. Thus, the Jacobian is rather easy to remember. It would not be difficult to extend this result to additional variables if the occasion arose.

In every case our use of the iterated integral has taken advantage of the fact that the integrand we wished for the final integral could best be obtained as an integral of something more elementary. If you think of problems in terms of adding up a large number of increments, not all in a line, you will find that the matter of setting up the mathematical formulation of the problem will often be very much eased by adding in one direction at a time and thus obtaining iterated sums, later to be refined into iterated integrals.

EXERCISES

1. (a) Using the relations $r = (x^2 + y^2)^{1/2}$ and $\theta = \arctan(y/x)$, evaluate the Jacobian $\partial(r, \theta)/\partial(x, y)$.
   (b) Using the result of part (a) show that $r\, dr\, d\theta = dx\, dy$.
   (c) Evaluate the integral $\int_0^{\pi/2} \int_0^{4/(\sin\theta + \cos\theta)} r\, dr\, d\theta$.
   (d) Sketch the area implied by the integral of part (c).
   (e) Find the integral equivalent to that of part (c) expressed in terms of rectangular coordinates.
   (f) Evaluate the integral of part (e) and compare your result with that of part (c).

2. Use an iterated integral to find the area inside one loop of $r = 5 \sin 3\theta$.

3. A disk is made in the shape of a limacon with the equation $r = 3 - 2 \sin \theta$. The density of the disk is $10r$ grams per square unit of area for points on the disk at a distance r units from the pole. Find the mass of the disk.

4. In each of the following evaluate $\iint f(x, y)\, dx\, dy$ where the integration is taken over the indicated region. Perform this evaluation by changing to polar coordinates first and then integrating.

   (a) $f(x, y) = (1 - x^2 - y^2)^5$ and the region is the disk $x^2 + y^2 \le 2$.
   (b) $f(x, y) = y/(x^2 + y^2)^{1/2}$ and the region is the ring $1 \le x^2 + y^2 \le 9$.
   (c) $f(x, y) = x^3 + xy^2$ and the region is that portion of the disk $x^2 + y^2 \le 16$ for which x is negative and y is positive.

5. (a) Use an iterated integral to find the area inside the circle $x^2 + y^2 - 4x = 0$.
   (b) Convert this equation to polar coordinates and use an iterated integral to find the area.

6. (a) Set up the iterated integral for the area inside $r = f(\theta)$ between the radii $\theta = \alpha$ and $\theta = \beta$.
   (b) Evaluate the inner integral and show that you then have the integral for area in polar coordinates which we derived in Chapters II and III.

7. (a) Evaluate the integral $\int_0^3 \int_{-x}^x (x^2 - y^2)dy\, dx$.
   (b) Find the Jacobian if $x = u + v$ and $y = u - v$.
   (c) Show that the relations in part (b) describe the situation in which the u-axis bisects the first quadrant and the v-axis bisects the fourth quadrant.
   (d) Show that the units on the u and v axes are $1/\sqrt{2}$ times the units on the x and y axes.
   (e) Show that the area implied by the limits of the integral of part (a) describe an area in the first quadrant of the u-v coordinate system for which $u + v \le 3$.
   (f) Find the integral in terms of u and v which corresponds to the integral of part (a). [Note the order $dy\, dx$ in the integral of part (a)].
   (g) Evaluate the integral of part (f). Show that you have the same value you obtained in part (a).

8. Cylindrical coordinates in three dimensions are related to the rectangular coordinates by the relations $x = r \cos\theta$, $y = r \sin\theta$, and $z = z$.

   (a) Show that $x^2 + y^2 = r^2$.
   (b) Evaluate the Jacobian $\partial(x, y, z)/\partial(r, \theta, z)$.
   (c) Obtain a triple iterated integral for the volume of the sphere

   $$x^2 + y^2 + z^2 = r^2 + z^2 = 25.$$

   (d) Evaluate the integral of part (c).

9. Spherical coordinates in three dimensions are related to the rectangular coordinates by the relations $x = r \sin\phi \cos\theta$, $y = r \sin\phi \sin\theta$, and $z = r \cos\phi$.

   (a) Show that $x^2 + y^2 + z^2 = r^2$.
   (b) Evaluate the Jacobian $\partial(x, y, z)/\partial(r, \phi, \theta)$.
   (c) Obtain a triple iterated integral for the volume of the sphere of radius 5 with center at the origin.
   (d) Evaluate the integral of part (c).
   (e) Obtain a triple iterated integral for the mass of the sphere if the density at any point is $3r$ where r is the distance from the center of the sphere.
   (f) Evaluate the integral of part (e).
   (g) Find the mean value of the density of the sphere using the results of parts (d) and (f).

IX.8 Trigonometric Series

Trigonometry appears to pervade many areas of the calculus, and the subject of infinite series is no exception. There are many problems that are periodic in nature. The period may be that of an alternating current, the seasons, or some other phenomenom. The power series which we considered earlier are hardly suited to such periodic performance, but the trigonometric functions are naturally periodic. We will concentrate on series involving sines and cosines, for these do not have points of discontinuity as do the other four trigonometric functions. Furthermore, we will consider that all of our functions have a period of 2π. It is always possible to determine a ratio between any given frequency and 2π, and then to convert units in either direction. For convenience we will also assume that we are to use either the interval $[-\pi, \pi]$ or $[0, 2\pi]$. The generalization to other intervals of length 2π is not hard to make. To further simplify our discussion we will consider only one series, and that will be the series

$$f(x) = \frac{a_0}{2} + a_1 \cos x + b_1 \sin x + a_2 \cos 2x + b_2 \sin 2x + \cdots$$

$$= \frac{a_0}{2} + \sum_{j=1}^{\infty} [a_j \cos jx + b_j \sin jx] \qquad (IX.8.1)$$

where $f(x)$ is the function we would like to represent as a trigonometric series. We must now evaluate the coefficients. You may wonder why we selected the constant term for special treatment in that we divided it by two. We shall evaluate the coefficients first, and the reason for using $(a_0/2)$ will become apparent as we proceed. This is one case in which we have looked ahead and then doctored up our starting expression in a way that will simplify the final results.

In order to simplify the discussion of evaluating the coefficients, we first prove the following theorem.

Theorem 8.1. *If m and n are non-negative integers and c is a real number, then*

(i) $\displaystyle\int_c^{c+2\pi} \sin mx \sin nx \, dx = 0 \qquad (if\ m \neq n)$

(ii) $\displaystyle\int_c^{c+2\pi} \cos mx \cos nx \, dx = 0 \qquad (if\ m \neq n)$

(iii) $\displaystyle\int_c^{c+2\pi} \sin mx \cos nx \, dx = 0$

(iv) $\displaystyle\int_c^{c+2\pi} \sin^2 mx \, dx = \int_c^{c+2\pi} \cos^2 mx \, dx = \pi \qquad (if\ m \neq 0)$

PROOF. Let y_m be either $\sin mx$ or $\cos mx$. Then

$$y_m'' = -m^2 y_m \quad \text{or} \quad y_m'' + m^2 y_m = 0$$

If we define y_n in a similar manner, we have the two relations

$$y_m'' + m^2 y_m = 0 \quad \text{and} \quad y_n'' + n^2 y_n = 0.$$

Upon multiplying the first of these equations by y_n and the second by y_m and then subtracting, we have

$$[y_m'' y_n - y_m y_n''] + (m^2 - n^2) y_m y_n = 0. \qquad \text{(IX.8.2)}$$

By straight differentiation, we can show that

$$D_x[y_m' y_n - y_m y_n'] = (y_m'' y_n + y_m' y_n') - (y_m' y_n' + y_m y_n'') = y_m'' y_n - y_m y_n''. \qquad \text{(IX.8.3)}$$

Substituting the results of (IX.8.3) into (IX.8.2) and multiplying by dx we have

$$d(y_m' y_n - y_m y_n') + (m^2 - n^2) y_m y_n \, dx = 0. \qquad \text{(IX.8.4)}$$

If we integrate (IX.8.4) over the interval $[c, c + 2\pi]$, we obtain

$$\int_c^{c+2\pi} d(y_m' y_n - y_m y_n') + (m^2 - n^2) \int_c^{c+2\pi} y_m y_n \, dx = 0 \qquad \text{(IX.8.5)}$$

or

$$(m^2 - n^2) \int_c^{c+2\pi} y_m y_n \, dx = 0,$$

since

$$\int_c^{c+2\pi} d(y_m' y_n - y_m y_n') = [y_m' y_n - y_m y_n']_{(x=c+2\pi)} - [y_m' y_n - y_m y_n']_{(x=c)}$$

and both y_m and y_n have the same values at $x = c$ and at $x = c + 2\pi$. If $m \neq n$, we have at once the fact that

$$\int_c^{c+2\pi} y_m y_n \, dx = 0,$$

and this proves parts (i), (ii), and (iii) of the theorem with the exception of the single case in part (iii) for which $m = n \neq 0$. However, if $m = n \neq 0$, part (iii) reduces to

$$\int_c^{c+2\pi} \sin mx \cos mx \, dx = \frac{1}{2} \int_c^{c+2\pi} \sin 2 mx \, dx.$$

This is an example of part (iii) in which m is replaced by $2m$ and n is replaced by 0, and hence we have taken care of the one gap in part (iii) left by the former proof.

For part (iv) we only need to notice that

$$\sin^2 mx = \frac{1}{2}(1 - \cos 2mx) \quad \text{and} \quad \cos^2 mx = \frac{1}{2}(1 + \cos 2mx),$$

where we consider both integrals by considering

$$\frac{1}{2} \int_c^{c+2\pi} (1 \pm \cos 2mx) dx = \frac{1}{2} \int_c^{c+2\pi} dx \pm \frac{1}{2} \int_c^{c+2\pi} \cos 2mx \, dx = \frac{1}{2}(2\pi) = \pi,$$

since the integral of $\cos 2mx$ is zero by either part (ii) or part (iii). □

Having this much background, although its use is not yet apparent, we will now proceed to obtain the coefficients of (IX.8.1). We must make the assumption that we can integrate the series term by term and obtain the integral of the entire series, just as we would with an expression with a finite number of terms. We will comment on this assumption near the end of the section. The key point here is to realize that Theorem 8.1 indicates that many of the integrals involving the sine and cosine may have a zero value. Hence we may be able to get rid of many terms by integration. Suppose we multiply (IX.8.1) by cos mx, and then integrate over an interval of length 2π. We shall let the c of Theorem 8.1 have a value of zero here, but it should be clear that we could use any other real value just as well. We then obtain, using our assumption about term-wise integration,

$$\int_0^{2\pi} f(x)\cos mx \, dx = \frac{a_0}{2} \int_0^{2\pi} \cos mx \, dx + \sum_{j=1}^{\infty} \left[a_j \int_0^{2\pi} \cos mx \cos jx \, dx \right.$$

$$\left. + b_j \int_0^{2\pi} \cos mx \sin jx \, dx \right].$$

If m is a positive integer, each of the integrals on the right side of this equation has a zero value except for the single integral

$$\int_0^{2\pi} \cos mx \cos jx \, dx$$

in which j has the specific value $j = m$. Therefore, of the infinite number of terms on the right side, only one is non-zero. However, in this case, we know from Theorem 8.1 that the value of this integral is π, and hence we have the equation

$$\int_0^{2\pi} f(x) \cos mx \, dx = a_m \pi.$$

Solving for a_m we have

$$a_m = \frac{1}{\pi} \int_0^{2\pi} f(x) \cos mx \, dx.$$

This was a rather nice way of isolating almost half the coefficients, one at a time, and then evaluating them. Since this worked so well, it might be worth while proceeding further. Suppose we had integrated (IX.8.1) just as it stands (or in other words performed the last operation with $m = 0$). Then we would have all of the integrals in the summations vanishing and our resulting equation would be

$$\int_0^{2\pi} f(x)dx = \frac{a_0}{2} \int_0^{2\pi} dx = \frac{a_0}{2} (2\pi) = \pi a_0,$$

and from this we would obtain

$$a_0 = \frac{1}{\pi} \int_0^{2\pi} f(x)dx.$$

Note that this gives us the same formula we had above, but permits us to include the case $m = 0$, or alternatively to permit the one formula to take care of the evaluation of all of the a_m's. Now you should be able to see the ulterior motive in using the denominator of two in the constant term. [We might well suspect that the first person to do this obtained a different formula for the constant term, and at some later time observed that he could make this nice simplification.] It is now time to turn our attention to the matter of determining the values for the b's. Our method has worked so well, however, that this poses little problem. We merely multiply by sin mx, and lo and behold we find that we obtain for the b's the formula

$$b_m = \frac{1}{\pi} \int_0^{2\pi} f(x) \sin mx \, dx.$$

We have now obtained the coefficients for the series (IX.8.1), provided, of course, we can perform the indicated integrations. We will consider an example at this point by way of illustration.

EXAMPLE 8.1. Let $f(x) = 1$ on the interval $[0, \pi)$ and $f(x) = 0$ on the interval $[\pi, 2\pi)$. Expand this function in a trigonometric series.

Solution. We can determine the coefficients of the trigonometric series by evaluating

$$a_m = \frac{1}{\pi} \int_0^{2\pi} f(x) \cos mx \, dx = \frac{1}{\pi} \left[\int_0^{\pi} f(x) \cos mx \, dx + \int_{\pi}^{2\pi} f(x) \cos mx \, dx \right]$$

$$= \frac{1}{\pi} \left[\int_0^{\pi} (1) \cos mx \, dx + \int_{\pi}^{2\pi} (0) \cos mx \, dx \right]$$

$$= \frac{1}{\pi} \int_0^{\pi} \cos mx \, dx.$$

Similarly

$$b_m = \frac{1}{\pi} \int_0^{2\pi} f(x) \sin mx \, dx = \frac{1}{\pi} \int_0^{\pi} \sin mx \, dx,$$

since $f(x) = 1$ on the interval $[0, \pi)$ and $f(x) = 0$ on $[\pi, 2\pi)$. Note that we have taken the integral over the interval $[0, 2\pi)$ and expressed it as the sum of two integrals in order that we would use the values of $f(x)$ as given. In the case of the second integral, the integrand is zero, and hence the integral vanishes. Now if $m \neq 0$, since m is an integer,

$$a_m = \frac{1}{m\pi} [\sin m\pi - \sin 0] = 0.$$

Similarly,

$$b_m = \frac{1}{m\pi} [1 - \cos m\pi].$$

If m is even, $b_m = 0$, but if m is odd $b_m = 2/(m\pi)$.

Finally, we have to take care of the case of a_0. Here we have $a_0 = (1/\pi) \int_0^\pi dx = 1$. Combining these results we have the series

$$f(x) = \frac{1}{2} + \frac{2}{\pi} \sin x + \frac{2}{3\pi} \sin 3x + \frac{2}{5\pi} \sin 5x + \cdots$$

$$= \frac{1}{2} + \sum_{k=1}^{\infty} \frac{1}{\pi(2k-1)} \sin(2k-1)x.$$

Here we have an expansion for a function that we could not have expanded in a Taylor's expansion, for the derivatives are zero where they exist and there are points where the derivatives fail to exist. We could apparently expand it in a trigonometric series. There is still the question of integration of the individual terms, but remember we are taking that on faith for the moment. In order to verify that this does seem to approach the rather weird function we proposed we have drawn a graph of $f(x)$ and of the curves

$$y = \frac{1}{2} + \frac{2}{\pi} \sin x,$$

and

$$y = \frac{1}{2} + \frac{2}{\pi} \sin x + \frac{2}{3\pi} \sin 3x + \frac{2}{5\pi} \sin 5x$$

in Figure IX.7a, and of $f(x)$ and the curve

$$y = \frac{1}{2} + \frac{2}{\pi} \sin x + \frac{2}{3\pi} \sin 3x + \frac{2}{5\pi} \sin 5x + \frac{2}{7\pi} \sin 7x + \frac{2}{9\pi} \sin 9x$$

in Figure IX.7b. Note the way in which the graphs of the partial sums approach the graph of the function as the number of terms in the partial sums increase. If you stop to think about it, it is rather remarkable that we can use a sum of continuous functions and approximate a discontinuous function as closely as this graph indicates. At the points of discontinuity, the partial sums, as representatives of the trigonometric series, appear to split the gap. As the number of terms increases the graph seems to overshoot slightly in attempting to get where it is supposed to be as quickly as possible.

The trigonometric series were used by Lagrange. However, they bear the name of *Joseph Fourier* (1768–1830) who used them in a study of the problem of heat flow at a time somewhat later than the work of Lagrange. They are now known as *Fourier series*.

The problem of determining whether the integration we have done in a term by term manner is valid in the case of an infinite series has been rather thoroughly investigated, and it can be shown that the resulting series is valid provided the series

$$\sum_{j=1}^{\infty} (a_j^2 + b_j^2)$$

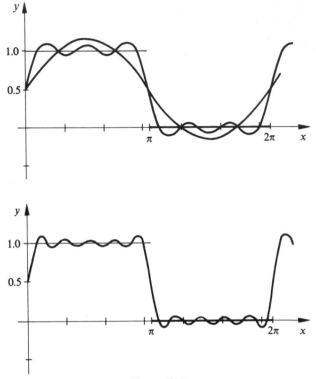

Figure IX.7

is convergent. A check in the case of our illustration would indicate that this is
the case. We will not endeavour to prove this result here, but the matter of
manipulation of any series is something that should always be questioned
before the results are accepted. (Our example of Section I.1 should illustrate this.) The Fourier series are helpful not only in cases in which there is a
cyclic repetition, but also in many cases in which the Taylor's expansion
fails to exist, for frequently a Fourier expansion will exist when the Taylor's
expansion does not. In general, the Fourier expansion will exist whenever the
function is bounded and is continuous except at a finite number of points in
the interval of investigation. While some slight generality is possible beyond
this, it is apparent that this covers a very large number of functions. This
expansion is a superior choice if there are any discontinuities, for the derivative is non-existent at the points of discontinuity and the Taylor's series
fails to exist.

The success of this type of expansion arises from the fact that the coefficients are determined by integration (which has less demanding hypotheses)
rather than by differentiation. The fact that the products of the trigonometric
functions yield integrals that vanish is, of course, the key to the ease with
which the coefficients can be determined. There are many other sets of

functions which have this same characteristic. They are called *orthogonal*
sets, and the trigonometric functions form the first such set with which we
have come in contact. We will not see others in this book, but you should keep
them in mind. They are important in many applications including the preparation of problems for computation on computers.

EXERCISES

1. Expand $f(x) = x$ in a Fourier series over the interval $[0, 2\pi]$.

2. Expand $f(x) = \sin 3x$ in a Fourier series over the interval $[0, 2\pi]$.

3. Expand $f(x) = \sin(x/2)$ in a Fourier series over the interval $[0, 2\pi]$.

4. (a) If $f(x)$ is defined as $f(x) = x$ over the interval $[0, \pi]$ and as $f(x) = 2\pi - x$ over
    the interval $[\pi, 2\pi]$, expand $f(x)$ in a Fourier series over the interval $[0, 2\pi]$.
   (b) Sketch $f(x)$ over this interval.
   (c) Let $x = 0$ and find the series of constants that will converge to $f(0) = 0$.
   (d) Assuming that the series of constants of part (c) does converge to zero, as it
    should, show that

$$\frac{\pi^2}{8} = \sum_{k=0}^{\infty} \frac{1}{(2k + 1)^2}.$$

   (e) Write a computer program or use a calculator and use the first 50 terms of the
    series of part (d) to approximate π. If you have a computer available, do the
    same thing for 500 terms.
   (f) Let $x = \pi$ and find the series of constants that will converge to $f(\pi) = \pi$.
   (g) Show that the result of part (f) agrees with the result of part (c).

5. (a) If $f(x)$ is defined as $f(x) = 0$ over $[0, \pi/2)$, and $f(x) = 1$ over $[\pi/2, \pi)$, $f(x) = 0$
    over $[\pi, 3\pi/2)$ and $f(x) = 1$ over $[3\pi/2, 2\pi)$, the function is sometimes referred to
    as a square wave. Expand $f(x)$ in a Fourier series over the interval $[0, 2\pi]$.
   (b) Sketch the function $f(x)$, and using the periodic nature of this function sketch
    the extensions of this function in order that you have a sketch covering the
    interval $[-4\pi, 4\pi]$.
   (c) Examine your sketch and determine whether $f(x)$ is an even function, an odd
    function, or neither.
   (d) Use the result of part (c) to explain the presence or absence of the sine terms and
    the cosine terms in part (a).

6. Expand the function $f(x) = e^{-x}$ in a Fourier series over the interval $[0, 2\pi]$.

7. (a) If $f(x)$ is an odd function over the interval $[-\pi, \pi]$ show that all of the a_m's in
    the Fourier series will be zero, and the b_m's could be obtained by integrating
    over the interval $[0, \pi]$ and then doubling the result.
   (b) If $f(x)$ is an even function over the interval $[-\pi, \pi]$ show that all of the b_m's in
    the Fourier series will be zero, and the a_m's could be obtained by integrating
    over the interval $[0, \pi]$ and then doubling the result.
   (c) If a function $f(x)$ is defined over the interval $[0, \pi]$, show that it is possible to
    so define the function over the interval $[-\pi, 0]$ that you can obtain a Fourier
    expansion of the function over the interval $[-\pi, \pi]$ that will consist only of sine
    functions or one that will consist only of cosine functions.

8. (a) If $f(x) = x$ over the interval $[0, \pi]$, obtain a Fourier expansion over this interval that consists only of sine functions.
   (b) Obtain a Fourier expansion of $f(x)$ that contains only cosine functions.
   (c) Let $g(x) = 0$ on the interval $[-\pi, 0]$ and $g(x) = x$ on the interval $[0, \pi]$, expand $g(x)$ in a Fourier series over the interval $[-\pi, \pi]$.
   (d) Show that the expansion of part (c) is the average of the expansions of parts (a) and (b). Explain why this should be the case.

9. (a) Expand the function $f(x) = \sin^2 4x$ over the interval $[-\pi, \pi]$.
   (b) Check your answer by use of identities.

10. (a) Expand $f(x) = \cos^3 x$ over the interval $[-\pi, \pi]$.
    (b) Check your answer by use of identities.

11. (a) If $f(x) = 0$ for $x < 0$ and $f(x) = e^x$ for $x \geq 0$, expand $f(x)$ over the interval $[-\pi, \pi]$.
    (b) Find the series for $f(0)$ obtained by using the result of part (a).
    (c) To what value does the series of part (b) appear to converge? Would you have expected this value?
    (d) Find the series for $f(\pi)$.
    (e) To what value does the series of part (d) appear to converge? Would you have expected this value? (Draw a sketch of the curve $y = f(x)$.)

12. (a) If x is in the interval $[-4, 4]$, show that $y = \pi x/4$ is in the interval $[-\pi, \pi]$.
    (b) Use the information of part (a) to expand $f(x)$ in a Fourier series over the interval $[-4, 4]$ if $f(x) = |x|$. Do this by first changing to the variable y for the expansion and then changing back to x.

13. Verify the integration results of Theorem 8.1 by actually performing the integrations involved.

IX.9 Uses and Mis-uses of Series

Infinite series are useful for the purpose of evaluating functions. They are also useful for defining functions. Just as we defined the gamma function as an improper integral, so there are functions for which the basic definition is an infinite series. (See Exercise IX.5.14). In fact, any series of functions which converges represents a function. There is the natural question concerning whether we can perform the various operations with series that we can perform with the more elementary functions. The answer in many cases is yes, but the problem of proving in advance that the results will converge, and moreover converge to the correct value, is not inconsiderable. We have attempted to prove a few theorems in that direction in this chapter. For the majority of series with which one is apt to come in contact in applications, things go well.

However, one should always check to make certain that the series resulting from the various manipulations is convergent. If it is, there is a good chance the results of the manipulations are correct ones. There is, of course, a great deal of theory concerning the various operations and the conditions under which they can be performed. This will undoubtedly be covered if you take a course in *advanced calculus* under one of the various titles that such a course may have. As far as we are concerned here, we have shown that if a series is absolutely convergent a rearrangement of terms does not affect the convergence, or the limit. We have stated, without proof, that if a power series is dominated (in the sense of the comparison test) by a series of constants which is absolutely convergent, then the series can be integrated or differentiated term by term and the result will be the one which you would like to have— namely a correct one. The sum of two convergent series and product of two absolutely convergent series are convergent.

We have also found that it is possible to obtain different types of series of functions. As you might suspect, we have only begun to tap the surface in this regard. The power series is the oldest from the standpoint of mathematical history, and perhaps it is the most natural. The trigonometric series, however, was not long delayed in making its appearance, for it fits so well in those situations which have a periodic nature. There is the further fact, which is of greater importance than might appear at first glance, that the coefficients of the trigonometric series were found by integration rather than differentiation, and it is frequently possible to integrate functions which may not have derivatives. There are many other orthogonal functions with the property we utilized so effectively in obtaining the coefficients of the Fourier series, that is the property that the integrals of the products of distinct terms in the set of functions over the specified interval is zero. Such functions have assumed a great deal of importance in many applications since they permit just exactly the type of manipulation we performed in Section IX.8.

We have indicated that series can be used for the evaluation of functions. They can also be used to aid in the evaluation of integrals which do not yield to formal techniques as indicated in the Exercises of Section IX.5. If one can express an integrand as an infinite series and if the criterion for term-wise integration is satisfied, we have a means for evaluating the resulting integral. Additional examples are given in the Exercises at the end of this section. This method was not overly appealing in the pre-computer days, but now it presents relatively little difficulty. This is not to imply that a great deal of effort has not gone into finding even better ways of evaluating such integrals, but the series technique is certainly one that is available if others do not work. In the next chapter we will find that the series, particularly the Taylor's series, is of great assistance in determining potential error in a variety of numerical techniques. We will be using it as an essential tool in evaluating the error for methods of differentiation and integration which are very popular in many modern computer programs designed to perform the basic operations of integration and differentiation.

It should always be borne in mind that in general a series is of value only when it converges. (There are exceptions, as in the case of asymptotic series, but these are not likely to appear in the less sophisticated applications.) Therefore, you should always be certain that you are operating within the interval of convergence of a series, and it is even better if it is within the interval of absolute convergence.

We will close with one further illustration of the usefulness of infinite series.

EXAMPLE 9.1. Evaluate $\int_0^a e^{-x^2}\, dx$.

Solution. We have often indicated that we cannot evaluate this integral by formal means. However, we do have a Maclaurin's expansion for e^y, and we could use this if we replaced y by $-x^2$. Since the series for e^y converges for all values of y, we have no fear that we can select a value of x for which the resulting composite series will fail to converge. Since

$$e^y = \sum_{k=0}^{\infty} \frac{y^k}{k!},$$

composition would give us $e^{-x^2} = \sum_{k=0}^{\infty} ((-x^2)^k/k!) = \sum_{k=0}^{\infty} ((-1)^k x^{2k}/k!)$. If we integrate this result, we will have $\int_0^a e^{-x^2}\, dx = \sum_{k=0}^{\infty} ((-1)^k/k!) \int_0^a x^{2k}\, dx$ or $\int_0^a e^{-x^2}\, dx = \sum_{k=0}^{\infty} ((-1)^k a^{2k+1}/(2k+1)k!)$. If in particular we wish to evaluate $\int_0^1 e^{-x^2}\, dx$ we would have

$$\int_0^1 e^{-x^2}\, dx = 1 - \frac{1}{3} + \frac{1}{10} - \frac{1}{42} + \frac{1}{216} - \frac{1}{1320} + \frac{1}{9360} - \cdots = 0.74736.$$

Since this is an alternating series and since we have reached a point from which the absolute values of the terms are monotonic decreasing, we know that the error in stopping at the point indicated is no greater than the first term omitted. Since the first omitted term in our computation above would only change the 6 to a 5, we are certain that our result is sufficiently close for the majority of computations. Observe that we have checked on the convergence of the series we used, and also checked on the accuracy of the result. It is always easy to *assume* that something will work, and then find that your assumption was not warranted in the case at hand.

EXERCISES

1. (a) Expand $f(x) = x^5 - 7x^4 + 2x^2 - 3x + 17$ in a Taylor's expansion about the value $x = -2$.
   (b) Simplify your result and check the accuracy of your work.

2. (a) Expand $f(x) = \ln x$ in a Taylor' expansion about $x = 10$.
   (b) Find the interval of convergence of your series.
   (c) At what point do the coefficients become monotonic decreasing in absolute value?

3. (a) Expand $f(x) = (\sin x)/x$ in a Maclaurin's series. Note that it would be possible to obtain this series from the series for $\sin x$ without differentiating $f(x)$.
   (b) Obtain the value of $\int_{0.1}^{1} ((\sin x)/x)dx$ with an accuracy of four decimal places.

4. (a) Expand $f(x) = \cos x$ in a Maclaurin's series.
   (b) Obtain a series for $\cos \sqrt[3]{x}$.
   (c) Use series to evaluate $\int_{0}^{0.125} \cos \sqrt[3]{x}\, dx$ to four decimal places.
   (d) See whether you can evaluate this integral directly. If so, compare your result with that of part (c).

5. (a) Expand $f(x) = e^x$ in a Maclaurin's expansion.
   (b) Obtain the expansion of $x^2 e^{-x^2}$.
   (c) Evaluate $\int_{-1}^{1} x^2 e^{-x^2}\, dx$ with an error no greater than 10^{-5}.

6. (a) Expand the function $f(x) = |x - \pi|$ in a Fourier series over the interval $[0, 2\pi]$.
   (b) Find the area bounded by $y = f(x)$ and the x-axis over the given interval by integrating the series of part (a).
   (c) Find the area involved in part (b) by geometry.

7. (a) Use the Maclaurin's expansions for e^x and $\sin x$ and obtain the series for $e^{\sin x}$ by composition.
   (b) Use any facts you know about each of the series and about composition to determine the interval of convergence for your result of part (a).

8. (a) Using the Maclaurin's expansion for $\sin x$, find the expansion for $\sin^2 x$.
   (b) Using the Maclaurin's expansion for $\cos x$, find the expansion for $\cos^2 x$.
   (c) Add the series of parts (a) and (b).

9. Let $g(x)$ be the function given by the fact that $g(x) = -1$ on the interval $[0, \pi/2)$, $g(x) = 0$ on $[\pi/2, \pi)$ and on $(3\pi/2, 2\pi)$, and $g(x) = 2$ on the interval $[\pi, 3\pi/2)$.

   (a) Expand $g(x)$ in a Fourier series.
   (b) Sketch the graph of $g(x)$.
   (c) Can you determine what values the series will ascribe to $g(x)$ at $x = 0, \pi/2$, and $3\pi/2$?
   (d) What value would seem logical for assignment at such points.
   (e) Does this assignment of values that you think is reasonable seem to be the one selected by the Fourier series?

10. (a) If $f(x)$ is expanded in a Fourier series such that

$$f(x) = (a_0/2) + \sum_{j=1}^{\infty} (a_j \cos jx + b_j \sin jx),$$

show that the series is absolutely convergent for each value of x

$$\text{if } \sum_{j=1}^{n} (|a_j| + |b_j|) \quad \text{converges.}$$

(b) If the condition of part (a) is fulfilled show that the series can be integrated and differentiated term by term.

CHAPTER X

Numerical Methods

X.1 Introduction

The use of arithmetic to obtain at least approximate answers to very complicated problems pre-dated the development of the calculus by Newton and Leibniz. The amount of arithmetic done by some of the early astronomers as they attempted to explain the movement of the planets would bring anguish to any person today who might attempt to duplicate their computations, even with the aid of a hand calculator. The introduction of integration and differentiation, however, added a new impetus to the search for methods which would permit more efficient and more accurate approximations for results which even the new mathematics could not provide in a formal sense. We have frequently referred to the fact that formulas fail to exist for the evaluation of much used integrals. Problems such as this are the object of the continuing search for improved numerical methods. In fact, the development of the modern high speed digital computer was hastened by the need for just such computation in time of war. It is our purpose in this chapter to explore some of the more commonly used techniques for differentiation and integration using numerical methods. In many cases the results will be aided by the use of a hand calculator or a computer, although the techniques can be used in hand computation, just as they have been used for decades, and in many cases centuries.

Our development of the numerical methods must cover several points. In the first case we must know where the given numbers may be expected to originate. In some instances we may have a function given and we can then obtain values for the function at such points as we may wish. In other cases we may be given data. If we have the latter case there is the question whether the data points are points of a regular partition or not. If they are, the computation will usually be somewhat more straightforward. If they are not, it might well require the development of special formulas. While such special formulas generally exist, we will not endeavor to obtain them here. Rather we will assume that we could obtain a function by interpolation or regression based upon the given data. With the function thus derived we would be able to get such points as we would need.

Our second consideration concerns methods of obtaining the numerical approximations of the results we seek. It is often possible to use a series, such as a Taylor's series or a Fourier series. Of course, we could use only a finite number of terms, but if the series is convergent we can take a sufficient number of terms that the error will be small. However, this brings in the matter of how many terms will be required. If the number of terms required should be 500,000, the problem would take a substantial amount of time even on a computer. If such a large number of terms were required, there would also be a resulting error due to the fact that each term can be expected to have some roundoff error, and the sum of the roundoff errors may not cancel out. Thus, the result might be suspect. Therefore, it would be well to try to find some method which uses fewer terms in the computation.

A third consideration has to do with accuracy. If the result is to be useful, it must be sufficiently accurate for the purpose at hand. We have already mentioned the problem of accumulated roundoff error in cases in which we have a large number of computations involved. There are other considerations that can introduce errors in a computer, such as that involved in evaluating the expression $(1 - \cos \theta)$ for very small values of θ. Since $\cos \theta$ in this case would be very close to one, and since the computer retains only a finite number of digits, say ten in a particular situation, we could have a value of θ sufficiently close to zero that the first eleven decimal places of $\cos \theta$ would be nines. In this instance the difference $(1 - \cos \theta)$ would appear to be zero although it would, in fact, not be zero. This would make a great difference if we were to divide by this expression. We will not concern ourselves with this problem in this chapter, but it is well to keep it in mind. Such problems are dealt with in courses in Numerical Analysis. However, accuracy does not depend entirely upon problems related to the computer. We must consider whether the method we are using would give the correct answer even if there were no computational errors. If we were to use a series and use only the first 100 terms, the result would be in error by the amount represented by the infinite number of terms we have failed to use. Such an error is called a *truncation error*. It is often possible to find some upper bound for the truncation error, and we will do this where possible. While this does not establish the precise error, it does indicate whether the results are worth obtaining in the first place.

The questions we have raised are not intended to be discouraging, but rather to give a note of caution. The methods given here are widely used, and are generally very accurate. However, it is well to check in each instance to insure that the result you obtain will meet the requirements of the problem at hand.

EXERCISES

C1. Let $f(x) = x^2/(1 - \cos x)$.

    (a) Evaluate $f(x)$ for $x = 2^{-1}, 2^{-2}, 2^{-3}, \ldots, 2^{-60}$.
    (b) Examine your results in part (a) and determine whether they seem to converge.
    (c) Show that $\lim_{x \to 0} f(x) = 2$.
    (d) Do the results of part (a) agree with the result of part (c)?
    (e) Show that $(1 - \cos x) = 2 \sin^2(x/2)$.
    (f) Replacing the denominator of $f(x)$ with the results of part (e) repeat part (a).
    (g) Compare the roundoff errors of part (f) with those of part (a).

C2. (a) Compute $\sin x$ using the first four non-zero terms of the Maclaurin's series for $x = 0, 0.1, 0.2, 0.3, \ldots, 1.5$.
    (b) Compare the result of your computation with the actual value of $\sin x$ for each of these values of x.
    (c) The primary error in the evaluation of part (a) is truncation error. Show why this error increases for larger values of x.

3. Let $f(n) = \sum_{k=1}^{n} ((-1)^{k+1}/k)$.

    (a) How many terms would be required in order that two successive values of $f(n)$ would differ by less than 0.00001.
    (b) How many divisions would be required to obtain the result for the number of terms you determined would be necessary in answer to part (a)?
    (c) How much roundoff error could be introduced in the computation of part (a) if you used a computer with an accuracy of 6 decimal places?
    (d) Indicate the possible error contribution due to roundoff error and the possible error contribution due to truncation error.

X.2 Numerical Computation of Derivatives

If we are asked to compute the derivative of a function by numerical methods, our first thoughts would probably return to the differential quotient. It is certainly true that $DQf(x, x + \Delta x)$ *should* give us a good approximation of the derivative for small values of Δx. There is a serious question of accuracy, however, for if we make Δx very small we find ourselves dividing the difference of two nearly equal numbers by a very small number. Division by 0.00001, for instance, would have the effect of multiplying any errors in the numerator by 100,000. As a consequence it would appear that we should use very small values of Δx in order that the differential quotient would have a value close to that of the derivative, but on the other hand if we use a small value of Δx, we magnify the computational errors. After this rather discouraging start, we will examine the process of obtaining a numerical approximation of the derivative through the use of the Taylor's expansion. In the Taylor's expansion we had an error term, and this might give us some clue concerning a bound for the error of computation. If we replace Δx by h, something frequently done by the numerical analyst, we then have

$$f(x + h) = f(x) + hf'(x) + \frac{h^2}{2!} f''(c) \qquad (X.2.1)$$

where c is some value in the interval $(x, x + h)$. Since we are seeking the derivative, $f'(x)$, we can solve (X.2.1) for $f'(x)$ and obtain

$$f'(x) = \frac{f(x + h) - f(x)}{h} - \frac{h}{2} f''(c). \qquad (X.2.2)$$

This shows us that the differential quotient $DQf(x, x + h)$ is in error by an amount $[-(h/2)f''(c)]$. We cannot expect to know $f''(x)$, for this would require knowledge of the second derivative which in turn would require knowledge of the first derivative. If we had known the first derivative we would not have had to go through this computation in the first place. However, we may have some idea of an upper bound for the values of $f''(x)$ in the interval $(x, x + h)$. If we do have a reason to know such an upper bound, we then can estimate an upper bound for the error in our numerical computation of $f'(x)$. Furthermore, we can frequently determine the sign of the error, and thus we would know whether our result was too large or too small.

If we examine the statement of (X.2.2) closely, we see that if we use a small value of h, we then have a smaller possible error in theory. That is to say that the truncation error obtained by using only the differential quotient will be small. This does not give any indication of the magnitude of the roundoff error, and that could be large as we inferred in the preceding paragraph.

EXAMPLE 2.1. Use numerical methods to approximate the value of $f'(3)$ if $f(x) = x^3$.

Solution. If we use a value $h = 0.01$, we would then consider $f'(3) = ((f(3.01) - f(3))/0.01) - (0.01/2)f''(c)$ where c is in the interval $(3, 3.01)$. We then have $f'(3) = ((27.270901 - 27)/0.01) - 0.005f''(c) = 27.0901 - 0.005f''(c)$. In the ordinary course of events we would not have written $0.005f''(c)$ and as a result we would have had an approximation. However, in this instance we have continued to carry this term in order to emphasize its presence. We have the question of possible error in our result of 27.0901. (Of course we know that the error is 0.0901 in this case. That is we have an error of 0.3337%.) If we had not known the correct result, however, we might have approximated $f'(3.01)$ by the same method, and obtained 27.2707. It would then be clear that the derivative has increased by an amount which can be approximated by $27.2707 - 27.0901 = 0.1806$ while x has increased by 0.01. Therefore the approximate rate of increase of the derivative is

(0.1806/0.01) = 18.06. This would serve as an approximation for the second derivative in the interval, and consequently we would have an error of (−0.005)(18.06) = (−0.0903). While this computation does not give an exact upper bound, it gives a result which would generally be very close to the upper bound. In this case the approximate bound differed from the actual error by only 0.0002. Note also that the indicated error is in the correct direction.

There is no reason why we could not have used a negative value for h. The results would be very close to those obtained here. In that case the approximation of the second derivative would be slightly low, but the value would be sufficiently close to the correct value of 18 that we would have reason to trust our numerical method as being sufficiently good to take care of most requirements.

The example indicates that our method will give us a good approximation, but the question should have come to mind whether we might not be able to do something better. Such a question cannot always be answered in the affirmative, but it is always worth asking. In this case we might also have tried to compute the differential quotient $DQf(3, 2.99)$ where we are using a value of -0.01 for h. We would have found $DQf(3, 2.99) = 26.9101$. In one instance we found a value too large and in this instance a value too small. It might be worth taking the average. If we did this we would have obtained 27.0001, a value that is obviously much closer to the correct value of 27 in this case. While this *seems* like a good idea, it is one we should check out, for a single instance does not constitute an argument for using this method. This suggests that we wish to try

$$f'(x) \doteq \frac{DQf(x, x + h) + DQf(x, x - h)}{2} = \frac{f(x + h) - f(x - h)}{2h}. \quad (X.2.3)$$

We have indicated that this is an approximation, of course, and we leave the algebra used in obtaining the last expression of this relation to you. Our concern now centers on the error we would have if we were to use this result in computing $f'(x)$. We can again use Taylor's expansion. Here we will need to use

$$f(x + h) = f(x) + hf'(x) + \frac{h^2}{2!} f''(x) + \frac{h^3}{3!} f'''(c_1)$$

and

$$f(x - h) = f(x) - hf'(x) + \frac{h^2}{2!} f''(x) - \frac{h^3}{3!} f'''(c_2).$$

Subtraction of these results gives

$$f(x + h) - f(x - h) = 2hf'(x) + \frac{h^3}{6} [f'''(c_1) + f'''(c_2)].$$

Upon solving this last result for $f'(x)$, we obtain

$$f'(x) = \frac{f(x+h) - f(x-h)}{2h} - \frac{h^2}{6}\left[\frac{f'''(c_1) + f'''(c_2)}{2}\right]. \quad (\text{X.2.4})$$

Since the term in the brackets is the average of two third derivatives in the interval $(x - h, x + h)$, we know that there must be some number c in this interval such that the average is equal to $f'''(c)$ if the third derivative is continuous in the interval. This follows from the intermediate value theorem. If the third derivative is not continuous, we might still be able to make some estimate which would give us at least an estimate of the size of the error. Therefore, we can write

$$f'(x) = \frac{f(x+h) - f(x-h)}{2h} - \frac{h^2}{6} f'''(c)$$

$$= DQf(x - h, x + h) - \frac{h^2}{6} f'''(c). \quad (\text{X.2.5})$$

EXAMPLE 2.2. Use (X.2.5) to approximate $f'(x)$ in Example 2.1.

Solution. Again using $h = 0.01$, we have

$$f'(x) \doteq \frac{f(3.01) - f(2.99)}{2(0.01)}.$$

This gives

$$f'(x) \doteq \frac{27.270901 - 26.730899}{0.02} = \frac{0.540002}{0.02} = 27.0001.$$

As we can see, this is very much better than our preceding result, and in fact is just the average we had obtained in our discussion above. We can now approximate the error term at least in part. Since $h = 0.01$, we would have the error term

$$-\frac{(0.01)^2}{6} f'''(c) = 0.00001667 f'''(c).$$

It would seem that this must be much better. Of course it is possible that the third derivative would be large, but it would seem improbable that it is large enough to cause a great deal of difficulty. In this instance we know that the third derivative is 6, and hence the result should be changed by -0.0001. In general we can only expect to obtain an approximation of the error, but even that would appear to be reassuring here.

It is apparent that the method used in Example 2.2 is superior to that of Example 2.1. It is rather easy to see why this should be so from a look at the graph of $y = f(x)$ in Figure X.1. We see that $DQf(x - h, x + h)$ is the

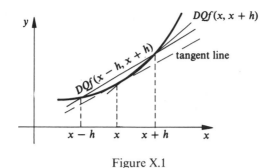

Figure X.1

slope of the line segment joining $(x - h, f(x - h))$ and $(x + h, f(x + h))$. This slope is certainly more likely to be near the slope of the line tangent to the curve than either the slope given by $DQf(x, x + h)$ or the slope given by $DQf(x - h, x)$. This sketch merely bears out the development of (X.2.5) indicating that we can expect a smaller error if we take the average slope over the longer interval provided the longer interval has the point $(x, f(x))$ as a midpoint. We should also note that the computation in this case is likely to be more accurate, for we will have a larger difference in the numerator and a larger denominator. This will reduce the likelihood of significant roundoff error. Thus, (X.2.5) is superior from many points of view.

One might ask whether it is possible to find more accurate methods than these. We were able to gain in accuracy by using two Taylor's expansions instead of one. Could we gain further by using more? The answer is yes. However, the relative gain in accuracy may be overshadowed by the need for additional computation and the resulting possible increase in roundoff error. With this in mind we will content ourselves with the result of (X.2.5) except for an exercise designed to give you an opportunity to test your powers of finding additional formulas.

EXERCISES

1. Let $f(x) = x^2$.

   (a) Find $f'(3)$ if $h = 0.1$ using both (X.2.2) and (X.2.5).
   (b) Show that the error term gives the actual error in each instance.
   (c) Show that (X.2.5) will give the derivative of $f(x)$ for any value of x without error.

2. (a) Find $f'(2)$ if $f(x) = x^3$ using (X.2.2) and $h = 0.05$.
   (b) Find $f'(2)$ using (X.2.5) and $h = 0.05$.
   (c) Show that the absolute value of the error in computing $f'(x)$ is a monotonic increasing function of the absolute value of h using (X.2.2).
   (d) Show that the error in computing $f'(x)$ using (X.2.5) is independent of the value of x.

3. Find $f'(2)$ using (X.2.2) and (X.2.5) in each of the following cases. Also find an upper bound for the error term in each case if possible.
   (a) $f(x) = x^3 - 3x$ using $h = 0.2$.
   (b) $f(x) = (x^2 + 2)^2 - 5$ using $h = 0.1$.
   (c) $f(x) = e^x$ using $h = 0.05$.
   (d) $f(x) = \sqrt{2x}$ using $h = 0.41$.
   (e) $f(x) = x^4$ using $h = 0.2$.
   (f) $f(x) = x^4$ using $h = -0.2$.

4. Use the values in the tables of Appendix C to evaluate $f'(x)$ in each of the following cases.
   (a) $f(x) = \sin x$ at $x = 0.25$ radians using the smallest h you can use while using only values in the tables.
   (b) $f(x) = \tan x$ at $x = 43°$ using the smallest h you can use with the information given in the tables. Also show that this derivative in terms of degrees is a good approximation by developing the relation between the derivative with respect to degrees and the derivative with respect to radian measure.
   (c) $f(x) = \cos x$ at $x = 0.60$ radians.
   (d) $f(x) = \ln x$ at $x = 2.8$.
   (e) $f(x) = \ln x$ at $x = 3.0$.
   (f) $f(x) = e^x$ at $x = 4.2$.
   (g) $f(x) = e^{-x}$ at $x = 5.5$.
   (h) $f(x) = \Gamma(x)$ at $x = 1.24$.

5. In each of the following cases find a value of h for (X.2.2) and for (X.2.5) which would give $f'(x)$ with an error no greater than 0.0001 in absolute value.
   (a) $f(x) = \sin x$ at $x = -1$.
   (b) $f(x) = x^4$ at $x = -2.1$.
   (c) $f(x) = e^x$ at $x = -1.3$.
   (d) $f(x) = x^{-3}$ at $x = 2.3$.

6. (a) Use the Taylor's expansions for $f(x - 2h)$, $f(x - h)$, $f(x)$, $f(x + h)$, and $f(x + 2h)$ and obtain a formula for $f'(x)$ in the manner of this section. Your formula should have non-zero coefficients for at least four of the five function values.
   (b) Use your formula to evaluate $f'(-2)$ if $f(x) = x^5$ and $h = 0.2$.
   (c) Obtain an expression for the error in your formula of part (a).
   (d) Show that the derivative given by your formula will have a zero error if $f(x) = 1$.
   (e) Use your result of part (d) to aid in showing that the sum of the coefficients in part (a) must be zero.

7. (a) In Section III.4 we developed formulas $(a^h - 1)/h$ and $(a^h - a^{-h})/2h$. Review the development of these formulas and show that our derivation was a correct one.
   (b) Show that these formulas are respectively equivalent to (X.2.2) and (X.2.5).
   (c) Use these results with $a = 2$ and $h = 2^{-10}$ to approximate $\ln 2$. [Note that the use of this value of h is equivalent to taking ten successive square roots.]
   (d) Show that your results come within the error bound you would establish using the results of this section.

C8. (a) Write a computer program and run it for determining the derivative of $\tan x$ at $x = \pi/4$ using $h = 2^{-n}$ for all positive integers n not greater than 60.

(b) Examine your results in part (a) and determine when the roundoff error appears to make the results less reliable.

(c) Find out the number of significant digits which your computer uses and see if you can correlate this information with your observations in part (b).

(d) Does the roundoff error appear to affect the use of (X.2.2) faster than the results of (X.2.5)?

(e) Give a reason why your answer to part (d) should be the one you would expect.

S9. (a) Obtain a table of values for the GNP (gross national product) for the last five years for which it is available.

(b) Use (X.2.2) with the smallest value of h (probably one year) that the information of part (a) will permit to find the derivative of the GNP for the middle year of your table.

(c) Find the marginal GNP based on the middle year of your table.

(d) Compare the results in parts (b) and (c).

(e) Why should these results compare in the way in which they do?

(f) Is there any way in which you could put error bounds on your results in parts (b) and (c)?

SB10. (a) Find a table in a recent *World Almanac* which provides annual data for some community for population growth.

(b) Use this data to find the rate of growth for the next to the last year for which such information is available. Use both (X.2.2) and (X.2.5).

(c) Compare the two results you obtain in part (b). Which do you think is more accurate? Why?

(d) What could you say about any errors in determining the rate of growth in part (b).

(e) Would the rate of growth you have obtained appear to explain the observed population over a five year period? (In other words, is the rate of growth relatively constant?)

B11. (a) Obtain a table for weight as a function of age for a person of your sex and for the ten year period starting with your present age.

(b) Find the expected rate of growth one year from now if your weight fits the expected data from this chart.

(c) How accurate is the computation of part (b) as a projection for your weight of growth one year hence?

P12. (a) Make a table for the function $w(n)$ if $w(n)$ represents the atomic weight of chemical element number n in the periodic table.

(b) If we think of $w(n)$ as a continuous function of n whose values coincide with the periodic table at the integer points, find $w'(n)$ by numerical methods for $n = 12$.

(c) Use interpolation involving five points to find a polynomial which is correct for $w(10), w(11), w(12), w(13), w(14)$.

(d) Take the derivative of this polynomial and compare your result with those of part (b).

(e) If you restrict your attention to the halides, find the value of the derivative corresponding to chlorine.

X.3 The Trapezoidal Rule

We now turn our attention to the problem of numerical integration. We will start with the simplest case in the development of a series of results known as the *Newton–Cotes formulas*. These are all derived from the Lagrangian interpolation formulas, and are developed under the assumption that the points upon which the interpolation is based are the points of a regular partition of the interval of integration. Thus, if we wished to integrate over the interval $[a, b]$, we would use a regular partition, $P[a, b]$, and assume that the value of the function is known at the points $(x_j, f(x_j))$. The case in which the partition consists of only two points gives rise to the particular Newton–Cotes formula known as the *trapezoidal rule*. Thus, we are considering that we have but two points, $(a, f(a))$ and $(b, f(b))$. In this instance it is not difficult to proceed directly from these two points, but in order to ease the computation in the development of some of the later results we are going to take advantage of some of the arbitrariness of the coordinate systems we use. In order that you have an opportunity to become accustomed to this simplification for later use we will take time to discuss it before proceeding with the problem at hand. We will use a coordinate system such that $x = 0$ at the midpoint of the interval $[a, b]$. We will also consider that the common distance between the successive points of the regular partition, $P[a, b]$, is the distance h. With two points we have the interval $[-h/2, h/2]$ in the new coordinate system we have selected. Remember that the choice of origin was arbitrary in the first place, and the value of the integral is not affected by a translation of the origin. This was shown in Corollary 3.5 of Theorem III.3.3. Thus we wish to evaluate

$$\int_{-h/2}^{h/2} g(x)dx$$

where $g(x)$ is the translated function. Consequently $g(-h/2) = f(a)$, $g(0) = f((a + b)/2)$, $g(h/2) = f(b)$. By Lagrangian interpolation we have the approximation

$$y(x) = \frac{(x - h/2)}{(-h/2 - h/2)} g(-h/2) + \frac{(x - (-h/2))}{(h/2) - (-h/2)} g(h/2)$$

$$= \frac{1}{h} [-(x - h/2)g(-h/2) + (x + h/2)g(h/2)]. \tag{X.3.1}$$

The function $y(x)$ is an approximation of $g(x)$ and has the same values as $g(x)$ at the points $(-h/2, g(-h/2))$ and $(h/2, g(h/2))$. We obtain an approximation for the integral by integrating $y(x)$ instead of $g(x)$, and hence we have

$$
\begin{aligned}
\int_{-h/2}^{h/2} g(x)dx &\doteq \int_{-h/2}^{h/2} y(x)dx \\
&\doteq \frac{1}{h}\int_{-h/2}^{h/2}\left[-\left(x-\frac{h}{2}\right)g\left(-\frac{h}{2}\right)+\left(x+\frac{h}{2}\right)g\left(\frac{h}{2}\right)\right]dx \\
&\doteq \frac{2}{h}\int_{0}^{h/2}\left[\frac{h}{2}g\left(\frac{-h}{2}\right)+\frac{h}{2}g\left(\frac{h}{2}\right)\right]dx \\
&\doteq \frac{h}{2}\left[g\left(\frac{-h}{2}\right)+g\left(\frac{h}{2}\right)\right]=\frac{h}{2}\left[f(a)+f(b)\right]. \qquad \text{(X.3.2)}
\end{aligned}
$$

Here we see evidence of the wisdom of our choice of coordinates. Since the portions of the integrand which involve an odd power of x (in this case x to the first power) are odd functions and the interval of integration is symmetric with respect to the origin, Corollary 8.2 of Theorem VI.8.4 states that these portions of the integrand will contribute nothing to the final result. The same corollary assures us that the integrals of the even functions over the same interval can be evaluated by doubling the result obtained by integrating over the right hand half interval $[0, h/2]$. Since h is a constant, $g(-h/2)$ and $g(h/2)$ are both constants. The integration reflected that fact. (The value of the integral is not altered, but the evaluation is considerably easier when we make use of these labor-saving devices.) We now find that the approximate value of the integral is the length of the interval multiplied by the average of the values of the function at the two ends. If you think of area, this is exactly the formula for the area of a trapezoid, and this explains the name. Note that this agrees with our discussion of the trapezoidal method in Section II.6.

EXAMPLE 3.1. Use the trapezoidal rule to approximate the value of $\int_0^{\pi/2} \sin x \, dx$.

Solution. Here the end points of the interval are $(0, \sin 0)$ and $(\pi/2, \sin \pi/2)$, or equivalently $(0, 0)$ and $(\pi/2, 1)$. Since the interval is of length $(\pi/2) - 0 = \pi/2$, we have $h = \pi/2$. Therefore, the approximation yields

$$
\int_0^{\pi/2} \sin x \, dx \doteq \frac{\pi}{4}[0 + 1] = \frac{\pi}{4} \doteq 0.7854.
$$

If we evaluate the integral formally (as we can in this case) we see that the value of the integral is one. We have an error of about 0.215 in this case, and this method does not appear to work well. However, we note that the value of h is rather large. Perhaps it would help if we were to use a smaller value of h. Probably the most convenient calculation occurs if we divide the interval by some integer. For convenience we will consider both the case of dividing it by two and by three. Therefore, we have in the two cases

$$\int_0^{\pi/2} \sin x \, dx = \int_0^{\pi/4} \sin x \, dx + \int_{\pi/4}^{\pi/2} \sin x \, dx$$

$$\doteq \frac{\pi}{8}\left[0 + \frac{\sqrt{2}}{2}\right] + \frac{\pi}{8}\left[\frac{\sqrt{2}}{2} + 1\right] = \frac{\pi}{8}[\sqrt{2} + 1] \doteq 0.948$$

and

$$\int_0^{\pi/2} \sin x \, dx = \int_0^{\pi/6} \sin x \, dx + \int_{\pi/6}^{\pi/3} \sin x \, dx + \int_{\pi/3}^{\pi/2} \sin x \, dx$$

$$\doteq \frac{\pi}{12}\left[0 + \frac{1}{2}\right] + \frac{\pi}{12}\left[\frac{1}{2} + \frac{\sqrt{3}}{2}\right] + \frac{\pi}{12}\left[\frac{\sqrt{3}}{2} + 1\right]$$

$$= \frac{\pi}{12}[2 + \sqrt{3}] \doteq 0.97705.$$

Example 3.1 gives food for thought in at least two directions. First it would appear that we can improve the accuracy of our result by increasing the number of partition points and using the trapezoidal rule over each of the subintervals. This procedure will be used with sufficient frequency that it would pay us to investigate whether we can find a simpler formula for the case with more than a single subinterval in the partition. We will continue to use the regular partition, and we will continue to denote the length of each subinterval by h. We can now approximate the integral by adding the integrals over the subintervals. Thus, we would have

$$\int_a^b f(x)dx \doteq \sum_{k=1}^n \frac{h}{2}[f(x_{k-1}) + f(x_k)].$$

If we factor out the common factor $h/2$, we can rewrite this as

$$\int_a^b f(x)dx \doteq \frac{h}{2}[f(x_0) + 2f(x_1) + 2f(x_2) + \cdots + 2f(x_{n-1}) + f(x_n)]$$

$$\doteq h[\tfrac{1}{2}f(x_0) + f(x_1) + f(x_2) + \cdots + f(x_{n-1}) + \tfrac{1}{2}f(x_n)].$$

(X.3.3)

The value of h in this case could be computed by noting that $h = (b - a)/n$. In Example 3.1 we used $n = 2$ and $n = 3$. We could illustrate the use of this result in case $n = 4$ by noting the sketch in Figure X.2. Here we have four trapezoids. The right hand base of the first trapezoid is also the left hand base of the second. Therefore, we double the result of the computation of this single base, since we will use it in both instances. We would perform a similar computation for the other common bases. Figure X.2 also illustrates intuitively the fact that we can expect more accurate results using $n = 4$ than using $n = 1$ in this case.

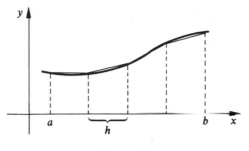

Figure X.2

A second thought that should occur from the work of Example 3.1 would involve the size of the computational error in evaluating the integral. We will now attempt to explain the size of the error in this example. In order to do this, we will assume that $f(x)$ is sufficiently well behaved that the second derivative exists and is continuous in the interval $[-h/2, h/2]$. We return to the result given in (X.3.2) and modify this by replacing each appearance of the function with its corresponding Maclaurin expansion including the remainder term. We will include in our equation an error term, E, in order that we allow for any discrepancies which occur in this evaluation. Thus we have

$$\int_{-h/2}^{h/2} \left[f(0) + xf'(0) + \frac{x^2}{2} f''(z) \right] dx = \frac{h}{2} \left[f(0) - \frac{h}{2} f'(0) + \frac{h^2}{8} f''(c_1) \right.$$

$$\left. + f(0) + \frac{h}{2} f'(0) + \frac{h^2}{8} f''(c_2) \right] + E.$$

$$(X.3.4)$$

Note that z is within the interval $(-h/2, h/2)$ and will vary with the choice of x. Hence, we are assured that there will be some number c_3 in $(-h/2, h/2)$ such that

$$\int_{-h/2}^{h/2} \frac{x^2}{2} f''(z) dx = f''(c_3) \frac{x^3}{6} \Big|_{-h/2}^{h/2} = f''(c_3) \left[\frac{h^3}{48} - \left(\frac{h^3}{-48} \right) \right] = f''(c_3) \frac{h^3}{24}$$

by the mean value theorem for integrals. This follows from our assumption that $f''(x)$ is continuous in this interval. The error terms in the use of $f(-h/2)$ and $f(h/2)$ include constants c_1, and c_2 respectively, constants which are unknown except that we are assured that they are in the interval $(-h/2, h/2)$. Therefore, (X.3.4) becomes

$$hf(0) + \frac{h^3}{24} f''(c_3) = hf(0) + \frac{h^3}{8} \left[\frac{f''(c_1) + f''(c_2)}{2} \right] + E \quad (X.3.5)$$

by direct computation. Under the assumption that the second derivative is continuous in the interval in question, there is some value c in $(-h/2, h/2)$ such that we can replace each of the constants c_1, c_2, and c_3 by c and still have a correct equation. Hence from (X.3.5) we can obtain

$$\frac{h^3}{24} f''(c) = \frac{h^3}{16} [f''(c) + f''(c)] + E = \frac{h^3}{8} f''(c) + E.$$

From this we have

$$E = \left(\frac{h^3}{24} - \frac{h^3}{8}\right) f''(c) = -\frac{h^3}{12} f''(c). \tag{X.3.6}$$

Therefore, the term which we must add to our approximation to make the result correct should be $-(h^3/12)f''(c)$, where c is some value that we have little prospect of ascertaining but for the fact that it is a point in $(-h/2, h/2)$. For purposes of establishing a bound, however, we can frequently find the largest absolute value that this term could have, and then we would know that our approximation is wrong by no more than this bound.

It is very nice to have the error term, but let us see whether this would explain the discrepancies that we found in Example 3.1. In this case we had $f(x) = \sin x$, and hence $f''(x) = -\sin x$. The error would be $-(h^3/12)f''(x) = -(\pi^3/96)(-\sin c) = +(31.0063/96)\sin c = 0.323 \sin c$. We have made use, of course, of the fact that $h = \pi/2$ in this case. Noting that c is in $(0, \pi/2)$, we see that $\sin c$ is in $(0, 1)$ and therefore an error which does not exceed $0.323(1) = 0.323$ should not be surprising. We could have anticipated in advance, then, that the error might be intolerable in this case. The correct value for the integral was indeed in the interval $[0.7854, 1.1084]$. Now if we proceed to the next case, we find that for the two divisions we have a possible error of

$$-\frac{1}{12}\left(\frac{\pi}{4}\right)^3 (-\sin c_1) = \frac{-\pi^3}{768} (-\sin c_1) = 0.04037 \sin c_1$$

and

$$-\frac{1}{12}\left(\frac{\pi}{4}\right)^3 (-\sin c_2) = -\frac{\pi^3}{768} (-\sin c_2) = 0.04037 \sin c_2$$

respectively where c_1 is in $(0, \pi/4)$, c_2 is in $(\pi/4, \pi/2)$ and $h = \pi/4$. Although we could improve our bound by using the separate intervals, it is convenient to simply note that in both instances we have an upper bound of one for the sine, whence we have a bound of $(2)(0.04037) = 0.08074$ for the possible error in the computation. This is a substantial improvement over the preceding case. In both instances our error bounds indicate that our computations were at least as close as we had a right to expect. Having gotten this far, let us try the case of three divisions. In this case we might as well combine the three terms in the computation, since we will again make the assumption of the worst possible case and we will not try to make use of the fact that we have three distinct intervals. Therefore, since $h = \pi/6$, we will have as a bound $|E| \leq (3)(\pi^3/2592)(1) \doteq 0.03589$. Our result was well within this bound, and again we are able to determine a bound for the error in our approximation. It is true that we could improve this bound by making use of the fact that $\sin x$ is smaller on the first two intervals, and there are occasions where one would wish to do this. In general, however, this does not help a great deal. With many functions we do not know enough about the possible values of the second derivative to do much more than make a rather conservative estimate of the maximum size the second derivative can assume in the interval under consideration. Therefore, the procedure we have used is the one used most frequently in applications.

From the method we used in the determination of error in the case of three divisions of the interval it would seem plausible that some form of the error term for an expression such as that of (X.3.3) would be possible. Indeed this is the case. We assumed we had a regular partition, $P[a, b]$, having n intervals, and hence $h = (b - a)/n$. Also we had n applications of the trapezoidal rule, and therefore n distinct error terms, whence the resulting error would be

$$E = \frac{-h^3}{12} \sum_{j=1}^{n} f''(c_j) = \frac{-h^2(b - a)}{12n} \sum_{j=1}^{n} f''(c_j)$$

where c_j is in (x_{j-1}, x_j). Then

$$|E| \leq \frac{h^2}{12} |b - a| [\max |f''(x)|]. \tag{X.3.7}$$

The maximum is taken, of course, over (a, b), and will give a pessimistic value for the error. However, it does give an absolute bound for the error.

You will remember that in the case of differentiation we had a significant problem in the selection of the value for h, for if we made h large we had a larger inherent mathematical error, but if we made h small we increased the likelihood of a round-off error of significant size. Let us take a look at the consequences of making h small here. We note that in no case does h appear in a denominator in the process of numerical integration, at least with the procedures we have used. The fact that h appears only in the numerator would indicate that we can always improve our results by further diminishing the value of h. This is almost true. (You will note that mathematicians are rather cautious individuals.) The decrease of h causes a corresponding increase in the number of intervals, and hence in the number of terms that must be added. If we had no round-off error in any of our calculations, we would then be assured without question that the decrease of h would always give improvement, unless perchance the bound is already zero. On the other hand, since there is likely to be some round-off present in each of the terms of (X.3.3), an increase in the number of terms increases the possibility that the round-off error may accumulate and cause some degradation of accuracy. One can always hope, of course, that some of the errors will be positive and some negative and the net effect would be small. On the other hand, there is no guarantee that this will always be the case. (In computers there is frequently no round-off procedure, but merely the dropping of all digits after some given number of decimal places. Hence the errors would be in the same direction. This can cause even greater difficulties with an ever increasing n.) Usually it is adequate to simply keep an extra digit or two in each number during the computation with the thought that the round-off error will be in the additional digits, and when they are dropped the result is at least accurate to the number of digits retained. The fact that the error term includes the square of h is significant for if $h = 0.01$, then $h^2 = 0.0001$. This may well be sufficient for our needs. If we apply the rule a single time, of course, we have h^3 in the error term, and this would be even better, but we usually need to integrate over an interval larger than 0.01.

One may well wonder what we are doing with such computations when errors seem to be an ever present evil. There are many instances in which we will have a function for which no formal integration is possible. Therefore an approximation is not only desirable, but probably essential if we are to handle the problem at hand. Furthermore, it is very seldom in actual practice that one would need an exact number. It is true that you count out exactly twelve items if you want a dozen items on a given purchase, but in measuring the length of your room you are not able to measure the room length more accurately than some rather small fraction of an inch. This does not worry you even when you are considering the purchase of wall-to-wall carpeting for the room. In the same way there is some lower bound for the possible resolution of the numbers used in the vast majority of applications. It *is* usually necessary to be certain that you are sufficiently accurate, but not that you are exact. Thus, approximate methods have their place. At times they represent the only means for even getting close to the results we may need.

EXERCISES

1. Use the trapezoidal rule to evaluate each of the following integrals using the number of divisions indicated. Find a bound for the error in each case. Check your result by evaluating the integral where possible.

   (a) $\int_1^2 (1/x)dx$ using one interval
   (b) $\int_1^2 (1/x)dx$ using three intervals
   (c) $\int_1^2 (1/x)dx$ using ten intervals
   (d) $\int_4^8 (1/x)dx$ using four intervals
   (e) $\int_0^1 (1/(1 + x^2))dx$ using one interval

(f) $\int_0^1 (1/(1 + x^2))dx$ using two intervals

(g) $\int_0^1 (1/(1 + x^2))dx$ using four intervals

(h) $\int_0^1 (1/(1 + x^3))dx$ using two intervals

2. In each of the following cases find the value of h, and consequently the number of intervals required, to obtain an approximation for the value of the integral by means of the trapezoidal rule if the error is to be be no greater than 0.001.

(a) $\int_1^2 (1/x)dx$

(b) $\int_0^{\pi/4} \tan x \, dx$

(c) $\int_0^{\pi/2} \sin 2x \, dx$

(d) $\int_0^1 e^{-2x} \, dx$

(e) $\int_{\pi/6}^{\pi/2} ((\sin x)/x)dx$

(f) $\int_0^{0.5} (1/\sqrt{1 - x^2})dx$

(g) $\int_1^3 (x^3 - 3x + 2)dx$

(h) $\int_1^2 \ln x \, dx$

3. (a) Set up the integral which would give the circumference of the ellipse $4x^2 + 9y^2 = 36$. This can involve obtaining the length in the first quadrant and multiplying by four.

(b) Use the trapezoidal rule with six intervals to evaluate the length of arc in the first quadrant from the point where $x = 0$ to $x = 2.4$.

(c) Find an upper bound for the error in your approximation.

This integral is an example of an *elliptic integral of the second kind*. No formula exists for its evaluation, and consequently numerical methods are required.

4. (a) Set up an integral to obtain the length of the closed curve with equations $x = 5 \cos t, y = 3 \sin t$.

(b) Approximate this length using the trapezoidal rule with 12 intervals.

(c) Find an upper bound for the error in your computation.

5. The *complete elliptic integral of the first kind* can be written

$$\int_0^{\pi/2} \frac{dx}{(1 - k^2 \sin^2 x)^{1/2}} \quad \text{where } 0 < k^2 < 1.$$

(a) Approximate the value of this integral using 3 intervals if $k = 0.5$.

(b) Find an upper bound for the error in your computation.

6. (a) Use the trapezoidal rule to approximate the length of the limacon $r = 3 - 2 \cos \theta$ using 12 intervals.

(b) Find an upper bound for the error in part (a).

7. You wish to evaluate the integral $\int_a^b f(x)dx$ using trapezoidal methods.

(a) Show that your approximation will be too large if $y = f(x)$ is concave upward in the interval $[a, b]$.

(b) Show that your approximation will be too small if $y = f(x)$ is concave downward in the interval $[a, b]$.

(c) Show that your approximation will be without error if the curvature is zero throughout $[a, b]$.

C8. (a) Write a program to evaluate the elliptic integral of the first kind using the trapezoidal rule. Your program should accomodate any given value of k and should use 20 subintervals

   (b) Use your program of part (a) to make a table of complete elliptic integrals of the first kind for $k = 0, 0.1, 0.2, 0.3, 0.4, 0.5, 0.6,$ and 0.7.

   (c) Find an upper bound for the error in each of your computations of part (b).

   (d) Would the error bound be greater for $k = 0.9$ than for $k = 0.1$? Why?

C9. (a) Write a program to evaluate the integral $\int_1^a (1/x)dx$ for any value of $a > 1$ using the trapezoidal rule with n subintervals.

   (b) Determine the number of subintervals that would be required to insure that the error bound in part (a) is less than 10^{-3} when $a = 2$.

   (c) Use your program to make a table of natural logarithms for $a = 1.00, 1.01, 1.02, 1.03, \ldots, 2.00$ with an accuracy of at least three decimal places.

10. (a) Show that the sum of the coefficients of the function values in the trapezoidal rule is the length of the interval involved.

   (b) Show that the trapezoidal rule gives exact results if $f(x)$ is a linear polynomial.

   (c) Show that the trapezoidal rule plus the error term gives exact results if $f(x)$ is a quadratic polynomial.

X.4 The Newton–Cotes Formulas

In the last section we considered a general procedure for obtaining the Newton-Cotes formulas, and their error terms, although we discussed only the special case of a single interval. We will now use two intervals (or three points). The generalization of this procedure should then be apparent. We start with two intervals, each of length h, and we again use the device of translating axes in such a way that the midpoint of the interval of integration is the zero point of the domain. We will integrate over the interval $(-h, h)$, and will use the three points with coordinates $(-h, f(-h))$, $(0, f(0))$, and $(h, f(h))$. (We could again use the semantic device of relating $f(a)$ and $g(-h)$, but the relation should be clear enough at this point that we can use $f(-h)$ without misunderstanding.) Using these three points our Lagrangian interpolation formula becomes

$$y(x) = \frac{x(x-h)}{(-h)(-2h)} f(-h) + \frac{(x+h)(x-h)}{(h)(-h)} f(0) + \frac{(x+h)x}{(2h)(h)} f(h)$$

$$= \frac{1}{h^2} [\tfrac{1}{2}(x^2 - xh)f(-h) - (x^2 - h^2)f(0) + \tfrac{1}{2}(x^2 + xh)f(h)].$$

$$(X.4.1)$$

As before, $y(x)$ is the approximation for the actual function $f(x)$, and in this case we have done some of the algebraic simplification rather than writing out all terms *ad nauseam* as we did in (X.3.1). We now proceed to integrate. As before we will utilize the symmetry of the interval of integration with respect to $x = 0$. Therefore, the terms which are odd functions of x will give zero values. We have

$$\int_{-h}^{h} f(x)dx \doteq \int_{-h}^{h} y(x)dx$$

$$\doteq \frac{1}{h^2} \left[\frac{h^3}{3} f(-h) - \left(\frac{2h^3}{3} - 2h^3 \right) f(0) + \frac{h^3}{3} f(h) \right]$$

$$\doteq \frac{h}{3} [f(-h) + 4f(0) + f(h)]. \tag{X.4.2}$$

This result is known as *Simpson's rule*, and was named for *Thomas Simpson* (1710–1761). It is probably the most frequently used of the various methods for numerical integration, and was known to the Greeks as the *prismoidal formula*, for it can be developed readily as a general formula for obtaining the volume of a large variety of solids commonly called "prismoids".

Having learned our lesson in the last section to the effect that there can be, and probably are, errors inherent in numerical computations, we will investigate the error term before giving an illustration of the use of Simpson's rule. We follow the same procedures as before, and again use the Maclaurin's expansion with the remainder term in each case. We have the formula for the numerical procedure in (X.4.2) and we need only substitute the Maclaurin's expansions. Since we have $f(-h)$ and $f(h)$ with equal coefficients, we would expect all of the terms involving h to an odd power would disappear in the expression $[f(-h) + 4f(0) + f(h)]$. Because of the symmetry of the interval of integration about $x = 0$, the odd powers will contribute nothing to the integral. In view of the presence of an additional term in the approximation formula, it would seem reasonable that we should be able to go beyond the term involving the second derivative. Therefore, we will include terms through the fourth derivative. The reasoning for this is all included in the foregoing statements, but it might be worth a quick review. We desire to use terms beyond that involving the second derivative. The terms involving the third derivative also include x^3 in the integrand and h^3 in the expansions in the formula. Since these are of odd degree, they will contribute nothing to the result on either side. Therefore, if we expect to have anything left with which to calculate an error term, we must expect to have to include the term involving the fourth derivative. With this line of thought in mind, we proceed.

$$\int_{-h}^{h} \left[f(0) + xf'(0) + \frac{x^2}{2} f''(0) + \frac{x^3}{6} f'''(0) + \frac{x^4}{24} f^{iv}(z) \right] dx$$

$$= \frac{h}{3} \left[f(0) - hf'(0) + \frac{h^2}{2} f''(0) - \frac{h^3}{6} f'''(0) + \frac{h^4}{24} f^{iv}(c_1) + 4f(0) \right.$$

$$\left. + f(0) + hf'(0) + \frac{h^2}{2} f''(0) + \frac{h^3}{6} f'''(0) + \frac{h^4}{24} f^{iv}(c_2) \right] + E$$

$$= \frac{h}{3} \left[6f(0) + h^2 f''(0) + \frac{h^4}{24} (f^{iv}(c_1) + f^{iv}(c_2)) \right] + E.$$

Therefore,

$$2hf(0) + \frac{h^3}{3} f''(0) + \frac{h^5}{60} f^{iv}(c_3) = 2hf(0) + \frac{h^3}{3} f''(0) + \frac{h^5}{36} f^{iv}(c_4) + E,$$

$$(X.4.3)$$

where c_3 is obtained as in (X.3.5) and c_4 is the value such that $f^{iv}(c_4)$ is the average of $f^{iv}(c_2)$ and $f^{iv}(c_3)$. From (X.4.3) we obtain the error term as

$$E = \frac{h^5}{60} f^{iv}(c) - \frac{h^5}{36} f^{iv}(c) = -\frac{h^5}{90} f^{iv}(c) \qquad (X.4.4)$$

where c is some value in $(-h, h)$. We have assumed that the fourth derivative is continuous in this interval. If it were not, we would have difficulty establishing a realistic bound for the error.

We should be aware of the fact that we may not always know the function $f(x)$ precisely in some applications, and it is very probable that we would know little concerning the value of the fourth derivative in the interval. Under some circumstances, however, it is possible to find some upper bound for the value of the fourth derivative, and then we have an upper bound for the error using Simpson's rule. Furthermore, the bound is generally going to be rather encouraging. If we were to use $h = 0.1$, then the error would be bounded by $-((0.1)^5/90)f^{iv}(c) = -0.0000001111111f^{iv}(c)$. Thus, the error would be small even if the value of the fourth derivative happened to be moderately large.

EXAMPLE 4.1. Evaluate $\int_0^{\pi/2} \sin x \, dx$ by Simpson's rule and find a bound for the error.

Solution. Since the length of the interval is $2h$, we then have $h = (1/2)(\pi/2 - 0) = \pi/4$. Therefore, the approximation is given by

$$\int_0^{\pi/2} \sin x \, dx \doteq \frac{\pi}{12} \left[\sin 0 + 4 \sin \frac{\pi}{4} + \sin \frac{\pi}{2} \right] = \frac{\pi}{12} [2\sqrt{2} + 1] \doteq 1.0028.$$

This result is fairly close to the correct value of one. We see that a single application of Simpson's rule gives a better result than we obtained with three applications of the trapezoidal rule. If we did not know the correct result, we could not have made this deduction without the use of the error term. Therefore we should investigate the error bound. Here we have, in a manner similar to that of the last section

$$E = \frac{-h^5}{90} f^{iv}(c) = -\frac{\pi^5}{(90)(4^5)} \sin c = -\frac{306.0197}{92160} \sin c \doteq -0.00332 \sin c.$$

Since c is in $(0, \pi/2)$, $\sin c$ is positive and the required correction is negative but does not exceed 0.00332 in absolute value. Our result is certainly within the interval $[1.00228, 1.00228 - 0.00332] = [1.00228, 0.99896]$.

The next question we face concerns whether we could subdivide the interval of integration here, as we did with the trapezoidal rule in order to achieve even greater accuracy. For two applications of Simpson's rule we would have $x = \pi/4$ as the upper end of the first evaluation and the lower end of the second. This gives us

$$\int_0^{\pi/2} \sin x \, dx \doteq \frac{\pi}{24}\left[\sin 0 + 4 \sin \frac{\pi}{8} + 2 \sin \frac{\pi}{4} + 4 \sin\frac{3\pi}{8} + \sin \frac{\pi}{2}\right]$$

$$= 1.0001346. \tag{X.4.5}$$

We note that the value $\sin(\pi/4)$ appears once as the upper end of the first application of Simpson's rule and once as the lower end of the second application. This would lead us to the development of the more general formula

$$\int_a^b f(x)dx \doteq \frac{h}{3}\left[f(x_0) + 4f(x_1) + 2f(x_2)\right.$$

$$\left. + 4f(x_3) + \cdots + 4f(x_{n-1}) + f(x_n)\right], \tag{X.4.6}$$

where n must be an even number. Since $b - a = nh$ we have $h = ((b - a)/n)$, as before, but note that it takes two h's for each application of Simpson's rule. Therefore, in our evaluation of the error bound we only have $(n/2)$ applications of Simpson's rule and the error is

$$E = - \left(\frac{n}{2}\right)\left(\frac{b - a}{n}\right)\frac{h^4}{90} f^{iv}(c) = - \frac{h^4(b - a)}{180} f^{iv}(c). \tag{X.4.7}$$

The error is bounded by

$$|E| \leq \frac{h^4(b - a)}{180} [\max|f^{iv}(c)|]. \tag{X.4.8}$$

Again the maximum is taken for values of c in the interval (a, b). Returning now to (X.4.5), the error bound is given by

$$|E| \leq \frac{\pi^4(\pi/2)}{8^4(180)} (1) \doteq \frac{306.0197}{1474560} \doteq 0.0002075.$$

You will note again that this bound is larger than the actual error. That is the virtue of having an error bound, for it does give us the assurance that we are in no worse shape than the bound would indicate. Whether you would need to use more divisions here is entirely a matter of the accuracy you would require for the situation at hand.

The procedure we have used in this section and in the preceding section indicates the general method for obtaining any Newton–Cotes integration formula. (These are frequently called *quadrature formulas*, but they are for the purpose of obtaining integrals, so don't let the change in terminology fool you.) If we had used three subintervals, that is four points, and obtained the corresponding Newton–Cotes formula, we would have found that we would

still have used the fourth derivative for the error term, since the terms of odd
degree in the Maclaurin's expansion contribute nothing, but the term with x^4
cannot be ignored. In similar fashion, the use of five or six points would
involve an error term which includes the sixth derivative. Using the Newton–
Cotes formula several times, thus diminishing the size of h, provides a
theoretical advantage in the computation of the integral, for you note that
we have no values of h in the denominator at any point. The advantages
gained by using formulas of higher order (that is more subintervals) is
generally not sufficient to recommend their use over that of using the lower
ordered formulas with several applications to cover the interval. The use of
Simpson's rule twice, as in the case of (X.4.5), will be easier to implement and
about as accurate as the use of a single expression involving five points or
four subintervals.

There are other numerical techniques for obtaining integrals, but the
Newton–Cotes formula will suffice for our purposes. Note that if you know
a sufficient number of values of the function in question, and if these values are
taken at the points of a regular partition, you can obtain the value of the
integral with a great deal of accuracy. It would be possible to handle the
case in which the partition is not regular, but this would require the use of
the Lagrangian formula in essentially the same manner as it was used
here for regular partitions.

EXERCISES

1. (a) Use Simpson's rule to evaluate $\int_1^2 (1/x)dx$ using three points.
    (b) Find a bound for the error in part (a).
    (c) Use Simpson's rule with five points to evaluate the integral of part (a).
    (d) Find an error bound for your computation in part (c).
    (e) Determine the maximum size for h, and hence the number of points required,
         if we are required to compute the value of the integral of part (a) accurate to
         the eighth decimal place.

2. (a) Use Simpson's rule to evaluate $\int_0^1 (1/(x^2 + 1))dx$ using three points.
    (b) Find a bound for the error in part (a).
    (c) Use Simpson's rule with five points to evaluate the integral of part (a).
    (d) Find an error bound for your computation in part (c).
    (e) Determine the maximum size for h, and hence the number of points required,
         if we are required to compute the value of the integral of part (a) accurate to
         the eighth decimal place.

3. A function $f(t)$ is given by the following table of values.

t	0	1	2	3	4	5
$f(t)$	1.24	1.68	1.90	1.96	1.60	1.38

    (a) Evaluate $f'(2)$ as accurately as you can.
    (b) Evaluate $\int_0^5 f(t)dt$ using the trapezoidal rule five times.

(c) Evaluate the integral of part (b) using Simpson's rule for the first four subintervals and using the trapezoidal rule for the fifth subinterval.

(d) Evaluate the integral of part (b) using Simpson's rule for the first two and last two subintervals and using the trapezoidal rule for the center subinterval.

(e) Sketch the graph of $f(t)$.

(f) Which of the three integrations, parts (b), (c), and (d), do you think is most accurate? Why?

4. (a) Use the table in Appendix C to evaluate $\int_{1.0}^{1.4} \Gamma(x)dx$ using the trapezoidal rule.

(b) Evaluate the integral of part (a) using Simpson's rule.

(c) With the data given in Appendix C what is the most accurate value you can obtain for the integral of part (a) with the formulas we have developed to date?

(d) Find the mean value of $\Gamma(x)$ over the interval $[1, 1.4]$.

5. (a) Find the circumference of the ellipse $4x^2 + 9y^2 = 36$ using one application of Simpson's rule for the integral with respect to x for each portion of the circumference between $x = -2$ and $x = 2$ and one application for each integral with respect to y for the remainder of the circumference.

(b) Find an error bound for your computation of part (a).

(c) Double the number of applications of Simpson's rule and reevaluate the circumference of part (a).

(d) Find a bound for your error in part (c).

6. (a) Use Simpson's rule with one application to find an approximation for the complete elliptic integral of the first kind if $k = 0.5$.

(b) Find a bound for your error.

(c) Compare your results in parts (a) and (b) with the corresponding results using the trapezoidal rule.

7. (a) Obtain the Newton–Cotes quadrature formula for three subintervals. If you use the techniques of Sections X.3 and X.4 this will involve integration over the interval $[-3h/2, 3h/2]$.

(b) Obtain the error term for part (a). Note the discussion in this section concerning the number of terms of the Maclaurin's expansion you will need to use.

(c) Apply the results of part (a) (known as the *three-eighths rule*) to approximate the integral of Exercise 1.

(d) Obtain an upper bound for the error of part (c).

(e) Apply the results of part (a) to approximate the integral of Exercise 2.

(f) Obtain an upper bound for the error of part (e).

8. (a) Find an upper bound for the error in evaluating the integral $\int_0^1 f(x)dx$ by one application of Simpson's rule in terms of the fourth derivative $f^{iv}(c)$.

(b) Find an upper bound for the error in the approximation of the integral of part (a) using one application of the three-eighths rule.

(c) Find an upper bound for the error in the approximation of the integral of part (a) using one application of the trapezoidal rule.

(d) Compare the error bounds found in parts (a), (b), and (c) and determine which is smallest, if possible.

(e) Which of the error bounds is largest?

9. (a) Find values of A and B such that $\int_a^b f(x)dx = Af(a) + Bf(b)$ for any constants a and b and for both the functions $f(x) = 1$ and $f(x) = x$.
   (b) Simplify your result of part (a) if you let $h = b - a$.
   (c) Find constants A, B, and C such that $\int_{-h}^h f(x)dx = Af(-h) + Bf(0) + Cf(h)$ for any constant h and for $f(x) = 1$, $f(x) = x$, $f(x) = x^2$.
   (d) Find values of A, B, C, and D such that

   $$\int_{-h}^{2h} f(x)dx = Af(-h) + Bf(0) + C(h) + Df(2h)$$

   for any constant h and for $f(x) = 1$, $f(x) = x$, $f(x) = x^2$, and $f(x) = x^3$.

S10. (a) Obtain figures for the GNP (gross national product) for the last five years.
   (b) Use Simpson's rule twice to obtain the mean value of the GNP over this period.
   (c) Compare your result of part (b) with the average of the five years.
   (d) Is the additional accuracy obtained using Simpson's rule warranted in this case for the purposes for which GNP is normally used?

C11. (a) What value of h would be required to approximate $\int_0^1 e^{-x^2} dx$ with an error no greater than 10^{-5} if you are to use Simpson's rule?
   (b) Write a program that will approximate the integral of part (a) with the desired accuracy.
   (c) Run your program and obtain the result.

C12. Write a program and make a table for the values of arc sin a for values of a from 0 to 0.8 at intervals of 0.02 by using Simpson's rule to approximate

$$\int_0^a \frac{1}{\sqrt{1 - x^2}}\, dx.$$

You are to obtain the results with four decimal place accuracy.

X.5 Richardson Extrapolation

In view of the increased use of the high speed computer in solving problems which require extensive integration, much work has been done attempting to improve the available numerical techniques. One could spend several years studying such attempts and trying to determine where each technique might be most useful. We will content ourselves here with the introduction of one method for improving the accuracy of already available methods. This method appeared in 1927 and is due to *L. F. Richardson*. While the method will work for any Newton–Cotes formula, as well as for formulas we have not discussed here, we will demonstrate the method in the trapezoidal case. Let us consider the problem of obtaining the integral $\int_a^b f(x)dx$ using the trapezoidal rule and a regular partition of $[a, b]$ involving $(n + 1)$ points or n intervals. In this case we would have $h = (b - a)/n$, and therefore

$$\int_a^b f(x)dx = h\left[\frac{f(a)}{2} + \sum_{k=1}^{n-1} f(a + kh) + \frac{f(b)}{2}\right] - n\left[\frac{(b - a)^3}{12n^3}\right]f''(c_n).$$

$$\text{(X.5.1)}$$

Observe the replacement of h in the error term by $(b - a)/n$. If we define I_n to be

$$I_n = h\left[\frac{f(a)}{2} + \sum_{k=1}^{n-1} f(a + kh) + \frac{f(b)}{2}\right],$$

we can rewrite (X.5.1) as

$$\int_a^b f(x)dx = I_n - \frac{(b - a)^3}{12n^2} f''(c_n) \tag{X.5.2}$$

where c_n is a number in the interval (a, b) and is the particular number required to make (X.5.2) correct. We have used the subscript n in I_n and c_n to indicate that these are the values associated with the use of n intervals in the numerical integration.

Since no restrictions have been placed upon n except that it be a positive integer, we could use two different values and obtain two relations similar to (X.5.2). Thus, we could write

$$\int_a^b f(x)dx = I_m - \frac{(b - a)^3}{12m^2} f''(c_m)$$
$$\int_a^b f(x)dx = I_n - \frac{(b - a)^3}{12n^2} f''(c_n). \tag{X.5.3}$$

The question now arises whether it might be possible to improve the accuracy by removing the effect of the error terms in (X.5.3). We have no knowledge of the actual value of c_m and c_n other than the fact that each of these is in the interval (a, b). However, for many functions it would be reasonable to expect that the values of $f''(c_m)$ and $f''(c_n)$ will not differ by a *significant* amount. If this be the case, we can eliminate a major portion of the error between the two equations of (X.5.3). Note that we are operating here on ground that is not completely solid for we are assuming certain characteristics hold for the function involved which will make this manipulation plausible. Despite this question concerning the validity of the work involved, we will plunge ahead. Multiplying the first of these equations by m^2 and the second by n^2 and then subtracting, we get

$$(m^2 - n^2)\int_a^b f(x)dx \doteq m^2 I_m - n^2 I_n.$$

The notation for approximation has been used here, for we are no longer certain that we have equality between the two sides of this relation. Since the object of our concern is the value of the integral, it is appropriate to solve for this value and obtain

$$\int_a^b f(x)dx \doteq \frac{m^2 I_m - n^2 I_n}{m^2 - n^2}. \tag{X.5.4}$$

We have no value for an error term in this approximation, but it would seem reasonable that the error using (X.5.4) would be less than the error that would have occured as a result of using (X.5.1).

While the approximation given in (X.5.4) looks promising, the matter of ease of computation cannot be ignored. If $m > n$, we would need more evaluations of $f(x)$ to compute I_m than to compute I_n, and we might well need $(m + n)$ evaluations. (Observe that the value of $f(a)$ and of $f(b)$ occur in both of these computations, and therefore there would be no need for re-evaluating these.) If, however, we were to let m be a multiple of n, say $m = kn$ for some positive integer k, then we would be able to use each of the function values obtained for I_n in the computation of I_m. In this case (X.5.4) reduces to

$$\int_a^b f(x)dx \doteq \frac{k^2 I_{kn} - I_n}{(k^2 - 1)}. \tag{X.5.5}$$

In particular, if $k = 2$, we have

$$\int_a^b f(x)dx \doteq \frac{4I_{2n} - I_n}{3}.$$

In order to keep things simple, let us investigate this result when $n = 1$. Furthermore, we could let $h = (b - a)/2n$, and then h would be the interval to be used in the evaluation of I_{2n}. The interval for I_n would be $(2h)$. Note that this will diminish the number of fractions later on. From this we have

$$\int_a^b f(x)dx \doteq \frac{4}{3} h \left[\frac{f(a)}{2} + f(m) + \frac{f(b)}{2} \right] - \frac{1}{3} (2h) \left[\frac{f(a)}{2} + \frac{f(b)}{2} \right]$$

$$\doteq \frac{h}{3} [(2 - 1)f(a) + 4f(m) + (2 - 1)f(b)]$$

$$= \frac{h}{3} [f(a) + 4f(m) + f(b)].$$

We have used m to represent the midpoint, or $m = (a + b)/2$. This result is precisely the result of the last section, for we observe that we now have Simpson's rule as a special case of Richardson extrapolation of the trapezoidal rule.

It is true that we do not have an error term for the approximation of (X.5.5), but we could obtain one by the method of the last two sections. There are other methods for arriving at the error term, but we will not introduce them here. There would have been no difficulty in establishing a result for $n = 2$ and any positive integer n, but there would have been more terms to keep track of. The result would again have been the same one developed in (X.4.6). Since we know the error term for Simpson's rule, we can observe that the Richardson extrapolation has improved the situation in this case. It is apparent that we could have used other values for k and obtained other formulas for approximating the integral. In most cases we would have to establish the error term, if needed, for we would not have the results ready-made as in this case.

EXAMPLE 5.1. Using the results of Section X.3 apply Richardson extrapolation to obtain a better approximation to the value of

$$\int_0^{\pi/2} \sin x \, dx.$$

Solution. In Section X.3 we found $I_1 = \pi/4 = 0.785398$. We also found $I_2 = (\pi/8)(\sqrt{2} + 1) = 0.948059$. If we use (X.5.5) with $k = 2$, we have $(4(0.948059) - (0.785398))/3 = 1.002279$. Example 4.1 gave the result 1.002280. The difference here is due to the round-off error resulting from rounding all results to six decimal places. Note that the result of the extrapolation in this case was too large despite the fact that the values of I_1 and I_2 were too small. This result also verifies our observation that Richardson extrapolation on the trapezoidal rule produces Simpson's rule. If we had used I_1 and I_3 we would have obtained the result 1.0010049, obviously much closer to the correct result than our previous numerical results. It is also worth noting that the use of I_2 and I_3 produces a result of 1.00023995, a result which is closer than any of the above to the correct result.

You may question the matter of introducing this new development if we are only going to re-derive Simpson's rule. However, this method is rather general, and we could use it to arrive at an extension of Simpson's rule. In this case we would observe that a suitable redefinition of the symbol I_n would yield

$$\int_a^b f(x)dx = I_n - \frac{(b-a)^5}{180n^4} f^{\mathrm{iv}}(c_n).$$

The result corresponding to (X.5.4) is then

$$\int_a^b f(x)dx \doteq \frac{m^4 I_m - n^4 I_n}{m^4 - n^4}.$$

Again, if we let $m = kn$, we have

$$\int_a^b f(x)dx \doteq \frac{k^4 I_{kn} - I_n}{k^4 - 1}. \tag{X.5.6}$$

If $k = 2$, this becomes

$$\int_a^b f(x)dx \doteq \frac{16}{15} I_{2n} - \frac{1}{15} I_n.$$

It is an interesting exercise to demonstrate that this result would be the Newton–Cotes formula based upon $2n$ intervals in the regular partition of $[a, b]$.

There are many formulas for obtaining approximations of integrals by numerical methods. In fact, the problem of evaluating integrals with reasonable accuracy was one of the driving forces in making the digital computer desirable and in encouraging its development. The Newton–Cotes formulas give us an insight into methods of development of such algorithms, and the Richardson extrapolation indicates one of the many ways in which these results can be extended to give greater accuracy without greatly increasing the amount of work involved. You might question whether our emphasis on economy of work is essential in view of the high speed available in modern computers. There are really two main reasons for desiring an *efficient* algorithm. The first has to do with the time involved, for many functions are exceedingly complex and the short time required to evaluate such a function each time multiplied by the number of times such evaluations may be needed becomes significant. Therefore, a reduction in the number of evaluations does save time (and with computers saving time is saving money). Furthermore, each operation can be expected to introduce some round-off error, and the greater the number of operations, the greater the accumulated round-off error can become. Hence, an efficient algorithm can also represent a gain in accuracy.

EXERCISES

1. (a) Use the results of I_1 and I_3 obtained in Section 3 and check the work using these two evaluations in Example 5.1.
   (b) Use the results of I_2 and I_3 obtained in Section 3 and check the result given for the Richardson extrapolation in Example 5.1.
   (c) Evaluate I_4 for the integral of Example 5.1.
   (d) Using I_4 with I_1 obtain the Richardson extrapolation.
   (e) Using I_4 with I_2 obtain the Richardson extrapolation.
   (f) Using I_4 with I_3 obtain the Richardson extrapolation.

2. (a) Use the trapezoidal rule to evaluate I_2 for the integral

$$\int_0^1 \frac{1}{1 + x^2}\, dx$$

   (b) Evaluate I_6 using the trapezoidal rule.
   (c) Use Richardson extrapolation with $k = 3$ on the results of parts (a) and (b).
   (d) Since you know the correct result in this instance, determine the extent of the improvement in accuracy (if any) obtained by using the extrapolation compared with the results of I_2 and I_6.

3. (a) Use Simpson's rule to evaluate I_2 for the integral of Exercise 2.
   (b) Evaluate I_6 using Simpson's rule.
   (c) Use Richardson extrapolation with $k = 3$ on the results of parts (a) and (b).
   (d) Determine the extent of the improvement in accuracy produced by the extrapolation (if any) when compared with your results of parts (a) and (b).

4. (a) Use the trapezoidal rule to evaluate I_2 for the integral $\int_1^2 (1/x)dx$.
   (b) Use the trapezoidal rule to evaluate I_4 for the integral of part (a).
   (c) Use Richardson extrapolation on the results of parts (a) and (b).
   (d) Use the trapezoidal rule to evaluate I_8 for the integral of part (a).
   (e) Compare the result in part (c) with that in part (d).
   (f) Use the result of parts (b) and (d) to obtain a Richardson extrapolation for the value of the integral of part (a).
   (g) Indicate the decrease in percentage error in your evaluation of the integral of part (a) (if any) as you went from part (a) to part (b) to part (c) to part (d) and finally to part (f).

5. (a) Show that the result in (X.5.4) is based upon the degree of h in the error term of the Newton–Cotes formula and not on the value of the derivative involved.
   (b) Given the fact that the error term for the three-eighths rule has a factor of h^5, show that the Richardson extrapolation for the three-eighths rule would have the same coefficients of I_m and I_n as the Richardson extrapolation for Simpson's rule.
   (c) If the values of I_m and I_n are equal, show that the Richardson extrapolation will give a result which is equal to I_m and I_n.

6. (a) Using the argument of Section X.4, show that the error term of any Newton–Cotes formula can be expected to have an odd power of h.
   (b) Elaborating on this same argument show that a Richardson extrapolation can be expected to have an error term in which the exponent of h and the order of the derivative involved are each increased by two.
   (c) Show that the information of part (b) together with the information of part (b) of Exercise 5 is sufficient to provide the formula for the Richardson extrapolation involving Richardson extrapolations.

C7. (a) Write a program that will approximate the value of an integral using the trapezoidal rule for n intervals and for kn intervals, and will then perform a Richardson extrapolation to obtain an improved approximation.
   (b) Run your program using $n = 10$ and $k = 2$ to obtain an approximation for the value of

$$\frac{\sqrt{2}}{\pi} \int_0^x e^{-t^2/2}\, dt \quad \text{for } x = 0.05, 0.10, 0.15, \ldots, 3.00.$$

   For each value of x print x, I_{10}, I_{20}, and the extrapolated result. This is a very important result in statistics.

C8. (a) Write a program that will approximate the value of the integral of Exercise 7 using Simpson's rule for n intervals and for kn intervals, and will then perform a Richardson extrapolation to obtain an improved approximation.
   (b) Run your program using $n = 6$ and $k - 2$ to obtain a table similar to the one requested in Exercise 7.
   (c) Compare the tables of Exercise 7 and part (b) of this exercise. Indicate which values you feel to be more accurate.

C9. (a) Write a program using Simpson's rule to evaluate I_n, I_{kn}, and I_{2kn} which will then perform a Richardson extrapolation on I_n and I_{kn}, will perform a Richardson extrapolation on I_{kn} and I_{2kn}, and will finally perform a Richardson extrapolation on the Richardson extrapolations. Remember that you can expect the power of h in the error to increase by 2 in each of the Richardson extrapolations when compared with the error term of the approximations on which the extrapolations are performed.

(b) Use this program to make a table similar to that requested in part (b) of Exercise 7 using $n = 4$ and $k = 2$.

(c) Compare your results of part (b) with the results of the preceding two exercises.

M10. (a) Show that the trapezoidal rule evaluates a Riemann integral with no error provided the integrand is a linear polynomial.

(b) Show that Richardson extrapolation performed on the trapezoidal formulas will yield perfect results if the integrand is a polynomial of degree no greater than three.

(c) Show that the result of a Richardson extrapolation performed on applications of Simpson's rule would be exact if the integrand is a polynomial of degree no greater than five.

X.6 A Variation of Integration

In the last three sections we dealt with the problem of evaluating an integral when we had some means of evaluating the integrand at points of our own choosing. As a result of the fundamental Theorem of the calculus, the derivative and the integral are essentially inverse operations, and therefore we are called upon to integrate when we are asked to find a function having a given derivative. We shall face this problem again in this section. Let us suppose we have been told that $D_x y = f(x, y)$, or that the derivative of y with respect to x is some function which involves both x and the same y of which we know only the derivative at the moment. The object of this problem, as you have probably guessed, is to find an expression for y in terms of x. An equation of this variety is called a *differential equation*. We cannot obtain the solution function by numerical methods, for numerical procedures only produce numbers. We will have to be content, at least for the moment, with obtaining what amounts to a table of values of y in terms of the values of x. If this table is fairly complete, we can then use interpolation techniques to find the intermediate values, and with our methods for numerical differentiation and integration we can carry out the operations of the calculus as may be required. We realize, of course, that doing things numerically exacts a price for it is not possible to obtain exact results, but we have found that in general we can get good approximations. We expect to obtain only approximations here.

In considering the problem posed above we stop to remind ourselves that a derivative is a rate of change. Hence $D_x y$ is the rate at which y is changing for a unit change in x. Euler recognized this, and observed that if we decided to start with the particular value $y = y_0$ at the time when $x = x_0$, we could then obtain an approximation for the value of $y(x_1)$ corresponding to $x = x_1 = x_0 + h$ for some increment h. The mean value theorem tells us that $y(x_1) = y(x_0) + (x_1 - x_0)y'(c) = y(x_0) + hy'(c)$ where c is some value in the interval (x_0, x_1). We know the value of $y(x_0)$ and it is presumed that we have decided what value we would choose for h. Therefore, we know the value of x_1. But we do not know the value of c. We might do as we did in using this theorem to obtain approximations and replace c by x_0, realizing full well that we would have just an approximation. This, at least, would be better than doing nothing. Therefore, we write

$$y_1(x_1) = y(x_0) + hy'(x_0) = y(x_0) + hf(x_0, y_0), \qquad \text{(X.6.1)}$$

since $y'(x) = f(x, y)$ by the relation which is the source of our entire problem. You will note that we can calculate every term on the right side of this equation. We have written $y_1(x_1)$ to represent the approximation to $y(x_1)$ that we obtain using this equation. It is apparent that $y_1(x_1)$ is not necessarily equal to $y(x_1)$.

We prefer not to leave an approximation unless we feel we are unable to improve the result, or alternatively that the present approximation is sufficiently good for our purposes. Since we know nothing about the possible error here, we rather obviously ought to be unhappy with this result, whether we are or not. The lack of accuracy came here when we replaced $y'(c)$ by $y'(x_0)$. It is true that if h is small, then c is close to x_0, but there is no reason to suspect that $y'(x_0) = y'(c)$. (If the derivative were constant, we would not have been bothered by this problem in the first place.) We might do better to replace $y'(c)$ by the average of $y'(x_0)$ and $y'(x_1)$. This sounds good until we start to calculate the value of these two derivatives. We already have $y'(x_0)$, but we would have to know $y(x_1)$ to calculate $y'(x) = f(x, y)$ at $x = x_1$. At first we knew nothing about $y(x_1)$, but now we at least have an approximation for this value. Therefore, we could use

$$y_2(x_1) = y(x_0) + h\frac{[f(x_0, y_0) + f(x_1, y_1(x_1))]}{2} \qquad \text{(X.6.2)}$$

where we are using $f(x_1, y_1(x_1))$ as an approximation for $f(x_1, y(x_1)) = y'(x_1)$. This will give us a second approximation to the value we seek, namely $y_2(x_1)$, and it would seem that this should be closer to the correct value than $y_1(x_1)$. If applying (X.6.2) improves the value, then why not apply it again? In fact we could continue to apply it as long as there is any further modification of the resulting approximation of $y(x_1)$. This would suggest using the iteration

$$y_{k+1}(x_1) = y(x_0) + h\frac{[f(x_0, y_0) + f(x_1, y_k(x_1))]}{2}. \qquad \text{(X.6.3)}$$

When we finally reach the point where $y_{k+1}(x_1)$ is very close to the value of $y_k(x_1)$, little additional improvement is possible by further iteration, and we would be well advised to stop.

We have developed a scheme for finding the value of $y(x_1)$, or rather an approximation thereto. We can use this value to find $y(x_2)$, etc. Before we contemplate continuing, however, let us go back long enough to see what error we may have incurred along the way. We will be concerned here only with the error in (X.6.3), for this is the expression we propose to use until further modification is no longer possible. Utilizing the fact that if we are given $y'(x)$ we can obtain $y(x)$ by taking the anti-derivative, and the anti-derivative is related to the integral via the fundamental theorem, we can consider the integral

$$\int_{x_0}^{x_1} y'(x)dx.$$

However, using the trapezoidal rule, we can evaluate this, and we can also obtain a *bound* for the error term. Thus

$$\int_{x_0}^{x_1} y'(x)dx = y(x_1) - y(x_0) = h\left[\frac{y'(x_0) + y'(x_1)}{2}\right] - \frac{h^3}{12} y'''(c). \quad (X.6.4)$$

We have obviously used the fact that $x_1 - x_0 = h$. It should be pointed out that the error term here involves the second derivative of the integrand, and that is the third derivative of $y(x)$. We can rewrite (X.6.4) as

$$y(x_1) = y(x_0) + \frac{h}{2}[f(x_0, y_0) + f(x_1, y(x_1))] - \frac{h^3}{12} D_x^2 f(x, y(x)). \quad (X.6.5)$$

For sufficiently small values of h, the error will be small, unless we hit the hopefully unusual case in which the second derivative of $f(x, y)$ is large.

We can summarize our discussion to this point. Our procedure for obtaining the values of $y(x)$ corresponding to the points x_0, x_1, x_2, etc., is to obtain $y(x_1)$ as accurately as possible, then using this as a new starting point advance to $y(x_2)$, etc. We make the best first approximation for each new value of $y(x)$ possible by calculating

$$y_1(x_{j+1}) = y(x_j) + hf(x_j, y(x_j)). \quad (X.6.6)$$

We then proceed to obtain improved values $y_{k+1}(x_{j+1})$ by the relation

$$y_{k+1}(x_{j+1}) = y(x_j) + \frac{h}{2}[f(x_j, y(x_j)) + f(x_{j+1}, y_k(x_{j+1}))]. \quad (X.6.7)$$

By the time this iteration settles down, that is by the time the sequence $y_1(x_{j+1}), y_2(x_{j+1}), y_3(x_{j+1}), \ldots$ converges, we should have an approximation to $y(x_{j+1})$ which differs from the correct value by no more than $(h^3/12)y'''(c)$ for some point $x = c$ in the interval (x_j, x_{j+1}). In this type of evaluation we

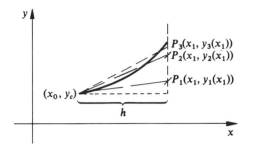

Figure X.3

are using two relations, the first for the purpose of *predicting* a starting approximation and the second for *correcting* the prediction. Such a method is called a *predictor-corrector* method. The situation here is illustrated schematically in Figure X.3. The segment terminating at P_1 is the line tangent to the curve at (x_0, y_0). The short segment through this P_1 indicates the slope which would be calculated using $D_x y = f(x_1, y_1(x_1))$. The slope of the line segment terminating at P_2 is the segment whose slope is the average of the slope at (x_0, y_0) and the slope at P_1. The slope at the point P_2 is indicated and the segment terminating at P_3 is then drawn with a slope obtained by repeating the averaging process. It is intuitively rather clear that in a situation such as that portrayed here the method should converge toward the correct value. Our development has shown, of course, that we cannot expect perfection, however. This method does permit us to determine a table of values for y corresponding to values of x, but even the first value can have a small error. The second value could well have its own error in addition to that which it had accumulated from the first error, etc. This is not to intimate such solutions have no value, but one should watch the solution carefully. If one were doing this on a computer, it might be well to perform the solution for two different values of h and compare results. The extent to which the results agree may indicate the amount of comfort that one could derive with respect to accuracy.

   You remember that we rather arbitrarily started with a beginning point (x_0, y_0). Any time that one wishes to reverse the derivative process, it is likely that one will have an arbitrary constant. Hence, it becomes necessary to know some additional fact to be able to determine the value of the arbitrary constant. Here we start from a specific *initial point*. We do not have the counterpart of the indefinite integral here, and so we must have sufficient information to make the answer explicit. The problem we have discussed is that of solving a differential equation. We will have more to say about differential equations in the next chapter, but we have here a technique which can be used to obtain numerical solutions if we have no better way of obtaining them. Before bringing this section to a close, we should consider an Example to illustrate the many things we have said herein.

EXAMPLE 6.1. Find an approximation for $y(3)$ if $y'(x) = x^2 - y$ and $y(1) = 1$. Use $h = 0.2$ for this solution.

Solution. Here we have $f(x, y) = x^2 - y$, and our predictor and corrector relations become respectively $y_1(x_1) = y(x_0) + 0.2(x_0^2 - y(x_0))$ and $y_{j+1}(x_1) = y(x_0) + 0.1[(x_0^2 - y(x_0)) + (x_1^2 - y_j(x_1))]$. We note that the expression $(x_0^2 - y(x_0))$ is used many times and we could reduce our work by evaluating this expression just one time. We start the calculations as follows

$$y_1(1.2) = 1 + 0.2(1^2 - 1) = 1$$
$$y_2(1.2) = 1 + 0.1[(1^2 - 1) + (1.2^2 - 1)] = 1.044$$
$$y_3(1.2) = 1 + 0.1[(1^2 - 1) + (1.2^2 - 1.044)] = 1.0396.$$

Note that we obtain additional digits with each iteration in this particular case. The results of this computation can be summarized in the following table of results. We have indicated by y_1, y_2, etc., in the table the values corresponding to $y_1(x_j)$, $y_2(x_j)$, etc. Hence these values indicate the successive approximations to the value requested for the given row. The last value of y obtained by iteration, namely y_6, is the one used as the true value of y in the next step.

x	y	y_1	y_2	y_3	y_4	y_5	y_6
1.0	1.0						
1.2		1.00000	1.04400	1.03960	1.04004	1.04000	1.04000
1.4		1.12000	1.16400	1.15960	1.16004	1.16000	1.16000
1.6		1.32001	1.36401	1.35961	1.36005	1.36000	1.36000
1.8		1.60001	1.64401	1.63961	1.64005	1.64001	1.64001
2.0		1.96001	2.00401	1.99961	2.00005	2.00001	2.00001
2.2		2.40001	2.44401	2.43961	2.44005	2.44001	2.44001
2.4		2.92002	2.96402	2.95962	2.96006	2.96001	2.96002
2.6		3.52002	3.56402	3.55962	3.56006	3.56001	3.56002
2.8		4.20002	4.24402	4.23962	4.24006	4.24001	4.24002
3.0		4.96002	5.00402	4.99962	5.00006	5.00001	5.00002

We see that our approximation for $y(3) = 5.00002$. The true value, as you might have guessed, is exactly 5 in this case. The computations were carried out to eight decimal places, and the final results were rounded to the five places given in this table. Note that there was some increase in accuracy from y_4 to y_5 in each instance, but very little, if any, in going from y_5 to y_6. Hence, this was a good place to stop. Using more decimal places is not apt to assist us here, for there is an inherent inaccuracy resulting from the fact that we have used the trapezoidal rule with the error which this involves. (This suggests that we might consider using Simpson's rule or some other formula with a smaller error. We will not pursue this at this point, but it is an idea that has merit.) Note also that the error is tending to increase slowly. Since each new value of y has a small error, the next value will be based on slightly erroneous information, and hence can well have a slightly larger error. This error is called the *cumulative error*.

We have not seriously considered the problem of convergence in our solution as yet. Note that the succession of $y_j(x_1)$ form a sequence, and it is this sequence which will ultimately yield our approximation for $y(x_1)$. In the vast majority of cases this method will converge. It is not likely that you will see a case in which it does not converge, unless the problem at hand is of such magnitude that you would have to call upon information from some tome in numerical analysis. The problems of stability, as it is sometimes called, and of cumulative error can be very serious in some situations, and should not be ignored. However, the full discussion of these topics is one that we will not undertake here. These comments are simply to alert you to the fact that we have not considered all of the questions that are essential to a full study of the numerical techniques for the solution of differential equations. We have, however, pointed out that such methods exist, and that they can be useful. Note in our example that we were able to get more than sufficient accuracy for most purposes with a value of h as large as 0.2.

EXERCISES

1. Obtain an approximation for $y(2)$ if $y'(x) = x^3 - y$, $y(1) = 0$, and $h = 0.5$.

2. Obtain the approximation for $y(2)$ in Exercise 1 if $h = 0.2$. Compare the accuracy in the two cases.

3. (a) Obtain an approximation for $y(3)$ in Example 6.1 in this Section using $h = 0.5$ and using $h = 1$.
   (b) Compare the results with those given in the solution and with each other.
   (c) How many iterations should be taken with these values of h before there appears to be little further chance for improvement?

4. (a) Expand $y(x)$ in Taylor's expansion if $y'(x) = x^2 - y$ and $y(1) = 1$.
   (b) Use this result to obtain an approximation for $y_1(x)$ and then proceed with the corrector equation, using $h = 0.5$.
   (c) Obtain $y(2)$ and compare the accuracy with the other evaluations you have for $y(2)$. [*Hint*: Since $y'(x) = x^2 - y$, then $y''(x) = 2x - y'(x)$, $y'''(x) = 2 - y''(x)$, etc., and by substitution of each expression, where appropriate, in the next term it is possible to obtain a series.]

5. Find an approximation for $y(3)$ if $y'(x) = y/2x$ and $y(1) = 1$. Use different values of h until you are satisfied that you have as much accuracy as you can get.

6. Make a table of values of $y(x)$ in Exercise 5, starting with $x = 1$ and then proceeding by intervals of 0.25 until you have $y(4)$. From your table find $y(x)$ as a regression function.

C7. You are asked to solve the differential equation

$$y'(x) = (x^3 y + y \sin x)/(x^2 + y^2).$$

(a) You have the initial point $(1, -1)$. Use numerical methods to find the value of $y(1.8)$ using $h = 0.1$.
(b) Solve this same problem using $h = 0.2$ and using $h = 0.01$.
(c) How far do they agree? Do you know anything about the relative accuracy of these results? If so, how much do you know and how do you know it?

P8. The current in a given circuit at time t seconds is governed by the differential equation $D_t i = 10t - LRi$ where $L = 0.1$ henries and $R = 30$ ohms. If the initial current is zero, what is the current when $t = 4$ seconds? Use $h = 0.5$.

S9. The marginal revenue for a certain product is given by the equation $MR(n) = 14n - 0.05n^2 + 0.003R$ in view of the fact that increased revenue improves the ability to obtain financing and therefore cuts certain overhead costs. If the revenue is $100 when 10 items are handled, what would be the revenue when 15 items are handled. [*Hint:* Although one thinks of a number of items as being a discrete quantity that must be an integer, it is frequently necessary to consider that n may be a continuous variable—a concept that is not quite so bizarre when one thinks of a continuous production process.]

B10. The rate of change of the percentage of those infected in an epidemic can be assumed to be approximately $D_t p = p(1 - p)e^{-0.2t}$ where p is the percentage infected and t is measured in weeks. If 5 percent are infected initially, how many are infected two weeks later, if this law holds. Let h represent a half week.

M11. Consider the differential equation $y''(x) = 4y$. This can be expressed as a system of two equations

$$y'(x) = z(x)$$
$$z'(x) = 4y(x).$$

Starting with the fact that when $x = 0$, $y'(x) = 1$ and $y''(x) = 0$, find a solution of a nature similar to that used in this section. Use your method to evaluate $y(1)$. [*Hint:* Can you find $y'(x)$ for the next value of x and then find the next value of $y(x)$?]

X.7 Numerical Problems and Answers

There are many applications requiring solutions for which the necessary operations cannot be carried out by formal means. If one is aware of the operations involved, however, it is frequently possible to obtain approximations to the solutions with sufficient accuracy for the purposes at hand. There are also instances in which large problems, too large to be feasible by hand techniques, are handled using modern high-speed digital computers. If within the large problems there is the need for a differentiation, an integration, or even the solution of a differential equation, it is not very efficient to require the computer to stop work in order that we can solve the particular calculus problem. Furthermore, the computer knows the function only as a set of coordinates or data points, and therefore we would not have the function in the nice form we have dealt with in general. For this reason, numerical methods are important, and they are becoming increasingly important. This is not to say that the formal methods will not remain as a center of attention for those using mathematics but rather that one cannot reasonably focus attention on either approach and avoid the other. There is also the very important matter of knowing something about the accuracy of the result. As you have discovered, it takes quite a bit of mathematics to establish bounds for the error, even if the bounds are not always satisfactory. It will be the lot of many who are now studying mathematics to find themselves concerned with problems such as those suggested herein, whether their main field of endeavor be mathematics or not. Thus a knowledge of how such procedures are handled together with the capabilities of the techniques and the accuracy of the results will be of increasing value to persons in many disciplines.

Numerical methods are always slightly disappointing, for they represent one portion of mathematics where we feel reasonably certain we do not have exactly the correct answer. We have only an approximation. That approximation may, however, be worth much more than the correct answer to some other question. In order that the result you get will be of sufficient accuracy, you must first know what accuracy is required. Then you must examine all the methods you have available and determine whether any of them will meet your requirements. It is wise not to do more than is required, for you increase the opportunity for human arithmetic error, or machine roundoff error. Therefore, the choice of h can be significant. If you cannot handle a given problem by formal means, do not give up. Look for a suitable numerical technique. If you can do it by formal methods, however, you will probably find less work and greater accuracy. Which direction should you take in handling a given problem? That depends on the problem and on the use to be made of the solution. If you can find a formal technique, this is probably optimal. If you do not find one after a reasonable search (exhaustive in some cases), then look for the numerical approach. It will often suffice.

EXERCISES

1. Compare the problems that occur in the choice of h in differentiation, integration, and in the solution of a differential equation.

2. You are required to obtain the value of the integral

$$\int_a^b f(x)dx$$

correct to five decimal places. You know that $f^{(n)}(x)$ is no larger than n in absolute value in the interval (a, b). What method would you use, and what value would you pick for h? Why? [Note that one possible method of solution consists of solving the differential equation $F'(x) = f(x)$ with $F(a) = 0$. Here you would want $F(b)$.]

3. You need to solve the differential equation $y'(x) = f(x, y)$ and you know that the successive values of $y(x)$ obtained by iteration will converge. What value of h should be used to permit you to start with $x = -1$ and find the approximation for $x = 2$ correct to two decimal places if you know that $|y'''(x)| \leq 5$ for each x in $(-1, 2)$? Neglect round-off error.

4. It is customary to use either the trapezoidal rule or to use a Newton–Cotes formula with an even number of subintervals (an odd number of points) if a Newton–Cotes formula is to be used for integration. Considering the form of the error term, can you explain this preference? [*Hint*: As an example we noted that both Simpson's rule and the three-eighths rule have an error term involving h^5 and $f^{iv}(x)$].

5. Evaluate

$$\int_0^1 \frac{dx}{x^2 + 1}$$

as a differential equation using $h = 0.2$, and compare this result with your earlier evaluations of this same integral.

6. (a) Using a table of local temperatures, integrate and find the number of degree days in the past month using each of the methods for numerical integration.
   (b) Find the mean temperature by dividing by the number of days.
   (c) Find the average temperature by taking an average in the arithmetic sense.
   (d) Do all of your results agree? If not, why do you think they differ?
   (e) Which do you think would be more accurate?

P7. In the case of an expanding gas at constant temperature the work done in the expansion is given by

$$\int_{v_0}^{v_1} p \, dv.$$

An experiment has been arranged such that the pressure is taken when the volume is 10, 11, 12, 13, 14, and 15 cubic inches, or in other words at regular intervals with respect to volume. The following table is produced as a result of these observations.

v	10	11	12	13	14	15
p	94	92	90	89	88	86

where p is given in pounds per square inch and v is measured in cubic inches. Find the work done in this expansion.

M8. Using the method mentioned in Exercise X.6.11, find $y(1)$ if $y(x)$ satisfies the differential equation $y''(x) + 2y'(x) + 5y(x) = 3$ if $y(0) = 0$ and $y'(0) = 0$. Use $h = 0.2$. [*Hint*: Solve for $y''(x)$, and then use the substitution used in Exercise 6.11.]

CHAPTER XI

Differential Equations

XI.1 Separation of Variables

In the last chapter we dealt with a rather simple case of the more general differential equation $y'(x) = f(x, y)$. Since the function $f(x, y)$ can be rather complicated, it should not be surprising that this differential equation, simple in appearance though it is, can be difficult to solve in some instances. We could have anticipated this, for the solution of a differential equation can be expected to involve taking the anti-differential, or effectively integrating, and we have discovered that integration can be somewhat less than straightforward on occasion. With this introduction, we might look forward to a variety of methods for solving differential equations, each one suitable under certain circumstances. Each method may apply in only special cases, but taken together the methods will cover the majority of cases appearing in routine types of applications.

In this chapter we will consider a few of these methods of solution. While they may be few in number, they will be found to cover a rather large number of the cases you are apt to see in your elementary work. We will start with probably the simplest method of all. This will cover cases in which it is possible to transform the given equation in such a way that we have two differentials which are equal. We can then take the anti-differentials and know that these must differ by at most an additive constant. Of course, the process of taking the anti-differentials is precisely the process of taking the indefinite integral.

Before we proceed with an example we should remember that by definition we have $dy = D_x y \, dx$. Therefore $y'(x) = f(x, y)$ is equivalent to $dy = f(x, y)dx$. The method of separating variables, then, is merely the method whereby we would transform $dy = f(x, y)dx$ into an equation of the form

$g(y)dy = h(x)dx$ through the use of algebra, provided this is possible. In this case we would have the anti-differential of $g(y)dy$ differing from the anti-differential of $h(x)dx$ by at most an additive constant. Having outlined the method, we will now proceed with an Example.

EXAMPLE 1.1. Solve the differential equation $y'(x) = x^2(y + 1)$.

Solution. By the discussion above we can write $dy = x^2(y + 1)dx$. Also following the discussion above we should now see whether we can *separate the variables*. This means getting only one variable on one side of the equal sign and only one variable on the other side. Therefore we try to get all of the terms involving y on the side with dy and all of the terms involving x on the side with dx. This is possible in this case if we divide both sides by $(y + 1)$. Hence we have

$$\frac{dy}{y + 1} = x^2\, dx$$

or

$$d[\ln(y + 1)] = d\left[\frac{x^3}{3}\right].$$

Since the differentials are equal, the functions can differ by at most an additive constant, and consequently we have

$$\ln(y + 1) = \frac{x^3}{3} + c,$$

or

$$y + 1 = e^{(x^3/3)}e^c = Ce^{(x^3/3)},$$

whence

$$y = Ce^{(x^3/3)} - 1.$$

This last simplification was not done solely to provide a confusion factor. It was suggested by the fact that we usually like to see y expressed as a function of x in some explicit way. The constant, C, is merely e^c. If we can evaluate e^c we can evaluate C and vice versa. If we knew, for instance, that $y(0) = 17$, we would then know that $17 = Ce^0 - 1 = C - 1$, or $C = 18$. In this case we would have the equation

$$y = 18e^{(x^3/3)} - 1.$$

Note that knowledge of the rate of change does not tell you your present position unless you know where you started from. This is very much like the constant of integration for the indefinite integral, as discussed in Chapter V and again in Section X.6.

This is an example of a differential equation in which the variables are separable. It simply means that we can employ some algebra, either directly or with subterfuge as the case may require, and obtain two expressions which are equal and which have the fine property that each involves only a single variable. It would not have been possible to obtain the answer easily if we had only written $dy = x^2(y + 1)dx$ in the example above. When trying to find the anti-differential of the right side we would have been left without any knowledge of how to handle the y which appears there. We know that y is some function of x, but the object of our search is to find which function. We would be left in a quandary when asked to integrate an expression which involves the answer we seek. Since the method of separation of variables requires the smallest amount of manipulation, one of the first things you should do if you are asked to solve any differential equation of the first order (that is having only first derivatives present) is to determine whether the variables can be separated.

We have illustrated this method with a problem that permitted the separation of variables by simple algebra. The algebra is not always so simple, and sometimes a bit of calculus will help.

EXAMPLE 1.2. Solve $D_x y = 4x^3 y^2 - (y/x)$.

Solution. It is not at all apparent that this can be solved by separating variables. We could express this in terms of differentials, and note, upon clearing fractions, that we have $x\, dy = 4x^4 y^2\, dx - y\, dx$. While this doesn't seem to help, there is a slight glimmer of hope for the very alert. Note that we might write this as $x\, dy + y\, dx = 4x^4 y^2\, dx$ or $d(xy) = 4x^2(xy)^2\, dx$. In this case we can separate variables, but we are not separating the x and y, but rather x and (xy). Thus, we have

$$(xy)^{-2}\, d(xy) = 4x^2\, dx.$$

From this we obtain

$$-(xy)^{-1} = 4x^3/3 + C$$

or

$$y = \frac{3}{Kx - 4x^4}$$

where we have replaced the constant $(-3C)$ by K, for convenience. If you didn't see this particular possibility for separating variables before you read through the discussion, don't feel sorry for yourself. This is one of those combinations that is far from obvious but if you do see it, by all means use it. The moral of this example is that it pays to keep one's eyes open, for the person who does so may find a way to get out of using a harder method.

The method of separation of variables is generally confined to the case of first order differential equations, but there are other examples, and they occur too frequently to be ignored, in which this method, or a slight variation thereof, can be applied.

EXAMPLE 1.3. Solve the equation $D_x^2 y = e^y/2$.

Solution. It is convenient here to make a substitution, and we will let

$$u = D_x y,$$

but we will only do it on the left hand side of the equation. Thus, we have

$$2D_x u = e^y.$$

This doesn't appear to offer any help at all, for we have only introduced another variable. Note, of course, that since u is the first derivative, the second derivative is $D_x u$. Now comes the demonstration that we have looked ahead. We multiply both sides of this equation by

$$u = D_x y,$$

using the right hand side of this on the right side of the given equation and the left side on the left side of the given equation. Thus,

$$2u D_x u = e^y D_x y,$$

or in differential form, after multiplying through by dx,

$$2u(du) = e^y \, dy.$$

This can be integrated. Hence we have

$$u^2 = e^y + C_1.$$

This is equivalent to

$$u = D_x y = (e^y + C_1)^{1/2}$$

where we have arbitrarily chosen the positive square root. There might have been compelling reasons for selecting the negative square root, of course. We now have the equation in its differential form as

$$\frac{dy}{(e^y + C_1)^{1/2}} = dx.$$

Thus, except for the integration yet to be performed, we have been able to solve this second order equation by the device of multiplying both sides by a variable that made integration possible. Since this is a second order equation, it is not surprising that there should be two constants involved. In case you have not finished the integration of the above equation by this time, we suggest you let

$$z = (e^y + C_1)^{1/2},$$

and then obtain the successive relations

$$z^2 = e^y + C_1,$$
$$e^y = z^2 - C_1,$$
$$y = \ln(z^2 - C_1),$$
$$dy = 2z \, dz / (z^2 - C_1),$$

and upon substitution the left side becomes $2 \, dz/(z^2 - C_1)$. This can be factored as the difference of two squares if C_1 is positive or handled as the differential of the inverse tangent if C_1 is negative. The constant here has been labeled C_1 in order that the constant arising from the second integration can be labeled C_2 and the two can be distinguished in the final solution.

It is true that a rather small percentage of differential equations are such that it is possible to separate variables, but this is a most valuable technique when it is applicable. By all means keep it in mind, and use it whenever possible. It will save a great deal of work. For those cases in which we cannot use this method, there are further sections detailing other methods. The fact that there is a large number of different methods for solving differential equations should not be surprising in view of the many techniques we had to consider in attempting to find formal means for integration, a direct counterpart of solving differential equations.

EXERCISES

1. Solve each of the following differential equations. Find the value of the constant if sufficient information is given.

   (a) $y'(x) = xy$ and $y(1) = 2$
   (b) $y'(x) = y + x^2y$ and $y(0) = 2$
   (c) $D_x y = x(y^2 - y)$
   (d) $y'(x) = x^2y^2 + x^2 + y^2 + 1$ if $y(0) = 0$

2. Solve the following differential equations, finding the value of the constant when sufficient information is given.

   (a) $y'(x) = \cos^2 y$ and $y(2) = 0$.
   (b) $y^2 x D_x y = x - 1$ and $y(1) = 3$.
   (c) $y'(x) = 1 - x - y + xy$ and $y(-1) = 4$.
   (d) $2x^2 yy' = y^2 + 1$

3. Solve the following differential equations, finding the value of the constant if sufficient information is given.

   (a) $x \, dy - y \, dx = 0$ and $y = 3$ when $x = 1$.
   (b) $x \, dy + y \, dx = 0$ and $y(3) = 4$
   (c) $y \, dy - x \, dx = 0$ and $y(3) = 5$.
   (d) $y \, dy + x \, dx = 0$ and $y(3) = 5$.

4. (a) Solve the differential equation $y + xy' = e^{xy}$ if $y(2) = 1$.
   (b) Solve the differential equation $y' \sin x = (1 - 4y^2)^{1/2}$.
   (c) Solve the differential equation $dx + dy = x(x + y)^4 \, dx$.

5. (a) Complete the solution of Example 1.3 if $C_1 > 0$. [$Hint$: Let $C_1 = k^2$.]
   (b) Complete the solution of Example 1.3 if $C_1 < 0$. [$Hint$: Let $C_1 = -k^2$.]
   (c) Complete the solution of Example 1.3 if $C_1 = 0$.

6. (a) Solve $D_x^2 y + k^2 y = 0$ where k is a constant.
   (b) Solve $D_x^2 y - k^2 y = 0$ where k is a constant.
   (c) Solve $D_x^2 y = 0$.

P7. Van't Hoff discovered the law of reaction $D_T(\ln k) = Q/RT^2$ relating the "constant" of a chemical reaction k, the temperature, T degrees Kelvin, the amount of heat freed or absorbed, Q, and the universal gas constant, R. If we assume that Q and R do not vary with T, find k as a function of T. It is known that Q and R change very slightly with respect to T if they change at all, and hence this result is a very good approximation of the situation which determines the value of k.

SB8. In a certain epidemic it is known that the proportion of the population infected at time t, $p(t)$, is determined by the relation $D_t p(t) = kp(t)[1 - 2p(t)]$ where k is a constant. It is also known that $p(t)$ is always in the interval $0 < p(t) < 0.5$.
   (a) Find a formula for $p(t)$ in terms of $p(0)$, the proportion when $t = 0$.
   (b) Under what condition will $p(t)$ increase and when will it decrease?
   (c) What proportion of the population can expect to escape infection?
   (d) What effect does the limitation of $p(t)$ to $p(t) < 0.5$ have on your solution of the various parts of this exercise?
   (e) Do you think it would have been realistic to consider this relation as a model of an epidemic if $p(t) > 0.5$?

SB9. In a certain animal population which lives in such isolation that the only source of population control is the presence of a plentiful food supply, we will denote the population at time t by $p(t)$. The average number of births per thousand population is a constant b and the average number of deaths is a constant d. It is then understood that the population is governed by the differential equation $p'(t) = [b - d - kp(t)]p(t)$ where k is a constant determined by the rate of lessening of the food supply as the population increases.
   (a) Find $p(t)$ as a function of time.
   (b) Sketch $p(t)$ as a function of time.
   (c) Determine the limiting population size, if any.

SB10. In a simple chemical reaction a molecule of A will combine with a molecule of B to create a molecule of X. If we start with an amount a of A and an amount b of B, and if it is assumed that the amount of X is initially zero, then at time t we will have $x(t)$ of X, $a - x(t)$ of A and $b - x(t)$ of B. The rate of formation of X is governed by the equation $x'(t) = k[a - x(t)][b - x(t)]$ where k is a constant dependent upon the specific reaction involved, the temperature, pressure, etc.
   (a) Find x as a function of t.
   (b) Find t as a function of x.
   (c) Does it make a difference whether $a > b$, $a = b$, or $a < b$? If so, what difference does it make?

PB11. A substance dissolves in water at a rate which is effectively proportional to the product of the amount remaining undissolved and the difference between the present level of concentration and a saturated concentration.
   (a) Set up the differential equation which describes the situation of this Exercise.
   (b) Find the concentration as a function of time.
   (c) Find the amount dissolved as a function of time.
   (d) If it is known that a saturated solution has 80 grams of a solute dissolved in one liter, and if it is also known that when 80 grams are placed in one liter of pure water 10 grams will dissolve in the first hour, how much would be dissolved in the first four hours?
   (e) How much would be dissolved in the first ten hours?
   (f) How long would it take to dissolve 75 grams?

S12. Domar's simplest growth model assumes three relations.
   (i) The actual rate of flow of savings, $S(t)$, at any time t is proportional to the income flow, $Y(t)$, at the same time. Thus $S(t) = kY(t)$.
   (ii) The intended rate of investment flow, $I(t)$, is proportional to $D_t Y(t)$, the rate at which the rate of income flow is changing.
   (iii) In the ideal situation the flow of savings should equal the intended rate of interest flow.
   (a) Obtain the differential equation which combines these three relations.
   (b) Find $Y(t)$ as a function of t.

XI.2 First Order Linear Equations and Companions

As you know by this time, the mathematician is one who leans heavily on precedents, particularly those that work. Having found a method for solving some differential equations, we will now attempt to see whether we can make others fit in the same mold. It will be the purpose of this section to consider a class of differential equations for which we can do that, at least in part. In the process we shall take our first look at *linear* differential equations. You will hear the word *linear* used a great deal by mathematicians. Since the linear case, involving as it does the case in which the exponents used all have the value one, is generally the simplest case it is usually the one that we know something about. When we get to the non-linear cases, the situation is frequently rather spotty, and we are able to solve only certain specific types. All of these thoughts conspire to bring us to the linear equations as the first ones we face after handling those in which we can separate variables.

Definition 1.2. A differential equation is said to be a *linear differential equation* if and only if each term containing either the dependent variable or one of its derivatives contains just one such term as a factor, and that only to the first power.

With this definition the general first order linear equation must appear as

$$a(x)D_x y + b(x)y = c(x) \qquad\qquad (XI.2.1)$$

where $a(x)$, $b(x)$ and $c(x)$ are functions of x alone. (They may be constant functions, of course.) Note that we are only permitted the presence of y and its first derivative in the case of the first order equation, and these cannot appear in the same term as factors, nor can either appear except to the first power. If $a(x)$ can never be zero in the interval in which we are interested, we can then divide both sides of the equation by $a(x)$ and obtain

$$D_x y + p(x)y = q(x) \qquad\qquad (XI.2.2)$$

where $p(x) = b(x)/a(x)$ and $q(x) = c(x)/a(x)$. It is this form with which we shall work. If $a(x)$ can be zero in the interval in question, this will pose problems that are better reserved for a course in differential equations.

Having now put the problem in perspective, let us proceed with an illustration of our general method.

EXAMPLE 2.1. Solve the differential equation $D_x y + y \tan x = \sec x$.

Solution. It is apparent from a close investigation of this equation (and a close investigation should be given) that it is not possible without some sort of trickery to separate the variables here, and the particular trickery is not at all obvious. Therefore, we will attempt to solve this in two parts. We will pick the first part in such a way that we know we can get a solution to the problem we pose and then we will hope that we can build on the first solution to obtain the solution for the stated problem. The first part consists of forgetting about $\sec x$. If we ignore, for the time being at least, the right hand side of this equation we have only to solve

$$D_x y + y \tan x = 0.$$

We *can* separate variables here! We have

$$\frac{dy}{y} = -\tan x \, dx,$$

and integration yields

$$\ln y = -\ln \sec x + C = \ln(K \cos x).$$

We have made use of the fact that $-\ln \sec x = \ln \cos x$, since $\sec x$ is the reciprocal of $\cos x$. We have also written $C = \ln K$. It is as easy to evaluate K as to evaluate C. Since we have two equal logarithms, we know their arguments are equal, and we have

$$y = K \cos x$$

as a solution to our easier problem. This is not a solution to the main problem, but we had said that we were going to do this in two parts. Since we know that $y = K \cos x$ will, if substituted in our original equation, make $(D_x y + y \tan x)$ vanish, perhaps we could help things a bit by seeing what would happen if we permitted K to become a variable for a moment. In order to make K appear a bit more like a variable, let us then see what would happen if we tried

$$y = v \cos x,$$

where v is the variable replacing K. It would be ideal if we would then be able to find some way of finding a value of v that would make $y = v \cos x$ the solution of the original equation.

On substitution of $y = v \cos x$ in the equation $D_x y + y \tan x = \sec x$, we obtain

$$[-v \sin x + (D_x v)\cos x] + (v \cos x)\tan x = \sec x,$$

or

$$-v \sin x + (D_x v)\cos x + v \sin x = \sec x,$$

since $(\cos x)(\tan x) = \sin x$. This now reduces to the equation

$$D_x v = \sec^2 x$$

and this is again an equation we can solve by separation of variables! We now have $v = \tan x + C_1$, where we have affixed the subscript "1" to distinguish this constant from the preceding constant C. Looking back, our solution to this equation was

$$y = v \cos x = (\tan x + C_1)\cos x = \sin x + C_1 \cos x.$$

If we try this to see whether it works, we will obtain

$$[\cos x - C_1 \sin x] + [\sin x + C_1 \cos x]\tan x \overset{?}{=} \sec x.$$

We have placed the question mark over the equal sign until we are certain we have an equality. Algebra and trigonometry give us

$$\cos x - C_1 \sin x + \sin^2 x/\cos x + C_1 \sin x = [\cos^2 x + \sin^2 x]/\cos x$$
$$= 1/\cos x$$
$$= \sec x,$$

and we see that we do have a solution.

Upon analyzing our work we see that the method we tried first involved the replacement of $q(x)$ in our general equation by zero in order to obtain what is called the homogeneous equation. [A function, $f(x)$, is called *homogeneous of degree n* provided $f(kx) = k^n f(x)$. An equation is homogeneous provided the equation relates two homogeneous functions of degree n. The linear differential equation with no non-zero term failing to include y or one of its derivatives as a factor is homogeneous, for if we replace y by ky, we would find that we could divide the value k into all terms of the equation and then return to the original equation.] We can solve the homogeneous equation by separating variables, for we have in general

$$\frac{dy}{y} + p(x)dx = 0. \qquad (XI.2.3)$$

The solution of this homogeneous equation will have a constant of integration and our next step will be to permit this constant to vary. Observe that the term $[yp(x)]$ is not appreciably altered by this change. On the other hand, the term involving the derivative will have two parts, one of which will be the portion of the derivative resulting from the variation in v as though x were held constant, and the other part resulting from the variation in x as though v were held constant. (Investigate the formula for the derivative of a product, and note that this is the case.) The portion of this derivative in which the v appears to have been constant will with the result of substitution in $yp(x)$ give a zero contribution to the left hand side. That is the way we obtained this part of the solution. Hence, there will remain on the left hand side only the part which involves the rate of change of v. Therefore, the situation which appeared in this example was not an accident after all. To be a little more formal in our consideration, the homogeneous equation $D_x y + yp(x) = 0$ will yield $\ln y + \int p(x)dx = 0$, or

$$y = Ke^{-\int p(x)dx}. \qquad (XI.2.4)$$

Upon substitution in the general linear equation, we will replace y by $y = vf(x)$, where $f(x)$ represents the exponential term. Hence, we know that $f'(x) + f(x)p(x) = 0$ for this is the basis on which we obtained the function

$$f(x) = e^{-\int p(x)dx}.$$

Now using $y = vf(x)$ in the original equation, we have

$$[vf'(x) + v'f(x)] + vf(x)p(x) = q(x)$$

or

$$v'f(x) + v[f'(x) + f(x)p(x)] = v'f(x) = q(x)$$

whence we have $dv = (q(x)/f(x))dx$, and we have separated variables. Of course it would be possible to replace $f(x)$ with its value and obtain a formula which could be memorized, but in the interests of being able to recall the result when you need it, and also in the interest of having this method available for other situations later on, we suggest that you merely remember the two steps and derive the solution by means of integration.

At this point you can begin to see why we enthused over linear equations in the beginning of the section. First order linear differential equations represent a class of equations which we know how to solve. Having this success behind us, we can now look for ways of converting equations which are almost, but not quite, linear to corresponding equations which are linear. Since we are only scratching the surface of things here, we will consider only one case of this variety. This is an equation that was first mentioned by *James Bernoulli* (1654–1705), a member of a very prolific mathematical family. The dates of his life indicate that this was discovered very soon after the initial development of the calculus in the 1670's. This particular equation, called appropriately enough the *Bernoulli equation*, is of the form

$$D_x y + y p(x) = y^n q(x).$$

It is apparent at a glance that it is *almost* linear, but the y^n on the right hand side is the cruncher here. We will assume that y is not always zero. (If it were, the entire equation would be satisfied all of the time, for y would be a constant and all terms would vanish.) Therefore, we will divide both sides of the equation by y^n, reserving for special consideration, if necessary, any points at which y momentarily takes on a zero value. We obtain

$$y^{-n} D_x y + y^{-n+1} p(x) = q(x).$$

But we suddenly notice that the derivative of y^{-n+1} is $(1 - n) y^{-n} D_x y$. This is fine if $n \neq 1$, but if $n = 1$ we could have moved the term on the right side to the left side and we would have been able to separate variables initially. Now, we can write

$$\frac{1}{1 - n} D_x(y^{-n+1}) + y^{-n+1} p(x) = q(x),$$

and it is apparent that we have a linear equation with the independent variable being y^{-n+1}. It is probably convenient to make a substitution of the form $u = y^{-n+1}$, and then have

$$D_x u + (1 - n) u p(x) = (1 - n) q(x).$$

EXAMPLE 2.2. Solve the equation $D_x y + y \tan x = y^3 \sec x$.

Solution. In this case we have a Bernoulli equation with $n = 3$. Instead of using the derivation above, we will start from scratch in the way in which we developed the above explanation. We will multiply by y^{-3}, and then obtain

$$y^{-3} D_x y + y^{-2} \tan x = \sec x,$$

or

$$-\frac{1}{2} D_x(y^{-2}) + y^{-2} \tan x = \sec x.$$

This is equivalent to

$$D_x(y^{-2}) + y^{-2}(-2 \tan x) = -2 \sec x.$$

However, this is a linear equation in y^{-2}, and we can solve this in a manner similar to that above. In order to make it slightly more obvious, we will let $u = y^{-2}$, and then solve

$$D_x u + u(-2 \tan x) = -2 \sec x. \qquad (XI.2.5)$$

Step one would have us solve

$$D_x u + u(-2 \tan x) = 0,$$

obtaining

$$\frac{du}{u} = 2 \tan x \, dx.$$

Since

$$\ln u = 2 \ln \sec x + C = \ln K \sec^2 x,$$

we can write

$$u = K \sec^2 x$$

by using methods similar to those in Example 2.1. In step two we let K become a variable, and substitute

$$u = v \sec^2 x$$

in (IX.2.5). Hence,

$$2v \sec x(\sec x \tan x) + v' \sec^2 x + v \sec^2 x(-2 \tan x) = -2 \sec x,$$

or

$$v' \sec^2 x = -2 \sec x.$$

From this we have $v' = -2 \cos x$, and $v = -2 \sin x + C$. Therefore

$$u = (-2 \sin x + C)\sec^2 x = -2 \sec x \tan x + C \sec^2 x$$

or

$$y^{-2} = -2 \sec x \tan x + C \sec^2 x.$$

While this is not the most appetizing expression if we must express y as a function of x it is not difficult and it does give us the result we sought.

Each of the examples we have considered so far has left us with an undetermined constant of integration. This has provided us with a *general solution* of the differential equation. In many cases we will have sufficient information to permit us to determine a specific value for this constant. In such a case we have a *particular solution* of the differential equation.

EXAMPLE 2.3. A 200 pound man jumps from an airplane. At the instant his parachute opens he is falling at a rate of 160 feet per second. The parachute provides air resistance equivalent to a number of pounds determined by multiplying the falling speed by 10. How fast will the man be falling 5 seconds after the parachute opens?

Solution. Here we have a downward force due to gravity counteracted in part by an upward force due to the effect of the parachute. The result is a force producing an observed rate of fall. The downward force due to gravity is 200 pounds. The upward force is $10v$ pounds where v is the velocity in feet per second. Finally, the observed force is $(200/32)v'$ where v' is the derivative of velocity, or the acceleration, and this is multiplied by the mass of $200/g$. Hence we have the equation

$$\frac{200}{32} v' = 200 - 10v \quad \text{or} \quad v' + 1.6v = 32.$$

This is a linear equation. We first solve the corresponding homogeneous equation

$$v' + 1.6v = 0,$$

and in this case we can separate variables. Our solution will be

$$v - ke^{-1.6t}$$

where k is the constant of integration. We will now replace k with a variable, u, and upon substitution in the given equation we will have

$$u' = 32e^{1.6t}.$$

This will give us

$$u = 20e^{1.6t} + k_2.$$

Upon substitution we now have the general solution

$$v = [20e^{1.6t} + k_2]e^{-1.6t} = 20 + k_2 e^{-1.6t}.$$

It is now time to use the fact that the initial velocity is known. If we let $t = 0$ represent the time at which the parachute opens, we have $v(0) = 160$. Hence substitution will give us

$$160 = 20 + k_2 e^0 = 20 + k_2.$$

Therefore $k_2 = 140$, and the particular equation which describes the man's fall is

$$v = 20 + 140e^{-1.6t}.$$

When $t = 5$ seconds we have

$$v(5) = 20 + 140e^{-8} = 20.05$$

feet per second.

In this section we have found that it is possible in every (?) case to solve the linear first order differential equation. We place the question mark here, for the solution is dependent upon your ability to perform the necessary integrations. If the integration is possible the solution of the differential equation follows at once. We first solve the homogenous differential equation obtained by replacing every term not involving the dependent variable by zero. Having this solution, we replace the constant of integration with a variable, and then solve the resulting differential equation for that variable. This second solution involves nothing more arduous than solving an equation in which we can separate variables. It is this general technique which we will exploit in the more general case involving linear equations of order higher than the first. We also discovered one particular type of non-linear equation which we can make linear by a suitable modification. This is another good demonstration of the old adage that "it pays to be observant." The matter of evaluating the constants, if we have information given relating the variables, is just as it was in the earlier case, and this needs no further discussion.

EXERCISES

1. Find the solution for each of the following differential equations:

   (a) $xy' + y = x^3$
   (b) $y' + 3y = 7$
   (c) $y' + 2xy = e^{-x^2}$
   (d) $y' - (2xy/(x^2 + 4)) = 1$

2. Find the solution for each of the following differential equations:

   (a) $xy' + y = y^2 \ln x$
   (b) $y' + y = xy^3$
   (c) $xy' - y - x^2 \sin x = 0$
   (d) $xy' + xy^2 - y = 0$
   (e) $y' + y \cot x = y^2 \sin x$

3. Find the solution for each of the following differential equations which satisfies the given condition:

   (a) $y' + xy = x^3$ if $y(0) = 1$
   (b) $y' - y = e^x$ if $y(0) = 1$
   (c) $2y' \cos x - y \sin x = y^3$ if $y(0) = 1$
   (d) $y' + y/x = x^3$ if $y(2) = 3$

4. Find the solution of each of the following:

   (a) $(\tan \theta)D_\theta r = r + \tan^2 \theta$
   (b) $xD_x y + y(1 - 2y \ln x) = 0$
   (c) $xy' + 3y = (\sin x)/x^2$ if $y(\pi/2) = 1$
   (d) $y' + xy = x/y^3$ if $y(0) = 2$

5. (a) Solve $D_x y + 2y = e^{-2x}$
   (b) Solve $D_x u + 2u = y$ where y is the solution of part (a).
   (c) Show that the solution of part (b) is also the solution of

$$D_x^2 u + 4D_x u + 4u = e^{-2x}.$$

   (d) Show that we could write symbolically $(D_x + 2)y = e^{-2x}$ for part (a) and $(D_x + 2)u = y$ for part (b).
   (e) Show that symbolically $(D_x + 2)(D_x + 2)u = (D_x + 2)y = e^{-2x}$ combines the results of part (d).
   (f) Show that symbolically $(D_x + 2)(D_x + 2)u = (D_x^2 + 4D_x + 4)u = e^{-2x}$ gives the equation of part (c).
   (g) Explain the way in which the various parts of this exercise relate the equations of parts (a) and (b) with the equation of part (c).

6. (a) Show that we can write symbolically

$$(D_x^2 - D_x - 6)y = D_x^2 y - D_x y - 6y = x \text{ as } (D_x - 3)(D_x + 2)y = x.$$

   (b) Solve $(D_x - 3)u = D_x u - 3u = x$ for u as a function of x.
   (c) Solve $(D_x + 2)y = D_x y + 2y = u$ for y as a function of x if u is the solution of part (b)
   (d) Show that your solution of part (c) is also a solution of part (a).

SB7. The population, $p(t)$, of a certain country is presumed to grow at the rate indicated by the differential equation $p'(t) = 0.02p(t) + 10^5 e^{t/500}$. The first term on the right side of this equation represents a normal rate of growth of 2% and the second term could represent an increase due to immigration.

   (a) If $t = 0$ in 1950, and if $p(0) - 200{,}000{,}000$, find the population of this country in 1960.
   (b) Find the population in 1970.

B8. A certain culture of bacteria having an initial population of 10,000 will double its population in 4 days.

   (a) Find the number of bacteria present as a function of time if B bacteria are extracted from the culture at noon each day.
   (b) What should be the value of B if it is desired to maintain a constant bacterial population?
   (c) For what values of B would the bacteria population ultimately become extinct?

P9. A body whose temperature is 180°F is immersed in a liquid which is maintained at a constant temperature of 60°F. The rate of change of the temperature of the body is proportional to the difference between the temperature of the body and the temperature of the surrounding liquid.

   (a) Find the temperature of the body as a function of time.
   (b) If the temperature is reduced to 120°F in the first minute, how long will it take to reduce the temperature to 90°F?
   (c) How long will it take to reduce the temperature to 70°F?

PB10. A tank containing 1000 gallons of water was polluted when someone accidentally dumped 500 pounds of salt into the tank. The tank is being purged by letting the brine run out at the rate of 4 gallons per minute while fresh water is running in at the same rate. It is assumed that the mixture is in constant motion, and therefore the concentration is uniform throughout.

    (a) Find the number of pounds of salt, x, in the water at any time as a function of the time.

    (b) How long will it take to get the concentration down to an acceptable 0.02 pounds of salt per gallon of water?

S11. In attempting to analyze the tendency to seek a goal, Anderson in his paper in *Psychological Review* in 1962 modified Miller's Theory of conflict and assumed that $x'(t) = G(x) - F(x)$ where $x(t)$ is the distance from the desired goal, $G(x)$ is the avoidance gradient, and $F(x)$ is the approach gradient. It is further assumed that $G(x) = a[x(0) - x] + b$ and $F(x) = c[x(0) - x] + d$ where a is a punishment parameter, c is a reward parameter, and b and d are constants associated with a given situation.

    (a) Find $x(t)$ as a function of time.

    (b) Under what conditions would this theory indicate the existence of an equilibrium? [When would $x(t)$ be a constant function?]

    (c) If an equilibrium exists, what is its value?

XI.3 Homogeneous Linear Differential Equations

We appeared to simplify the problem of solving the linear differential equation of the first order by considering the solution in two steps. First we considered the homogeneous equation obtained by replacing the non-homogeneous term by zero. In a second step we replaced the constant of integration by a variable, and then determined the variable in such a way that we had a solution of the given equation. We will now turn our attention to the solution of linear differential equations of order higher than the first, and we will again use the same two steps. In this section we will consider only the first step, that is the solution of the homogeneous linear differential equation. In the next section we will consider the second step. Thus, we will consider the solution of the equation

$$D_x^n y + a_1 D_x^{n-1} y + a_2 D_x^{n-2} y + \cdots + a_{n-1} D_x y + a_n y = 0. \quad \text{(XI.3.1)}$$

In order to further simplify the problem, we will restrict our attention to the case in which all of the coefficients of (XI.3.1) are constants. Since we arc considering only constant coefficients, we would either have all coefficients zero, and hence no equation, or else we would have a derivative of highest order having a non-zero coefficient. Since this non-zero coefficient is a constant, by assumption, we could divide by this constant. Hence, there is no restriction in assuming that the leading coefficient is one. While we have restricted this equation rather severely with the assumptions we have made, you will find that there are many applications in which this equation is sufficient. For a more complete treatment of differential equations you could always consult a book whose sole purpose is the discussion of such equations.

Despite the limitations we have imposed (XI.3.1) looks rather formidable. One approach would be that of trying to *guess* a solution. We remember that the exponential function appeared frequently in the linear equations of the last section, and consequently this might be worth a try. With this motivation, we will try using the exponential function $y = e^{mx}$, and determining whether it will work by substituting this in the equation. In order to illustrate the technique, we will proceed with an example.

EXAMPLE 3.1. Find the general solution of $D_x^3 y + 6D_x^2 y + 11D_x y + 6y = 0$.

Solution. Following our discussion, we will let $y = e^{mx}$. Upon substitution we find

$$m^3 e^{mx} + 6m^2 e^{mx} + 11m e^{mx} + 6e^{mx} = 0.$$

We know that no finite values of m and x will permit e^{mx} to be anything other than positive if m and x are real. Even if they are complex there is no combination that would permit $e^{mx} = 0$. Therefore, we can divide both sides by e^{mx} and obtain $m^3 + 6m^2 + 11m + 6 = 0$. At this stage we have not found that e^{mx} will necessarily work, but we have found that if e^{mx} is to be a solution, then m must be chosen such that $m^3 + 6m^2 + 11m + 6 = 0$. This is an equation of the type we have seen before. We see that m will have to be negative in this case if we are to have a solution, and further if m is an integer, m must be a divisor of 6, for m divides every term but the constant, and must therefore divide the constant term. Hence, we only have (-1), (-2), (-3), and (-6) to try. We simply try each of these in turn and find that the first three are roots. (Note that if this had not worked, we might have used Newton's method to obtain a sufficiently good approximation of the roots.) Here we found e^{-x}, e^{-2x}, and e^{-3x} to be candidates for the solution of our given equation. Upon substitution each one of them does, in fact, satisfy the equation.

In view of the form of the given equation, that is homogeneous, a constant multiplied by any solution is also a solution, for in the differentiation we simply have the same constant multiplier for each term. (Since a constant times zero is zero, we would still have a solution.) There is no reason to believe that we should use the same constant in each case. Consequently, we are now in a position to try $c_1 e^{-x}$, $c_2 e^{-2x}$, and $c_3 e^{-3x}$. Each of these satisfy the equation. Since the derivative of a sum is the sum of the derivatives, we could even be so bold as to try the sum of these three terms. Thus, we could try $y = c_1 e^{-x} + c_2 e^{-2x} + c_3 e^{-3x}$. This is also a solution. Notice that we have three constants to be determined by outside conditions, and the equation is third order, implying ultimately three integrations, hence three constants of integration. Therefore, it would seem that we have come out just right. In fact, this is the most general solution of this equation.

We have now exposed the method we will be using here. We merely substitute e^{mx} for y and then find an *auxiliary equation* which is a polynomial equation. In this case the auxiliary equation was $m^3 + 6m^2 + 11m + 6 = 0$. We obtain the solutions of this equation and then have the information to write out the terms of the general solution of the homogeneous equation. It would seem that everything is taken care of. However, there are two cases that can cause difficulties in what would otherwise be about as perfect a situation as one might find. The first of these concerns the possibility of complex roots for the auxiliary equation. You might suspect that we would avoid these like the plague, but in point of fact these are among the most sought after of the possibilities for certain types of applications. If we have real coefficients for the equation (XI.3.1), we then have real coefficients for the auxiliary equation, and the complex roots must occur in pairs of complex conjugate numbers. Thus, if $(a + ib)$ is a root of the auxiliary equation, we also have $(a - ib)$ as a root. This would mean that we would have as part of the solution of the differential equation $C_1 e^{(a+ib)x} + C_2 e^{(a-ib)x}$ if we follow the procedure illustrated in Example 3.1. However, referring to the Euler relation in Chapter IX we see that

$$
\begin{aligned}
C_1 e^{(a+ib)x} + C_2 e^{(a-ib)x} &= e^{ax}[C_1 e^{ibx} + C_2 e^{-ibx}] \\
&= e^{ax}[C_1(\cos bx + i \sin bx) \\
&\quad + C_2(\cos bx - i \sin bx)] \\
&= e^{ax}[(C_1 + C_2)\cos bx + (iC_1 - iC_2)\sin bx].
\end{aligned}
$$

Since $(C_1 + C_2)$ is a constant, and $(iC_1 - iC_2)$ is also a constant, we could rename these constants c_1 and c_2 respectively and have

$$
C_1 e^{(a+ib)x} + C_2 e^{(a-ib)x} = e^{ax}[c_1 \cos bx + c_2 \sin bx].
$$

You may wonder whether one or both of the constants c_1 and c_2 are complex. They could be, of course, but then any of the constants in our solutions could be. On the other hand, if C_1 and C_2 were complex conjugates you would find that c_1 and c_2 are both real numbers. Therefore, if we have complex conjugate roots of the auxiliary equation, we know that we will have an exponential term involving the real part of these complex roots and that this exponential term is multiplied by a linear combination of sine and cosine terms involving the imaginary part of the complex conjugate roots.

EXAMPLE 3.2. Find the solution of $D_x^3 y - 3D_x^2 y + D_x y + 5y = 0$.

Solution. If we let $y = e^{mx}$, as before, we obtain the auxiliary equation $m^3 - 3m^2 + m + 5 = 0$. As before, if there is an integral root, it must be $(1), (-1), (5)$, or (-5), for it must divide (5). We find that (-1) satisfies this equation, and therefore $(m - (-1)) = (m + 1)$ is a factor of the left hand side. Upon dividing we obtain

$$
(m + 1)(m^2 - 4m + 5) = 0.
$$

We know that $m = -1$ is one solution, but if we consider any values for which $m \neq -1$, we then have $m^2 - 4m + 5 = 0$. This can be written

$$(m - 2)^2 + 1 = (m - 2)^2 - i^2 = 0.$$

We now factor the left hand side as the difference of two squares and obtain the roots by setting each factor in turn equal to zero. Thus, we have for roots (-1), $(2 - i)$ and $(2 + i)$. (We could have used the quadratic formula but it is sometimes easier to complete the square.) Here are the complex conjugates we mentioned above. Our solution would now be of the type

$$
\begin{aligned}
y &= c_1 e^{-x} + C_2 e^{(2-i)x} + C_3 e^{(2+i)x} \\
&= c_1 e^{-x} + e^{2x}[C_2 e^{-ix} + C_3 e^{ix}] \\
&= c_1 e^{ix} + e^{2x}[C_2(\cos x - i \sin x) + C_3(\cos x + i \sin x)] \\
&= c_1 e^{-x} + e^{2x}[(C_2 + C_3)\cos x + (-iC_2 + iC_3)\sin x] \\
&= c_1 e^{-x} + e^{2x}(c_2 \cos x + c_3 \sin x).
\end{aligned}
$$

The procedure we have followed is precisely that which we indicated in the foregoing discussion. The remarkable thing about this result is that if you substitute this in the original equation, it checks! Again we note the presence of the requisite number of constants of integration, and this gives us a feeling that we probably have the complete story here.

Before tackling the second possible complication, that of multiple roots, let us pause long enough to note that the presence of complex roots is apt to give us functions of the form $e^{ax}f(x)$, where, at least so far, the $f(x)$ have been sines and cosines. If we have to deal with many terms like this it would be very nice to have an easier way of handling such a product. Among other things it would ease the problem of checking, for you noted above in the checking we had to (or would have had to if we had done it) obtain the third derivative of something that didn't look overly pleasant in the original. In order to simplify this kind of computation, we will consider the following two Theorems.

Theorem 3.1. *If a is a constant and $f(x)$ is differentiable, then $D_x[e^{ax}f(x)] = e^{ax}[D_x f(x) + af(x)] = e^{ax}[D_x + a]f(x)$, where $[D_x + a]$ is to be interpreted as an operator that will perform the indicated operations of differentiation and adding to the result of multiplying by a on any functions appearing immediately to the right of the operator.*

PROOF. If we proceed to differentiate the product $e^{ax}f(x)$ in a straightforward manner, we have

$$
\begin{aligned}
D_x[e^{ax}f(x)] &= e^{ax}D_x f(x) + f(x)[ae^{ax}] \\
&= e^{ax}[D_x f(x) + af(x)] = e^{ax}[D_x + a]f(x).
\end{aligned}
$$

The final step merely restates the result using operator notation.   □

Theorem 3.2. *If n is a positive integer, a is a constant, and $f(x)$ is a function for which the n-th derivative exists, then*

$$D_x^n[e^{ax}f(x)] = e^{ax}[D_x + a]^n f(x)$$

where $[D_x + a]^n$ is to be obtained by performing the formal algebra of expanding $[D_x + a]^n$ by the binomial theorem where the exponents of D_x are interpreted as the order of the derivative. The entire expression is to be considered as an operator, operating on $f(x)$ as indicated.

PROOF. We will use mathematical induction. We have already shown in Theorem 3.1 that this result holds for $n = 1$. If k is a value of n for which this holds, we are then assuming that it is true that

$$D_x^k[e^{ax}f(x)] = e^{ax}[D_x + a]^k f(x).$$

Now

$$\begin{aligned}
D_x^{k+1}[e^{ax}f(x)] &= D_x[D_x^k(e^{ax}f(x)] \\
&= D_x[e^{ax}[(D_x + a)^k f(x)] \\
&= e^{ax}(D_x + a)[(D_x + a)^k f(x)] \\
&= e^{ax}(D_x + a)^{k+1} f(x).
\end{aligned}$$

We have used the fact that the $(k + 1)$-st derivative is the derivative of the k-th derivative, the fact that when differentiating we can use Theorem 3.1, and finally the fact that if we differentiate $[(D_x + a)^k f(x)]$ and then add $a[(D_x + a)^k f(x)]$ to the result, we would obtain $(D_x + a)^{k+1} f(x)$ in a fashion very similar to the fashion in which we derived the formula for the binomial expansion by mathematical induction.   □

EXAMPLE 3.3.   Substitute $y = e^{2x}(c_2 \cos x + c_3 \sin x)$ in the expression $y''' - 3y'' + y' + 5y$.

Solution. Here we have, upon letting $f(x) = c_2 \cos x + c_3 \sin x$ for the moment,

$$\begin{aligned}
D_x^3[e^{2x}f(x)] &- 3D_x^2[e^{2x}f(x)] + D_x[e^{2x}f(x)] + 5e^{2x}f(x) \\
&= e^{2x}[(D_x + 2)^3 f(x) - 3(D_x + 2)^2 f(x) \\
&\quad + (D_x + 2)f(x) + 5f(x)] \\
&= e^{2x}[D_x^3 + 3D_x^2 + D_x + 3]f(x) \\
&= e^{2x}[D_x^3 + 3D_x^2 + D_x + 3](c_2 \cos x + c_3 \sin x) \\
&= e^{2x}[(c_2 \sin x - c_3 \cos x) + 3(-c_2 \cos x - c_3 \sin x) \\
&\qquad\qquad + (-c_2 \sin x + c_3 \cos x) \\
&\qquad\qquad + 3(c_2 \cos x + c_3 \sin x)] = 0.
\end{aligned}$$

This provides us with a check for the trigonometric portion of the result in Example 3.2. Note that we replaced $f(x)$ in the next to the last step, but it was easier to write $f(x)$ in each of the early stages. Also observe our use of Theorem 3.2. This theorem will frequently ease the problem of differentiation very much, as illustrated here.

We are now ready to face the problem of the multiple roots. We will first do this in an example. Then we will discuss the implications.

EXAMPLE 3.4. Solve $y^{iv} - 6y''' + 12y'' - 8y' = 0$.

Solution. Here the auxiliary equation is

$$m^4 - 6m^3 + 12m^2 - 8m = 0,$$

and we have an obvious root of $m = 0$. Thus, we can consider the factorization

$$m(m^3 - 6m^2 + 12m - 8) = 0,$$

and the other roots must be roots of

$$m^3 - 6m^2 + 12m - 8 = 0.$$

If we have an integral root, it must divide 8, and it must also be even, for we note that the last three terms are even, and therefore m^3 must be even also. Upon trying $m = 2$ we find that this works, and therefore we now have

$$m(m - 2)(m^2 - 4m + 4) = m(m - 2)^3 = 0.$$

Upon setting each of the four factors equal to zero, the roots are 0, 2, 2, and 2. This would seem to yield a solution of the form

$$y = c_1 e^{0x} + c_2 e^{2x} + c_3 e^{2x} + c_4 e^{2x} = c_1 + (c_2 + c_3 + c_4)e^{2x}.$$

However, there are only two distinct constants in this solution, for the coefficient of e^{2x} is in reality a single constant. Thus, it would appear that we do not have the complete solution at this point. It would also seem reasonable that the culprit is the three-fold appearance of (2) as a root. Perhaps we should replace the coefficient of e^{2x} with a variable. It can do no harm, and might give us the additional constants we need. (We replaced a constant with a variable in the first order case.) Therefore, consider what would happen if we were to try

$$y = e^{2x}f(x).$$

The best way to determine whether this is a solution is to put it back into the given equation and see what happens. Hence, we have

$$[D_x^4 - 6D_x^3 + 12D_x^2 - 8D_x](e^{2x}f(x)) = 0$$

upon writing this equation in operator form. We are now in a position to use Theorem 3.2, and obtain

$$e^{2x}[(D_x + 2)^4 - 6(D_x + 2)^3 + 12(D_x + 2)^2 - 8(D_x + 2)]f(x) = 0.$$

Upon expansion and combination of terms, this becomes

$$e^{2x}[D_x^4 + 2D_x^3]f(x) = 0.$$

We note that if $f(x)$ is any function for which the third derivative is always zero, this will be satisfied. Thus we obtain in successive steps

$$f'''(x) = 0,$$
$$f''(x) = C_1,$$
$$f'(x) = C_1 x + C_2,$$
$$f(x) = (C_1/2)x^2 + C_2 x + C_3.$$

Since we were considering the portion of the solution involving e^{2x}, we can with a slight renaming of constants, obtain what appears to be a full solution, $y = c_1 + e^{2x}(c_2 x^2 + c_3 x + c_4)$. Note that we do have four constants of integration, and each of these is multiplying a term which differs essentially from each of the others. If we check this solution, we will find that it satisfies the given equation.

Example 3.4 has illustrated a technique which we can use in the case of multiple roots of the auxiliary equation. One can generalize, of course, and prove a theorem which states that if m_1 is a root of the auxiliary equation of multiplicity k, the corresponding part of the solution will be $e^{m_1 x}$ multiplied by the most general possible polynomial in x of degree $(k - 1)$. The method we have developed here would tell us the same thing, however, and does not require that we remember all the details of a theorem which is important when the need arises, but which is not used as often as some of our other results.

EXAMPLE 3.5. Find the solution of $y'' - 3y' - 10y = 0$ for which $y(1) = 3$ and $y'(1) = 1$.

Solution. If we make the substitution $y = e^{mx}$, the given equation becomes

$$e^{mx}(m^2 - 3m - 10) = 0$$

and we have the auxiliary equation

$$m^2 - 3m - 10 = (m - 5)(m + 2) = 0.$$

Therefore we have $m = 5$ and $m = -2$ as solutions of the auxiliary equation and the general solution of the given equation is

$$y = c_1 e^{5x} + c_2 e^{-2x}.$$

If we use the given conditions to obtain the particular solution requested here, we would have

$$y(1) = c_1 e^5 + c_2 e^{-2} = 3$$

and

$$y'(1) = 5c_1 e^5 - 2c_2 e^{-2} = 1.$$

We can solve these equations simultaneously and obtain $c_1 = e^{-5}$ and $c_2 = 2e^2$. It should be clear that these values for the two constants of integration, and only these constants, will fulfill the additional conditions given in the example. Therefore, the solution which fulfills these conditions is $y(x) = e^{-5}e^{5x} + 2e^2e^{-2x} = e^{5(x-1)} + 2e^{-2(x-1)}$. Since there are two constants of integration in the case of a differential equation of second order, it is reasonable that two independent conditions would have to be known in order to obtain specific values for the constants. These may be conditions known when $x = 0$ (sometimes called *initial conditions*) or they may be conditions known for other values of x, as in this Example.

In this section we have considered all cases for the solution of the linear differential equation which is homogeneous and which has constant coefficients. The problem of finding the solution of the auxiliary equation is one that can be handled by factoring, if one can find the factors, by Newton's method if the roots are real, and for which you will have to do some research otherwise. It may help to know that it has been proven that each polynomial with coefficients no worse than complex numbers can be factored into a number of linear factors equal to its degree. (It is true, of course, that some of the linear factors may involve complex coefficients.) In other words, the equation has a solution if you can find it. Consequently, you are not going on a wild goose chase in attempting to solve such an equation.

EXERCISES

1. Find the solution of each of the following differential equations:

   (a) $y'' + 2y' = 0$
   (b) $y'' - 3y' + 2y = 0$
   (c) $y'' - 4y = 0$
   (d) $y''' + y'' - 6y' = 0$

2. Find the solution of each of the following equations:

   (a) $6y'' - 11y' + 4y = 0$
   (b) $y'' + 2y' - y = 0$
   (c) $y'' - 3y = 0$
   (d) $2y'' - 5y' + 2y = 0$

3. Find the solution of each of the following equations:

   (a) $y'' + 4y = 0$
   (b) $y'' + 2y' + 2y = 0$
   (c) $y'' - 6y' + 13y = 0$
   (d) $y'' + 3y = 0$
   (e) $y'' - 4y' + 6y = 0$

4. Find the solution of each of the following equations:

(a) $y''' - 4y'' + 5y' = 0$
(b) $y''' + 6y'' + 10y' = 0$
(c) $y''' - 8y = 0$
(d) $y^{iv} - 16y = 0$

5. Find the solution of each of the following equations:

(a) $y'' - 4y' + 4y = 0$
(b) $y''' + 6y'' + 9y' = 0$
(c) $y''' - 3y'' + 3y' - y = 0$
(d) $y^v - 2y^{iv} = 0$
(e) $y^{iv} - 8y'' + 16y = 0$

6. Find the solution for each of the following equations which satisfies the given conditions:

(a) $y'' = 0$ if $y(1) = 2$ and $y'(1) = -1$
(b) $y'' - 4y' + 4y = 0$ if $y(0) = 1$ and $y'(0) = 1$
(c) $y'' - 2y' + 5y = 0$ if $y(0) = 2$ and $y'(0) = 4$
(d) $y'' - 4y' + 20 = 0$ if $y(\pi/2) = 1$ and $y'(\pi/2) = 0$

7. Evaluate each of the following:

(a) $D_x(e^{-2x} \sin x)$
(b) $D_x^3(e^{-2x} \cos 3x)$
(c) $D^4(e^x x^3)$
(d) $(D_x - 2)^3 e^{2x} x^5$

8. (a) Show that $(D_x + a)^2 e^{-ax} f(x) = e^{-ax} f''(x)$
(b) Show that $(D_x + a)^n e^{-ax} x^n = e^{-ax} n!$
(c) Carry out the steps necessary to prove that

$$(D_x + a)(D_x + a)^k f(x) = (D_x + a)^{k+1} f(x)$$

(d) Evaluate $D_x^2 e^{3x} \tan 2x$

P9. A certain electric circuit (Fig. XI.1) consists of a capacitance of 10^{-4} farads, a resistance of 120 ohms, and an inductance of 1 henry, as indicated in the accompanying figure. The forces involved are all additive, and all have the same sign. The electromotive force (emf) resulting from a charge on the capacitor is given by $(1/C)q$ where C is the capacitance in farads and q is the charge. The emf due to the resistance is given by Ri where R is the resistance in ohms and i is the current in amperes. The emf due to the inductance is given by $LD_t i$ where L is the inductance

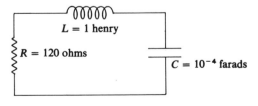

Figure XI.1

in henries. We also know that the current i is the rate of change of q, the charge on the capacitance.

(a) Find the sum of the emfs for this circuit.
(b) Since there is no external force operating on this circuit, the sum of the emfs must be zero. Obtain a differential equation which involves the charge, q, and the time, t.
(c) Find the general solution of the differential equation of part (b).
(d) Find the particular solution if $q = 5$ and $i = 0$ initially.
(e) Using the equation of part (d) find the current flowing in the circuit when $t = 1/120$ second.

S10. (a) If the rate of change of income is proportional to the income, Y, find a first order differential equation involving Y and time, t.
(b) If the rate of change of debt, D, is also proportional to income, find a first order differential equation involving D, Y, and t.
(c) Differentiate the equation of part (b) and with the aid of the equation in part (a) eliminate the term involving Y' in order to obtain a second order linear homogeneous differential equation involving D and t.
(d) Find the general solution of part (c), thus obtaining debt as a function of time.
(e) By differentiating your solution of part (d) obtain Y as a function of time. This will involve the results of parts (a) and (b).
(f) Show that the ratio of debt to income will approach a constant as time increases under the assumptions of this exercise.

S11. (a) Repeat part (b) of Exercise 10 under the assumption that the rate of change of debt is inversely proportional to income.
(b) Using the method outlined for Exercise 10 find the debt to income ratio under the assumption of part (a).
(c) Show that the debt to income ratio of part (b) will approach zero as time goes on under the assumption of this exercise.

B12. For purposes of detecting motion, we each have within the head a semicircular canal filled with a viscous fluid called the endolymph. Projecting into this canal is a gelatinous partition known as the *cupula*. If one is put in a chair that is rotated, the fluid rotates with the rest of the body, but when the rotation ceases the fluid continues to move long enough to give displacement to the cupula. This displacement and the subsequent return of the cupula to its equilibrium position gives one a sensation of continuing rotation. If we denote the displacement of the cupula (in this case an angular displacement since one end is fixed) by x, it has been theorized, and subsequent work has borne out the theory, that x satisfies the differential equation $AD_t^2 x + BD_t x + Cx = 0$ during the brief period in which the cupula is going from its position of maximum displacement (approximately 0.05 seconds after rotation stopped) back to equilibrium. Experiments have shown that the values $A = 1$, $B = 10$, and $C = 1$ yield results which agree with observation.

(a) Assuming $x(0) = 1$ mm and $x'(0) = +2$ mm/sec, find x as a function of time (in seconds) and note which part of the solution of the equation appears to be more significant.

(b) If there is some value of x, say x_{min} ,, which is the smallest x that can produce any sensation, what is the *extinction time*, that is the time required for displacement to get down to x_{min}? Use $x_{min} = 0.1$ mm.

(c) What would be the extinction time if $x_{min} = 0.05$ mm?

XI.4 Non-Homogeneous Linear Equations

In Section XI.2 we solved the first order linear differential equation by first solving the corresponding homogeneous equation and then letting the constant of integration become a variable. We were able to find the particular variable that would make the result satisfy the given equation. We suggested in Section XI.3 that we could do the same thing for linear equations of higher order. Consequently we devoted Section XI.3 to the first of these two steps, namely finding the solution of the homogeneous equation. We will now proceed to the second step, and will make use of the solutions for the homogeneous equations.

Assuming we are given a non-homogeneous linear differential equation with constant coefficients we will obtain the general solution of the corresponding homogeneous equation by the methods of the last section. We will then replace each of the constants of integration by a function in the manner of Section XI.2. If there are two constants, we will have two functions, if three constants we will have three functions, etc. In order to determine two functions, we expect to require two conditions or equations, and for three functions we expect three conditions. This can be extended to any larger number by rather obvious means. One condition is certainly not under our control, for we will most certainly be required to obtain a set of functions, however many are required, such that the given equation is satisfied. There seem to be no other real requirements inherent in the problem, and consequently we can invoke the other restrictions as we may wish in order to simplify the computation. Keep this in mind, for we do have some spare conditions sitting around that we can call on as we wish. Remember also that we are limited to a specific number of such conditions. The choices we will make for using these conditions are not the only ones that could be made, but they certainly expedite the solution, and that is the reason we make them. We will start with an example.

EXAMPLE 4.1. Solve $y'' - 2y' + 5y = 8e^x \tan 2x$.

Solution. We start by solving the equation

$$y'' - 2y' + 5y = 0.$$

The auxiliary equation is

$$m^2 - 2m + 5 = 0.$$

The roots are $(1 + 2i)$ and $(1 - 2i)$, and hence the solution of the homogeneous equation is

$$y_c = e^x(a \cos 2x + b \sin 2x),$$

where we have used a and b as the arbitrary constants rather than the subscripted c's. The reason for the subscript c on y, that is y_c, is that this portion of the final solution is often called the *complementary solution*. We will see this again later on. So far nothing is new. We now start the second step by considering $y = e^x(P \cos 2x + Q \sin 2x)$ where P and Q are both considered to be functions of x. We will not write the full $P(x)$ and $Q(x)$ for this would increase the apparent complexity and we do not envisage that you will have trouble with the simplified notation at this point. Upon substitution of this expression in the equation we have

$$[D_x^2 - 2D_x + 5](e^x(P \cos 2x + Q \sin 2x))$$
$$= e^x[D_x^2 + 4](P \cos 2x + Q \sin 2x) = 8e^x \tan x.$$

Since e^x cannot be zero, we have the equation

$$[D_x^2 + 4](P \cos 2x + Q \sin 2x) = 8 \tan 2x$$

to solve. This means that we need the second derivative of

$$(P \cos 2x + Q \sin 2x).$$

Since both P and Q are functions of x, $P \cos 2x$ and $Q \sin 2x$ are both products. The first derivative then contains four terms as follows.

$$D_x(P \cos 2x + Q \sin 2x)$$
$$= -2P \sin 2x + 2Q \cos 2x + P' \cos 2x + Q' \sin 2x.$$

Now we have a sinking feeling, for if we have this many terms from the first derivative, what will we have from the second—and we need the second derivative. But remember that we had one spare condition we could use since we have the two functions, P and Q, to determine. If we use the one condition that is at our disposal to require $P' \cos 2x + Q' \sin 2x = 0$, we would then be rid of the derivatives of P and Q, and would not in the next differentiation have any second derivatives of these two functions appearing. With the use of the one spare condition in this fashion, we now have

$$D_x(P \cos 2x + Q \sin 2x) = -2P \sin 2x + 2Q \cos 2x$$

and it is this we must differentiate in order to obtain the second derivative. Performing the second differentiation, we have

$$D_x^2(P \cos x + Q \sin x)$$
$$= -4P \cos 2x - 4Q \sin 2x - 2P' \sin 2x + 2Q' \cos 2x.$$

Therefore

$$(D_x^2 + 4)[P \cos 2x + Q \sin 2x]$$
$$= (-4P \cos 2x - 4Q \sin 2x - 2P' \sin 2x$$
$$+ 2Q' \cos 2x) + (4P \cos 2x + 4Q \sin 2x)$$
$$= -2P' \sin 2x + 2Q' \cos 2x = 8 \tan 2x.$$

This last result stems from the fact that we are required to obtain a solution for the given equation. This is one condition that must be met. At this point we have two equations, the one we imposed for convenience and the one required in order that we have a solution. These two, rewritten so that they appear together, are

$$P' \cos 2x + Q' \sin 2x = 0$$

and

$$-2P' \sin 2x + 2Q' \cos 2x = 8 \tan 2x.$$

If we solve for P' and Q', we obtain

$$P' = -4 \sin^2 2x/\cos 2x$$

and

$$Q' = 4 \sin 2x.$$

After integrating we have

$$P = 2 \sin 2x - 2 \ln|\sec 2x + \tan 2x| + c_1$$

and

$$Q = -2 \cos 2x + c_2.$$

Upon substituting these in the expression for y, we obtain

$$\begin{aligned} y &= e^x[(2 \sin 2x - 2 \ln|\sec 2x + \tan 2x| + c_1) \\ &\quad \times \cos 2x + (-2 \cos 2x + c_2)\sin 2x] \\ &= e^x[c_1 \cos 2x + c_2 \sin 2x - 2(\cos 2x) \\ &\quad \times \ln|\sec 2x + \tan 2x|]. \end{aligned}$$

Note the appearance of the complementary solution here, as indicated by the presence of the two constants of integration. The other portion, often called *a particular integral*, can be obtained without keeping any constants of integration if you remember that it must be combined with the complementary solution. The term *particular integral* merely denotes that we have a solution, but not the general solution. Since the difference between any two particular solutions would be a solution of the homogeneous equation, we can *complete* the particular solution by adding the most general solution of the homogeneous equation, namely the *complementary solution*. (Compare this use of the term *particular integral* with that of Section XI.2.)

We obtained the solution of the non-homogeneous equation. We obtained two equations in the derivatives of P and Q instead of the functions themselves. This is the pattern which we can expect to follow, a pattern not unlike that of the first order case. We replace each constant of the complementary solution by a function, and then differentiate as required in order to be able to substitute the modified expressions in the given equation. We simply get rid of the terms involving the derivatives of our parameters by letting the terms which involve them remain zero for all x. Since we only need to do this for the first $(n - 1)$ derivatives in the n-th order equation, and since in this case we would have n functions and $(n - 1)$ spare conditions at our disposal, this is always possible. Requiring that the equation be satisfied will give us the n-th equation, and we will have n equations in the first derivatives of the n functions. There will be no second derivatives for we have removed the first derivatives at each stage of the procedure before differentiating further. We can solve this system of n equations in the n unknown first derivatives, and then, provided we can integrate the first derivatives, we have the solution to our original problem. The method is not altered in any way by the presence of complex or multiple roots of the homogeneous equation, for by the time we have solved the homogeneous equation we have the complementary solution with the n constants of integration.

There are easier ways of solving the non-homogeneous equation in some cases, but this method has a general application. We, shall discuss one of the easier ways in the next section, but in case of doubt you can be assured that this method, generally known as *the method of variation of parameters* will work, provided you can perform the necessary integration, and of course provided you can solve the corresponding homogeneous equation.

EXERCISES

1. Find the general solution of each of the following equations:
   (a) $y'' + 3y' + 2y = 6$
   (b) $y'' + 3y' + 2y = 12e^x$
   (c) $y'' + 3y' + 2y = \cos x$
   (d) $y'' + 3y' + 2y = x^2$

2. Find the general solution of each of the following:
   (a) $y'' + 2y' + y = x^2 e^{-x}$
   (b) $y'' - 2y' - 8y = 3xe^x - 15e^{-x}$
   (c) $y'' - y' = x - \cos x$
   (d) $y'' + y = \sec^3 x$

3. Find the general solution of each of the following:
   (a) $y''' + y' = e^x$
   (b) $y''' - y' = e^{-x} + x$
   (c) $y''' + y' = \sin x$
   (d) $y''' - 3y'' + 3y' - y = e^x$

4. Find the general solution of each of the following:
   (a) $y'' - 2y' - 8y = 9xe^x + 10e^{-x}$
   (b) $y''' - y' = e^{2x} \sin^2 x$
   (c) $y^{iv} - 16y = e^{2x} + e^{-2x} + \sin 2x + \cos 2x$
   (d) $y''' + y'' - 2y' - 2y = 6e^{-x}$

5. (a) In the solution of Example 4.1 replace the equation $P' \cos 2x + Q' \sin 2x = 0$ with the equation $P' \cos 2x + Q' \sin 2x = 8$ making any other alterations which are indicated by this change.
   (b) Find the solution using the equation of part (a).
   (c) Show that this result is equivalent to the result obtained in Example 4.1.

6. (a) If both $y_1(x)$ and $y_2(x)$ are solution functions of a linear non-homogeneous differential equation, show that $[y_2(x) - y_1(x)]$ is a solution of the corresponding homogeneous differential equation.
   (b) Use the result of part (a) to show that the general solution of the non-homogeneous equation can be obtained by taking *any* solution and adding to it the complementary solution.

7. (a) Show that the method of this section will work in the case of first order linear differential equations if the coefficients of all terms involving the dependent variable and its derivatives are constants.
   (b) Use this method to find the solution of $y' + 3y = \sin x$.
   (c) Use this method to find the solution of $y' + 3y = e^{-3x} \sin x$.
   (d) Use this method to find the solution of $y' + 3y = e^{3x} \sin x$.

P8. An electric circuit contains a capacitance of 10^{-4} farads, a resistance of 120 ohms, an inductance of 1 henry, and a voltage source of $12 \sin 120\pi t$ volts. The current is initially zero amperes.

   (a) Find a differential equation which relates the charge, q, to time.
   (b) Find a solution of the equation of part (a), and hence q as a function of time.
   (c) If the initial charge on the capacitance was zero, find the current when $t = 1/120$.
   (d) If the initial charge on the cpacitance was 5, find the current when $t = 1/120$.

S9. (a) If the rate of change of income is a constant plus a term which is proportional to income, obtain a differential equation for income in terms of time.
   (b) If the rate of change of debt is proportional to income, obtain a second order differential equation relating debt and time using the result of part (a).
   (c) Solve your equation of part (b) to obtain the debt as a function of time.
   (d) Using your results for parts (a), (b), and (c) obtain income as a function of time.
   (e) Show that the debt to income ratio approaches a constant as time increases.

B10. We have two species, a host species, H, and a parasite species, P. Since the presence of a large number of P is detrimental to H and the lessening of the number of H is detrimental to the increase of P, it may be postulated that the number h of H and p of P vary according to the equations

$$D_t h = 0.7h - 0.002hp$$
$$D_t p = 0.4p - 0.001hp.$$

   (a) Show that $h = 400$ and $p = 350$ is an equilibrium population. That is, show that if we start with this number, we will continue to have this number for all time unless the conditions implicit in the equations are altered.
   (b) If $h = 400 + \Delta h$ and $p = 350 + \Delta p$, and if Δh and Δp are sufficiently small that we can ignore the product $\Delta h \Delta p$, obtain a differential equation of second order relating Δh and the time, t.
   (c) Find Δh as a function of t by solving the equation of part (b).
   (d) Using the information of parts (a), (b), and (c) find Δp as a function of t.

XI.5 Other Techniques

As we mentioned in the last section there are methods for handling certain non-homogeneous differential equations which might be considered easier than the method of variation of parameters. The particular method in which we will be interested here is that of "undetermined coefficients". That is a high sounding phrase that hides the fact that we are making an "educated" guess, and hoping we can cover up and find the solution. Thus, if we wished to solve the differential equation

$$y'' + y' - 2y = 2x^3,$$

we might guess that y would have a term involving x^3 in it. However, since we have to contend with both the first and second derivatives of y we could then expect to have to handle terms involving x^2 and x as well, but since we have to involve these terms, and that means taking their derivatives, it is likely that we will need a constant. Thus, we have now made a guess that at least one solution of this equation might have the form

$$y = Ax^3 + Bx^2 + Cx + E.$$

Here A, B, C, and E are constants to be determined, hence the name for the method. (Note that we skipped D, for D is often used to represent derivatives and there might well be confusion if D were also used as a coefficient.) Since the ultimate test of any solution is obtained by answering the question "Does it work?", we will substitute and see. We then have

$$D_x^2(Ax^3 + Bx^2 + Cx + E) + D_x(Ax^3 + Bx^2 + Cx + E)$$
$$-2(Ax^3 + Bx^2 + Cx + E) = 2x^3,$$

or

$$(6Ax + 2B) + (3Ax^2 + 2Bx + C) - (2Ax^3 + 2Bx^2 + 2Cx + 2E) = 2x^3.$$

As we did in handling integration by partial fractions we write down four equations involving the four unknown coefficients, obtaining

$$
\begin{aligned}
-2A &= 2, \\
3A - 2B &= 0, \\
6A + 2B - 2C &= 0, \\
2B + C - 2E &= 0,
\end{aligned}
$$

and we have $A = -1$, $B = -3/2$, $C = -9/2$, and $E = -15/4$. Thus, we have for one solution

$$y = -x^3 - \frac{3x^2}{2} - \frac{9x}{2} - \frac{15}{4}.$$

This is only one solution, but the entire solution can be obtained by augmenting this with the complementary solution. Thus the complete solution in this case would be

$$y = -x^3 - \frac{3x^2}{2} - \frac{9x}{2} - \frac{15}{4} + c_1 e^x + c_2 e^{-2x}.$$

This is the same solution you would have obtained by variation of parameters. It is suggested that you check this statement by finding the solution using variation of parameters.

The difficulty with this method is that one has to make a guess that includes all of the terms that can possibly appear in the particular integral. The penalty for not picking enough terms is the consequent inability to solve for the coefficients. There is no penalty for picking too many, so long as you include the right ones, except for the excess amount of work you have caused yourself. Frequently you can decide, as we did here, which terms should appear. However, this is not always the case. If the right hand side had been tan x, then we would have been in great trouble from the start, for as we consider taking the derivatives of tan x, and then the derivatives of the derivatives of tan x, etc., we would obtain $\sec^2 x$, a term involving $\sec^2 x \tan x$, etc., and the degree is increasing rather than decreasing. This not only does not look promising, but it is *not* promising, for if the right hand side had been tan x, we would not have been able to solve the equation by this method. We would have had to resort to the method of variation of parameters. In fact, a requirement for the type of reasoning we have indulged in here is that the successive derivatives will involve only a finite number of different functions. Otherwise, we would continually be in the position of having to include additional functions and this would lead to an infinite series. Upon a cursory examination of the functions we have been dealing with in this book, we find that if the right hand side includes an exponential term of the form e^{mx}, a sine, a cosine, or a positive integral power of x, we can hope to use the method of undetermined coefficients. In fact, we can use this method if we have any of these as individual terms or if we have a product of two or more of these. In all other cases we must resort to the method of Section XI.4.

Having made a hasty investigation of the conditions under which we can expect this method to work, we now take a closer look at the type of function we ought to try. In the majority of cases, if the method of undetermined coefficients will work at all, it is sufficient to take the function which makes the equation non-homogeneous and use this function together with every type of function that can be obtained from it by differentiation. It is not necessary, of course, to use repetitions. However, there is one case in which this is not adequate. We will consider the case in which this is sufficient in the following Example, and then consider the other case later.

EXAMPLE 5.1. Solve $y'' + 4y' + 3y = e^{2x} \sin x + x^2$.

Solution. The complementary solution is $y_c = c_1 e^{-x} + c_2 e^{-3x}$. We keep this for the record and add it to the particular integral later. For the particular integral we anticipate that we will need $e^{2x} \sin x$, x^2 and the various functions which can be obtained by differentiating these two. Thus, we have $e^{2x} \sin x$ and $e^{2x} \cos x$ in order to take care of the $e^{2x} \sin x$ and its derivatives, and then have x^2, x, and 1 in order to take care of x^2 and its derivatives. Note that we have not worried about the coefficients that these terms should have, but merely the type of terms involved. Our particular integral will supposedly be a linear combination of these terms, therefore we will try

$$y = Ae^{2x} \sin x + Be^{2x} \cos x + Cx^2 + Ex + F.$$

Upon substitution we have

$$(14A - 8B)e^{2x} \sin x + (8A + 14B)e^{2x} \cos x + 3Cx^2 + (3E + 8C)x$$
$$+ (3F + 4E + 2C) = e^{2x} \sin x + x^2.$$

Equating corresponding coefficients yields

$$\begin{aligned}
14A - 8B &= 1, \\
8A + 14B &= 0, \\
3C &= 1, \\
8C + 3E &= 0, \\
2C + 4E + 3F &= 0.
\end{aligned}$$

This will give the solution $A = 14/260$, $B = -8/260$, $C = 1/3$, $E = -8/9$, and $F = 26/27$. Placing these coefficients in the value of y above will give us a particular integral. We will combine this with the complementary solution obtained at the beginning of our solution and obtain the complete solution

$$y = e^{2x}\left[\frac{14}{260} \sin x - \frac{8}{260} \cos x\right] + \frac{x^2}{3} - \frac{8x}{9} + \frac{26}{27} + c_1 e^{-x} + c_2 e^{-3x}.$$

The method here is a copy of the one we used in the first illustration in this section but for the fact that we had more terms involved. Notice that frequently the system of equations which we must solve will partition nicely into separate systems, as in this case where the first two equations involved only A and B, and the last three involved only C, E, and F. The non-integral numbers that appear are not at all unusual if you are dealing with real, live applications, for nature seldom seems to cooperate in permitting us to confine ourselves to integers. Never feel that you must have the wrong answer if you get non-integral or irrational numbers. There is really nothing wrong with non-integers except for the inconvenience of handling them.

The suggestion was made earlier that there are times when we would run into difficulty in cases where trouble might not appear at first glance. It is now time to consider just such a case.

EXAMPLE 5.2. Find the complete solution of the equation $y'' - 4y' + 3y = 4e^x - 1$.

Solution. As before, we obtain the complementary solution $y_c = c_1 e^x + c_2 e^{3x}$. Upon looking at the non-homogeneous portion of the equation, we see that we will want the function e^x and 1 and their derivatives. Since the derivative of e^x is e^x, this does not require additional functions and since the derivative of 1 is 0, we have no additional terms in the second case either. We are all set to try

$$y = Ae^x + B(1) = Ae^x + B.$$

If you are alert, you may see trouble brewing, but we will proceed, trying to ignore it at this point, and obtain

$$(Ae^x) - 4(Ae^x) + 3(Ae^x + B) = 4e^x - 1 \quad \text{or} \quad 3B = 4e^x - 1.$$

It isn't difficult to let $B = -1/3$ and thus secure the negative one on the right side, but the exponential terms just disappeared from the left side of the equation. It would not be possible to let

$$B = 4e^x/3 - 1/3,$$

for we have assumed throughout that B is a constant, and to consider B otherwise would require reconsidering each differentiation. Therefore we are left with the task of finding some way of getting around this dilemma.

You might have felt that this was coming, for you could have noted that Ae^x is also part of the complementary solution, and the complementary solution produces zero on the right hand side. Perhaps we had better try variation of parameters after all. We can keep our value of $B = -1/3$ and only concern ourselves with trying $e^x f(x)$. We now have

$$(D_x^2 - 4D_x + 3)e^x f(x) = e^x(D_x^2 - 2D_x)f(x) = e^x,$$

in order to secure the exponential part of the non-homogeneous solution. This means that we would have

$$(D_x^2 - 2D_x)f(x) = 1.$$

This can easily be done if

$$D_x f(x) = -1/2,$$

for the second derivative of $f(x)$ would be zero. Hence, we would have $f(x) = -x/2$, and our complete solution would be

$$y = -\frac{xe^x}{2} - \frac{1}{3} + c_1 e^x + c_2 e^{3x}.$$

We note that we could have obtained a constant coefficient corresponding to our use of A if we had started with $y = Cxe^x + B$ instead of with $Ae^x + B$. If we had done this, we would have obtained on substitution

$$(Cxe^x + 2Ce^x) - 4(Cxe^x + Ce^x) + 3(Cxe^x + B) = e^x - 1$$

or

$$-2Ce^x + 3B = e^x - 1.$$

From this we obtain easily $C = -1/2$ and $B = -1/3$, thus giving us the result above.

All of this is fine, but it would appear that at crucial points we still have to go back to the method of variation of parameters. However, we note that by multiplying the offending function by x before using the method of undetermined coefficients, we were able to continue to use undetermined coefficients. In general we can show that if the conflicting term (in this case e^x) in the complementary function was due to a root of the auxiliary equation of multiplicity k, we should multiply all terms that relate to the conflicting term by x^k. Thus, if we had $\sin 2x$ in the non-homogeneous portion of the equation by itself and had $(c_1 + c_2 x + c_3 x^2)\sin 2x$ in the complementary equation, indicating a triple root in the auxiliary equation, we would then try $Ax^3 \sin 2x$ in implementing the method of undetermined coefficients. If you do not remember all of the rules we have indicated here, you can always return to the method of variation of parameters. However, if the method of undetermined coefficients can be easily employed, it will probably involve a little less work.

EXAMPLE 5.3. Find the complete solution of $y^{iv} + 4y'' = 48x - 16 \cos 2x$.

Solution. The auxiliary equation is $m^4 + 4m^2 = 0$, and consequently the complementary solution is $y_c = c_1 \sin 2x + c_2 \cos 2x + c_3 x + c_4$. If we look at the term involving $48x$ we see that this will indicate terms of the form Ax and B for the method of undetermined coefficients. This results from the fact that we must consider the function x and its derivative. However, we see that the complementary solution has a term involving x and a term involving a constant. These terms came as a result of the double root $m = 0$ in the auxiliary equation. Therefore we should multiply $Ax + B$ by x^2 in order to obtain a suitable candidate for the particular solution using the method of undetermined coefficients. Therefore this portion of the trial function should be $Ax^3 + Bx^2$. Similarly if we examine $\cos 2x$ and its derivatives we see the desirability of including $C \cos 2x + E \sin 2x$ in the trial solution. This also duplicates a portion of the complementary solution. However, the root of the auxiliary equation that caused these terms appeared as a single root. Therefore, we only need to multiply by the first power of x, and we will include in the trial solution the expression

$$Cx \cos 2x + Ex \sin 2x.$$

Hence the trial solution is $y = Ax^3 + Bx^2 + Cx \cos 2x + Ex \sin 2x$. Upon substitution we now have

$$(D_x^4 + 4D_x^2)(Ax^3 + Bx^2 + Cx \cos 2x + Ex \sin 2x)$$
$$= 24Ax + 2B + 16C \sin 2x - 16E \cos 2x.$$

This should equal the given expression of $48x - 16 \cos 2x$. This will happen if

$$
\begin{aligned}
24A &&&= 48 \\
&2B &&= 0 \\
&&16C &= 0 \\
&&-16E &= -16.
\end{aligned}
$$

This clearly indicates that $A = 2$, $B = 0$, $C = 0$, and $E = 1$. Our general solution, then, is $y = 2x^3 + x \sin 2x + c_1 \sin 2x + c_2 \cos 2x + c_3 x + c_4$.

Example 5.3 makes it clear that we can consider each component of the trial solution for the method of undetermined coefficients by considering it separately in its relationship to the solution of the auxiliary equation and to its counterparts in the complementary solution. Be on the alert for this case, for it can appear just often enough to make one wonder whether the method of undetermined coefficients is to be trusted. However, the method of undetermined coefficients often requires less work, and it should not be ignored as a possibility, despite its limitations.

EXERCISES

1. Find the general solution of each of the following:

   (a) $y'' + 3y' + 2y = 6$
   (b) $y'' + 3y' + 2y = \sin x$
   (c) $y'' + 3y' + 2y = e^x$
   (d) $y'' + 3y' + 2y = e^{-x}$

2. Find the general solution of each of the following:

   (a) $y'' - 3y' + 2y = 2xe^{3x} + 3 \sin x$
   (b) $y'' - 3y' = 2e^{2x} \sin x$
   (c) $y''' - 2y'' = x^2 + x + 1 + xe^{2x}$
   (d) $y'' + 4y = 8 \sin 2x$

3. Find the general solution of each of the following:

   (a) $y'' + 4y' + 13y = \sin 3x$
   (b) $y'' + 4y' + 13y = e^{-2x}$
   (c) $y'' + 4y' + 13y = e^{-2x} \sin 3x$
   (d) $y'' + 4y' + 13y = e^{-2x}(\sin 3x + 2 \cos 3x)$

4. Find the general solution of each of the following:

   (a) $y''' + y'' = 2e^{-x} + 96x^2$
   (b) $y''' + y'' = 2e^{2x} + 8 \sin 2x$
   (c) $y^{iv} + 8y'' + 16y = \sin 2x - 4 \cos 2x$

5. Find the solution for each of the following:

   (a) $y'' - 5y' - 6y = e^{2x}$ if $y(0) = 2$ and $y'(0) = 1$
   (b) $y''' + 4y'' + 4y' = 9e^x + 4e^{-x}$ if $y(0) = 1$, $y'(0) = 2$, and $y''(0) = 3$
   (c) $y'' + 16y = 15 \sin x$ if $y(\pi/2) = 1$ and $y'(\pi/2) = -1$.

6. In each of the following cases show why the method of undetermined coefficients cannot be expected to work:

   (a) $y'' + y' = x^{-1}$
   (b) $y'' + 3y' + 2y = \sec 2x$
   (c) $y''' - 4y = \ln 2x$
   (d) $y'' + 3y' - 4y = \ln \sec x$

7. Show that any linear differential equation with constant coefficients that can be solved by the method of undetermined coefficients can be solved by variation of parameters, but the converse of this statement would not be correct.

CHAPTER XII

Epilogue

XII.1 Whither Have We Come?

Whether or not you have read every word in this tome, you probably realize that you have gained at least to some extent by the exposure. A large number of problems of an *elementary* nature are now within your grasp, and you begin to see relationships which are capable of mathematical analysis which were hitherto unobserved. You have seen (and conquered?) such concepts as limit, integral, derivative, series, etc., and you have been exposed to problems in which each of these concepts is important. Furthermore, you have investigated means whereby you can obtain functions if you are given data. Thus, you should have some feel for the elementary problem. But what is the meaning of *elementary*, and why do we continue to emphasize this word? Elementary, of course, can mean different things to different people, but in general it means those items which do not require a great amount of background. If this be so, what kind of problems would fail to be elementary? We certainly have acquired quite some background during our perusal of the preceding chapters. It is the sole purpose of this chapter to explore the future from this particular point of view. We will not attempt to introduce additional mathematics herein, and therefore you can relax. We are only concerned with talking about some of the problems which you may face in the future and about places where you might find the information to permit you to handle such problems, provided of course you do not have that information at hand now.

Before embarking on a discussion of the areas you may wish (or need) to explore further, it would not be fair to omit a commendation for those who have been faithful in reading to this point. The mathematics which you have learned has involved many concepts, not all of them obvious at first glance.

The method of approaching and pursuing a problem has required a second look, for you have had to look behind the formulas to see why they work— and when they work. You have had to learn to be ever-observant so that you could employ any of the seemingly myriad of ideas we have introduced as they happen to be appropriate. You have, indeed, come a long way, and you should have a background which will make further steps much easier.

XII.2 More Variables

The first thing we might well observe as we contemplate problems that are likely to arise in practice is that very few problems depend upon a single quantity. For the purpose of introducing the calculus, we have handled primarily the problem in which each function depends upon one and only one variable. We have given passing notice to the fact that other cases exist in our discussion of partial derivatives and iterated integrals, but we have at best used our intuition, and we have found cases in which intuition can be dangerous. Therefore, one of the first things that you are apt to require is a consideration of methods whereby you can handle problems involving functions of many variables in a manner similar to that which we have used for functions of one variable. Needless to say, there are more details to consider in such situations and there are frequently more hypotheses for the various Theorems we would prove. There are many theorems that are similar to those we have discussed here, and hence it would not all look like foreign territory.

Before seriously considering the calculus of functions of many variables, it is necessary to consider an enlarged counterpart of the analytic geometry. We must deal with coordinates that can be written $(1, 3, 4, -3, 2)$. There is the question of what can be done with such as these. We rather obviously would find it difficult to draw axes depicting five coordinates, for this would require—if we amplify that with which we have become familiar—that we have five axes and that each one would be perpendicular to each of the other four. It is possible to consider algebraically the counterpart of coordinate geometry which would permit us to handle just such situations. These variables could, in applications, represent five different commodities, five different measurements, or any one of several quintuples of items. It should not surprise you that there are many facts which we would need to learn about such systems before we would be in a position to apply the calculus without fear of running into trouble. In general it is a little dangerous to trust intuition in dealing with what might be called five dimensional spaces. (It is probably true that our intuition hasn't been there, either.) Thus, a course in *linear algebra*, which covers such items, would be desirable if you find yourself facing multivariable problems. It would be worth pointing out that the largest application in the social sciences about which the author has heard involves nearly ten million dimensions (or unknowns). This is one for the computer but it would require a considerable knowledge of mathematical theory to design a method of attack.

With these considerations in mind, you may wish at some time to explore both the areas of linear algebra and of the *multi-variable calculus* (sometimes referred to as *advanced calculus*, although advanced calculus can mean different things to different people). The main thrust of these courses will be to provide the background for dealing with problems involving a very large number of variables in a somewhat expeditious manner. The course in linear algebra could include an introduction to *matrices* (the plural of matrix). These are used for various types of transformations, and they can also be used in such problems as the solution of large systems of linear equations. The linear algebra will introduce *vector spaces* and matrices. The Fourier series mentioned in Chapter IX can be considered as an illustration of the use of an infinite dimensional function space, or a vector space involving an infinite number of functions. The multi-variable calculus will extend the concept of directional derivative to multi-dimensional spaces. It will also consider both the iterated and the multiple integral, indicating where these concepts give the same result and where they do not.

An extension of the calculus (in more advanced courses we usually talk about *analysis*) would involve a further investigation of what happens when we make more general use of the complex numbers rather than just sporadic use. The calculus over the complex plane (or Argand diagram) has many facets which we made no effort to explore in this book. Those wishing to explore this area would probably be interested in a course in *the functions of a complex variable*. In view of the fact that the complex numbers themselves involve two real numbers it may not be surprising that you might be asked to take some work in multi-variable calculus first.

XII.3 Probability and its Consequences

There are few disciplines that do not make a great deal of use of statistics. Statistics develops from the study of probability or chance. This is a very old subject, for gambling is nothing new, and the attempt to determine strategies for improving one's chances at the gaming table were considered long before Blaise Pascal was asked by his benefactor to provide assistance during the 17th century. The comparison of what has happened with what might have happened under certain assumptions can be very helpful in determining whether the assumptions were probably incorrect. The use of such items as *mean* and *standard deviation* provide a means of standardization such that one probability distribution can be compared with another. Since different conditions can affect the probability of an occurence, the type of distribution of results, etc., there is a great deal that needs to be investigated carefully here before one can have any feeling of complacency about the results. The phrase "figures don't lie, but liars figure" has been used on many occasions to explain how the same data can be used to develop two apparently conflicting interpretations of the meaning of the data. Which of the two is correct can only be determined by making certain that the hypotheses are satisfied and that the reasoning from the hypotheses is correct. It is this area that is explored by the probabilist and statistician in courses in *probability* and *statistics*. Since one can either deal with a relatively small finite number of data points or with so many points that one is more apt to use continuous functions, you can expect to see the use of discrete methods involving primarily linear algebra, and the use of the calculus of several variables.

A more recent development in the area of applications has concerned the *theory of games*, which attempts to assess the probability of success of various strategies and then determine the optimal strategy. This portion of mathematics makes heavy use of linear algebra as well as probability, statistics, and calculus. There is the closely related subject of *linear programming*. This has nothing to do with computers, but does attempt to find optimal procedures for certain types of problems arising in business, management, and other areas. The solution of a problem in linear programming will generally rely on linear algebra, and in particular on the theory of matrices.

XII.4 Other Topics

Other areas which may be of use to you include *numerical analysis*, a subject which attempts to go beyond the material considered in Chapters VIII and X to find ways of obtaining results in many areas of mathematics with as much accuracy as possible and with the smallest number of computations possible. This requires an analysis of the accuracy that can be expected, and usually focuses on methods by which it is possible to obtain results with the aid of digital computers. It is impossible to restrict the background that numerical analysis calls upon. A great deal of use is made of Taylor's theorem, as we have seen, but work in matrix algebra, linear algebra, orthogonal polynomials, etc., is all heavily used.

Much can be done with the area of *logic*, and you might find it very helpful to explore this field. In addition to the mathematical or symbolic logic there is *Boolean Algebra*, named after *George Boole*, an Irish mathematician (1815–1863). The Boolean algebra has many uses, but the two valued Boolean algebra used heavily in the design of switching circuits, such as in computers and dial telephone switching systems is probably most used.

Number Theory is a subject with a long history. The development of material in this area is very much alive today. There are surprising applications of this subject in areas relating to energy distribution, biology, etc. It is a fascinating subject in and of itself. Among other interesting results are the conditions under which an integer can be expressed as the sum of the squares of two other integers, work concerning greatest common divisors, etc. One of the fascinating things about this subject is the number of problems which are easily stated but for which no solution is yet known. The solution of any of these unsolved problems could lead to mathematical fame in a hurry.

You have had experience with Euclidean geometry, but have you studied the *non-Euclidean geometries*, the geometries having only a finite number of points, or the *projective geometries*? Starting from geometry, but now very much a discipline of its own, is the subject of *topology*. The fact that at any moment of time there must be at least one point on the earth such that the wind is not blowing is an immediate consequence of the fixed point theorem of topology. Also included is the study of one-sided surfaces, the method by which a man who wears both a coat and a vest can take off the vest without first removing the coat, etc.

You will find upon inspection of the catalogue of any college or university a number of other courses listed, and on glancing through any mathematics library you will see many titles not mentioned here. Our purpose here has been to indicate that there are still many subjects in mathematics which are of interest and which have application. There are also subjects which you may well need for a given application. The fact that you have gotten this far indicates that you have good reason to believe that *you* might be able to handle some of these other courses, despite their sometimes exotic names. If you decide to take such courses, good luck in your venture. If not, you may have a better appreciation for what these courses contain and some knowledge of where to go if you need more mathematics at a later time.

Appendix A. Trigonometry

There are many equivalent methods for defining the six *trigonometric ratios.* We shall use a method which starts with the unit circle (the circle of radius one with center at the origin) shown in Figure A.1. If we take a ray OP with its initial point at the origin and making an angle θ with the positive x-axis, we have a unique point Q determined by the intersection of the circle and the ray. (Angles will be considered positive if measured in the counterclockwise direction and negative if measured in the clockwise direction.) The coordinates of Q will be defined as the *cosine of* θ and the *sine of* θ respectively. Using the customary abbreviations we have $x = \cos \theta$, $y = \sin \theta$, or $(\cos \theta, \sin \theta)$ as the coordinates of Q. We define the remaining four trigonometric functions by the following relations when they exist (if a denominator is zero for a particular value of θ, the corresponding function is undefined for that value of θ):

$$\text{tangent } \theta = \tan \theta = \frac{\sin \theta}{\cos \theta};$$

$$\text{cotangent } \theta = \cot \theta = \frac{\cos \theta}{\sin \theta};$$

$$\text{secant } \theta = \sec \theta = \frac{1}{\cos \theta};$$

$$\text{cosecant } \theta = \csc \theta = \frac{1}{\sin \theta}.$$

There are several relations involving these functions which are true for all of the values of θ for which the specific functions exist. By the Pythagorean Theorem it follows at once that

$$\sin^2 \theta + \cos^2 \theta = 1, \tag{A.1}$$

643

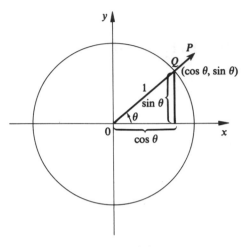

Figure A.1

ℓ where $\sin^2 \theta$ is a notation used for $(\sin \theta)^2$. [This convention of writing $(\sin \theta)^n = \sin^n \theta$ should never be used for $n = -1$, since $\sin^{-1}\theta$ has the meaning of inverse sine in some books and papers, and hence to use this notation instead of $(\sin \theta)^{-1}$ would be ambiguous.] If we divide (A.1) by $\cos^2 \theta$ or by $\sin^2 \theta$ and use the definitions above, we have

$$\tan^2 \theta + 1 = \sec^2 \theta, \tag{A.2}$$

and

$$1 + \cot^2 \theta = \csc^2 \theta, \tag{A.3}$$

respectively.

Let us now look at the relation of the trigonometric functions of negative angles to those with corresponding positive angles. Since the angles θ and $(-\theta)$ have the same absolute magnitude (Figure A.2), the triangles OQP and OQP' are congruent, for $OP = OP'$ and $OQ = OQ'$. Therefore $\cos \theta = OQ = \cos(-\theta)$ and $\sin(-\theta) = -QP = -\sin \theta$. Hence we can write

$$\begin{aligned} \sin(-\theta) &= -\sin \theta, \\ \cos(-\theta) &= \cos \theta. \end{aligned} \tag{A.4}$$

Using these results in the definitions we have

$$\tan(-\theta) = \frac{\sin(-\theta)}{\cos(-\theta)} = \frac{-\sin \theta}{\cos \theta} = -\tan \theta,$$

$$\cot(-\theta) = -\cot \theta,$$

$$\sec(-\theta) = \sec \theta,$$

$$\csc(-\theta) = -\csc \theta.$$

We now consider the relation between the sine and the cosine. Two angles are *complementary* if their sum is a right angle. (Note the spelling. The first

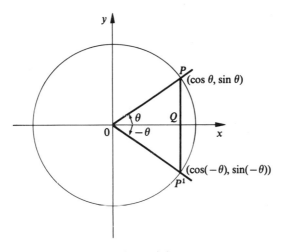

Figure A.2

"e" is related to the fact that this word descends from "complete". This is not equivalent to the word complimentary which denotes appreciation of one form or another.) Let $\bar{\alpha}$ be the complement of α. Hence $\alpha + \bar{\alpha}$ is a right angle. The geometric fact that it is a right angle does not impose conditions on the units of measurement, but for convenience we will use *radian measure* (see Section III.5) rather than *degrees*. Remember that 2π radians measure the same angle as 360°, and consequently

$$90° = \frac{\pi}{2} \text{ radians.}$$

Now

$$\alpha + \bar{\alpha} = \frac{\pi}{2}$$

or

$$\bar{\alpha} = \frac{\pi}{2} - \alpha.$$

Observe in Figure A.3 that triangle OPR is congruent to triangle OQS, and hence

$$\cos \bar{\alpha} = QS = PR = \sin \alpha$$
$$\sin \bar{\alpha} = OS = OR = \cos \alpha.$$

Rewriting these relations

$$\cos \bar{\alpha} = \sin \alpha$$
$$\sin \bar{\alpha} = \cos \alpha$$

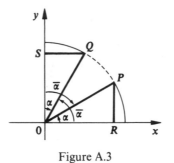

Figure A.3

we see that the sine of the complement of the angle is equal to the cosine of the angle. Note the underlined letters. Similarly we have

$$\tan \bar{\alpha} = \frac{\sin \bar{\alpha}}{\cos \bar{\alpha}} = \frac{\cos \alpha}{\sin \alpha} = \cot \alpha$$

$$\cot \bar{\alpha} = \frac{\cos \bar{\alpha}}{\sin \bar{\alpha}} = \frac{\sin \alpha}{\cos \alpha} = \tan \alpha$$

$$\sec \bar{\alpha} = \frac{1}{\cos \bar{\alpha}} = \frac{1}{\sin \alpha} = \csc \alpha$$

$$\csc \bar{\alpha} = \frac{1}{\sin \bar{\alpha}} = \frac{1}{\cos \alpha} = \sec \alpha.$$

While the drawing in Figure A.3 pictured α as an acute angle, the same construction is possible for any angle α provided only that

$$\alpha + \bar{\alpha} = \frac{\pi}{2}, \quad \text{or} \quad \bar{\alpha} = \frac{\pi}{2} - \alpha.$$

It may be helpful to summarize this paragraph with the following six relations:

$$\sin\left(\frac{\pi}{2} - \alpha\right) = \cos \alpha$$

$$\cos\left(\frac{\pi}{2} - \alpha\right) = \sin \alpha$$

$$\tan\left(\frac{\pi}{2} - \alpha\right) = \cot \alpha$$

$$\cot\left(\frac{\pi}{2} - \alpha\right) = \tan \alpha \tag{A.5}$$

$$\sec\left(\frac{\pi}{2} - \alpha\right) = \csc \alpha$$

$$\csc\left(\frac{\pi}{2} - \alpha\right) = \sec \alpha.$$

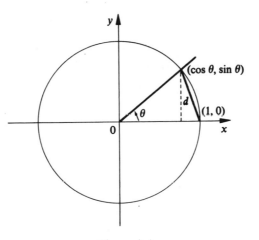

Figure A.4

Now if we consider the chord of length d in Figure A.4 that joins the points $(1, 0)$ and $(\cos \theta, \sin \theta)$ on the unit circle, we have

$$d^2 = (1 - \cos \theta)^2 + \sin^2 \theta$$
$$= 1 - 2 \cos \theta + (\cos^2 \theta + \sin^2 \theta) = 2 - 2 \cos \theta.$$

Thus the square of this chord is two minus twice the cosine of the subtended angle. We now turn our attention to the case in which we have two angles, α and β, both starting from the positive x-axis as shown in Figure A.5. The two terminal sides intersect the unit circle at $P : (\cos \alpha, \sin \alpha)$ and $Q : (\cos \beta, \sin \beta)$ respectively. By the result above we know that $d^2 = 2 - 2 \cos (\alpha - \beta)$. On the other hand, by the Pythagorean theorem we have the result

$$d^2 = (\cos \beta - \cos \alpha)^2 + (\sin \beta - \sin \alpha)^2$$
$$= (\cos^2 \beta + \sin^2 \beta) - 2(\cos \alpha \cos \beta + \sin \alpha \sin \beta) + (\cos^2 \alpha + \sin^2 \alpha)$$
$$= 2 - 2(\cos \alpha \cos \beta + \sin \alpha \sin \beta).$$

By equating the values of d^2 we have

$$\cos(\alpha - \beta) = \cos \alpha \cos \beta + \sin \alpha \sin \beta. \tag{A.6}$$

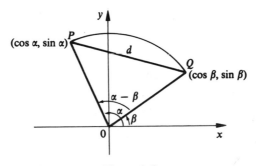

Figure A.5

Note that the result depends only on the coordinates of the points involved and is not restricted to any particular pair of angles.

Equation (A.6) refers only to the cosine of the difference of two angles, but we have all of the information necessary to expand this. Thus,

$$\begin{aligned}\cos(\alpha + \beta) = \cos[\alpha - (-\beta)] &= \cos\alpha\cos(-\beta) + \sin\alpha\sin(-\beta)\\ &= \cos\alpha\cos\beta + (\sin\alpha)(-\sin\beta)\\ &= \cos\alpha\cos\beta - \sin\alpha\sin\beta\end{aligned}$$

and

$$\begin{aligned}\sin(\alpha + \beta) &= \cos\left[\frac{\pi}{2} - (\alpha + \beta)\right]\\ &= \cos\left[\left(\frac{\pi}{2} - \alpha\right) - \beta\right]\\ &= \cos\left(\frac{\pi}{2} - \alpha\right)\cos\beta + \sin\left(\frac{\pi}{2} - \alpha\right)\sin\beta\\ &= \sin\alpha\cos\beta + \cos\alpha\sin\beta.\end{aligned}$$

Also

$$\begin{aligned}\sin(\alpha - \beta) &= \sin[\alpha + (-\beta)]\\ &= \sin\alpha\cos(-\beta) + \cos\alpha\sin(-\beta)\\ &= \sin\alpha\cos\beta + (\cos\alpha)(-\sin\beta)\\ &= \sin\alpha\cos\beta - \cos\alpha\sin\beta.\end{aligned}$$

The four results so far obtained are frequently summarized as

$$\begin{aligned}\sin(\alpha \pm \beta) &= \sin\alpha\cos\beta \pm \cos\alpha\sin\beta\\ \cos(\alpha \pm \beta) &= \cos\alpha\cos\beta \mp \sin\alpha\sin\beta\end{aligned} \tag{A.7}$$

where it is understood that we use either the upper algebraic sign throughout the equation or the lower algebraic sign throughout the equation. The results (A.7) can be used to obtain results for the tangent as well. Thus

$$\begin{aligned}\tan(\alpha \pm \beta) &= \frac{\sin(\alpha \pm \beta)}{\cos(\alpha \pm \beta)}\\[2mm] &= \frac{\sin\alpha\cos\beta \pm \cos\alpha\sin\beta}{\cos\alpha\cos\beta \mp \sin\alpha\sin\beta}\\[2mm] &= \frac{\dfrac{\sin\alpha\cos\beta}{\cos\alpha\cos\beta} \pm \dfrac{\cos\alpha\sin\beta}{\cos\alpha\cos\beta}}{\dfrac{\cos\alpha\cos\beta}{\cos\alpha\cos\beta} \mp \dfrac{\sin\alpha\sin\beta}{\cos\alpha\cos\beta}}\\[2mm] &= \frac{\tan\alpha \pm \tan\beta}{1 \mp \tan\alpha\tan\beta}.\end{aligned}$$

The number of steps was increased in order to return the result to the tangent functions of α and β from the sine and cosine functions.

We will consider two more items in this brief survey of trigonometry. The first involves a special case of the results obtained so far and the second is numerical in nature. If $\alpha = \beta$, we then obtain

$$\sin 2\alpha = \sin(\alpha + \alpha) = \sin \alpha \cos \alpha + \cos \alpha \sin \alpha = 2 \sin \alpha \cos \alpha.$$

$$\cos 2\alpha = \cos(\alpha + \alpha) = \cos \alpha \cos \alpha - \sin \alpha \sin \alpha = \cos^2 \alpha - \sin^2 \alpha$$

$$\tan 2\alpha = \frac{\tan \alpha + \tan \alpha}{1 - \tan \alpha \tan \alpha} = \frac{2 \tan \alpha}{1 - \tan^2 \alpha}.$$

With the use of the relation $\sin^2 \alpha + \cos^2 \alpha = 1$, we can now write

(1) $$\sin 2\alpha = 2 \sin \alpha \cos \alpha$$

(2a) $$\cos 2\alpha = \cos^2 \alpha - \sin^2 \alpha$$

(2b) $$\cos 2\alpha = 2 \cos^2 \alpha - 1 \qquad\qquad\text{(A.8)}$$

(2c) $$\cos 2\alpha = 1 - 2 \sin^2 \alpha$$

(3) $$\tan 2\alpha = \frac{2 \tan \alpha}{1 - \tan^2 \alpha}.$$

It is also convenient to use (2b) and (2c) in a different form, namely

$$\cos^2 \alpha = \frac{1}{2}(1 + \cos 2\alpha)$$

$$\text{(A.9)}$$

$$\sin^2 \alpha = \frac{1}{2}(1 - \cos 2\alpha).$$

If we let $\alpha = \beta/2$, we have

$$\cos \frac{\beta}{2} = \pm \sqrt{\frac{1 + \cos \beta}{1}}$$

$$\text{(A.10)}$$

$$\sin \frac{\beta}{2} = \pm \sqrt{\frac{1 - \cos \beta}{2}}$$

where the choice of sign depends on the quadrant in which $(\beta/2)$ is located.

An alternate, though less general, method for obtaining the "half-angle" formulas is based on an isosceles triangle. In Figure A.6 let $AB = AC = 1$ unit and let BD be perpendicular to AC. Then $BD = \sin \alpha$, $AD = \cos \alpha$, and $DC = 1 - \cos \alpha$. Since triangle BDC is a right triangle, the hypotenuse

$$BC = \sqrt{\sin^2 \alpha + (1 - \cos \alpha)^2}$$

or

$$BC = \sqrt{2 - 2 \cos \alpha}.$$

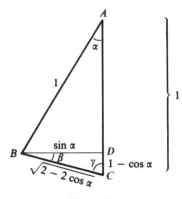

Figure A.6

Since the sum of the angles of a triangle is a straight angle and the sum of the acute angles of a right triangle is a right angle, we have $\alpha + 2\gamma = \pi$ and $\beta + \gamma = \pi/2$ whence $\beta = \alpha/2$. Therefore, using triangle BCD we have

$$\sin \beta = \sin \frac{\alpha}{2} = \frac{1 - \cos \alpha}{\sqrt{2(1 - \cos \alpha)}} = \sqrt{\frac{1 - \cos \alpha}{2}}$$

$$\cos \beta = \cos \frac{\alpha}{2} = \frac{\sin \alpha}{\sqrt{2(1 - \cos \alpha)}} = \sqrt{\frac{1 - \cos^2 \alpha}{2(1 - \cos \alpha)}} = \sqrt{\frac{1 + \cos \alpha}{2}}$$

$$\tan \beta = \tan \frac{\alpha}{2} = \frac{1 - \cos \alpha}{\sin \alpha} = \csc \alpha - \cot \alpha.$$

Note that α must be acute for this construction, but it may assist in remembering the half-angle formulas.

Finally, we wish to consider the sine of each of the five angles

$$0, \frac{\pi}{6}, \frac{\pi}{4}, \frac{\pi}{3}, \quad \text{and} \quad \frac{\pi}{2}.$$

These are special only to the extent that we can use rather simple geometry to obtain the required values. We note at once that $\sin 0 = 0$ and

$$\sin \frac{\pi}{2} = 1$$

from Figure A.7.

If $\angle ROP = \pi/4$, then $OR = RP$ because the $\angle OPR + \angle ROP + \angle ORP = \pi$, $\angle ORP = \pi/2$, and $\angle ROP = \pi/4$, hence $\angle OPR = \pi/4$ and the triangle is isosceles. Here $\sin(\pi/4) = \cos(\pi/4)$ and since $\sin^2(\pi/4) + \cos^2(\pi/4) = 1$, $\sin^2(\pi/4) = 1/2$ or $\sin(\pi/4) = \frac{1}{2}\sqrt{2}$.

If $\angle SOQ = \pi/6$, and $\angle SOQ' = \pi/6$, we have by elementary geometry an equilateral triangle OQQ'. Since $OQ = 1$, each side of this triangle is of

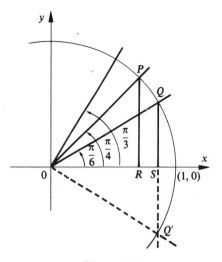

Figure A.7

length one. Since S bisects QQ', $QS = 1/2$ and by the Pythagorean theorem $OS = \sqrt{1 - (1/4)} = (1/2)\sqrt{3}$.

We now have $\sin(\pi/6) = 1/2$ and $\sin(\pi/3) = \cos(\pi/6) = (1/2)\sqrt{3}$. This set of values can easily be summarized in the table

α	0	$\dfrac{\pi}{6}$	$\dfrac{\pi}{4}$	$\dfrac{\pi}{3}$	$\dfrac{\pi}{2}$	
$\sin \alpha$	$\dfrac{1}{2}\sqrt{0}$	$\dfrac{1}{2}\sqrt{1}$	$\dfrac{1}{2}\sqrt{2}$	$\dfrac{1}{2}\sqrt{3}$	$\dfrac{1}{2}\sqrt{4}$	(A.12)

With these values and with the relations we have developed, we can obtain each of the functions for each of these angles for which the particular function exists.

EXERCISES

1. In each of the following sketch the angle in the unit circle which has the given value of its trigonometric function. Find the other five trigonometric functions of the angle if they exist. Assume the angles are in the first quadrant unless stated otherwise.

(a) $\sin \theta = 3/4$
(b) $\tan \theta = 5/13$
(c) $\sec \theta = 2$
(d) $\cos \theta = 8/17$
(e) $\csc \theta = -5/3$ in the third quadrant
(f) $\cot \theta = 2$ in the third quadrant
(g) $\sin \theta = -0.6$ in the fourth quadrant
(h) $\cos \theta = -0.8$ in the second quadrant

2. If θ is an angle in the second quadrant such that $\sin \theta = 0.6$, use the relations of this appendix to find each of the following:

   (a) $\tan \theta$
   (b) $\sin 2\theta$
   (c) $\tan 2\theta$
   (d) $\cos 2\theta$
   (e) $\sec \theta$
   (f) $\sin(\theta/2)$
   (g) $\cos(\theta/2)$
   (h) $\tan(\theta/2)$

3. Using the relations of this Appendix find each of the following formulas:

   (a) $\sin 3\theta = \sin(2\theta + \theta)$
   (b) $\cos 3\theta$
   (c) $\tan 3\theta$
   (d) $\sin 4\theta$
   (e) $\cos 5\theta$
   (f) $\sin(\alpha + \beta) + \sin(\alpha - \beta)$
   (g) $\cos(\alpha - \beta) - \cos(\alpha + \beta)$
   (h) $\cos(\alpha + \beta) + \cos(\alpha - \beta)$

4. Using the half angle formulas and the information given concerning the values of the special angles, find each of the following values:

   (a) $\sin(\pi/8)$
   (b) $\tan(5\pi/12)$
   (c) $\cos(11\pi/12)$
   (d) $\sin(-\pi/8)$
   (e) $\cos(\pi/12)$
   (f) $\sin(7\pi/12)$
   (g) $\tan(5\pi/8)$
   (h) $\cos(-\pi/12)$

5. Using the formulas for the sum and difference of angles and the short table of values given in this appendix, find each of the following:

   (a) $\sin(\pi/12)$
   (b) $\tan(7\pi/12)$
   (c) $\tan(2\pi/3)$
   (d) $\cos(5\pi/6)$
   (e) $\cos(2\pi/3)$
   (f) $\sin(7\pi/12)$

6. (a) Express $\sin 6x - \sin 2x$ as a product of trigonometric functions using the fact that $6x = 4x + 2x$ and $2x = 4x - 2x$.
   (b) Express $\cos 5x - \cos 3x$ as a product of trigonometric functions.
   (c) Express $\sin 3x + \sin x$ as a product of trigonometric functions.
   (d) Express $\cos 7x + \cos x$ as a product of trigonometric functions.

7. (a) Express $\cos 6x \sin x$ as a sum or difference of trigonometric functions through the use of the expressions for $\sin(6x + x)$ and $\sin(6x - x)$.

(b) Express $\cos 5x \cos x$ as a sum or difference of trigonometric functions using $\cos(5x + x)$ and $\cos(5x - x)$.

(c) Express $\sin 7x \sin 3x$ as a sum or difference of trigonometric functions.

8. (a) Show that $3 \sin 2x + 4 \cos 2x = 5 \sin(2x + \theta)$ where $\tan \theta = 4/3$. Note that $3^2 + 4^2 = 5^2$.

(b) Express $\sin 3x + \cos 3x$ in the form $[A \sin(3x + \theta)]$ and indicate how you could find the values for both A and θ.

9. Find $\sin(\alpha + \beta)$, $\sin(\alpha - \beta)$, $\cos(\alpha + \beta)$, $\cos(\alpha - \beta)$, $\tan(\alpha + \beta)$, and $\tan(\alpha - \beta)$ in each of the following situations. Assume angles are in the first quadrant unless stated otherwise.

(a) $\tan \alpha = 1/2$, $\tan \beta = 1/3$
(b) $\sin \alpha = 3/5$, $\cos \beta = 12/13$
(c) $\sin \alpha = 4/5$, $\cos \beta = 8/17$
(d) $\tan \alpha = 3/4$ (quad III), $\sin \beta = 4/5$
(e) $\cos \alpha = 3/5$ (quad IV), $\sin \beta = 7/25$ (quad II)
(f) $\sin \alpha = 24/25$, $\tan \alpha = 15/8$

Appendix B. Analytic Geometry

B.1 Introduction

As explained in Section I.8, there are many different coordinate systems. We shall concentrate on two of them, the rectangular coordinate system and the polar coordinate system. The first five sections of this appendix will discuss the *rectangular*, or *Cartesian*, *coordinate system*. The *polar coordinate system* will be discussed in the major portion of the remaining three sections. Throughout this appendix our discussion will be restricted to coordinate systems in the plane.

We start with the usual pair of mutually perpendicular axes, the horizontal one labeled the x-axis, and the vertical one labeled the y-axis. This is shown in Figure B.1. On the x-axis the positive direction is generally taken toward the right, and this will be the case here. In similar fashion the positive y-direction will be upward. These positive directions are designated by the arrowheads. The point of crossing of the axes is called the *origin*. This is the point for which both the x and y coordinates have the value zero. The one-to-one correspondence between points and coordinates is illustrated by the points $(2, 3)$, and $(-1, 2)$ in Figure B.1. The first coordinate in each case is the *x-coordinate* or *abscissa*, and represents the distance measured along the x-axis to a vertical line which will pass through the point. The second coordinate is the *y-coordinate* or *ordinate*, and this represents the distance along the y-axis to a horizontal line that will also pass through the point. The four regions into which the plane is divided by the two axes are called *quadrants*, and are denoted by number, starting with the quadrant in which both coordinates are positive as the first quadrant, and then moving counterclockwise to the second, third, and fourth quadrants. These are indicated by the Roman numerals in Figure B.1.

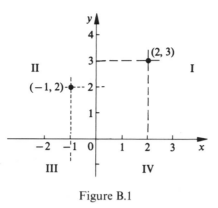

Figure B.1

In geometry, we are frequently concerned with the distance between points, and with the angle between lines. The distance is very easily obtained since we have only three possible cases to consider. The two points in question may determine a horizontal line, a vertical line, or some line which is neither horizontal nor vertical. If the two points determine a horizontal line, the y-coordinates for the two points will be equal, and it only takes a glance at the graph as shown in Figure B.2 to find out that if we take the difference of the two x-coordinates we have the length of the segment joining the two points. If we consider the segment to have a direction, we wish to subtract the starting coordinate from the ending or terminal coordinate. In this case, a positive distance would be toward the right. In some cases we do not worry about direction in the case of distance, but we shall be concerned about direction in other cases. A similar consideration would hold for the situation in which two points determine a vertical segment. If the two points determine some segment which is neither vertical nor horizontal, we can draw a vertical line through one end point and a horizontal line through the other. The point of intersection of these two lines will, with the two given points, form a right triangle. The length of the segment in question is the

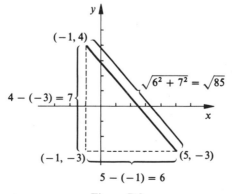

Figure B.2

length of the hypotenuse. Hence, by the Pythagorean Theorem we are able to ascertain that the length of the hypotenuse (the segment in question) is the square root of the sum of the squares of the length of the horizontal and vertical segments. In the case shown in Figure B.2, the horizontal segment has length $(5 - (-1)) = 6$ and the vertical segment has length $(4 - (-3)) = 7$. These were the lengths found using the method for horizontal and vertical segments. Therefore, the distance from $(-1, 4)$ to $(5, -3)$ is $\sqrt{6^2 + 7^2} = \sqrt{85}$. In general it is preferable to use the Pythagorean theorem to obtain the length of a segment (as we have done here) than to add another formula to those which you must remember.

As for the direction of a segment, we denote this by a ratio, following the methods used by surveyors in very early times. They considered the case in which a road might *rise* two feet while going thirty-five feet in the horizontal direction, or having a *run* of 35 feet. They would then say that the slope was 2/35, with the *rise over the run*. This means we will divide the vertical component of a segment by the horizontal component. In this case we will be concerned with the directions of the components involved. Since we wish the direction of a line we will take two points on the line, such as A and B, and we will divide the vertical distance from A to B by the horizontal distance from A to B. This quotient we will call the *slope*. Mathematicians have adopted the letter m to represent slope and we will use this designation here. Note that we have been very careful to concern ourselves with the directions of the horizontal and vertical segments in our determination of slope. Note the part the signs play in determining the ratios in Figure B.3. In the first case we have gone from A to B and in this case the y-coordinate has changed by an amount (-5) whereas the x-coordinate has changed by an amount $(+4)$. Hence the slope is $(-5)/(+4) = -5/4$. In the case of the segment from C to D we have, in similar fashion, the slope $(+3)/(+2) = 3/2$. If we had gone from B to A or from D to C we would not have altered the slope, for in that case we would have reversed the directions of both the numerator and the denominator as far as signs are concerned. It makes no

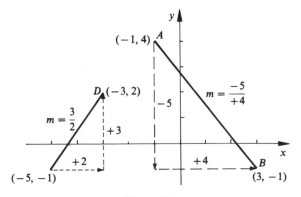

Figure B.3

difference which direction we select, but we must obtain the lengths of the corresponding x and y segments for the same direction along the given segment, that is both segments going from the A coordinates to the B coordinates, or vice versa.

If we have a horizontal line, there is no change in the y-coordinates as we go along any segment of the line, and therefore the fraction indicating the slope will have a zero numerator. Since the segment is presumed to have some length, it is evident that the slope would be zero in this case. If the segment is vertical, we have a non-zero numerator but a zero denominator, and in this case the slope fails to exist.

It has probably already occurred to you that the slope of the line, being the vertical distance over the horizontal, is precisely the tangent of the angle made by the segment in question with the positive x-axis. Thus, if we wished the angle that the line AB made with the x-axis, we would only have to find the angle which has $-5/4$ for its tangent, an angle of approximately $128°39'35''$. We could also have obtained the angle $-51°20'25''$, of course, for this would merely be the same line where we are proceeding in the opposite direction along the line. The fact that the slope of the line is the tangent of the angle made with the positive x-axis gives a very convenient means for finding the angle formed by two lines. We can obtain the slope of each line, or equivalently the tangent of the angle that the line makes with the positive x-axis for each line as shown in Figure B.4, and then use the information we have from trigonometry for the tangent of the difference of two angles. We note that line a has slope m_1 and makes an angle θ_1 with the positive x-axis, while line b has slope m_2 and makes an angle θ_2 with the positive x-axis. In order to better illustrate the angles involved, we have drawn a dashed line through the point of intersection parallel to the positive x-axis, and have indicated angles θ_1' and θ_2' which are equal to θ_1 and θ_2 respectively. From trigonometry we know that

$$\tan(\theta_2 - \theta_1) = \frac{\tan \theta_2 - \tan \theta_1}{1 + \tan \theta_2 \tan \theta_1}.$$

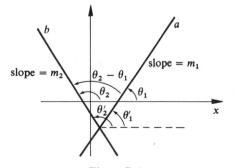

Figure B.4

Since $\tan \theta_1 = m_1$ and $\tan \theta_2 = m_2$, this becomes

$$\tan(\theta_2 - \theta_1) = \frac{m_2 - m_1}{1 + m_2 m_1}.$$

Thus, we have a formula for the tangent of the angle between two lines. We note that the only possible difficulty here occurs when $(1 + m_2 m_1) = 0$, for then the tangent fails to exist, but this only happens when $(\theta_2 - \theta_1)$ is a right angle. Hence, the condition for perpendicularity of the two lines is obtained as $m_2 m_1 = -1$, or $m_2 = -1/m_1$. Restating this, two lines are perpendicular if their slopes are negative reciprocals of each other. Since $\tan 0 = 0$, we note that if two lines are parallel (or make the same angle with the positive x-axis) then the slopes are equal.

EXERCISES

1. (a) Plot the points $A(2, 4)$, $B(-1, 4)$, $C(-3, -2)$, and $D(0, 1)$.
   (b) Find the slopes of AB, AC, AD, BC, BD, and CD.
   (c) Which pairs of segments of part (b) (if any) are parallel?
   (d) Which pairs of segments of part (b) (if any) are perpendicular?
   (e) Find the area of the quadrilateral $ABCD$.
   (f) Find the lengths of each of the segments of part (b).
   (g) Find the area of triangle ABC.
   (h) Find the area of triangle BCD.
   (i) Find the tangents of each of the angles of triangle ABC.

2. (a) Find an equation which determines all points (x, y) which are twice as far from the point $(2, 1)$ as from the point $(-1, 2)$.
   (b) Find the slope of the segment joining $(2, 1)$ and $(-1, 2)$.
   (c) Find the length of the segment joining $(2, 1)$ and $(-1, 2)$.
   (d) Find the midpoint of the segment of parts (b) and (c).
   (e) Find the slope of the line perpendicular to the segment of the line of parts (b) and (c).

3. (a) What is the slope of the segment joining $(-1, 5)$ and $(3, -2)$.
   (b) Find the coordinates of some point P such that the segment joining P and $(-1, 5)$ is perpendicular to the segment of part (a).
   (c) Find the length of the segment joining $(-1, 5)$ to P and the length of the segment of part (a).
   (d) Find the area of the triangle with vertices at P, $(-1, 5)$, and $(3, -2)$.
   (e) Find the length of the segment joining P and $(3, -2)$.
   (f) Determine whether the triangle of part (d) is a right triangle, and if it is, check by using the Pythagorean theorem.

4. (a) Determine four points by giving their coordinates such that the four points are the vertices of a square and such that none of the six segments determined by the four points are horizontal or vertical.
   (b) Use slopes to demonstrate that you have right angles at each of the vertices of the figure of part (a).

(c) Find the length of each of the six segments determined by the four points of part (a). Show that the sides of your figure are all equal in length.

(d) Show that the diagonals of your figure are $\sqrt{2}$ times the length of the sides of the figure.

(e) Find the tangents of the angles between the sides of the figure and the diagonals, and show that each tangent is equal to one, hence the angles are each $(\pi/4)$.

5. (a) Show that the point $(2, 3)$ is at the intersection of the vertical line $x = 2$ and the horizontal line $y = 3$.

(b) Show that the relation of part (a) applies for any set of rectangular coordinates.

B.2 Lines

We know from geometry that a *line* is determined if we are given a point on the line and the direction of the line or if we are given two points. We will start with the former case and develop the latter. Note in Figure B.5 we have the point (x_1, y_1), and we can think of the point (x, y) as being any other point on the line. If we can obtain a description which will fit each such point (x, y), we will have given the conditions which will distinguish between those points on the line and those not on the line. Since, as shown in Figure B.5, the slope of the segment joining (x, y) and (x_1, y_1) is $(y - y_1)/(x - x_1)$, and since it is presumed that the direction, hence the slope, is given, (we will call it m), we have the relationship

$$\frac{y - y_1}{x - x_1} = m. \tag{B.2.1}$$

It follows that

$$y = m(x - x_2) + y_1. \tag{B.2.2}$$

We now have an equation describing the line for which the coordinates of a point on the line and also the slope are known. If the point on the line is on the x-axis, say $(0, b)$ where b is called the *y-intercept*, then substitution gives us

$$y = mx + b. \tag{B.2.3}$$

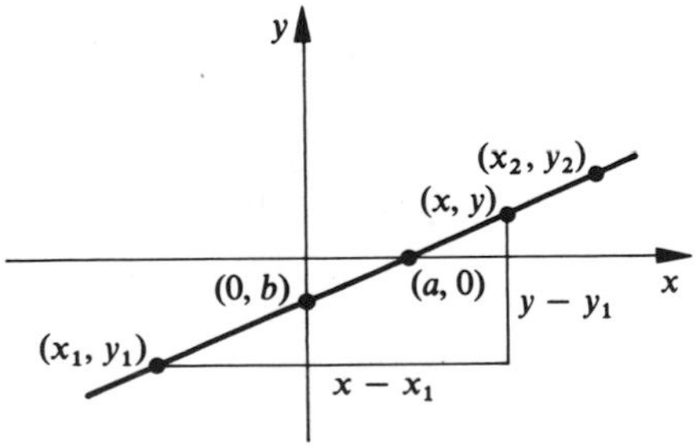

Figure B.5

This is frequently called the *slope-intercept* form of the line. Either (B.2.1) or (B.2.2) is called the *point slope* form of the line. If we have two points given, say (x_1, y_1) and (x_2, y_2), these two points determine the slope $(y_2 - y_1)/(x_2 - x_1)$ of the segment. Using the point-slope form we have

$$\frac{(y - y_1)}{(x - x_1)} = \frac{(y_2 - y_1)}{(x_2 - x_1)}, \tag{B.2.4}$$

a form commonly called the *two point* form.

If instead of being given sufficient geometric facts to determine the line, and hence the equation, we are given the equation, we can work in reverse order, for we can solve for y, knowing from (B.2.3) that the coefficient of x is the slope and the constant term is the point at which the line crosses the y-axis. If the slope is zero, that is the x term does not appear, the line is horizontal, and if the solution for y would call for dividing by zero, the line must be vertical. This, of course, is based upon the presumption that all variables appear only to the first power, for otherwise we do not have the equation of a line. Equations of lines have at most one x term, one y term, and one constant, with all the variables appearing just to the first power. For this reason any equation such that variables appear only to the first power is called *linear*.

We discussed the angle between two lines in Section B.1, and we can obtain the direction of a line by means of the slope. We can also determine whether two lines are parallel by seeing whether the slopes are equal. The question concerning the distance from a point to a line has not been answered, however. This distance would be measured along a line through the point perpendicular to the given line. Consider the line $ax + by = c$ and the point (x_1, y_1). We will carry out the computation in two parts. We will first find a method for determining how far a line is from the origin, and then we will note that the distance from (x_1, y_1) to $ax + by = c$ is the same as the distance from the line $ax + by = c$ to the parallel line $ax + by = ax_1 + by_1$ which passes through the point (x_1, y_1). Our first concern, therefore, will be the distance between the origin and $ax + by = c$. In Figure B.6 we have shown the line, and we know that the line crosses the x and y axes at c/a and c/b respectively. Let d denote the distance from the origin to the line measured along the segment perpendicular to the line. Observe that $\triangle OAB$ is similar to $\triangle DOB$, and hence the sides of the two triangles are proportional. Thus we have

$$\frac{DO}{BO} = \frac{OA}{AB}$$

which requires

$$\frac{d}{c/b} = \frac{c/a}{\left[\left(\frac{c}{a}\right)^2 + \left(\frac{c}{b}\right)^2\right]^{1/2}} = \frac{c/a}{\frac{c}{ab}\sqrt{a^2 + b^2}}$$

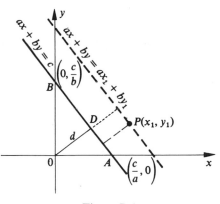

Figure B.6

or

$$d = \frac{c}{\sqrt{a^2 + b^2}}. \tag{B.2.5}$$

Now the distance from the line $ax + by = ax_1 + by_1$ to the origin is then

$$d_1 = \frac{ax_1 + by_1}{\sqrt{a^2 + b^2}}$$

since the constant term in this case is not c but is $(ax_1 + by_1)$. The distance from (x_1, y_1) to the line, then, is the distance $d_1 - d$ or

$$\frac{ax_1 + by_1 - c}{\sqrt{a^2 + b^2}}. \tag{B.2.6}$$

The form

$$\frac{ax + by - c}{\sqrt{a^2 + b^2}} = 0$$

is called the *polar form* of the straight line. It is frequently used to obtain the distance from a point to a line.

EXAMPLE 2.1. Sketch the line having the equation $5x - 12y = 26$. Also find the distance between this line and the origin and the distance between this line and the point $(3, 2)$.

Solution. In order to sketch this line we can first solve for y. Thus we obtain $y = (5/12)x - (26/12)$. From this we see that the line has a y-intercept of $(-26/12)$ and a slope of $(5/12)$. We first indicate the y-intercept in Figure B.7 as the point where the line intersects the y-axis. We next obtain the line with the correct slope by going one unit to the right and then permitting y

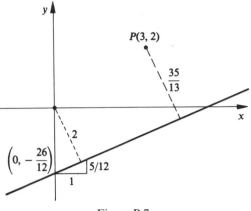

Figure B.7

to change by the amount of the slope, that is 5/12. This is also shown in Figure B.7.

We can find the distance from the origin through the use of (B.2.5). In this case we have $d = 26/\sqrt{5^2 + 12^2} = 26/13 = 2$ units. This distance is also indicated in the figure. Now to find the distance from the point $(3, 2)$, we can use (B.2.6) and we will have the distance

$$\frac{(5)(3) - (12)(2) - 26}{13} = \frac{-35}{13}.$$

Therefore there is a distance of 2 units from the origin and a distance of (35/13) units from the point $(3, 2)$ to the line. If one wishes to make the sign of the radical in the denominator the same as the sign of the constant term, a positive sign would indicate that the point in question and the origin are on the same side of the line with which we are concerned. A negative sign in this instance would indicate that the point and the origin are on opposite sides of the line. If we wished to obtain this additional information in this case, we would have had to make the denominator negative since the constant term appeared in the numerator with a negative sign. In this case our distance would have been positive, indicating that $(3, 2)$ and the origin are on the same side of the line, as indicated in Figure B.7.

EXERCISES

1. Find the equation of each of the lines determined by the given point and the given slope. Sketch each of the lines.

   (a) $(2, 3)$ and slope -2
   (b) $(-1, 2)$ and slope $2/3$
   (c) $(1, -1)$ and slope 0

(d) $(3, 2)$ and non-existent slope
(e) $(-2, -3)$ and slope 3
(f) $(0, 2)$ and slope $-2/3$
(g) $(3, 0)$ and slope $-7/3$
(h) $(0, 0)$ and slope -2

2. Find the equation of the lines determined by each pair of points. Sketch the lines.

   (a) $(2, 3)$ and $(1, -1)$
   (b) $(-2, 3)$ and $(3, -2)$
   (c) $(2, -1)$ and $(-2, -1)$
   (d) $(-2, -1)$ and $(-2, 1)$
   (e) $(2, 0)$ and $(0, 3)$
   (f) $(4, 0)$ and $(0, -5)$
   (g) $(a, 0)$ and $(0, b)$
   (h) $(0, 0)$ and $(2, -2)$

3. Graph each of the following lines:

   (a) $2x - 3y = 6$
   (b) $3x + y = -7$
   (c) $2x + 5y + 10 = 0$
   (d) $11x - 5y - 13 = 0$
   (e) $4x = 17$
   (f) $2y + 5 = 0$
   (g) $x = y$
   (h) $x + y = 0$

4. (a) Find the slope of $3x - 4y = 7$ and sketch the line.
   (b) Find the slope of $2x + y = 5$ and sketch the line.
   (c) Find the point of intersection of the lines of parts (a) and (b).
   (d) Find the angle formed by going counterclockwise from the line of part (a) to the line of part (b).
   (e) Find a point other than the point of part (c) which is the same distance from the line of part (a) as from the line of part (b).
   (f) Find the equation of the line joining the point of part (c) to the point of part (e).
   (g) Show that the line of part (f) bisects one of the two angles formed by the lines of parts (a) and (b).
   (h) Find the equation of the line through the point of intersection of part (c) which is perpendicular to the line of part (f).
   (i) Show that the line of part (h) bisects the other one of the two angles formed by the lines of parts (a) and (b).

5. (a) Find the distance from $(1, -2)$ to the line $2x - 3y = 13$.
   (b) Find the equation of the line through $(1, -2)$ perpendicular to the line $2x - 3y = 13$.
   (c) Find the point of intersection of the line of part (b) and the line $2x - 3y = 13$.
   (d) Find the distance between $(1, -2)$ and the point of part (c).
   (e) Compare the result of part (d) with that of part (a) and explain their equality or lack of equality.

6. Find the equation of the perpendicular bisector of the segments joining each of the following pairs of points and sketch the graph of the segment and the perpendicular bisector.

   (a) $(2, 3)$ and $(5, -1)$
   (b) $(-2, -4)$ and $(6, 0)$
   (c) $(-2, 3)$ and $(3, -1)$
   (d) $(0, 0)$ and $(2, -4)$
   (e) $(2, 2)$ and $(6, 2)$
   (f) $(-1, -5)$ and $(-1, -9)$
   (g) $(1. 1)$ and $(-1, -1)$
   (h) $(-2, -3)$ and $(-4, -6)$

B.3   Circles

The *circle* is defined as the path of all points located at a given distance from a fixed point called the *center* of the circle. The given distance is called the *radius*. Thus, if we have the point (x_1, y_1) and the distance r, we would have all points on the circle (x, y) described by the relation

$$(x - x_1)^2 + (y - y_1)^2 = r^2, \qquad (B.3.1)$$

where we have squared both sides of the distance formula in order to avoid square roots. If the center were located at the origin, this would reduce to

$$x^2 + y^2 = r^2. \qquad (B.3.2)$$

Note that in (B.3.1) we had essentially moved the circle from the position with center at the origin, as given in (B.3.2) to the position in which the center is at (x_1, y_1) by replacing x by $(x - x_1)$ or in other words starting all x-measurements from the value x_1, and similarly replacing y by $(y - y_1)$ or starting all y-measurements from y_1. When we have moved the curve in question either horizontally or vertically, (or some of each) but without any rotation or distortion, we say that we have *translated* the curve. This is equivalent to considering the curve relative to a new origin, which is generally not drawn in. It is possible to consider all circles as though they were of the form $x^2 + y^2 = r^2$, with the provision that there may be a *translation*.

If we are given the equation of a circle, we can locate it on the graph by reversing our procedures. Thus, if we are given the equation

$$x^2 + y^2 + 4x - 6y = 12,$$

we can write

$$(x^2 + 4x + 4) + (y^2 - 6y + 9) = 12 + 4 + 9$$

or

$$(x + 2)^2 + (y - 3)^2 = 25.$$

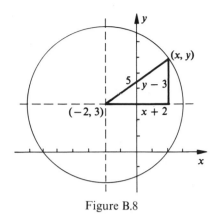

Figure B.8

This is apparently a circle, and the radius is 5. Since the expression $(y - 3)$ is zero when $y = 3$, the y-coordinate of the center must be 3, a fact which could easily be deduced by comparing with (B.3.1). When $(x + 2)$ is zero, we have $x = -2$, and we note that in fact $(x + 2) = (x - (-2))$, whence the x-coordinate of the center is -2. Therefore, the center is located at $(-2, 3)$ and we can easily draw the circle as shown in Figure B.8. Since (B.3.1) represents the most general case of the circle, we see that the most general equation of the circle in expanded form will contain both x^2 and y^2, and they must appear with equal coefficients. It is not possible to have a term here containing the product (xy). Therefore, if we have a circle we will always have the type of equation given above in which we can complete the square and the above procedure can be followed.

EXERCISES

1. Find the equation of the circles having the center and the radius as indicated:

    (a) $(-2, 3)$ with radius 4
    (b) $(5, 2)$ with radius 5
    (c) $(0, 4)$ with radius 4
    (d) $(-2, -4)$ with radius 3
    (e) $(1, -1)$ with radius 5
    (f) $(3, 3)$ with radius 3
    (g) $(-5, 0)$ with radius 5
    (h) $(-3, -4)$ with radius 5

2. Graph each of the following circles.

    (a) $x^2 + y^2 - 9 = 0$
    (b) $x^2 + y^2 - 4x + 12y + 4 = 0$
    (c) $x^2 + y^2 - 10x + 26y = 0$
    (d) $x^2 + y^2 - 4x + 6y - 13 = 0$
    (e) $x^2 + y^2 + 5x - 7y = 25$
    (f) $2x^2 + 2y^2 - 5x + 6y - 50 = 0$

3. (a) Find the points of intersection of the circles $x^2 + y^2 - 2x - 4 = 0$ and $x^2 + y^2 - 6x - 7 = 0$.
   (b) Find the equation of the line through the two points found in part (a). This line is called the *radical axis* of the two circles.
   (c) Subtract the second equation of part (a) from the first in order to obtain a first degree equation.
   (d) Compare your result in part (b) with your result in part (c).
   (e) Explain the similarity or lack of similarity of the two results compared in part (d).

4. Find the radical axis for each of the following pairs of circles. Sketch the circles and the radical axis in each case.

   (a) $x^2 + y^2 = 25$ and $x^2 + y^2 - 6x - 4y - 12 = 0$
   (b) $x^2 + y^2 - 4x = 0$ and $x^2 + y^2 - 4y = 0$
   (c) $x^2 + y^2 - 2x + 2y - 14 = 0$ and $4x^2 + 4y^2 - 4x - 4y - 14 = 0$

5. (a) Find the two points of intersection of $x^2 + y^2 - 4x - 6y = 24$ and $x^2 + y^2 - 8y = 0$.
   (b) Find the perpendicular bisector of the segment joining the two points of part (a).
   (c) Find the equation of the line through the centers of the two circles of part (a).
   (d) Compare the results in parts (b) and (c).

6. An equation such as $x^2 + y^2 + 16 = 0$ is the equation of what is sometimes called an *imaginary circle*. Explain what is meant by this term. Also indicate how you can be sure that there are no pairs of coordinates that satisfy this equation.

7. Any line with equation $y = mx + 2$ passes through the point $(0, 2)$.
   (a) Find m such that the line through $(0, 2)$ intersects the circle $x^2 + y^2 + 4x + 6y - 12 = 0$ in just one point.
   (b) Show that the two values of m you obtain in part a give two lines, each of which is tangent to the circle.
   (c) Sketch the graph of the circle and the two lines of part (b) to show that these lines are tangent to the circle.

B.4 The Conic Sections

We will now turn our attention to a family of curves which have one property in common. In each case there is a fixed point called the *focus* and a fixed line which does not pass through the focus. We call the line the *directrix*. The curves which we will discuss have the property that for each point on the curve the ratio of the distance from the focus to the distance from the directrix will be constant. The constant will vary from curve to curve, but it will be a constant for any given curve. This constant is called the *eccentricity*. Since we can always translate a curve we will derive the equations of these curves only for the locations which are most convenient from the standpoint of algebra. In the first case we will start with an eccentricity of one. The

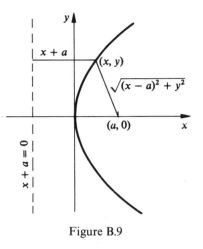

Figure B.9

directrix will have the equation $x + a = 0$ and the focus will be located at coordinates $(a, 0)$. This is shown in Figure B.9. The directrix is vertical, and the focus is on the x-axis. Since the eccentricity is one, we desire the equation which will describe all points whose location is such that the points are equally far from the directrix and from the focus. If (x, y) is such a point, the distance from the directrix, since this is measured horizontally, is $(x - (-a)) = (x + a)$. (Note the advantage we gain by picking the directrix as a vertical line.) The distance from the focus is $[(x - a)^2 + y^2]^{1/2}$, using the distance formula from Section B.1. Since the requirement for acceptable points is that these two distances be equal, we have

$$[(x - a)^2 + y^2]^{1/2} = (x + a).$$

Squaring both sides gives

$$x^2 - 2ax + a^2 + y^2 = x^2 + 2ax + a^2.$$

Now the merits of our selection of directrix and focus appear, for upon simplifying, we have

$$y^2 = 4ax. \tag{B.4.1}$$

We note at once that if $a > 0$, we cannot have any negative values of x, for this would require that $y^2 < 0$. This is not possible for real values of y. Furthermore, we note that for any acceptable value of x we have two values of y, one positive and one negative, except for the case $x = 0$ in which we have $y = +0$ and $y = -0$, and hence a single point. This curve is called the *parabola*. In this case we were given the fact that

$$\frac{\text{focal distance}}{\text{distance from directrix}} = \text{eccentricity} = 1. \tag{B.4.2}$$

The point on the parabola nearest the directrix, namely $(0, 0)$, is called the *vertex*. If the vertex is moved by translation to (x_1, y_1) we have

$$(y - y_1)^2 = 4a(x - x_1). \qquad (B.4.3)$$

It is possible to have the parabola open along the y-axis instead of along the x-axis. In this case the role of the two variables is reversed, giving us

$$(x - x_1)^2 = 4a(y - y_1).$$

After multiplication we find that we have a term involving x^2 or y^2 but not both. All other terms are of first degree. This is a clue for recognizing a parabola, and it is also a clue for determining which parabola we have. Suppose we are given the equation

$$x^2 + 4x + 8y = 0.$$

Upon completing the square in a manner similar to that used for the circle, we have

$$(x + 2)^2 = -8y + 4 = -8(y - (1/2)).$$

Thus we know that the vertex is at the point $(-2, 1/2)$, and the value of a is -2. Furthermore, the axis of the parabola is vertical, and since a is negative, the focus must be below the vertex. This parabola is drawn in Figure B.10. We have also shown the focus and directrix.

It might be interesting to note that the headlights on a car have a reflector in the form of a paraboloid of revolution with the filament located at the focus. This causes, except for stray reflections, all light rays to reflect parallel to the axis of the parabola, and hence to remain concentrated as they shine down the highway. There are many other uses of the parabola. Consequently, this is a curve one sees often in applications.

We now proceed to the case in which the eccentricity is a positive number other than one. Again we select a convenient location on the axes for the

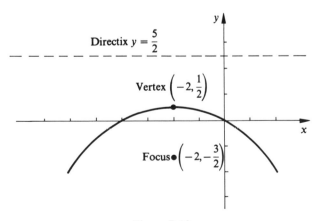

Figure B.10

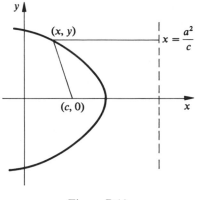

Figure B.11

purpose of simplifying the derivation since translation will permit us to use other locations as we see fit. This time we select as focus the point $(c, 0)$ and as directrix the line $x = a^2/c$. This is shown in Figure B.11. We will let our eccentricity be c/a. Now if (x, y) is a general point, we have the distance from the directrix given by $((a^2/c) - x)$ and the focal distance is $[(x - c)^2 + y^2]^{1/2}$. Thus, we have

$$\frac{[(x - c)^2 + y^2]^{1/2}}{\left(\dfrac{a^2}{c} - x\right)} = \frac{c}{a}.$$

Upon simplification, we obtain

$$a[(x - c)^2 + y^2]^{1/2} = (a^2 - cx).$$

Squaring both sides, we have

$$a^2x^2 - 2a^2xc + a^2c^2 + a^2y^2 = a^4 - 2a^2xc + c^2x^2.$$

This reduces to

$$(a^2 - c^2)x^2 + a^2y^2 = a^2(a^2 - c^2). \tag{B.4.4}$$

Note, incidentally, that whether we used $((a^2/c) - x)$ or $(x - (a^2c))$ made no difference, for we squared this term in our derivation, and both expressions have the same value for their square.

We now come to a moment of decision. We have drawn Figure B.11 as though $c < a$, but we have said nothing that would fail if $c > a$. We observe that if $c > a$, the focus would be further from the origin than the directrix. The point (x, y) would be changed, too, of course, to correspond to the change in relative position of the focus and directrix. Note that if the eccentricity were one, that is $c = a$, then the equation would reduce to $a^2y^2 = 0$, or equivalently we would have only the x-axis. In this case, however, using the assumptions for Figure B.11, the focus would be a point on the directrix, and this would be the only set of points obeying the conditions of our problem. (This would be a *degenerate parabola*.) We will, then, ignore the case where $c = a$, and take in turn the other two cases. If $c < a$, then $(a^2 - c^2) > 0$, and we can replace this cumbersome expression by b^2, knowing that b is a real number. Hence, we have

$$\frac{x^2}{a^2} + \frac{y^2}{b^2} = 1. \tag{B.4.5}$$

This is the standard form for the *ellipse*. The focus is located at $(c, 0)$, but since $a^2 = b^2 + c^2$, this can be written $(\sqrt{a^2 - b^2}, 0)$. In view of the *symmetry* of this curve with respect to both the x-axis and the y-axis as evidenced by the fact that if (x, y) is a point on the curve, so are $(x, -y)$, $(-x, y)$ and $(-x, -y)$, we see that there should be another focus and another directrix located symmetrically with respect to the y-axis. The ellipse has the interesting property that the sum of the distance from any point on the ellipse to one focus plus the distance from the point to the other focus is a constant, and is equal to $2a$. The distance a is called the *semi-major axis* and the distance b is called the *semi-minor axis*. The ends of the major axes are called the *vertices*.

As indicated before, we can obtain any ellipse with horizontal and vertical axes from (B.4.5) by translation, with the understanding (not difficult to see) that if the denominator of y^2 in standard form is larger, the ellipse looks like an egg standing on end. Consider the equation

$$4x^2 + 9y^2 - 16x + 18y - 11 = 0.$$

We can complete the square and obtain

$$4(x^2 - 4x + 4) + 9(y^2 + 2y + 1) = 11 + 4(4) + 9(1) = 36,$$

whence we have

$$\frac{(x - 2)^2}{9} + \frac{(y + 1)^2}{4} = 1.$$

This is an ellipse with center not at the origin, but at the point $(2, -1)$, and having a semi-major axis of 3 and a semi-minor axis of 2. The distance from the center to the focus is $\sqrt{9 - 4} = \sqrt{5}$. The graph is shown in Figure B.12. Note that if you start at the center, draw in the axes of the ellipse and find the ends of the major and minor axes, it is rather easy to draw this figure. Note the location of the foci (plural of focus) and the right triangle that can be used to establish the location of a focus. A similar construction would hold for an ellipse with a vertical major axis. We obtain the ellipse provided we have both x^2 and y^2, and they have coefficients of the same sign. The circle can be considered a special case of the ellipse in which the two foci have come together and coincided. Here, since $c = 0$, we would have an eccentricity of zero.

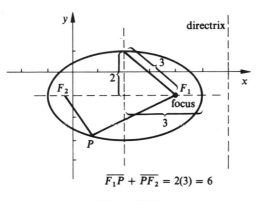

$$\overline{F_1 P} + \overline{PF_2} = 2(3) = 6$$

Figure B.12

We now take up the case in which $c > a$. Here we have $c^2 - a^2 > 0$, and we could replace $c^2 - a^2$ by b^2 to obtain, upon simplification,

$$\frac{x^2}{a^2} - \frac{y^2}{b^2} = 1. \tag{B.4.6}$$

In this case we note that $c^2 = a^2 + b^2$, and our eccentricity is greater than one. This yields an *hyperbola* having a *semi-transverse axis* of a and a *semi-conjugate axis* of b. The semi-transverse axis is determined here not by size, but by the fact that it is the square root of the denominator of the positive term when the equation is in the standard form indicated by (B.4.6). The ends of the transverse axis are called the *vertices*. We will now consider the equation

$$4x^2 - 9y^2 - 16x - 18y - 29 = 0.$$

If we complete the squares, as before, we obtain

$$4(x^2 - 4x + 4) - 9(y^2 + 2y + 1) = 29 + 4(4) - 9(1) = 36$$

or

$$\frac{(x - 2)^2}{9} - \frac{(y + 1)^2}{4} = 1.$$

As shown in Figure B.13, this gives us an hyperbola with center at $(2, -1)$ and with a horizontal axis. The semi-transverse axis is 3 and the semi-conjugate axis is 2. Since we obtain c^2 by $c^2 = 3^2 + 2^2$, we would construct the right triangle somewhat differently from the method used in the case of the ellipse. Here we have made a rectangle using the points at the ends of the semi-axes, and have drawn the diagonals of the rectangle. These are the *asymptotes* of the hyperbola. The hyperbola approaches them as it gets further from its center. Note the placement of the foci, as indicated by the arc of the circle. As for the asymptotes, we might observe that if we

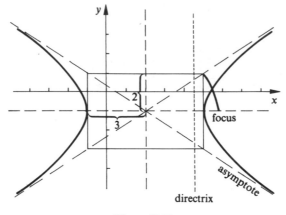

Figure B.13

were to divide both sides by x^2 (and this is all right, since we will be con-sidering what happens as x increases in size), we will obtain

$$\frac{1}{a^2} - \frac{y^2}{b^2x^2} = \frac{1}{x^2}.$$

As x increases the right hand side assumes insignificant proportions. Hence, we would have essentially the equation

$$\frac{1}{a^2} - \frac{y^2}{b^2x^2} = 0.$$

Upon factoring and clearing fractions, we have $(bx - ay)(bx + ay) = 0$. Each of these factors gives a line through the center, and these are the two asymptotes. We have used the standard form here, but the translation re-quired for our examples would easily yield the two equations

$$2(x - 2) - 3(y + 1) = 0$$

and

$$2(x - 2) + 3(y + 1) = 0$$

as the equations of the asymptotes. Since for all values the product

$$(bx - ay)(bx + ay)$$

would have to be positive, this would tell us that although the curve gets closer to these lines, the points on the curve must always be such that this product is positive, and hence in the region shown in Figure B.13.

The three curves, the parabola, the ellipse with its special case of the circle, and the hyperbola were known to the Greeks to be related to the *cone*. This can be shown by taking a right circular cone with its two parts (or *nappes*), and cutting the cone with a plane as in Figure B.14. If the plane cuts the cone perpendicular to the axis we have the circle. If the plane is

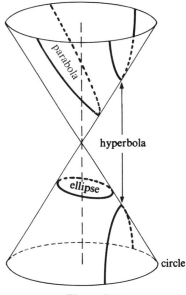

Figure B.14

such that it cuts completely across one of the nappes, but is not perpendicular to the axis, we would have an ellipse which is not a circle. If the plane is parallel to the edge of the cone (technically an element of the cone) then we see that the plane would never go completely through one nappe, nor would it ever cut the other nappe. This would give a parabola. If the plane were to cut both nappes, then we would have the hyperbola. If you are very energetic, you might take some paper and make a cone, being sure to make both nappes, and then draw the curve of intersection which a plane would make with the cone to illustrate these curves, called the *conic sections*.

Note from the standpoint of the equations that if we have either x^2 or y^2 but not both, the equation can be reduced to a parabola. If we have both x^2 and y^2 and if their coefficients have the same sign, we have an ellipse. If the coefficients are equal we have a circle. If we have opposite signs for the coefficients, we have the hyperbola. This is all based on the absence of a term involving the product (xy). If such a product term is present, then these rules are no longer valid, and we will see how to handle this in the last section of this appendix.

EXERCISES

1. Find the equation of the parabolas having the indicated properties.
   (a) Focus $(2, 0)$ and directrix $x + 2 = 0$
   (b) Focus $(2, 3)$ and directrix $y = 5$
   (c) Focus $(-2, 1)$ and directrix $x + 8 = 0$
   (d) Focus $(3, 3)$ and vertex $(3, 2)$

(e) Focus $(0, 2)$ and directrix $y - 4 = 0$
(f) Vertex $(2, 3)$ and directrix $x + 1 = 0$
(g) Focus $(1, -1)$ and vertex $(-1, -1)$

2. Sketch the graph of each of the following:
   (a) $x^2 = 8y$
   (b) $y^2 = 6x$
   (c) $x^2 - 6x - 2y + 1 = 0$
   (d) $4y^2 + 4x - 20y + 25 = 0$
   (e) $2x^2 + 4x - 5y + 7 = 0$
   (f) $x^2 + 2x - 6y - 17 = 0$

3. (a) Find the points at which the line $x = a$ intersects the parabola $y^2 = 4ax$.
   (b) Find the distance between the two points found in part (a). The segment between these points is called the *latus rectum* of the parabola.
   (c) Find the distance of each of the points of part (a) from the focus and also find the distance from the directrix.
   (d) Show that the points at the end of the latus rectum satisfy the definition of the parabola.

4. Find the equation of the ellipses having the indicated properties:
   (a) Center at $(0, 0)$, focus at $(0, 3)$ and semi-major axis of length 5.
   (b) Focus at $(3, 1)$, center at $(3, -1)$ and semi-major axis 4.
   (c) Foci at $(3, 1)$ and $(3, 5)$ with semi-minor axis 4.
   (d) Vertices $(0, 4)$ and $(0, -2)$ with focus at $(0, 3)$.

5. Sketch the graph of each of the following:
   (a) $x^2 + 4y^2 = 16$
   (b) $9x^2 + y^2 = 9$
   (c) $4x^2 + 5y^2 - 80y + 300 = 0$
   (d) $2x^2 + 3y^2 - 4x - 12y + 14 = 0$
   (e) $9x^2 + 4y^2 - 36x + 8y + 4 = 0$
   (f) $25x^2 + 16y^2 + 150x - 96y = 31$

6. Find the equation of the curve and sketch the curve such that the distance from $(0, 3)$ to a point (x, y) on the curve plus the distance from (x, y) to $(8, 3)$ is always equal to 10.

7. Find the equation of the hyperbolas having the indicated properties.
   (a) The foci are $(5, 0)$ and $(-5, 0)$ and a vertex at $(3, 0)$
   (b) A focus at $(0, 10)$ and vertices at $(0, 6)$ and $(0, -6)$
   (c) Foci at $(2, 1)$ and $(-2, 1)$ with semi-transverse axis of 1
   (d) Focus at $(4, 4)$ with asymptotes $y = x + 4$ and $y = -x + 4$.

8. Sketch the graph of each of the following:
   (a) $x^2 - 4y^2 - 4 = 0$
   (b) $x^2 - 4y^2 + 4 = 0$
   (c) $x^2 - 2y^2 - 6x - 4y + 7 = 0$

(d) $4x^2 - 9y^2 - 8x - 36y = 68$
(e) $x^2 - 4y^2 - 2x - 12y + 8 = 0$
(f) $4x^2 - 9y^2 - 16x - 54y - 101 = 0$

9. Sketch each of the following curves:
(a) $4x^2 + 4y - 4x - 3 = 0$
(b) $4x^2 + 4y^2 - 4x + 4y - 7 = 0$
(c) $4x^2 + y^2 - 4x + 4y - 11 = 0$
(d) $4x^2 - y^2 - 4x + 4y - 19 = 0$
(e) $4x^2 - y^2 - 4x + 4y + 13 = 0$

B.5 More Complicated Curves

We have discussed the problem of drawing the graphs of curves determined by equations of first and second degree (provided there is no xy term in the second degree equations) for the case involving rectangular coordinates. We will now take a rather brief look at curves having somewhat more complicated equations. (We will reserve the case of the second degree curves with the xy term for the last section of this appendix.) We will not attempt a detailed analysis as we did in our treatment of the conic sections, but we will look for certain helpful hints that can be determined from the equation. Then with a bit of sleuthing we will attempt to see where the curve would have to go. Of course, it is always possible to plot a large number of points and then draw the curve, but this is tedious, and it is usually sufficient to have a rather decent idea of the general shape of the curve. We will proceed via the method of looking at examples.

EXAMPLE 5.1. Sketch the graph of $y = x/(x^2 + 1)$.

Solution. First we will find where this curve crosses the axes. We note that if $x = 0$, then $y = 0$, and similarly if $y = 0$, then $x = 0$. Consequently the only point at which the graph crosses either axis is the origin. Secondly, we will note that if the point (x, y) is a point on the graph (that is if we substitute coordinates (x, y) in the given equation, the equality holds), then $(-x, -y)$ is also a point on the graph by direct substitution. Thus, if we can draw half of the curve, the other half can be obtained by drawing line segments from any point on the curve to the origin, extending them an equal length on the other side of the origin, and thus obtaining another point on the curve. We also note that if $x > 0$, then $y > 0$, and if $x < 0$, then $y < 0$. This latter information can be deduced by merely noting that if $x > 0$, then $y > 0$, for the symmetry with respect to the origin would then yield the information concerning negative values. We also note that as x gets larger, either in the positive or negative direction—hence, if x gets larger in absolute value—then y gets closer to zero, for the denominator will be larger

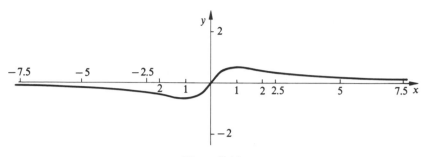

Figure B.15

than the numerator in absolute value. Thus, the curve approaches the x-axis both on the right and on the left as $|x|$ increases, and the x-axis is an asymptote. Finally, we observe that since

$$x^2 y - x + y = 0,$$

we have

$$x = \frac{1 \pm \sqrt{1 - 4y^2}}{2y}$$

indicating that we should not have $1 - 4y^2 < 0$, since we must take the square root of this quantity. Hence $|y| \leq 1/2$. When $|y| = 1/2$, we have $|x| = 1$. Hence we have the points $(1, 1/2)$ and $(-1, -1/2)$ as the highest and lowest points on the graph, respectively. With all of this information, we have sketched the curve in Figure B.15. Note that no points are used in the second and fourth quadrants, for we found that x and y must have the same sign. We also found that the only time that the curve intersected an axis was at the origin, and that the x-axis is an asymptote. Thus, with a minimum of point plotting, we have drawn the curve, known in mathematical literature as the *serpentine curve*.

EXAMPLE 5.2. Sketch the graph of $y = x^4 - x^2$.

Solution. Here we note that if $y = 0$, we have $x^2(x^2 - 1) = 0$ and $x = 0$, $x = 1$, or $x = -1$. Thus, the curve crosses the x-axis at three points, namely $(0, 0)$, $(1, 0)$, and $(-1, 0)$. If $x = 0$, then $y = 0$ and we have not added any new intercepts to the list. If (x, y) is a point on the curve, then $(-x, y)$ is also a point on the curve, and hence we have symmetry with respect to the y-axis. (That is to say, if we drew the right hand half of the curve and put a mirror on the y-axis, the reflection of the right hand half in the mirror would give us an exact picture of the left hand half.) This means that we only need to draw one half of the curve, and we can then use reflection. We have indicated this in Figure B.16 by drawing the left hand side in with dashes rather than as a solid curve. In this case we have no asymptotes, but we

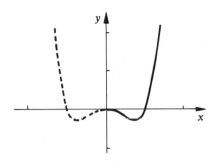

Figure B.16

note that since $y = x^2(x + 1)(x - 1)$, $y > 0$ provided $x > 1$ or $x < -1$, for then either four factors are positive, or exactly two are positive and the remainder negative. On the other hand, in the interval $(-1, 1)$ on the x-axis, we see that we have $(x + 1) > 0$, but $(x - 1) < 0$, and so $y < 0$ in this interval. Hence, we have the graph sketched in Figure B.16. Note again that we have been looking for pertinent information, and then combining the facts we have found to draw a sketch of the curve in question.

In the case of curves defined by equations of an algebraic nature, we will, if possible, find the intercepts, whether there is any symmetry (and there frequently is none), whether there are any areas that we can easily rule out as not containing the graph (such as we have done by indicating when $y > 0$ and when $y < 0$), and by determining any asymptotes that we can find easily. If there are values of x for which y is not defined, one should be alert for the possible existence of an asymptote. All in all, there are no direct rules that can be applied in all cases, but there are certain things we can look for. These are frequently helpful, perhaps in combination with some plotted points, in sketching the graph.

We now turn our attention to a non-algebraic case, just to illustrate that the same type of reasoning will be of assistance in this case also.

EXAMPLE 5.3. Sketch the graph of $y = \tan x$.

Solution. We note that y fails to exist if x is an odd integer multiplied by $\pi/2$. Since y fails to exist at these points due to the fact that as x approaches one of these values from either side the tangent of x increases in absolute value without limit we find that we have an asymptote at each such point. We also observe that if $x = n\pi$, when n is an integer, then $y = 0$, whence we have an infinite number of intercepts, each separated from its nearest neighbor by π units. It is also true, since $\tan(-x) = -\tan x$, that this curve is symmetric with respect to the origin, as was the curve in Example 5.1. Finally, we note that if x is in the interval $(0, \pi/2)$ $(\pi, 3\pi/2)$, $(2\pi, 5\pi/2)$, $(-\pi, \pi/2)$, etc., then $y > 0$ and for the intervening intervals $y < 0$. In other words, $y > 0$ for any interval corresponding to angles in the first and third

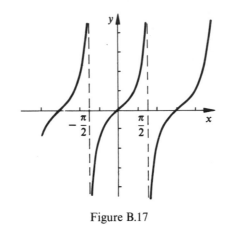

Figure B.17

quadrants, and $y < 0$ for angles found in the second and fourth quadrants. The periodic nature of $y = \tan x$ would also indicate that we should expect repetition every 2π units along the x-axis. The sketch is shown in Figure B.17. Again we have taken advantage of the information we could glean concerning intercepts, symmetry, areas in which the curve could not exist, and asymptotes. These are always good things to check, and with the information thus obtained, it is usually much easier to sketch the curves.

EXERCISES

1. Sketch the curve $x^2y - 2x^2 - 16y = 0$

2. Sketch the curve $y + x^2y = 1$

3. Sketch the curve $y = x^{2/3}$

4. Sketch the curve $y = \cosh x$. [$\cosh x = (e^x + e^{-x})/2$.]

5. Sketch the curve $x^3 - x^2y + y = 0$

6. Sketch the curve $y = x(x - 1)(x - 2)$

7. Sketch the curve $y^2 = x(x - 1)(x - 2)$

8. Sketch the curve $y = 2x + (1/x)$

9. Sketch the curve $y = x^3 - 4x^2 + 4x$

10. Sketch the curve $y = (x^2 - 1)/x$

11. Sketch the curve $y = (x^2 - 1)/(x^2 - 4)$

12. Sketch the curve $y = x^{3/2}$

13. Sketch the curve $y = x^4 - 2$

14. Sketch the curve $y^2 = x^4 - 16$

15. Sketch the curve $y = x^2/(x - 1)^2$

16. Sketch the curve $y = x^3/(x + 1)^2$

17. Sketch the curve $y^2 = x^4$

18. Sketch the curve $x^3 + y^3 = 6xy$

19. Sketch the curve $x^2y + 4y - 16x = 0$

20. Sketch the curve $y = x(x - 1)^2(x - 2)^3(x - 3)^4$

21. Sketch the curve $y = x(x - 1)(x - 2)(x - 3)$

22. Sketch the curve $y^2 = x(x - 1)(x - 2)(x - 3)$

23. Sketch the curve $y = x^2 + (1/x^2)$

24. Sketch the curve $y^2 = x^2 + (1/x^2)$

25. Sketch the curve $y = \sin x + \cos x$

26. Sketch the curve $y = x + \sin 2x$

27. Sketch the curve $y = \tan x - \sec x$

28. Sketch the curve $y = e^{-x} \cos x$

29. Sketch the curve $y = \ln x$

30. Sketch the curve $y = x + \ln x$

B.6 Polar Coordinates

There are many instances in which information is better expressed in terms of a coordinate system other than the rectangular coordinate system. In particular the *polar coordinate system* is often used. In polar coordinates we replace the origin with a *pole*, and from this pole we draw a ray, usually toward the right. This ray represents the positive portion of the *polar axis*. The length of the *radius vector*, denoted by r, is measured along this ray with positive distances to the right and negative distances to the left. The point thus plotted is then moved as though the entire plane were rotated through a central angle, θ, about the pole. We denote the polar coordinates of the point by the ordered pair (r, θ). For problems relating to the calculus θ is usually measured in radians, but the angle is a geometric entity and can be measured in such units as seem appropriate to the occasion. The plotting of points is illustrated in Figure B.18, where we have indicated points with positive angles, negative angles, and negative measurements for the radius vector. In polar coordinates it is possible to describe a single point with more than one pair of coordinates. For instance, the point $(2, -(\pi/4))$ in Figure B.18 could equally well be described by $(-2, (3\pi/4))$ or $(2, (7\pi/4))$.

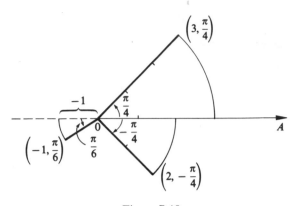

Figure B.18

The coordinates might be important in the functions which determine such points, but geometrically these all give the same point. This is confusing at times, but convenient at other times. By contrast, you remember we had a one-to-one correspondence between points and coordinates in the case of the rectangular coordinate system.

It is frequently convenient to establish a relationship between the two coordinate systems we are using in this Appendix. In order to do this, we will place the pole over the origin and place the polar axis along the positive x-axis. We note in Figure B.19 that for any point, $P(x, y) = P(r, \theta)$, we have a triangle for which the x distance is the horizontal side, the y distance is the vertical side, the r distance is the hypotenuse, and θ is the angle with vertex at the pole. Since this is a right triangle, we have the relations

$$x = r \cos \theta$$
$$y = r \sin \theta \tag{B.6.1}$$

and

$$r = x^2 + y^2$$
$$\theta = \text{arc tan}(y/x). \tag{B.6.2}$$

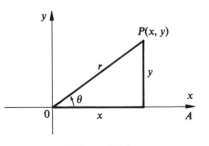

Figure B.19

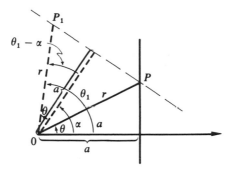

Figure B.20

These relations enable us to obtain the polar coordinates of a point for which rectangular coordinates are given and vice versa. It is necessary, of course, to be aware of the coordinate system in which one is working. If it is not clear whether polar or rectangular coordinates are being used, a specific statement should be made to remove any doubt. As a rule if the coordinates are given in the polar system, we will express our angles in fractions of π, and this will serve as a tip-off. In this section we will be using polar coordinates unless we specifically state otherwise.

We now turn to the equation of a line. We start with the line $x = a$, or in polar coordinates $r \cos \theta = a$. We could have obtained the same equation by noting that if we are given the vertical line, crossing the polar axis at a distance a from the pole, then an arbitrary point on the line would have coordinates determined by $r = a \sec \theta$. But this is equivalent to $r \cos \theta = a$. If we note the construction in Figure B.20, and then note the dashed line obtained by rotating the figure through the angle α, we see that we are now measuring the acute angle of the triangle by the difference $(\theta_1 - \alpha)$, where subscripts are used to distinguish the points P and P_1. This gives the equation $r_1 \cos(\theta_1 - \alpha) = a$. The subtraction of a constant angle from θ is much like translation in rectangular coordinates, for it changes the effective direction from which the angle is measured. This last equation can be expanded, and we will have

$$r_1 \cos \theta_1 \cos \alpha + r_1 \sin \theta_1 \sin \alpha = a,$$

or in rectangular coordinates $x \cos \alpha + y \sin \alpha = a$. Since the sum of the squares of the coefficients of x and of y is one, this is much like the polar form we had for rectangular coordinates, and we note that a is the distance from the pole (or origin). In case the line passes through the pole, we need only specify the angle θ and we have the equation $\theta = c$ where c is a constant.

Later we will have more to say about replacing r by $(r - b)$, but in general this is a much more difficult situation to handle than translation in rectangular coordinates. The problem results from the fact that the measurement would be diminished (or increased) in different directions dependent upon the value of θ used for the point in question. The rotation of graphs is very straighforward in polar coordinates, however, and we will use this again later on.

EXERCISES

1. Plot each of the following points:

   (a) $(2, \pi/4)$
   (b) $(0, \pi)$
   (c) $(-1, \pi/4)$
   (d) $(3, 11\pi/6)$
   (e) $(-3, 13\pi/6)$
   (f) $(2, -5\pi)$
   (g) $(-3, 9\pi/2)$
   (h) $(2, -3\pi/4)$

2. (a) Find the rectangular coordinates corresponding to $(2, \pi/3)$ in polar coordinates.
   (b) Find the polar coordinates corresponding to $(-3, 3\sqrt{3})$ in rectangular coordinates.
   (c) Find the rectangular coordinates corresponding to $(-3, \pi/2)$ in polar coordinates.
   (d) Find the polar coordinates corresponding to $(0, 0)$ in rectangular coordinates.

3. Sketch the lines

   (a) $r \cos \theta = 3$
   (b) $r \cos \theta = -2$
   (c) $r = 4 \sec \theta$
   (d) $r = -2 \sec \theta$
   (e) $r \sin \theta = 2$
   (f) $r \sin \theta = -1$
   (g) $r = - \csc \theta$
   (h) $r = 2 \csc \theta$

4. Sketch the lines

   (a) $r \cos (\theta - (\pi/3)) = 2$
   (b) $r \cos (\theta + (\pi/6)) = -1$
   (c) $r \cos(\theta - (5\pi/6)) = 2$
   (d) $r \cos(\theta - (15\pi/4)) = -4$

5. Sketch each of the following lines

   (a) $\theta = 0$
   (b) $\theta = \pi$
   (c) $\theta = 7\pi/6$
   (d) $\theta = -17\pi/6$

6. Find the equations for the lines described in each of the following situations:

   (a) The line parallel to the polar axis and 4 units below it.
   (b) The line through the point $(2, \pi/3)$ which makes an angle of $5\pi/6$ with the polar axis, and which proceeds from upper left toward lower right on your graph.
   (c) The line perpendicular to the polar axis 3 units to the left of the pole.
   (d) The line which passes through the points $(2, 0)$ and $(4, \pi/2)$.

7. (a) Convert $r \cos \theta = 4$ to rectangular coordinates
   (b) Convert $r \cos(\theta - (\pi/4)) = \sqrt{2}$ to rectangular coordinates
   (c) Convert $x + \sqrt{3}y = 5$ to polar coordinates.
   (d) Convert $\theta = \pi/6$ to rectangular coordinates.
   (e) Convert $2x - 3y = 0$ to polar coordinates.

B.7 Some Common Curves in Polar Coordinates

After the line, the circle is probably the simplest curve, and we will start our discussion of polar graphs with the circle. While it is possible to consider a *circle* placed in any position on the polar coordinate system, there is a distinct simplification if we restrict our attention to those circles for which the pole is either a point on the circle or else the pole is the center of the circle. In our brief description we will so restrict our discussion.

If the center is at the pole, it is only necessary to specify the radius, and we have the equation $r = c$ where c is a constant. This is shown in Figure B.21. Let us now consider the equation of the circle which has its diameter along the polar axis, which passes through the pole, and which has diameter of length a. This is also shown in Figure B.21. If we select a point P with coordinates (r, θ) on the circle we can join P to the two ends of the diameter and have a right angle with vertex at P since this angle is inscribed in a semi-circle. Therefore, we have a right triangle with an acute angle θ, with the adjacent side of length r and the hypotenuse of length a. Hence, using the definition of the cosine, we have $r = a \cos \theta$. If we were to rotate this circle through an angle α about the pole, we would have the equation $r = a \cos(\theta - \alpha)$, and if the angle $\alpha = \pi/2$, this would become

$$r = a \cos(\theta - (\pi/2)) = a \cos((\pi/2) - \theta) = a \sin \theta,$$

since the cosine is an even function. This gives the equation of the circle of diameter a which is tangent to the polar axis at the pole, and which is above the polar axis. You should take the time to consider what happens to the radius vector, r, as θ makes a complete revolution. In the equation $r = a \cos \theta$ you will note that as θ goes from $-\pi/2$ to $\pi/2$ we obtain a complete circle. As θ goes from $\pi/2$ to $3\pi/2$ we will repeat the same circle, for all values

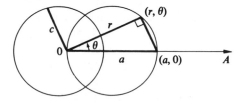

Figure B.21

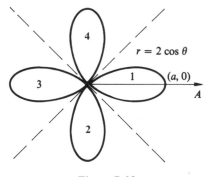

Figure B.22

of r will be negative in this latter half revolution. Therefore, each point is plotted twice if θ makes a complete revolution. We have a half revolution for which $r \geq 0$ and a half revolution for which $r \leq 0$.

We now consider the case of the equation $r = a \cos n\theta$. Here we see that when θ is $(1/n)$-th as big as it was formerly, we obtain the same value of r we obtained earlier. Thus, we have squeezed our circle of diameter a by pressing in from both sides in such a way that the squeezed curve occupies a sector with angular width π/n instead of the former width π. We have done nothing to reduce the maximum length of the radius vector, and consequently we have a curve that looks much like the petal of a flower. This is shown in Figure B.22. This is fine as far as it goes, but we must ask the question concerning what happens as θ makes a complete revolution. You will remember that in the case of the circle we had a *petal* in the form of a circle that occupied a sector of angular width π. In the next sector of angular width π we had the same circle, but with negative values for the radius vector, and for this reason, the second circle was plotted on top of the first. In our squeezed situation, we can then expect that we will have $2n$ sectors (if n is an integer), each of width π/n, and in the sector which straddles the polar axis, that is the one including the value $\theta = 0$, we will have positive values for the radius vector. The values of r will alternate from positive to negative and back to positive as we progress from sector to sector. If n is an even integer, we will find that we then fill in each sector, for the sectors in which $r < 0$ neatly fit in between the sectors in which $r > 0$. On the other hand, if n is an odd integer, we find that the sectors in which $r < 0$ are so placed that the petals retrace those formed when $r > 0$. Hence, in case n is an odd integer, we have $2n$ petals, but these appear as n petals, each of which is traced twice.

This is what happened in the case $n = 1$, or in the case of the circle. If n is not an integer, then we have overlapping petals, but a finite number of them if n is rational, whereas we will find that we never come back to retrace a petal already drawn if n is irrational. For purposes of illustration we have indicated the order in which the petals appear in the graph of $r = a \cos 2\theta$ in Figure B.22. Observe that the effect of multiplying a variable by a constant

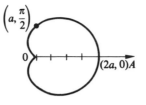

Figure B.23

in an equation is one of squeezing the space that is occupied as a result of the movement of that variable. If you would re-examine the equation of the ellipse in rectangular coordinates, you would find that this could have been obtained from the circle by the same type of squeezing effect where we squeeze by different ratios along the two axes.

We now come to the case in which we replace r by $(r - a)$. As we mentioned before, this produces weird results since the movement is in the direction in which r happens to be measured at the moment. Let us consider the equation $r - a = a \cos \theta$, or $r = a + a \cos \theta$. This can also be written $r = a(1 + \cos \theta)$. We observe that the coefficient of a can no longer be negative in this case, and hence we have lost the re-tracing of the curve that we had in the circle equation. In the right half plane we find that the length of the radius vector is more than a (and is $2a$ on the polar axis). However, in the left half plane we find that $r < a$, and $r = 0$ when $\theta = \pi$. The resulting curve is drawn in Figure B.23. This curve is called the *cardioid*, and from the heart shape it is rather easy to see why. If we attempted to investigate $r = b + a \cos \theta$, obtained by re-placing r by $(r - b)$ in the circle equation, we would have had two cases depending on the relative values of a and b. If $|b| > |a|$, we see that r is always of one sign and we have a figure approaching the heart shaped figure of the cardioid, although it will not reach the pole at any point. This gives us the *limacon without loop*. On the other hand, if $|b| < |a|$, we have a sector in which r is negative, giving us a petal with an extreme radius vector of $|a| - |b|$ in absolute value. The larger part of the curve would have a petal of maximum length $|b| + |a|$. This curve is called the *limacon with loop*. These two curves are shown in Figures B.24a and B.24b respectively. Note the effect of replacing r by what would constitute a translation modification in rectangular co-ordinates, or would constitute an angular translation (or rotation) if applied

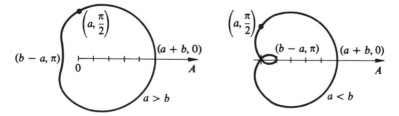

Figure B.24

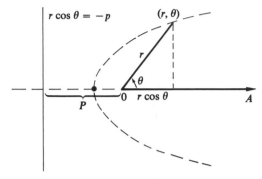

Figure B.25

to θ, is dependent upon the direction of the radius vector and therefore each point *translates* in a different direction. It follows, of course, that we can obtain the rotated *rose curve*, *cardioids*, and *limacons* by replacing θ by $(\theta - \alpha)$, as before.

We now take a look at the conic sections, but only for the case in which a focus is at the pole. We will consider that the directrix is the line $r \cos \theta = -p$ where p is positive in Figure B.25. This is a vertical line p units to the left of the pole. Let us further assume that we have an eccentricity, e, and that $e \neq 0$. Now the distance from the focus to the point on the conic is merely r, for the focus is at the pole. On the other hand, the distance from the point to the directrix is the distance p from the directrix to the pole plus the distance $r \cos \theta$ by which the point is further right than is the pole, or focus. Thus the distance from the point to the directrix is given by $(p + r \cos \theta)$. Using the relation (B.4.2) we have

$$\frac{r}{p + r \cos \theta} = e.$$

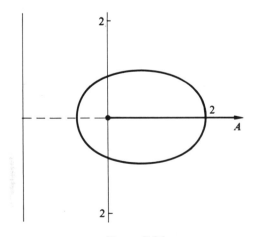

Figure B.26

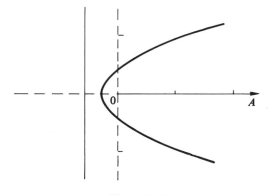

Figure B.27

With some algebraic manipulations this becomes

$$r = \frac{ep}{1 - e \cos \theta}. \tag{B.7.1}$$

Note that if $e < 1$ the denominator cannot fail to be positive and hence there is no angle for which a value of r fails to exist, as shown in Figure B.26. This, you will remember, is the case of the *ellipse*. If $e = 1$, there is but one value where no finite length exists for the radius vector. This is the case of the *parabola*, Figure B.27. The direction for which r fails to exist will be the direction of the open end of the parabola or the axis of the parabola. If $e > 1$, we have two values of θ in each revolution for which r fails to exist, as shown in Figure B.28. These two values divide the revolution into two intervals. In one of these intervals $r > 0$ and in the other $r < 0$. This gives two distinct portions of the curve, both appearing on the same side of the pole (or focus). This gives the *hyperbola*. We repeat that p is the distance from the focus to the directrix. Rotation is again possible using an angular translation, as before.

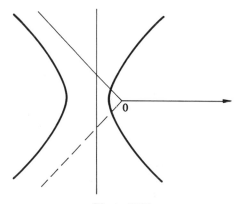

Figure B.28

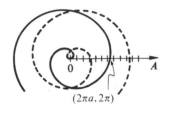

$(2\pi a, 2\pi)$

Figure B.29

We now come to a type of curve which we did not attempt to describe in rectangular coordinates. This curve is the *spiral*. We are usually interested in one of two spirals. In the spiral, of course, one would expect to see the radius vector increasing in magnitude as the angle increases, and on each revolution the radius vector would be larger than on the preceding revolution. Hence, it should not be surprising to note that the equation $r = a\theta$ would produce such a figure. This is called the *spiral of Archimedes*, and is shown in Figure B.29. In this spiral we see that if we start with $\theta = 0$, and let θ increase, r increases by an amount $2a\pi$ with each revolution. This is shown by the solid line. On the other hand if we start with $\theta = 0$ and let θ decrease, then r will be negative. The absolute value of r will increase but r will remain negative. This half of the spiral is shown by the dashed curve. Note the points of intersection of the two spirals. These are points at which they coincide geometrically, but at which they do not have the same coordinates, for on one curve both r and θ are positive while on the other curve they are both negative.

The final curve we will consider in this section is the *logarithmic spiral*. This curve has the equation

$$r = e^{a\theta}.$$

Note here that r is always positive, but if $\theta > 0$ we have $r > 1$ whereas if $\theta < 0$ we have $r < 1$. Drawing a radius vector here does not produce equal segments between successive intersections with the spiral as in the case of the spiral of Archimedes, but rather as we recede from the pole the intervals between intersections are greater. This is shown in Figure B.30. Rather interestingly this spiral is the one favored by the lowly snail, for the simple

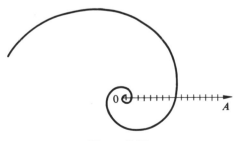

Figure B.30

reason that it is possible to have more snail with less shell using this spiral. If you find a snail shell with a significant variation from this pattern, it will indicate a significant perturbation in the growth cycle of the snail, and not a voluntary act on the snail's part.

EXERCISES

1. Sketch each of the following curves:

(a) $r = 3 \cos \theta$
(b) $r = 2 \sin \theta$
(c) $r = -3 \sin \theta$
(d) $r = - \cos \theta$
(e) $r = 3 \cos \theta + 4 \sin \theta$
(f) $r = 4 \sin(\theta - (\pi/6))$
(g) $r = \sin 2\theta$
(h) $r = \cos 3\theta$

2. Sketch each of the following curves:

(a) $r = -2 \sin 5\theta$
(b) $r = -3 \cos 4\theta$
(c) $r = 2 \cos(4\theta/3)$
(d) $r = 3 \sin(3\theta/2)$
(e) $r = 2 \cos 4(\theta - (\pi/6))$
(f) $r = 3 \sin(4\theta - (\pi/6))$

3. Sketch each of the following curves:

(a) $r = 2 - 2 \cos \theta$
(b) $r = 1 + \sin \theta$
(c) $r = 2 \cos \theta - 2$
(d) $r = 3 + \sin \theta$
(e) $r = 2 + 2 \sin \theta$
(f) $r = 2 - \sin \theta$

4. Sketch each of the following:

(a) $r = 2 + 3 \sin \theta$
(b) $r = 2 - 3 \sin \theta$
(c) $r = 1 - 4 \cos \theta$
(d) $r = 3 + 2 \sin \theta$
(e) $r = 3 - 2 \sin \theta$
(f) $r = 2 - 5 \sin \theta$

5. Sketch each of the following:

(a) $r = 2/(1 + \cos \theta)$
(b) $r = 3/(1 - 2 \cos \theta)$
(c) $r = 5/(1 - 0.4 \sin \theta)$
(d) $r = 3/(2 - \sin \theta)$
(e) $r = 14/(3 + 3 \sin \theta)$
(f) $r = 5/(2 + 3 \sin \theta)$

6. Sketch each of the following:

   (a) $r = 0.5\theta$
   (b) $r = 2\theta$
   (c) $r = -\theta$
   (d) $r = e^{0.5\theta}$
   (e) $r = 2e^{0.2\theta}$
   (f) $r = 3e^{-\theta}$

7. (a) Find the polar equation of the parabola having its focus at the pole and having a directrix parallel to the polar axis and 2 units above the polar axis. Sketch the curve.

   (b) Find the polar equation of the ellipse for which one focus is at the pole and the corresponding directrix is parallel to the polar axis and 3 units below if the eccentricity is 0.5. Sketch the curve.

   (c) Find the polar equation of the conic section for which the eccentricity is 2 if one focus is at the pole and the corresponding directrix is perpendicular to the polar axis and three units to the left of the pole. Sketch the curve.

8. (a) Convert the polar equation $r = 3 \cos \theta$ to rectangular coordinates.

   (b) Convert the polar equation $r(1 + 2 \cos \theta) = 3$ to rectangular coordinates.

   (c) Convert the rectangular equation $0.5 \ln(x^2 + y^2) = \text{arc } \tan(y/x)$ to polar coordinates and sketch the curve.

9. (a) Sketch the spiral $r = 3\theta$.

   (b) Find all of the points at which the portion of this curve for which r is positive intersect the branch for which r is negative.

   (c) Show the relation of the coordinates of the points of intersection of part (b).

B.8 Rotation of Axes

Throughout the discussion of polar coordinates you were reminded that it is possible to rotate the curve by the mere device of replacing θ by $(\theta - \alpha)$, where α is the angle of rotation. We should point out here that if $\alpha > 0$, this is tantamount to rotating the point for which θ would have been zero in a positive direction to the point where $\theta = \alpha$. We would have obtained the same effect by keeping the curve in the same position but rotating the polar axis in the clockwise (or negative) direction, provided we returned later to straighten up the picture by again making the polar axis horizontal. Therefore, to rotate the curve in the positive direction, we effectively rotated the axis in the negative direction. Conversely, if we would desire to rotate the axis in the positive direction, we should then replace θ by $(\theta + \alpha)$. It is not our sole purpose here to discuss directions of rotation, but it is suggested that you draw sketches and satisfy yourself that what has been said is correct. Our concern here will be to see the effect of rotation on the rectangular coordinates of a point. In other words, if we were to start with the rectangular axes, x and y, indicated in Figure B.31, and were to consider axes designated

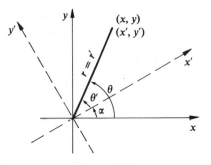

Figure B.31

by x' and y' which have effectively been rotated through the angle α, we would like to know the relation between the coordinates (x, y) on the given system and the coordinates (x', y') on the system that has been rotated. If we think of replacing those pairs of coordinates which represent the same point by their polar equivalents (r, θ) and (r', θ'), we can then find the result very easily, for we see that

$$\theta' = \theta - \alpha, \quad \text{or} \quad \theta = \theta' + \alpha, \quad \text{and} \quad r' = r.$$

Therefore

$$x' = r' \cos \theta' = r \cos(\theta - \alpha)$$
$$= r \cos \theta \cos \alpha + r \sin \theta \sin \alpha \qquad (B.8.1)$$
$$= x \cos \alpha + y \sin \alpha,$$

and

$$y' = r' \sin \theta' = r \sin(\theta - \alpha)$$
$$= r \sin \theta \cos \alpha - r \cos \theta \sin \alpha$$
$$= y \cos \alpha - x \sin \alpha.$$

These relations permit us to replace (x', y') with expressions involving x and y namely $(x \cos \alpha + y \sin \alpha, -x \sin \alpha + y \cos \alpha)$. In similar fashion, we can show that (x, y) would be replaced by

$$(x' \cos \alpha - y' \sin \alpha, x' \sin \alpha + y' \cos \alpha).$$

Thus we have the relations between the rectangular coordinates in the two axis systems. The two systems share the same origin, but one is rotated with respect to the other.

We will close this appendix with an illustration which makes use of the result we have just derived. Consider the problem of drawing the graph of the curve determined by the equation

$$36x^2 - 24xy + 29y^2 - 120x - 10y - 55 = 0.$$

This curve has some of the attributes of the conic sections in rectangular coordinates, but we have not yet handled a case in which we have an xy term present. Let us replace x and y by $(x' \cos \alpha - y' \sin \alpha)$ and $(x' \sin \alpha + y' \cos \alpha)$, respectively, as indicated by our development of the rotation relations. We will then see whether we can choose a value of α such that we might make the xy term—or rather its new counterpart of $x'y'$—disappear. The algebra looks formidable, but not quite impossible. Hence, we obtain

$$36(x' \cos \alpha - y' \sin \alpha)^2 - 24(x' \cos \alpha - y' \sin \alpha)(x' \sin \alpha + y' \cos \alpha)$$
$$+ 29(x' \sin \alpha + y' \cos \alpha)^2 - 120(x' \cos \alpha - y' \sin \alpha)$$
$$- 10(x' \sin \alpha + y' \cos \alpha) - 55 = 0.$$

Since we are interested in seeing whether we can eliminate the $(x'y')$ term, we need only look at the portion of this expanded result which would comprise the coefficient of $(x'y')$. Thus, we would look at the middle term that results from the first term of the original expression, the middle term resulting from the third term of the original expression, and the two terms that would involve $(x'y')$ resulting from the multiplication of the second term. This coefficient of $(x'y')$ will be

$$-72 \sin \alpha \cos \alpha - 24 \cos^2 \alpha + 24 \sin^2 \alpha + 58 \sin \alpha \cos \alpha.$$

If we now attempt to find the value of α which will make this zero, we might find it more convenient to divide the two sides of the equation in which we are interested by $\cos^2 \alpha$ for this will reduce the complications to just the tangent function, and it is probably easier to cope with only one function at a time. We know that we are not dividing by zero, for if $\cos \alpha = 0$, then we have $\alpha = \pi/2$, and the result would only be to rotate through a right angle, hardly something that would cause a major alteration in the equation of a conic section that has not already been considered in Section 4. Upon dividing by two this gives us the equation $12 \tan^2 \alpha - 7 \tan \alpha - 12 = 0$. This can be factored and we obtain $(4 \tan \alpha + 3)(3 \tan \alpha - 4) = 0$, whence we have $\tan \alpha$ equal $-3/4$ or $4/3$. Note that these are negative reciprocals, and therefore the two choices for α differ by exactly a right angle. This choice only affects which axis will be the x'-axis and which the y'-axis. We will arbitrarily select the angle having the positive tangent, and we can then find the required sine and cosine by drawing the triangle shown in Figure B.32. From this triangle we see that $\sin \alpha = 4/5$ and $\cos \alpha = 3/5$. We can

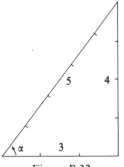

Figure B.32

either go back to our very long equation and try to insure that we make no wrong substitutions, or we can start all over again, but this time with numbers. Most people seem to consider it easier to use numbers than the expressions for the trigonometric functions. We will do the re-substitution, for this also permits us to check our work from the beginning and insure that we have really found the angle α which will produce a zero coefficient for the $(x'y')$ term.

We now return to the original equation and use the substitutions

$$x = \frac{3x' - 4y'}{5}$$

$$y = \frac{4x' + 3y'}{5}.$$

Upon substituting we have

$$36\left(\frac{3x' - 4y'}{5}\right)^2 - 24\left(\frac{3x' - 4y'}{5}\right)\left(\frac{4x' + 3y'}{5}\right) + 29\left(\frac{4x' + 3y'}{5}\right)^2$$
$$- 120\left(\frac{3x' - 4y'}{5}\right) - 10\left(\frac{4x' + 3y'}{5}\right) - 55 = 0.$$

After multiplying and collecting terms this reduces to

$$20(x')^2 + 45(y')^2 - 80(x') + 90(y') - 55 = 0.$$

Before dividing all terms by 5, let's pause long enough to note that $20 + 45 = 36 + 29$, or in other words that the sum of the coefficients of x^2 and y^2 is equal to the sum of the coefficients of $(x')^2$ and $(y')^2$. If you are very daring you might show that this will always work. If we now divide this last result by 5, we obtain $4(x')^2 + 9(y')^2 - 16x' + 18y' - 11 = 0$. This is precisely the same equation we considered in Section 4 but for the fact that the curve is relative to the x' and y' axes instead of the x and y axes. Thus, we would obtain the sketch shown in Figure B.33. Note that we have indicated

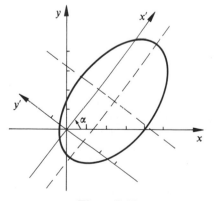

Figure B.33

the angle α here. Since we know the size of angle α geometrically in view of
the work shown in Figure B.32, we should have no problem in making a
careful drawing.

In general if we have any second degree equation in rectangular coordinates, we will obtain a conic section (or in certain cases degenerate conic
sections such as circles with imaginary radii). If there is an xy term present,
we would have to rotate axes to recognize the form of the conic section,
but this can always be done in the manner illustrated here.

EXERCISES

1. Rotate axes to remove the xy term and sketch each of the following. Show both
   the original and the rotated set of axes.

   (a) $x^2 - xy = 1$
   (b) $2xy + y^2 = 4$
   (c) $x^2 + 3xy - y^2 = 1$
   (d) $x^2 + xy + y^2 = 4$

2. Rotate axes to remove the xy term and sketch each of the following. Show both the
   original and the rotated set of axes.

   (a) $9x^2 + 24xy + 16y^2 + 80x - 60y = 0$
   (b) $9x^2 + 4xy + 6y^2 + 12x + 36y + 44 = 0$
   (c) $x^2 - 10xy + y^2 + x + y + 1 = 0$
   (d) $73x^2 - 72xy + 52y^2 + 30x + 40y - 75 = 0$
   (e) $16x^2 - 24xy + 9y^2 - 5x - 90y + 25 = 0$
   (f) $10x^2 - 12xy + 10y^2 - 16\sqrt{2}x + 16\sqrt{2}y = 16$

3. (a) Rotate the axes to remove the xy term in the equation $xy = 2$, and sketch the
       curve.
   (b) Show that no rotation of axes will alter the equation $x^2 + y^2 = 25$. Sketch the
       curve and explain why rotation leaves this equation unchanged.
   (c) Rotate the axes through an angle of $\pi/4$ for the equation $x^2 + y^2 + 4x - 4y = 4$.
       Sketch the curve and explain the changes that the rotation makes in the equation
       by the actual geometry of the graph.

4. (a) If we start with the equation $Ax^2 + Bxy + Cy^2 + Dx + Ey + F = 0$, show
       that it is always possible to find an angle through which you can rotate the axes
       and remove the xy term.
   (b) Show that any rotation of axes will leave unchanged the value of $A + C$, the
       sum of the coefficients of the squared terms.
   (c) Show that any rotation of axes will leave unchanged the value of $B^2 - 4AC$, the
       discriminant of the equation giving the conic section.
   (d) Show that the conic section is an ellipse, a parabola, or a hyperbola (or a degen-
       erate case of one of these) if the discriminant is negative, zero or positive respec-
       tively. [*Hint*: Use the standard form for each of the conics combined with the
       result of part (c).]
   (e) Show that any rotation of axes will leave unchanged the value of $D^2 + E^2$.

Appendix C. Tables

Table 1a Trigonometric Functions (Radian measure)

(radians)	Sin x	Cos x	Tan x	Cot x	Sec x	Csc x
0.00000	0.00000	1.00000	0.00000	*******	1.00000	*******
0.05000	0.04998	0.99875	0.05004	19.98333	1.00125	20.00834
0.10000	0.09983	0.99500	0.10033	9.96664	1.00502	10.01669
0.15000	0.14944	0.98877	0.15114	6.61659	1.01136	6.69173
0.20000	0.19867	0.98007	0.20271	4.93315	1.02034	5.03349
0.25000	0.24740	0.96891	0.25534	3.91632	1.03209	4.04197
0.30000	0.29552	0.95534	0.30934	3.23273	1.04675	3.38386
0.35000	0.34290	0.93937	0.36503	2.73951	1.06454	2.91632
0.40000	0.38942	0.92106	0.42279	2.36522	1.08570	2.56793
0.45000	0.43497	0.90045	0.48306	2.07016	1.11056	2.29903
0.50000	0.47943	0.87758	0.54630	1.83049	1.13949	2.08583
0.55000	0.52269	0.85252	0.61311	1.63104	1.17299	1.91319
0.60000	0.56464	0.82534	0.68414	1.46170	1.21163	1.77103
0.65000	0.60519	0.79608	0.76020	1.31544	1.25615	1.65238
0.70000	0.64422	0.76484	0.84229	1.18724	1.30746	1.55227
0.75000	0.68164	0.73169	0.93160	1.07343	1.36670	1.46705
0.80000	0.71736	0.69671	1.02964	0.97121	1.43532	1.39401
0.85000	0.75128	0.65998	1.13833	0.87848	1.51519	1.33106
0.90000	0.78333	0.62161	1.26016	0.79335	1.60873	1.27661
0.95000	0.81342	0.58168	1.39838	0.71511	1.71915	1.22938
1.00000	0.84147	0.54030	1.55741	0.64209	1.85082	1.18840
1.05000	0.86742	0.49757	1.74332	0.57362	2.00976	1.15284
1.10000	0.89121	0.45360	1.96476	0.50897	2.20460	1.12207
1.15000	0.91276	0.40849	2.23450	0.44753	2.44806	1.09557
1.20000	0.93204	0.36236	2.57215	0.38878	2.75970	1.07292
1.25000	0.94898	0.31532	3.00957	0.33227	3.17136	1.05376
1.30000	0.96356	0.26750	3.60210	0.27762	3.73833	1.03782
1.35000	0.97572	0.21901	4.45522	0.22446	4.56607	1.02488
1.40000	0.98545	0.16997	5.79788	0.17248	5.88349	1.01477
1.45000	0.99271	0.12050	8.23809	0.12139	8.29856	1.00734
1.50000	0.99749	0.07074	14.10142	0.07091	14.13683	1.00251
1.55000	0.99978	0.02079	48.07848	0.02080	48.08888	1.00022
1.60000	0.99957	-0.02919	-34.23252	-0.02920	-34.24713	1.00043
1.65000	0.99687	-0.07911	-12.59925	-0.07936	-12.63888	1.00314
1.70000	0.99166	-0.12883	-7.69659	-0.12992	-7.76128	1.00841
1.75000	0.98399	-0.17824	-5.52037	-0.18114	-5.61021	1.01627
1.80000	0.97385	-0.22719	-4.28625	-0.23329	-4.40136	1.02685
1.85000	0.96128	-0.27558	-3.48805	-0.28668	-3.62857	1.04028
1.90000	0.94630	-0.32328	-2.92709	-0.34163	-3.09319	1.05675
1.95000	0.92896	-0.37017	-2.50947	-0.39848	-2.70137	1.07647
2.00000	0.90930	-0.41614	-2.18503	-0.45765	-2.40299	1.09975

Table 1b Trigonometric Functions (Degrees)

Degrees	Sin	Cos	Tan	Cot	Sec	Csc
0	0.00000	1.00000	0.00000	*******	1.00000	*******
1	0.01745	0.99985	0.01746	57.28994	1.00015	57.29867
2	0.03490	0.99939	0.03492	28.63624	1.00061	28.65370
3	0.05234	0.99863	0.05241	19.08113	1.00137	19.10732
4	0.06976	0.99756	0.06993	14.30066	1.00244	14.33558
5	0.08716	0.99619	0.08749	11.43005	1.00382	11.47371
6	0.10453	0.99452	0.10510	9.51436	1.00551	9.56677
7	0.12187	0.99255	0.12278	8.14434	1.00751	8.20551
8	0.13917	0.99027	0.14054	7.11537	1.00983	7.18529
9	0.15643	0.98769	0.15838	6.31375	1.01247	6.39245
10	0.17365	0.98481	0.17633	5.67128	1.01543	5.75877
11	0.19081	0.98163	0.19438	5.14455	1.01872	5.24084
12	0.20791	0.97815	0.21256	4.70463	1.02234	4.80973
13	0.22495	0.97437	0.23087	4.33147	1.02630	4.44541
14	0.24192	0.97030	0.24933	4.01078	1.03061	4.13356
15	0.25882	0.96593	0.26795	3.73205	1.03528	3.86370
16	0.27564	0.96126	0.28675	3.48741	1.04030	3.62795
17	0.29237	0.95630	0.30573	3.27085	1.04569	3.42030
18	0.30902	0.95106	0.32492	3.07768	1.05146	3.23607
19	0.32557	0.94552	0.34433	2.90421	1.05762	3.07155
20	0.34202	0.93969	0.36397	2.74748	1.06418	2.92380
21	0.35837	0.93358	0.38386	2.60509	1.07115	2.79043
22	0.37461	0.92718	0.40403	2.47509	1.07853	2.66947
23	0.39073	0.92050	0.42447	2.35585	1.08636	2.55930
24	0.40674	0.91355	0.44523	2.24604	1.09464	2.45859
25	0.42262	0.90631	0.46631	2.14451	1.10338	2.36620
26	0.43837	0.89879	0.48773	2.05030	1.11260	2.28117
27	0.45399	0.89101	0.50953	1.96261	1.12233	2.20269
28	0.46947	0.88295	0.53171	1.88073	1.13257	2.13005
29	0.48481	0.87462	0.55431	1.80405	1.14335	2.06266
30	0.50000	0.86603	0.57735	1.73205	1.15470	2.00000
31	9.51504	0.85717	0.60086	1.66428	1.16663	1.94160
32	0.52992	0.84805	0.62487	1.60033	1.17918	1.88708
33	0.54464	0.83867	0.64941	1.53986	1.19236	1.83608
34	0.55919	0.82904	0.67451	1.48256	1.20622	1.78829
35	0.57358	0.81915	0.70021	1.42815	1.22077	1.74345
36	0.58779	0.80902	0.72654	1.37638	1.23607	1.70130
37	0.60182	0.79864	0.75355	1.32704	1.25214	1.66164
38	0.61566	0.78801	0.78129	1.27994	1.26902	1.62427
39	0.62932	0.77715	0.80978	1.23490	1.28676	1.58902
40	0.64279	0.76604	0.83910	1.19175	1.30541	1.55572
41	0.65606	0.75471	0.86929	1.15037	1.32501	1.52425
42	0.66913	0.74314	0.90040	1.11061	1.34563	1.49448
43	0.68200	0.73135	0.93252	1.07237	1.36733	1.46628
44	0.69466	0.71934	0.96569	1.03553	1.39016	1.43956
45	0.70711	0.70711	1.00000	1.00000	1.41421	1.41421

Table 2 Exponentials and Logarithms

x	e^x	e^{-x}	Ln x
0.0	1.00000	1.00000	*******
0.1	1.10517	0.90484	-2.30258
0.2	1.22140	0.81873	-1.60943
0.3	1.34986	0.74082	-1.20396
0.4	1.49182	0.67032	-0.91628
0.5	1.64872	0.60653	-0.69314
0.6	1.82212	0.54881	-0.51082
0.7	2.01375	0.49659	-0.35666
0.8	2.22554	0.44933	-0.22313
0.9	2.45960	0.40657	-0.10535
1.0	2.71828	0.36788	0.00000
1.1	3.00417	0.33287	0.09531
1.2	3.32012	0.30119	0.18232
1.3	3.66930	0.27253	0.26236
1.4	4.05520	0.24660	0.33647
1.5	4.48169	0.22313	0.40547
1.6	4.95303	0.20190	0.47000
1.7	5.47395	0.18268	0.53063
1.8	6.04956	0.16530	0.58779
1.9	6.68589	0.14957	0.64185
2.0	7.38906	0.13534	0.69315
2.1	8.16617	0.12246	0.74194
2.2	9.02501	0.11080	0.78846
2.3	9.97418	0.10026	0.83291
2.4	11.02318	0.09072	0.87547
2.5	12.18249	0.08208	0.91629
2.6	13.46374	0.07427	0.95551
2.7	14.87973	0.06721	0.99325
2.8	16.44465	0.06081	1.02962
2.9	18.17415	0.05502	1.06471
3.0	20.08554	0.04979	1.09861
3.2	24.53253	0.04076	1.16315
3.4	29.96410	0.03337	1.22378
3.6	36.59823	0.02732	1.28093
3.8	44.70118	0.02237	1.33500
4.0	54.59815	0.01832	1.38629
4.2	66.68633	0.01500	1.43508
4.4	81.45087	0.01228	1.48160
4.6	99.48432	0.01005	1.52606
4.8	121.51042	0.00823	1.56862
5.0	148.41316	0.00674	1.60944
5.5	244.69193	0.00409	1.70475
6.0	403.42879	0.00248	1.79176
6.5	665.14163	0.00150	1.87180
7.0	1096.63316	0.00091	1.94591
7.5	1808.04242	0.00055	2.01490
8.0	2980.95801	0.00034	2.07944
8.5	4914.76885	0.00020	2.14007
9.0	8103.08396	0.00012	2.19722
9.5	13359.72680	0.00007	2.25129
10.0	22026.46582	0.00005	2.30259

Table 3 Gamma Functions

X	GAMMA(X)	X	GAMMA(X)
1.00	1.0000	1.52	0.8870
1.04	0.9784	1.56	0.8896
1.08	0.9597	1.60	0.8935
1.12	0.9436	1.64	0.8986
1.16	0.9298	1.68	0.9050
1.20	0.9182	1.72	0.9126
1.24	0.9085	1.76	0.9214
1.28	0.9007	1.80	0.9314
1.32	0.8946	1.84	0.9426
1.36	0.8902	1.88	0.9551
1.40	0.8873	1.92	0.9688
1.44	0.8858	1.96	0.9837
1.48	0.8857	2.00	1.0000

Appendix D. **FORTRAN** Language

D.1 Introduction

The **FORTRAN** language is one of many used today in communicating with computers. It is one of the more frequently used languages, and consequently the great majority of computers have compilers which permit the use of **FORTRAN**. We will discuss here a sufficient number of items concerning **FORTRAN** to permit you to write computer programs in this language for any problem which should arise in a calculus course. It should be emphasized, however, that there are additional features of **FORTRAN** which we will not attempt to discuss in this brief account. The **FORTRAN** language is a specific language with its own syntax in which we can write instructions which we wish the computer to follow. Any problem for which we desire a computer solution can be written in this language. The instructions are then entered into the computer together with a *translator* or *compiler* which will translate the **FORTRAN** instructions into the language of ones and zeros built into the machine. The entry of this program is most frequently performed by first punching the instructions into 80 column cards, sometimes known as IBM cards, one instruction per card, and then causing these cards to be read by the card reader of the computer. Instructions concerning procedures to be used should be obtained from your computing center, both as to the manner in which the program should be presented to the center, and concerning the use of any equipment you are required to use in preparing the program for the computer. One other item of information which you will need to obtain from your computing center will concern the *device codes* to be used with the *input* and *output devices*. While the manner of writing instructions is standardized, the device codes and procedures for submitting programs depend upon the computer and the computing center. It is for this reason that we must ask

699

you to obtain certain information locally. We will include enough information here, however, to enable you to write a *source program*, or set of instructions to perform the calculations in which you are interested.

FORTRAN distinguishes between *integers* and *real numbers*. The computer will fail to handle correctly integers larger in absolute value than some fixed number which is dependent upon the computer involved. Furthermore, if all of the operands in any computation are integers, then the result will be an integer. This gives rise to such peculiar arithmetic as $9/10 = 0$, since the largest integer in the quotient obtained when dividing 9 by 10 is 0. *Integer constants* are distinguished by the fact that they are numbers, written in the decimal system, which do not contain a decimal point. By contrast the real numbers are written with a decimal point. Thus 9 is an integer, but 9.0 is a real number insofar as FORTRAN is concerned. The majority of programming is done by writing statements involving variables. A *variable* consists of a name of not more than six characters (five for some computers), the first of which must be a letter of the alphabet and the remaining characters must be either letters or digits. If the initial letter of the symbol is I, J, K, L, M, or N, the variable is considered to be an integer. Otherwise the variable is considered to be a real number. Thus we can write both constants and variables, and we can also write integers and real numbers. Note that 175 and ITEM would represent integers while 47.3 and DATA would represent real numbers.

EXERCISES

1. Identify each of the following as an integer or a real number when interpreted by FORTRAN:

   (a) -45
   (b) 0
   (c) 3.1415962
   (d) 4978
   (e) -39876
   (f) 34
   (g) 28
   (h) 57.25
   (i) -5.678
   (j) 42.00

2. Identify each of the following variables as being integers or real numbers when interpreted by FORTRAN:

   (a) NUMBER
   (b) KOUNT
   (c) SUM
   (d) JJJJJJ
   (e) KZZZZ
   (f) COUNT
   (g) TOTAL

(h) ITEM
(i) ABCDEF
(j) LITTLE

3. Which of the following are not legitimate FORTRAN variable names, and why?

(a) ABCDEFG
(b) 2ABKL
(c) A23B4
(d) R2D2
(e) 2453
(f) ABC,DE
(g) ABKL2
(h) NO-GO
(i) SUM.
(j) FREE

4. Give the result of each of the following integer computations as done by FORTRAN:

(a) 23/4
(b) 15/3
(c) 99/100
(d) −13/4
(e) 19/3
(f) 14/8

D.2 Arithmetic Statements

Our next concern is to be able to do some arithmetic. We use the arithmetic operations $+$, $-$, $*$, $/$, and $**$ to represent *addition, subtraction, multiplication, division*, and *raising to a power* respectively. We *cannot* represent multiplication by simply placing two symbols next to each other. We must always insert $*$ between the two symbols if the computer is expected to multiply them. We are free to use *parentheses* as often as we wish, but we must remember to provide a closing parenthesis to match each opening parenthesis. Each arithmetic FORTRAN statement consists of a single variable followed by an equal sign and the expression indicating the arithmetic to be performed. Such a statement instructs the computer to do the arithmetic indicated on the right side of " $=$ ", and then to assign this resulting value to the symbol on the left side. Thus, in the statement

$$X = (-B + SQRT(B**2 - 4*A*C))/(2*A) \qquad (D.2.1)$$

the computer expects to take the values previously given it for A, B, and C, to compute $B^2 - 4AC$, take the square root of this quantity, add the result to $(-B)$, and then divide the value thus obtained by the product of two and A. The final result is then to be assigned to X so that X will have the value thus obtained by the time this statement has been executed. The values

Table D.1

ABS(E)	The absolute value of E.
ALOG(E)	The natural logarithm of E (that is to the base e).
ATAN(E)	The arc tangent of E. The result will be in radians.
COS(E)	The cosine of E. The angle will be considered to be in radians.
EXP(E)	The exponential function of E. This is e^E.
INT(E)	The result of ignoring any decimal fractional part of E. This is the greatest integer function if E is positive.
SGN(E)	The signum function. It has a value of $+1$ if $E > 0$, 0 if $E = 0$, and -1 if $E < 0$.
SIN(E)	The sine of E. The angle will be considered to be in radians.
SQRT(E)	The square root of E.

of A, B, and C have not been altered, but any previous value which X might have had has now been lost. Note that this procedure performed the instructions in the order indicated by the parentheses. Note also that the sequence followed is the one we would desire, namely the completion of the multiplication prior to addition or subtraction.

In (D.2.1) we saw a FORTRAN statement which required a square root. We saw that this could be done with the function notation SQRT(E) where E is an expression. There are several commonly used functions available to us in FORTRAN. Different compilers (or versions) of FORTRAN may have different sets of functions, but one is almost certain to find those of Table D.1. In the functions listed in Table D.1 it is usually necessary that E be a real number, a real variable, or an arithmetic expression which will give a result which is a real number as opposed to an integer.

These functions can be used as required in any *assignment statement*. An assignment statement is one involving an " = " sign. They cannot appear on the left side. With these functions and the arithmetic operations we can construct statements that will instruct the computer to perform any computation that we may need to have performed. While these statements are not sufficient to give complete instructions to a computer, they will go far in meeting our needs. We will consider another very important set of instructions in the next section.

Before leaving the matter of assignment statements, we should consider the manner in which such statements can be used. Consider the following set of statements as following one another in a *program* designed to inform the computer of our desires.

$$X = 10$$
$$Y = 15$$
$$X = 3*X + Y**2$$
$$A = SQRT(X - 155) + (X - Y)/X$$

Here we have a succession of statements. The first one assigns the value 10 to the variable X and the second one assigns the value 15 to the variable Y. In the third statement we take the current values of X and Y, that is 10 and 15, and we calculate the value $3(10) + (15)^2 = 255$. Having determined that the expression to the right of " $=$ " has the value 255, the third statement goes on to assign this value to X. Consequently the variable X now has the value 255 and there is no further record of the earlier use of 10 as a value since we did not arrange to save this value by the use of some other variable. The fourth statement uses the now current values of the variables X and Y, namely 255 and 15, the latter not having been changed to this point. The computation of the right side of the fourth statement is now carried out, and since $(255 - 155) = 100$, the square root will be 10. Also in the second term of this expression we will have $(255 - 15)/255 = 0.941176$. Consequently we now have 10.941176 as the value of the expression on the right, and this will be assigned as the new value for the variable A. This process would continue throughout the program with the purpose of assigning values to variables in such a way that ultimately some assignment will result in obtaining the answer to what might be a very large problem.

It is the task of the programmer to determine the computations that will be required, to make the necessary assignments throughout the program, and finally to arrange to have the results made available through some form of communication between the machine and the outside world. It is essential to remember that each variable can have only a single value at any one time. Any record of prior values held by the variable will be lost. Therefore, if the preservation of some value is needed, that value must be *stored* in some other variable prior to the changing of the variable originally having the value in question.

EXERCISES

1. What is the value assigned in each of the following statements, and to which variable is the value assigned?

   (a) X = 5*7 − 4/2
   (b) I = 45/7 − 2
   (c) SUM = 2.3 − 4.65 + 3*5.2
   (d) ANSWER = 3**2 − (9 + 4/2)*2
   (e) QUEST = (3*3 + 4*4)**0.5

2. Give the value of each variable involved in the following set of statements at the termination of the computation indicated.

   A = 12/5
   B = 2*3
   K = SQRT(34.)
   K = K + INT(A + B)
   A = K − B
   B = A**2

3. Give the value of the variable ANSWER in each of the following cases.

   (a) ANSWER = ABS(E) if E = −42.3
   (b) ANSWER = ALOG(4. −2. + .718281828)
   (c) ANSWER = ATAN(ANSWER/ANSWER) where it is known that
       ANSWER ≠ 0
   (d) ANSWER = COS(0.0) + SIN(0.0)
   (e) ANSWER = (1.0) + INT(2.3) − SGN(−23.9)
   (f) ANSWER = SQRT(14 − 5)

4. In each of the following determine whether it is an acceptable FORTRAN name or
   indicate why it is not.

   (a) MATH
   (b) WHICH
   (c) CALCULUS
   (d) COMP.
   (e) ZZZZ
   (f) ENGLISH
   (g) B2345
   (h) 3Z425
   (i) NEW$
   (j) A*B

5. In each of the following statements determine whether there is an error, and if there
   is, indicate a method for correcting the error, if possible.

   (a) X + Y = Z
   (b) ANS ≑ (A + B)(C − D)
   (c) XK = SIN ((A*B)
   (d) RESULT = FIRST*SECOND − THIRD/FOURTH**FIFTH
   (e) KKK = NEW + −OLD
   (f) COUNT = SIN(K)
   (g) ANSWER = SIN(ALOG(B)

D.3 Control Statements

The computer will execute the successive statements in a program in the
order in which they are written unless instructed to do otherwise. Therefore, in
writing a program you will need to be certain to put the statements in an
order which is logically correct for the satisfactory completion of your problem. There are times, however, when one wishes to return to an earlier part of
the program and use a statement a second time or perhaps to jump over some
statements and return to them later. This can be done in two ways. The first
method for doing this is simply to tell the computer to proceed to a specific
statement regardless of the status of the computation. This can be done by
the simple statement GO TO. However, it is apparent that this statement by
itself is not adequate, for there would have to be some method for determining

the statement to which the computer should next address its attention. This is accomplished by giving numbers to those statements to which we need to make reference. The *statement numbers* must be integers, and cannot be greater than five digit numbers. Thus, the smallest possible statement number is 1 and the largest possible statement number is 99999. There should be no commas or spaces in such numbers. Now it is possible to write the statement

GO TO 245

and the computer would proceed to the statement numbered 245 at such time as it came upon the statement just given. This makes it possible to write a program involving many computations but having only a small number of statements. The statements can be used many times by transferring control back to earlier statements in the program.

The GO TO statement has a major shortcoming if this is the only method for transferring control. It demands that the control be transferred every time we come to the statement, and this would probably have us in an *infinite loop*, that is one from which we could never exit. For this reason we would like to have some means for transferring control from one statement to another only if certain conditions were fulfilled. This can be done through the use of an IF statement. There are two constructions for the IF statement. The first one is of the form

IF (E) m_1, m_2, m_3

where E is an expression that when evaluated will be positive, negative, or zero. If E is positive the statement instructs the computer to proceed to the statement with the number indicated by m_3. If the expression is negative the computer is to go to the statement with the number m_1. If E is zero the computer is to go to the statement numbered m_2. If you think of the number line and the fact that in the customary diagram of the number line the negative numbers are on the left, zero is in the middle, and the positive numbers are on the right, you will find this helpful, for in this case the computer goes to the left number, m_1, if E is negative, to the middle one, m_2, if E is zero, and to the right number, m_3, if E is positive. Thus, the statement

IF (A*B + SQRT(A + B)) 230, 253, 125

would cause the computer to go to the statement with number 230 for the next assignment if the expression (A*B + SQRT(A + B)) is negative, to the statement numbered 253 if this expression is zero, and to the statement numbered 125 if the expression is positive. This requires that one be able to formulate the condition that will cause the decision in such form that it will cause the expression to become negative, zero, or positive at the correct times to insure the correct performance of the program. If one wishes only two choices, there is no reason why two of the three statement numbers cannot be the same number. However, they must be written with the three numbers separated by commas as we have indicated in the sample statements.

Table D.2

Relation	Meaning
.EQ.	Is equal to
.GE.	Is greater than or equal to
.LE.	Is less than or equal to
.NE.	Is not equal to
.GT.	Is greater than
.LT.	Is less than

There is a second form of the IF statement, called the *logical* IF statement. This form of the IF statement is not included in all FORTRAN compilers and you should check to make certain that it is available to you before using it. In this form of the statement there are two possibilities, for a condition is either true or false. We might write

IF (SUM.GE.OLD.OR.NEW.EQ.NEXT) A = B

In this case we are instructing the computer that if SUM is greater than or equal to OLD or if NEW is equal to NEXT, then replace the former value of A with the value currently held by B. Otherwise the "A = B" should be ignored and the program continues to the next statement. The *relations* that are available to use are given in Table D.2. These relations can be combined through the use of the *logical connectives* given in Table D.3. The general form of the logical IF consists of IF followed by a relation or a set of relations connected with appropriate connectives enclosed in parentheses and followed by a statement that is to be performed if the logical expression is true. In every case the next statement is performed unless the relation is true and the statement involved is a GO TO.

We can use the information we have gained so far to write a short program that will add up the first 500 integers. The program might be written

```
          SUM = 0
          NUMBER = 0
100       NUMBER = NUMBER + 1
          SUM = SUM + NUMBER
          IF (NUMBER.LT.500) GO TO 100
```

Table D.3

Connective	Meaning
.OR.	True if either relation is true or if both are true
.AND.	True only if both relations are true
.NOT.	Negates the truth or falsity of the relation
.XOR.	True only if one relation is true and the other one is false
.EQV.	True only if both relations are true or if both relations are false

In this case we initially start both **SUM** and **NUMBER** with the value zero. We then increase **NUMBER** by one and add number to **SUM**. If **NUMBER** has reached 500 we go to the next set of statements. Otherwise we increment **NUMBER** again and add the new value of **NUMBER** to **SUM**. It would have been possible to replace the **IF** statement by the statement

$$\text{IF (NUMBER} - 500) \ 100, \ 200, \ 200$$

provided the next statement is numbered 200. In this instance

$$\text{NUMBER} - 500$$

will be negative if **NUMBER** is less than 500. Only in this case do we wish to transfer control back to statement 100.

EXERCISES

1. Does every statement have to have a statement number? If not, which statements must have statement numbers?

2. What are the limitations on numbers that can be used as statement numbers?

3. In the quadratic equation $A*X**2 + B*X + C = 0$, the roots are real and distinct if the discriminant $(B*B - 4*A*C)$ is positive, the roots are real and equal if the discriminant is zero, and the roots are complex if the discriminant is negative. The case of real, distinct roots is handled in statement number 100, the case of equal roots in statement number 200, and the case of complex roots in statement 300. Write a single statement that will cause the program to go to the appropriate statement for the case at hand.

4. (a) The majority of **FORTRAN** compilers require that a statement following a **GO TO** statement or an arithmetic **IF** statement have a statement number. Why is this necessary?
   (b) Why is it not necessary that the statement following a logical **IF** statement have a statement number?

5. (a) Make a table that has in the first two columns the four possible cases involving the logical expressions **A** and **B**, such as both true or both false.
   (b) Append to this table the state resulting from **A.OR.B** for each of the four cases.
   (c) Append the state resulting from **A.AND.B** in a manner similar to that of part (b).
   (d) Append the states for the other connectives listed in Table D.3 in a manner similar to that of parts (b) and (c).

6. (a) Where will command be transferred as a result of the statement

$$\text{IF (SQRT(N*N)} - \text{N)20,30,40}$$

   (b) Where will command be transferred as a result of the statement

$$\text{IF (SQRT(A*A)} - \text{A)20,30,40}$$

   (c) Since the computer treats the number **A** as a binary number with a finite number of binary places together with the binary equivalent of the decimal point, and since the computer only retains a finite number of places, it is possible that **SQRT(A*A)** will not be *exactly* **A**. How might this affect your answer in part (b)?

(d) If you know that the discrepancy mentioned in part (c) is not more than one millionth, how might you modify the statement of part (b) to achieve the result that you think you should have?

7. (a) What simple logical expression is equivalent to .NOT.(A.OR.B)?
   (b) What simple logical expression is equivalent to .NOT.(A.EQ.B)?
   (c) What simple logical expression is equivalent to .NOT.(A.GT.B)?
   (d) What simple logical expression is equivalent to .NOT.(A.XOR.B)?

8. The program is able to make decisions through the use of the IF statement. State all of the options available to you for expressing the conditions on which a decision is to be made.

9. What will be the value of X at the conclusion of the following sequence of instructions?

```
      A = 3.5
      B = 4
      C = 6
      K = 2.3
   5  IF(A*B − 10)12,13,14
  14  B = B − 1
      GO TO 5
  13  X = 57
      GO TO 24
  12  X = K*C/B
  24  X = X − 2
```

D.4 Input/Output Statements

We have now learned how to write arithmetic statements and how to control the order in which statements are executed. Before we put all of these things together in an illustrative program, we should pause to indicate a means whereby we can enter information into the computer and by which we can obtain information from the computer. In general we will enter information with a READ statement. The READ statement will appear as

$$\text{READ(INPUT, 78)ITEM1, ITEM2, DATA1, DATA2, DATA3} \quad \text{(D.4.1)}$$

with as many symbols as necessary following the right parenthesis. The word "READ" is self-explanatory. However, the majority of computers have several devices from which they can receive information, and some instruction must be given concerning the device from which information is to appear. This information will be given in the form of a positive integer which we have here labeled INPUT. For instance, if the device code for the card reader were "2", we might have the sequence of instructions

```
      INPUT = 2
      READ(INPUT, 78)ITEM1, ITEM2, DATA1, DATA2, DATA3
```

and this would be equivalent to the single statement

READ(2,78) ITEM1, ITEM2, DATA1, DATA2, DATA3

You will need to obtain through a representative of your computing center the particular device code that you should use to indicate the input device that you will be using. This number depends in general upon the particular computer system being used, and may also be influenced by local practice. This number is known as the *input device code*.

The number 78 in (D.4.1) must match the statement number of a **FORMAT** statement. This requires that there be a statement numbered 78 in your program, and this statement must start with the word **FORMAT**. The statement of (D.4.1) is instructed to read five items, the first two of which are integers and the last three are real numbers. In fact the statement will probably read a long string of digits with the possibility of decimal points appearing in the real numbers. The program must then have instructions concerning how many of those digits are to be considered as the digits of **ITEM1**, and which digits are to be so considered. Similarly it is necessary to know which digits are to constitute **ITEM2**. Additional information must be supplied to locate the numbers **DATA1, DATA2,** and **DATA3**. The only thing that is certain is that the five numbers referred to in (D.4.1) must appear in the order in which they are given. It is the purpose of the **FORMAT** statement to indicate where this data is to be found. If we find in the program the statement

78 FORMAT(I5, 4X, I8, 2F10.4, F12.8)   (D.4.2)

we know that this is the particular **FORMAT** statement that corresponds to the **READ** statement (D.4.1). This informs us that the first integer to be found, namely **ITEM1**, is to occupy the first five places in the string of characters that are read by the **READ** statement. This is indicated by **I5**, with the **I** standing for *integer* and the **5** indicating that five spaces are to be used. The next four spaces are to be ignored, as indicated by **4X**. The **X** indicates that the space is to be considered as though it were blank. The next number, in this case **ITEM2**, is to be an eight digit integer as indicated by **I8**. The description **F10.4** indicates that we have a *real number* which occupies ten spaces, the last four of which will be considered as decimal places *unless* a decimal point actually appears in the string of characters within these ten spaces. In the latter case the number would be read as it appeared, namely using the decimal point that actually appeared in the string of characters. The **2** in front of **F10.4** indicates that there are to be two numbers with this description. These numbers, of course, would be **DATA1** and **DATA2** from left to right. Finally **DATA3** would be described by **F12.8**. This indicates that twelve spaces are to be used, and in case a decimal point fails to appear in this set of twelve characters the last eight digits are to be considered as following the decimal point. Incidentally, the **F** remains from an earlier designation of *floating point* numbers in lieu of *real numbers*.

```
    1         2         3         4         5         6         7         8
12345 678901234 5678 9012345678 9012345678 9012 345678901234567890123456789012345678 90
ITEM1     ITEM2---DATA1-----DATA2-----DATA3-------
```

Figure D.1

We can summarize the last paragraph in case we have statement (D.4.2) in the program by pointing out that the numbers would appear on the card as indicated in Figure D.1. The "—" indicates additional spaces associated with the preceding entry, and the blanks indicate spaces that will not be read by the statements of (D.4.1) and (D.4.2). It should be relatively easy to produce READ statements and FORMAT statements for other arrangements of data. It is often helpful to prepare a *layout* as shown in Figure D.1, for then one can count the number of places required for each entry in the FORMAT statement, and can insure that these entries match the corresponding entries in the READ statement. The one thing you must remember is that *all* spaces beginning with the first must be accounted for with the exception that it is not necessary to include spaces at the right hand end of the line if they are not to be involved in placing additional data.

Instructions for writing numbers follow a pattern similar to the one for reading numbers. The statement used might be

$$\text{WRITE (IOUT, 95) NUMBER, ANS1, ANS2, ANS3} \qquad \text{(D.4.3)}$$

where IOUT refers to an *output device code*. The output device code is very similar to the input device code discussed earlier. As indicated above, you can obtain the list of the available output devices and their corresponding output device codes from your local computer center. The number 95 is a statement number referring to a FORMAT statement just as in the case of the input statements. The symbols NUMBER, ANS1, ANS2 and ANS3 denote the quantities to be written. The FORMAT statement serves a purpose here similar to that in the READ statement. It will indicate how we want the information delivered to us. Note that we must provide for an integer and three real numbers in that order. However, we have the privilege of spacing the numbers so they can be more easily read, inserting labels as we may wish, and a choice of forms in which the real numbers can be written. A sample FORMAT statement might be

95 FORMAT(1X, 'THERE ARE', I5,' ITEMS WITH RESULTS',
    2F12.5, E18.6)   (D.4.4)

or

95 FORMAT(1X, 9HTHERE ARE, I5, 19H ITEMS WITH RESULTS,
    2F12.5, E18.6)   (D.4.5)

If this is to be printed on a printer, it is probable that some means will be used to control the number of lines by which the paper moves up between successive lines of print. The insertion of the 1X in this case is to cover a wide variety of situations. If the printer requires such information, this will provide it in the form of an instruction to print on the next line. If your output device does not require such information, this merely means that all of your subsequent output via the **FORMAT** statement will start in the second available space, counting from the left. The characters contained between the apostrophes in (D.4.4.) will be printed (or punched) just as they are. Thus, the first available space will have a T in it, the second one an H, etc., and the ninth place will have the E of ARE. Some computers do not permit the use of the apostrophe in the manner of (D.4.4). In such a case one omits the apostrophes and precedes the phrase "$THERE\ ARE$" with **9H** where the 9 indicates the number of characters, including the space, that are to be printed or punched. The H is the initial letter of Hollerith, the name of the man who initiated some of the current processes for use of such cards. This is shown in (D.4.5). One of these two methods will provide you with the capability for printing the words which will make the results much easier to interpret. The particular method used will have to be ascertained by checking with your computing center, for it will depend on the compiler available to you. After the expression $THERE\ ARE$ the value of **NUMBER** will be placed. The value that will be printed is the value which the variable **NUMBER** represents at the time that the **WRITE** statement is executed. Remember that **NUMBER** may have many different values during the execution of the program, and only the last one will be available. This value is to occupy the five spaces in positions 10-14. As before, the I5 is a specification indicating that an integer is to be placed in this space and that it is to occupy five spaces. Following this number, and beginning in space 15, we will find ITEMS WITH RESULTS. Note that we left a space in front of $ITEMS$, for had we not done so the I of $ITEMS$ would have followed the number preceding it without a space. The underlining here is inserted merely to indicate the positions that will be printed with this specification. Note that the spaces are included, and are counted in D.4.5). The first two answers would then be printed, using 12 spaces each. The last five of the spaces would follow the decimal point, and the decimal point would be inserted in the seventh of the 12 spaces. This leaves 6 spaces for the sign and the integral part of the number combined. Finally, the third real number would be printed out in a form indicating a number between zero and one and giving a power of ten by which this number is to be multiplied. We would find Avogadro's number printed out in this specification as $0.606000E\ 24$, indicating that we have 0.606 multiplied by 10^{24}. The space between the E and the 24 is reserved for a negative sign in case the exponent is negative. You will find that this exponential form, designated by the E, is very handy for very large and very small numbers.

　　We will now put some of these concepts together in an example to illustrate the information that we have given to this point.

EXAMPLE 4.1. Write a **FORTRAN** program which will cause the computer to read a set of data cards, each card having one real number punched into the first ten columns of the card (the decimal point is included as part of the number), which will then obtain the average of the data and will also obtain the square root of the average of the squares of the data items. The number of cards present is not known, but it is known that no card has a number larger than 90000.

Solution. Our job is to write a program that will add up the various numbers, keep track of the number of data items, and also add up the squares of these numbers. After the last card has been read, it will then be possible to divide the sum of the numbers and the sum of the squares by the number of cards and then to take the square root of the average of the squares. We must then write the results, for they will do us no good if they remain in the computer. A description of the various steps involved is given in Figure D.2. To the right of the description is a flowchart indicating the logic of the problem in a pictorial manner and to the right of the flowchart we find the FORTRAN program. A flowchart is *very* important. It is difficult to keep the logic straight for a program of any size. The flowchart gives a pictorial representation of this logic. (See the discussion of flowcharts in Section I.9.) A good flowchart will be of inestimable assistance in tracking down errors in logic if the program fails to perform as desired. You should *never* try to write a program without first drawing a flowchart as indicated in Figure D.2. Note that the program is written using only capital letters. The keypunch for punching the computer cards or the typewriter terminal for entering a program will ordinarily be restricted to capital letters, and you might as well get used to this.

After indicating START in the flowchart, we have step (a) in which we place zeros in the variables we are going to use to insure that our results reflect only our data and do not include someone else's leftovers. As indicated, we will need three registers, and we have chosen to call these SUM, SUMSQ, and KOUNT. The use of SUM should be apparent, as is the use of SUMSQ to indicate the sum of the squares. The word KOUNT is spelled with the initial letter K to insure that FORTRAN will treat it as an integer. The SUM and SUMSQ are to be treated as real numbers.

In step (b) we read a card. We have chosen to call the number on this card DATA. This number will change many times during the execution of the program, for it will always represent a single number, namely the last one read in. Therefore we must be certain that we do everything we need to with one value of DATA before we read another card and replace the former value. It could well happen, however, that we would not wish to use the information read, for we might have finished reading all of the cards given us. If someone else has entered computer cards after ours, the machine would have no way of knowing that we were through. For this reason we will use an artificial device for informing the machine that it has completed our particular problem. We will simply insert a final card with the number 99999.9 punched in it. By the information given us we know that this could not be a legitimate data card, and therefore this should be a signal to the computer that no more

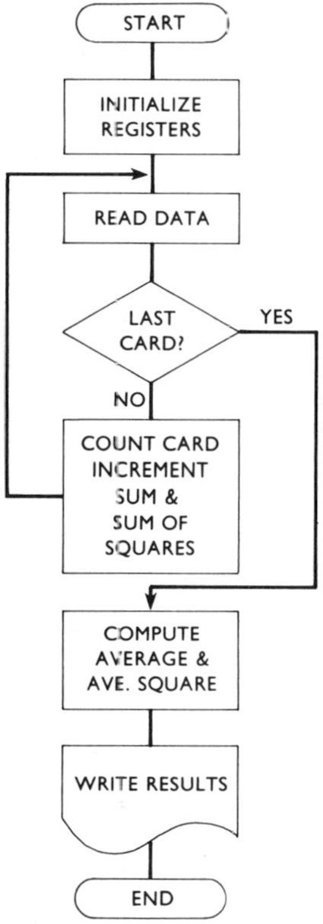

```
       SUM = 0
       SUMSQ = 0
       KOUNT = 0
       ICARD = 2
       IOUT = 5
  5    READ (ICARD, 10) DATA
 10    FORMAT (F10.3)

       IF(DATA-95000.) 20, 20, 30

 20    KOUNT = KOUNT + 1
       SUM = SUM + DATA
       SUMSQ = SUMSQ + DATA**2
       GO TO 5

 30    AVE = SUM/KOUNT
       AVSQ = SQRT(SUMSQ/KOUNT)

       WRITE (IOUT, 40) KOUNT, AVE, AVSQ
 40    FORMAT(1X,I5,' CARDS', 10X, 2F12.6)

       END
```

Figure D.2

a. Place zero in each register

b. Read in a number from a card.

c. Is this the last card? If not increment registers. Otherwise compute final results.

d. Compute final answers and print results.

e. The end of the program. This is an instruction to the compiler to finish translation into 1's and 0's used by the computer.

information should be accumulated. With this in mind, we have inserted an IF statement to permit us to answer the question of part (c) of the problem. This particular statement will take the last number read from a card and subtract 95000. The number 95000 is written with a decimal point, for some computers will not handle integers this large, but they will handle this as a real number. If this last number, which was read, and which is the current number to be named DATA, was one of the numbers from the data cards, the expression in the IF statement would be negative, for none of the numbers in the given data exceeded 90000. In this case we would wish to go ahead with our computation and then read another card. However, when our fictitious card is read, the result will be $99999.9 - 95000 = 4999.9$ and will be positive, indicating that we should not use this information, but should complete the required calculations. Thus, if the result is negative we have informed the computer that it should proceed to statement 20, and process this data in accordance with our instructions. If the difference is positive, the execution is to proceed immediately with statement 30 which will finish the computation and print out the results. It is necessary to include a statement number to be used if the difference is zero, even though we know it cannot be zero. Since the zero case cannot arise, it makes little difference whether we indicate that it should go to statement 20 or to statement 30, but we must indicate some *existing* statement to which it could go. In the case of statement 20, the program instructs the computer to add one to the number already in KOUNT, to add the number read from the card to the number already in SUM and to add the square of the number DATA to the number already accumulated in SUMSQ. After doing this, we indicate, just as the flowchart directs, that the program should go to statement 5 which is the one that requires the reading of another card.

If the program is directed to statement 30, as will be the case when the card containing 99999.9 is read, the average is calculated and given the name AVE. In one step we will divide the sum of the squares of the number by the number of such numbers and then take the square root of the quotient, denoting this result by AVSQ. The next statement directs the computer to write out the results, and finally the last statement indicates that we have come to the end of the program by so stating with the FORTRAN statement END. It may be necessary to place a statement which reads CALL EXIT before the END statement, but this again should be in the set of instructions you receive from your computing center.

Notice that we have used ICARD and IOUT as device codes for the READ and WRITE statements respectively. In this program we have given ICARD the value 2 and IOUT the value 5. You would need to replace these with the correct device codes for the equipment you are using. Statements 10 and 40 are the FORMAT statements for READ and WRITE respectively. While it is not necessary to assign the device code and put the FORMAT statement near the corresponding READ and WRITE statements, it is sometimes convenient to do so.

This program is now ready to be punched into cards and *read* into the computer. **FORTRAN** statements are punched starting with the *7th column* of the card. The *statement number*, if there is one, must appear somewhere in the first five columns of the card. If you wish to aid the understanding of the program by inserting comments for your edification with the assurance that the computer will ignore them, you need only place a *C* in the *first column* of a card you wish the computer to ignore. Finally, the computer will ignore the last eight columns (73–80 of an 80 column card) in reading a **FORTRAN** program. Therefore do *not* extend your statements beyond column number 72. If you need to continue any statements, this can be done by continuing in column 7 of the next card while placing a "1" in column 6 and insuring that the continuation card does not have a statement number. Note that our discussion of the location of statements applies only to the source program you have written, such as the one in Figure D.2. As for data, you can place that in any location that you may wish in a card provided you give an appropriate description in the **FORMAT** statement. This permits you to use all 80 columns for data if you so desire.

With the information given you have enough information to write a program to perform any calculation. The appropriate use of the **IF** statement will permit you to implement any decisions that have to be made en route to the result. If you find the **FORMAT** statements confusing, you only need realize that you can read all numbers in as real numbers, inserting a decimal point in each number. You can also print all numbers out as real numbers (if you use only symbols that represent real numbers), and furthermore, you can adopt some standard **FORMAT** such as **F15.7** which would allow you to use numbers with as many as six integer digits and seven decimal digits regardless of the sign of the number. That should certainly be sufficient for the vast majority of your problems. If that is not enough use **E16.6** since the exponential notation removes all restrictions except those imposed by the computer itself.

EXERCISES

1. (a) What is the purpose of a device code?
   (b) What is the purpose of the **FORMAT** statement?
   (c) What is the purpose of the **1X** at the beginning of the expression following the word **FORMAT** in the **FORMAT** statement?

2. Which of the following are legitimate expressions for use in a **FORMAT** statement and what do they indicate?

   (a) **F12.3**
   (b) **I8**
   (c) **I9.4**
   (d) **E18.8**

(e) F14
(f) F10.12
(g) 7X
(h) X5.2

3. If K = −423 and X = 234.718 and if IOUT is the device code for a line printer, state precisely what the output would look like for each of the following parts. [Be sure to include spacing.]

(a)   WRITE (IOUT, 47) K, X, X
    47   FORMAT(1X, 'K HAS THE VALUE', I6, F8.2, E16.4)
(b)   WRITE (IOUT, 5) X, K
     5   FORMAT (1X, F5.0, 7HFOR THE, I4, 9H-TH VALUE)
(c)   WRITE (IOUT, 234) X, K, X
   234   FORMAT ('THE VALUE OF', F12.6, 'IS THE', I6, '-TH VALUE OF',
           F6.1)

4. The first ten columns of a card have the ten digits 0002349876. What value or values would the variables have after the READ statement in each of the following cases?

(a)   READ(ICARD, 14) X
    14   FORMAT(F9.3)
(b)   READ(ICARD, 23) X, K
    23   FORMAT (F7.2, I3)
(c)   READ (ICARD, 37)A
    37   FORMAT (2X, F6.4)
(d)   READ (ICARD, 49) K, L
    49   FORMAT(I4, 3X, I3)

5. (a) How do you indicate that you are inserting a comment in a FORTRAN program?
   (b) Why would you wish to use a comment in a program?
   (c) How many comments should you use in a program?

6. You wish to write a program which would instruct the computer to take the average of the square roots of the first thousand positive integers.

   (a) Draw a flowchart for this program.
   (b) Write a program that will carry out this computation.
   (c) Run your program, doing any debugging that may be necessary.

7. You are to write a program that will read a card having four numbers, the first two of which are in columns 1–8 and 9–16 and the second pair being in columns 31–40 and 41–50. Each of the numbers is to have two decimal places. The first pair on each card is to be considered as the coordinates of a point, and the second pair the coordinates of a second point. Your program is to read a card, compute the length of the segment joining the two points and the slope of the segment joining the two points, and then print out the coordinates of the two points and the length and slope you have calculated. Draw a flowchart and write and run the program.

D.5 Subscripts

You have been exposed to sufficient information thus far to permit handling any situation that may arise as far as programming is concerned. However, there are additional features of **FORTRAN** which make the task of programming certain problems much easier. We will cover just a few of these features, the ones that are most commonly used in beginning programming, in the next few pages.

As you have often noted in mathematics, it is convenient to use subscripts rather than try to give each variable a distinct name. This is particularly true when we wish to go through a rather long list of items. In view of the fact that the computer is not able to half-space the printer as a rule, it follows that we must use some modification of the method of writing small numbers on a line slightly below the character itself for denoting subscripts. The method used is that of writing the *subscript* immediately following the variable name, but enclosing the subscript in parentheses. Thus, we would write $X(1), X(2), X(3), \ldots, X(50)$ if we wished to indicate a single array having fifty subscripted variables. It is clear that the subscript must be an integer, and consequently we would use an integer variable in indicating a variable subscript. Thus, we would write $X(ISUB)$ where **ISUB** is the subscript and the **I**, as before, indicates that we have an integer. All of this is very nice, but we must not forget that the computer is not able to look ahead and see the range of values we intended to use for subscripts. Therefore, we must precede the first use of a subscripted variable with information indicating the maximum value the subscript is to be permitted to have for each variable. This is done with a **DIMENSION** statement. If we were to use three subscripted variables, **X**, **DATA**, and **NUMBER** in a given program, and if the maximum subscript we would use for **X** is fifty, the maximum subscript for **DATA** is one hundred, and the maximum subscript for **NUMBER** is twenty, then we would use a statement such as

$$\text{DIMENSION } X(50), \text{ DATA}(100), \text{ NUMBER}(20) \qquad (D.5.1)$$

This statement must come at the very beginning of your program in order that the compiler may know how many values it must arrange for. Observe that we can give the dimension of several dimensioned *arrays* in a single statement. Also observe that we separate the various dimensioned variables by commas, but we do not put any punctuation at the end of the statement.

It is possible to use more than a single subscript in **FORTRAN**. The number of subscripts you may use will depend upon your compiler, and this number is additional information you may want to get from your computing center. If you wish to use more than a single subscript, you must again indicate this in the **DIMENSION** statement. Thus, we could use **DIMENSION** $Y(25, 30)$ if we wished to use the doubly subscripted variable **Y** with the first subscript never being larger than 25 and the second never larger than 30. If we were to use this array, we could later refer to $Y(K, L)$ where **K** and **L** are integers and K should have a positive value no larger than 25 and L should have a positive value no larger than *30*.

The use of subscripts permits the use of a large number of variables without having to devise distinct names for each variable. It also permits relatively easy storage of a relatively large number of values for so much of the time the program is running as one may require. The use of subscripts in a program is eased by the fact that we can use such expressions as X(I), X(I + 4), and X(2*L − 5). In each case the subscript, when evaluated, must be in the interval indicated in the corresponding DIMENSION statement. If arithmetic is to be performed in the subscript, it is mandatory that the variable term be the first term, and that it be modified only by multiplying by a constant and adding or subtracting a constant. If there is multiplication, the constant must precede the variable.

EXERCISES

1. Which of the following are valid expressions in FORTRAN?

   (a) A(X)
   (b) DATA(KSUB + 4)
   (c) XYZ(KSUB*4)
   (d) ALGEBRA(K)
   (e) MATH(4*M − 34)
   (f) TERM(34)
   (g) NUMBER(4 + I)
   (h) TOTAL(3*J)
   (i) SUM(I/J)
   (j) NEXT(I5 − 3)

2. Which of the following are valid in FORTRAN?

   (a) NUMBER(2*X, 3*Y)
   (b) X(345, 432)
   (c) Y(2*K − 1, L + 1)
   (d) DATA(2*K, 3*1 + 2, 4*M)
   (e) ITEM(K + 1, L − 2, M − 3)
   (f) Z(I, I, I)

3. What information is conveyed by each of the following statements?

   (a) DIMENSION X(23, 35), Y(15)
   (b) DIMENSION ITEM(100), JOB(25, 40)
   (c) DIMENSION DATA(10, 20, 20)

4. (a) Why is the DIMENSION statement necessary if you are going to use subscripted variables?
   (b) Where does the DIMENSION statement come in the program? Do you have any choice in where you put this statement?

5. List all of the possible types of subscripts if you denote constants by c_1 and c_2 and denote variables by v_1 and v_2, using as many constants and variables as may be permitted. [*Hint*: $v_1 + c_1$ is a valid form for a subscript.]

6. Write a program that will use an array with 100 variables and which will assign the number I to the I-th variable. Have the computer print the numbers on ten lines, with each line having ten equally spaced numbers.

D.6 Do Loops

As you observed in the program of Section D.4, we had a loop in the flow-chart and a consequent loop in the program. We implemented this by sending the execution back to statement 5 at regular intervals. We are often in the position of wanting to execute a *loop* a given number of times. While this can be done by using some symbol, such as **KOUNT**, in conjunction with an **IF** statement to keep track of the number of times we have gone around the loop, a somewhat more direct method has been made available in the **FORTRAN** language. This involves the use of the **DO** statement. The **DO** statement itself is the first statement of the loop. The last statement of the loop must be given a statement number. The statement number given the last statement of the loop must follow the word **DO** in the **DO** statement. If we were to write

$$\text{DO 67 INDEX} = \text{IBEG, IEND} \qquad \text{(D.6.1)}$$

we would indicate that the statement (D.6.1) is the first statement of the loop and statement 67 is the last statement of the loop. We further indicate that we are using the integer variable **INDEX** to keep track of the number of times we have gone around the loop, that **INDEX** is to start with the value given by **IBEG**, which can be either an integer variable or a positive integer, and that we will go around the loop for the last time when any further incrementing of **INDEX** would give us a value larger than **IEND**. Normally we do not place a comma after the location indicated by **IEND** in (D.6.1), nor do we place anything else after **IEND**, particularly if we wish to let **INDEX** take on the successive integer values, **IBEG, IBEG + 1, IBEG + 2**, etc. We would stop with **IEND** in our statement. However if we wished to increment **INDEX** by something other than one, we would include the desired increment as additional information. Thus

$$\text{DO 67 INDEX} = 13, 26, 2$$

would cause the program to go through the loop first with **INDEX = 13**, and upon completing the execution of statement 67 we would proceed with **INDEX** $= 13 + 2 = 15$, going again through statement 67, and successively using the values *17, 19, 21, 23*, and *25*. Since *27* would be larger than *26*, the loop would not be executed with **INDEX** = *27*. If the value of **INDEX** is equal to the terminating value, the execution will take place.

EXAMPLE 6.1. Write a program that will find the sum of the first 100 integers and will also find the sum of the squares of the first 100 integers.

Solution. In this case we will give a program first which would accomplish the required computation, and then we will explain it. You should realize that in general you will need to draw the flowchart and go through the procedures we used in Section D.4.

```
      SUM = 0
      SUMSQ = 0
      DO 50 K = 1, 100
      SUM = SUM + K
 50   SUMSQ = SUMSQ + K*K
      WRITE (IOUT, 60) SUM, SUMSQ
 60   FORMAT (1X, 'THE SUM IS', F10.0, 10X, 'THE SUM OF THE
         SQUARES IS', F25.0)
      END
```

In this case we have made certain that both SUM and SUMSQ start with the value zero. We then have the DO loop in which K starts with one and continues with increments of one to one hundred. For each value of K we have added that value to SUM and have added the square of that value to SUMSQ, since both are completed by the time we have finished executing statement number 50. We have assumed that the computer being used will permit adding an integer to a real number. As soon as the loop has been performed 100 times and K has been incremented to 101, a number larger than the largest value for which we will carry out the loop, the program proceeds to the first statement following the loop, that is the statement involving WRITE. The use of the WRITE statement follows our instructions concerning input/output statements, and should be easily followed. Finally we conclude with an END statement.

It should be clear that the DO loop can be a very convenient programming device in that it eases the task of programming loops. However, there is one pitfall we must avoid. The performance of the DO loop is jeopardized if the final statement of the loop, the numbered statement referred to in the DO statement itself, causes the program to branch somewhere else. Thus, this final statement cannot be a GO TO or an IF statement. If it appears that this is necessary and there is no convenient method for avoiding this impasse, we can simply add one more statement to the loop and place the relevant statement number on this additional statement. This is obviously permissible if this additional statement is there but does nothing. Such a statement consists of the single word CONTINUE. In other words, the final statement of the DO loop starting with DO 57 I = 1, 10 might well be

57 CONTINUE

It should be realized that **CONTINUE** is more than six letters in length, but this is a special case and the computer will accept this word.

EXERCISE

1. You find the statement DO 234 KEY = INIT, LAST, JIG

   (a) What is the purpose of 234?
   (b) What is the role of **KEY**?
   (c) What is the role of **INIT**?
   (d) What is the role of **LAST**?
   (e) What is the role of **JIG**?
   (f) What parts of this statement (if any) could be omitted and still have a valid statement?

2. Is it possible to use other programming and accomplish the same thing without using a **DO** loop? If so, how would you do it?   .

3. (a) What is the purpose of the word **CONTINUE** in a **FORTRAN** program?
   (b) Is the word **CONTINUE** invalid because it has more than six letters? Explain your answer.
   (c) What statements are *not* permitted at the end of a **DO** loop?

4. Write a **DO** loop for the purpose of computing $N!$ for a given value of N.

5. Write a program using a **DO** loop that would assign to X(I) the value I*I for each number in an array of 50 numbers.

D.7 Function Statements

One other bit of **FORTRAN** that may be helpful in the calculus involves the **FUNCTION** statement. If we write at the beginning of the program a function statement, (in case there is a **DIMENSION** statement we would place the function statement immediately after the **DIMENSION** statement) we have the luxury of letting the computer calculate function values with a minimum of effort on our part. For instance, we might write

$$F(X) = 1/(SIN(X) + COS(X)) \qquad (D.7.1)$$

and later write F(.4). The program will supply the value of $1/(\sin .4 + \cos .4)$. We can include as many function statements as we may wish, but each must have its own function name, and they must all come at the beginning of the program, subject only to the fact that they must follow any **DIMENSION** statements. If we wish to use a function of more than one variable, we would do this by using a notation such as F(X, Y). We will use these concepts in the following example.

EXAMPLE 7.1. We are given two numbers, $a = .4$ and $b = 1.8$, and we are to divide the interval from a to b inclusive into 100 equal subintervals. If we denote the 101 points by x_k where k has values $1, 2, 3, \ldots, 101$ we will have $x_1 = a$ and $x_{101} = b$. We also wish to consider 100 values t_k where t_k is the midpoint of the interval from x_{k-1} to x_k. This means we will have values t_k for $k = 2, 3, 4, \ldots, 101$. Next we wish to obtain the sum of 100 terms, each of the form $\sin t_k [x_k^3 - x_{k-1}^3]$. Write a **FORTRAN** program that will perform this computation.

Solution. We need 101 values of a variable we will designate by X and we need 100 values of a variable called T. However, if we follow the terminology given us, the largest subscript for T will be 101, and therefore we must make provision for this subscript in our **DIMENSION** statement. We also see that we will have to evaluate both $\sin x$ and x^3 for values of t_k and x_k respectively on 100 occasions. It would be convenient to simplify the reference to these functions through the use of function statements. With these thoughts for easing the programming involved, let us write the description, the flowchart, and thence the program as we did before. You will find this in Figure D.3. Note that we started with the **DIMENSION** statement, as we are required to do. We next inserted the function statements, for these must come next. The next thing that must be done is the determination of the values of the x_k and t_k. The determination must be in that order since we need the end points of the interval to determine the midpoint. There are, of course, alternatives because of the very regular arrangements of the x's in this case, but we have chosen the obvious method. After we have the points determined, we are now in a position to compute the value of the 100 summands, and to add them as we go along. This is done in the **DO** loop which terminates with statement 20. Finally we have given instructions for writing out the results.

We have used two **DO** loops. It is important to observe that the first one must either terminate before the second one begins, as it does in this case, or else the second one must be completely included in the first one. Thus if the statement number 10 referred to as the end of the loop in the first of the two loops had not come before the statement that begins **DO** 20, then statement 10 must not appear before statement 20. Observe the use of comments to assist in understanding the program.

EXAMPLE 7.2. Read a number N which represents the number of data items we are to process and then read in this many data items. Sort the numbers to put them in an ascending order such that each number in the sequence is at least as large as those preceding it. Write the resulting sequence.

Solution. We have written out a description and drawn a flowchart to indicate the logic involved in Figure D.4. In this case we read the number N first, as directed. Note in the program that we have used a **DIMENSION** statement that allocates 1000 locations for data entries. This does not mean

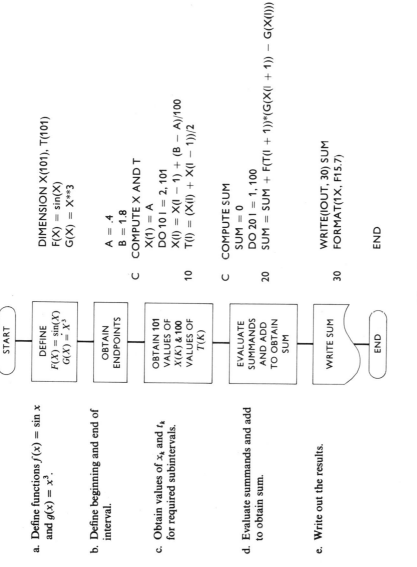

START

DEFINE
$F(X) = \sin(X)$
$G(X) = X^3$

OBTAIN
ENDPOINTS

OBTAIN 101
VALUES OF
$X(K)$ & 100
VALUES OF
$T(K)$

EVALUATE
SUMMANDS
AND ADD
TO OBTAIN
SUM

WRITE SUM

END

a. Define functions $f(x) = \sin x$ and $g(x) = x^3$.

b. Define beginning and end of interval.

c. Obtain values of x_k and t_k for required subintervals.

d. Evaluate summands and add to obtain sum.

e. Write out the results.

```
        DIMENSION X(101), T(101)
        F(X) = sin(X)
        G(X) = X**3

        A = .4
        B = 1.8
C       COMPUTE X AND T
        X(1) = A
        DO 10 I = 2, 101
        X(I) = X(I - 1) + (B - A)/100
10      T(I) = (X(I) + X(I - 1))/2

C       COMPUTE SUM
        SUM = 0
        DO 20 I = 1, 100
20      SUM = SUM + F(T(I + 1))*(G(X(I + 1)) - G(X(I)))

30      WRITE(IOUT, 30) SUM
        FORMAT(1X, F15.7)

        END
```

Figure D.3

723

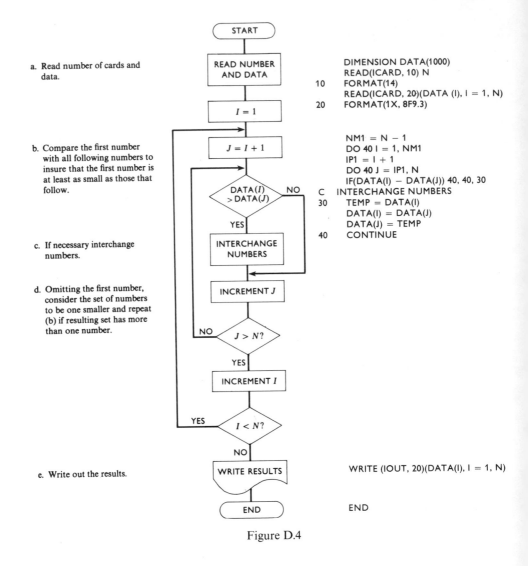

a. Read number of cards and data.

b. Compare the first number with all following numbers to insure that the first number is at least as small as those that follow.

c. If necessary interchange numbers.

d. Omitting the first number, consider the set of numbers to be one smaller and repeat (b) if resulting set has more than one number.

e. Write out the results.

```
                    DIMENSION DATA(1000)
                    READ(ICARD, 10) N
         10         FORMAT(14)
                    READ(ICARD, 20)(DATA (I), I = 1, N)
         20         FORMAT(1X, 8F9.3)


                    NM1 = N − 1
                    DO 40 I = 1, NM1
                    IP1 = I + 1
                    DO 40 J = IP1, N
                    IF(DATA(I) − DATA(J)) 40, 40, 30
         C    INTERCHANGE NUMBERS
         30         TEMP = DATA(I)
                    DATA(I) = DATA(J)
                    DATA(J) = TEMP
         40         CONTINUE




                    WRITE (IOUT, 20)(DATA(I), I = 1, N)


                    END
```

Figure D.4

that we must use 1000 entries, but rather that we are permitting ourselves that many. After reading in N with the appropriate **FORMAT** statement for reading an integer, we proceed to read the data. In this case it is assumed we are ignoring the first column of the card and then putting 8 numbers to a card, allocating 9 columns for each number. In the **READ** statement we have indicated we wish to read **DATA(1)**, and then follow it with **DATA(2)**, proceeding until we have read **DATA(N)**. Note that the notation used is very similar to that in the **DO** statement. This is often called an *implied* **DO** *loop*. If there are more than 8 numbers, the program will cause the reader to read successive cards, each having the same **FORMAT**, until the total number of data items requested have been read. The questions raised by the diamonds in the flowchart are answered in this case by means of two DO loops. Note that one is inside the other and that both end with statement *40*. In the first instance we wish to take our earlier entry for comparison, and since this could not be the *N*-th number we must end the DO loop with the entry $N - 1$. Since we are not permitted to put such an expression in as a limit for a DO loop, we have written NM1 (standing for *N minus 1*) $= N - 1$ in order to have a single variable that can serve as the limit for the DO loop. We wish to compare this with a second number that will come some time after the first in sequence, and hence we start the inner DO loop at IP1 (*I plus 1*). We then make the comparison in the IF statement. If DATA(I) is larger than DATA(J), we will go to statement *30* and in this case we will interchange the numbers. Note that we must temporarily save the value of DATA(I) before we replace DATA(I) with DATA(J) or we would lose DATA(I). If on the other hand the numbers are already in order, we merely want to go to the next comparison. We do this by sending the program to statement *40*, a CONTINUE statement, which causes no computing to occur, but does give us the end of the DO loops. Note that we can end two DO loops with the same statement. The operation here will have a value of I determined by the outer DO loop (the first DO statement we come to) and then for each value of I we will run through an entire list of values for J starting with the value (I + 1) and going to the value N. After we have completed the outer DO loop, which means that we have run the entire range of the inner DO loop (N − 1) times, we will proceed to the WRITE statement. You will note that in this case we used the same FORMAT we had used for the second READ statement. The presence of the 1X in the FORMAT statement makes this suitable for a printer, in case that is what we are using, and there is no reason why many input and output statements cannot refer to the same FORMAT statement. The implied DO loop for the WRITE statement is similar to that for the READ statement.

The use of the DO loops in this example makes it slightly more difficult to relate the description, the flowchart, and the program, for we do many things at once with a DO loop which would require separate description in the flowchart. The logic is the same, however.

EXERCISES

1. (a) What are the advantages of using a FUNCTION statement?
   (b) Where should a FUNCTION statement appear if there is no DIMENSION statement in the program?
   (c) Where should a FUNCTION statement appear if there is a DIMENSION statement in the program?
   (d) How many FUNCTION statements can there be in a single program?

2. Explain in detail all of the computation that takes place as a result of the single statement labeled **20** in Example 7.2.

3. Why is it necessary to introduce the variable **TEMP** in Example 7.2?

4. Write a function statement that would use the function $f(x) = x^3 + 2x^2 + 3x + 4$.

5. Write a program that will use a function statement corresponding to the function $f(x) = x^3 + 3x^2 - 4x + 2$ and will then, using a **DO** loop, print out a table giving the values of x from -2 to $+4$ at intervals of 0.5 and for each value giving the value of $f(x)$.

6. Write a program similar to Example 7.2 that will sort numbers in such a way that the largest number is first and successive numbers are smaller.

7. Write a program that will make a list of factorials for all non-negative integers which are not greater than 20.

D.8 Summary

In this brief account we have given a sufficient amount of information to permit you to write programs to do any computation that may be required in the calculus. If you wish to write more complicated programs, you should consult the technical manuals on **FORTRAN** supplied by the manufacturer of the computer you are using, or consulting one of many books available on the **FORTRAN** language.

After the program has been written, you will try to run it. You may find that you have one or more *errors* (or *bugs*) in your program due to errors in typing, incorrect punctuation or spelling, or due to a misunderstanding of some feature of the **FORTRAN** language. Such errors, often called *syntax errors*, are usually accompanied by diagnostic messages from the computer which will assist you in locating the errors. Do not be dismayed by such errors, for every programmer has bugs in programs at frequent intervals. Correct the errors and try again. Eventually you will have a program that will "*run*". Do not relax at this point, for you should check the results to make certain that the program has done what you wished it to do. The fact that a program runs merely means that the computer is able to make an interpretation of each of your statements, and consequently it is able to do something. It is your task to make certain that that something is what you wanted done. Do this by using somewhat simple data initially. If it works for simple data, it is probable that it will work equally well for the more cumbersome numbers you really wanted to use in the first place. Here the problem is likely to be in the logic. If you have followed the examples which contained a verbal description of the problem, then a flowchart, and finally the coding, you can use this to check the logic. It is advised that you use this method or one very much like it. If you have tried all of the things you can think of, ask someone for assistance. However, be certain that you understand why your program did not work properly and why the suggestions do work (if they do). Also check carefully to be certain that the program is the way you wrote it. It is very easy to get cards out of order, and this can cause statements to be performed in the wrong order.

Finally, the only way to learn to program is to write programs. While this may seem trite, it is very true. You will not hurt the computer, so do not worry on that score. Therefore, go ahead. Be sure you understand the problem, that you have a method which will lead to a solution, and then accept the challenge of instructing the computer concerning the steps to be taken to realize the solution. Do not give up when you have a bug, but persist, and you will ultimately have a successful program.

Appendix E. BASIC Language

E.1 Introduction

The **BASIC** language is one of many *computer languages* in common use. It is frequently available in educational institutions and was designed to be particularly easy to use. This language is generally used on a computer equipped with terminals, and you will need to find out from your computing center the procedures necessary to sign on the *terminals*. Once you have signed on you will type your instructions to the computer in the **BASIC** language, and you will find that the computer almost seems to talk to you in response to your typing. There are some systems on which this is not true, but if you are working on an *interactive* system, you will find this statement to be true.

It is not our purpose here to give all of the details of the **BASIC** language. Rather we will concentrate on giving sufficient information that you will be able to write any program likely to be assigned in the course of your study of the calculus. At some later time you may wish to explore additional features of the language, and this can be done through using any one of a number of books currently on the market. Since there are many versions (or dialects) of **BASIC**, be certain to get a book that is recommended by personnel in your computing center. The information given here is valid for the great majority of versions of the **BASIC** language, and this is another reason for restricting ourselves to those portions of the language which are most used.

One of the first things that we need to know in using any computer language is the method whereby we enter numbers or variables. In **BASIC** we can refer to a number just as we would ordinarily write it. Thus 23.4 is entered as 23.4. If we wish to use a *variable*, we can do this by using a designation consisting of a *single letter of the alphabet* or a designation consisting of *a letter followed by a single digit.* If the variable is denoted by two characters, the first one must be a letter of the alphabet and the second one must be a numeric digit. No variable can be designated by more than two characters. If we wish to enter more than one number on a line, we can do this by inserting commas between the numbers. All of this information indicates that the process of writing mathematical statements in **BASIC** should not be considered as being much more than writing statements in algebra.

EXERCISES

1. Which of the following are valid designations for variables in **BASIC**?
   (a) A
   (b) D2
   (c) 23
   (d) Y9
   (e) NO
   (f) 3A
   (g) KK
   (h) X
   (i) DATA
   (j) 2A

2. Through what device are programs ordinarily entered into the computer when using the **BASIC** language?

E.2 Arithmetic Statements

Our next concern is to be able to do arithmetic. We use the arithmetic operations $+$, $-$, $*$, $/$, and $\uparrow$ to represent *addition, subtraction, multiplication, division,* and *raising to a power* respectively. (For some computers $**$ is used in lieu of $\uparrow$.) We cannot represent multiplication by simply placing two symbols next to each other, but must always insert $*$ between the two symbols if the computer is expected to multiply them. We are free to use *parentheses* as often as we wish, but we must remember to provide a closing parenthesis to match each opening parenthesis. Each arithmetic **BASIC** statement consists of the work **LET** followed by a single variable, an equal sign, and then the expression containing the arithmetic to be performed. Such a statement instructs the computer to do the arithmetic indicated on the right side of " $=$ ", and to assign this resulting value to the symbol on the left side. Thus, since this statement assigns a new value to the variable on the left side of the " $=$ " sign, it is often called an *assignment* statement. The word **LET** is optional in some systems, but is mandatory on others. In the statement

$$\text{LET } X = (-B + \text{SQR}(B \uparrow 2 - 4*A*C))/(2*A) \qquad (E.2.1)$$

Table E.1

ABS(E)	The absolute value of the expression E.
ATN(E)	The arc tangent of E measured in radians.
COS(E)	The cosine of E where E is measured in radians.
EXP(E)	The exponential function of E. [That is e^E.]
INT(E)	The greatest integer in E.
LOG(E)	The natural logarithm of E.
LOG10(E)	The common logarithm of E.
SGN(E)	The signum function of E. This is $+1$ if E is positive, 0 if E is zero, and -1 if E is negative.
SIN(E)	The sine of E where E is measured in radians.
SQR(E)	The square root of E.
TAN(E)	The tangent of E where E is measured in radians.

the computer expects to take the values previously given it for A, B, and C, to compute $B^2 - 4AC$, then take the square root of this quantity, add the result to $(-B)$, and finally divide the value thus obtained by the product of 2 and A. The final result is then to be assigned to X so that X will have the value thus obtained after this statement has been exececuted. The values of A, B, and C have not been altered, but any previous value which X might have had has now been lost. Note that this procedure follows the instructions given via the grouping indicated by parentheses. Note also that the sequence followed is the one we would desire, namely the completion of the multiplication prior to addition or subtraction.

We noted the fact that we were to take the square root in (E.2.1) through the use of SQR(E) where E was an expression that would yield a number at the time we executed the statement in question. Needless to say, it is very convenient to be able to take a square root so easily. There are several other functions that are available to us in BASIC. Those which are likely to be most useful to you are listed in Table E.1.

Through the use of the arithmetic operations and with the aid of the functions of Table E.1, we can write assignment statements that will cause the computer to calculate just about anything we may require. A program usually consists of a succession of such statements, many of them using results found in preceding statements. We need a few more bits of information before we are ready to write complete programs, but the type of statement that will be of assistance should already be apparent.

EXERCISES

1. What is the value of X in each of the following situations?

    (a) LET X = 5*7 − 4/2
    (b) LET X = 45/7 + 2.3
    (c) LET X = 3↑3 − 4↑0.5
    (d) LET X = A2 − A*2 if $A = 4$ and $A2 = 7$.

2. What is the value of K in each of the following?

   (a) LET K = SIN(3.14159/6)
   (b) LET K = SQR(25) − 3.2
   (c) LET K = K + 1 if $K = 4$ to begin with.

3. Each of the following statements has at least one error. Find all of the errors, and indicate a correct statement that would do what you think was intended (if possible).

   (a) LET X + Y = 3
   (b) LET D = (A − B)(2 + A)
   (c) LET X = SIN(X))
   (d) LET A = B+ −C
   (e) LET X = SIN(LOG(B)

E.3 Control Statements

Ordinarily we write a sequence of **BASIC** statements and expect the computer to execute the first one, then the second statement, etc. There are occasions, however, when we would like to reuse some statement or we would like to jump ahead depending on whether a given result is larger or smaller than some given number. In order to instruct the computer to do this, we must have some way of indicating that we wish to change the sequential nature of program execution. To facilitate reference to any statement, every statement in **BASIC** must be numbered, and the numbers must appear in *ascending order*. All *statement numbers* must be positive numbers with larger numbers denoting statements that appear later in the program. The numbers are placed to the left of the statement and on the same line as the statement. We can use a statement **GO TO 40** to instruct the computer to leave the normal sequence and execute statement number **40** next. Since all statements are numbered, all you need to do is to refer to the appropriate number in order to refer to the given statement. The **GO TO** statement does not give the computer permission to do anything but go to statement **40**. On the other hand, if we write **IF (A*B) > = C, THEN GO TO 40**, we are telling the computer to compare the present value of **A*B** with the value of **C** and if the former is either greater than or equal to the value of the latter, the next statement executed would be statement *40*. There are six relations that can be used in the **IF** statement. There are =, <, >, < =, > =, < >, and they are used respectively to make the transfer if the *expressions are equal*, if the *first is smaller than the second*, if the *first is larger than the second*, if the *first is not larger than the second*, if the *first is not smaller than the second*, and when the *first is not equal to the second*. In case the relation following the word **IF** is not true the program will proceed automatically to the next statement.

It is also possible to follow the word **THEN** in the **IF** statement with any other **BASIC** statement. For instance we might have

$$\text{100 IF A} > = \text{B THEN LET A} = 5$$

or perhaps

$$\text{100 IF A} > = \text{B THEN IF B} < \text{C THEN B} = \text{SQR(A*C)}.$$

In each of these cases the statement following **THEN** will be executed only if the relation following the **IF** is a true relation. In both instances control will then pass to the next statement. However, in the first case the program would replace the value of A with 5 if and only if $A > B$ or $A = B$. In the second case B would be replaced by the square root of A times C if and only if $A > B$ or $A = B$ and in either case it is also true that B is less than C.

EXERCISES

1. What will be the value of X at the conclusion of the following sequence of instructions?

```
10 LET A = 3.5
20 LET B = 4
30 LET C = 6
40 LET K = 2.3
50 IF A*B < = 10 THEN GO TO 80
60 LET B = B − 1
70 GO TO 50
80 LET X = K*C/B
90 LET X = X − 2
```

2. Does every statement have to have a statement number? If not, which statements do not have to have such a number?

3. If $A = 3$ and $B = 4$, describe the result of each of the following statements.

   (a) IF 2*A > B THEN TO TO 30
   (b) IF A < > B THEN A = 5
   (c) IF SQR(A*B) < A THEN GO TO 450
   (d) IF B**3 > = 20 THEN GO TO 200

4. Why do you think an **IF** statement is referred to as a *conditional transfer* statement whereas a **GO TO** statement is referred to as an *unconditional transfer* statement?

5. Would it be possible for a **BASIC** program to have line number (or statement number) 30 before line number 20? Explain.

E.4 Input/Output Statements

We have now learned how to write assignment statements and how to control the order in which the statements are executed. Before we put all these things together in an illustrative program, we pause to indicate a means whereby we can enter information into the computer and can obtain information from the computer. In BASIC we have two ways of assigning values to variables other than by using the LET statement. We can use the command READ followed by a list of variables, each variable separated from its predecessor by a comma. In this case we must also use a DATA statement. The DATA statement (or statements) can be as numerous as we care to make them and can occur any place in the program. The DATA statement is followed by one or more numbers, with the numbers separated by commas. There should not be a comma after the last number in the DATA statement. The instruction READ will cause the BASIC program to seek out the DATA statements in the order in which they occur and to assign to each variable in turn a number from a DATA statement starting with the first DATA number in the program. Using this method, one inserts the data to be used as part of the program at the time it is written. In case there are more variables to be read by a READ statement than there are numbers in the DATA statement, an ERROR message will be printed out. It is the responsibility of the programmer to see to it that there is sufficient data for the number of variables which are to be read. There is one way out of this dilemma. If you wish to use some data more than once, you can insert the command consisting of the single word RESTORE. This will cause the first READ statement following RESTORE to go back to the first item of data given in a DATA statement in the program just as though no READ statement had preceded the RESTORE command.

One frequently wants to insert information after the execution of the program has been initiated. In this case we would use the command INPUT followed by a list of variables we wish to enter at the time the program is run. The INPUT statement is put at the appropriate place in the program, and upon execution of the program the typewriter will type a question mark indicating it is expecting you to put in data at that time. Data is entered in a BASIC program by typing in as many numbers as one requires on a single line with the numbers separated by commas. We illustrate these input commands by the following examples:

$$\text{INPUT A, B, C4} \qquad\qquad \text{(E.4.1)}$$

whereupon during execution the program will type

$$?$$

and we will respond with three numbers such as

$$25.5, \ 37, \ -.004$$

with the numbers separated by commas. If we type in fewer than three numbers the program will respond with another question mark and will await the next number until it has the three numbers it requires. If we had used the command READ, we would have had to include a DATA statement. Thus we would have

READ A, B, C4

DATA 24.5, 37, −.004, 598

and in this case, if this is the first occurrence of a READ and DATA statement in the program we would assign *24.5* to *A*, *37* to *B*, and *−.004* to *C4*. Since the DATA statement included a fourth number and this was not used by the READ statement, the number *598* would be the first one read by the next READ statement. If the next READ statement is preceded by a RESTORE command, the first item read by the next READ statement would be *24.5*. This again presumes that this is the first DATA statement in the program, for the first READ will start with the first item in the first DATA statement.

In order to write results we use the command PRINT. This instruction is to be followed by the list of variables to be printed. If the variables are separated by *commas* the information will be printed in five *columns* of 14 spaces each. If the items are separated by *semicolons*, the program will cause the data to be separated by single spaces in most versions of BASIC. In the PRINT command, it is also possible to insert expressions we wish printed. Thus, if we had the command

PRINT A, 24 + 5, "THE VALUE OF X IS", X   (E.4.3)

we would have printed in the first column the value of A, in the second column the number *29*, then the words THE VALUE OF X IS and finally the value of X. Since the literal expression will occupy more than one column of fourteen spaces in this case, the value of X will be located in what would ordinarily be the fifth column across the page (that is starting in the *57*th space). Note that we can insert labels by the simple device of including the material we wish printed between quotation marks.

We will now put some of these concepts together in an example to illustrate the information we have given to this point.

EXAMPLE 4.1. Write a BASIC program which will cause the computer to ask for numbers to be typed in, which will then obtain the average of the data and will also obtain the square root of the average of the squares of the data items. The number of items we will wish to process is not known at the time we write the program, but it is known that no number is larger than *90000*.

Solution. Our job is to write a program that will add up the various numbers, keep track of the number of data items, and also add up the squares of these numbers. After the last item has been given to the computer, it will then be possible to divide the sum of the numbers and the sum of the squares by the number of items used and then to take the square root of the average of the squares. We must then write out the results, for they will do us no good if they remain in the computer. A description of the various steps involved is given in Figure E.1. To the right of the description is a

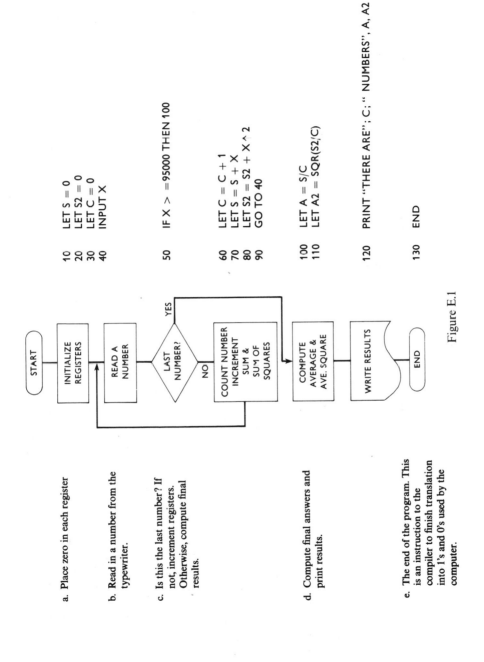

```
10    LET S = 0
20    LET S2 = 0
30    LET C = 0
40    INPUT X

50    IF X > =95000 THEN 100

60    LET C = C + 1
70    LET S = S + X
80    LET S2 = S2 + X ^ 2
90    GO TO 40

100   LET A = S/C
110   LET A2 = SQR(S2/C)

120   PRINT "THERE ARE"; C; " NUMBERS", A, A2

130   END
```

a. Place zero in each register

b. Read in a number from the typewriter.

c. Is this the last number? If not, increment registers. Otherwise, compute final results.

d. Compute final answers and print results.

e. The end of the program. This is an instruction to the compiler to finish translation into 1's and 0's used by the computer.

Figure E.1

735

flowchart indicating the logic of the problem in a pictorial manner, and to the right of the flowchart we find the BASIC program. Note that the program is written using only capital letters. The typewriter terminal for entering the program will often be restricted to capital letters and you might as well get used to this fact.

After indicating START in the flowchart, we have step (a) in which we set the variables we are going to use to a zero value to insure that our results reflect only our data and do not include someone else's leftovers. As indicated above, we will need three registers, and we have chosen to call these S, S2, and C. The symbol S stands for *sum*, S2 for *sum of 2nd powers* and C for *count*. It is frequently helpful to use initial letters to aid in remembering what the variable is to represent. You must remember, however, that you are limited to a single letter or a letter followed by a digit.

In step (b) we read an input item from the terminal. We have given this value the label X. This variable will assume many values during the execution of the program, for it will always represent a single number, namely the last one entered. Therefore, we must be certain that we do everything we need to with one value of X before we read another number and replace the former value. We must have some method for letting the program know when we are through supplying information. We can make use of the information given, and enter the number *99999* after the last item we wish to use. We do not wish to use *99999* as a regular entry, however, and for this reason we have included the IF statement. If the number just introduced is not smaller than *95000*, the program is directed to statement *100* for its next instruction. Otherwise it will continue with the next instruction, in this case number *60*. This permits us to answer the question of part (c) of the problem. Thus, if we have one of the numbers we wish to process, the program will proceed to statement *60* and increment C, add X to S and add X↑2 to S2 whereupon it will return to statement *40* to obtain another item. Note the way in which we have instructed the computer to use the old value of S and then add X to it after which we assign this new value to S.

If the program is directed to statement *100*, as will be the case when we type in *99999*, the average is calculated and given the name A. In one step we will divide the sum of the squares of the numbers by the number of such numbers and then take the square root of the quotient, denoting this result by A2. The next statement directs the computer to write out the results, and finally the last statement indicates that we have come to the end of the program by so stating with the BASIC statement END. After the program has been entered, you can instruct the computer to execute the statement by typing in the word RUN. When all of the information has been entered, and you have finally indicated the end of the data by typing in *99999*, the terminal will then type out the number of numbers, C, the word NUMBERS preceded by one space since we have inserted a semicolon here and a space between the apostrophe and the N, the value of the average A and the square root of the average of the squares A2. To put this program in the computer you will need to obtain the attention of the computer at the terminal you are using. You will need instructions from your computing center for this task. You will then use the procedures given you to indicate that you have a program written in BASIC. Either you must type out a number for each line or else the computer will type out the line number and wait for you to complete the line, depending upon the system being used. If you make a mistake in typing, retype the number of the incorrect line, and then retype the line correctly. Upon typing the word END you indicate to the computer that you are at the end of your program. The word RUN will cause the program to be executed. If you wish to insert remarks for your own information which the BASIC program should ignore, you can do this by preceding such remarks with the three letters REM. Anything following REM and on the same line will be ignored, but this will be typed out in any listing of your program to remind you of that which you intended to do at that point.

EXERCISES

1. What is the difference between using a comma and using a semicolon to separate items following the command PRINT?

2. How does the program indicate that it has come to an INPUT statement and it is waiting for you to type in data?

3. What statement *must* occur in the program if you use a READ statement?

4. If you wish to use data in a set of DATA statements twice, how can you accomplish this?

5. If you wish to use the second item of data in a DATA statement twice, but did not wish to repeat the first number, how might you accomplish this?

6. (a) How would you enter a remark into a program?
   (b) For what purpose would you want a remark?
   (c) How often should remarks be used in a program?

7. You are asked to find the average of the square roots of the first thousand positive integers.

   (a) Draw a flow chart.
   (b) Write and run the program.

8. You have a list of coordinates of points (each involving a pair of numbers). Write a program which will find the distance between successive points and will also find for each set of four consecutive points the angle formed by the line through the first pair of points and the line through the second pair of points. [*Hint*: Be sure to draw a flow chart. Consider the case in which one of the lines might be vertical.]

E.5 Subscripts

With the material presented to this point you have enough information to write a program to take care of the majority of problems likely to appear in the calculus. The IF statement will permit you to implement any decisions that have to be made enroute to the result. If you decide you would like to read on, however, there are additional items in the BASIC language that will simplify some of your programming problems. The first of these has to do with the BASIC equivalent of mathematical subscripts. It is frequently convenient to use *subscripts* to distinguish variables such as $x_0, x_1, \ldots, x_{50}$. Since the majority of computers do not have the capability of writing small numbers below the normal line of print, this particular notation is not available to us here. We can, however, use the same idea and enclose these numbers in parentheses following the symbol. Thus we would have X(0), X(1), X(2), ..., X(50). The subscript can be a number or an expression. If the subscript is not an integer, the program will use the largest integer which does not exceed the value of the subscript at the time of execution. Ordinarily we would have little reason to use other than integers for the subscripts, but X(4.73) would be treated as X(4) in the BASIC language. If we indicate that we are using subscripts by the simple device of just writing them in the program, the BASIC compiler will set aside ten spaces for items with subscripts up to ten, and we can use subscripts up to a maximum of ten without difficulty. If we try to go above this number we will receive an ERROR message when we try to run the program. If we wish to have subscripts larger than ten, we must so inform the program by a statement such as DIM X(50). In order to indicate that we wish to use subscripts larger than ten with several variables we can use several DIM statements or we can use a single DIM statement separating the items to be dimensioned by commas. Thus we could write

$$\text{DIM } X(50), D(40), Y(5, 8) \tag{E.5.1}$$

where we observe that no punctuation follows the last entry in the line. Notice also that we can use two subscripts if we like. If we do not use the DIM statement, the. two subscripted variables would have maximum subscripts of Y(10, 10). By inserting the statement DIM Y(5, 8) we have indicated we will not need so much space and the space will be saved for other purposes. The DIM statement, if used, *must* precede the first reference to any variable which appears in the statement, and for this reason it is customary to place it at the beginning of the program.

EXERCISES

1. When is a DIM statement necessary in a BASIC program?

2. What information is imparted to the computer by the statement

   100 DIM X(300), A2(15,20), B3(7)

3. A segment of a **BASIC** program includes the following sequence of statements:

```
100 I = 1
110 X(I) = 2*I
120 IF I < 10 THEN I = I +1
130 IF I < = 10 THEN GO TO 110
```

(a) What is the value of X(3) at the conclusion of this sequence?
(b) What is the value of X(5.8) at the end of this sequence?
(c) Is a DIM statement necessary for the array X insofar as it is used in this sequence? Why?

4. A segment of a **BASIC** program includes the following sequence of statements

```
50 I = 1
60 J =1
70 A(I, J) = 2*I + J
80 I = I + 1
90 IF I < 20 THEN GO TO 70
```

(a) If this sequence is not to produce an error, would a   statement be necessary? If so, how much do you know about the requirements for the DIM statement for this program?
(b) What elements of the array A would have values assigned as a result of this sequence?
(c) What would be the value of A(13, 1)?

5. In each of the following statements indicate whether it is a valid statement and if it is not indicate any necessary corrections.

(a) 120 DIM A(25), A(24, 12)
(b) 130 DIM B(5), C(2, 200)
(c) 130 LET A(I) = A(I, J) + 4
(d) 140 LET A(I) = A + 43.25

E.6 For-Next Loops

The program of Section 4 had a loop in the flowchart with a consequent loop in the program. This was implemented by sending the control of the program back to line *40* through the use of a GO TO statement at line *90*. We also used a counter, C, and an IF statement to count the number of times we went through the loop and to provide a means for getting out of the loop. Any loop can be executed in a manner somewhat similar to this. However, the use of loops in programs is so frequent that we have a somewhat more direct method for controlling a loop. This is done through the use of two statements, the first of which starts with FOR and the second one starts with NEXT. If we wish to add the first hundred numbers, add their squares, and add their cubes, we could do so with a segment of program as follows:

```
100 FOR K = 1 TO 100
110 S = S + K
120 S2 = S2 + K*K
130 S3 = S3 + K↑3
140 NEXT K
```

In this segment we start with a statement which tells the computer that K is to be used as an *index* for the "FOR-NEXT" loop. The loop is to be started using the value K = 1. The computer is to complete all steps following this "FOR" statement down to the "NEXT" statement. When the program comes to the statement NEXT K, it proceeds, as instructed, to the next value of K, namely 2. The program then proceeds again through statements on lines *110, 120*, and *130* using K = 2 and then increments K to the value *3*. This continues until we have completed this set of three statements for the value K = *100*. At this point the program continues with the statement that would follow NEXT K. This provides us with an easy method for setting up the necessary procedures to go through a loop *100* times.

In the preceding paragraph we indicated a program segment that would let K start with the value *1* and which would increase K by 1 after each completion of the loop. There are times when we might wish to use some other *increment*. This can be done by replacing line *100* of our previous segment by

100 FOR K = 1 TO 100 STEP 0.5

In this case the action would be similar, but K would take on the successive values 1, 1.5, 2.0, 2.5, 3.0, . . . , 99.5, 100. It is also possible to start the loop index at a value which has to be computed, to terminate the loop with a computed value, and to use a STEP which is a computed value. For instance, a statement such as

100 FOR K = SQR(X) TO X*Y STEP Z

would be permissible if X, Y, and Z had been assigned values before coming to line *100*. This gives a great deal of flexibility in handling loops.

It is possible to have one loop contained within a second loop. The one requirement is that any FOR-NEXT loop which starts after the FOR of another loop must end before the NEXT of the other loop. In other words, such loops can be *nested*, but the one that starts later in the program cannot include within it the NEXT statement of one that had started earlier. There is no limit to the number of lines which can appear between the FOR and the NEXT other than the limit which the computer places on the maximum size of your program. The use of this loop should depend only upon the logic of the program, for it is the purpose of the FOR-NEXT loop, as it is of the other features of BASIC programming that the program provides the means for carrying out the computation required to successfully solve the problem at hand.

EXERCISES

1. Assuming that there is a line **NEXT K** appropriately placed, what does each of the following instruct the computer to do?

   (a) **50 FOR K = 7 TO 23**
   (b) **55 FOR K = SQR(9) TO 41 STEP 5**
   (c) **60 FOR K = 100 TO 0 STEP −1**
   (d) **70 FOR K = 100 TO 0 STEP −3**

2. If you have statements which start **FOR I, FOR J,** and **FOR K** in that order, in what order would you place the **NEXT I, NEXT J,** and **NEXT K** statements?

3. You find the statement **FOR I = J TO K STEP L** in a **BASIC** program.

   (a) What is the role that I plays in this loop?
   (b) What is the role that J plays in this loop?
   (c) What is the role that K plays in this loop?
   (d) What is the role that L plays in this loop?
   (e) If L is positive would you expect K to be larger than J or smaller than J?

4. Write a **FOR-NEXT** loop that would perform the computation for each of the following problems.

   (a) Add all of the even numbers starting with 6 and ending with 100.
   (b) Add all of the multiples of 5 starting with 100 and ending with 1000.
   (c) Multiply all of the positive integers which are not larger than 14.
   (d) Add the square roots of the integers from 2 to 18.

5. Is there any program which you can write using a **FOR-NEXT** loop which you could not write if you did not use **FOR** and **NEXT**? Explain your answer by indicating a loop that you could not handle without **FOR** and **NEXT** or by indicating how you could handle all such loops.

E.7 Function Statements

One other bit of **BASIC** that may be helpful for programming problems in the calculus involves use of function statements. If we insert the statement

$$\text{DEF FNF (X)} = 1/(\text{SIN(X)} + \text{COS(X)}) \tag{E.7.1}$$

and later write FNF(0.4) the program will supply the value of

$$1/(\sin 0.4 + \cos 0.4).$$

We can supply several function statements by simply using **DEF FN*( )** where the asterisk is to be replaced by the letter that will designate the particular function, and each function must use a different letter. We place within the parentheses the argument (or arguments) of the function. If we use more than one argument, we separate the arguments by commas. We will use these concepts in the following Example:

EXAMPLE 7.1. We are given two numbers, $a = 0.4$ and $b = 1.8$, and we are to divide the interval from a to b inclusive into 100 equal subintervals. If we denote the 101 points by x_k where k has values $1, 2, 3, \ldots, 101$ we will have $x_1 = a$ and $x_{101} = b$. We also wish to consider 100 values t_k, where t_k is the midpoint of the interval *from* x_{k-1} to x_k. This means we will have values t_k for $k = 2, 3, 4, \ldots, 101$. Next we wish to obtain the sum of 100 terms, each of the form $\sin t_k [x_k^3 - x_{k-1}^3]$. Write a BASIC program that will perform this computation.

Solution. We need *101* values of a variable we will call X and we need *100* values of a variable called T. However, if we follow the terminology given us, the largest subscript for T will be *101*, and therefore we must make provisions for this subscript in our DIM statement. We also see that we will have to evaluate both $\sin x$ and x^3 for values of t_k and x_k respectively on *100* occasions. It would be convenient to simplify the reference to these functions through the use of function statements. With these thoughts for easing the programming involved, let us write out the description, the flowchart and thence the program as we did before. You will find this in Figure E.2. Note that we started with the DIM statement to insure that we did not use any subscripted variables before we had made adequate provisions for them. The DIM statement is necessary since we will use subscripts larger than ten. We next inserted the function definitions to be certain we did not overlook them. The next thing to be done is the determination of the values of x_k and t_k. The determination must be in that order since we need the end points of the interval to determine the midpoint. There are, of course, alternatives due to the regular arrangements of the x's in this case, but we have chosen the obvious method. After we have the points determined, we are now in a position to compute the value of the *100* summands, and to add them as we go along. This is done in the loop involving statements *130, 140,* and *150.* Finally we have given instructions to write out the results.

Note that we have used two FOR-NEXT loops. As we stated before, the first one must either terminate before the second one begins, as it does in this case, or else the second one must be completely included in the first one. Thus, the NEXT statement from the first loop must either come before the FOR statement for the second loop, or it must come after the NEXT statement for the second loop. Also note that we used the READ statement. The values of A and B will be the two items in the DATA statement in the order in which they appear. We assign the value *0.4* to A and the value *1.8* to B. Observe the use of remarks to assist in understanding the program. We will further illustrate these concepts in Example 7.2.

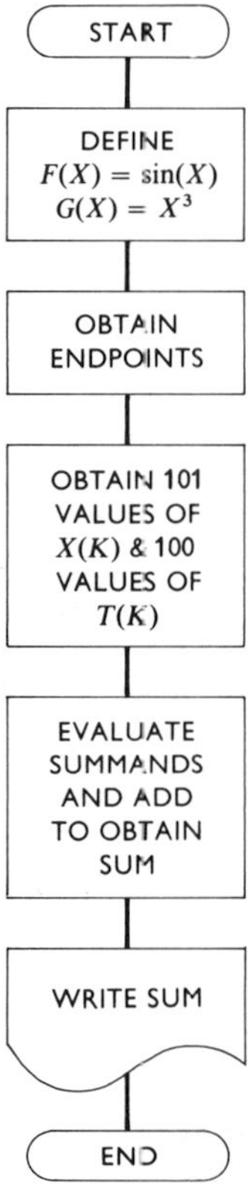

```
10    DIM X(101), T(101)
20    DEF FNF(X) = sin(X)
30    DEF FNG(X) = X↑3

40    READ A, B

50    REM COMPUTE X AND T
60    LET X(1) = A
70    FOR I = 2 TO 101
80    LET X(I) = X(I − 1) + (B − A)/100
90    LET T(I) = (X(I) + X(I − 1))/2
100   NEXT I

110   REM COMPUTE SUM
120   LET S = 0
130   FOR I = 1 TO 100
140   LET S = S + FNF(T(I + 1))*(FNG(X(I + 1)) − FNG(X(I)))
150   NEXT I

160   PRINT S

170   DATA .4, 1.8

180   END
```

Figure E.2

START

DEFINE
$F(X) = \sin(X)$
$G(X) = X^3$

OBTAIN
ENDPOINTS

OBTAIN 101
VALUES OF
$X(K)$ & 100
VALUES OF
$T(K)$

EVALUATE
SUMMANDS
AND ADD
TO OBTAIN
SUM

WRITE SUM

END

a. Define functions $f(x) = \sin x$ and $g(x) = x^3$.

b. Define beginning and end of interval.

c. Obtain values of x_k and t_k for required subintervals.

d. Evaluate summands and add to obtain sum.

e. Write out the results.

743

EXAMPLE 7.2. Type in a number N which represents the number of data items we wish to process and then type in N data items. Sort the numbers to put them in ascending order such that each number in the sequence is as least as large as those preceding it. Write the resulting sequence.

Solution. In Figure E.3, we have written out a description and drawn a flowchart to indicate the logic involved. In this case we type in N first, as directed. Note in the program we have used a DIM statement that allocates *200* locations for data entries. This does not mean that we must use *200* entries, but rather than we are permitting ourselves that many. After obtaining N from the terminal, we proceed to obtain the data. The question raised by the diamonds in the flowchart is answered in this case by means

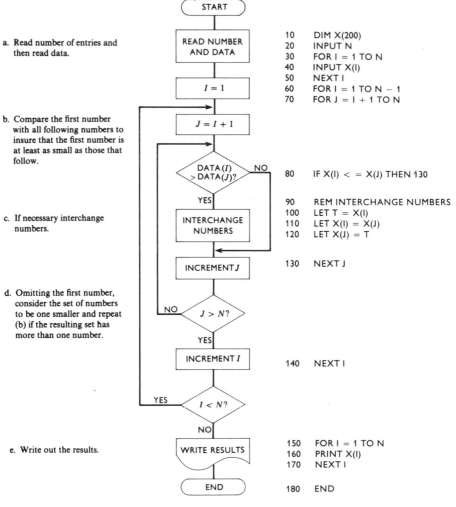

a. Read number of entries and then read data.

```
10   DIM X(200)
20   INPUT N
30   FOR I = 1 TO N
40   INPUT X(I)
50   NEXT I
60   FOR I = 1 TO N − 1
70   FOR J = I + 1 TO N
```

b. Compare the first number with all following numbers to insure that the first number is at least as small as those that follow.

```
80   IF X(I) < = X(J) THEN 130
```

c. If necessary interchange numbers.

```
90   REM INTERCHANGE NUMBERS
100  LET T = X(I)
110  LET X(I) = X(J)
120  LET X(J) = T
```

```
130  NEXT J
```

d. Omitting the first number, consider the set of numbers to be one smaller and repeat (b) if the resulting set has more than one number.

```
140  NEXT I
```

e. Write out the results.

```
150  FOR I = 1 TO N
160  PRINT X(I)
170  NEXT I
```

```
180  END
```

Figure E.3

of two **FOR-NEXT** loops. Note that one is inside the other. In the first instance we wish to take our earlier entry for comparison, and since this could not be the N-th number we must end the loop with the subscript $(N - 1)$. We wish to compare this with a second number that will come some time after the first one in the sequence, and hence we start the inner loop with $(I + 1)$. We make the comparison in the **IF** statement. If X(I) is larger than X(J), we will go to statement *100* and interchange the numbers. Note that we must temporarily save the value of X(I) before we replace X(I) with X(J) or we would lose X(I). If the numbers are already in order, we merely want to go to the next comparison. We do this by sending the program to statement *130* which causes the value of J to be incremented, unless it is already N in which case we increment I and then start the loop with J all over again. After we have completed the outer loop, which means we have run through the entire range of the inner loop $N - 1$ times, we proceed to the loop at statement *150*. This loop consisting of three statements will cause the resulting list to be typed.

The use of the **FOR-NEXT** loop in this Example makes it slightly more difficult to relate the description, the flowchart, and the program, for we do many things at once. The logic is the same, however, and you should check this out.

EXERCISES

1. What is the purpose of a function statement?

2. Where does a function statement appear in the program if there is no **DIM** statement? Where does it appear if there is a **DIM** statement?

3. Explain in words the various actions that are to take place as a result of statement *140* of Example 7.1.

4. (a) If you wanted to repeat Example 7.1 with $f(x) = \tan x$ and $g(x) = e^x$, what changes would be necessary in the program given in Figure E.2?
   (b) What changes would be required if we wished to use *200* intervals instead of *100* intervals?

5. Why is it necessary to introduce the variable T in Figure E.3?

6. Draw a flowchart and write a program that will give a table of factorials for all positive integers less than 26.

7. Draw a flowchart and write a program that will make a table of values of the function $f(x) = x^3 + 2x^2 - 4x - 3$ for all integer values of x from $x = -5$ to $x = 7$. You might see how few statements you need for this program.

8. Draw a flowchart and write a program that will make a table of values for $\sin x$, $\cos x$, and $\tan x$ for $0°, 1°, 2°, \ldots, 90°$. Remember to convert degrees to radians if you use the functions provided by the **BASIC** language.

E.8 Summary

As we stated in the beginning of this Appendix, there was no intent to give
a complete description of the BASIC language. However, we have included
the information necessary to write a program for any problem which you
are likely to face in this course. It is far better to start with this small amount
of information and write programs that will do the job than to find yourself
with much more information concerning the language, but in a quandary
over which procedure to use. The only way to learn to write programs is to
write them. If you have *errors* (or *bugs*) in your program due to an incorrect
use of BASIC, there will be diagnostic messages on the terminal and you can
use these to help locate the errors. Once the program *runs*, you should check
carefully to determine that it is doing what you wished it to do. If it does not,
one of the best things you can do is check the logic with the aid of your
flowchart to make certain that your instructions were logically correct. It
is also helpful to insert additional PRINT statements to determine the value
of various variables at different points in the program. If you use simple
numbers for test data, you can check to see whether you and the computer
agree. This will frequently be very helpful. In extreme cases, ask someone
for aid. You won't be the first one to use a computer who has had to ask
for help. Remember that there are very few programs that are written that
do not have errors the first time through. Keep at it.

In writing a program it is strongly suggested that you follow the procedure
given in the Examples of this Appendix. Write out the logical flow of the
problem and draw a flowchart. It then becomes relatively easy to write the
instructions in the BASIC language. It is not necessary to be fancy in the
way you do things. There are several distinct programs which will give a
correct result for any simple problem.

It may also help if you keep listings of your programs. In Example 7.1,
we would only have had to change the instructions defining FNF(X) and
FNG(X), and change the DATA for A, and B to handle a large variety of
problems. It might even be worthwhile to replace the number *100* by N
throughout the program so that a value of N could be inserted when the
program is run to provide greater flexibility. If you intend to use a value
N larger than *101*, you will have to increase the dimension allocated for X
and T through a modification of the DIM statement. You will find many
programs that can be used more than once provided you build in the flexibility in the manner suggested here.

Answers to Selected Exercises

Chapter I

Section I.2

1. Integer: $2, 0, -5$; Rational: all but π and $\sqrt{3}$; Real: all of the numbers.

5. Let x be the multiplicative inverse of 0 and show that this leads to a contradiction.

7. $1, 2, 3$.

M17. All properties are satisfied. & is zero and $+$ is one.

M19. No. \$ has no additive inverse.

Section I.3

2. $2.8, 3, -9/20, 22/7, 22/7, -3/4, -\sqrt{10}, 0.001, \sqrt[3]{3}$.

4. (a) $\{-2, -\sqrt{5/2}, 0, \sqrt{3}, 2.8, \pi, 22/7\}$.

6. No.

10. No, if $b \geq 0$ and $d \geq 0$.

12. (a) Yes; (b) no.

M18. No.

Section I.4

1. (a) $1 + 5i$; (b) $-3 - 5i$; (c) $2 - 8i$; (d) $6 - 8i$.

3. (a) $(-46 + 9i)/169$; (b) $(1 + 18i)/5$; (c) $1 + 1.5i$; (d) $9.6 + 9.2i$;
   (e) $\sqrt{3}/2 - i/2$.

5. (a) -2; (b) $2 + i$; (c) $(7 + i)/10$.

7. (a) $5(\cos \theta + i \sin \theta)$ where $\tan \theta = 4/3$, 1st quadrant; (d) $4(\cos \pi/2 + i \sin \pi/2)$.

9. (b) $4(\cos \pi/3 + i \sin \pi/3)$; (d) $13(\cos \theta + i \sin \theta)$ where $\tan \theta = -5/12$, 2nd quadrant.

15. $-a - bi$.

17. 2.

21. $x = r \cos \theta$, $y = r \sin \theta$, $r = \sqrt{x^2 + y^2}$, $\theta = $ arc tan $y/x = $ angle whose tangent has the value (y/x).

M25. No.

Section I.5

2. $\sqrt{5}$, $\sqrt{10}$, $\sqrt{13}$, $\sqrt{2}$, $\sqrt{2}$, 13, 17, 5, $\sqrt{34}$.

5. (a) If z is real, then $1 < z < 5$. If z is complex, z is within the circle of radius 2 with center at $z = 3$.
    (c) If z is real either $z \le -5$ or else $z \ge -1$. If z is complex, z is outside of the circle of radius 2 with center at $z = -3$.

7. (a) $x > 2 \cup x < -8$;   (e) all values of x;   (h) $(-2 \le x < 0) \cup (-5 < x \le -3)$.

8. (a) $|x - 2| < 4$;   (i) not possible; (j) $|x - 2| > 0$.

9. (b) $4(\cos 5\pi/3 + i \sin 5\pi/3)$;   (e) $\sqrt{313}(\cos 42°43' + i \sin 42°43')$

10. (b) $\sqrt{3} + i$;   (d) $-1.665 + 3.637i$;   (g) $3.920 - 0.795i$.

12. (c) $-1 < x < 3$;   (e) $x > 1$ or $x < -4$;   (g) 3.

15. (b) i: (d) $-32 - 32i$.

17. (b) $x = 0$.

21. All points more than 2 units from the origin.

23. $8 - \sqrt{59} \le t \le 8 + \sqrt{59}$.

Section I.6

1. (a) and (c).

3. $x = 2$; $x > 2$ and $x < 0$; 0; $-1/2$; $1/(1 - \sqrt{2})$; $(y + 2)/y$.

5. $x \ge 0$; 14; 78; $169 - \sqrt{13}$; no; no; 0; 0; no.

7. 2; 2; 2.

9. 10; 17; $x^2 + x - 2$.

12. Yes; domain $x \ge 2$, range $x \ge 0$; yes; $f^{-1}(x) = x^2 + 2$ with the domain restricted to non-negative real numbers.

14. No; no.

16. Yes; yes.

18. 2; yes; $(2x + 1)/(x - 1)$.

21. $(7x + 5)/(4x - 3)$.

23. Increasing if $x > 0$ and decreasing if $x < 0$.

25. Yes; both.

Section I.7

1. 0.3846.

3. The result should be in the interval (2.615, 2.814).

5. The result should be in the interval (1.16, 1.199).

7. No; 0; 5.

C14. 3.535534.

Section I.8

1.

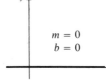

$(-3, 0)$ $(0, 2)$

$(-1, -3)$

3. (a)

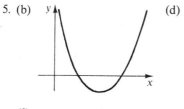

$m = -\frac{2}{3}$
$b = 2$

(c)

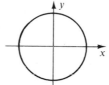

$m = 2$
$b = \frac{3}{2}$

·(e)

$m = 0$
$b = 0$

5. (b)

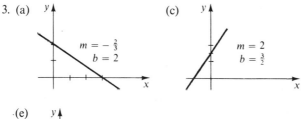

(d)

(f)

7. (a) $2x - y = 1$;   (c) $2x + 3y = 6$;   (e) $7x + 3y = 0$.

9. (a)

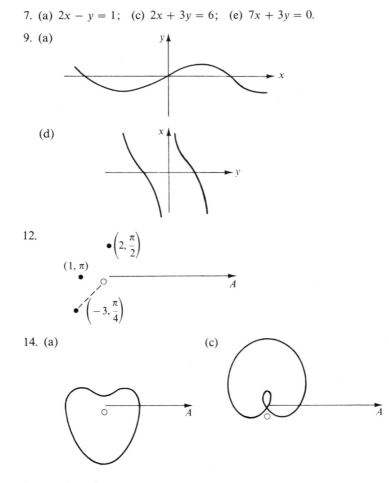

(d)

12.

14. (a)   (c)

17. $r = 2 \cos \theta$.

19. $y = mx + 3; m = 1, 5; 1 < m < 5$.

Section I.10

1. (a) 12;   (b) 135;   (d) 91;   (g) 620;   (i) 15.16

2. (c) $\sum_{k=1}^{99} (-1)^{k+1} k$;   (f) $\sum_{k=0}^{8} (-2)^k$.

9. (a) 35;   (c) 479,001,600.

Section I.11

16. $\cos 3\theta = \cos^3 \theta - 3 \cos \theta \sin^2 \theta$, $\sin 3\theta = 3 \cos^2 \theta \sin \theta - \sin^3 \theta$.

22. $64x^6 - 576x^5 + 2160x^4 - 4320x^3 + 4860x^2 - 2916x + 729$.

Chapter II

Section II.1

1. $2123.22.

3. $\sum_{k=1}^{100} p(k)[n(k) - n(k - 1)]$.

6. 55π.

Section II.2

2. $\{-1, -1/2, 0, 1/2, 1, 3/2, 2, 5/2, 3\}$, $1/2$, 17.

4. $P = \{-4, -11/3, -10/3, -3, -8/3, -7/3, -2\}$,
   $T = \{-23/6, -7/2, -19/6, -17/6, -5/2, -13/6\}$

15. 2.642.

16. $\{0.4, 1, 1.2, 1.5, 2\}$, $\{0, 0.4, 1, 1.2, 1.5\}$, 3.627, 1.853, yes.

18. 7.391. 19. 0.786. 22. 0.6.

Section II.3

1. (a) 1814.5; (c) 2234.89; (e) 1443.88.

4. 40.625

6. (a) -2.25; (b) -0.625.

10. (a) $\{-3, -1.5, 1.5, 4\}$; (b) 0.2155, -0.4941.

C17. 25 intervals 283.74 & 251.59; 50 intervals 276.02 & 259.93; 75 intervals 273.40 & 262.68; 100 intervals 272.08 & 264.03; 1000 intervals 268.48 & 267.68.

C19. Both upper and lower sums = 4.18879 correct to five decimal places.

Section II.4

1. 204.375, 76.219.

3. 32, 4.

6. (b) -5.6; (c) -4.8; (d) -4.4, -4.2; (e) -4.1, -4.05.

C12.

n	L	M	R
10	2.74352	3.28076	3.92432
20	2.98204	3.26355	3.57255
30	3.06980	3.26050	3.46349
40	3.11527	3.25945	3.41054
50	3.14307	3.25896	3.37928
60	3.16181	3.25870	3.35866
70	3.17531	3.25854	3.34403
80	3.18548	3.25843	3.32466
90	3.19343	3.25836	3.32466
100	3.19981	3.25831	3.31792

S14. (c) \$20,077.50, \$19.977.50;   (d) \$100.13.

P17. 28.27 lbs.

P19. (d) 2,964,000 lbs, 2,184,000 lbs.

B22. 25.4, 12.7.

Section II.5

1. (a) 8;   (c) does not exist; 1.

3. $a^3/3$.

5. 3.

7. $g(b) - g(a)$.

9. (a) 14/3, 0;   (b) does not exist.

12. (b) $\sum_{k=1}^{n} t_k^2 [x_k - x_{k-1}]$;   (c) $\int_0^3 x^2 \, dx$;   (d) 9.

C14. 25 intervals 6.89368, 6.19334; 50 intervals 6.71438, 6.36424; 75 intervals 6.65524, 6.42182; 100 intervals 6.62579, 6.45073.

P16. (c) $t_k^2 [(0.2x_k + 0.004x_k^2) - (0.2x_{k-1} + 0.004x_{k-1}^2)]$;
     (e) $\int_0^{12} x^2 \, d(0.2x + 0.004x^2)$.

B21. $5\pi c(10^{-5}) \, \text{cm}^3/\text{sec}$.

S24. (a) $(3 + 0.1t)(18000 + 360t + 4t^2)$;
     (c) $\int_0^{10} (3 + 0.1t)(18000 + 360t + 4t^2) d(365t)$.

Section II.6

1. 1; 8; 27; 64.

3. 0.693.

4. (b) $\{0, 0.079, 0.112, 0.137, \ldots\}$

6. (a) 1.875, 0.9375;   (b) 30, 15.

C14. 25 intervals 3.2122, 3.0522, 3.1443; 100 interval 3.1604, 3.1204, 3.1419; 500 intervals 3.1455, 3.1375, 3.1416.

Chapter III

Section III.1

1. (a) 15;   (c) $3.5(\sqrt{2} - 1)$;   (e) 0;   (g) 1/15.

3. (b) $\sqrt{2}$.

6. (a) 64/5;   (c) 5/9;   (e) $27\sqrt{3}/5$.

8. (b) $2a^3/3$;   (d) 0.

Section III.2

1. (a) 15; (c) 1016/7; (e) -2; (g) 384/7.

3. (a) $-2432/3$; (c) $(6a^6 - 7a^4 - 6a^2)/4$; (e) $-286/3$.

5. 1421.5.

7. (a) 16/3; (d) 16/3, 32/3.

9. (b) $\sum_{k=1}^{n} \pi(3t_k/5)^2[x_k - x_{k-1}]$; (c) $(9\pi/25) \int_0^5 x^2\, dx$; (d) 15π.

12. (a) 3/8 & 3; (c) $\pi/3$ & $2\pi/3$; (e) 3 & 4.

14. (c) 20.

16. $625\pi/4$.

S19. \$415,797.67.

Section III.3

1. (a) 3; (c) $1 + 2/\sqrt{3}$; (e) no value; (g) any value of c.

3. (a) 99/160; (c) 4/7; (e) $(128 - \sqrt{2}/16)/7$; (g) $243(2\sqrt[3]{2} - 1)/4$.

5. (a) 96/5; (c) $3(2^{-7/6} - 1)/7$; (e) 0; (g) 78.

7. (a) 14/3; (b) 14/9.

8. (a) $\pi/3$ & $\pi/6$; (c) $\pi/2$ & 0; (e) $\pi^2/16$ & 0; (g) $\pi^2/9$ & $\pi^2/72$.

11. 625π.

P13. (a) Density $= 16.5 - 2.5x/1400$; (b) $1.68(10^{23})$ kg.

P14. $8.67(10^{22})$ kg.

BS16. 101.27.

P18. (a) 4; (b) 3/4; (c) 64/15.

Section III.4

1. (a) $e - 1/e$; (c) 0.3181; (e) 8.091; (g) 36.859.

3. (a) 7.0212; (c) -1.8436; (e) 52.102.

5. (a) $\cosh x$; (b) $\sinh x$; (e) 6.5771.

7. 1.1149.

BP12. (a) -0.08664; (c) 0.01953 milliroentgens.

BP13. (a) 18.358 gms; (e) 11.98 minutes.

S15. \$780,037, \$761,483, \$762,022, \$764,436, \$767,162.

Section III.5

1. (a) 1;   (d) 0.1172;   (e) 1/4;   (g) $3\sqrt{3}/4$.

3. (a) $-19/8$;   (c) $\sqrt{3}/2 + 4\pi/3$;   (e) $(\cos 2 - \cos 2e)/2$;   (g) $3/\pi$.

5. (a) $(e^2 - e^{-2})/2 + 3\sqrt{3}/\pi$;   (c) -1;   (e) $\cos 1 - \cos 4$.

7. (a) 1;   (b) 1.

9. $2/\pi$.

Section III.6

1. (a) 20;   (c) 24768/7;   (e) $2(e^{3/2} - e^{1/2})$;   (g) $1 - e^{-3}$.

3. (a) 0;   (c) $(\cosh 4 - \cosh 2)/4$;   (e) $2(e^2 - e^{-6})$.

5. (b) $4 \sinh 3$;   (c) $2\pi(6 + \sinh 6)$;   (d) $(2 \sinh 3)/3$.

7. $(1 - e^{-6})/2$.

9. (b) 96/5, 64/5;   (d) $512\pi/7, 384\pi/7$;   (f) $576\pi/5, 704\pi/5$.

11. $\pi/8$.

13. (a) 81/4;   (b) 9/4;   (c) 243/10;   (d) 27/10.

P15. (a) 729/7;   (b) 81/7;   (c) $9\sqrt{7}/7$.

P17. 3/40 kilowatts.

B19. (a) $(\ln 2)/r$;   (c) $0.12/(1 - e^{-0.3})$.

S.21 221199 tons.

S23. $300, $900.

S25. $183.43, $2498.57, $1578.32.

Chapter IV

Section IV.1

1. (a) 9;   (c) 7;   (e) 6.01;   (g) 6.0001;   (i) 4;   (k) 5.9;   (m) 5.999.

2. (a) $-1/[x(x + \Delta x)]$;   (c) $-1 + 2x + \Delta x$;   (e) $3x^2 + 3x\Delta x + (\Delta x)^2$;   (g) 0;   (i) $8x + 4\Delta x - 12$.

B13. 22025, 294.83, 31.95, 17.18, 12.97, 11.66, 11.07, 10.52.

Section IV.2

1. (a) 11;   (c) 11/4;   (e) 46;   (g) 8.

3. (a) $3x^2 - 6x - 9$;   (b) $-1, 3$.

5. (a) $y = (-a/b)x + (-c/a)$, (b) $-a/b$.

7. (b) $-1/x^2$.

9. (b) 3; (c) 1/3.

11. (a) $3x^2/2 + 2x - 10$; (b) $3x + 2$.

13. (b) $y = x/6 + 3/2$; (c) $(10, 19/6)$.

15. (a) $x_0^3/3 + 3x_0^2/2 + x_0$; (b) $x_0^2 + 3x_0 + 1$.

SB18. (a) $p'(t) = kp(t)[1 - p(t)]$.

S20. 38, 42, 34 + $0.04n_0$.

P22. (a) 8.88 ft/sec^2; (b) $t < 17.32$ secs, $t > 17.32$ secs.

P24. (a) $96 - 32t$ ft/sec; (b) 3 secs; (c) 656 ft; (d) 204.9 ft/sec.

Section IV.3

1. (a) $6x - 5$; (c) $8x - 12$; (e) $x/2 - 1/3$; (g) $3x^2 - 6x + 7 + 4/x^2$.

3. (b) $2a$; (c) $-1/2a$.

5. (c) $-1, 1$.

7. (a) $(1, -2), (-1, 2)$; (c) $(0, 0)$.

P10. (a) gt ft/sec; (b) g ft/sec^2; (d) no.

Section IV.4

1. (a) $8x - 7 + 3/2\sqrt{x}$; (c) $3x^2$; (e) $28x - 84$; (g) $-3x^{-4}$; (j) $9x^2$.

3. (a) $-1 < x < 0, x > 1$; (c) $-1, 0, 1$.

5. (a) $2x_0 x - y = x_0^2$. ·

7. (a) $y = 300x - 2000$; (b) 940; (c) 941.192.

9. (a) $2\pi r$ units^2/unit.

P12. (a) $-2GMm/r^3$.

S14. $3p^2/(10000 - 1.5p^2)$.

Section IV.5

1. (a) $28x^6 - 15x^4 - 50/x^{11}$; (c) $3\sqrt{x}/2\sqrt{5} - 7/2\sqrt{5x} - 15/2x^{7/2}$;
   (e) $\sum_{k=0}^{20} k^2 x^{k-1}$; (g) $4\pi x^2$; (i) $(2x - 3)(32x^2 - 24x + 12)$.

2. (b) $2x + 3x^{-2} - 12x^{-5} + 7x^{-8}$; (d) $x^2(3 + 7x^4)$; (h) $(3x^2 - x^6)/(1 + x^4)^2$.

9. $-1, 0, 1$.

10. (a) $-1/(x + 1)^2$.

P15. (a) 160 ft/sec; (c) 0 ft/sec^3; (e) 16.307 secs; (g) $[0, 16.307)$.

P17. Decreasing $3/4$ lbs/in^2/sec.

S20. (a) $1.2 - 2n + 0.3n^2$; (c) when $n < 10$.

B21. (b) -474.34; (d) -0.265.

Chapter V

Section V.1

1. (a) $4e^{-3x}x^{-3/7}/7$; (c) $2x(\sec 2x - \csc 2x)^2$; (e) $2x \sin 4x$;
   (g) $2x(3 \sin 2x - 2)^3$.

2. (b) $(\tan x)/2x^{1/2}$; (d) 7; (f) x^{-4}; (h) 0; (j) $2x^{-1/2}$.

4. (a) $8x^{17/12}/17 + C$; (c) $-\cos x^2 + C$; (e) $-2 \cos(x/2) + C$;
   (g) $\sin x^3 + C$.

5. (a) $(e^9 - e)/2$; (c) $(1 - e^{-12})/6$.

7. $f(x) = 0.4x^{5/2} - \cos x - x^3/3 + 13 + \cos 3 - 18\sqrt{3}/5$.

9. (a) Increasing $-2 < x < 0, x > 2$, decreasing $x < -2, 0 < x < 2$;
   (c) increasing $x > 0$.

11. $\int_a^x \tan t \, dt^3$, the interval (a, x) must not include an odd multiple of $(\pi/2)$.

14. (a) $0.4(x^{2.5} - 1)$.

P17. $s = 256 + 8t - 16t^2$.

S19. $14,442.27.

B21. 59673.

Section V.2

2. (a) $3e^{3x} - 4 \cos 4x + 12.5x^{11.5}$; (c) $-6(3^{-2x})(\ln 3)(\cos 3x) - 3(3^{-2x})(\sin 3x)$;
   (e) $e^{2x}(2x^{-3} - 3x^{-4})$; (g) $\sqrt{x}[\cos(x/2) + \sin(x/2)/x]/2$;
   (i) $\sec^2 x - 2 \sec 2x \tan 2x$; (k) $-e^{-2x}(2 \cos x + \sin x)$.

4. (a) $-(1/3) \cos 3x + C$; (c) $(\tan 5x)/5 + C$; (e) $(\sec x)/2 + C$;
   (g) $e^{2x}/2$.

6. (b) $(\tan 3x)/3 + (\cot 4x)/4 + C$; (d) $(\sec 3x)/3 + C$.

7. (a) $1/2$; (c) $2\sqrt{3}/2\pi - 1/6$; (e) $11421/40$.

9. (a) $x + 2y = \pi/6 + \sqrt{3}$; (b) $y = 1$.

12. (a) $e^{-x}(1 - x)$; (f) $1 - 11e^{-10}$.

14. (a) 0; (b) 0; (c) 0.

15. π.

SB17. (a) $0.025P(0)\exp(0.025t)$;   (c) 2.5%.

SP19. $(R/\ln t)(1 - t^{-x})$.

Section V.3

1. (a) 3.5;   (c) 3;   (e) 4.

2. (a) $(3e^{3x} - 2\sin 2x)dx$;   (c) $0.5\exp(-x/2)(3\sin 2x + 5\cos 2x)dx$;
   (e) $(3\sqrt{x}/2 - \cos x)dx$;   (g) $(4\sec 4x \tan 4x + 2\csc 2x \cot 2x)dx$.

3. (b) 1.02;   (d) 0.85;   (f) 3.977;   (h) $1 + 2\pi/90$.

4. (a) Yes;   (c) no.

5. (a) $2x\,dx$.

7. (a) $4\pi r^2\,dr$.

10. $2(37\sqrt{37} - 10\sqrt{10})/27$.

12. $\sinh 3$.

14. $\pi(577\sqrt{577} - 1)/54$.

17. 2π.

19. 1.

P22. 25.6π, no.

Section V.4

1. (a) $(1/2, -1/4)$min;   (c) $(-1, -1)$min, $(0, 0)$max, $(1, -1)$min;
   (e) $(-1, 4)$max, $(1, -4)$min;   (g) none.

3. (a) min at $(-2, -60)$, $(2.577, -0.385)$, max at $(1.423, 0.385)$, $(4, 6)$, infl at $(2, 0)$;
   (c) min at $(0,0)$, $(3, 0.448)$, max at $(-3, 180.77)$, $(2, 4/e)$, infl at $(0.586, 0.1910)$, $(3.414, 0.3835)$;
   (e) max at $(-3, 17.086)$, $(3.3.050)$, min at $(0, 1)$;
   (g) min at $(0, 0)$, max at $(7, \sqrt{7})$.

5. (a) $2\sec^2 x \tan x$;   (c) $n(n - 1)x^{n-2}$;   (e) $2/x^3$;   (g) $\csc^3 x + \csc x \cot^2 x$.

7. (a) $2, 0$;   (c) $24, 0$.

9. min at $(0, 0)$.

P12. At $t = 3\pi/2$ location $= -10$, velocity $= 0$, acceleration $= 10$.

P14. 5.

B15. (c) $2r_0/3$.

S17. (a) $1000 + 8.95N$;   (c) $50,000,000/N + 447,500 + 0.375N$.

Section V.5

1. (a) -1.4656;   (c) -3.1038;   (e) -0.5671.

'3. (a) 2.37144;   (c) 0.2581.

5. $(0.6662, 0.7862)$.

7. 1.8955.

9. $4.4934, 7.7253, 10.9041$.

11. $149.12.

Section V.6

1. (a) $e^{\sin x} \cos x$;   (c) $3(x^2 - 4x)^2(2x - 4)$;   (d) $3 \tan^2 x \sec^2 x$;
   (g) $3 \sin^2[\exp(x^2)]\cos[\exp(x^2)]2x(\exp x^2)$.

3. (a) $(12 + 48x^2 + 16x^4)\exp(x^2)$;   (c) $-6(x^3 - x^{-3})^{-3}(x^2 + x^{-4})dx$;
   (e) $24(1 + 10x^2 + 5x^4)/(1 - x^2)^5$.

5. (a) $0.94e = 2.555$;   (c) 1.05;   (d) 1.

6. (a) $y = 1$;   (c) $2\pi x + 3y = 3\sqrt{3}$;   (e) $ex - y = \cosh 1$.

7. (a) $-(1/3)\cos(x^3) + C$;   (c) $-\exp(-x^2)/2 + C$.

8. (a) $(\sin^3 x)/3 + C$;   (c) $\exp(\sin x) + C$;   (e) $(x^2 - 4)^{3/2}/3 + C$;
   (g) $-\cos(e^x) + C$.

9. (a) $2^t(1 + t \ln 2)$;   (c) slope $= (2t_0 - 1)/[2^{t_0}(1 + t_0 \ln 2)]$.

11. (b) $12:21.6$ P.M.

13. (a) $-5/72\pi$ in/min;   (c) increasing.

15. (a) $(15t - 15 \sin t, 15 - 15 \cos t)$;   (c) t is an odd multiple of π.

P17. (a) $-9/100\pi$ ft/min.

S19. $409.60, $170.67.

Section V.7

1. (a) $(3x^2y + 3)/(2y + 12y^2 - x^3)$;   (c) $-(3x^2 + 2xy)/(x^2 + 3y^2)$;
   (e) $[3x^2 - 2xy \sec^2(x^2y)]/[x^2 \sec^2(x^2y) - 3y^2]$.

2. (b) $2x/(1 + x^4)$;   (d) $1/(17 - 8x + x^2)$;
   (f) $2x/[(x^2 + 3)(x^4 + 6x^2 + 8)^{1/2}]$;   (h) $3(\text{arc sec } x)^2/[x(x^2 - 1)^{1/2}]$.

3. (a) $\pi/4$;   (c) $\text{arc sec}(x + 3) + C$;   (e) $\text{arc sin } e^x + C$.

5. $x + 2y = 1$.

7. $K = 12/145\sqrt{145}, R = 145\sqrt{145}/12$.

9. $-48/(5y^2 + 16)^{3/2}$

11. (a) $2x_0^3/(x_0^4 + 1)^{3/2}$;   (b) $(1, 1)$ and $(-1, -1)$.

S13. (a) $P(D_n Q) + Q(D_n P)$.

Section V.8

1. (a) $\cot x$;   (c) $(3x^2 - 4)/(x^3 - 4x + 2)$;   (e) $\tan x$;   (g) $4/(x + 5)$.

3. (a) $2 + \ln 81$;   (d) $219/8 + 27 \ln 2$;   (f) $\ln \sin x + C$;
   (h) $-\ln(\cos 3x - 4)/3 + C$.

4. (a) $-24(1 - x)^{-5}$;   (c) $1/2$;   (e) $-2.6058(1 - x)^{-4}$;   (g) $\ln(17/2)$.

5. (a) -0.05;   (c) 0;   (e) $-\ln 2 - \pi\sqrt{3}/60$.

6. (b) 0.653.

7. 1.317.

9. $\ln(10!)$;   (c) $\ln(30!/3^{29})$.

SB11. (b) 31436 years.

SB12. 101 years.

B14. (c) 35.00 years.

Chapter VI

Section VI.1

1. (a) $2x \ln 3x + 2 - x \exp(-x^2/2)$;   (c) $4^x \ln 4$;
   (e) $3x^2 \text{ arc } \sec(x^2 - 3x) + (2x^3 - 3x^2)/[x - 3)(x^4 - 6x^3 - 9x^2 - 1)^{1/2}]$;
   (g) $-4 \cot 4x$.

3. (a) $0.25 \ln(4/3)$;   (c) 0.2595;   (e) $1 - \cos 1$;   (g) 2.

5. (a) 40.212;   (c) 49.763;   (3) 16π.

7. 1.317.

9. 0.866.

12. 2.5 radians/hr.

13. 7.746 ft.

Section VI.2

1. (b) $e^{3x} \sin 4x(3 + 4 \cos 4x)$;   (d) $x^{\sin^2 x}(2 \sin x \cos x \ln x + \sin^2 x/x)$;
   (f) $-3/(x - 1)^2$;   (h) $4^{\sin x} \cos x \ln 4 + \pi x^{\pi - 1}$.

3. (a) $1/(x \ln x \ln \ln x)$;   (c) $1/(x \ln x \ln \ln x \ln \ln \ln x)$.

5. (b) 1.763.

7. $(2.128, 1.898)$.

9. $[(\cos t)^{\cos t - t^3}][t^3 \tan t - \sin t - (\sin t + 3t^2)\ln \cos t]/[(\sin t)^{\tan t}][1 + \sec^2 t \ln \sin t]$.

Section VI.3

1. (a) $(-2/3)\cos^3 x + C$;   (c) $(\tan^5 2x)/10 - (\tan^3 2x)/6 + (\tan 2x)/2 - x + C$;
   (e) $(1/6)\cos^3 x^2 - (1/2)\cos x^2 + C$;   (g) $5\pi/32$.

2. (a) $\ln(\sec x + \tan x) - \sin x + C$;   (c) $(1/2)\tan 2x + (1/6)\tan^3 2x + C$;
   (e) $(1/3)\sin^3 x + C$;   (g) $1/6$.

3. (a) $-(\cos 7x)/14 - (\cos 3x)/6 + C$;   (c) $\tan 2x/2 - x + C$;
   (e) $(\sin 11x)/22 + (\sin 3x)/6 + C$;   (g) $(\cos^7 4x)/28 - (\cos^5 4x)/20 + C$.

5. $1/2$.

7. $16\pi - 24\sqrt{3}$.

9. $\pi/2$.

11. 10π.

13. (a) 0;   (d) 0.

Section VI.4

1. (a) -3.720;   (c) $-3(1 - x)^{2/3}(5x^2 + 6x + 9)/40 + C$;
   (e) $-(2x - 3)^{1/2}(54x^2 + 408x + 544)/405 + C$;   (g) $(1 + x^2)^{1/2} + C$.

2. (a) $(1 + x^2)^{3/2}/3 + C$;   (c) $x^3/3 - (1/3)\ln(x^3 + 1) + C$;
   (e) $x + 1 + 8(x + 1)^{1/2} + 16\ln(\sqrt{x + 1} - 2) + C$;
   (g) $2(21x + 106)(3x - 5)^{1/2}/27 + C$.

3. (a) $-(1 - x^2)^{1/2} + \arcsin x + C$;
   (c) $2x^4 - 8x^3/3 - 2x^2 + x - 2\ln(2x - 1) + C$;
   (e) $-x^2/2 - 4\ln(3 - x) + C$;   (g) $e^{2x}/2 - 2e^x - 3\ln(e^x + 2) + C$.

5. (a) $52/3$;   (c) $38/3 - 14\ln 3$;   (e) $1311/28$.

6. $2 - 2\,\text{arc sec}\sqrt{10} + 2\,\text{arc sec}\sqrt{5} = 1.7162$.

9. (a) $-\ln(e^{-x} + 1)$;   (b) $\pi/12$.

S10. \$64.

Section VI.5

1. (a) $\ln(x + \sqrt{x^2 + 9}) + C$;   (c) $(1/6)\ln[(3 + x)/3 - x)] + C$;
   (e) $\sqrt{x^2 + 9} + C$;
   (g) $(3645/16)\text{arc sin}(x/3) + \sqrt{9 - x^2}(2187x - 180x^3 + 8x^5)/16 + C$.

2. (a) 18.017;   (c) 1.988;   (e) 4π;   (g) 1.080

3. (a) 0.2125;   (c) 0.2079;   (e) 0.309;   (g) 1.218.

4. (a) 0.2838;   (c) 0.6423;
   (e) $(-6/5)\ln[(10 - x)/(10x - x^2)^{1/2}] - (1/2)\ln(10x - x^2) + C$.

5. (b) $\arctan x + C$.

6. (b) $25\pi/3\sqrt{3} - (63/2)$.

7. (a) $\arctan e^x + C$;   (c) $(1/2a)\ln\dfrac{x - a}{x + a} + C$;   (e) $(1/3)\text{arc sin}\dfrac{3x - 2}{2} + C$.

9. (a) $25\pi/3 + 25\sqrt{3}/2$.

11. (a) $15/2 - 4\ln 4$.

13. $4a/3\pi$.

15. (a) 0.2986;   (c) (4.0792, 0.0774).

Section VI.6

1. (a) $(-0.25)[\csc 2x \cot 2x - \ln(\csc 2x - \cot 2x)] + C$;
   (c) $-e^{-x}(x^3 + 3x^2 + 6x + 6) + C$;   (e) $(x^{11}/11)\ln x - x^{11}/121 + C$;
   (g) $\sqrt{3}/2 - \pi/6$.

2. (b) $-x^2\cos x + 2x\sin x + 2\cos x + C$;
   (d) $(x^2/2)\text{arc csc } x + (1/2)\sqrt{x^2 - 1} + C$.

3. (b) $(x^{101}/101)(\ln x - 1/101) + C$;   (d) $x \text{ arc cos } x - (1 - x^2)^{1/2} + C$;
   (f) $x \text{ arc cot } x + (1/2)\ln(x^2 + 1) + C$;   (g) $x \text{ arc csc } x + \ln(x + \sqrt{x^2 - 1}) + C$.

4. (b) $(-1/33)(4\sin 4x \sin 7x + 7\cos 4x \cos 7x) + C$.

5. (a) 0.2173;   (c) 0.1090;   (e) 0.2804

6. (b) 0.1535;   (d) $2\pi/3 - \sqrt{3}/2$.

8. (a) If $n = 3$, the result is $e^x(x^3 - 3x^2 + 6x - 6)$.

9. (a) $\dfrac{e^{ax}}{a^2 + b^2}(a\cos bx + b\sin bx) + C$.

12. (a) 12.1718.

13. (a) 0.1278;   (c) 0.7774;   (e) 0.6504.

Section VI.7

1. (b) $x^4/4 + x^3/3 + 7x^2/2 + 13x + (243/5)\ln(x - 3) + (32/5)\ln(x + 2) + C$;
   (d) $(1/2a)\ln[(x - a)/(x + a)] + C$;
   (f) $(1/4)\ln[(x - 1)/(x + 1)] - (1/2)\text{arc tan } x + C$.

2. (a) $3\ln[x/(x + 1)] + 3/x - 3/2x^2 + 2/3x^3 + C$;
   (c) $(1/\sqrt{3})[\text{arc tan}(2x - 1)/\sqrt{3} + \text{arc tan}(2x + 1)/\sqrt{3}] + C$;
   (e) $(1/27)[-3/2(x + 1)^2 - 2/(x + 1) - 1/(x - 2) + \ln(x + 1)(x - 2)] + C$.

3. (b) $(1/2)[\text{arc tan}(x - 1)/2 - \text{arc tan}(x + 1)/2] + C$;
   (d) $1/(x + 1) + \ln(x - 2) + C$;   (f) $\ln[(x^2 - 1)/x] + C$.

4. (a) $(1/4)\ln[(2x + 3)^2/(4x^2 + 4x + 17)] - (1/8)\text{arc tan}[(2x + 1)/4] + C$;
   (c) $(1/4)\ln[(x - 4)/x] + (x + 2)/x^2 + C$;   (e) $\ln[(x^3 - 1)/(x^3 + 1)] + C$.

5. (b) $30\ln x - 24\ln(x - 1) - 16\ln(x + 1) + 13\ln(x - 2) - 3\ln(x + 2) + C$;
   (d) 0.4089;   (f) 0.05699.

7. 0.4463.

8. 0.4604.

10. $(1/4)\sec^3 x \tan x + (3/8)\sec x \tan x + (3/8)\ln(\sec x + \tan x) + C$;

BP11. $x = -(kI/2p) + [K(1 + C \exp(-Kt))/(1 - C \exp(-K/t))$
      where $K = (4kI + k^2I^2)^{1/2}$ and C is the constant of integration.

S12. (a) 4,622,000,000 people.

Section VI.8

1. (a) $(-1/11)\sin^4 x \cos^7 x - (4/99)\sin^2 x \cos^7 x - (8/693)\cos^7 x + C$;
   (c) $(1/9)\sin^9 x - (2/7)\sin^7 x + (1/5)\sin^5 x + C$; (e) 16/1155; (g) 16/35.

2. (b) $\pi/16$; (d) 2/35; (f) $e^{2x}(2 \sin 3x - 3 \cos 3x)/13 + C$;

   (h) $e^{x^2}(x^4 - x^2 + 2)/2 + C$.

3. (c) $-2/3$.

4. $21\pi/64 - 2/3$.

6. (a) 54332/315; (c) 1.3224; (d) $1323\pi/1024$; (f) 2.

7. (a) 1280π; (c) 28π; (e) 0; (g) $3^9\pi/16$.

10. (b) $\pi^2/2$.

Section VI.9

1. (a) $(1 - \cos x)/\sin x + \ln(1 + \cos x) + C$; (c) $2 \sec x - 2 \tan x + 2x + C$.

2. (b) $8x(\text{arc sec } x)x^{1/3}[4/3x + 1/(x \text{ arc sec } x(x^2 - 1)^{1/2})] + 2x + C$.

3. (a) $(\tan^2 2x)/4 + C$;
   (c) $(560x - 256 \sin 4x + 128 \sin^3 4x/3 + 56 \sin 8x + \sin 16x)/2^{19} + C$.

4. (b) 0.06451; (d) 0.00286.

5. (a) $e^{3x}(11 \sin 5x - 7 \cos 5x)/34 + C$;
   (d) $e^{x/2}(2 \ln x - 4/x - 8/x^2 - 32/x^3) - 96 \int e^{x/2}x^{-4}\, dx$.

6. (b) 0; (d) 128/1300075.

7. (a) $(\tan^4 x)/4 + (\tan^3 x)/3 + \ln \sec x + x + C$;
   (c) $x + (5/8)\ln(\sec x + \tan x) + (7/8)\sec x \tan x + 2 \tan x + (1/3)\tan^3 x$
   $+ (1/4)\sec^3 x \tan x + C$.

8. (b) $(1/14)\cos^7 2x - (1/10)\cos^5 2x + C$; (d) $(1/4)\tan^4 x + C$;
   (f) $(1/10)\cos 5x + (1/26)\cos 13x + C$.

9. (a) $x^2/2 + \ln([(x^2 - 1)/x] + C$;
   (c) $x^5/5 - x^3/3 + x + \ln[x^2/(x^2 + 1)] - \text{arc tan } x + C$.

11. 8.

12. 640π.

S15. $105.00.

B16. $P(t) = (1 + b)P(0)/[1 + b \exp(-kt)]$

Chapter VII

Section VII.1

1. (c) $(0, 1]$.

2. (b) 0.35.

Section VII.2

1. (a) 1; (c) 2.

2. (b) -5; (d) 0.

5. (a) Yes, 0.

6. $2x - 1 - 1/x^2$.

9. (a) No.

11. (a) 0.

B12. (c) $T = 300, N = 200$.

Section VII.3

1. (a) 4; (c) -1; (e) 2/27.

2. (b) 0; (d) 0; (f) ∞.

5. (b) ∞; (d) 3/2; (f) 0.

Section VII.4

8. (c) No.

B10. (a) Yes.

Section VII.5

1. (a) $\sqrt{x}$.

2. (a) $[-\pi/2, \pi/2]$.

5. (c) Yes, if x is the number of radians.

6. (b) None; (d) $n\pi/2$ for any integer n; (f) $x \le -2$; (h) $\pm 1, \pm 2$.

7. (a) No; (d) yes.

8. (b) Each interval that does not include $x = 0$.

9. (e) 15; (f) 35.

10. (b) Yes; (e) no.

SB12. (a) Yes; (c) no.

Section VII.6

1. (a) 1;   (c) 1/6;   (e) 2;   (g) 0.

3. (a) Does not exist;   (c) 192;   (f) 0;   (h) 1/2.

4. (b) Does not exist;   (d) does not exist;   (f) 0.

5. (a) $\pi/2$;   (c) $\pi/2$;   (f) ∞;   (h) 1;   (j) 0.

6. (a) No.

7. (a) 2.25;   (c) 2.48832.

S10. (b) P.

Section VII.7

1. (a) $\pi/2$;   (c) $\pi/6 - (1/3)\text{arc sin}(5/3)$;   (e) $\pi/6$;   (g) 1.

2. (b) Does not exist;   (d) does not exist;   (f) 1.

3. (b) 6.7798;   (d) does not exist;   (f) 0;   (h) 2.

4. (b) Does not exist;   (d) 1;   (g) does not exist.

5. (b) $-1 < n < 0$;   (d) -1.

6. (b) $n < -1$.

8. (c) -0.9453087204.

10. (b) Does not exist;   (c) π.

11. (c) $\pi^2/2$.

12. (c) $\pi^2/2$.

P14. (a) $0.693k$;   (c) does not exist.

SB15. (a) 32.69 years;   (c) 24.9 years vs. 9.6 years.

P16. (a) 58.78 watt-seconds.

Chapter VIII

Section VIII.1

1. $x^2 + 7x - 3$.

3. $x^3 - 3x + 2$.

5. $x^4/30 + x^2/6 - 1/5$.

7. (b) $a = 7.1009, b = -0.5068$.

9. $19.4912(2^{-t/2}) - 4.4735(2^{-t}) + 2.9823(2^{-3t/2})$.

Section VIII.2

1. (a) $(-13x^3 - 144x^2 + 433x + 3924)/840$;   (c) $-1.25F^{iv}(c)$.

2. (a) $x^3 + 12$.

4. (a) $1 - 9x^2/2\pi^2$;   (c) $-0.07177f'''(c)$.

5. (a) $(-13x^3 + 27x^2 + 154x + 906)/6$ if 1940 is equivalent to -1.

6. (c) 2,784,000,000 megawatts.

7. (a) 0.7600;   (e) 0.0006 actual error vs. 0.0009 as an upper bound for error.

9. (b) $(h/3)(y_1 + 4y_2 + y_3)$;   (d) $(2h^2 - 24)/3$.

11. (a) $2x, 2y$;   (c) $3x^2 - 6xy + 3y^2, 3x^2 + 1 + 6xy - 3y^2$;

Section VIII.3

2. (a) $-0.4x + 0.6$;   (c) 1.2.

4. (a) $-x + 7.5$;   (e) 0.

5. (a) $0.7392x^2 - 2.4973x + 3.0161$;   (b) 0.0001.

7. (a) $-0.25482x^2 + 1.09589x - 0.00063$.

8. (c) $0.94869 \exp(1.10259x)$.

9. (c) $19.79312 \exp(-0.98255x)$.

11. (a) $2x, 2y$;   (c) $3x^2 - 6xy + 3y^2, 3x^2 + 1 + 6xy - 3y^2$;
    (e) $y\cos(xy), x\cos(xy)$;   (g) $\cos(x + y), \cos(x + y)$.

13. (a) $y + z, x + z, x + y$;   (c) $yz\cos(xyz) + e^y \ln z, xz\cos(xyz) + xe^y \ln z$,
    $xy\cos(xyz) + (x/z)e^y$.

Section VIII.4

1. (a) $3.1435x + 0.04166$.

3. (b) $1.9410x^2 + 3.5506x - 0.3353$.

5. (a) $-6x - 5$.

SB7. (a) $199.436x + 17935.988$ using 1965 as a base year and giving units in
thousands;   (c) $17908 \exp(0.01108t)$.

BC9. (a) $19.914 \exp(0.25635t)$; $3563 - 6745593 \exp(-0.41874t)$ if $c = 3563$.

C11. (a) $0.399801x^2 + 4.20855x + 1.96811$.

Section VIII.5

1. (a) $2x\sin y + y^2 \sin x, x^2 \cos y - 2y\cos x$;   (c) $67/3, 75/4$;
   (e) $2r\sin 2\theta, 2r^2 \cos 2\theta$.

3. (b) $3x - 4y + 6z - 41 = 0$;   (d) $x + 4y - 2z = 3 - \pi/2$.

4. (a) $(3\,dx + 6\,dy)/(x + 2y)$;   (c) $(2x + 3y)dx + (3x - 6y^2)dy$;
   (e) $\cos 2\theta\,dr - 2r\sin 2\theta\,d\theta$.

5. (a) 3.86;   (c) -54.51;   (e) 0.48956.

8. (a) $(2xy + y^2 - ye^{xy})(2t - 2) + (x^2 + 2xy - xe^{xy})(-3)$;
   (c) $2(e^t + 1)/(3y - 2) + (6x + 9)(e^{-t} + 1)/(3y - 2)^2$.

10. (c) 3π;   (d) 8.

S11. (d) If $L/C < 3$.

S12. (c) $Y = (I + G + a - bc)/(1 - b + bd)$.

BP13. (a) 28985.7 calories;   (c) 211.45 calories.

Section VIII.6

2. (a) $3x^2\cos\theta - 3y^2\sin\theta$, arc $\tan(-9/4)$;
   (c) $(2y\cos\theta - 2x\sin\theta)/(x + y)^2$, arc $\tan(-2/3)$.

4. (a) Min at $(1, 1)$, Max at $(-1, -1)$;   (c) Min at $(1, 1), (1, -1), (-1, 1)$, and $(-1, -1)$.

P6. (a) Decrease 5%.

S7. (a) $w = \$1,661,800$, $p = \$1,285,500$ produces a minimum.

S8. (c) $c = 109.75$, $w = 101.56$.

S9. (e) 1/2, Maximum.

S10. (b) $500 + 30x$;   (d) 1044 units;   (f) $x = 0, t = 31$.

Chapter IX

Section IX.1

1. (a) 0;   (c) 0;   (e) ∞.

2. (b) $\pi/2$;   (f) 0.

4. (a) 1/2, 3/4, 7/8, 15/16, 31/32;   (d) no.

5. (c) 1, 2/3;   (e) no.

6. (d) $\sqrt{30}$.

SB10. (a) The first two terms are B and $0.63212A + 0.36788B$, for $k = 1$.

P12. (b) No;   (c) yes.

Section IX.2

1. (a) Converges, 3/2;   (d) diverges.

2. (a) Converges, 2;   (c) diverges.

3. (a) Yes; (c) no; (e) yes.

4. (a) $1 + 2 + 2 + 4/3 + 2/3 + 4/15 + \cdots$.

5. (b) 26149/9900.

6. (c) $e^k/(1 - e^k)$.

7. 56 feet.

C15. (a) 5.18738; (d) no.

C16. (a) $4, 11, 31, 83$, etc.; (c) e.

Section IX.3

1. (a) Convergent; (d) divergent; (g) divergent.

3. (a) All values; (c) $x = 1$; e) $|x| < 1$.

4. (a) All values; (c) $|x| \leq 1$; (e) $1 < x < 2$; (g) $1 < x \leq 3$.

5. (b) $1 + 2x + 2x^2 + 4x^3/3 + 2x^4/3$;
   (e) $1 - x^2/2 + x^4/24 - x^6/720 + x^8/40320$; (g) 1.

B7. (a) $c_0 \sum_{j=0}^{k} e^{-Tj}$.

B8. (b) 0.2117 units.

Section IX.4

1. (a) $x - x^3/6 + x^5/120 - x^7/5040 + x^9/362880 + \cdots$, converges for all x;
   (e) $x + x^3/3 + 2x^5/15 + 17x^7/315 + 1097x^9/45360 + \cdots$;
   (h) $1 + x^2 + x^4/2 + x^6/6 + x^8/24 + x^{10}/120 + \cdots$, converges for all x.

3. (a) $x - x^2/2 + x^3/3 - x^4/4 + x^5/5 - \cdots$, converges for $-1 < x \leq 1$;

   (c) $\sum_{k=1}^{\infty} \frac{2x^{2k-1}}{2k - 1}$, converges for $|x| < 1$; (e) 1.6904.

5. (a) $a^n + na^{n-1}bx + n(n - 1)a^{n-2}b^2x^2/2 + n(n - 1)(n - 2)a^{n-3}b^3x^3/6 + \cdots$;
   (d) 5.

6. (a) $x + x^3/6 + x^5/120 + x^7/5040 + x^9/362880 + \cdots$.

9. (a) $1 - x^3/3 + x^5/5 - x^7/7 + x^9/9 - \cdots$.

C10. (b) 10.

M12. (b) $-e \leq x < e$.

Section IX.5

1. (a) 500,000; (c) 0.321751.

3. (d) 1.0986123.

4. (e) 0.523599.

5. (c) 0.461281.

7. (b) 0.22413.

8. (b) 3.104379

12. (a) $\ln \sqrt{2} + \pi i/4$;   (c) $\ln 2 + 5\pi i/3, \ln 2 + 3\pi i/2$.

13. (a) $2\sqrt{2} \exp(-\pi i/4)$.

P14. Converges for all values of x.

Section IX.6

1. (a) 26.55;   (c) 2016;   (e) $(\pi + 1)/8$.

2. $4\pi/9$.

4. (c) $32\pi/3$.

5. (a) 32π.

7. (a) $4\pi abc/3$;   (d) $abc/8\pi$.

9. (a) 30.78 lbs.

S10. 2500.

BP12. 0.02016.

Section IX.7

1. (a) $1/r$;   (c) 8;   (e) $\int_0^4 \int_0^{4-y} dx\, dy$,

3. 300π grams.

5. (b) $r = 4\cos\theta, 4\pi$.

7. (a) 27; (b) -2.

9. (b) $r^2 \sin\varphi$;   (d) $500\pi/3$;   (f) 1875π.

Section IX.8

1. $\pi - \sum_{n=1}^{\infty} \frac{2}{n} \sin nx$

3. $2/\pi - (2/\pi) \sum_{n=1}^{\infty} (\cos nx)/(4n^2 - 1)$.

5. $1/2 - (4/\pi) \sum_{k=1}^{\infty} [(\sin(4k - 2)]/(4k - 2)$.

8. (b) $\pi/2 - (4/\pi) \sum_{k=1}^{\infty} \dfrac{\cos(2k - 1)x}{(2k - 1)^2}$.

10. (a) $(3/4)\cos x + (1/4)\cos 3x$.

11. (a) $(e^{\pi} - 1)/2\pi + (1/\pi) \sum_{j=1}^{\infty} (e^{\pi} \cos \pi j - 1)(\cos jx - j \sin jx)/(j^2 + 1)$.

Section IX.9

1. (a) $(x + 2)^5 - 17(x + 2)^4 + 96(x + 2)^3 - 246(x + 2)^2 + 293(x + 2) - 113$.

2. (b) $0 < x \le 20$.

3. (b) 0.8461.

4. (b) $1 - x^{2/3}/2 + x^{4/3}/24 - x^3/720 + x^{8/3}/40320 - \cdots$.

5. (c) 0.37894.

6. (c) π^2.

7. (a) $1 + x + x^2/2 - x^4/8 - x^5/15 - x^6/240 + x^7/90 + \cdots$.

9. (a) $1/4 - (3/\pi)\cos x + (1/\pi)\cos 3x - (3\pi/5)\cos 5x + \cdots$
$- (3/\pi)\sin x + (1/\pi)\sin 2x - (1/\pi)\sin 3x - (3/5\pi)\sin 5x + \cdots$.

Chapter X

Section X.1

3. (a) 1000,000; (c) 0.05.

Section X.2

1. (a) 6.1, 6.

3. (a) $10.24 \pm 1.32, 9.04 \pm 0.04$; (c) $7.5769 \pm 0.1942, 7.3921 \pm 0.00324$;
(e) $37.128 \pm 5.808, 32.32 \pm 0.352$.

5. (a) 0.0002, 0.02449; (c) 0.00733, 0.0469.

7. (c) 0.69338, 0.6931475.

P12. (b) 1.995; (d) 2.00075.

Section X.3

1. (a) 0.75; (d) 0.69702; (g) 0.78279.

2. (b) 0.06; (d) 0.054; (f) 0.088.

3. (b) 6.4804 (c) 0.0059

4. (a) $\displaystyle\int_0^{2\pi} \sqrt{25 \sin^2 t + 9 \cos^2 t}\, dt$.

5. (a) 1.6858.

6. (a) 21.0085.

Section X.4

1. (a) 0.69444; (c) 0.69325; (e) 0.0165.

3. (a) 0.157; (c) 8.5567

4. (b) 0.3706.

6. (a) 1.6836.

7. (c) 0.69375;   (e) 0.78462.

8. (b) 0.000017 $f^{iv}(c)$.

9. (a) $A = B = (b - a)/2$.

C11. 0.0974.

Section X.5

1. (a) 1.00101;   (d) 1.00057;   (f) 1.00007.

2. (c) 0.785395.

3. (c) 0.78542375.

4. (b) 0.697024;   (f) 0.6931545.

Section X.6

1. 2.8.

3. (a) 5.

4. (a) $1 + x^2$.

5. 1.73354 for $h = 0.25$.

C7. -0.39354.

S9. $937.22.

B10. 0.215.

Section X.7

3. $h < 0.089$.

5. 0.78373.

P7. 449 inch-lbs.

Chapter XI

Section XI.1

1. (a) $2 \exp[(x^2 - 1)/2]$;   (c) $1/[1 - C \exp(x^2/2)]$.

3. (a) $y = 3x$;   (c) $y^2 = x^2 + 16$.

4. (b) $\arcsin 2y = 2 \ln|\csc x - \cot x| + C$.

6. (a) $y = c_1 \sin kx + c_2 \cos kx$;   (c) $y = c_1 x + c_2$.

SB8. (a) $1/[C \exp(-kt) + 2]$.

SB10. (b) $t = C + [1/(a - b)k]\ln[(a - x)/(b - x)]$ if $a \neq b$, $t = C - 1/k(a - x)$ if $a = b$.

PB11. (b) $S - 1/(kVt + C)$; (d) 29.1 grams.

S12. (b) $Y = c_1 \exp(\sqrt{k/m}t) + c_2 \exp(-\sqrt{k/m}t)$.

Section XI.2

1. (a) $x^3/4 + C/x$; (c) $(x + c)\exp(-x^2)$.

2. (b) $y^{-2} = x + 0.5 + ce^{2x}$; (d) $y = 2x/(x^2 + c)$.

3. (a) $x^2 - 2 + 3\exp(-x^2/2)$; (c) $y^{-2} = \sin x + C\cos x$.

4. (b) $x/[c - 2(\ln x)^2]$; (d) $4y^4 = 1 + c\exp(2x^2)$.

5. (a) $y = e^{-2x}(x + c_1)$.

6. (b) $u = -x/3 - 1/9 + c_1 e^{3x}$.

SB7. (b) 300,870,572.

B8. (b) 1892.

P9. (b) 2 minutes.

PB10. (a) $500\exp(-t/250)$.

S11. (b) If $a = c$ and $b = d$.

Section XI.3

1. (a) $c_1 + c_2 e^{-2x}$; (c) $c_1 e^{2x} + c_2 e^{-2x}$.

3. (a) $c_1 \sin 2x + c_2 \cos 2x$; (c) $e^{3x}(c_1 \cos 2x + c_2 \sin 2x)$.

5. (a) $e^{2x}(c_1 + c_2 x)$; (c) $e^x(c_1 + c_2 x + c_3 x^2)$:
   (e) $e^{2x}(c_1 + c_2 x) + e^{-2x}(c_3 + c_4 x)$.

7. (a) $e^{-2x}(\cos x - 2\sin x)$; (c) $e^x(x^3 + 12x^2 + 36x + 24)$.

P9. (b) $D_t^2 q + 120 D_t q + 10^4 q = 0$; (e) 3.79 amps.

S10. (c) $D'' - kD' = 0$; $Y = Cke^{kt}$.

BP12. (b) 24.72 seconds.

Section XI.4

1. (a) $3 + c_1 e^{-x} + c_2 e^{-2x}$; (c) $(3\sin x + \cos x)/10 + c_1 e^{-x} + c_2 e^{-2x}$.

3. (a) $e^x/2 + c_1 + c_2 \sin x + c_3 \cos x$; (c) $-(x\sin x)/2 + c_1 + c_2 \sin x + c_3 \cos x$.

4. (a) $-xe^x - 2e^{-x} + c_1 e^{4x} + c_2 e^{-2x}$.

7. (a) $(3 \sin x - \cos x)/10 + ce^{-3x}$;   (c) $-e^{-3x}(\cos x + c)$.

P8. (c) 0.0168 amps.

S9. (a) $Y' - kY = C$;   (d) $Y = \sqrt{k}(c_1 e^{t\sqrt{k}} - c_2 e^{-t\sqrt{k}})$.

B10. (d) $2.36(c_1 e^{-0.2\sqrt{7}t} - c_2 e^{-0.2\sqrt{7}t})$.

Section XI.5

1. (a) $3 + c_1 e^{-x} + c_2 e^{-2x}$;   (c) $e^x/6 + c_1 e^{-x} + c_2 e^{-2x}$.

3. (a) $(\sin 3x - 3 \cos 3x)/40 + e^{-2x}(c_1 \cos 3x + c_2 \sin 3x)$;
   (c) $(-xe^{-2x}/6)\cos 3x + e^{-2x}(c_1 \cos 3x + c_2 \sin 3x)$.

5. (a) $-e^{2x}/12 + 34e^{-x}/21 + 13e^{6x}/28$;   (c) $\sin x - (\sin 4x)/4$.

Appendix A

1. (a) $\cos \theta = \sqrt{7}/4$, $\tan \theta = 3/\sqrt{7}$;   (d) $\sin \theta = 15/17$, $\tan \theta = 15/8$;
   (g) $\cos \theta = 0.8$, $\tan \theta = -0.75$.

3. (a) $3 \sin \theta \cos^2 \theta - \sin^3 \theta$;   (d) $4 \sin \theta \cos^3 \theta - 4 \sin^3 \theta \cos \theta$;   (g) $2 \sin \alpha \sin \beta$.

4. (b) $2 + \sqrt{3}$;   (d) $-\sqrt{2 - \sqrt{2}}/2$;   (g) $-(1 + \sqrt{2})$.

5. (b) $-2 - \sqrt{3}$;   (d) $-\sqrt{3}/2$;   (f) $\sqrt{6}/4 + \sqrt{2}/4$.

7. (a) $(\sin 7x - \sin 5x)/2$.

9. (b) 56/65, 16/65, 33/65, 63/65, 56/33, 16/33.

Appendix B

Section B.1

1. (b) 0, 6/5, 3/2, 3, -3, 1;   (d) None;   (f) 3, $\sqrt{61}$, $\sqrt{13}$, $2\sqrt{10}$, $\sqrt{10}$, $3\sqrt{2}$.

3. (a) $-7/4$;   (c) $\sqrt{65}$;   (e) $\sqrt{130}$.

Section B.2

1. (a) $2x + y = 7$;   (c) $y = -1$;   (e) $3x - y + 3 = 0$;   (h) $2x + y = 0$.

3. (a)   (c)   (f)

5. (a) $5\sqrt{13}$;   (c) $(29/13, -37/13)$.

6. (a) $6x - 8y = 13$;   (d) $2x - y = 4$;   (g) $x = y$.

Section B.3

1. (a) $x^2 + y^2 + 4x - 6y - 3 = 0$;   (c) $x^2 + y^2 - 8y = 0$;
   (f) $x^2 + y^2 - 6x - 6y + 9 = 0$.

3. (b) $4x + 3 = 0$.

5. (a) $(-4, 4), (-2.4, 7.2)$;   (c) $x + 2y = 8$.

7. (a) 0 or $-20/21$.

Section B.4

1. (a) $y^2 = 8x$;   (c) $y^2 - 12x - 2y = 59$;   (f) $y^2 - 6x - 6y + 12 = 0$.

3. (a) $(a, 2a), (a, -2a)$;   (c) $2a$.

4. (a) $25x^2 + 16y^2 = 400$;   (c) $5(x - 3)^2 + 4(y - 3)^2 = 80$.

6. $9(x - 4)^2 + 25(y - 3)^2 = 225$.

7. (a) $x^2/9 - y^2/16 = 1$;   (d) $x^2 - (y - 4)^2 = 8$.

Section B.5

1.   6.

12.   20.

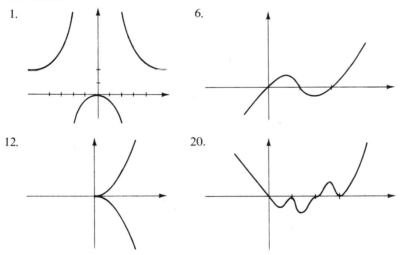

Section B.6

2. (a) $(1, \sqrt{3})$;   (c) $(0, -3)$.

6. (a) $r \sin \theta = -4$;   (c) $r \cos(\theta - \pi/3) = 2$.

7. (a) $x = 4$;   (c) $r \cos(\theta - \pi/3) = 5/2$;   (e) $\theta = \arctan(2/3)$.

Section B.7

1. (a)   (h)

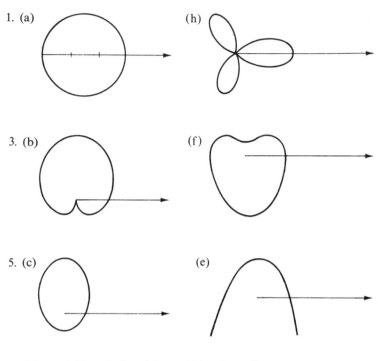

3. (b)   (f)

5. (c)   (e)

7. (a) $r = 2/(1 + \sin \theta)$;   (c) $r = 6/(1 - 2 \cos \theta)$.

8. (b) $3x^2 - y^2 - 12x + 9 = 0$.

9. (b) $\theta = (n + 0.5)\pi$ for any integer n.

Section B.8

1. (b) $(1 + \sqrt{5})^2 x^2 - 4y^2 = 8(1 + \sqrt{5})$;   (c) $y^2 - x^2 = 6\sqrt{13}/13$.

3. (a) $x^2 - y^2 = 4$;   (c) $x^2 + (y - 2\sqrt{2})^2 = 12$.

Appendix D

Section D.1

1. (a) Integer;   (d) integer;   (h) real;   (j) real.

2. (b) Integer;   (g) real;   (j) integer.

3. (a) Not legitimate;   (d) legitimate;   (h) not legitimate.

4. (a) 5;   (c) 0;   (f) 1.

Section D.2

1. (a) 33; (d) -13.

3. (a) 42.3; (d) 1.

4. (b) Acceptable; (d) contains period; (h) starts with a digit.

5. (a) Arithmetic operation left of " $=$ "; (g) parentheses do not match.

Section D.3

1. No.

3. IF (B*B $-$ 4*A*C) 300,200,100

5. (a)

A	B	A .AND. B
T	T	T
T	F	F
F	T	F
F	F	F

6. (a) 30 if N is a square.

7. (b) A .NE. B; (c) A .LE. B.

9. 4.9.

Section D.4

2. (a) Legitimate, a real number occupying 12 spaces with 3 following the decimal point; (f) illegal, more decimal places than total spaces for the number.

4. (a) 234.987; (d) $K = 2, L = 876$.

5. (a) Use "C" in the first column.

Section D.5

1. (a) Not valid; (e) valid; (g) not valid.

2. (c) Valid; (e) valid.

3. (a) X will be an array having a range of 1 to 23 for the first subscript and 1 to 35 for the second subscript. Y will be an array having a single subscript with a range of 1 to 15.

Section D.6

2. Yes.

3. (b) No.

Section D.7

4. F(X) = ((X + 2) * X + 3) * X + 4.

Appendix E

Section E.1

1. (a) Valid;   (e) not valid;   (i) not valid.

Section E.2

1. (a) 33;   (c) 7.

2. (b) 1.8.

3. (b) LET D = (A − B) * (2 + A);   (e) LET X = SIN(LOG(B)).

Section E.3

1. 4.9.

2. Yes.

3. (b) A = 5, B = 4;   (d) control is transferred to line 200.

5. No.

Section E.4

2. By printing a question mark.

4. Use the command RESTORE and then READ from the beginning to the desired data item.

6. (a) Start the line with the command REM.

Section E.5

3. (a) 6;   (c) No.

5. (a) Not valid;   (d) valid, although it may not do what you expect.

Section E.6

1. (b) Perform all statements in the loop down to NEXT K for the values K = 3, 8, 13, 18, 23, 28, 33, and 38.

4. (a) 100 LET S = 0
       110 FOR J = 6 TO 100 STEP 2
       120 LET S = S + J
       130 NEXT J

5. No.

Section E.7

4. (a) Replace lines 20 and 30 with DEF FNF(X) = TAN(X) and FNG(X) = EXP(X) respectively.

Index

Page numbers in *italics* refer to Figures.